ELEMENTARY
Algebra Within Reach

SIXTH EDITION

ELEMENTARY
Algebra Within Reach

SIXTH EDITION

Ron Larson
The Pennsylvania State University
The Behrend College

With the assistance of Kimberly Nolting
Hillsborough Community College

Australia • Brazil • Japan • Korea • Mexico • Singapore • Spain • United Kingdom • United States

**Elementary Algebra Within Reach,
Sixth Edition, International Edition**

Ron Larson

Senior Publisher: Charlie Van Wagner
Senior Acquiring Sponsoring Editor: Marc Bove
Associate Development Editor: Stefanie Beeck
Assistant Editor: Lauren Crosby
Senior Editorial Assistant: Jennifer Cordoba
Media Editor: Bryon Spencer
Senior Market Development Manager: Danae April
Content Project Manager: Jill Quinn
Manufacturing Planner: Doug Bertke
Rights Acquisition Specialist: Shalice Shah-Caldwell
Text and Cover Designer: Larson Texts, Inc.
Cover Image: Vlade Shestakov/Shutterstock.com
Compositor: Larson Texts, Inc.

© 2014, 2010, 2005 Brooks/Cole, Cengage Learning

CENGAGE and CENGAGE LEARNING are registered trademarks of Cengage Learning, Inc., within the United States and certain other jurisdictions.

ALL RIGHTS RESERVED. No part of this work covered by the copyright herein may be reproduced, transmitted, stored, or used in any form or by any means graphic, electronic, or mechanical, including but not limited to photocopying, recording, scanning, digitizing, taping, Web distribution, information networks, or information storage and retrieval systems, except as permitted under Section 107 or 108 of the 1976 United States Copyright Act or applicable copyright law of another jurisdiction, without the prior written permission of the publisher.

> For permission to use material from this text or product,
> submit all requests online at www.cengage.com/permissions.
> Further permissions questions can be emailed to
> permissionrequest@cengage.com.

International Edition:
ISBN-13: 978-1-285-16033-7
ISBN-10: 1-285-16033-9

Cengage Learning International Offices

Asia
www.cengageasia.com
tel: (65) 6410 1200

Australia/New Zealand
www.cengage.com.au
tel: (61) 3 9685 4111

Brazil
www.cengage.com.br
tel: (55) 11 3665 9900

India
www.cengage.com.in
tel: (91) 11 4364 1111

Latin America
www.cengage.com.mx
tel: (52) 55 1500 6000

UK/Europe/Middle East/Africa
www.cengage.co.uk
tel: (44) 0 1264 332 424

Represented in Canada by Nelson Education, Ltd.
www.nelson.com
tel: (416) 752 9100 / (800) 668 0671

Cengage Learning is a leading provider of customized learning solutions with office locations around the globe, including Singapore, the United Kingdom, Australia, Mexico, Brazil and Japan. Locate your local office at www.cengage.com/global

For product information and free companion resources:
www.cengage.com/international
Visit your local office: www.cengage.com/global

Printed in the United States of America
1 2 3 4 5 6 7 16 15 14 13 12

Contents

1 ▶ THE REAL NUMBER SYSTEM — 1
- 1.1 Real Numbers — 2
- 1.2 Adding and Subtracting Integers — 10
- 1.3 Multiplying and Dividing Integers — 18
- **Mid-Chapter Quiz** — 28
- 1.4 Operations with Rational Numbers — 30
- 1.5 Exponents and Properties of Real Numbers — 40
- **Chapter Summary** — 48
- **Review Exercises** — 50
- **Chapter Test** — 54

2 ▶ FUNDAMENTALS OF ALGEBRA — 55
- 2.1 Writing and Evaluating Algebraic Expressions — 56
- 2.2 Simplifying Algebraic Expressions — 64
- **Mid-Chapter Quiz** — 74
- 2.3 Algebra and Problem Solving — 76
- 2.4 Introduction to Equations — 86
- **Chapter Summary** — 94
- **Review Exercises** — 96
- **Chapter Test** — 100

3 ▶ LINEAR EQUATIONS AND PROBLEM SOLVING — 101
- 3.1 Solving Linear Equations — 102
- 3.2 Equations that Reduce to Linear Form — 110
- 3.3 Problem Solving with Percents — 118
- **Mid-Chapter Quiz** — 126
- 3.4 Ratios and Proportions — 128
- 3.5 Geometric and Scientific Applications — 136
- 3.6 Linear Inequalities — 144
- **Chapter Summary** — 152
- **Review Exercises** — 154
- **Chapter Test** — 157
- **Cumulative Test: Chapters 1–3** — 158

4 ▶ EQUATIONS AND INEQUALITIES IN TWO VARIABLES — 159
- 4.1 Ordered Pairs and Graphs — 160
- 4.2 Graphs of Equations in Two Variables — 168
- 4.3 Slope and Graphs of Linear Equations — 176
- **Mid-Chapter Quiz** — 184
- 4.4 Equations of Lines — 186
- 4.5 Graphs of Linear Inequalities — 194
- **Chapter Summary** — 202
- **Review Exercises** — 204
- **Chapter Test** — 208

5 ▶ EXPONENTS AND POLYNOMIALS 209

- 5.1 Negative Exponents and Scientific Notation 210
- 5.2 Adding and Subtracting Polynomials 218
- **Mid-Chapter Quiz** 226
- 5.3 Multiplying Polynomials: Special Products 228
- 5.4 Dividing Polynomials 238
- **Chapter Summary** 246
- **Review Exercises** 248
- **Chapter Test** 252

6 ▶ FACTORING AND SOLVING EQUATIONS 253

- 6.1 Factoring Polynomials with Common Factors 254
- 6.2 Factoring Trinomials 262
- 6.3 More About Factoring Trinomials 270
- **Mid-Chapter Quiz** 278
- 6.4 Factoring Polynomials with Special Forms 280
- 6.5 Solving Quadratic Equations by Factoring 288
- **Chapter Summary** 296
- **Review Exercises** 298
- **Chapter Test** 301
- **Cumulative Test: Chapters 4−6** 302

7 ▶ RATIONAL EXPRESSIONS AND EQUATIONS 303

- 7.1 Simplifying Rational Expressions 304
- 7.2 Multiplying and Dividing Rational Expressions 312
- 7.3 Adding and Subtracting Rational Expressions 320
- **Mid-Chapter Quiz** 328
- 7.4 Complex Fractions 330
- 7.5 Rational Equations and Applications 338
- **Chapter Summary** 346
- **Review Exercises** 348
- **Chapter Test** 352

8 ▶ SYSTEMS OF LINEAR EQUATIONS AND INEQUALITIES 353

- 8.1 Solving Systems of Equations by Graphing 354
- 8.2 Solving Systems of Equations by Substitution 362
- 8.3 Solving Systems of Equations by Elimination 370
- **Mid-Chapter Quiz** 378
- 8.4 Applications of Systems of Linear Equations 380
- 8.5 Systems of Linear Inequalities 388
- **Chapter Summary** 396
- **Review Exercises** 398
- **Chapter Test** 402

9 ▶ RADICAL EXPRESSIONS AND EQUATIONS 403

9.1	Roots and Radicals	404
9.2	Simplifying Radicals	412
	Mid-Chapter Quiz	422
9.3	Operations with Radical Expressions	424
9.4	Radical Equations and Applications	432
	Chapter Summary	440
	Review Exercises	442
	Chapter Test	446
	Cumulative Test: Chapters 7–9	447

10 ▶ QUADRATIC EQUATIONS AND FUNCTIONS 449

10.1	Solving by the Square Root Property	450
10.2	Solving by Completing the Square	458
10.3	Solving by the Quadratic Formula	466
	Mid-Chapter Quiz	474
10.4	Graphing Quadratic Equations	476
10.5	Applications of Quadratic Equations	484
10.6	Complex Numbers	492
10.7	Relations, Functions, and Graphs	500
	Chapter Summary	508
	Review Exercises	510
	Chapter Test	514

APPENDICES

Appendix A Introduction to Graphing Calculators (web)*

Appendix B Further Concepts in Geometry (web)*
        B.1   Exploring Congruence and Similarity
        B.2   Angles

Appendix C Further Concepts in Statistics (web)*

Answers to Odd-Numbered Exercises	A1
Index of Applications	A35
Index	A37

*Available at the text-specific website *www.cengagebrain.com*

Preface

Welcome to *Elementary Algebra Within Reach*, Sixth Edition. I am proud to present to you this new edition. As with all editions, I have been able to incorporate many useful comments from you, our user. And, while much has changed with this revision, you will still find what you expect—a pedagogically sound, mathematically precise, and comprehensive textbook.

I'm very excited about this edition. As I was writing, I kept one thought in mind—provide students what they need to learn algebra *within reach*. As you study from this book, you should notice right away that something is different. I've structured the book so that examples and exercises are on the same page—*within reach*. I am also offering something brand new with this edition: a companion website at **AlgebraWithinReach.com**. This site offers many resources that will help you as you study algebra. All of these resources are just a click away—*within reach*.

My goal for every edition of this textbook is to provide students with the tools that they need to master algebra. I hope that you find the changes in this edition, together with AlgebraWithinReach.com, will accomplish just that.

New To This Edition

REVISED Exercises Within Reach

The exercise sets have been carefully and extensively reviewed to ensure they are relevant and cover all topics suggested by our users. Additionally, the exercises have been completely restructured. Exercises now appear on the *same* page and immediately follow a corresponding example. There is no need to flip back and forth from example to exercise. The end-of-section exercises focus on mastery of conceptual understanding. View and listen to worked-out solutions at AlgebraWithinReach.com.

NEW Data Spreadsheets

Download editable spreadsheets from AlgebraWithinReach.com, and use this data to solve exercises.

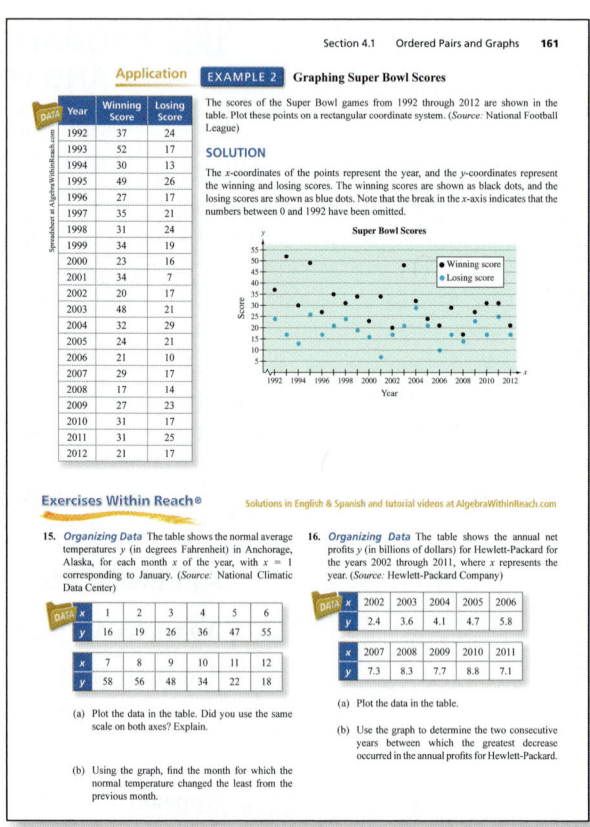

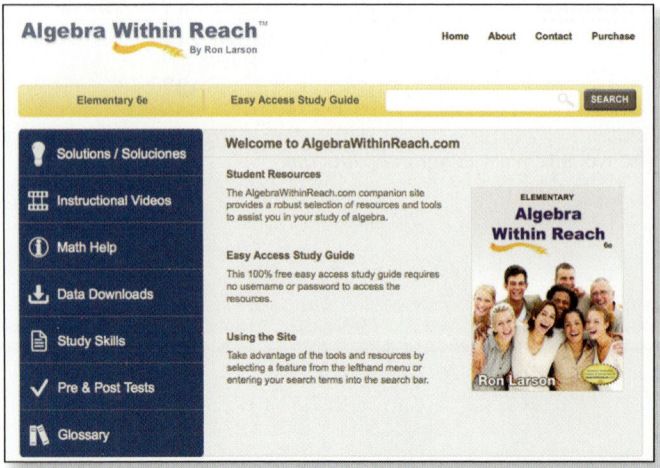

NEW AlgebraWithinReach.com

This companion website offers multiple tools and resources to supplement your learning. Access to these features is free. View and listen to worked-out solutions of thousands of exercises in English or Spanish, download data sets, take diagnostic tests, watch lesson videos and much more.

Preface   ix

NEW Concept Summary

This simple review of important concepts appears at the end of every section. Each Concept Summary reviews *What*, *How*, and *Why*—what concepts you studied, how to apply the concepts, and why the concepts are important. The Concept Summary includes four exercises to check your understanding.

NEW Math Helps

Additional instruction is available for every example and many exercises at AlgebraWithinReach.com. Just click on *Math Help*.

REVISED Section Objectives

A bulleted list of learning objectives provides you the opportunity to preview what will be presented in the upcoming section.

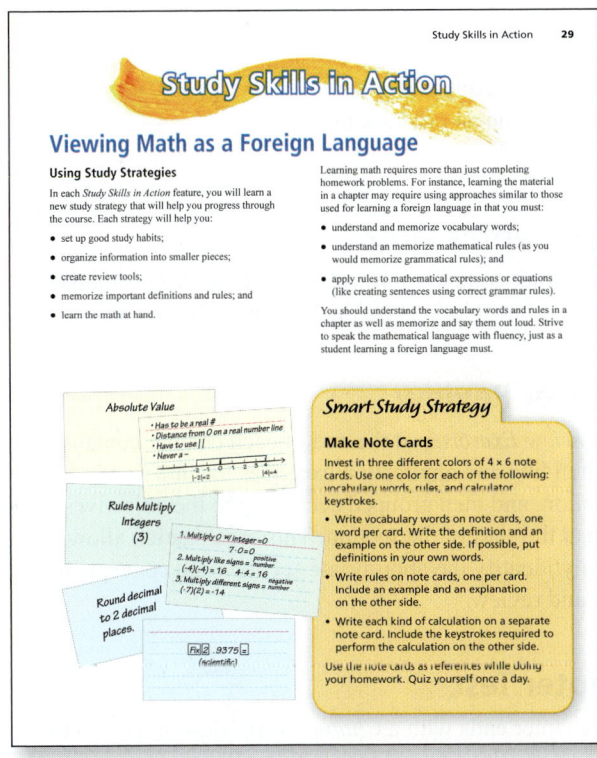

REVISED Study Skills in Action

Each chapter presents a study skill essential to success in mathematics. Read and apply these throughout the course. Print them out at AlgebraWithinReach.com to keep them as reminders to develop strong study skills.

REVISED Applications

A wide variety of real-life applications are integrated throughout the text in examples and exercises. These applications demonstrate the relevance of algebra in the real world. Many of these applications use current, real data.

REVISED Chapter Summaries

The *Chapter Summary* now includes explanations and examples of the objectives taught in the chapter. Review exercises that cover these objectives are listed to check your understanding of the material.

Trusted Features

Examples

Each example has been carefully chosen to illustrate a particular mathematical concept or problem-solving technique. The examples cover a wide variety of problems and are titled for easy reference. Many examples include detailed, step-by-step solutions with side comments, which explain the key steps of the solution process.

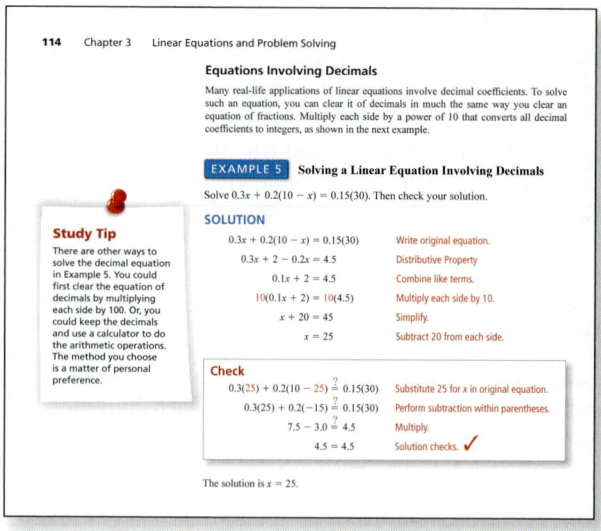

Study Tips

Study Tips offer students specific point-of-use suggestions for studying algebra, as well as pointing out common errors and discussing alternative solution methods. They appear in the margins.

Technology Tips

Point-of-use instructions for using graphing calculators or software appear in the margins as *Technology Tips*. These features encourage the use of graphing technology as a tool for visualization of mathematical concepts, for verification of other solution methods, and for help with computations.

Cumulative Review

Each exercise set (except in Chapter 1) is followed by *Cumulative Review* exercises that cover concepts from previous sections. This serves as a review and also a way to connect old concepts with new concepts.

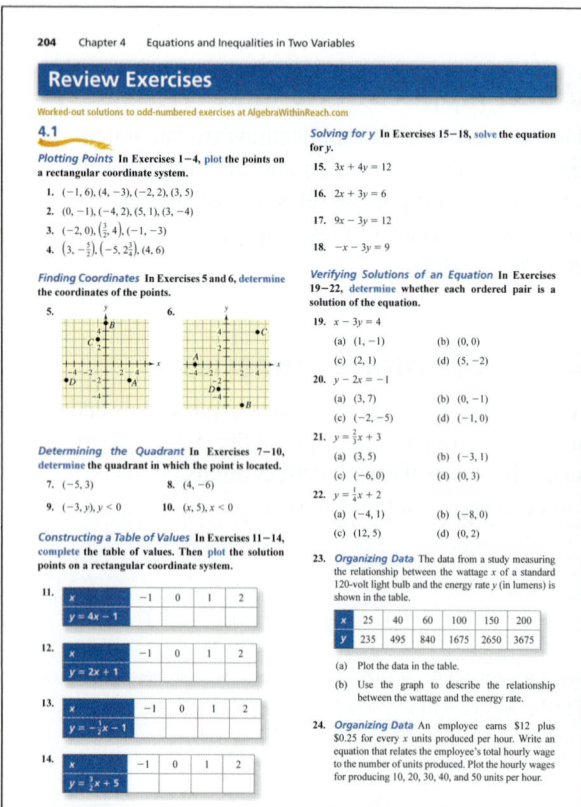

Mid-Chapter Quiz

Each chapter contains a *Mid-Chapter Quiz*. View and listen to worked-out solutions at AlgebraWithinReach.com.

Chapter Review

The *Review Exercises* at the end of each chapter contain skill-building and application exercises that are first ordered by section, and then grouped according to the objectives stated at the start of the section. This organization allows you to easily identify the appropriate sections and concepts for study and review.

Chapter Test

Each chapter ends with a *Chapter Test*. View and listen to worked-out solutions at AlgebraWithinReach.com.

Cumulative Test

The *Cumulative Tests* that follow Chapters 3, 6, and 9 provide a comprehensive self-assessment tool that helps you check your mastery of previously covered material. View and listen to worked-out solutions at AlgebraWithinReach.com.

Supplements

Student

Student Solutions Manual

ISBN 978-1-285-41974-9

Author: Ron Larson

The Student Solutions Manual provides detailed, step-by-step solutions to all odd-numbered problems in both the section exercise sets and review exercises. It also contains detailed, step-by-step solutions to all Mid-Chapter Quiz, Chapter Test, and Cumulative Test questions.

Student Workbook

ISBN 978-1-285-41975-6

Author: Maria H. Andersen, Muskegon Community College

Get a head start! The Student Workbook contains assessments, activities and worksheets for classroom discussions, in-class activities, and group work.

Instructor

Complete Solutions Manual

ISBN 978-1-285-16006-1

Author: Ron Larson

The Complete Solutions Manual provides detailed step-by-step solutions to all problems in the text. It contains Chapter and Final Exam test forms with answer keys as well as individual test items and answers for Chapters 1–10.

Instructor's Resource Binder

ISBN 978-0-538-73675-6

Author: Maria H. Andersen, Muskegon Community College

The Instructor's Resource Binder contains uniquely designed Teaching Guides, which include instruction tips, examples, activities, worksheets, overheads, and assessments with answers to accompany them.

ISBN: 978-0-538-73807-1

Instant feedback and ease of use are just two reasons why *WebAssign* is the most widely used homework system in higher education. *WebAssign*'s homework delivery system allows you to assign, collect, grade, and record homework assignments via the web. And now this proven system has been enhanced to include a multimedia eBook, video examples, and problem-specific tutorials. *Enhanced WebAssign* is more than a homework system—it is a complete learning system for math students.

PowerLecture with Examview

ISBN: 978-1-285-41982-4

Author: Ron Larson

This supplement provides the instructor with dynamic media tools for teaching. Create, deliver and customize tests (both print and online) in minutes with *Examview® Computerized Testing Featuring Algorithmic Equations*. Easily build solution sets for homework or exams using *Solution Builder*'s online solution manual. Microsoft® Powerpoint® lecture slides including *all* examples from the text, figures from the book, and easy-to-use PDF testbanks, in electronic format, are also included on this DVD-ROM.

Solution Builder

This online instructor database offers complete worked-out solutions to all exercises in the text, allowing you to create customized, secure solutions printouts (in PDF format) matched exactly to the problems you assign in class. For more information, visit www.cengage.com/solutionbuilder.

Acknowledgements

I would like to thank the many people who have helped me revise the various editions of this text. Their encouragement, criticisms, and suggestions have been invaluable.

Reviewers

Tom Anthony, *Central Piedmont Community College*
Tina Cannon, *Chattanooga State Technical Community College*
LeAnne Conaway, *Harrisburg Area Community College and Penn State University*
Mary Deas, *Johnson County Community College*
Jeremiah Gilbert, *San Bernadino Valley College*
Jason Pallett, *Metropolitan Community College-Longview*
Laurence Small, *L.A. Pierce College*
Dr. Azar Raiszadeh, *Chattanooga State Technical Community College*
Patrick Ward, *Illinois Central College*

My thanks to Robert Hostetler, The Behrend College, The Pennsylvania State University, David Heyd, The Behrend College, The Pennsylvania State University, and Patrick Kelly, Mercyhurst University, for their significant contributions to previous editions of this text.

I would also like to thank the staff of Larson Texts, Inc., who assisted in preparing the manuscript, rendering the art package, and typesetting and proofreading the pages and the supplements.

On a personal level, I am grateful to my spouse, Deanna Gilbert Larson, for her love, patience, and support. Also, a special thanks goes to R. Scott O'Neil.

If you have suggestions for improving this text, please feel free to write to me. Over the past two decades I have received many useful comments from both instructors and students, and I value these comments very much.

Ron Larson
Professor of Mathematics
Penn State University
www.RonLarson.com

1 The Real Number System

- **1.1** Real Numbers
- **1.2** Adding and Subtracting Integers
- **1.3** Multiplying and Dividing Integers
- **1.4** Operations with Rational Numbers
- **1.5** Exponents and Properties of Real Numbers

MASTERY IS WITHIN REACH!

"When I trained to be a tutor, I learned how note cards can be used to help review and memorize math concepts. I started using them in my own math class, and it really helps me, especially when I am caught with short amounts of time on campus. I don't have to pull out all my books to study. I just keep my note cards in my backpack."

Chris
Computer Science

See page 29 for suggestions about using note cards.

1.1 Real Numbers

▶ Classify numbers and plot them on a real number line.
▶ Use the real number line and inequality symbols to order real numbers.
▶ Find the absolute value of a number.

Classifying Real Numbers and the Real Number Line

The numbers you use in everyday life are called **real numbers**. They are classified into different categories, as shown at the right.

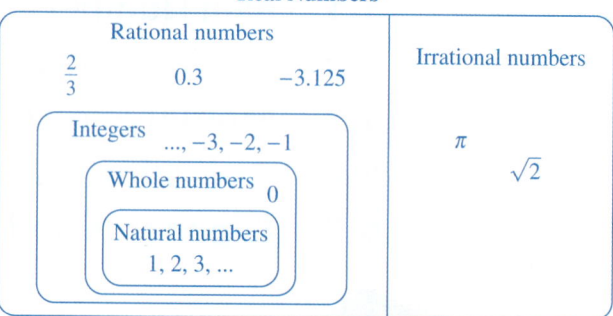

Study Tip

In *decimal form*, you can recognize rational numbers as decimals that terminate

$\frac{1}{2} = 0.5$ or $\frac{3}{8} = 0.375$

or repeat

$\frac{4}{3} = 1.\overline{3}$ or $\frac{2}{11} = 0.\overline{18}$.

Irrational numbers are represented by decimals that neither terminate nor repeat, as in

$\sqrt{2} = 1.414213562\ldots$

or

$\pi = 3.141592654\ldots$.

EXAMPLE 1 Classifying Real Numbers

Which of the numbers in the following set are (a) natural numbers, (b) integers, (c) rational numbers, and (d) irrational numbers?

$$\left\{\frac{1}{2}, -1, 0, 4, -\frac{5}{8}, \frac{4}{2}, -\frac{3}{1}, 0.86, \sqrt{2}, \sqrt{9}\right\}$$

SOLUTION

a. Natural numbers: $\left\{4, \frac{4}{2} = 2, \sqrt{9} = 3\right\}$

b. Integers: $\left\{-1, 0, 4, \frac{4}{2} = 2, -\frac{3}{1} = -3, \sqrt{9} = 3\right\}$

c. Rational numbers: $\left\{\frac{1}{2}, -1, 0, 4, -\frac{5}{8}, \frac{4}{2}, -\frac{3}{1}, 0.86, \sqrt{9} = 3\right\}$

d. Irrational number: $\left\{\sqrt{2}\right\}$

Exercises Within Reach®

Solutions in English & Spanish and tutorial videos at AlgebraWithinReach.com

Classifying Real Numbers In Exercises 1–4, *determine* which of the numbers in the set are (a) natural numbers, (b) integers, (c) rational numbers, and (d) irrational numbers.

1. $\left\{-3, 20, \pi, -\frac{3}{2}, \frac{9}{3}, 4.5, -\sqrt{3}\right\}$

2. $\left\{\frac{1}{5}, \sqrt{5}, -\frac{24}{3}, -42, -4.5, 10, -\pi\right\}$

3. $\left\{\sqrt{7}, -\sqrt{25}, -\frac{5}{1}, 9.4, 0, -12, \frac{7}{14}\right\}$

4. $\left\{\frac{6}{1}, -\frac{6}{18}, -\sqrt{11}, \sqrt{36}, -1, -9.98, 22\right\}$

Section 1.1 Real Numbers 3

The diagram used to represent real numbers is called the **real number line**.

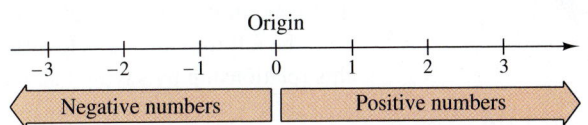

Each point on the real number line corresponds to exactly one real number, and each real number corresponds to exactly one point on the real number line.

EXAMPLE 2 Plotting Real Numbers

Plot each number on the real number line.

a. $-\dfrac{1}{2}$ b. 2 c. $-\dfrac{3}{2}$ d. 1

SOLUTION

a.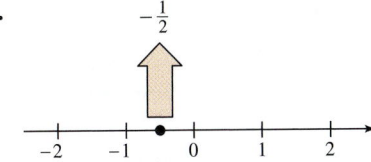

b.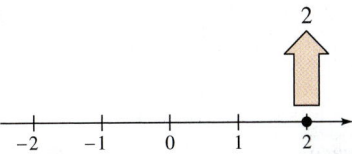

c.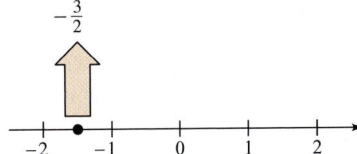

d.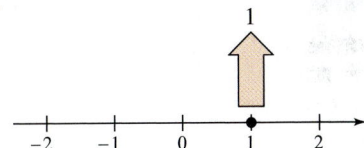

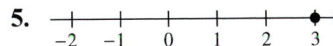

Solutions in English & Spanish and tutorial videos at AlgebraWithinReach.com

Identifying the Number In Exercises 5–10, identify the real number that corresponds to the point plotted on the real number line.

5.   6.   7.

8.   9.   10.

Plotting Real Numbers In Exercises 11–16, plot the numbers on the real number line.

11. $\dfrac{5}{2}, 5$ 12. $-\dfrac{3}{2}, -3$ 13. $-6, -7, -3$

14. $9, 6, 10$ 15. $-0.8, 1.2, 1.8$ 16. $-1.4, -0.3, 0.8$

Chapter 1   The Real Number System

Study Tip
The symbols < and > are called **inequality symbols**.

Ordering Real Numbers

The real number line provides you with a way of comparing any two real numbers. If a is to the left of b, then a is **less than** b, which is written as $a < b$. You can also describe this relationship by saying that b is **greater than** a, or $b > a$.

a is less than b or b is greater than a

EXAMPLE 3   **Ordering Real Numbers**

a. **Ordering Integers**

$3 < 5$ (3 is less than 5.) $-5 < -3$ (-5 is less than -3.)

b. **Ordering Decimals**

$-3.1 < 2.8$ (-3.1 is less than 2.8.) $-1.90 < -1.09$ (-1.90 is less than -1.09.)

c. **Ordering Fractions**

To order two fractions, you can write both fractions with the same denominator, or you can rewrite both fractions in decimal form. Here are two examples.

$$\frac{1}{3} = \frac{4}{12} \quad \text{and} \quad \frac{1}{4} = \frac{3}{12} \quad \Rightarrow \quad \frac{1}{3} > \frac{1}{4}$$

$$\frac{11}{131} \approx 0.084 \quad \text{and} \quad \frac{19}{209} \approx 0.091 \quad \Rightarrow \quad \frac{11}{131} < \frac{19}{209}$$

The symbol $\approx$ means "is approximately equal to."

Exercises Within Reach®   Solutions in English & Spanish and tutorial videos at AlgebraWithinReach.com

Ordering Real Numbers   In Exercises 17–24, **plot** each real number as a point on the real number line and **place** the correct inequality symbol (< or >) between the real numbers.

17. 3 ☐ −4   18. 6 ☐ −2   19. 3 ☐ 9   20. −4 ☐ −7

21. −4.6 ☐ 1.5   22. 28.60 ☐ −3.75   23. −6.58 ☐ −7.66   24. 20.156 ☐ 54.235

Ordering Fractions   In Exercises 25–28, **order** the fractions by (a) writing both fractions with the same denominator and (b) rewriting both fractions in decimal form.

25. $\frac{9}{16}$ ☐ $\frac{5}{8}$ 26. $-\frac{3}{8}$ ☐ $-\frac{5}{4}$ 27. $-\frac{7}{3}$ ☐ $-\frac{5}{2}$ 28. $\frac{3}{4}$ ☐ $\frac{5}{6}$

Section 1.1 Real Numbers 5

Application **EXAMPLE 4** **Comparing Profits**

The bar graph shows the profits for a company from 2008 through 2013.

a. Compare the profit for 2008 with the profit for 2009.

b. Write a statement that summarizes the trend in the company's profits.

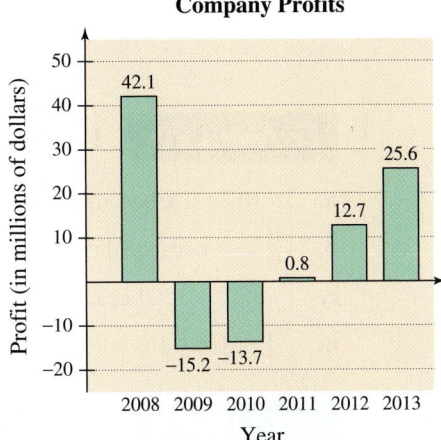

SOLUTION

a. The profit for 2008 was greater than the profit for 2009.

$$\$42.1 > -\$15.2$$

Profit for 2008 > Profit for 2009

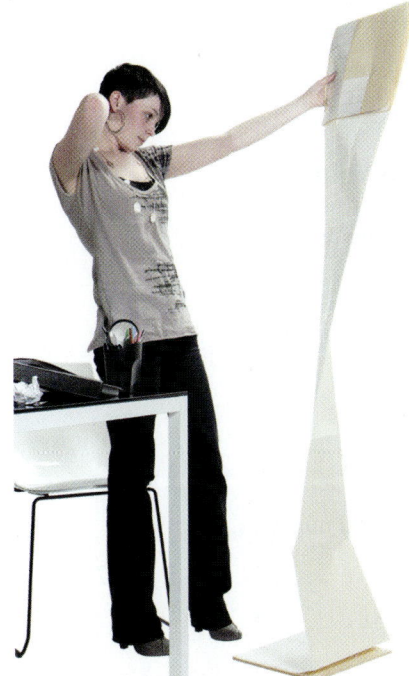

A negative profit is called a *loss*. However, when comparing profits and losses, it is better to list all the figures as profits. This makes comparisons easier.

b. Here is one possible summary statement.

"The company's profits had a big fall from 2008 to 2009. After that, profits have been increasing, but as of 2013, they have still not reached the level of profit in 2008."

Exercises Within Reach®

Solutions in English & Spanish and tutorial videos at AlgebraWithinReach.com

29. Miniature Golf The table shows your scores relative to par in six consecutive rounds of miniature golf.

(a) In golf, the lowest score wins. Which was your best score?

(b) Describe the trend in your scores.

Round	Score
1	4
2	1
3	0
4	−2
5	−3
6	−5

30. Temperature The line graph shows the temperature (in degrees Celsius) of a cut of meat moved from a freezer to a refrigerator to thaw.

(a) Compare the temperatures at hour 0 and hour 1.

(b) Write a statement that describes the trend in the temperature over the 6-hour period.

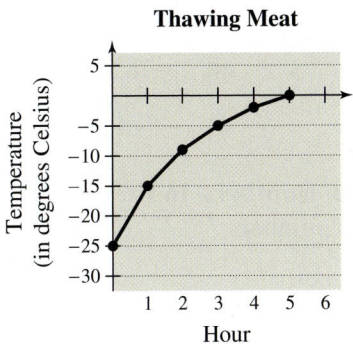

YG/Shutterstock.com

Finding Absolute Value

For any real number, the distance between the number and 0 on the real number line is its **absolute value**. A pair of vertical bars, | |, is used to denote absolute value. For instance, the absolute value of -8 is 8.

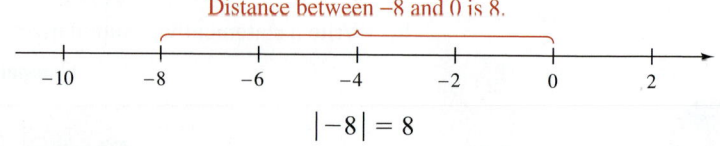

$|-8| = 8$

Study Tip

Two real numbers are **opposites** of each other if they lie the same distance from, but on opposite sides of, zero. For example, -2 is the opposite of 2, and 4 is the opposite of -4.

EXAMPLE 5 Evaluating an Absolute Value

a. $|-10| = 10$, because the distance between -10 and 0 is 10.

b. $\left|\frac{3}{4}\right| = \frac{3}{4}$, because the distance between $\frac{3}{4}$ and 0 is $\frac{3}{4}$.

c. $|-3.2| = 3.2$, because the distance between -3.2 and 0 is 3.2.

d. $-|-6| = -(6) = -6$

EXAMPLE 6 Comparing Real Numbers

Place the correct symbol ($<$, $>$, or $=$) between the real numbers.

a. $|-9|$ ▢ $|9|$ b. $|-3|$ ▢ 5 c. 0 ▢ $|-7|$
d. -4 ▢ $-|-4|$ e. $|12|$ ▢ $|-15|$ f. 2 ▢ $-|-2|$

SOLUTION

a. $|-9| = |9|$, because $|-9| = 9$ and $|9| = 9$.

b. $|-3| < 5$, because $|-3| = 3$ and 3 is less than 5.

c. $0 < |-7|$, because $|-7| = 7$ and 0 is less than 7.

d. $-4 = -|-4|$, because $-|-4| = -4$ and -4 is equal to -4.

e. $|12| < |-15|$, because $|12| = 12$, $|-15| = 15$, and 12 is less than 15.

f. $2 > -|-2|$, because $-|-2| = -2$ and 2 is greater than -2.

Exercises Within Reach®

Solutions in English & Spanish and tutorial videos at AlgebraWithinReach.com

Using Absolute Value In Exercises 31–34, **find** the distance between a and zero on the real number line.

31. $a = 2$ 32. $a = 5$ 33. $a = -8$ 34. $a = -17$

Evaluating an Absolute Value In Exercises 35–46, **evaluate** the expression.

35. $|10|$ 36. $|1|$ 37. $|-3|$ 38. $|-19|$

39. $|-3.4|$ 40. $|-16.2|$ 41. $\left|-\frac{7}{2}\right|$ 42. $\left|-\frac{9}{16}\right|$

43. $-|-23.6|$ 44. $-|-0.08|$ 45. $|0|$ 46. $|\pi|$

Comparing Real Numbers In Exercises 47–52, **place** the correct symbol ($<$, $>$, or $=$) between the real numbers.

47. $|-16|$ ▢ $|16|$ 48. $|525|$ ▢ $|-525|$ 49. $|-4|$ ▢ $|3|$

50. $|16|$ ▢ $|-25|$ 51. $-|-48.5|$ ▢ $|-48.5|$ 52. $|-\pi|$ ▢ $-|-2\pi|$

Application

EXAMPLE 7 Explaining Elevation

You are visiting Death Valley, California. Your GPS receiver lists your elevation as -282 feet. However, an outdoor sign states that the elevation is $+282$ feet below sea level. How can you explain the difference in signs?

SOLUTION

GPS receivers usually display elevations above sea level as positive numbers and elevations below sea level as negative numbers. If you want to talk about the elevation "below sea level," take the absolute value of the actual elevation.

$$|-282| = 282$$

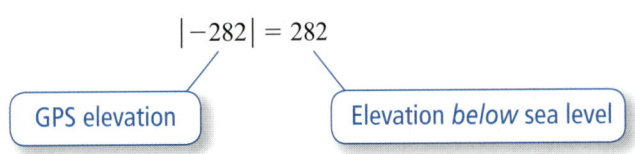

GPS elevation   Elevation *below* sea level

Exercises Within Reach®

Solutions in English & Spanish and tutorial videos at AlgebraWithinReach.com

53. Oceanography A whale is hunting at an elevation of -456 meters relative to sea level. How many meters below sea level is the whale?

54. Falcon A soaring falcon dives to the surface of a lake to catch a fish. The change in the falcon's elevation is -192 feet. What was the falcon's elevation relative to the lake before the dive?

55. Checking Account What payment amount should you enter for check number 143? Explain.

Check number	Date	Transaction description	Payment amount	Deposit amount	Account balance
142	1/11	Art text	78.99		-78.99 23.24
	1/12	Deposit		100.00	+100.00 123.24
143	1/14	Art supplies			-49.12 74.12
	1/16	Deposit			+150.00 224.12

56. Checking Account What deposit amount should you enter for January 16th? To find the amount, can you use the same method you used in Exercise 55? Do you *need* to use the same method you used in Exercise 55? Explain.

Concept Summary: Ordering Real Numbers

What

When you are asked to order two **real numbers**, the goal is to determine which of the two numbers is greater.

EXAMPLE

Order $-\frac{6}{3}$ and $-\frac{5}{2}$.

How

You can use the **real number line** to order two real numbers. For example, to order two **fractions**, rewrite them with the same denominator, or rewrite them as decimals. Then plot each number on a number line.

EXAMPLE

$-\frac{6}{3} = -2, \; -\frac{5}{2} = -2.5$

$-\frac{6}{3} > -\frac{5}{2}$

Why

There are many situations in which you need to order real numbers. For instance, to determine the standings at a golf tournament, you order the scores of the golfers.

Exercises Within Reach®

Worked-out solutions to odd-numbered exercises at AlgebraWithinReach.com

Concept Summary Check

57. *Using a Number Line* Explain how the number line above shows that $-2 > -2.5$.

58. *Using a Number Line* Describe the meaning of the expression $-\frac{6}{3} > -\frac{5}{2}$.

59. *Ordering Methods* Which method for ordering fractions is shown in the solution above?

60. *Rewriting Fractions* Describe another way to rewrite and order $-\frac{6}{3}$ and $-\frac{5}{2}$.

Extra Practice

Plotting Real Numbers In Exercises 61–64, **plot** the numbers on the real number line.

61. $\frac{5}{2}, \pi, -1, -|-3|$

62. $3.7, \frac{16}{3}, -|-1.9|, -\frac{1}{2}$

63. $-5, \frac{7}{3}, |-3|, 0, -|4.5|$

64. $|-2.3|, 3.2, -2.3, -|3.2|$

Distance on the Real Number Line In Exercises 65–70, **find** all real numbers whose distance from a is given by d.

65. $a = 8, d = 12$

66. $a = 6, d = 7$

67. $a = 21.3, d = 6$

68. $a = 42.5, d = 7$

69. $a = -2, d = 3.5$

70. $a = -7, d = 7.2$

Identifying Real Numbers In Exercises 71–76, **give** three examples that satisfy the given conditions.

71. A real number that is not a rational number

72. A real number that is not an irrational number

73. An integer that is a rational number

74. A rational number that is not a negative number

75. A real number that is not a positive rational number

76. An integer that is not a whole number

77. **Determining a Solution Set** Describe the real numbers n that satisfy the equation $n + |-n| = 2n$.

78. **Determining a Solution Set** Describe the real numbers n that satisfy the equation $n + |-n| = 0$.

79. **Volcanoes** The *summit elevation* of a volcano is the elevation of the top of the volcano relative to sea level. The table shows the summit elevations of several volcanoes.

Volcano	Summit Elevation
Kīlauea	1277 m
Lo`ihi	−969 m
Mauna Loa	4170 m
Ruby	−230 m
Anatahan	790 m

(a) Which of the volcanoes have summits below sea level?

(b) Which volcano summit is closest to sea level?

(c) Which volcano summit is farthest from sea level?

80. **Scuba Diving** The positions relative to sea level of two scuba divers are shown.

(a) Which position is represented by the greater integer?

(b) Which integer has the greater absolute value?

(c) What does the absolute value of each integer represent in the context of the problem?

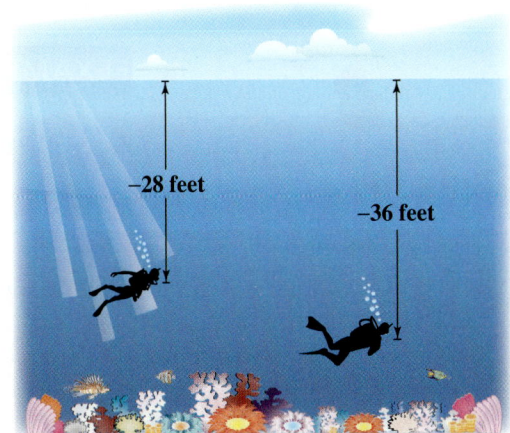

Explaining Concepts

81. Explain the difference between plotting the numbers 4 and −4 on the real number line.

82. How many numbers are three units from 0 on the real number line? Explain your answer.

83. **Writing** Explain why $\frac{8}{4}$ is a natural number, but $\frac{7}{4}$ is not.

84. **Number Sense** Which real number lies farther from 0 on the real number line, −15 or 10? Explain.

85. **Number Sense** Which real number lies farther from −4 on the real number line, 3 or −10? Explain.

86. **Precision** Which real number is smaller, $\frac{3}{8}$ or 0.37? Explain.

True or False? In Exercises 87–92, **decide** whether the statement is true or false. **Justify** your answer.

87. The absolute value of any real number is always positive.

88. The absolute value of a number is equal to the absolute value of its opposite.

89. The absolute value of a rational number is a rational number.

90. A given real number corresponds to exactly one point on the real number line.

91. The opposite of a positive number is a negative number.

92. Every rational number is an integer.

1.2 Adding and Subtracting Integers

▶ Add integers using a number line.
▶ Add integers with like signs and with unlike signs.
▶ Subtract integers with like signs and with unlike signs.

Adding Integers Using a Number Line

To find the sum $a + b$ using a number line, start at 0. Move right or left a units depending on whether a is positive or negative. From that position, move right or left b units depending on whether b is positive or negative. The final position is the **sum**.

EXAMPLE 1 Adding Integers Using a Number Line

a. Like signs: $5 + 2 = 7$

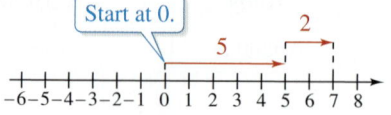

b. Like signs: $-3 + (-5) = -8$

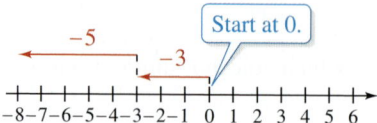

c. Unlike signs: $-5 + 2 = -3$

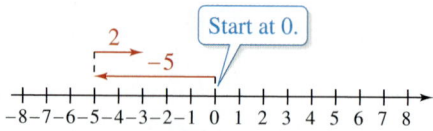

d. Unlike signs: $7 + (-3) = 4$

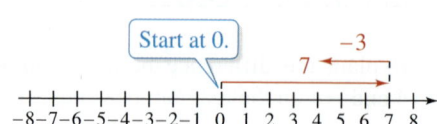

Study Tip

The sum of a number and its opposite is 0. For instance,

$-4 + 4 = 0.$

Exercises Within Reach®

Solutions in English & Spanish and tutorial videos at AlgebraWithinReach.com

Adding Integers Using a Number Line In Exercises 1–12, **find** the sum and **demonstrate** the addition on the real number line.

1. $2 + 7$
2. $3 + 9$
3. $-8 + (-3)$
4. $-4 + (-7)$
5. $10 + (-3)$
6. $14 + (-8)$
7. $-6 + 4$
8. $-12 + 5$
9. $3 + (-9)$
10. $-2 + 7$
11. $-5 + 9$
12. $5 + (-8)$

Application EXAMPLE 2 Finding the Total Yardage

The yards gained or lost by a football team on its four downs are shown.

1st Down: +3 yards
2nd Down: −4 yards
3rd Down: +7 yards
4th Down: +4 yards

Did the team earn a new first down? (*Note:* In football, a team earns a new first down when it gains 10 or more yards in 4 downs.)

SOLUTION

You can use a number line to find the total yards gained.

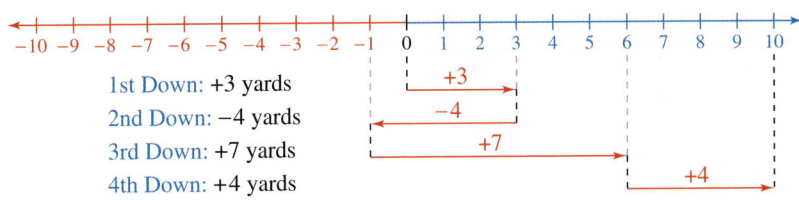

The team gained 10 yards total. So, it did earn a new 1st down.

$$3 + (-4) + 7 + 4 = 10$$

Exercises Within Reach®

Solutions in English & Spanish and tutorial videos at AlgebraWithinReach.com

13. **Highway Speed** While driving on a highway, you set the cruise control at the speed limit. You change your speed by the following amounts for traffic.

    −7 miles per hour
    +6 miles per hour
    −3 miles per hour
    +8 miles per hour

    Are you traveling within 5 miles per hour of the speed limit after these changes?

14. **Game Score** Your team's score in a game is the sum of the scores of the four players on your team.

    Player 1: +7 points
    Player 2: −1 point
    Player 3: +5 points
    Player 4: −4 points

    Is your team's score greater than the opposing team's score of 8 points?

15. **Temperature Change** When you left for class in the morning, the temperature was 25°C. By the time class ended, the temperature had increased by 4°. While you studied, the temperature increased by 3°. During your soccer practice, the temperature decreased by 9°. What was the temperature after your soccer practice?

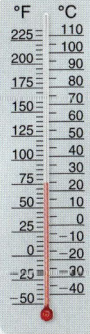

16. **Temperature Change** When you left for class in the morning, the temperature was 40°F. By the time class ended, the temperature had increased by 13°. While you studied, the temperature decreased by 5°. During your club meeting, the temperature decreased by 6°. What was the temperature after your club meeting?

Adding Integers Algebraically

1. **To add two integers with like signs**, add their absolute values and attach the common sign to the result.

2. **To add two integers with unlike signs**, subtract the lesser absolute value from the greater absolute value and attach the sign of the integer with the greater absolute value.

EXAMPLE 3 Adding Integers

a. Like Signs: $-18 + (-62) = -(|-18| + |-62|)$
$= -(18 + 62)$
$= -80$

b. Unlike Signs: $22 + (-17) = |22| - |-17|$
$= 22 - 17$
$= 5$

c. Unlike Signs: $-84 + 14 = -(|-84| - |14|)$
$= -(84 - 14)$
$= -70$

There are different ways to add three or more integers. You can use the carrying algorithm with a vertical format with nonnegative integers, as shown below, or you can add them two at a time, as illustrated in Example 4.

$$\begin{array}{r} {\scriptstyle 1\ 1} \\ 1\ 4\ 8 \\ 6\ 2 \\ +\ 5\ 3\ 6 \\ \hline 7\ 4\ 6 \end{array}$$

Vertical carrying algorithm
$148 + 62 + 536 = 746$

Exercises Within Reach®

Solutions in English & Spanish and tutorial videos at AlgebraWithinReach.com

Adding Integers In Exercises 17–46, find the sum.

17. $6 + 10$
18. $8 + 3$
19. $14 + (-14)$
20. $10 + (-10)$
21. $-45 + 45$
22. $-23 + 23$
23. $-14 + (-13)$
24. $-20 + (-19)$
25. $-23 + (-4)$
26. $-32 + (-16)$
27. $18 + (-12)$
28. $34 + (-16)$
29. $-75 + 100$
30. $-54 + 68$
31. $9 + (-14)$
32. $18 + (-26)$

33. $\begin{array}{r} 110 \\ 45 \\ +\ 208 \\ \hline \end{array}$

34. $\begin{array}{r} 44 \\ 115 \\ +\ 380 \\ \hline \end{array}$

35. $\begin{array}{r} 250 \\ 354 \\ +\ 122 \\ \hline \end{array}$

36. $\begin{array}{r} 275 \\ 416 \\ +\ 316 \\ \hline \end{array}$

37. $10 + (-6) + 34$
38. $7 + (-4) + 1$
39. $-15 + (-3) + 8$
40. $-82 + (-36) + 82$
41. $9 + (-18) + 4$
42. $2 + (-51) + 13$
43. $803 + (-104) + (-613) + 214$
44. $4365 + (-2145) + (-1873) + 40{,}084$
45. $312 + (-564) + 119 + (-100)$
46. $1200 + (-1300) + 62 + (-275)$

Section 1.2 Adding and Subtracting Integers

Application

EXAMPLE 4 Finding Your Account Balance

At the beginning of a month, your account balance was $28. During the month, you deposited $60 and withdrew $40. What was your balance at the end of the month?

ANY BANK
12345 Main Street
Anytown, NY 01234

CHECKING ACCOUNT STATEMENT

Statement period: 01/01/2013 - 02/01/2013
Account No.: 00005-123-456-7

Date	Description	Ref.	Withdrawals	Deposits	Balance
01/01	Beginning Balance				$28.00
01/01	Deposit			$60.00	?
01/01	Withdrawal		$40.00		?
01/01	Ending Balance				?

SOLUTION

$$\$28 + \$60 + (-\$40) = (\$28 + \$60) + (-\$40)$$
$$= \$88 + (-\$40)$$
$$= \$48 \qquad \text{Balance}$$

Your balance at the end of the month was $48.

Exercises Within Reach®

Solutions in English & Spanish and tutorial videos at AlgebraWithinReach.com

47. *Finding Your Account Balance* At the beginning of a month, your account balance was $46. During the month, you deposited $552 and wrote a check for $489. What was your balance at the end of the month?

Date	Description	Withdrawals	Deposits	Balance
09/01	Beginning Balance			$46.00
09/08	Deposit		$552.00	?
09/18	Check #321	$489.00		?
09/30	Ending Balance			?

48. *Finding Your Account Balance* At the beginning of a month, your checking account balance was $89. During the month, you deposited $120 and withdrew $108. What was your balance at the end of the month?

Date	Description	Withdrawals	Deposits	Balance
10/01	Beginning Balance			$89.00
10/09	Deposit		$120.00	?
10/22	Withdrawal	$108.00		?
10/31	Ending Balance			?

49. *Fishing Depth* A fisherman drops a line 56 feet below the surface of the water. The fisherman raises the line 18 feet, then 16 feet more. Finally, the fisherman drops the line 45 feet and catches a fish. How far below the surface does the fisherman catch the fish?

50. *Forensics* A forensic archaeologist finds a human skull 220 centimeters below ground. The archaeologist had already found a leg bone 75 centimeters above the skull and a hand bone 36 centimeters below the leg bone. At what depth was the hand bone?

Apollofoto/Shutterstock.com; Iurii Davydov/Shutterstock.com

Study Tip

Be sure you understand that the terminology of subtraction is not the same as that used for negative numbers. For instance, −5 is read as "negative 5," but 8 − 5 is read as "8 subtract 5" or "8 minus 5."

Subtracting Integers Algebraically

To subtract one integer from another, add the opposite of the integer being subtracted to the other integer. The result is called the **difference** of the two integers.

EXAMPLE 5 Subtracting Integers

a. $3 - 8 = 3 + (-8) = -5$ Add the opposite of 8.

b. $10 - (-13) = 10 + 13 = 23$ Add the opposite of −13.

c. $-5 - 12 = -5 + (-12) = -17$ Add the opposite of 12.

For subtraction problems involving two nonnegative integers, you can use the borrowing algorithm shown below.

Vertical borrowing algorithm

$415 - 276 = 139$

EXAMPLE 6 Evaluating Expressions

a.
$$-13 - 7 + 11 - (-4) = -13 + (-7) + 11 + 4$$ Add opposites.
$$= -20 + 11 + 4$$ Add −13 and −7.
$$= -9 + 4$$ Add −20 and 11.
$$= -5$$ Add.

b.
$$-1 - 3 - 4 + 6 = -1 + (-3) + (-4) + 6$$ Add opposites.
$$= -4 + (-4) + 6$$ Add −1 and −3.
$$= -8 + 6$$ Add −4 and −4.
$$= -2$$ Add.

Exercises Within Reach®

Solutions in English & Spanish and tutorial videos at AlgebraWithinReach.com

Subtracting Integers In Exercises 51−82, find the difference.

51. $21 - 18$
52. $47 - 12$
53. $51 - 25$
54. $37 - 37$
55. $1 - (-4)$
56. $7 - (-8)$
57. $15 - (-10)$
58. $8 - (-31)$
59. $18 - (-18)$
60. $62 - (-28)$
61. $27 - 57$
62. $18 - 32$
63. $61 - 85$
64. $53 - 74$
65. $22 - 131$
66. $48 - 222$
67. $2 - 11$
68. $3 - 15$
69. $13 - 24$
70. $26 - 34$
71. $-135 - (-114)$
72. $-63 - (-8)$
73. $-4 - (-4)$
74. $-942 - (-942)$
75. $-10 - (-4)$
76. $-12 - (-7)$
77. $-71 - 32$
78. $-84 - 55$
79. $-210 - 400$
80. $-120 - 142$
81. $-110 - (-30)$
82. $-2500 - (-600)$

Evaluating an Expression In Exercises 83−88, evaluate the expression.

83. $-3 + 2 - 20 + 9$
84. $-1 + 3 - (-4) + 10$
85. $12 - 6 + 3 - (-8)$
86. $6 + 7 - 12 - 5$
87. $-(-5) + 7 - 18 + 4$
88. $-15 - (-2) + 4 - 6$

Section 1.2 Adding and Subtracting Integers

Application

EXAMPLE 7 Finding a Temperature

The temperature in Minneapolis, Minnesota, at 4 P.M. was 15°F. By midnight, the temperature had decreased by 18°. What was the temperature in Minneapolis at midnight?

SOLUTION

To find the temperature at midnight, subtract 18 from 15.

$$15 - 18 = 15 + (-18)$$
$$= -3$$

The temperature in Minneapolis at midnight was -3°F.

Minnehaha Falls in Minneapolis has a 53-foot drop. It is located near the confluence of Minnehaha Creek and the Mississippi.

EXAMPLE 8 Using a Calculator

Evaluate each expression using a calculator.

a. $-4 - 5$ b. $2 - (-3) + 9$

SOLUTION

	Keystrokes	Display	
a.	4 [+/−] [−] 5 [=]	-9	Scientific
	[(−)] 4 [−] 5 [ENTER]	-9	Graphing

	Keystrokes	Display	
b.	2 [−] [(] 3 [+/−] [)] [+] 9 [=]	14	Scientific
	2 [−] [(] [(−)] 3 [)] [+] 9 [ENTER]	14	Graphing

Exercises Within Reach®

Solutions in English & Spanish and tutorial videos at AlgebraWithinReach.com

89. Altitude An airplane flying at a cruising altitude of 31,000 feet is instructed to descend as shown in the diagram below. How many feet must the airplane descend?

31,000 ft   24,000 ft

Not drawn to scale

90. Profit A telephone company lost $650,000 during the first half of the year. By the end of the year, the company had an overall profit of $362,000. What was the company's profit during the second half of the year?

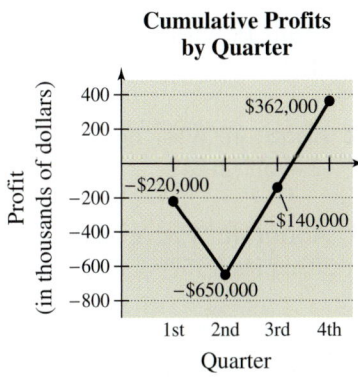

Cumulative Profits by Quarter

📱 **Using a Calculator** In Exercises 91–96, **write** the keystrokes used to evaluate the expression with a calculator (either scientific or graphing). Then **evaluate** the expression.

91. $-3 - 7$

92. $9 - (-2)$

93. $6 + 5 - (-7)$

94. $4 - 3 - (-9)$

95. $-6 + (-2) - 5$

96. $-3 + (-7) - 9$

16 Chapter 1 The Real Number System

> ### Concept Summary: Adding and Subtracting Integers
>
> **What**
> When you are asked to find the **sum** of two integers, you add the integers.
>
> **EXAMPLE**
> Evaluate the equation
> $-6 + 3$.
>
> **How**
> To find the sum of two integers,
> - use a number line.
> - add the integers algebraically.
>
> **EXAMPLE**
> $$-6 + 3 = -(|-6| - |3|)$$
> $$= -(6 - 3)$$
> $$= -3$$
>
> **Why**
> When you know how to add any two integers, you can subtract any two integers by adding the **opposite**.

Exercises Within Reach®

Worked-out solutions to odd-numbered exercises at AlgebraWithinReach.com

Concept Summary Check

97. *Explaining the Process* In the first step of the solution above, why is subtraction used between the absolute value expressions? Why is there a negative sign in front of the expression in parentheses?

98. *Using a Number Line* Describe the process for adding -6 and 3 using a number line.

99. *Adding Integers* Explain how to add two integers with like signs.

100. *Subtracting Integers* State the rule for subtracting integers.

Extra Practice

101. *Think About It* What number must be added to 10 to obtain -5?

102. *Think About It* What number must be added to 36 to obtain -12?

103. *Think About It* What number must be subtracted from -12 to obtain 24?

104. *Think About It* What number must be subtracted from -20 to obtain 15?

105. *Trail Elevation* The North Bass Trail in the Grand Canyon starts at Swamp Point and descends 244 meters to Muav Saddle. The trail continues on to Shinumo Creek, which is 1433 meters lower than Swamp Point, and then descends another 182 meters to the Colorado River. How much lower than Muav Saddle is the Colorado River?

106. *Trail Elevation* The North Kaibab Trail in the Grand Canyon descends 439 meters from the North Kaibab trailhead to the Supai Tunnel. The trail continues on to Roaring Springs, which is 921 meters lower than the North Kaibab trailhead, and then descends another 457 meters to Ribbon Falls. How much lower than the Supai Tunnel is Ribbon Falls?

The Grand Canyon, located in Arizona, is 277 river miles long and 1 mile deep on average.

John Burcham/National Geographic/Getty Images

107. *Stock Values* On Monday, you purchased $500 worth of stock. The values of the stock during the remainder of the week are shown in the bar graph.

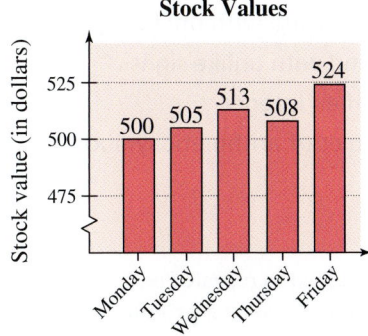

(a) Use the graph to complete the table.

Day	Daily Gain or Loss
Tuesday	
Wednesday	
Thursday	
Friday	

(b) What was the total change in the value of the stock from trading on Thursday and Friday?

(c) Find the sum of the daily gains and losses. Interpret the result in the context of the problem. How could you determine this sum from the graph?

108. *Profits* The bar graph shows the yearly revenues and expenses of a company for 2010 through 2013.

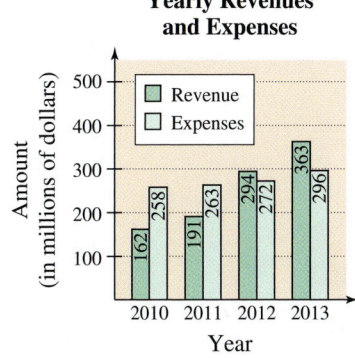

(a) Use the graph to complete the table for the yearly profit, using Profit = Revenue − Expenses.

Year	Profit
2010	
2011	
2012	
2013	

(b) Find the total profit for the years 2010 through 2013. How is the total profit related to the total revenue and total expenses for these years?

(c) From 2010 through 2013, between which two consecutive years was the change in profits the greatest?

Explaining Concepts

Adding Integers Using a Number Line In Exercises 109 and 110, an addition problem is shown visually on the real number line. (a) *Write* the addition problem and find the sum. (b) *State* the rule for adding integers that is demonstrated.

109.

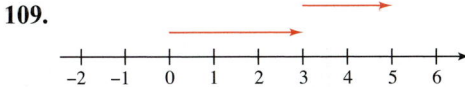

110.

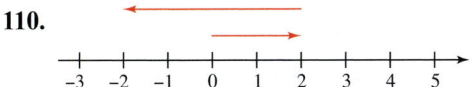

111. *Writing* Explain why the sum of two negative integers is a negative integer.

112. *Writing* When is the sum of a positive integer and a negative integer a positive integer?

113. *Writing* Is is possible that the sum of two positive integers is a negative integer? Explain.

114. *Writing* Is it possible that the difference of two negative integers is a positive integer? Explain.

1.3 Multiplying and Dividing Integers

▶ Multiplying integers with like signs and with unlike signs.
▶ Divide integers with like signs and with unlike signs.
▶ Find factors and prime factors of an integer.
▶ Represent the definitions and rules of arithmetic symbolically.

Study Tip

To find the product of more than two numbers, first find the product of their absolute values. If the number of negative factors is even, then the product is positive. If the number of negative factors is odd, then the product is negative.

Multiplying Integers

Multiplication of two integers can be described as repeated addition or subtraction. The result of multiplying one number by another is called a **product**.

1. The product of an integer and zero is 0.
2. The product of two integers with like signs is positive.
3. The product of two integers with unlike signs is negative.

EXAMPLE 1 Multiplying Integers

a. $4(10) = 40$ (Positive) · (positive) = positive
b. $-6 \cdot 9 = -54$ (Negative) · (positive) = negative
c. $-5(-7) = 35$ (Negative) · (negative) = positive
d. $3(-12) = -36$ (Positive) · (negative) = negative
e. $-12 \cdot 0 = 0$ (Negative) · (zero) = zero
f. $-2(8)(-3)(-1) = -(2 \cdot 8 \cdot 3 \cdot 1)$ Odd number of negative factors
                $= -48$ Answer is negative.

To multiply two integers having two or more digits, use the vertical multiplication algorithm shown below. The sign of the product is determined by the usual multiplication rule.

Vertical multiplication algorithm

$47(23) = 1081$

$$\begin{array}{r} 47 \\ \times\ 23 \\ \hline 141 \\ 94 \\ \hline 1081 \end{array}$$

141   Multiply 3 times 47.
94   Multiply 2 times 47.
1081 Add columns.

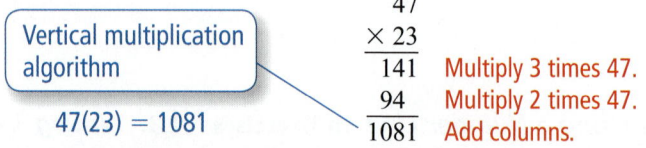

Solutions in English & Spanish and tutorial videos at AlgebraWithinReach.com

Multiplying Integers In Exercises 1–24, **find** the product.

1. 5×7
2. 8×3
3. $0 \cdot 4$
4. $-12 \cdot 0$
5. $2(-16)$
6. $8(-7)$
7. $-9(4)$
8. $-6(5)$
9. $230(-3)$
10. $175(-2)$
11. $-7(-13)$
12. $-40(-4)$
13. $-200(-8)$
14. $-150(-4)$
15. $3(-5)(6)$
16. $4(2)(-6)$
17. $7(-3)(-1)$
18. $-2(5)(-3)$
19. $-2(-3)(-5)$
20. $-10(-4)(-2)$
21. $|(-3)4|$
22. $|8(-9)|$
23. $|3(-5)(6)|$
24. $|8(-3)(5)|$

Multiplying Integers Vertically In Exercises 25–32, use the vertical multiplication algorithm to **find** the product.

25. 26×13
26. -14×24
27. $75(-63)$
28. $-72(866)$
29. $-13(-20)$
30. $-11(-24)$
31. $-21(-429)$
32. $-14(-585)$

Section 1.3 Multiplying and Dividing Integers 19

Application   EXAMPLE 2   Geometry: Finding the Volume of a Box

Find the volume of the rectangular box.

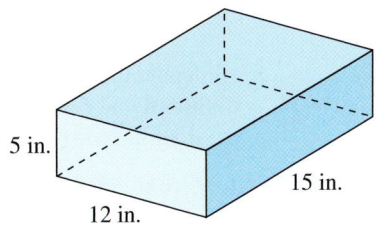

Jets typically provide about 30 cubic feet of air per passenger in coach and about 50 cubic feet per passenger in first class.

SOLUTION

To find the volume, multiply the length, width, and height of the box.

Volume = (Length) • (Width) • (Height)
= (15 inches) • (12 inches) • (5 inches)
= 900 cubic inches

So, the box has a volume of 900 cubic inches.

Exercises Within Reach®

Solutions in English & Spanish and tutorial videos at AlgebraWithinReach.com

33. *Geometry* Find the area of the football field.

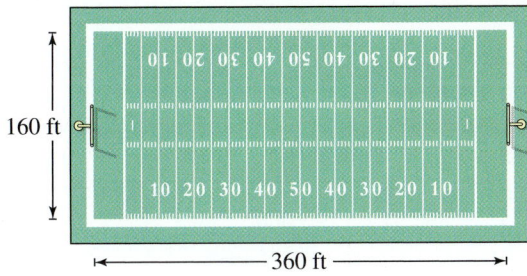

34. *Geometry* Find the area of the city park.

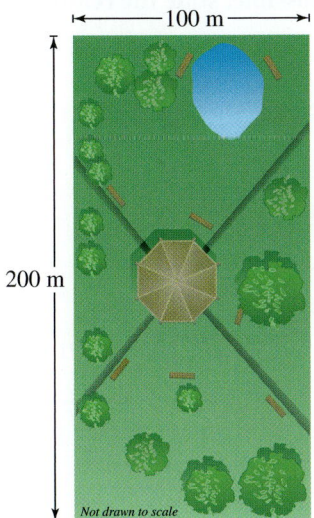

35. *Geometry* Find the volume of the rectangular shipping box.

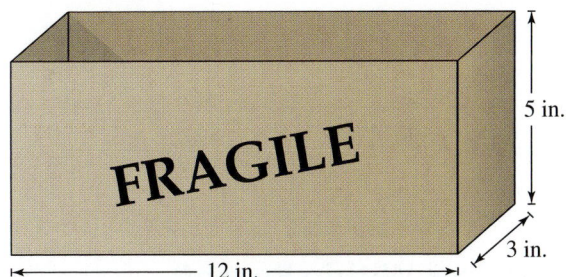

36. *Geometry* Find the volume of the rectangular shipping box.

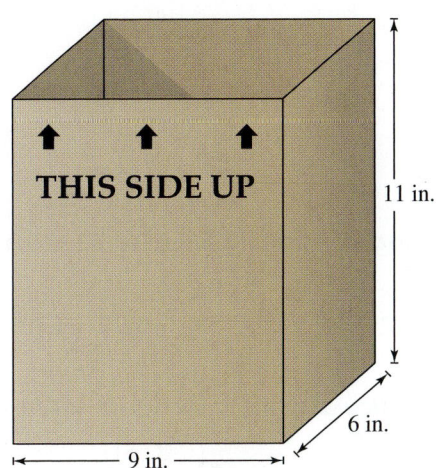

Dmitriy Shironosov/Shutterstock.com

Dividing Integers

The result of dividing one integer by another is called the **quotient** of the integers. Division is denoted by the symbol ÷, or by /, or by a horizontal line. For instance,

$$30 \div 6, \quad 30/6, \quad \text{and} \quad \frac{30}{6}$$

all denote the quotient of 30 and 6, which is 5. Using the form $30 \div 6$, 30 is called the **dividend** and 6 is the **divisor**. In the forms 30/6 and $\frac{30}{6}$, 30 is the **numerator** and 6 is the **denominator**.

1. Zero divided by a nonzero integer is 0, whereas a nonzero integer divided by zero is undefined.
2. The quotient of two nonzero integers with like signs is positive.
3. The quotient of two nonzero integers with unlike signs is negative.

EXAMPLE 3 **Dividing Integers**

a. $\dfrac{-42}{-6} = 7$ because $-42 = 7(-6)$.

b. $36 \div (-9) = -4$ because $36 = (-4)(-9)$.

c. $0 \div (-13) = 0$ because $0 = (0)(-13)$.

d. $-105 \div 7 = -15$ because $-105 = (-15)(7)$.

e. $-97 \div 0$ is undefined.

When dividing large numbers, the long division algorithm can be used. For instance, the long division algorithm below shows that $351 \div 13 = 27$.

$$\begin{array}{r} 27 \\ 13\overline{)351} \\ \underline{26} \\ 91 \\ \underline{91} \\ 0 \end{array}$$

Long division algorithm

$351 \div 13 = 27$

Exercises Within Reach®

Solutions in English & Spanish and tutorial videos at AlgebraWithinReach.com

Dividing Integers In Exercises 37–52, perform the division, if possible. If not possible, state the reason.

37. $27 \div 9$
38. $35 \div 7$
39. $72 \div (-12)$
40. $54 \div (-9)$
41. $-28 \div 4$
42. $-108 \div 9$
43. $-56 \div (-8)$
44. $-68 \div (-4)$
45. $8 \div 0$
46. $17 \div 0$
47. $0 \div 8$
48. $0 \div 17$
49. $\dfrac{-81}{-3}$
50. $\dfrac{-125}{-25}$
51. $\dfrac{-28}{4}$
52. $\dfrac{72}{-12}$

Using Long Division In Exercises 53–60, use the long division algorithm to find the quotient.

53. $1440 \div 45$
54. $936 \div 52$
55. $1440 \div (-9)$
56. $936 \div (-8)$
57. $-1312 \div 16$
58. $-5152 \div 23$
59. $-9268 \div (-28)$
60. $-6804 \div (-36)$

Using a Calculator In Exercises 61–64, use a calculator to find the quotient.

61. $\dfrac{44{,}290}{515}$
62. $\dfrac{33{,}511}{47}$
63. $\dfrac{169{,}290}{-162}$
64. $\dfrac{-1{,}027{,}500}{250}$

Application — EXAMPLE 4 Finding an Average Gain in Stock Prices

On Monday, you bought $500 worth of stock in a company. During the rest of the week, you recorded the gains and losses in your stock's value.

Tuesday	Wednesday	Thursday	Friday
Gained $15	Lost $18	Lost $23	Gained $10

a. What was the value of the stock at the close of Wednesday?

b. What was the value of the stock at the end of the week?

c. What would the total loss have been if Thursday's loss had occurred on each of the four days?

d. What was the average daily gain (or loss) for the four days recorded?

SOLUTION

a. The value at the close of Wednesday was

$$500 + 15 - 18 = \$497.$$

b. The value of the stock at the end of the week was

$$500 + 15 - 18 - 23 + 10 = \$484.$$

c. The loss on Thursday was $23. If this loss had occurred each day, the total loss would have been

$$4(23) = \$92.$$

d. To find the average daily gain (or loss), add the gains and losses of the four days and divide by 4.

$$\text{Average} = \frac{15 + (-18) + (-23) + 10}{4} = \frac{-16}{4} = -4$$

This means that during the four days, the stock had an average loss of $4 per day.

Several million shares of stock are traded on the New York Stock Exchange each business day.

Exercises Within Reach®

Solutions in English & Spanish and tutorial videos at AlgebraWithinReach.com

65. *Average Speed* A space shuttle orbiting Earth travels about 45 miles in 9 seconds. What is the average speed of the space shuttle in miles per second?

66. *Sports* A football team gains a total of 20 yards after four plays.

(a) What is the average number of yards gained per play?

(b) The gains on the four plays are 8 yards, 4 yards, 2 yards, and 6 yards. Plot each of the gains and the average gain on the real number line.

(c) Find the difference between each gain and the average gain. Find the sum of these differences and give a possible explanation of the result.

67. *Exam Scores* A student has a total of 328 points after four 100-point exams.

(a) What is the average number of points scored per exam?

(b) The scores on the four exams are 87, 73, 77, and 91. Plot each of the scores and the average score on the real number line.

(c) Find the difference between each score and the average score. Find the sum of these differences and give a possible explanation of the result.

68. *Average Speed* A hiker jogs a mountain trail that is 6 miles long in 54 minutes. How many minutes does the hiker average per mile?

Factors and Prime Numbers

If a and b are positive integers, then a is a **factor** (or divisor) of b if and only if a divides evenly into b. For instance, 1, 2, 3, and 6 are all factors of 6.

The concept of factors allows you to classify positive integers into three groups: prime numbers, composite numbers, and the number 1.

1. An integer greater than 1 with no factors other than itself and 1 is called a **prime number**, or simply a prime.

2. An integer greater than 1 with more than two factors is called a **composite number**, or simply a composite.

Every composite number can be expressed as a unique product of prime factors.

Study Tip

A tree diagram is a nice way to record your work when you are factoring a composite number into prime numbers. For instance, the following tree diagram shows that $45 = 3 \cdot 3 \cdot 5$.

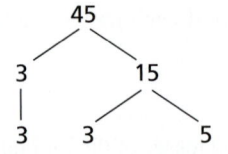

EXAMPLE 5   Prime Factorization

Write the prime factorization of each number.

a. 84   **b.** 78   **c.** 133   **d.** 43

SOLUTION

a. 2 is a recognized divisor of 84. So,
$$84 = 2 \cdot 42 = 2 \cdot 2 \cdot 21 = 2 \cdot 2 \cdot 3 \cdot 7.$$

b. 2 is a recognized divisor of 78. So,
$$78 = 2 \cdot 39 = 2 \cdot 3 \cdot 13.$$

c. If you do not recognize a divisor of 133, you can start by dividing any of the prime numbers 2, 3, 5, 7, 11, 13, etc., into 133. You will find 7 to be the first prime to divide into 133. So,
$$133 = 7 \cdot 19.$$

d. In this case, none of the primes less than 43 divides 43. So, 43 is prime.

Exercises Within Reach®   Solutions in English & Spanish and tutorial videos at AlgebraWithinReach.com

Classifying an Integer In Exercises 69–80, decide whether the number is prime or composite.

69. 2	70. 3	71. 4	72. 5
73. 7	74. 9	75. 12	76. 35
77. 240	78. 533	79. 643	80. 257

Prime Factorization In Exercises 81–98, write the prime factorization of the number.

81. 11	82. 13	83. 4
84. 6	85. 16	86. 27
87. 37	88. 29	89. 12
90. 52	91. 561	92. 245
93. 210	94. 525	95. 192
96. 264	97. 2535	98. 1521

Application

EXAMPLE 6 Finding Factors of Note Frequencies

From A220 to A440 on the piano, there are 12 semitones.

A, A#, B, C, C#, D, D#, E, F, F#, G, G#, A

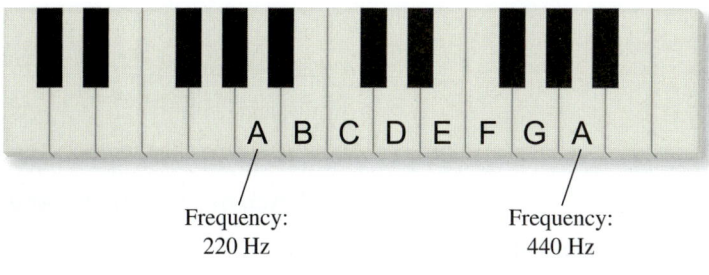

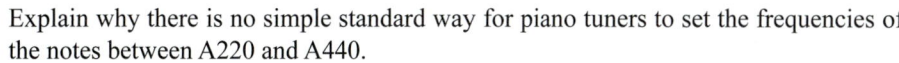

Frequency: 220 Hz   Frequency: 440 Hz

Explain why there is no simple standard way for piano tuners to set the frequencies of the notes between A220 and A440.

SOLUTION

The reason for this is a bit complicated. However, you can see the problem when you consider that piano tuners have two conflicting goals.

1. One goal is that from each note to the next higher note, you should have the same multiple. For instance, you want a number a such that

   (Frequency of A)(a) = (Frequency of A#)
   (Frequency of A#)(a) = (Frequency of B), ... and so on.

   The number that works is $a \approx 1.0594$.

2. A second goal is that you want the frequencies of the 11 notes between A220 and A440 to have as many common factors with 220 as possible. Notes whose frequencies have common factors harmonize.

When you try satisfying these two goals, you will see that they are conflicting.

Bartolomeo Cristofori (1655–1731) of Padua, Italy, is credited with the invention of the modern piano. He was employed by the famous Medici family.

Exercises Within Reach®

Solutions in English & Spanish and tutorial videos at AlgebraWithinReach.com

99. *Packaging* You are designing a shipping carton to hold 144 tomatoes. The carton will contain multiple layers of tomatoes arranged in rows and columns. One possible arrangement has 4 layers of 4 rows and 9 columns, as shown. Use the factors of 144 to determine the possible numbers of rows and columns when the carton has 6 layers.

100. *Inventory* You need to arrange 72 boxes on a skid with the same number of rows and columns in each layer. One possible arrangement has 6 layers of 3 rows and 4 columns, as shown. Use the factors of 72 to determine the possible numbers of rows and columns in an arrangement of 4 layers. Which of these arrangements do you prefer? Explain your reasoning.

Summary of Rules and Definitions

At its simplest level, algebra is a symbolic form of arithmetic. This arithmetic-algebra connection can be illustrated in the following way.

Arithmetic — Verbal rules and definitions
Algebra — Symbolic rules and definitions
Specific examples of rules and definitions

Arithmetic Summary

Definitions: Let a, b, and c be integers.

Definition — *Example*

1. Subtraction
$a - b = a + (-b)$ $5 - 7 = 5 + (-7)$

2. Multiplication: (a is a positive integer)
$a \cdot b = \underbrace{b + b + \cdots + b}_{a \text{ terms}}$ $3 \cdot 5 = 5 + 5 + 5$

3. Division: ($b \neq 0$)
$a \div b = c$ if and only if $a = c \cdot b$. $12 \div 4 = 3$ because $12 = 3 \cdot 4$.

4. Less than:
$a < b$ if there is a positive real number c such that $a + c = b$. $-2 < 1$ because $-2 + 3 = 1$.

5. Absolute value: $|a| = \begin{cases} a, & \text{if } a \geq 0 \\ -a, & \text{if } a < 0 \end{cases}$ $|-3| = -(-3) = 3$

6. Divisor:
a is a divisor of b if and only if there is an integer c such that $a \cdot c = b$.   7 is a divisor of 21 because $7 \cdot 3 = 21$.

Rules: Let a and b be integers.

Rule — *Example*

1. Addition
   (a) To add two integers with *like* signs, add their absolute values and attach the common sign to the result.
   $3 + 7 = |3| + |7| = 10$

   (b) To add two integers with *unlike* signs, subtract the smaller absolute value from the larger absolute value and attach the sign of the integer with the larger absolute value.
   $-5 + 8 = |8| - |-5|$
   $= 8 - 5$
   $= 3$

2. Multiplication:
   (a) $a \cdot 0 = 0 = 0 \cdot a$ $3 \cdot 0 = 0 = 0 \cdot 3$
   (b) Like signs: $a \cdot b > 0$ $(-2)(-5) = 10$
   (c) Unlike signs: $a \cdot b < 0$ $(2)(-5) = -10$

3. Division:
   (a) $\dfrac{0}{a} = 0, a \neq 0$ $\dfrac{0}{4} = 0$
   (b) $\dfrac{a}{0}$ is undefined. $\dfrac{6}{0}$ is undefined.
   (c) Like signs: $\dfrac{a}{b} > 0, b \neq 0$ $\dfrac{-2}{-3} = \dfrac{2}{3}$
   (d) Unlike signs: $\dfrac{a}{b} < 0, b \neq 0$ $\dfrac{-5}{7} = -\dfrac{5}{7}$

EXAMPLE 7 Using Rules and Definitions

a. Use the definition of subtraction to complete the statement.

$$4 - 9 = \boxed{}$$

b. Use the definition of multiplication to complete the statement.

$$6 + 6 + 6 + 6 = \boxed{}$$

c. Use the definition of absolute value to complete the statement.

$$|-9| = \boxed{}$$

d. Use the rule for adding integers with unlike signs to complete the statement.

$$-7 + 3 = \boxed{}$$

e. Use the rule for multiplying integers with unlike signs to complete the statement.

$$-9 \times 2 = \boxed{}$$

SOLUTION

a. $4 - 9 = 4 + (-9) = -5$

b. $6 + 6 + 6 + 6 = 4 \cdot 6 = 24$

c. $|-9| = -(-9) = 9$

d. $-7 + 3 = -(|-7| - |3|) = -4$

e. $-9 \times 2 = -18$

The word *algebra* comes from the Arabic language and much of its methods from Arabic/Islamic mathematics.

Exercises Within Reach®

Solutions in English & Spanish and tutorial videos at AlgebraWithinReach.com

Using a Rule or Definition In Exercises 101–104, complete the statement using the indicated definition or rule.

101. Definition of division: $12 \div 4 = \boxed{}$

102. Definition of absolute value: $|-8| = \boxed{}$

103. Rule for multiplying integers by 0:

$$6 \cdot 0 = \boxed{} = 0 \cdot 6$$

104. Rule for dividing integers with unlike signs:

$$\frac{30}{-10} = \boxed{}$$

Analyzing a Rule In Exercises 105 and 106, write an example and an algebraic description of the arithmetic rule.

105. The product of 1 and any real number is the real number itself.

106. Any nonzero real number divided by itself is 1.

Finding a Pattern In Exercises 107–110, complete the pattern. Decide which rule the pattern demonstrates.

107. $2(0) = 0$
 $1(0) = 0$
 $-1(0) = \boxed{}$
 $-2(0) = \boxed{}$

108. $0 \div 2 = 0$
 $0 \div 1 = 0$
 $0 \div (-1) = \boxed{}$
 $0 \div (-2) = \boxed{}$

109. $|2| = 2$
 $|1| = 1$
 $|0| = \boxed{}$
 $|-1| = \boxed{}$
 $|-2| = \boxed{}$

110. When $a + 2 = b$, $a < b$.
 When $a + 1 = b$, $a < b$.
 When $a + 0 = b$, $a \boxed{} b$.
 When $a + (-1) = b$, $a \boxed{} b$.
 When $a + (-2) = b$, $a \boxed{} b$.

Concept Summary: Multiplying and Dividing Integers

What
When you are asked to multiply or divide two integers a and b, the goal is to determine the correct sign of the **product** or **quotient** of the absolute values of the integers.

EXAMPLE

Find the product $-5 \cdot 8$.

Find the quotient $\dfrac{-40}{-8}$.

How
To determine the correct sign, use the rules for multiplying and dividing (nonzero) integers.

Like signs: $a \cdot b > 0, \dfrac{a}{b} > 0$

Unlike signs: $a \cdot b < 0, \dfrac{a}{b} < 0$

EXAMPLE

$-5 \cdot 8 = -40$

$\dfrac{-40}{-8} = 5$

Why
The sign is as important as the number in applications involving multiplying and dividing integers. In accounting, for instance, an incorrect sign could make a loss appear to be a profit.

Exercises Within Reach®

Worked-out solutions to odd-numbered exercises at AlgebraWithinReach.com

Concept Summary Check

111. *The Signs of the Factors* Do the integers multiplied in the example above have like signs or unlike signs?

112. *The Signs in a Quotient* Do the integers divided in the example above have like signs or unlike signs?

113. *A Sign of a Product* Is the product of -6 and -4 positive or negative? Explain.

114. *The Sign of a Quotient* Is the quotient of 12 and -4 positive or negative? Explain.

Extra Practice

Multiplying Integers In Exercises 115–126, **find** the product.

115. $-5(2)(-3)(-4)$

116. $-2(6)(-2)(-1)$

117. $4(-5)(2)(-1)$

118. $2(-3)(4)(-9)$

119. $-7(-2)(-5)(-3)(4)$

120. $-6(-2)(-3)(-4)(5)$

121. $-10(3)(7)(-2)(-6)$

122. $-10(4)(5)(-8)(-2)$

123. $4(-2)(3)(-4)(2)(-1)$

124. $5(-3)(2)(-2)(-1)(4)$

125. $8(-3)(2)(-4)(5)(0)$

126. $7(-2)(-1)(5)(0)(-9)$

Relatively Prime Numbers Two or more numbers that have no common prime factors are called *relatively prime*. In Exercises 127–138, **decide** whether the numbers are relatively prime.

127. 15, 28

128. 18, 49

129. 64, 232

130. 27, 36

131. 495, 784

132. 621, 1496

133. 63, 1375

134. 403, 899

135. 51, 85, 119

136. 24, 65, 161

137. 21, 20, 143

138. 84, 289, 325

139. Temperature Change The temperature measured by a weather balloon is decreasing approximately 3° for each 1000-foot increase in altitude. The balloon rises 8000 feet. What is the total temperature change?

140. Savings Plan A homeowner saves $250 per month for home improvements. After 2 years, how much money has the homeowner saved?

141. Stock Price The price per share of a technology stock drops $0.29 on each of four consecutive days, and then increases $0.32 on the fifth day. What is the total price change per share during the five days?

142. Loss Leaders To attract customers, a grocery store runs a sale on bananas. The bananas are *loss leaders*, which means the store loses money on the bananas but hopes to make it up on other items. The store sells 800 pounds of bananas at a loss of $0.26 per pound, and 1200 pounds of potatoes at a profit of $0.22 per pound. What is the store's profit?

143. Geometry The rectangular prism shown below has a volume of 4095 cubic feet. What is the height of the prism?

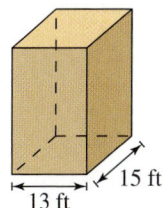

144. Geometry The rectangular prism shown below has a volume of 3456 cubic feet. What is the height of the prism?

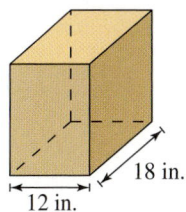

Explaining Concepts

145. Reasoning What is the only even prime number? Explain why there are no other even prime numbers.

146. Logic The number 1997 is not divisible by a prime number that is less than 45. Explain why this implies that 1997 is a prime number.

147. Number Sense Explain why the product of an even integer and any other integer is even. What can you conclude about the product of two odd integers?

148. Reasoning Explain how to check the result of a division problem.

149. Think About It A nonzero product has 25 factors, 17 of which are negative. What is the sign of the product? Explain.

150. Think About It A nonzero product has 25 factors, 16 of which are negative. What is the sign of the product? Explain.

151. Investigation Twin primes are prime numbers that differ by 2. For instance, 3 and 5 are twin primes. What are the other twin primes less than 100?

152. Investigation The **proper factors** of a number are all its factors less than the number itself. A number is **perfect** if the sum of its proper factors is equal to the number. A number is **abundant** if the sum of its proper factors is greater than the number. Which numbers less than 25 are perfect? Which are abundant? Try to find the first perfect number greater than 25.

153. Investigation The numbers 14, 15, and 16 are an example of three consecutive composite numbers. Is it possible to find ten consecutive composite numbers? If so, list an example. If not, explain why.

154. Think About It An integer n is divided by 2 and the quotient is an even integer. What does this tell you about n? Give an example.

155. The Sieve of Eratosthenes Write the integers from 1 through 100 in 10 lines of 10 numbers each.

(a) Cross out the number 1. Cross out all multiples of 2 other than 2 itself. Do the same for 3, 5, and 7.

(b) Of what type are the remaining numbers? Explain why this is the only type of number left.

Mid-Chapter Quiz: Sections 1.1–1.3

Solutions in English & Spanish and tutorial videos at AlgebraWithinReach.com

Take this quiz as you would take a quiz in class. After you are done, check your work against the answers in the back of the book.

In Exercises 1–4, plot each real number as a point on the real number line and place the correct inequality symbol (< or >) between the real numbers.

1. $\frac{3}{16}$ ▢ $\frac{3}{8}$
2. -2.5 ▢ -4
3. -7 ▢ 3
4. 2π ▢ 6

In Exercises 5 and 6, evaluate the expression.

5. $-|-0.75|$
6. $\left|-\frac{17}{19}\right|$

In Exercises 7 and 8, place the correct symbol (<, >, or =) between the real numbers.

7. $\left|\frac{7}{2}\right|$ ▢ $|-3.5|$
8. $\left|\frac{3}{4}\right|$ ▢ $-|0.75|$

9. Subtract -13 from -22.

10. Find the absolute value of the sum of -54 and 26.

In Exercises 11–22, evaluate the expression.

11. $52 + 47$
12. $-18 + (-35)$
13. $-15 - 12$
14. $-35 - (-10)$
15. $25 + (-75)$
16. $72 - 134$
17. $12 + (-6) - 8 + 10$
18. $-9 - 17 + 36 + (-15)$
19. $-6(10)$
20. $-7(-13)$
21. $\dfrac{-45}{-3}$
22. $\dfrac{-24}{6}$

In Exercises 23–26, decide whether the number is prime or composite. If it is composite, write its prime factorization.

23. 23
24. 91
25. 111
26. 144

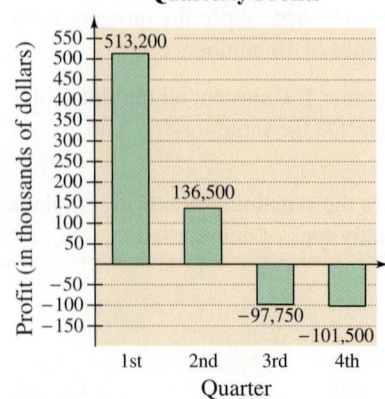
Quarterly Profits

Applications

27. An electronics manufacturer's quarterly profits are shown in the bar graph at the left. What is the manufacturer's total profit for the year?

28. A cord of wood is a pile 8 feet long, 4 feet wide, and 4 feet high. The volume of a rectangular solid is its length times its width times its height. Find the number of cubic feet in a cord of wood.

29. You cut a 90-foot rope into 6 pieces of equal length. What is the length of each piece?

30. At the beginning of a month, your account balance was $738. During the month, you withdrew $550, deposited $189, and paid a fee of $10. What was your balance at the end of the month?

Study Skills in Action

Viewing Math as a Foreign Language

Using Study Strategies

In each *Study Skills in Action* feature, you will learn a new study strategy that will help you progress through the course. Each strategy will help you:

- set up good study habits;
- organize information into smaller pieces;
- create review tools;
- memorize important definitions and rules; and
- learn the math at hand.

Learning math requires more than just completing homework problems. For instance, learning the material in a chapter may require using approaches similar to those used for learning a foreign language in that you must:

- understand and memorize vocabulary words;
- understand an memorize mathematical rules (as you would memorize grammatical rules); and
- apply rules to mathematical expressions or equations (like creating sentences using correct grammar rules).

You should understand the vocabulary words and rules in a chapter as well as memorize and say them out loud. Strive to speak the mathematical language with fluency, just as a student learning a foreign language must.

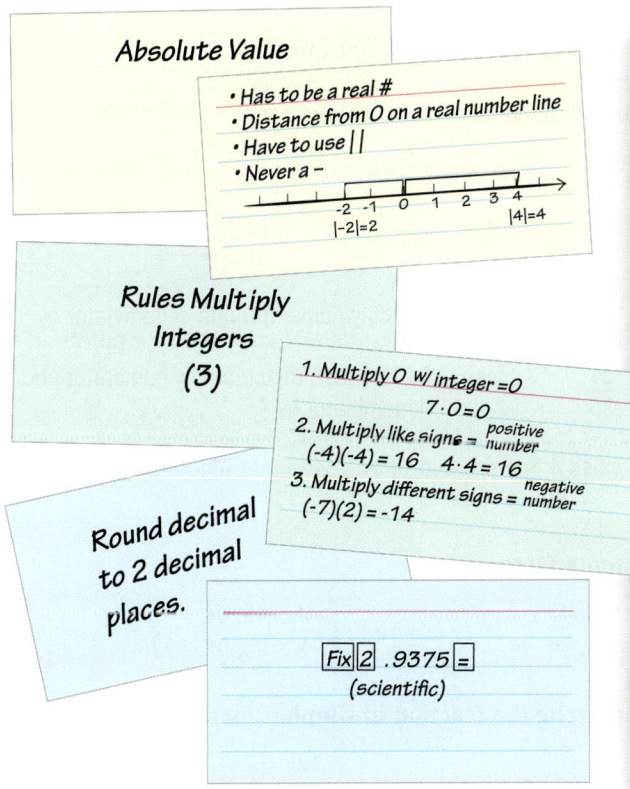

Smart Study Strategy

Make Note Cards

Invest in three different colors of 4 × 6 note cards. Use one color for each of the following: vocabulary words, rules, and calculator keystrokes.

- Write vocabulary words on note cards, one word per card. Write the definition and an example on the other side. If possible, put definitions in your own words.

- Write rules on note cards, one per card. Include an example and an explanation on the other side.

- Write each kind of calculation on a separate note card. Include the keystrokes required to perform the calculation on the other side.

Use the note cards as references while doing your homework. Quiz yourself once a day.

1.4 Operations with Rational Numbers

- Rewrite fractions as equivalent fractions.
- Add and subtract fractions.
- Multiply and divide fractions.
- Add, subtract, multiply, and divide decimals.

Rewriting Fractions

To write a fraction in simplest form, divide both the numerator and denominator by their greatest common factor (GCF).

EXAMPLE 1 Writing Fractions in Simplest Form

a. $\dfrac{18}{24} = \dfrac{2 \cdot 3 \cdot 3}{2 \cdot 2 \cdot 2 \cdot 3} = \dfrac{3}{4}$ b. $\dfrac{35}{21} = \dfrac{5 \cdot 7}{3 \cdot 7} = \dfrac{5}{3}$ c. $\dfrac{24}{72} = \dfrac{2 \cdot 2 \cdot 2 \cdot 3}{2 \cdot 2 \cdot 2 \cdot 3 \cdot 3} = \dfrac{1}{3}$

Divide out GCF of 6.   Divide out GCF of 7.   Divide out GCF of 24.

You can obtain an **equivalent fraction** by multiplying or dividing the numerator and denominator by the same nonzero number. For instance, 9/12 and 3/4 are equivalent.

EXAMPLE 2 Writing Equivalent Fractions

Write an equivalent fraction with the indicated denominator.

a. $\dfrac{2}{3} = \dfrac{\square}{15}$ b. $\dfrac{9}{15} = \dfrac{\square}{35}$

SOLUTION

a. $\dfrac{2}{3} = \dfrac{2 \cdot 5}{3 \cdot 5} = \dfrac{10}{15}$ Multiply numerator and denominator by 5.

b. $\dfrac{9}{15} = \dfrac{3 \cdot 3}{3 \cdot 5} = \dfrac{3 \cdot 7}{5 \cdot 7} = \dfrac{21}{35}$ Reduce first, then multiply numerator and denominator by 7.

$\dfrac{9}{12} = \dfrac{3 \cdot 3}{3 \cdot 4}$

$\dfrac{9}{12}$ and $\dfrac{3}{4}$ are equivalent.

Exercises Within Reach ®

Solutions in English & Spanish and tutorial videos at AlgebraWithinReach.com

Finding the GCF In Exercises 1–8, find the greatest common factor.

1. 5, 10
2. 3, 9
3. 20, 45
4. 48, 64
5. 45, 90
6. 27, 54
7. 18, 84, 90
8. 84, 98, 192

Writing a Fraction in Simplest Form In Exercises 9–16, write the fraction in simplest form.

9. $\dfrac{2}{4}$
10. $\dfrac{4}{16}$
11. $\dfrac{12}{15}$
12. $\dfrac{14}{35}$
13. $\dfrac{60}{192}$
14. $\dfrac{90}{225}$
15. $\dfrac{28}{350}$
16. $\dfrac{88}{154}$

Writing an Equivalent Fraction In Exercises 17–20, write an equivalent fraction with the indicated denominator.

17. $\dfrac{3}{8} = \dfrac{\square}{16}$
18. $\dfrac{4}{5} = \dfrac{\square}{15}$
19. $\dfrac{6}{15} = \dfrac{\square}{25}$
20. $\dfrac{21}{49} = \dfrac{\square}{28}$

Section 1.4 Operations with Rational Numbers   31

Application

EXAMPLE 3 **Comparing Fraction and Decimal Systems**

Why do humans continue to use fractions to represent numbers? Why don't they simply write all numbers in decimal form?

SOLUTION

The simple answer is that in the decimal (base 10) system, most fractions cannot be written in exact form using a finite number of digits. For instance, the decimal form of the fraction $\frac{1}{3}$ has infinitely many digits.

$$\frac{1}{3} = 0.333333\ldots$$

Rather than using base 10, if you use a base that has more factors, then it is possible to write more fractions in exact form using a finite number of digits. For instance, the ancient Babylonians used base 60 for their number system. This is still present today in the measurement of time—there are 60 minutes in an hour.

The ancient Babylonians used a base 60 number system.

Exercises Within Reach®

Solutions in English & Spanish and tutorial videos at AlgebraWithinReach.com

Writing the Decimal Form of a Fraction In Exercises 21–24, each figure is divided into regions of equal area. Write the fraction in simplest form that represents the shaded portion of the figure. Then use a calculator to find the decimal form of the fraction. Can the fraction be written in decimal form using a finite number of digits?

21.

22.

23.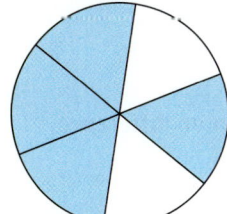

24.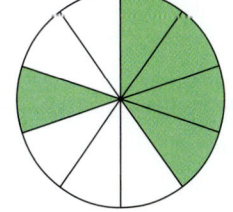

Adding and Subtracting Fractions

Let *a*, *b*, and *c* be integers with $c \neq 0$. To add or subtract the fractions a/c and b/c, use the following rules.

1. **With like denominators:**

$$\frac{a}{c} + \frac{b}{c} = \frac{a+b}{c} \quad \text{or} \quad \frac{a}{c} - \frac{b}{c} = \frac{a-b}{c}$$

2. **With unlike denominators:** Rewrite the fractions so that they have like denominators. Then use the rule for adding and subtracting fractions with like denominators.

Study Tip

To find a like denominator for two or more fractions, find the least common multiple (or LCM) of their denominators. For instance, the LCM of the denominators of 3/8 and 5/12 is 24.

EXAMPLE 4 Adding and Subtracting with Like Denominators

a. $\dfrac{3}{12} + \dfrac{4}{12} = \dfrac{3+4}{12} = \dfrac{7}{12}$

b. $\dfrac{7}{9} - \dfrac{2}{9} = \dfrac{7-2}{9} = \dfrac{5}{9}$

EXAMPLE 5 Adding and Subtracting with Unlike Denominators

a. $\dfrac{4}{5} + \dfrac{11}{15} = \dfrac{4(3)}{5(3)} + \dfrac{11}{15}$ LCM of 5 and 15 is 15.

   $= \dfrac{12}{15} + \dfrac{11}{15}$ Rewrite with like denominators.

   $= \dfrac{23}{15}$ Add numerators.

b. $1\dfrac{7}{9} - \dfrac{11}{12} = \dfrac{16}{9} - \dfrac{11}{12}$ Rewrite $1\frac{7}{9}$ as $\frac{16}{9}$.

   $= \dfrac{16(4)}{9(4)} - \dfrac{11(3)}{12(3)}$ LCM of 9 and 12 is 36.

   $= \dfrac{64}{36} - \dfrac{33}{36}$ Rewrite with like denominators.

   $= \dfrac{31}{36}$ Subtract numerators.

Exercises Within Reach®

Solutions in English & Spanish and tutorial videos at AlgebraWithinReach.com

Adding and Subtracting with Like Denominators In Exercises 25–32, find the sum or difference. Write the result in simplest form.

25. $\dfrac{7}{15} + \dfrac{1}{15}$ 26. $\dfrac{13}{35} + \dfrac{5}{35}$ 27. $\dfrac{3}{2} + \dfrac{5}{2}$ 28. $\dfrac{5}{6} + \dfrac{13}{6}$

29. $\dfrac{9}{16} - \dfrac{3}{16}$ 30. $\dfrac{15}{32} - \dfrac{7}{32}$ 31. $\dfrac{3}{4} - \dfrac{5}{4}$ 32. $\dfrac{7}{8} - \dfrac{9}{8}$

Adding and Subtracting with Unlike Denominators In Exercises 33–48, find the sum or difference. Write the result in simplest form.

33. $\dfrac{1}{2} + \dfrac{1}{3}$ 34. $\dfrac{3}{5} + \dfrac{1}{2}$ 35. $\dfrac{3}{16} + \dfrac{3}{8}$ 36. $\dfrac{2}{3} + \dfrac{4}{9}$

37. $\dfrac{5}{6} - \dfrac{1}{3}$ 38. $\dfrac{2}{3} - \dfrac{1}{6}$ 39. $\dfrac{3}{4} - \dfrac{2}{5}$ 40. $\dfrac{5}{6} - \dfrac{2}{7}$

41. $3\dfrac{1}{2} + 5\dfrac{2}{3}$ 42. $5\dfrac{3}{4} + 8\dfrac{1}{10}$ 43. $1\dfrac{3}{16} - 2\dfrac{1}{4}$ 44. $5\dfrac{7}{8} - 2\dfrac{1}{2}$

45. $15 - 20\dfrac{1}{4}$ 46. $6 - 3\dfrac{5}{8}$ 47. $-5\dfrac{1}{3} - 4\dfrac{5}{12}$ 48. $-2\dfrac{3}{4} - 3\dfrac{1}{5}$

Section 1.4   Operations with Rational Numbers   33

Application   **EXAMPLE 6**   Finding the Yardage for a Clothing Design

A designer uses $3\frac{1}{6}$ yards of material to make a skirt and $2\frac{3}{4}$ yards to make a shirt. Find the total amount of material required.

SOLUTION

$3\frac{1}{6} + 2\frac{3}{4} = \frac{19}{6} + \frac{11}{4}$ Write mixed numbers as fractions.

$= \frac{19(2)}{6(2)} + \frac{11(3)}{4(3)}$ LCM of 6 and 4 is 12.

$= \frac{38}{12} + \frac{33}{12}$ Rewrite with like denominators.

$= \frac{71}{12}$ Add numerators.

$= 5\frac{11}{12}$ Write fraction as mixed number.

The designer needs $5\frac{11}{12}$ (about 6) yards of material.

Exercises Within Reach®   Solutions in English & Spanish and tutorial videos at AlgebraWithinReach.com

49. Construction Project A sign near a construction site indicates what fraction of the work has been completed. At the beginnings of May and June, the fractions of work completed were $\frac{5}{16}$ and $\frac{2}{3}$, respectively. What fraction of the work was completed during the month of May?

50. Fund Drive During a fund drive, a charity has a display showing how close it is to reaching its goal. At the end of the first week, the display shows $\frac{1}{8}$ of the goal. At the end of the second week, the display shows $\frac{3}{5}$ of the goal. What fraction of the goal was gained during the second week?

51. Geometry Determine the unknown fractional part of the circle graph.

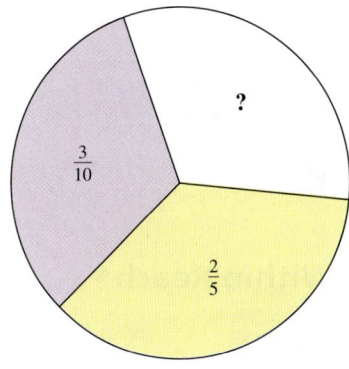

52. Geometry Determine the unknown fractional part of the circle graph.

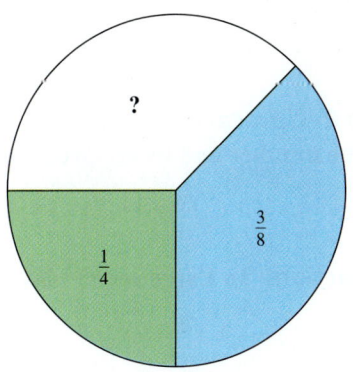

Multiplying and Dividing Fractions

Let a, b, c, and d be integers with $b \neq 0$ and $d \neq 0$. Then the product of $\dfrac{a}{b}$ and $\dfrac{c}{d}$ is

$$\frac{a}{b} \cdot \frac{c}{d} = \frac{a \cdot c}{b \cdot d}.$$ Multiply numerators and denominators.

Study Tip

When you use a scientific or graphing calculator to add, subtract, multiply, or divide fractions, your answer may appear in decimal form. An answer such as 0.583333333 is not as exact as $\dfrac{7}{12}$ and may introduce roundoff error. Refer to the user's manual for your calculator for instructions on displaying answers in fraction form.

EXAMPLE 7 Multiplying Fractions

$$\frac{5}{8} \cdot \frac{3}{2} = \frac{5(3)}{8(2)}$$ Multiply numerators and denominators.

$$= \frac{15}{16}$$ Simplify.

Let a, b, c, and d be integers with $b \neq 0$, $c \neq 0$, and $d \neq 0$. Then the quotient of $\dfrac{a}{b}$ and $\dfrac{c}{d}$ is

$$\frac{a}{b} \div \frac{c}{d} = \frac{a}{b} \cdot \frac{d}{c}.$$ Invert divisor and multiply.

EXAMPLE 8 Dividing Fractions

$$\frac{5}{8} \div \frac{20}{12} = \frac{5}{8} \cdot \frac{12}{20}$$ Invert divisor and multiply.

$$= \frac{(5)(12)}{(8)(20)}$$ Multiply numerators and denominators.

$$= \frac{(5)(3)(4)}{(8)(4)(5)}$$ Divide out common factors.

$$= \frac{3}{8}$$ Write in simplest form.

Exercises Within Reach®

Solutions in English & Spanish and tutorial videos at AlgebraWithinReach.com

Multiplying Fractions In Exercises 53–64, evaluate the expression. Write the result in simplest form.

53. $\dfrac{1}{2} \cdot \dfrac{3}{4}$
54. $\dfrac{3}{5} \cdot \dfrac{1}{2}$
55. $-\dfrac{2}{3} \cdot \dfrac{5}{7}$
56. $-\dfrac{5}{6} \cdot \dfrac{1}{2}$

57. $\dfrac{2}{3}\left(-\dfrac{9}{16}\right)$
58. $\dfrac{5}{3}\left(-\dfrac{3}{5}\right)$
59. $-\dfrac{3}{4}\left(-\dfrac{4}{9}\right)$
60. $-\dfrac{15}{16}\left(-\dfrac{12}{5}\right)$

61. $\dfrac{11}{12}\left(-\dfrac{9}{44}\right)$
62. $\dfrac{5}{12}\left(-\dfrac{6}{25}\right)$
63. $9\left(\dfrac{4}{15}\right)$
64. $24\left(\dfrac{7}{18}\right)$

Finding the Reciprocal In Exercises 65–68, find the reciprocal of the number. Show that the product of the number and its reciprocal is 1.

65. 7
66. 14
67. $\dfrac{4}{7}$
68. $-\dfrac{5}{9}$

Dividing Fractions In Exercises 69–80, evaluate the expression and write the result in simplest form.

69. $\dfrac{3}{8} \div \dfrac{3}{4}$
70. $\dfrac{5}{16} \div \dfrac{25}{8}$
71. $-\dfrac{5}{12} \div \dfrac{45}{32}$
72. $-\dfrac{16}{21} \div \dfrac{12}{27}$

73. $\dfrac{8}{3} \div \dfrac{8}{3}$
74. $\dfrac{5}{7} \div \dfrac{5}{7}$
75. $\dfrac{3}{5} \div \dfrac{7}{5}$
76. $\dfrac{7}{8} \div \dfrac{3}{8}$

77. $-\dfrac{5}{6} \div \left(-\dfrac{8}{10}\right)$
78. $-\dfrac{14}{15} \div \left(-\dfrac{24}{25}\right)$
79. $-10 \div \dfrac{1}{9}$
80. $-6 \div \dfrac{1}{3}$

Section 1.4 Operations with Rational Numbers 35

Application **EXAMPLE 9** **Finding the Number of Calories Burned**

You decide to take a tennis class. You burn about 400 calories per hour playing tennis. In one week, you played tennis for $\frac{3}{4}$ hour on Tuesday, 2 hours on Wednesday, and $1\frac{1}{2}$ hours on Thursday. How many total calories did you burn playing tennis during that week? What was your average number of calories burned per day playing tennis?

SOLUTION

The total number of calories you burned playing tennis during the week was

$$400\left(\frac{3}{4}\right) + 400(2) + 400\left(1\frac{1}{2}\right) = 300 + 800 + 600$$
$$= 1700 \text{ calories.}$$

The average number of calories burned per day was

$$\frac{1700 \text{ calories}}{3 \text{ days}} = 566\frac{2}{3} \text{ calories per day.}$$

Exercises Within Reach®

Solutions in English & Spanish and tutorial videos at AlgebraWithinReach.com

81. **Basketball** You burn about 600 calories per hour playing basketball. You played for $1\frac{1}{4}$ hours on Monday, $1\frac{1}{2}$ hours on Tuesday, and $\frac{5}{6}$ hour on Wednesday. How many total calories did you burn playing basketball on the three days? What was the average number of calories you burned per day playing basketball?

82. **Racewalking** You burn about 240 calories per hour racewalking. You racewalked for $\frac{3}{4}$ hour on Thursday, $1\frac{1}{6}$ hours on Friday, and $1\frac{4}{5}$ hours on Saturday. How many total calories did you burn racewalking on the three days? What was the average number of calories you burned per day racewalking?

83. **Skiing Time** You cross-country ski on a trail that is $5\frac{3}{5}$ miles long at an average rate of $9\frac{1}{3}$ miles per hour. How long does it take you to ski the entire trail?

84. **Walking Time** Your apartment is $\frac{3}{4}$ mile from the subway. You walk at the rate of $3\frac{1}{4}$ miles per hour. How long does it take you to walk to the subway?

Operations with Decimals

To round a decimal, use the following rules.

1. Determine the number of digits of accuracy you wish to keep. The digit in the last position you keep is called the **rounding digit**, and the digit in the first position you discard is called the **decision digit**.
2. If the decision digit is 5 or greater, round up by adding 1 to the rounding digit.
3. If the decision digit is 4 or less, round down by leaving the rounding digit unchanged.

EXAMPLE 10 Adding and Multiplying Decimals

a.
```
  1 1
  0.583
  1.06
+ 2.9104
  4.5534
```

b.
```
     -3.57    Two decimal places
   × 0.032    Three decimal places
      714
     1071
   -0.11424   Five decimal places
```

Study Tip

Rational numbers can be represented as terminating or repeating decimals. Here are some examples.

Terminating Decimals

$\frac{1}{4} = 0.25 \quad \frac{3}{8} = 0.375$

Repeating Decimals

$\frac{1}{6} = 0.1666\ldots$ or $0.1\overline{6}$

$\frac{8}{33} = 0.2424\ldots$ or $0.\overline{24}$

EXAMPLE 11 Dividing Decimals

$$\frac{1.483}{0.56} \implies \begin{array}{r} 2.648 \\ 56\overline{)148.300} \\ \underline{112} \\ 363 \\ \underline{336} \\ 270 \\ \underline{224} \\ 460 \\ \underline{448} \end{array}$$

Rounded to two decimal places, the answer is 2.65. This can be written as

$$\frac{1.483}{0.56} \approx 2.65.$$

Exercises Within Reach®

Solutions in English & Spanish and tutorial videos at AlgebraWithinReach.com

Adding and Subtracting Decimals In Exercises 85–96, *evaluate* the expression.

85. $12.33 + 14.76$
86. $6.983 + 241.5$
87. $0.287 + 1.65 + 2.1932$
88. $2.013 + 0.1145 + 1.12$
89. $132.1 + (-25.45)$
90. $408.9 + (-13.12)$
91. $8.673 - 2.55$
92. $5.1146 - 1.028$
93. $4.54 - 6.668$
94. $4.25 - 7.998$
95. $1.21 + 4.06 - 3.00$
96. $3.4 + 1.062 - 5.13$

Multiplying Decimals In Exercises 97–100, *evaluate* the expression. Round your answer to two decimal places.

97. $-6.3(9.05)$
98. $3.7(-14.8)$
99. $-0.05(-85.95)$
100. $-0.09(-0.45)$

Dividing Decimals In Exercises 101–104, *evaluate* the expression. Round your answer to two decimal places.

101. $4.69 \div 0.12$
102. $7.14 \div 0.94$
103. $1.062 \div (-2.1)$
104. $2.011 \div (-3.3)$

Section 1.4 Operations with Rational Numbers 37

EXAMPLE 12 Finding a Cell Phone Charge

A cellular phone company charges $5.35 for the first 200 text messages per month and $0.10 for each additional text message.

a. Find the cost of 263 text messages.

b. Can you save money by switching to a plan that allows unlimited text messages for $10 per month?

SOLUTION

a. You sent or received 63 text messages above 200. The cost of these is

($0.10)(63) = $6.30.

So, your total charge for the month is

$5.35 + $6.30 = $11.65.

b. If you continue to send and receive this number of text messages each month, you can save money by switching to a plan that allows unlimited texts messages for $10 per month.

Exercises Within Reach ®

Solutions in English & Spanish and tutorial videos at AlgebraWithinReach.com

105. Consumer Awareness At a convenience store, you buy two gallons of milk at $3.94 per gallon and three loaves of bread at $2.47 per loaf. You give the clerk a 20-dollar bill. How much change will you receive? (Assume there is no sales tax.)

107. Fuel Efficiency The sticker on a new car gives the fuel efficiency as 22.3 miles per gallon. The average cost of fuel is $3.479 per gallon. You expect to drive the car 12,000 miles per year. Find the expected annual fuel cost for the car.

108. Consumer Awareness The prices per gallon of regular unleaded gasoline at three service stations are $3.439, $3.479, and $3.589, respectively. Find the average price per gallon.

106. Stock Purchase You buy 200 shares of stock at $23.63 per share and 300 shares at $86.25 per share.

(a) Estimate the total cost of the stock.

(b) Use a calculator to find the total cost of the stock.

Concept Summary: Operations with Rational Numbers

What
As you study mathematics, you will be asked to add, subtract, multiply, or divide two fractions or two decimals.

EXAMPLE
Evaluate each expression.
a. $\frac{1}{2} + \frac{2}{3}$
b. $\frac{1}{2} \div \frac{2}{3}$
c. $2.2 + 10.03$
d. $0.2 \cdot 6.33$

How
To perform such operations with rational numbers, you can use the rules for fractions and the rules for decimals.

EXAMPLE
a. $\frac{1}{2} + \frac{2}{3} = \frac{3}{6} + \frac{4}{6} = \frac{7}{6}$
b. $\frac{1}{2} \div \frac{2}{3} = \frac{1}{2} \cdot \frac{3}{2} = \frac{3}{4}$

c. 2.2
$\underline{+10.03}$
12.23

d. 6.33
$\underline{\times 0.2}$
1.266

Why
When you know the rules for the operations with rational numbers, you will be able to add, subtract, multiply, or divide any two rational numbers. For instance, to determine the gas mileage for your car, you will be able to divide the number of miles you drove by the number of gallons you used.

Exercises Within Reach®
Worked-out solutions to odd-numbered exercises at AlgebraWithinReach.com

Concept Summary Check

109. *Using the Least Common Denominator* Explain how to use the least common denominator to add two fractions.

110. *Dividing Fractions* Describe how to divide two fractions.

111. *Adding Decimals* Describe how to add two decimals using a vertical format.

112. *Multiplying Decimals* When you multiply two decimals, how do you determine where to place the decimal point in the product?

Extra Practice

Geometry In Exercises 113 and 114, **determine** the unknown fractional part of the circle graph.

113.
114.

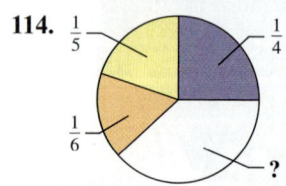

Operations with Fractions In Exercises 115–134, **evaluate** the expression and write the result in simplest form. If it is not possible, explain why.

115. $-\frac{23}{11} + \frac{12}{11}$
116. $-\frac{39}{23} + \frac{11}{23}$
117. $\frac{7}{10} + \left(-\frac{3}{10}\right)$
118. $\frac{11}{15} + \left(-\frac{2}{15}\right)$

119. $-\frac{1}{8} - \frac{1}{6}$
120. $-\frac{13}{8} - \frac{3}{4}$
121. $\frac{2}{5} + \frac{4}{5} + \frac{1}{5}$
122. $\frac{2}{9} + \frac{4}{9} + \frac{1}{9}$

123. $2\frac{3}{4} \cdot 3\frac{2}{3}$
124. $-8\frac{1}{2} \cdot 3\frac{2}{5}$
125. $-5\frac{2}{3} \cdot 4\frac{1}{2}$
126. $2\frac{4}{5} \cdot 6\frac{2}{3}$

127. $-\frac{3}{2}\left(-\frac{15}{16}\right)\left(\frac{12}{25}\right)$
128. $\frac{1}{2}\left(-\frac{4}{15}\right)\left(-\frac{5}{24}\right)$
129. $6\left(\frac{3}{4}\right)\left(\frac{2}{9}\right)$
130. $8\left(\frac{5}{12}\right)\left(\frac{3}{10}\right)$

131. $\frac{3}{5} \div 0$
132. $\frac{11}{13} \div 0$
133. $3\frac{3}{4} \div 1\frac{1}{2}$
134. $2\frac{4}{9} \div 5\frac{1}{3}$

Operations with Decimals In Exercises 135–138, **evaluate** the expression. Round your answer to two decimal places.

135. $-0.0005 - 2.01 + 0.111$
136. $-1.0012 - 3.25 + 0.2$
137. $-2.54(3.8)(6.55)$
138. $7.8(12.32)(-0.95)$

139. Consumer Awareness The prices of a 16-ounce bottle of soda at three different convenience stores are $1.09, $1.25, and $1.10, respectively. Find the average price for the bottle of soda.

140. Cooking You make 60 ounces of dough for breadsticks. Each breadstick requires $\frac{5}{4}$ ounces of dough. How many breadsticks can you make?

Explaining Concepts

141. Number Sense Is it true that the sum of two fractions of like signs is positive? If not, give an example that shows the statement is false.

142. Structure Does $\frac{2}{3} + \frac{3}{2} = \frac{(2+3)}{(3+2)} = 1$? Explain your answer.

143. Writing In your own words, describe the rule for determining the sign of the product of two fractions.

144. Precision Is it true that $\frac{2}{3} = 0.67$? Explain your answer.

145. Modeling Use the figure to determine how many one-fourths are in 3. Explain how to obtain the same result by division.

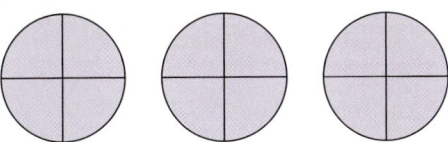

146. Modeling Use the figure to determine how many one-sixths are in $\frac{2}{3}$. Explain how to obtain the same result by division.

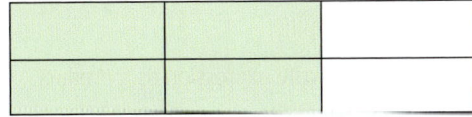

147. Investigation When using a calculator to perform operations with decimals, you should try to get in the habit of rounding your answers *only* after all the calculations are done. By rounding the answer at a preliminary stage, you can introduce unnecessary roundoff error. The dimensions of a box are $l = 5.24$, $w = 3.03$, and $h = 2.749$. Find the volume $l \cdot w \cdot h$, by multiplying the numbers and then rounding the answer to one decimal place. Now use a second method, first rounding each dimension to one decimal place and then multiplying the numbers. Compare your answers, and explain which of these techniques produces the more accurate answer.

True or False? In Exercises 148–153, decide whether the statement is true or false. Justify your answer.

148. The reciprocal of every nonzero integer is an integer.

149. The reciprocal of every nonzero rational number is a rational number.

150. The product of two nonzero rational numbers is a rational number.

151. The product of two positive rational numbers is greater than either factor.

152. If $u > v$, then $u - v > 0$.

153. If $u > 0$ and $v > 0$, then $u - v > 0$.

154. Estimation Use mental math to determine whether $\left(5\frac{3}{4}\right) \times \left(4\frac{1}{8}\right)$ is less than 20. Explain your reasoning.

155. Think About It Determine the placement of the 3, 4, 5, and 6 in the following addition problem so that you obtain the specified sum. Use each number only once.

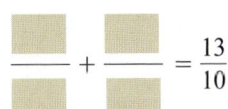

156. Think About It If the fractions represented by the points P and R are multiplied, what point on the number line best represents their product: M, S, N, P, or T? (*Source:* National Council of Teachers of Mathematics)

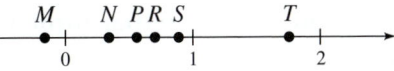

Reprinted with permission from Mathematics Teacher, © 1997, by the National Council of Teachers of Mathematics. All rights reserved.

1.5 Exponents and Properties of Real Numbers

▶ Write expressions in exponential form and evaluate exponential expressions.
▶ Evaluate expressions using the order of operations.
▶ Identify and use the properties of real numbers.

Exponents

Repeated multiplication can be described in exponential form.

Repeated Multiplication   *Exponential Form*
$\underbrace{7 \cdot 7 \cdot 7 \cdot 7}_{\text{4 factors of 7}}$ 7^4

In the exponential form, 7 is the **base** and it specifies the repeated factor. The number 4 is the **exponent** and it indicates how many times the base occurs as a factor. When you write the exponential form, you can say that you are raising 7 to the fourth **power**.

Study Tip

Keep in mind that an exponent applies only to the factor (number) directly preceding it. Parentheses are needed to include a negative sign or other factors as part of the base. Here is an example.

$(-5)^2 = (-5)(-5) = 25$
$-5^2 = -(5 \cdot 5) = -25$

EXAMPLE 1   Evaluating Exponential Expressions

a. $2^5 = 2 \cdot 2 \cdot 2 \cdot 2 \cdot 2$ Rewrite expression as a product.
      $= 32$ Simplify.

b. $\left(\dfrac{2}{3}\right)^4 = \dfrac{2}{3} \cdot \dfrac{2}{3} \cdot \dfrac{2}{3} \cdot \dfrac{2}{3}$ Rewrite expression as a product.

      $= \dfrac{2 \cdot 2 \cdot 2 \cdot 2}{3 \cdot 3 \cdot 3 \cdot 3}$ Multiply fractions.

      $= \dfrac{16}{81}$ Simplify.

EXAMPLE 2   Evaluating Exponential Expressions

a. $(-4)^3 = (-4)(-4)(-4)$ Rewrite expression as a product.
      $= -64$ Simplify.

b. $(-3)^4 = (-3)(-3)(-3)(-3)$ Rewrite expression as a product.
      $= 81$ Simplify.

c. $-3^4 = -(3 \cdot 3 \cdot 3 \cdot 3)$ Rewrite expression as a product.
      $= -81$ Simplify.

Exercises Within Reach ®

Solutions in English & Spanish and tutorial videos at AlgebraWithinReach.com

Writing an Expression in Exponential Form   In Exercises 1 and 2, **rewrite** the expression in exponential form.

1. $2 \cdot 2 \cdot 2 \cdot 2 \cdot 2 \cdot 2$
2. $4 \cdot 4 \cdot 4 \cdot 4 \cdot 4 \cdot 4$

Evaluating an Exponential Expression   In Exercises 3–14, **evaluate** the expression.

3. 3^2
4. 4^3
5. 2^6
6. 5^3
7. $\left(\dfrac{1}{4}\right)^3$
8. $\left(\dfrac{4}{5}\right)^4$
9. $(-5)^3$
10. $(-4)^2$
11. -4^2
12. $-(-6)^3$
13. $(-1.2)^3$
14. $(-1.5)^4$

Section 1.5   Exponents and Properties of Real Numbers   41

Application   **EXAMPLE 3**   **Finding the Amount a Truck Can Transport**

A truck can transport a load of motor oil that is 6 cases high, 6 cases wide, and 6 cases long. Each case contains 6 quarts of motor oil. How many quarts can the truck transport?

SOLUTION

A sketch can help you solve this problem.

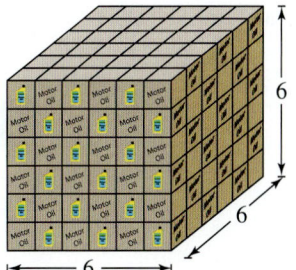

There are $6 \cdot 6 \cdot 6$ cases of motor oil, and each case contains 6 quarts. You can see that 6 occurs as a factor 4 times, which implies that the total number of quarts is

$$(6 \cdot 6 \cdot 6) \cdot 6 = 6^4 = 1296.$$

So, the truck can transport 1296 quarts of oil.

Exercises Within Reach®

Solutions in English & Spanish and tutorial videos at AlgebraWithinReach.com

15. Cereal A grocery store has a cereal display that is 8 boxes high, 8 boxes wide, and 8 boxes long. How many cereal boxes are in the display?

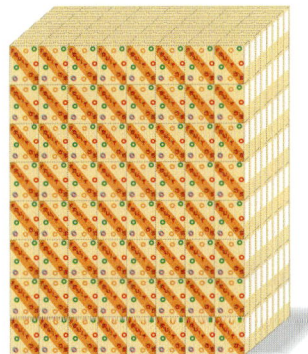

16. Propane A truck can transport a load of propane tanks that is 4 cases high, 4 cases wide, and 4 cases long. Each case contains 4 propane tanks. How many tanks can the truck transport?

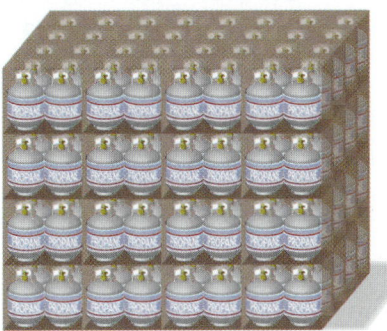

17. Exponential Communication You send an email to 5 people. Each person forwards the message to 5 other people, and so on. Complete the table to show the results of the first 6 stages. After what stage have more than 3900 people seen the message?

Stage	Emails sent, as a power	Emails sent
1	5^1	5
2	5^2	25
3		
4		
5		
6		

18. Exponential Communication You send an email to 3 people. Each person forwards the message to 3 other people, and so on. How many stages does it take for the total number of people who see the message to exceed the total number after Stage 4 in Exercise 17? Explain.

David Touchtone/Shutterstock.com

Order of Operations

Study Tip
You can remember the order of operations P, E, M, D, A, S using the pneumonic 'Please excuse my dear Aunt Sally.'

The accepted priorities for the **order of operations** are summarized below.

1. Perform operations inside symbols of grouping: **P**arentheses, brackets, or absolute value symbols, starting with the innermost symbols.
2. Evaluate all **E**xponential expressions.
3. Perform all **M**ultiplications and **D**ivisions from left to right.
4. Perform all **A**dditions and **S**ubtractions from left to right.

EXAMPLE 4 Using Order of Operations

a. $7 - [(5 \cdot 3) + 2^3] = 7 - [15 + 2^3]$ Multiply inside the parentheses.
$ = 7 - [15 + 8]$ Evaluate exponential expression.
$ = 7 - 23$ Add inside the brackets.
$ = -16$ Subtract.

b. $36 \div (3^2 \cdot 2) - 6 = 36 \div (9 \cdot 2) - 6$ Evaluate exponential expression.
$ = 36 \div 18 - 6$ Multiply inside the parentheses.
$ = 2 - 6$ Divide.
$ = -4$ Subtract.

c. $\dfrac{8}{3}\left(\dfrac{1}{6} + \dfrac{1}{4}\right) = \dfrac{8}{3}\left(\dfrac{2}{12} + \dfrac{3}{12}\right)$ Find common denominator.
$\phantom{\dfrac{8}{3}\left(\dfrac{1}{6} + \dfrac{1}{4}\right)} = \dfrac{8}{3}\left(\dfrac{5}{12}\right)$ Add inside the parentheses.
$\phantom{\dfrac{8}{3}\left(\dfrac{1}{6} + \dfrac{1}{4}\right)} = \dfrac{40}{36}$ Multiply fractions.
$\phantom{\dfrac{8}{3}\left(\dfrac{1}{6} + \dfrac{1}{4}\right)} = \dfrac{10}{9}$ Simplify.

Exercises Within Reach®

Solutions in English & Spanish and tutorial videos at AlgebraWithinReach.com

Using Order of Operations In Exercises 19–58, *evaluate* the expression. If it is not possible, state the reason. Write fractional answers in simplest form.

19. $4 - 6 + 10$
20. $8 + 9 - 12$
21. $5 - (8 - 15)$
22. $13 - (12 - 3)$
23. $15 + 3 \cdot 4$
24. $9 - 5 \cdot 2$
25. $25 - 32 \div 4$
26. $16 + 24 \div 8$
27. $(45 \div 10) \cdot 2$
28. $(38 \div 5) \cdot 4$
29. $(16 - 5) \div (3 - 5)$
30. $(19 - 4) \div (7 - 2)$
31. $(10 - 16) \cdot (20 - 26)$
32. $(14 - 17) \cdot (13 - 19)$
33. $17 - |2 - (6 + 5)|$
34. $125 - |10 - (25 \cdot 3)|$
35. $[360 - (8 + 12)] \div 5$
36. $[127 - (13 + 4)] \div 10$
37. $5 + (2^2 \cdot 3)$
38. $181 - (13 \cdot 3^2)$
39. $(-6)^2 - (48 \div 4^2)$
40. $(-3)^3 + (12 \div 2^2)$
41. $\left(3 \cdot \dfrac{5}{9}\right) + 1 - \dfrac{1}{3}$
42. $\dfrac{2}{3}\left(\dfrac{3}{4}\right) + 2 - \dfrac{3}{2}$
43. $18\left(\dfrac{1}{2} + \dfrac{2}{3}\right)$
44. $4\left(-\dfrac{2}{3} + \dfrac{4}{3}\right)$
45. $\dfrac{7}{25}\left(\dfrac{7}{16} - \dfrac{1}{8}\right)$
46. $\dfrac{3}{2}\left(\dfrac{2}{3} + \dfrac{1}{6}\right)$
47. $\dfrac{7}{3}\left(\dfrac{2}{3}\right) \div \dfrac{28}{15}$
48. $\dfrac{3}{8}\left(\dfrac{1}{5}\right) \div \dfrac{25}{32}$
49. $\dfrac{3 + [15 \div (-3)]}{16}$
50. $\dfrac{5 + [(-12) \div 4]}{24}$
51. $\dfrac{1 - 3^2}{-2}$
52. $\dfrac{2^2 + 4^2}{5}$
53. $\dfrac{7^2 - 4^2}{0}$
54. $\dfrac{0}{3^2 - 1^2}$
55. $\dfrac{0}{6^2 + 1}$
56. $\dfrac{3^3 + 1}{0}$
57. $\dfrac{5^2 + 12^2}{13}$
58. $\dfrac{8^2 - 2^3}{4}$

Section 1.5 Exponents and Properties of Real Numbers

Properties of Real Numbers

You are now ready for the symbolic versions of the properties that are true about operations with real numbers. These properties are referred to as **properties of real numbers**. The table shows a verbal description and an illustrative example for each property. Keep in mind that the letters a, b, and c represent real numbers, even though only rational numbers have been used to this point.

Properties of Real Numbers

Let a, b, and c be real numbers.

Property	*Example*

1. *Commutative Property of Addition:*

Two real numbers can be added in either order.

$a + b = b + a$ $\quad\quad\quad 3 + 5 = 5 + 3$

2. *Commutative Property of Multiplication:*

Two real numbers can be multiplied in either order.

$ab = ba$ $\quad\quad\quad 4 \cdot (-7) = -7 \cdot 4$

3. *Associative Property of Addition:*

When three real numbers are added, it makes no difference which two are added first.

$(a + b) + c = a + (b + c)$ $\quad\quad\quad (2 + 6) + 5 = 2 + (6 + 5)$

4. *Associative Property of Multiplication:*

When three real numbers are multiplied, it makes no difference which two are multiplied first.

$(ab)c = a(bc)$ $\quad\quad\quad (3 \cdot 5) \cdot 2 = 3 \cdot (5 \cdot 2)$

5. *Distributive Property:*

Multiplication distributes over addition.

$a(b + c) = ab + ac$ $\quad\quad\quad 3(8 + 5) = 3 \cdot 8 + 3 \cdot 5$
$(a + b)c = ac + bc$ $\quad\quad\quad (3 + 8)5 = 3 \cdot 5 + 8 \cdot 5$

6. *Additive Identity Property:*

The sum of zero and a real number equals the number itself.

$a + 0 = 0 + a = a$ $\quad\quad\quad 3 + 0 = 0 + 3 = 3$

7. *Multiplicative Identity Property:*

The product of 1 and a real number equals the number itself.

$a \cdot 1 = 1 \cdot a = a$ $\quad\quad\quad 4 \cdot 1 = 1 \cdot 4 = 4$

8. *Additive Inverse Property:*

The sum of a real number and its opposite is zero.

$a + (-a) = 0$ $\quad\quad\quad 3 + (-3) = 0$

9. *Multiplicative Inverse Property:*

The product of a nonzero real number and its reciprocal is 1.

$a \cdot \dfrac{1}{a} = 1, a \neq 0$ $\quad\quad\quad 8 \cdot \dfrac{1}{8} = 1$

Study Tip

One of the distinctive things about algebra is that its rules make sense. You do not have to accept them on "blind faith." Instead, you can learn the reasons that the rules work.

EXAMPLE 5 Identifying Properties of Real Numbers

Identify the property of real numbers illustrated by each statement.

a. $3(6 + 2) = 3 \cdot 6 + 3 \cdot 2$
b. $5 \cdot \frac{1}{5} = 1$
c. $7 + (5 + 4) = (7 + 5) + 4$
d. $(12 + 3) + 0 = 12 + 3$

SOLUTION

a. This statement illustrates the Distributive Property.
b. This statement illustrates the Multiplicative Inverse Property.
c. This statement illustrates the Associative Property of Addition.
d. This statement illustrates the Additive Identity Property.

EXAMPLE 6 Using Properties of Real Numbers

Complete each statement using the specified property of real numbers.

a. Commutative Property of Addition: $5 + 9 =$ ▢

b. Associative Property of Multiplication: $6(5 \cdot 13) =$ ▢

c. Distributive Property: $4 \cdot 3 + 4 \cdot 7 =$ ▢

SOLUTION

a. By the Commutative Property of Addition, you can write
$$5 + 9 = 9 + 5.$$

b. By the Associative Property of Multiplication, you can write
$$6(5 \cdot 13) = (6 \cdot 5)13.$$

c. By the Distributive Property, you can write
$$4 \cdot 3 + 4 \cdot 7 = 4(3 + 7).$$

Exercises Within Reach®

Solutions in English & Spanish and tutorial videos at AlgebraWithinReach.com

Identifying a Property of Real Numbers In Exercises 59–70, identify the property of real numbers illustrated by the statement.

59. $(10 + 3) + 2 = 10 + (3 + 2)$
60. $(32 + 8) + 5 = 32 + (8 + 5)$
61. $6(-3) = -3(6)$
62. $16 + 10 = 10 + 16$
63. $5 + 10 = 10 + 5$
64. $-2(8) = 8(-2)$
65. $6(3 + 13) = 6 \cdot 3 + 6 \cdot 13$
66. $(14 + 2)3 = 14 \cdot 3 + 2 \cdot 3$
67. $7(\frac{1}{7}) = 1$
68. $1 \cdot 4 = 4$
69. $0 + 15 = 15$
70. $25 + (-25) = 0$

Using a Property of Real Numbers In Exercises 71–74, complete the statement using the specified property of real numbers.

71. Commutative Property of Multiplication: $10(-3) =$ ▢

72. Distributive Property: $6(19 + 2) =$ ▢

73. Associative Property of Addition: $18 + (12 + 9) =$ ▢

74. Associative Property of Multiplication: $12(3 \cdot 4) =$ ▢

Section 1.5   Exponents and Properties of Real Numbers   45

Application

EXAMPLE 7   Finding the Area of a Billboard

You measure the width of a billboard and find that it is 60 feet. You are told that its height is 22 feet less than its width.

a. Write an expression for the area of the billboard.
b. Use the Distributive Property to rewrite the expression.
c. Find the area of the billboard.

SOLUTION

a. Begin by drawing and labeling a diagram. To find an expression for the area of the billboard, multiply the width by the height.

$$\text{Area} = \text{Width} \times \text{Height}$$
$$= 60(60 - 22)$$

b. To rewrite the expression $60(60 - 22)$ using the Distributive Property, distribute 60 over the subtraction.

$$60(60 - 22) = 60(60) - 60(22)$$

c. To find the area of the billboard, evaluate the expression in part (b) as follows.

$$60(60) - 60(22) = 3600 - 1320 \qquad \text{Multiply.}$$
$$= 2280 \qquad \text{Subtract.}$$

So, the area of the billboard is 2280 square feet.

Exercises Within Reach®

Solutions in English & Spanish and tutorial videos at AlgebraWithinReach.com

75. *Movie Screen* The width of a movie screen is 30 feet and its height is 8 feet less than its width.

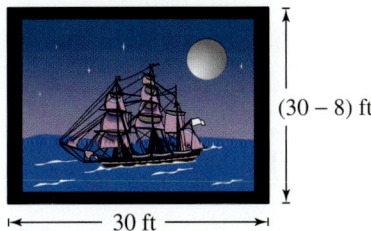

(a) Write an expression for the area of the movie screen.

(b) Use the Distributive Property to rewrite the expression.

(c) Find the area of the movie screen.

76. *Picture Frame* A picture frame is 36 inches wide and its height is 9 inches less than its width.

(a) Write an expression for the area of the picture frame.

(b) Use the Distributive Property to rewrite the expression.

(c) Find the area of the picture frame.

77. *Geometry* Consider the rectangle shown in the figure.

(a) Find the area of the rectangle by adding the areas of Regions I and II.

(b) Find the area of the rectangle by multiplying its length by its width.

(c) Explain how the results of parts (a) and (b) relate to the Distributive Property.

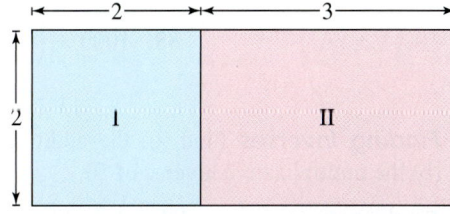

78. *Sales Tax* You purchase a sweater for $35.95. There is a 6% sales tax, which means that the total amount you must pay is $35.95 + 0.06(35.95)$.

(a) Use the Distributive Property to rewrite the expression.

(b) How much must you pay for the sweater including sales tax?

Concept Summary: Using the Order of Operations

What
To evaluate expressions consisting of grouping symbols, **exponents**, or more than one operation correctly, you need to use the established **order of operations**.

EXAMPLE
Evaluate
$(7 + 3)2^4 + 5$.

How
You can apply the order of operations as follows.
1. **P**arentheses
2. **E**xponents
3. **M**ultiplication and **D**ivision
4. **A**ddition and **S**ubtraction

EXAMPLE
$(7 + 3)2^4 + 5 = 10 \cdot 2^4 + 5$
$= 10 \cdot 16 + 5$
$= 160 + 5$
$= 165$

Why
When you use the order of operations, along with the **properties of real numbers**, you will be able to evaluate expressions correctly and efficiently.

Exercises Within Reach®

Worked-out solutions to odd-numbered exercises at AlgebraWithinReach.com

Concept Summary Check

79. **Understanding Exponents** Which part of the exponential expression 2^4 is the base? Which part is the exponent?

80. **Order of Operations** Explain why addition is the first operation performed in the solution steps above.

81. **Order of Operations** Explain how each step in the solution above relates to PEMDAS.

82. **Equivalent Expressions** Explain why the expressions $10 \cdot 16 + 5$ and $(10 \cdot 16) + 5$ are equivalent.

Extra Practice

Using Order of Operations In Exercises 83–86, **evaluate** the expression. If it is not possible, state the reason. Write fractional answers in simplest form.

83. $\dfrac{3 \cdot 6 - 4 \cdot 6}{5 + 1}$

84. $\dfrac{5 \cdot 3 + 5 \cdot 6}{7 - 2}$

85. $7 - \dfrac{4 + 6}{2^2 + 1} + 5$

86. $11 - \dfrac{3^3 - 30}{8 + 1} + 1$

Using a Calculator In Exercises 87–90, use a calculator to **evaluate** the expression. Round your answer to two decimal places.

87. $300\left(1 + \dfrac{0.1}{12}\right)^{24}$

88. $1000 \div \left(1 + \dfrac{0.09}{4}\right)^8$

89. $\dfrac{1.32 + 4(3.68)}{1.5}$

90. $\dfrac{4.19 - 7(2.27)}{14.8}$

91. **Finding Inverses** Find (a) the additive inverse and (b) the multiplicative inverse of 50.

92. **Finding Inverses** Find (a) the additive inverse and (b) the multiplicative inverse of -8.

93. **Geometry** Write and evaluate an expression for the perimeter of the triangle.

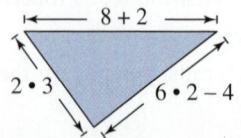

94. **Geometry** Find the area of the region.

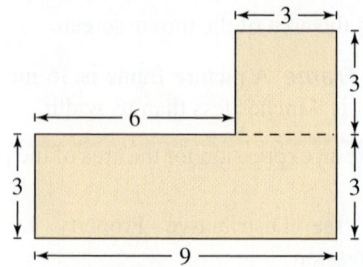

Explaining Concepts

95. *Writing* Are -6^2 and $(-6)^2$ equal? Explain.

96. *Writing* Are $2 \cdot 5^2$ and 10^2 equal? Explain.

Writing In Exercises 97–106, **explain** why the statement is true.

97. $4 \cdot 6^2 \neq 24^2$

98. $-3^2 \neq (-3)(-3)$

99. $4 - (6 - 2) \neq 4 - 6 - 2$

100. $\dfrac{8-6}{2} \neq 4 - 6$

101. $100 \div 2 \times 50 \neq 1$

102. $\dfrac{16}{2} \cdot 2 \neq 4$

103. $5(7 + 3) \neq 5(7) + 3$

104. $-7(5 - 2) \neq -7(5) - 7(2)$

105. $\dfrac{8}{0} \neq 0$

106. $5\left(\dfrac{1}{5}\right) \neq 0$

107. *Error Analysis* Describe and correct the error.
$$-9 + \dfrac{9 + 20}{3(5)} - (-3) = -9 + \dfrac{9}{3} + \dfrac{20}{5} - (-3)$$
$$= -9 + 3 + 4 - (-3)$$
$$= 1$$

108. *Error Analysis* Describe and correct the error.
$$7 - 3(8 + 1) - 15 = 4(8 + 1) - 15$$
$$= 4(9) - 15$$
$$= 36 - 15$$
$$= 21$$

109. *Matching* Match each expression in the first column with its value in the second column.

Expression	Value
$(6 + 2) \cdot (5 + 3)$	19
$(6 + 2) \cdot 5 + 3$	22
$6 + 2 \cdot 5 + 3$	64
$6 + 2 \cdot (5 + 3)$	43

110. *Determining Order of Operations* Using the established order of operations, which of the following expressions has a value of 72? For those that do not, decide whether you can insert parentheses into the expression so that its value is 72.

(a) $4 + 2^3 - 7$ (b) $4 + 8 \cdot 6$

(c) $93 - 25 - 4$ (d) $70 + 10 \div 5$

(e) $60 + 20 \div 2 + 32$ (f) $35 \cdot 2 + 2$

Geometry In Exercises 111 and 112, **find** the area of the shaded rectangle in two ways. Explain how the results are related to the Distributive Property.

111.
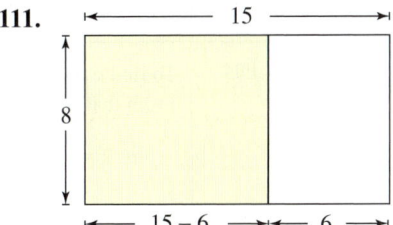

112.
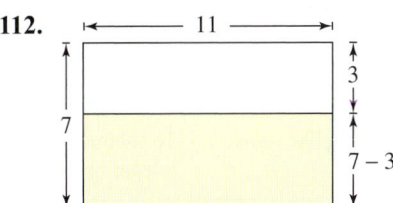

Think About It In Exercises 113 and 114, **determine** whether the order in which the two activities are performed is "commutative." That is, do you obtain the same result regardless of which activity is performed first?

113. (a) "Put on your socks."

(b) "Put on your shoes."

114. (a) "Weed the flower beds."

(b) "Mow the lawn."

1 Chapter Summary

	What did you learn?	Explanation and Examples	Review Exercises
1.1	Classify numbers and plot them on a real number line *(p. 2)*.	The real numbers include rational numbers, irrational numbers, integers, whole numbers, and natural numbers. Each point on the real number line corresponds to exactly one real number. The Real Number Line	1–10
	Use the real number line and inequality symbols to order real numbers *(p. 4)*.	a is less than b or b is greater than a	11–16
	Find the absolute value of a number *(p. 6)*.	The absolute value of a number is its distance from 0 on the real number line. Distance between -8 and 0 is 8. $|-8| = 8$	17–38
1.2	Add integers using a number line *(p. 10)*.	To find $a + b$, start at 0. Move a units along the number line. From that position, move b units. The final position is the sum. $-5 + 2 = -3$	39–42
	Add integers with like signs and with unlike signs *(p. 12)*.	To add two integers with like signs, add their absolute values and attach the common sign to the result. To add two integers with unlike signs, subtract the lesser absolute value from the greater absolute value and attach the sign of the integer with the greater absolute value.	43–54
	Subtract integers with like signs and with unlike signs *(p. 14)*.	To subtract one integer from another, add the opposite of the integer being subtracted to the other integer.	55–66
1.3	Multiply integers with like signs and with unlike signs *(p. 18)*.	1. The product of an integer and zero is 0. 2. The product of two integers with like signs is positive. 3. The product of two integers with unlike signs is negative.	67–80
	Divide integers with like signs and with unlike signs *(p. 20)*.	1. Zero divided by a nonzero integer is 0, whereas a nonzero integer divided by zero is undefined. 2. The quotient of two nonzero integers with like signs is positive. 3. The quotient of two nonzero integers with unlike signs is negative.	81–94
	Find factors and prime factors of an integer *(p. 22)*.	If a and b are positive integers, then a is a factor of b if and only if a divides evenly into b. The prime factorization of b is its representation as the product of all its prime factors.	95–106

Chapter Summary

	What did you learn?	Explanation and Examples	Review Exercises
1.3	Represent the definitions and rules of arithmetic symbolically (p. 24).	Arithmetic → Algebra Verbal rules and definitions → Symbolic rules and definitions Specific examples of rules and definitions →	107–110
	Rewrite fractions as equivalent fractions (p. 30).	To write a fraction in simplest form, divide both the numerator and denominator by the greatest common factor. You can obtain an equivalent fraction by multiplying the numerator and denominator by the same nonzero number.	111–124
	Add and subtract fractions (p. 32).	1. **With like denominators:** $$\frac{a}{c} + \frac{b}{c} = \frac{a+b}{c} \quad \text{or} \quad \frac{a}{c} - \frac{b}{c} = \frac{a-b}{c}$$ 2. **With unlike denominators:** Rewrite the fractions so that they have like denominators. Then use the rule for adding and subtracting fractions with like denominators.	125–140
1.4	Multiply and divide fractions (p. 34).	$\frac{a}{b} \cdot \frac{c}{d} = \frac{a \cdot c}{b \cdot d}$, where a, b, c, and d are integers with $b \neq 0$ and $d \neq 0$. $\frac{a}{b} \div \frac{c}{d} = \frac{a}{b} \cdot \frac{d}{c}$, where a, b, c, and d are integers with $b \neq 0$, $c \neq 0$, and $d \neq 0$.	141–156
	Add, subtract, multiply, and divide decimals (p. 36).	**Add:** 0.43 + 1.1 = 1.53 **Subtract:** 8.40 − 1.38 = 7.02 **Multiply:** 2.32 × 0.4 = 0.928 **Divide:** 28.8 ÷ 12 = 2.4	157–166
	Write expressions in exponential form and evaluate exponential expressions (p. 40).	Repeated Multiplication: $7 \cdot 7 \cdot 7 \cdot 7$ (4 factors of 7) Exponential Form: 7^4	167–176
	Evaluate expressions using the order of operations (p. 42).	1. Perform operations inside symbols of grouping: **P**arentheses, brackets, or absolute value symbols, starting with the innermost symbols. 2. Evaluate all **E**xponential expressions. 3. Perform all **M**ultiplications and **D**ivisions from left to right. 4. Perform all **A**dditions and **S**ubtractions from left to right.	177–202
1.5	Identify and use the properties of real numbers (p. 43).	Let a, b, and c be real numbers. *Commutative Property of Addition:* $a + b = b + a$ *Commutative Property of Multiplication:* $ab = ba$ *Associative Property of Addition:* $(a + b) + c = a + (b + c)$ *Associative Property of Multiplication:* $(ab)c = a(bc)$ *Distributive Property:* $a(b + c) = ab + ac \quad (a + b)c = ac + bc$ *Additive Identity Property:* $a + 0 = 0 + a = a$ *Multiplicative Identity Property:* $a \cdot 1 = 1 \cdot a = a$ *Additive Inverse Property:* $a + (-a) = 0$ *Multiplicative Inverse Property:* $a \cdot \frac{1}{a} = 1, a \neq 0$	203–215

Review Exercises

Worked-out solutions to odd-numbered exercises at AlgebraWithinReach.com

1.1

Classifying Real Numbers In Exercises 1–4, determine which of the numbers in the set are (a) natural numbers, (b) integers, (c) rational numbers, and (d) irrational numbers.

1. $\left\{-1, 4.5\frac{2}{5}, -\frac{1}{7}, \sqrt{4}, \sqrt{5}\right\}$

2. $\left\{10, -3, \frac{4}{5}, \pi, -3.16, -\frac{19}{11}\right\}$

3. $\left\{\frac{30}{2}, 2, -\sqrt{3}, 1.5, -\pi, -\frac{10}{7}\right\}$

4. $\left\{3.75, 33, \frac{2}{3}, \frac{1}{10}, -92, -\frac{\pi}{4}\right\}$

Plotting Real Numbers In Exercises 5–10, plot the numbers on the real number line.

5. $-3, 5$
6. $-8, 11$
7. $-6, \frac{5}{4}$
8. $-\frac{7}{2}, 9$
9. $-1, 0, \frac{1}{2}$
10. $-2, -\frac{1}{3}, 5$

Ordering Real Numbers In Exercises 11–16, plot each real number as a point on the real number line and place the correct inequality symbol (< or >) between the real numbers.

11. $-\frac{1}{10}$ ▢ 4
12. $\frac{25}{3}$ ▢ $\frac{5}{3}$
13. -3 ▢ -7
14. 10.6 ▢ -3.5
15. 5 ▢ $\frac{7}{2}$
16. $\frac{3}{8}$ ▢ $\frac{4}{9}$

Using Absolute Value In Exercises 17–20, find the distance between a and zero on the real number line.

17. $a = 152$
18. $a = -10.4$
19. $a = -\frac{7}{3}$
20. $a = \frac{2}{3}$

Evaluating an Absolute Value In Exercises 21–28, evaluate the expression.

21. $|-8.5|$
22. $|-9.6|$
23. $|3.4|$
24. $|5.98|$
25. $-|-6.2|$
26. $-\left|-\frac{7}{9}\right|$
27. $-\left|\frac{8}{5}\right|$
28. $-|4|$

Comparing Real Numbers In Exercises 29–34, place the correct symbol (<, >, or =) between the real numbers.

29. $|-84|$ ▢ $|84|$
30. $|-10|$ ▢ $|4|$
31. $\left|\frac{5}{2}\right|$ ▢ $\left|\frac{8}{9}\right|$
32. $-|-1.8|$ ▢ $|5.7|$
33. $\left|\frac{3}{10}\right|$ ▢ $-\left|\frac{4}{5}\right|$
34. $|2.3|$ ▢ $-|2.3|$

Distance on the Real Number Line In Exercises 35–38, find all real numbers whose distance from a is given by d.

35. $a = 5, d = 7$
36. $a = -1, d = 4$
37. $a = 2.6, d = 5$
38. $a = -3, d = 6.5$

1.2

Adding Integers Using a Number Line In Exercises 39–42, find the sum and demonstrate the addition on the real number line.

39. $4 + 3$
40. $15 + (-6)$
41. $-1 + (-4)$
42. $-6 + (-2)$

Adding Integers In Exercises 43–52, find the sum.

43. $16 + (-5)$
44. $25 + (-10)$
45. $-125 + 30$
46. $-54 + 12$
47. $-13 + (-76)$
48. $-24 + (-25)$
49. $-10 + 21 + (-6)$
50. $-23 + 4 + (-11)$
51. $-17 + (-3) + (-9)$
52. $-16 + (-2) + (-8)$

53. **Profit** A small software company had a profit of $95,000 in January, a loss of $64,400 in February, and a profit of $51,800 in March. What was the company's overall profit (or loss) for the three months?

54. *Account Balance* At the beginning of a month, your account balance was $3090. During the month, you withdrew $870 and $465, deposited $109, and earned $10.05 in interest. What was your balance at the end of the month?

Subtracting Integers In Exercises 55–64, find the difference.

55. $28 - 7$
56. $43 - 12$
57. $8 - 15$
58. $17 - 26$
59. $14 - (-19)$
60. $28 - (-4)$
61. $-18 - 4$
62. $-37 - 14$
63. $-12 - (-7) - 4$
64. $-26 - (-8) - (-10)$

65. *Account Balance* At the beginning of a month, your account balance was $1560. During the month, you withdrew $50, $255, and $490. What was your balance at the end of the month?

66. *Gasoline Prices* At the beginning of a month, gas cost $4.14 per gallon. During the month, the price increased by $0.05 and $0.02, decreased by $0.10, and then increased again by $0.07. How much did gas cost at the end of the month?

1.3

Multiplying Integers In Exercises 67–78, find the product.

67. $15 \cdot 3$
68. $21 \cdot 4$
69. $-3 \cdot 24$
70. $-2 \cdot 44$
71. $6(-8)$
72. $12(-5)$
73. $-5(-9)$
74. $-10(-81)$
75. $3(-6)(3)$
76. $15(-2)(7)$
77. $-4(-5)(-2)$
78. $-12(-2)(-6)$

79. *Savings Plan* You save $150 per month for 2 years. What is the total amount you have saved?

80. *Average Speed* A truck drives 65 miles per hour for 5 hours. How far has the truck traveled?

Dividing Integers In Exercises 81–92, perform the division, if possible. If not possible, state the reason.

81. $72 \div 8$
82. $63 \div 9$
83. $\dfrac{-72}{6}$
84. $\dfrac{-162}{9}$
85. $75 \div (-5)$
86. $48 \div (-4)$
87. $\dfrac{-52}{-4}$
88. $\dfrac{-64}{-4}$
89. $0 \div 815$
90. $0 \div 25$
91. $135 \div 0$
92. $26 \div 0$

93. *Average Speed* A commuter train travels 195 miles between two cities in 3 hours. What is the average speed of the train in miles per hour?

94. *Unit Price* At an auction, you buy a box of six glass canisters for a total of $78. All the canisters are of equal value. How much is each one worth?

Classifying an Integer In Exercises 95–100, decide whether the number is prime or composite.

95. 137
96. 296
97. 839
98. 909
99. 1764
100. 1847

Prime Factorization In Exercises 101–106, write the prime factorization of the number.

101. 264
102. 195
103. 378
104. 858
105. 1612
106. 1787

Using a Rule or Definition In Exercises 107–110, complete the statement using the indicated definition or rule.

107. Rule for multiplying integers with unlike signs:

$12 \times (-3) = $ ▢

108. Definition of multiplication:

$(-4) + (-4) + (-4) = $ ▢

109. Definition of absolute value:

$|-7| = $ ▢

110. Rule for adding integers with unlike signs:

$-9 + 5 = $ ▢

1.4

Finding the GCF In Exercises 111–116, find the greatest common factor.

111. 54, 90
112. 154, 220
113. 2, 6, 9
114. 8, 12, 24
115. 63, 84, 441
116. 99, 132, 253

Chapter 1 The Real Number System

Writing a Fraction in Simplest Form In Exercises 117–120, write the fraction in simplest form.

117. $\frac{3}{12}$
118. $\frac{15}{25}$
119. $\frac{30}{48}$
120. $\frac{126}{162}$

Writing an Equivalent Fraction In Exercises 121–124, write an equivalent fraction with the indicated denominator.

121. $\frac{2}{3} = \frac{\;\;\;}{15}$
122. $\frac{3}{7} = \frac{\;\;\;}{28}$
123. $\frac{6}{10} = \frac{\;\;\;}{25}$
124. $\frac{9}{12} = \frac{\;\;\;}{16}$

Adding and Subtracting Fractions In Exercises 125–138, find the sum or difference. Write the result in simplest form.

125. $\frac{3}{25} + \frac{7}{25}$
126. $\frac{9}{64} + \frac{7}{64}$
127. $\frac{27}{16} - \frac{15}{16}$
128. $-\frac{5}{12} + \frac{1}{12}$
129. $\frac{3}{8} + \frac{1}{2}$
130. $\frac{7}{12} + \frac{5}{18}$
131. $-\frac{5}{9} + \frac{2}{3}$
132. $\frac{7}{15} - \frac{2}{25}$
133. $-\frac{25}{32} + \left(-\frac{7}{24}\right)$
134. $-\frac{7}{8} - \frac{11}{12}$
135. $5 - \frac{15}{4}$
136. $\frac{12}{5} - 3$
137. $5\frac{3}{4} - 3\frac{5}{8}$
138. $-3\frac{7}{10} + 1\frac{1}{20}$

139. **Meteorology** The table shows the daily amounts of rainfall (in inches) during a five-day period. What was the total amount of rainfall for the five days?

Day	Mon	Tue	Wed	Thu	Fri
Rainfall (in inches)	$\frac{3}{8}$	$\frac{1}{2}$	$\frac{1}{8}$	$1\frac{1}{4}$	$\frac{1}{2}$

140. **Fuel Consumption** The morning and evening readings of the fuel gauge on a car were 7/8 and 1/3, respectively. What fraction of the tank of fuel was used that day?

Multiplying and Dividing Fractions In Exercises 141–154, evaluate the expression and write the result in simplest form. If it is not possible, explain why.

141. $\frac{5}{8} \cdot \frac{-2}{15}$
142. $\frac{3}{32} \cdot \frac{32}{3}$
143. $35\left(\frac{1}{35}\right)$
144. $-6\left(\frac{5}{36}\right)$
145. $\frac{3}{8}\left(-\frac{2}{27}\right)$
146. $-\frac{5}{12}\left(-\frac{4}{25}\right)$
147. $\frac{5}{14} \div \frac{15}{28}$
148. $-\frac{7}{10} \div \frac{4}{15}$
149. $-\frac{3}{4} \div \left(-\frac{7}{8}\right)$
150. $\frac{15}{32} \div \left(-\frac{5}{4}\right)$
151. $-\frac{5}{9} \div 0$
152. $0 \div \frac{1}{12}$
153. $-5 \cdot 0$
154. $0 \cdot \frac{1}{2}$

155. **Meteorology** During an eight-hour period, $6\frac{3}{4}$ inches of snow fell. What was the average rate of snowfall per hour?

156. **Sports** In three strokes on a golf course, you hit your ball a total distance of $64\frac{7}{8}$ meters. What is your average distance per stroke?

Operations with Decimals In Exercises 157–164, evaluate the expression. Round your answer to two decimal places.

157. $4.89 + 0.76$
158. $1.29 + 0.44$
159. $3.815 - 5.19$
160. $7.234 - 8.16$
161. $1.49(-0.5)$
162. $2.34(-1.2)$
163. $5.25 \div 0.25$
164. $10.18 \div 1.6$

165. **Consumer Awareness** An engagement ring is advertised for $299.99 plus $26.99 per month for 24 months. Find the total cost of the engagement ring.

166. **Consumer Awareness** A plasma television costs $599.99 plus $32.96 per month for 18 months. Find the total cost of the television.

1.5

Writing an Expression in Exponential Form In Exercises 167–170, rewrite the expression in exponential form.

167. $6 \cdot 6 \cdot 6 \cdot 6 \cdot 6$
168. $(-3) \cdot (-3) \cdot (-3)$
169. $\left(\frac{6}{7}\right) \cdot \left(\frac{6}{7}\right) \cdot \left(\frac{6}{7}\right) \cdot \left(\frac{6}{7}\right)$
170. $-[(3.3) \cdot (3.3)]$

Evaluating an Exponential Expression In Exercises 171–176, evaluate the expression.

171. 2^4
172. $(-6)^2$
173. $\left(-\frac{3}{4}\right)^3$
174. $\left(\frac{2}{3}\right)^2$
175. -7^2
176. $-(-3)^3$

Review Exercises

Using Order of Operations In Exercises 177–196, evaluate the expression. Write fractional answers in simplest form.

177. $12 - 2 \cdot 3$
178. $1 + 7 \cdot 3 - 10$
179. $18 \div 6 \cdot 7$
180. $3^2 \cdot 4 \div 2$
181. $20 + (8^2 \div 2)$
182. $(8 - 3) \div 15$
183. $240 - (4^2 \cdot 5)$
184. $5^2 - (625 \cdot 5^2)$
185. $3^2(5 - 2)^2$
186. $-5(10 - 7)^3$
187. $\frac{3}{4}\left(\frac{5}{6}\right) + 4$
188. $75 - 24 \div 2^3$
189. $122 - [45 - (32 + 8) - 23]$
190. $-58 - (48 - 12) - (-30 - 4)$
191. $\dfrac{6 \cdot 4 - 36}{4}$
192. $\dfrac{144}{2 \cdot 3 \cdot 3}$
193. $\dfrac{54 - 4 \cdot 3}{6}$
194. $\dfrac{3 \cdot 5 + 125}{10}$
195. $\dfrac{78 - |-78|}{5}$
196. $\dfrac{300}{15 - |-15|}$

Using a Calculator In Exercises 197–200, use a calculator to evaluate the expression. Round your answer to two decimal places.

197. $(5.8)^4 - (3.2)^5$
198. $\dfrac{(15.8)^3}{(2.3)^8}$
199. $\dfrac{3000}{(1.05)^{10}}$
200. $500\left(1 + \dfrac{0.07}{4}\right)^{40}$

201. **Depreciation** After 3 years, the value of a $25,000 car is given by $25{,}000\left(\frac{3}{4}\right)^3$.
   (a) What is the value of the car after 3 years?
   (b) How much has the car depreciated during the 3 years?

202. **Geometry** The volume of water in a hot tub is given by $V = 6^2 \cdot 3$ (see figure). How many cubic feet of water will the hot tub hold? Find the total weight of the water in the tub. (Use the fact that 1 cubic foot of water weighs 62.4 pounds.)

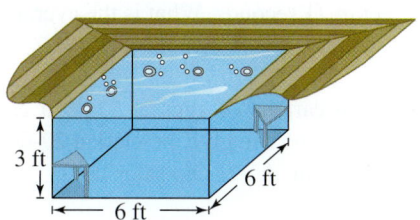

Identifying a Property of Real Numbers In Exercises 203–210, identify the property of real numbers illustrated by the statement.

203. $123 - 123 = 0$
204. $9 \cdot \frac{1}{9} = 1$
205. $14(3) = 3(14)$
206. $5(3 \cdot 8) = (5 \cdot 3)8$
207. $17 \cdot 1 = 17$
208. $10 + 6 = 6 + 10$
209. $-2(7 + 12) = (-2)7 + (-2)12$
210. $2 + (3 + 19) = (2 + 3) + 19$

Using a Property of Real Numbers In Exercises 211–214, complete the statement using the specified property of real numbers.

211. Additive Identity Property:
   $-16 + 0 = $

212. Distributive Property:
   $8(7 + 2) = $

213. Commutative Property of Addition:
   $24 + 1 = $

214. Associative Property of Multiplication:
   $8(5 \cdot 7) = $

215. **Geometry** Find the area of the shaded rectangle in two ways. Explain how the results are related to the Distributive Property.

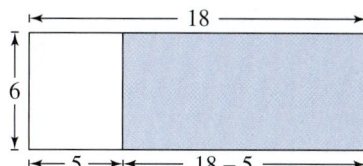

Chapter Test

Solutions in English & Spanish and tutorial videos at AlgebraWithinReach.com

Take this test as you would take a test in class. After you are done, check your work against the answers in the back of the book.

1. Which of the following are (a) natural numbers, (b) integers, (c) rational numbers, and (d) irrational numbers?

    $\left\{4, -6, \frac{1}{2}, 0, \pi, \frac{7}{9}\right\}$

2. Place the correct inequality symbol (< or >) between the real numbers.

    $-\frac{3}{5}$ ▢ $-|-2|$

In Exercises 3–20, evaluate the expression. Write fractional answers in simplest form.

3. $|-13|$
4. $-|-6.8|$
5. $16 + (-20)$
6. $-50 - (-60)$
7. $7 + |-3|$
8. $64 - (25 - 8)$
9. $-5(32)$
10. $\frac{-72}{-9}$
11. $\frac{15(-6)}{3}$
12. $-\frac{(-2)(5)}{10}$
13. $\frac{5}{6} - \frac{1}{8}$
14. $-\frac{9}{50}\left(-\frac{20}{27}\right)$
15. $\frac{7}{16} \div \frac{21}{28}$
16. $\frac{-8.1}{0.3}$
17. $-(0.8)^2$
18. $35 - (50 \div 5^2)$
19. $5(3 + 4)^2 - 10$
20. $18 - 7 \cdot 4 + 2^3$

In Exercises 21–24, identify the property of real numbers illustrated by the statement.

21. $3(4 + 6) = 3 \cdot 4 + 3 \cdot 6$
22. $5 \cdot \frac{1}{5} = 1$
23. $3 + (4 + 8) = (3 + 4) + 8$
24. $3(7 + 2) = (7 + 2)3$

25. Write the fraction $\frac{36}{162}$ in simplest form.

26. Write the prime factorization of 216.

27. An electric train travels 1218 feet in 21 seconds. What is the average speed of the train in feet per second?

28. At the grocery store, you buy five cartons of eggs at $1.49 a carton and two gallons of orange juice at $3.06 a gallon. You give the clerk a 20-dollar bill. How much change will you receive? (Assume there is no sales tax.)

Fundamentals of Algebra

- **2.1** Writing and Evaluating Algebraic Expressions
- **2.2** Simplifying Algebraic Expressions
- **2.3** Algebra and Problem Solving
- **2.4** Introduction to Equations

MASTERY IS WITHIN REACH!

"When I am in math class, I struggle with keeping all my notes neat. After class, I rewrite them so that I can review what we have covered and so I have a neat set for future reference. A friend of mine does this too."

Adishree
Psychology/film

See page 75 for suggestions about reworking your notes.

2.1 Writing and Evaluating Algebraic Expressions

▶ Define and identify terms, variables, and coefficients of algebraic expressions.
▶ Define exponential form and interpret exponential expressions.
▶ Evaluate algebraic expressions using real numbers.

Study Tip

The terms of an algebraic expression depend on the way the expression is written. Rewriting the expression can (and, in fact, usually does) change its terms. For instance, the expression $4 + x + 2$ has three terms, but the equivalent expression $6 + x$ has only two terms.

Variables and Algebraic Expressions

A collection of letters (**variables**) and real numbers (**constants**) combined by using addition, subtraction, multiplication, or division is an **algebraic expression**.
The **terms** of an algebraic expression are those parts that are separated by *addition*. For example, the expression $x^2 - 4x + 5$ has three terms: x^2, $-4x$, and 5. Note that $-4x$, rather than $4x$, is a term of $x^2 - 4x + 5$ because

$$x^2 - 4x + 5 = x^2 + (-4x) + 5.$$ To subtract, add the opposite.

For terms such as x^2, $-4x$, and 5, the numerical factor is called the **coefficient** of the term. Here, the coefficients are 1, -4, and 5.

EXAMPLE 1   Identifying the Terms of Expressions

Algebraic Expression	Terms
a. $x + 2$	$x, 2$
b. $3x + \frac{1}{2}$	$3x, \frac{1}{2}$
c. $2y - 5x - 7$	$2y, -5x, -7$

EXAMPLE 2   Identifying the Coefficients of Terms

Term	Coefficient	Comment
a. $-5x^2$	-5	Note that $-5x^2 = (-5)x^2$.
b. x^3	1	Note that $x^3 = 1 \cdot x^3$.
c. $-\dfrac{2x}{3}$	$-\dfrac{2}{3}$	Note that $-\dfrac{2x}{3} = -\dfrac{2}{3}(x)$.

Exercises Within Reach®   Solutions in English & Spanish and tutorial videos at AlgebraWithinReach.com

Identifying Variables In Exercises 1–4, identify the variable(s) in the expression.

1. $x + 3$ 2. $y - 1$ 3. $m + n$ 4. $a + b$

Identifying Terms In Exercises 5–10, identify the terms of the expression.

5. $4x + 3$ 6. $5 - 3t^2$ 7. $\frac{5}{3} - 3y^3$
8. $6x + \frac{2}{3}$ 9. $a^2 + 4ab + b^2$ 10. $x^2 + 18xy + y^2$

Identifying the Coefficient In Exercises 11–18, identify the coefficient of the term.

11. $14x$ 12. $25y$ 13. $-\frac{1}{3}y$ 14. $\frac{2}{3}n$
15. $\dfrac{2x}{5}$ 16. $-\dfrac{3x}{4}$ 17. $2\pi x^2$ 18. πt^4

Section 2.1 Writing and Evaluating Algebraic Expressions 57

Exponential Form

In general, for any positive integer n and any real number a, you have

$$a^n = \underbrace{a \cdot a \cdot a \cdots a}_{n \text{ factors}}.$$

This rule applies to factors that are variables as well as factors that are *algebraic expressions*.

Study Tip

Be sure you understand the difference between repeated addition

$$\underbrace{x + x + x + x}_{4 \text{ terms}} = 4x$$

and repeated multiplication

$$\underbrace{x \cdot x \cdot x \cdot x}_{4 \text{ factors}} = x^4.$$

EXAMPLE 3 Interpreting Exponential Expressions

a. $3^4 = 3 \cdot 3 \cdot 3 \cdot 3$

b. $3x^4 = 3 \cdot x \cdot x \cdot x \cdot x$

c. $(-3x)^4 = (-3x)(-3x)(-3x)(-3x)$
$= (-3)(-3)(-3)(-3) \cdot x \cdot x \cdot x \cdot x$

d. $(y+2)^3 = (y+2)(y+2)(y+2)$

e. $(5x)^2 y^3 = (5x)(5x) \cdot y \cdot y \cdot y$
$= 5 \cdot 5 \cdot x \cdot x \cdot y \cdot y \cdot y$

Be sure you understand the priorities for order of operations involving exponents. Here are some examples that tend to cause problems.

Expression	Correct Evaluation	Incorrect Evaluation
-3^2	$-(3 \cdot 3) = -9$	$(-3)(-3) = 9$
$(-3)^2$	$(-3)(-3) = 9$	$-(3 \cdot 3) = -9$
$3x^2$	$3 \cdot x \cdot x$	$(3x)(3x)$
$-3x^2$	$-3 \cdot x \cdot x$	$-(3x)(3x)$
$(-3x)^2$	$(-3x)(-3x)$	$-(3x)(3x)$

Exercises Within Reach®   Solutions in English & Spanish and tutorial videos at AlgebraWithinReach.com

Interpreting an Exponential Expression In Exercises 19–36, **expand** the expression as a product of factors.

19. y^5

20. $(-x)^6$

21. $2^2 x^4$

22. $(-5)^3 x^2$

23. $4y^2 z^3$

24. $3uv^4$

25. $(a^2)^3$

26. $(z^3)^3$

27. $-4x^3 \cdot x^4$

28. $a^2 y^2 \cdot y^3$

29. $-9(ab)^3$

30. $2(xz)^4$

31. $(x+y)^2$

32. $(s-t)^5$

33. $\left(\dfrac{a}{3s}\right)^4$

34. $\left(-\dfrac{2}{5x}\right)^3$

35. $[2(a-b)^3][2(a-b)^2]$

36. $[3(r+s)^2][3(r+s)]^2$

Evaluating Algebraic Expressions

In applications of algebra, you are often required to **evaluate** an algebraic expression. This means you are to find the value of an expression when its variables are replaced by real numbers.

EXAMPLE 4 Evaluating Algebraic Expressions

Evaluate each expression when $x = -3$ and $y = 5$.

a. $-x$
b. $x - y$
c. $3x + 2y$
d. $y - 2(x + y)$
e. $y^2 - 3y$

SOLUTION

a. When $x = -3$, the value of $-x$ is

$$-x = -(-3) \quad \text{Substitute } -3 \text{ for } x.$$
$$= 3. \quad \text{Simplify.}$$

b. When $x = -3$ and $y = 5$, the value of $x - y$ is

$$x - y = (-3) - 5 \quad \text{Substitute } -3 \text{ for } x \text{ and } 5 \text{ for } y.$$
$$= -8. \quad \text{Simplify.}$$

c. When $x = -3$ and $y = 5$, the value of $3x + 2y$ is

$$3x + 2y = 3(-3) + 2(5) \quad \text{Substitute } -3 \text{ for } x \text{ and } 5 \text{ for } y.$$
$$= -9 + 10 \quad \text{Multiply.}$$
$$= 1. \quad \text{Add.}$$

d. When $x = -3$ and $y = 5$, the value of $y - 2(x + y)$ is

$$y - 2(x + y) = 5 - 2[(-3) + 5] \quad \text{Substitute } -3 \text{ for } x \text{ and } 5 \text{ for } y.$$
$$= 5 - 2(2) \quad \text{Add.}$$
$$= 1. \quad \text{Simplify.}$$

e. When $y = 5$, the value of $y^2 - 3y$ is

$$y^2 - 3y = (5)^2 - 3(5) \quad \text{Substitute } 5 \text{ for } y.$$
$$= 25 - 15 \quad \text{Simplify.}$$
$$= 10. \quad \text{Subtract.}$$

Study Tip

As shown in parts (a), (b), (c), and (d) of Example 4, it is a good idea to use parentheses when substituting a negative number for a variable.

Exercises Within Reach®

Solutions in English & Spanish and tutorial videos at AlgebraWithinReach.com

Evaluating an Algebraic Expression In Exercises 37–44, evaluate the algebraic expression for the given values of the variable(s).

Expression *Values*

37. $2x - 1$ (a) $x = \frac{1}{2}$ (b) $x = -4$

38. $3x - 2$ (a) $x = \frac{4}{3}$ (b) $x = -1$

39. $2x^2 - 5$ (a) $x = 2$ (b) $x = 3$

40. $64 - 16t^2$ (a) $t = 2$ (b) $t = 3$

41. $3x - 2y$ (a) $x = 4, y = 3$
(b) $x = \frac{2}{3}, y = -1$

42. $10u - 3v$ (a) $u = 3, v = 10$
(b) $u = -2, v = \frac{4}{7}$

43. $x - 3(x - y)$ (a) $x = 3, y = 3$
(b) $x = 4, y = -4$

44. $-3x + 2(x + y)$ (a) $x = -2, y = 2$
(b) $x = 0, y = 5$

Section 2.1 Writing and Evaluating Algebraic Expressions 59

EXAMPLE 5 Evaluating Algebraic Expressions

a. When $y = -6$, the value of y^2 is
$$y^2 = (-6)^2 = 36.$$

b. When $y = -6$, the value of $-y^2$ is
$$-y^2 = -(y^2) = -(-6)^2 = -36.$$

c. When $x = 4$ and $y = -6$, the value of $y - x$ is
$$y - x = (-6) - 4 = -10.$$

d. When $x = 4$ and $y = -6$, the value of $|y - x|$ is
$$|y - x| = |(-6) - 4| = |-10| = 10.$$

e. When $x = 4$ and $y = -6$, the value of $|x - y|$ is
$$|x - y| = |4 - (-6)| = |4 + 6| = |10| = 10.$$

EXAMPLE 6 Evaluating an Algebraic Expression

When $x = -5$, $y = -2$, and $z = 3$, the value of $\dfrac{y + 2z}{5y - xz}$ is

$$\dfrac{y + 2z}{5y - xz} = \dfrac{-2 + 2(3)}{5(-2) - (-5)(3)} \quad \text{Substitute for } x, y, \text{ and } z.$$

$$= \dfrac{-2 + 6}{-10 - (-15)} \quad \text{Multiply.}$$

$$= \dfrac{-2 + 6}{-10 + 15} \quad \text{Add the opposite of } -15.$$

$$= \dfrac{4}{5}. \quad \text{Simplify.}$$

Exercises Within Reach®

Solutions in English & Spanish and tutorial videos at AlgebraWithinReach.com

Evaluating an Algebraic Expression In Exercises 45–54, evaluate the algebraic expression for the given values of the variables. If it is not possible, state the reason.

	Expression	Values		Expression	Values
45.	$b^2 - 4ab$	(a) $a = 2, b = -3$	46.	$a^2 + 2ab$	(a) $a = -2, b = 3$
		(b) $a = 6, b = -4$			(b) $a = 4, b = -2$
47.	$\|2x - 3y\|$	(a) $x = 2, y = 3$	48.	$y - \|-3x + y\|$	(a) $x = -2, y = -1$
		(b) $x = -1, y = 4$			(b) $x = 7, y = 3$
49.	$\dfrac{x - 2y}{x + 2y}$	(a) $x = 4, y = 2$	50.	$\dfrac{5x}{y - 3}$	(a) $x = 2, y = 4$
		(b) $x = 4, y = -2$			(b) $x = 2, y = 3$
51.	$\dfrac{-y}{x^2 + y^2}$	(a) $x = 0, y = 5$	52.	$\dfrac{2x - y}{y^2 + 1}$	(a) $x = 1, y = 2$
		(b) $x = 1, y = -3$			(b) $x = 1, y = 3$
53.	$(x + 2y)(-3x - z)$	(a) $x = 2, y = -1, z = -1$	54.	$\dfrac{yz - 3}{x + 2z}$	(a) $x = 0, y = -7, z = 3$
		(b) $x = -3, y = 2, z = -2$			(b) $x = -2, y = -3, z = 3$

Application

EXAMPLE 7 Converting Temperatures

There is one temperature that has the same degree measure in the Fahrenheit and Celsius scales. Use the formula $F = \frac{9}{5}C + 32$ to find this temperature.

SOLUTION

A spreadsheet is useful to answer this type of question. Enter several Celsius temperatures into Column A. Then program Column B to calculate corresponding Fahrenheit temperatures.

	A	B
1	Celsius	Fahrenheit
2	100	212
3	90	194
4	80	176
5	70	158
6	60	140
7	50	122
8	40	104
9	30	86
10	20	68
11	10	50
12	0	32
13	−10	14
14	−20	−4
15	−30	−22
16	−40	−40
17	−50	−58

(9/5)*A2+32

Inuvik is a town in the Northwest Territories of Canada. The population is about 3500. The average low temperature in January is about −32°C.

From the result, you can see that the temperature −40 degrees is the same on both scales.

Exercises Within Reach®

Solutions in English & Spanish and tutorial videos at AlgebraWithinReach.com

55. Savings The interest a savings account earns is given by $I = 850(0.095)t$, where I is the interest the account earns after t years. Use a spreadsheet to determine the interest the account earns after 8 years.

	A	B
1	Years	Interest
2	1	
3	2	
4	3	
5	4	
6	5	
7	6	
8	7	
9	8	

56. Traveling The distance a car travels is given by $d = 63t$, where d is the distance (in miles) the car travels after t hours. Use a spreadsheet to determine the distance the car travels after 7 hours.

	A	B
1	Hours	Distance
2	1	
3	2	
4	3	
5	4	
6	5	
7	6	
8	7	

Section 2.1 Writing and Evaluating Algebraic Expressions 61

Application **EXAMPLE 8** Writing an Algebraic Expression

You accept a part-time job for $9 per hour. The job offer states that you will be expected to work between 15 and 30 hours a week. Because you do not know how many hours you will work during a week, your total income for a week is unknown. Moreover, your income will probably *vary* from week to week. By representing the variable quantity (the number of hours worked) by the letter x, you can represent the weekly income by the following algebraic expression.

In the product $9x$, the number 9 is a *constant* and the letter x is a *variable*.

Application **EXAMPLE 9** Geometry: Writing an Algebraic Expression

Write an expression for the area of the rectangle shown at the left. Then evaluate the expression to find the area of the rectangle when $x = 7$.

SOLUTION

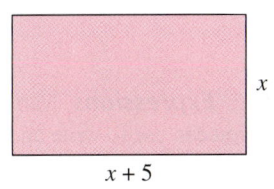

Area of a rectangle = Length • Width

$\qquad\qquad\qquad\quad = (x + 5) \cdot x$ Substitute.

To find the area of the rectangle when $x = 7$, substitute 7 for x in the expression for the area.

$(x + 5) \cdot x = (7 + 5) \cdot 7$ Substitute 7 for x.

$\qquad\qquad\;\, = 12 \cdot 7$ Add.

$\qquad\qquad\;\, = 84$ Multiply.

So, the area of the rectangle is 84 square units.

Exercises Within Reach ®

Solutions in English & Spanish and tutorial videos at AlgebraWithinReach.com

Writing an Algebraic Expression In Exercises 57–60, **write** an algebraic expression for the statement.

57. The income earned at $7.55 per hour for w hours
58. The cost for a family of n people to see a movie when the cost per person is $8.25
59. The cost of m pounds of meat when the cost per pound is $3.79
60. The total weight of x bags of fertilizer when each bag weighs 50 pounds

Geometry In Exercises 61–64, **write** an expression for the area of the figure. Then **evaluate** the expression for the given value(s) of the variable(s).

61. $n = 8$ 62. $x = 10, y = 3$ 63. $a = 5, b = 4$ 64. $x = 9$

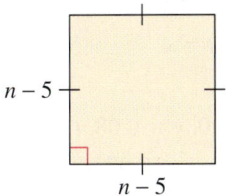 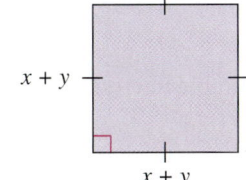 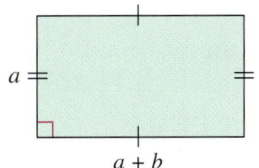 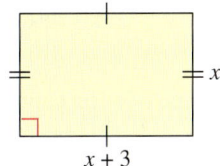

Concept Summary: Writing and Evaluating Algebraic Expressions

What
When solving a word problem, the goal is to translate the words and phrases into an **algebraic expression**. Then **evaluate** the expression.

EXAMPLE

You can earn $12 every hour you work. Write an expression that represents the amount of money you earn working x hours. How much money do you earn after working 8 hours?

How
When translating a word problem into an algebraic expression, represent unknown quantities with **variables** and known quantities with **constants**. Then to evaluate the expression, replace the variables with real numbers, and simplify.

EXAMPLE

Write: $12x$

Evaluate: $12x = 12(8) = 96$

Why
When given a real-life problem with unknown quantities, you can use algebra to write an expression to represent the situation.

Then you can find the value of the expression for different values of the variable. For example, you can use the expression $12x$ to find the amount of money you earn for any number of hours you work.

Exercises Within Reach®
Worked-out solutions to odd-numbered exercises at AlgebraWithinReach.com

Concept Summary Check

65. *Interpreting Expressions* Are the expressions -3^2 and $(-3)^2$ equivalent? Explain.

66. *Writing Algebraic Expressions* Do you always have to use x to represent an unknown value when writing an algebraic expression? Give an example of when you may want to use another letter.

67. *Writing Algebraic Expressions* Name four mathematical operations you can use to write an algebraic expression.

68. *Number Sense* What value of y would cause $3y + 2$ to equal 8? Explain.

Extra Practice

Identifying Variables In Exercises 69 and 70, identify the variable(s) in the expression.

69. $2^3 - k$

70. $3^2 + z$

Identifying Terms In Exercises 71 and 72, identify the terms of the expression.

71. $3(x + 5) + 10$

72. $\dfrac{6}{t} - 22$

Rewriting in Exponential Form In Exercises 73–76, rewrite the product in exponential form.

73. $-2 \cdot u \cdot u \cdot u \cdot u$

74. $\frac{1}{3} \cdot x \cdot x \cdot x \cdot x$

75. $-3 \cdot (x - y) \cdot (x - y) \cdot (-3) \cdot (-3)$

76. $(u - v) \cdot (u - v) \cdot 8 \cdot 8 \cdot 8 \cdot (u - v)$

Evaluating an Algebraic Expression In Exercises 77–80, evaluate the algebraic expression for the given values of the variable(s).

77. Area of a Triangle
$\frac{1}{2}bh$
(a) $b = 3, h = 5$
(b) $b = 2, h = 10$

78. Distance Traveled
rt
(a) $r = 50, t = 3.5$
(b) $r = 35, t = 4$

79. Volume of a Rectangular Prism
lwh
(a) $l = 4, w = 2, h = 9$
(b) $l = 100, w = 0.8, h = 4$

80. Simple Interest
Prt
(a) $P = 1000, r = 0.08, t = 3$
(b) $P = 500, r = 0.07, t = 5$

81. **Advertising** An advertisement for a new pair of basketball shoes claims that the shoes will help you jump 6 inches higher than without shoes.

   (a) Let x represent the height (in inches) jumped without shoes. Write an expression that represents the height of a jump while wearing the new shoes.

   (b) You can jump 23 inches without shoes. How high can you jump while wearing the new shoes?

   (c) Your friend can jump 20.5 inches without shoes. How high can she jump while wearing the new shoes?

82. **Distance** You are driving 60 miles per hour on the highway.

   (a) Write an expression that represents the distance you travel in t hours.

   (b) How far will you travel in 2.75 hours?

83. **Exploration** For any natural number n, the sum of the numbers $1, 2, 3, \ldots, n$ is equal to

   $$\frac{n(n+1)}{2}, \quad n \geq 1.$$

   Verify the formula for (a) $n = 3$, (b) $n = 6$, and (c) $n = 10$.

84. **Exploration** A convex polygon with n sides has

   $$\frac{n(n-3)}{2}, \quad n \geq 4$$

   diagonals. Verify the formula for (a) a square (two diagonals), (b) a pentagon (five diagonals), and (c) a hexagon (nine diagonals).

Explaining Concepts

85. **Identifying Terms** Is $3x$ a term of $4 - 3x$? Explain.

86. **Number Sense** Is it possible to evaluate the expression

    $$\frac{x+2}{y-3}$$

    when $x = 5$ and $y = 3$? Explain.

87. **Logic** Explain why the formulas in Exercises 83 and 84 will always yield natural numbers.

88. **Error Analysis** Describe and correct the error in evaluating $y - 2(x - y)$ for $x = 2$ and $y = -4$.

    $y - 2(x - y) = -4 - 2(2 - 4)$
    $= -4 - 2(-2)$
    $= -4 + 4$
    $= 0$

Cumulative Review

In Exercises 89–96, evaluate the expression.

89. $10 - (-7)$

90. $6 - 10 - (-12) + 3$

91. $-5 + 10 - (-9) - 4$

92. $-(-8) + 6 - 4 - 2$

93. $(-6)(-4)$

94. $\dfrac{-56}{7}$

95. $\dfrac{-144}{-12}$

96. $5(-7)$

In Exercises 97–100, identify the property of real numbers illustrated by the statement.

97. $3(4) = 4(3)$

98. $10 - 10 = 0$

99. $3(6 + 2) = 3 \cdot 6 + 3 \cdot 2$

100. $7 + (8 + 5) = (7 + 8) + 5$

2.2 Simplifying Algebraic Expressions

- Use the properties of algebra.
- Combine like terms of an algebraic expression.
- Simplify an algebraic expression by rewriting the terms.
- Use the Distributive Property to remove symbols of grouping.

Properties of Algebra

You can rewrite and simplify algebraic expressions using the properties of real numbers listed on page 43.

EXAMPLE 1 Applying Properties of Real Numbers

Use the indicated rule to complete each statement.

a. Additive Identity Property: $(x - 2) + \boxed{} = x - 2$

b. Commutative Property of Multiplication: $5(y + 6) = \boxed{}$

c. Associative Property of Addition: $(x^2 + 3) + 7 = \boxed{}$

d. Additive Inverse Property: $\boxed{} + 4x = 0$

SOLUTION

a. $(x - 2) + 0 = x - 2$ b. $5(y + 6) = (y + 6)5$

c. $(x^2 + 3) + 7 = x^2 + (3 + 7)$ d. $-4x + 4x = 0$

EXAMPLE 2 Using the Distributive Property

Use the Distributive Property to expand each expression.

a. $2(7 - x)$ b. $(10 - 2y)3$ c. $2x(x + 4y)$ d. $-(1 - 2y)$

SOLUTION

a. $2(7 - x) = 2 \cdot 7 - 2 \cdot x$
    $= 14 - 2x$

b. $(10 - 2y)3 = 10(3) - 2y(3)$
    $= 30 - 6y$

c. $2x(x + 4y) = 2x(x) + 2x(4y)$
    $= 2x^2 + 8xy$

d. $-(1 - 2y) = (-1)(1 - 2y)$
    $= (-1)(1) - (-1)(2y)$
    $= -1 + 2y$

Exercises Within Reach®

Solutions in English & Spanish and tutorial videos at AlgebraWithinReach.com

Applying Properties of Real Numbers In Exercises 1–4, **complete** the statement. Then **state** the property of algebra that you used.

1. $v \cdot 2 = \boxed{}$
2. $(2x - y)(-3) = -3\boxed{}$
3. $5(t - 2) = 5\left(\boxed{}\right) + 5\left(\boxed{}\right)$
4. $x(y + 4) = x\left(\boxed{}\right) + x\left(\boxed{}\right)$

Using the Distributive Property In Exercises 5–12, use the Distributive Property to **expand** the expression.

5. $2(16 + 8z)$
6. $3(7 - 4a)$
7. $-5(2x - y)$
8. $-3(11y - 6)$
9. $(x + 1)8$
10. $(r + 10)2$
11. $-6s(6s - 1)$
12. $-(u - v)$

Section 2.2   Simplifying Algebraic Expressions   65

Application   **EXAMPLE 3**   **Geometry: Visualizing the Distributive Property**

Write the area of each component of each figure. Then demonstrate the Distributive Property by writing the total area of each figure in two ways.

a. 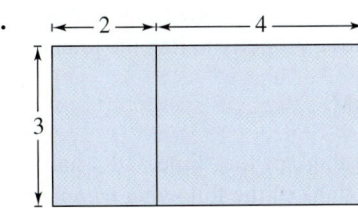   b.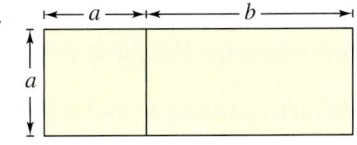

SOLUTION

a.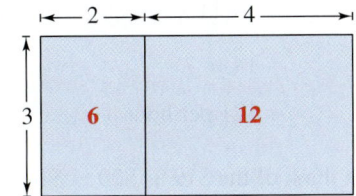

The total area is $3(2 + 4) = 3 \cdot 2 + 3 \cdot 4$
$= 6 + 12$
$= 18.$

b.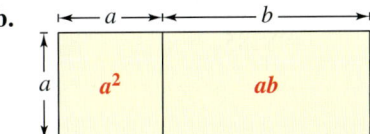

The total area is $a(a + b) = a \cdot a + a \cdot b$
$= a^2 + ab.$

Exercises Within Reach®   Solutions in English & Spanish and tutorial videos at AlgebraWithinReach.com

Visualizing the Distributive Property In Exercises 13–16, **write** the area of each component of the figure. Then **demonstrate** the Distributive Property by writing the total area of each figure in two ways.

13.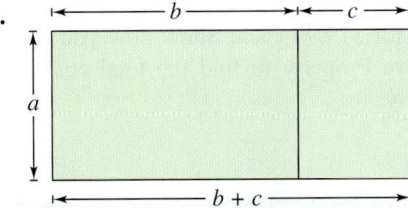

14.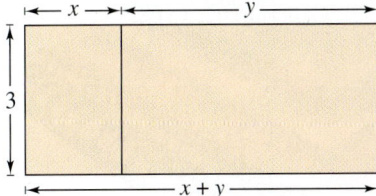

15.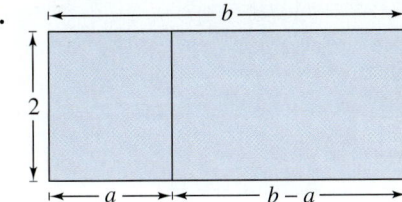

16.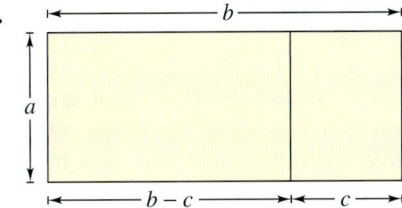

Application

EXAMPLE 4 Using Mental Math in Everyday Applications

a. You earn $14 per hour and time-and-a-half for overtime. Show how you can use the Distributive Property to find your overtime wage mentally.

b. You are buying 15 potted plants that cost $19 each. Show how you can use the Distributive Property to find the total cost mentally.

SOLUTION

a. You can think of the "time-and-a-half" as $1\frac{1}{2}$. So, using the Distributive Property, you can think of the following *without* writing anything on paper.

$$14\left(1\tfrac{1}{2}\right) = 14\left(1 + \tfrac{1}{2}\right)$$
$$= 14(1) + 14\left(\tfrac{1}{2}\right)$$
$$= 14 + 7$$
$$= \$21 \text{ per hour}$$

b. You can think of the $19 as $20 − $1. So, using the Distributive Property, you can think of the following *without* writing anything on paper.

$$15(19) = 15(20 - 1)$$
$$= 15(20) - 15(1)$$
$$= 300 - 15$$
$$= \$285$$

Exercises Within Reach®

Solutions in English & Spanish and tutorial videos at AlgebraWithinReach.com

17. Notebooks You are shopping for school supplies. You want to buy 8 notebooks for $1.25 each. Show how you can use the Distributive Property to find the total of the notebooks mentally.

18. Work You earn $12 per hour working at a grocery store. You receive a 10% raise in pay. Show how you can use the Distributive Property to find your new hourly pay rate.

19. Lighting You are installing solar lighting units along the driveway to your house. Each unit is on sale for $5. You purchase 18 units. Show how you can use the Distributive Property to find the total cost of the lighting units mentally.

20. Movies You are shopping for DVDs. You want to buy 7 DVDs for $19.99 each. Show how you can use the Distributive Property to find the total cost of the DVDs mentally.

Combining Like Terms

In an algebraic expression, two terms are said to be **like terms** if they are both constant terms or if they have the same variable factor(s). Factors such as x in $5x$ and ab in $6ab$ are called **variable factors**.

EXAMPLE 5 Identifying Like Terms

Expression	Like Terms
a. $5xy + 1 - xy$	$5xy$ and $-xy$
b. $12 - x^2 + 3x - 5$	12 and -5
c. $7x - 3 - 2x + 5$	$7x$ and $-2x$, -3 and 5

EXAMPLE 6 Combining Like Terms

a. $5x + 2x - 4 = (5 + 2)x - 4$ Distributive Property
$= 7x - 4$ Add.

b. $-5 + 8 + 7y - 5y = (-5 + 8) + (7 - 5)y$ Distributive Property
$= 3 + 2y$ Simplify.

c. $2y - 3x - 4x = 2y - x(3 + 4)$ Distributive Property
$= 2y - x(7)$ Add.
$= 2y - 7x$ Simplify.

d. $7x + 3y - 4x = 3y + 7x - 4x$ Commutative Property
$= 3y + (7x - 4x)$ Associative Property
$= 3y + (7 - 4)x$ Distributive Property
$= 3y + 3x$ Subtract.

e. $12a - 5 - 3a + 7 = 12a - 3a - 5 + 7$ Commutative Property
$= (12a - 3a) + (-5 + 7)$ Associative Property
$= (12 - 3)a + (-5 + 7)$ Distributive Property
$= 9a + 2$ Simplify.

Exercises Within Reach®
Solutions in English & Spanish and tutorial videos at AlgebraWithinReach.com

Identifying Like Terms In Exercises 21–24, **identify** the like terms.

21. $16t^3 + 4t - 5t + 3t^3$
22. $-\frac{1}{4}x^2 - 3x + \frac{3}{4}x^2 + x$
23. $4rs^2 - 5 - 2r^2s + 12rs^2 + 1$
24. $3 + 6x^2y + 2xy - 2 - 4x^2y$

Combining Like Terms In Exercises 25–32, **simplify** the expression by combining like terms.

25. $3y - 5y$
26. $-16x + 25x$
27. $x + 5 - 3x$
28. $7s + 3 - 3s$
29. $2x + 9x + 4$
30. $10x - 6 - 5x$
31. $5r + 6 - 2r + 1$
32. $2t - 4 + 8t + 9$

Simplifying Algebraic Expressions

To **simplify an algebraic expression** generally means to remove symbols of grouping and combine like terms.

EXAMPLE 7 Simplifying Algebraic Expressions

Simplify each expression.

a. $-3(-5x)$ b. $7(-x)$ c. $\dfrac{5x}{3} \cdot \dfrac{3}{5}$

d. $x^2(-2x^3)$ e. $(-2x)(4x)$ f. $(2rs)(r^2s)$

SOLUTION

a. $-3(-5x) = (-3)(-5)x$ Associative Property
$= 15x$ Multiply.

b. $7(-x) = 7(-1)(x)$ Coefficient of $-x$ is -1.
$= -7x$ Multiply.

c. $\dfrac{5x}{3} \cdot \dfrac{3}{5} = \left(\dfrac{5}{3} \cdot x\right) \cdot \dfrac{3}{5}$ Coefficient of $\dfrac{5x}{3}$ is $\dfrac{5}{3}$.
$= \left(\dfrac{5}{3} \cdot \dfrac{3}{5}\right) \cdot x$ Commutative and Associative Properties
$= 1 \cdot x$ Multiplicative Inverse
$= x$ Multiplicative Identity

d. $x^2(-2x^3) = (-2)(x^2 \cdot x^3)$ Commutative and Associative Properties
$= -2 \cdot x \cdot x \cdot x \cdot x \cdot x$ Repeated multiplication
$= -2x^5$ Exponential form

e. $(-2x)(4x) = (-2 \cdot 4)(x \cdot x)$ Commutative and Associative Properties
$= -8x^2$ Exponential form

f. $(2rs)(r^2s) = 2(r \cdot r^2)(s \cdot s)$ Commutative and Associative Properties
$= 2 \cdot r \cdot r \cdot r \cdot s \cdot s$ Repeated multiplication
$= 2r^3s^2$ Exponential form

Exercises Within Reach®

Solutions in English & Spanish and tutorial videos at AlgebraWithinReach.com

Simplifying an Algebraic Expression In Exercises 33–46, simplify the expression.

33. $2(6x)$ 34. $-7(5a)$

35. $-(4x)$ 36. $-(5t)$

37. $(-2x)(-3x)$ 38. $(-3y)(-4y)$

39. $(-5z)(2z^2)$ 40. $(10t)(-4t^2)$

41. $\dfrac{18a}{5} \cdot \dfrac{15}{6}$ 42. $\dfrac{5x}{8} \cdot \dfrac{16}{5}$

43. $\left(-\dfrac{3x^2}{2}\right)\left(\dfrac{4x}{18}\right)$ 44. $\left(\dfrac{4x}{3}\right)\left(\dfrac{3x}{16}\right)$

45. $(12xy^2)(-2x^3y^2)$ 46. $(7r^2s^3)(3rs)$

Symbols of Grouping

EXAMPLE 8 Removing Symbols of Grouping

Simplify each expression.

a. $-(3y + 5)$
b. $5x + (x - 7)2$
c. $-2(4x - 1) + 3x$
d. $5x - 2[4x + 3(x - 1)]$
e. $-7y + 3[2y - (3 - 2y)] - 5y + 4$

SOLUTION

a. $-(3y + 5) = -3y - 5$ — Distributive Property

b. $5x + (x - 7)2 = 5x + 2x - 14$ — Distributive Property
$= 7x - 14$ — Combine like terms.

c. $-2(4x - 1) + 3x = -8x + 2 + 3x$ — Distributive Property
$= -8x + 3x + 2$ — Commutative Property
$= -5x + 2$ — Combine like terms.

d. $5x - 2[4x + 3(x - 1)]$
$= 5x - 2[4x + 3x - 3]$ — Distributive Property
$= 5x - 2[7x - 3]$ — Combine like terms.
$= 5x - 14x + 6$ — Distributive Property
$= -9x + 6$ — Combine like terms.

e. $-7y + 3[2y - (3 - 2y)] - 5y + 4$
$= -7y + 3[2y - 3 + 2y] - 5y + 4$ — Distributive Property
$= -7y + 3[4y - 3] - 5y + 4$ — Combine like terms.
$= -7y + 12y - 9 - 5y + 4$ — Distributive Property
$= (-7y + 12y - 5y) + (-9 + 4)$ — Group like terms.
$= -5$ — Combine like terms.

Study Tip

When a parenthetical expression is preceded by a *plus* sign, you can remove the parentheses without changing the signs of the terms inside.

$3y + (-2y + 7)$
$= 3y - 2y + 7$

When a parenthetical expression is preceded by a *minus* sign, however, you must change the sign of each term to remove the parentheses.

$3y - (2y - 7)$
$= 3y - 2y + 7$

Remember that $-(2y - 7)$ is equal to $(-1)(2y - 7)$, and the Distributive Property can be used to "distribute the minus sign" to obtain $-2y + 7$.

Exercises Within Reach®

Solutions in English & Spanish and tutorial videos at AlgebraWithinReach.com

Removing Symbols of Grouping In Exercises 47–66, *simplify* the expression by removing symbols of grouping and combining like terms.

47. $2(x - 2) + 4$
48. $3(x - 5) - 2$
49. $6(2s - 1) + s + 4$
50. $(2x - 1)2 + x + 9$
51. $m - 3(m - 7)$
52. $8l - (3l - 7)$
53. $-6(2 - 3x) + 10(5 - x)$
54. $3(r - 2s) - 5(3r - 5s)$
55. $\frac{2}{3}(12x + 15) + 16$
56. $\frac{3}{8}(4 - y) - \frac{5}{2} + 10$
57. $3 - 2[6 + (4 - x)]$
58. $10x + 5[6 - (2x + 3)]$
59. $7x(2 - x) - 4x$
60. $-6x(x - 1) + x^2$
61. $4x^2 + x(5 - x) - 3$
62. $-z(z - 2) + 3z^2 + 5$
63. $-3t(4 - t) + t(t + 1)$
64. $-2x(x - 1) + x(3x - 2)$
65. $3t[4 - (t - 3)] + t(t + 5)$
66. $4y[5 - (y + 1)] + 3y(y + 1)$

Application EXAMPLE 9 Geometry: Writing and Simplifying a Formula

Write and simplify an expression for (a) the perimeter and (b) the area of the triangle.

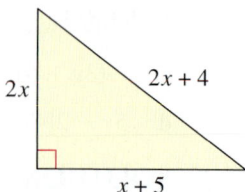

SOLUTION

a. Perimeter of a Triangle = Sum of the Three Sides

$= 2x + (2x + 4) + (x + 5)$ Substitute.
$= (2x + 2x + x) + (4 + 5)$ Group like terms.
$= 5x + 9$ Combine like terms.

b. Area of a Triangle $= \frac{1}{2} \cdot$ Base $\cdot$ Height

$= \frac{1}{2}(x + 5)(2x)$ Substitute.
$= \frac{1}{2}(2x)(x + 5)$ Commutative Property
$= x(x + 5)$ Multiply.
$= x^2 + 5x$ Distributive Property

Exercises Within Reach®

Solutions in English & Spanish and tutorial videos at AlgebraWithinReach.com

Geometry In Exercises 67 and 68, write and simplify expressions for (a) the perimeter and (b) the area of the rectangular sandboxes.

67.

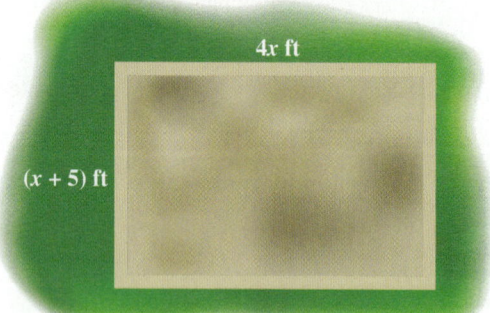

68.

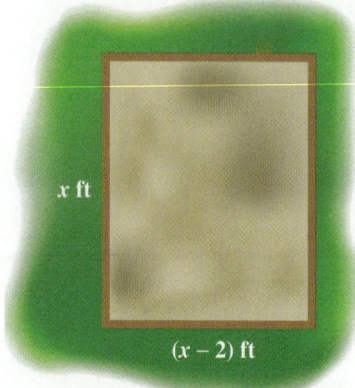

Komarov/Shutterstock.com

Section 2.2 Simplifying Algebraic Expressions 71

Application **EXAMPLE 10** **Geometry: Writing and Simplifying a Formula**

The formula for the area of a trapezoid is $A = \frac{1}{2}h(b_1 + b_2)$. Use this formula to write and simplify an expression for the area of the proposed trapezoidal park.

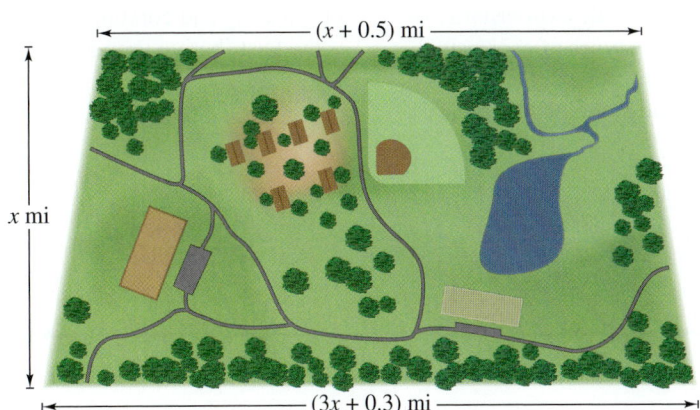

SOLUTION

Begin by assigning the following values.

$$h = x \text{ mi} \qquad b_1 = (x + 0.5) \text{ mi} \qquad b_2 = (3x + 0.3) \text{ mi}$$

Then use the formula to write an expression.

$$\begin{aligned}\tfrac{1}{2}h(b_1 + b_2) &= \tfrac{1}{2}x[(x + 0.5) + (3x + 0.3)] \\ &= \tfrac{1}{2}x[x + 0.5 + 3x + 0.3] \\ &= \tfrac{1}{2}x[4x + 0.8] \\ &= 2x^2 + 0.4x \end{aligned}$$

Exercises Within Reach®

Solutions in English & Spanish and tutorial videos at AlgebraWithinReach.com

Geometry In Exercises 69 and 70, use the formula for the area of a trapezoid to write and simplify an expression for the area of the trapezoidal house lot and park.

69.
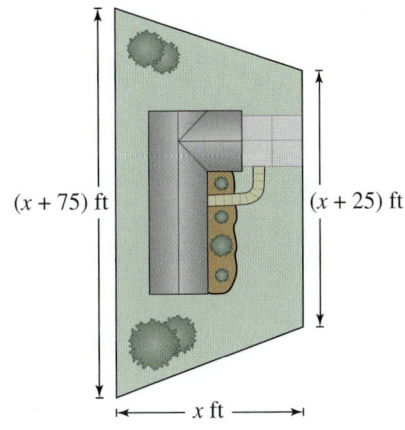

70.
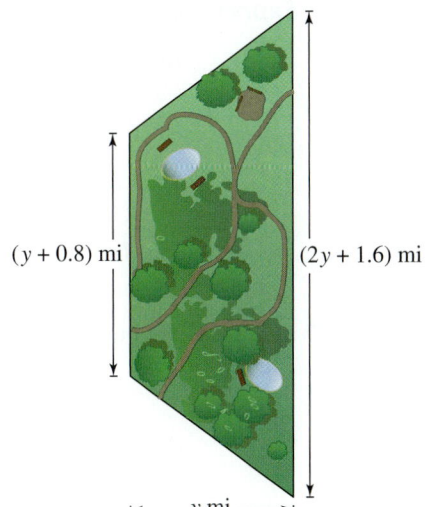

Concept Summary: Simplifying Algebraic Expressions

What
You can use properties of algebra to **simplify algebraic expressions**. To simplify these types of expressions usually means to remove symbols of grouping and combine **like terms**.

EXAMPLE
Simplify the expression $3x + 2(x + 4)$.

How
The main tool for removing symbols of grouping and combining like terms is the Distributive Property.

EXAMPLE
$3x + 2(4 + x)$
$= 3x + 8 + 2x$ Distributive Prop.
$= 3x + 2x + 8$ Comm. Prop.
$= x(3 + 2) + 8$ Distributive Prop.
$= 5x + 8$ Simplify.

Why
Simplifying an algebraic expression into a more usable form is one of the most frequently used skills in algebra.

Exercises Within Reach®
Worked-out solutions to odd-numbered exercises at **AlgebraWithinReach.com**

Concept Summary Check

71. *Describing Like Terms* In your own words, state the definition of like terms. Give an example of like terms and an example of unlike terms.

72. *Combining Like Terms* Describe how to combine like terms. Give an example of an expression that can be simplified by combining like terms.

73. *Writing* In your own words, describe the procedure for removing symbols of grouping.

74. *Writing* What does it mean to simplify an algebraic expression?

Extra Practice

Simplifying an Algebraic Expression In Exercises 75–86, simplify the expression.

75. $x^2 - 2xy + 4 + xy$

76. $r^2 + 3rs - 6 - rs$

77. $5z - 5 + 10z + 2z + 16$

78. $7x - 4x + 8 + 3x - 6$

79. $(7y^2)(-3y)$

80. $(-2t^3)(4t^2)$

81. $\left(\dfrac{2x}{5}\right)\left(\dfrac{4x}{8}\right)$

82. $\left(-\dfrac{6y^2}{7}\right)\left(-\dfrac{y}{6}\right)$

83. $-4(2 - 5x) + 3(x + 6)$

84. $5(x + 9) - 2(30 + 4x)$

85. $7 - 3[7 - (3 + x)]$

86. $2x[1 - (x - 4)] + x(x - 3)$

Geometry In Exercises 87 and 88, write and simplify an expression for the perimeter of the triangle.

87.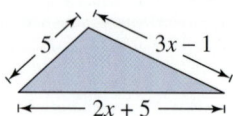

88.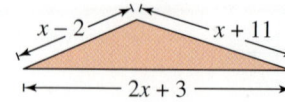

Geometry In Exercises 89 and 90, write and simplify expressions for (a) the perimeter and (b) the area of the rectangle.

89.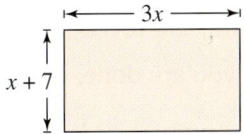

90.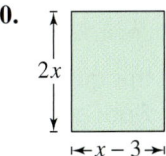

91. **Geometry** The remaining area of a square with side length x after a smaller square with side length y has been removed (see figure) is $(x + y)(x - y)$.

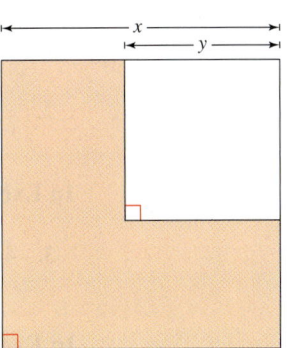

(a) Show that the remaining area can also be expressed as $x(x - y) + y(x - y)$, and give a geometric explanation for the area represented by each term in this expression.

(b) Find the remaining area of a square with side length 9 after a square with side length 5 has been removed.

Explaining Concepts

Writing In Exercises 92 and 93, explain why the two expressions are not like terms.

92. $\frac{1}{2}x^2y, \frac{5}{2}xy^2$

93. $-16x^2y^3, 7x^2y$

94. **Using Order of Operations** Does the expression $[x - (3 \cdot 4)] \div 5$ change when the parentheses are removed? Does it change when the brackets are removed? Explain.

95. **Writing** Discuss the difference between $(6x)^4$ and $6x^4$.

96. **Error Analysis** Describe and correct the error.

$$4x - 3(x - 1) = 4x - 3(x) - 3(1)$$
$$= 4x - 3x - 3$$
$$= x - 3$$

Cumulative Review

In Exercises 97–102, evaluate the expression.

97. $0 - (-12)$

98. $5 - 4 \div 2 + 6$

99. $-12 - 2 + |-3|$

100. $6 + 3(4 + 2)$

101. $\frac{5}{16} - \frac{3}{10}$

102. $\frac{9}{16} + 2\frac{3}{12}$

In Exercises 103 and 104, evaluate the algebraic expression for the given values of the variable.

103. $3x - 2$
(a) $x = 2$
(b) $x = -1$

104. $2x^2 + 3$
(a) $x = 3$
(b) $x = -4$

Mid-Chapter Quiz: Sections 2.1–2.2

Solutions in English & Spanish and tutorial videos at AlgebraWithinReach.com

Take this quiz as you would take a quiz in class. After you are done, check your work against the answers in the back of the book.

In Exercises 1 and 2, evaluate the algebraic expression for the given values of the variable(s). If it is not possible, state the reason.

1. $x^2 - 3x$ (a) $x = 3$ (b) $x = -2$

2. $\dfrac{x}{y-3}$ (a) $x = 5, y = 3$ (b) $x = 0, y = -1$

In Exercises 3 and 4, identify the terms of the expression and their coefficients.

3. $4x^2 - 2x$

4. $5x + 3y - z$

In Exercises 5 and 6, rewrite the product in exponential form.

5. $(-3y)(-3y)(-3y)(-3y)$

6. $2 \cdot (x - 3) \cdot (x - 3) \cdot 2 \cdot 2$

In Exercises 7–10, identify the property of algebra illustrated by the statement.

7. $-3(2y) = (-3 \cdot 2)y$

8. $(x + 2)y = xy + 2y$

9. $3y \cdot \dfrac{1}{3y} = 1, y \neq 0$

10. $x - x^2 + 2 = -x^2 + x + 2$

In Exercises 11 and 12, use the Distributive Property to expand the expression.

11. $2x(3x - 1)$

12. $-6(2y + 3y^2 - 6)$

In Exercises 13–20, simplify the expression.

13. $-4(-5y^2)$

14. $\dfrac{x}{3}\left(-\dfrac{3x}{5}\right)$

15. $(-3y)^2 y^3$

16. $\dfrac{2z^2}{3y} \cdot \dfrac{5z}{7}$

17. $y^2 - 3xy + y + 7xy$

18. $(2x + 2)3 + x - 10$

19. $5(a - 2b) + 3(a + b)$

20. $4x + 3[2 - 4(x + 6)]$

Applications

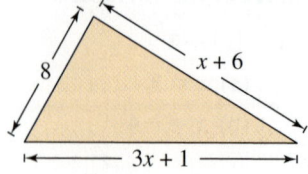

21. Write and simplify an expression for the perimeter of the triangle (see figure).

22. Your teacher divides your class of x students into 6 teams.

    (a) Write an expression representing the number of students on each team.

    (b) There are 30 students in your class. How many students are on each team?

Study Skills in Action

Absorbing Details Sequentially

Math is a sequential subject. Learning new math concepts successfully depends on how well you understand all the previous concepts. So, it is important to learn and remember concepts as they are encountered. One way to work through a section sequentially is by following these steps.

1 ▶ Work through an example. If you have trouble, consult your notes or seek help from a classmate or instructor.

2 ▶ Complete the exercises following the example.

3 ▶ If you get the exercises correct, move on to the next example. If not, make sure you understand your mistake(s) before you move on.

4 ▶ When you have finished working through all the examples in the section, take a short break of 5 to 10 minutes. This will give your brain time to process everything.

5 ▶ Start the exercises that follow the lesson.

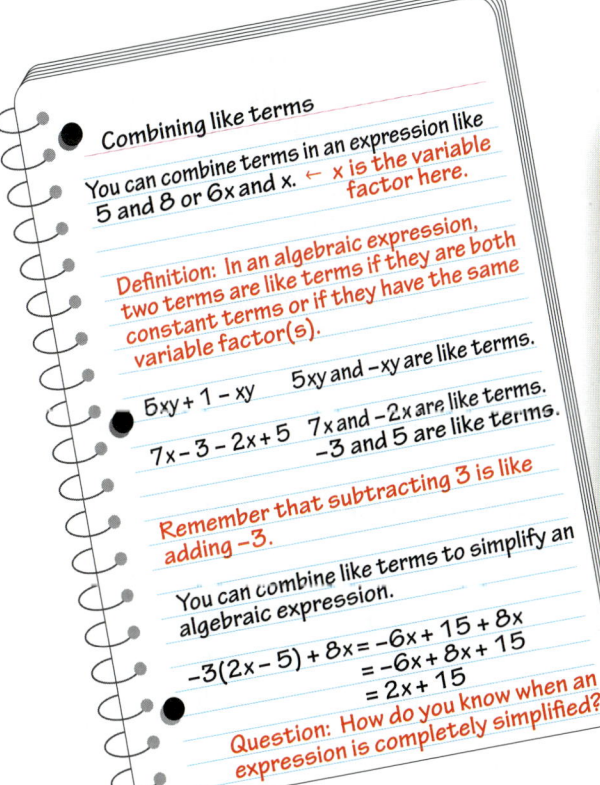

Smart Study Strategy

Rework Your Notes

It is almost impossible to write down in your notes all the detailed information you are taught in class. A good way to reinforce the concepts and put them into your long-term memory is to rework your notes. When you take notes, leave extra space on the pages. You can go back after class and fill in:

- important definitions and rules
- additional examples
- questions you have about the material

2.3 Algebra and Problem Solving

▶ Construct verbal mathematical models from written statements.
▶ Translate verbal phrases into algebraic expressions.
▶ Identify hidden operations when writing algebraic expressions.
▶ Use problem-solving strategies to solve application problems.

Constructing Verbal Models

Algebra is a problem-solving language that is used to solve real-life problems. It has four basic components, which tend to nest within each other.

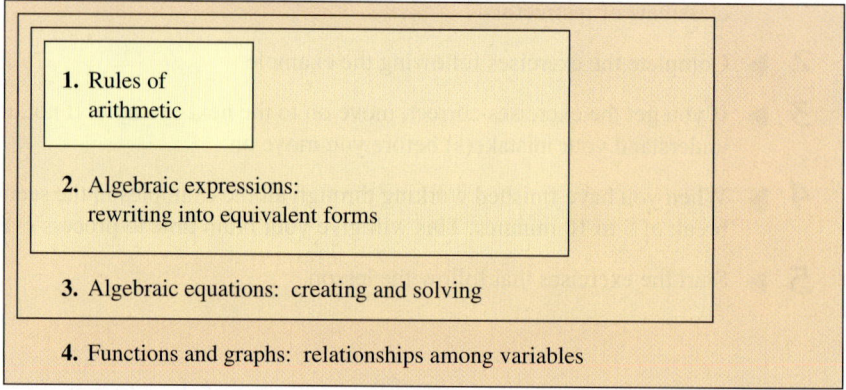

In the first two sections of this chapter, you studied techniques for rewriting and simplifying algebraic expressions. In this section, you will study ways to write algebraic expressions from written statements by first constructing a **verbal mathematical model**.

Take another look at Example 8 in Section 2.1 (page 61). In that example, you are paid $9 per hour and your weekly pay can be represented by the verbal model

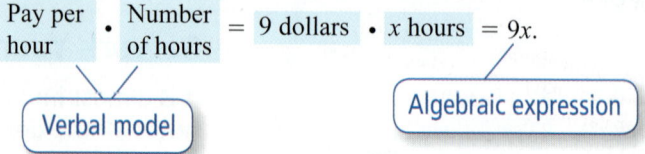

Exercises Within Reach ®   Solutions in English & Spanish and tutorial videos at AlgebraWithinReach.com

Constructing a Verbal Model In Exercises 1–6, **construct** a verbal model for the given situation.

1. You earn $10 per hour. How much do you earn for working x hours?

2. You bought x CDs online and y CDs from a store. How many total CDs did you buy?

3. You have 20 coupons. You use c coupons when you pay the bill. How many coupons do you have left?

4. You want to evenly divide 15 tickets between p people. How many tickets will each person receive?

5. A carton of eggs costs $2.89. How much will it cost to buy m cartons?

6. You have x dollars. How much money will you have after loaning $5 to a friend?

Section 2.3 Algebra and Problem Solving

Application — EXAMPLE 1 Writing an Algebraic Expression

You are paid 5 cents for each aluminum soda can and 3 cents for each plastic soda bottle you collect. Write an algebraic expression that represents the total weekly income (in cents) for this recycling activity.

Study Tip

In Example 1, notice that c is used to represent the number of cans and b is used to represent the number of bottles. When writing algebraic expressions, it is convenient to choose variables that can be identified with the unknown quantities.

SOLUTION

Before writing an algebraic expression for the weekly income, it is helpful to construct an informal verbal model. For instance, the following verbal model could be used.

$$\text{Pay per can} \cdot \text{Number of cans} + \text{Pay per bottle} \cdot \text{Number of bottles} \quad \text{Verbal model}$$

Note that the word *and* in the problem indicates addition. Because both the number of cans and the number of bottles can vary from week to week, you can use the two variables c and b, respectively, to write the following algebraic expression.

$$5 \text{ cents} \cdot c \text{ cans} + 3 \text{ cents} \cdot b \text{ bottles} = 5c + 3b \quad \text{Algebraic expression}$$

Exercises Within Reach®   Solutions in English & Spanish and tutorial videos at AlgebraWithinReach.com

7. **Money** A cash register contains d dimes. Write an algebraic expression that represents the total amount of money (in dollars).

8. **Fruit** A bag of apples costs $4.99. Write an algebraic expression that represents the total cost of b bags of apples.

9. **Money** A cash register contains d dimes and q quarters. Write an algebraic expression that represents the total amount of money (in dollars).

10. **Fruit** Apples cost $3.29 per pound and oranges cost $2.99 per pound. Write an algebraic expression that represents the total cost of buying a pounds of apples and g pounds of oranges.

Translating Phrases

Translating Phrases into Algebraic Expressions

Key Words and Phrases	Verbal Description	Expression
Addition: Sum, plus, greater than, increased by, more than, exceeds, total of	The sum of 6 and x Eight more than y	$6 + x$ $y + 8$
Subtraction: Difference, minus, less than, decreased by, subtracted from, reduced by	Five decreased by a Four less than z	$5 - a$ $z - 4$
Multiplication: Product, multiplied by, twice, times, percent of	Seven times x	$7x$
Division: Quotient, divided by, ratio, per	The ratio of x and 3	$\dfrac{x}{3}$

EXAMPLE 2 Translating Verbal Phrases into Algebraic Expressions

a. Three **less than** m

   $m - 3$ "Less than" indicates subtraction.

b. y **decreased by** 10

   $y - 10$ "Decreased by" indicates subtraction.

c. The **product** of 5 and x

   $5x$ "Product" indicates multiplication.

d. The **quotient** of n and 7

   $\dfrac{n}{7}$ "Quotient" indicates division.

Study Tip

Order is important when writing subtraction and division expressions. For instance, *three less than m* means $m - 3$, not $3 - m$, and the *quotient of n and 7* means $\dfrac{n}{7}$, not $\dfrac{7}{n}$.

Exercises Within Reach®

Solutions in English & Spanish and tutorial videos at AlgebraWithinReach.com

Translating a Verbal Phrase In Exercises 11−22, **translate** the verbal phrase into an algebraic expression.

11. x increased by 5
12. 17 more than y
13. b decreased by 25
14. k decreased by 7
15. Six less than g
16. Ten more than x
17. Twice h
18. The product of 30 and c
19. w divided by 3
20. d divided by 100
21. The ratio of x and 50
22. One-half of y

Section 2.3 Algebra and Problem Solving 79

EXAMPLE 3 **Translating Algebraic Expressions into Verbal Phrases**

Without using a variable, write a verbal description for each expression.

a. $x - 12$
b. $7(x + 12)$
c. $5 + \dfrac{x}{2}$
d. $\dfrac{5 + x}{2}$

SOLUTION

a. Algebraic expression: $x - 12$
   Operation: Subtraction
   Key phrase: Less than
   Verbal description: Twelve less than a number

b. Algebraic expression: $7(x + 12)$
   Operations: Multiplication, addition
   Key words: Times, sum
   Verbal description: Seven times the sum of a number and 12

c. Algebraic expression: $5 + \dfrac{x}{2}$
   Operations: Addition, division
   Key words: Plus, quotient
   Verbal description: Five plus the quotient of a number and 2

d. Algebraic expression: $\dfrac{5 + x}{2}$
   Operations: Addition, division
   Key words: Sum, divided by
   Verbal description: The sum of 5 and a number, all divided by 2

Exercises Within Reach ®

Solutions in English & Spanish and tutorial videos at AlgebraWithinReach.com

Translating an Algebraic Expression In Exercises 23–36, write a verbal description of the algebraic expression, without using a variable. (There is more than one correct answer.)

23. $x - 10$
24. $x + 9$
25. $3x + 2$
26. $4 - 7x$
27. $\frac{1}{2}x - 6$
28. $9 - \frac{1}{4}x$
29. $3(2 - x)$
30. $-10(t - 6)$
31. $\dfrac{t + 1}{2}$
32. $\dfrac{y - 3}{4}$
33. $\dfrac{1}{2} - \dfrac{t}{5}$
34. $\dfrac{1}{4} + \dfrac{y}{8}$
35. $x^2 + 5$
36. $x^3 - 1$

Verbal Models with Hidden Operations

Application

EXAMPLE 4 Discovering Hidden Operations

A cash register contains n nickels and d dimes. Write an algebraic expression for this amount of money in cents.

SOLUTION

The amount of money is a sum of products.

Verbal Model:	Value of nickel · Number of nickels + Value of dime · Number of dimes	
Labels:	Value of nickel = 5	(cents)
	Number of nickels = n	(nickels)
	Value of dime = 10	(cents)
	Number of dimes = d	(dimes)
Expression:	$5n + 10d$	(cents)

Study Tip

In Example 5, the final answer is listed in terms of miles. This unit is found as follows.

$$12 \, \frac{\text{miles}}{\text{hour}} \cdot t \, \text{hours}$$

Note that the hours "divide out," leaving miles as the unit of measure. This technique is called *unit analysis* and can be very helpful in determining the final unit of measure.

EXAMPLE 5 Discovering Hidden Operations

A person riding a bicycle travels at a constant rate of 12 miles per hour. Write an algebraic expression showing how far the person can ride in t hours.

SOLUTION

The distance traveled is a product.

Verbal Model:	Rate of travel · Time traveled	$(d = r \cdot t)$
Labels:	Rate of travel = 12	(miles per hour)
	Time traveled = t	(hours)
Expression:	$12t$	(miles)

Exercises Within Reach®

Solutions in English & Spanish and tutorial videos at AlgebraWithinReach.com

37. Sales Tax The sales tax on a purchase of L dollars is 6%. Write an algebraic expression that represents the total amount of sales tax. (*Hint:* Use the decimal form of 6%.)

38. Income Tax The state income tax on a gross income of I dollars in Pennsylvania is 3.07%. Write an algebraic expression that represents the total amount of income tax. (*Hint:* Use the decimal form of 3.07%.)

39. Rentals A movie rental costs $3 per day. A video game rental costs $4 per day. Write an algebraic expression that represents the total cost of renting m movies and v video games per day.

40. Frame The height of a rectangular picture frame is 1.5 times the width w. Write an algebraic expression that represents the perimeter of the picture frame.

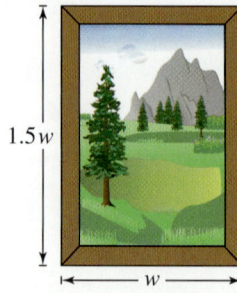

Application EXAMPLE 6 Discovering Hidden Operations

A person paid x dollars plus 6% sales tax for an automobile. Write an algebraic expression for the total cost of the automobile.

SOLUTION

The total cost is a sum.

Verbal Model: Cost of automobile $+$ Sales tax rate $\cdot$ Cost of automobile

Labels:
Sales tax rate $= 0.06$ (decimal form)
Cost of automobile $= x$ (dollars)

Expression: $x + 0.06x = (1 + 0.06)x$ (dollars)
$= 1.06x$

Application EXAMPLE 7 Discovering Hidden Operations

A truck travels 100 miles at an average speed of r miles per hour. Write an expression that represents the total travel time.

SOLUTION

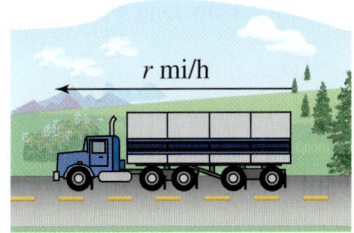

Verbal Model: $\dfrac{\text{Distance}}{\text{Rate}}$ $\left(t = \dfrac{d}{r}\right)$

Labels:
Distance $= 100$ (miles)
Rate $= r$ (miles per hour)

Expression: $\dfrac{100}{r}$ (hours)

Exercises Within Reach®

Solutions in English & Spanish and tutorial videos at AlgebraWithinReach.com

41. Hourly Wage An employee's hourly wage is $12.50 per hour plus $0.75 for each of the q units produced during the hour. Write an algebraic expression that represents the employee's total hourly earnings.

42. Camping A campground charges $15 for adults and $2 for children. Write an algebraic expression that represents the total camping fee for m adults and n children.

43. Mobile Device Applications for a cellular phone cost $0.99 each. Ringtones cost $1.99 each. Write an algebraic expression that represents the total cost of buying a applications and r ringtones.

44. Tickets You buy t tickets to a baseball game for a total of $45. Write an algebraic expression that represents the cost of each ticket.

Johann Helgason/Shutterstock.com; Vjom/Shutterstock.com

Additional Problem-Solving Strategies

> **Summary of Additional Problem-Solving Strategies**
>
> 1. **Guess, Check, and Revise** Guess a reasonable solution based on the given data. Check the guess, and revise it, if necessary. Continue guessing, checking, and revising until a correct solution is found.
>
> 2. **Make a Table/Look for a Pattern** Make a table using the data in the problem. Look for a number pattern. Then use the pattern to complete the table or find a solution.
>
> 3. **Draw a Diagram** Draw a diagram that shows the facts of the problem. Use the diagram to visualize the action of the problem. Use algebra to find a solution. Then check the solution against the facts.
>
> 4. **Solve a Simpler Problem** Construct a simpler problem that is similar to the original problem. Solve the simpler problem. Then use the same procedure to solve the original problem.

Application

EXAMPLE 8 Guess, Check, and Revise

You deposit $500 in an account that earns 6% interest compounded annually. The balance A in the account after t years is $A = 500(1 + 0.06)t$. How long will it take for your investment to double?

SOLUTION

You can solve this problem using a guess, check, and revise strategy. For instance, you might guess that it will take 10 years for your investment to double. The balance after 10 years is

$$A = 500(1 + 0.06)^{10} \approx \$895.42$$

Because the amount has not yet doubled, you increase your guess to 15 years.

$$A = 500(1 + 0.06)^{15} \approx \$1198.28$$

Because this amount is greater than double the investment, your next guess should be a number between 10 and 15. After trying several more numbers, you can determine that your balance will double in about 11.9 years.

	A Time, t (years)	B Balance, A (dollars)
1		
2	2	$561.80
3	4	$631.24
4	6	$709.26
5	8	$796.92
6	10	$895.42
7	12	$1,006.10

Another way to solve this problem is to use a spreadsheet to make a table.

Exercises Within Reach ®

Solutions in English & Spanish and tutorial videos at AlgebraWithinReach.com

Guess, Check, and Revise In Exercises 45–50, an expression for the balance in an account is given. Use a guess, check, and revise strategy to determine the time (in years) necessary for the investment of $1000 to double.

45. Interest rate: 7%
    $1000(1 + 0.07)^t$

46. Interest rate: 5%
    $1000(1 + 0.05)^t$

47. Interest rate: 6%
    $1000(1 + 0.06)^t$

48. Interest rate: 8%
    $1000(1 + 0.08)^t$

49. Interest rate: 6.5%
    $1000(1 + 0.065)^t$

50. Interest rate: 7.5%
    $1000(1 + 0.075)^t$

Section 2.3 Algebra and Problem Solving 83

Application   EXAMPLE 9   Draw a Diagram

The outer dimensions of a rectangular apartment are 25 feet by 40 feet. The combination living room, dining room, and kitchen areas occupy two-fifths of the apartment's area. Find the total area of the remaining rooms.

SOLUTION

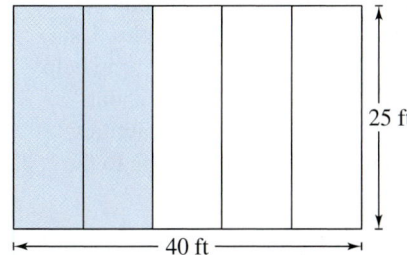

For this problem, it helps to draw a diagram, as shown at the left. From the figure, you can see that the total area of the apartment is

Area = (Length)(Width)
     = (40)(25)
     = 1000 square feet.

The area occupied by the living room, dining room, and kitchen is

$\frac{2}{5}(1000) = 400$ square feet.

This implies that the remaining rooms must have a total area of

$1000 - 400 = 600$ square feet.

Application   EXAMPLE 10   Solve a Simpler Problem

You are driving on an interstate highway at an average speed of 60 miles per hour. How far will you travel in $12\frac{1}{2}$ hours?

SOLUTION

One way to solve this problem is to use the formula that relates distance, rate, and time. Suppose, however, that you have forgotten the formula. To help you remember, you could solve some simpler problems.

- If you travel 60 miles per hour for 1 hour, you will travel 60 miles.
- If you travel 60 miles per hour for 2 hours, you will travel 120 miles.
- If you travel 60 miles per hour for 3 hours, you will travel 180 miles.

From the examples, it appears that you can find the total miles traveled by multiplying the rate by the time. So, if you travel 60 miles per hour for $12\frac{1}{2}$ hours, you will travel a distance of

$(60)(12.5) = 750$ miles.

Exercises Within Reach®

Solutions in English & Spanish and tutorial videos at AlgebraWithinReach.com

Drawing a Diagram In Exercises 51 and 52, draw figures satisfying the specified conditions.

51. The sides of a square have lengths of *a* centimeters. Draw the square. Draw the rectangle obtained by extending two parallel sides of the square 6 centimeters. Find expressions for the perimeter and area of each figure.

52. The dimensions of a rectangular lawn are 150 feet by 250 feet. The property owner buys a rectangular strip of land *x* feet wide along one 250-foot side of the lawn. Draw diagrams representing the lawn before and after the purchase. Write an expression for the area of each.

Solving a Simpler Problem In Exercises 53 and 54, solve the problem.

53. A bubble rises through water at a rate of about 1.15 feet per second. How far will the bubble rise in 5 seconds?

54. A train travels at an average speed of 50 miles per hour. How long will it take the train to travel 350 miles?

Concept Summary: Using Verbal Models to Write Algebraic Expressions

What
You can construct algebraic expressions from written statements by constructing **verbal mathematical models**.

EXAMPLE
A gallon of milk costs $3.29. Write an algebraic expression that represents the total cost of buying g gallons of milk.

How
Determine any known and unknown quantities from the written statement. Then construct a verbal model. Use the model to write an algebraic expression.

EXAMPLE

Cost per gallon	·	Number of gallons	=
$3.29	·	g gallons	=

$3.29 \cdot g = 3.29g$

Note that nowhere in the written statement does it say to multiply 3.29 by g. It is *implied* in the statement.

Why
Algebra is a problem-solving language that you can use to solve real-life problems. Some of these problems can consist of several known and unknown quantities. Using a verbal mathematical model can help you organize your thoughts about the known and unknown quantities, as well as the overall solution to the real-life problem.

Exercises Within Reach®
Worked-out solutions to odd-numbered exercises at AlgebraWithinReach.com

Concept Summary Check

55. Describing Strategies Describe other problem-solving strategies besides constructing verbal models that you can use to solve exercises in this chapter.

56. Assigning Variables Two unknown quantities in a verbal model are "Number of cherries" and "Number of strawberries." What variables would you use to represent these quantities. Explain.

57. Identifying Operations When constructing a verbal model from a written statement, what are some key words and phrases that indicate the four operations of arithmetic?

58. Writing What is a *hidden operation* in a verbal phrase? Explain how to identify hidden operations.

Extra Practice

Translating an Algebraic Expression In Exercises 59−62, write a verbal description of the algebraic expression, without using a variable. (There is more than one correct answer.)

59. $t(t + 16)$

60. $\frac{4}{5}(w + 10)$

61. $\frac{4}{x - 2}$

62. $x^2 - (x + 17)$

Translating a Verbal Phrase In Exercises 63−68, translate the verbal phrase into an algebraic expression. Simplify the expression.

63. x times the sum of x and 3

64. n times the difference of 6 and n

65. x minus the sum of 25 and x

66. The sum of 4 and x added to the sum of x and -8

67. The square of x decreased by the product of x and $2x$

68. The square of x added to the product of x and $x + 1$

69. **Supplies** Pens cost $0.25 each. Pencils cost $0.10 each. Write an algebraic expression that represents the total cost of buying p pens and n pencils.

70. **Gasoline** A consumer buys g gallons of gasoline for a total of d dollars. Write an algebraic expression that represents the price per gallon.

71. **Finding a Pattern** Describe the pattern and use your description to find the value of the expression when $n = 20$.

n	0	1	2	3	4	5
Value of expression	−1	1	3	5	7	9

72. **Finding a Pattern** Find values for a and b such that the expression $an + b$ yields the values in the table.

n	0	1	2	3	4	5
an + b	4	9	14	19	24	29

Finding Additional Information In Exercises 73 and 74, describe **what additional information is needed to solve the problem. (Do not solve the problem.)**

73. A family taking a Sunday drive through the country travels at an average speed of 45 miles per hour. How far have they traveled by 3:00 P.M.?

74. You paint a rectangular room that is twice as long as it is wide. One gallon of paint covers 100 square feet. How much money do you spend on paint?

Explaining Concepts

75. **Reasoning** Determine which verbal phrase(s) is (are) equivalent to the expression $n + 4$.

    (a) 4 more than n
    (b) the sum of n and 4
    (c) n less than 4
    (d) the ratio of n to 4
    (e) the total of 4 and n

76. **Number Sense** Determine whether order is important when translating each verbal phrase into an algebraic expression. Explain.

    (a) x increased by 10
    (b) 10 decreased by x
    (c) The product of x and 10
    (d) The quotient of x and 10

77. **Writing** Give two interpretations of "the quotient of 5 and a number times 3." Explain why $\dfrac{3n}{5}$ is not a possible interpretation.

78. **Writing** Give two interpretations of "the difference of 6 and a number divided by 3." Explain why $\dfrac{n-6}{3}$ is not a possible interpretation.

Cumulative Review

In Exercises 79–84, evaluate the expression.

79. $(-6)(-13)$
80. $|4(-6)(5)|$
81. $\left(-\dfrac{4}{3}\right)\left(-\dfrac{9}{16}\right)$
82. $\dfrac{7}{8} \div \dfrac{3}{16}$
83. $\left|-\dfrac{5}{9}\right| + 2$
84. $-7\dfrac{3}{5} - 3\dfrac{1}{2}$

In Exercises 85–88, identify the property of algebra illustrated by the statement.

85. $2a + b = b + 2a$
86. $-4x(1) = -4x$
87. $2(c - d) = 2c - 2d$
88. $-3y^3 + 3y^3 = 0$

2.4 Introduction to Equations

- Check whether a given value is a solution of an equation.
- Use properties of equality to form equivalent equations.
- Use a verbal model to write an algebraic equation.

Checking Solutions of Equations

An **equation** is a statement that two algebraic expressions are equal. For instance,

$$x = 3, \quad 5x - 2 = 8, \quad \frac{x}{4} = 7, \quad \text{and} \quad x^2 - 9 = 0$$

are equations. To **solve** an equation involving the variable x means to find all values of x for which the equation is true. Such values are called **solutions**. For instance, $x = 2$ is a solution of the equation $5x - 2 = 8$ because $5(2) - 2 = 8$ is a true statement. The solutions of an equation are said to **satisfy** the equation.

EXAMPLE 1   Checking Solutions of an Equation

Determine whether (a) $x = -2$ and (b) $x = 2$ are solutions of $x^2 - 5 = 4x + 7$.

SOLUTION

a.
$$x^2 - 5 = 4x + 7 \quad \text{Write original equation.}$$
$$(-2)^2 - 5 \stackrel{?}{=} 4(-2) + 7 \quad \text{Substitute } -2 \text{ for } x.$$
$$4 - 5 \stackrel{?}{=} -8 + 7 \quad \text{Simplify.}$$
$$-1 = -1 \quad \text{Solution checks. } \checkmark$$

b.
$$x^2 - 5 = 4x + 7 \quad \text{Write original equation.}$$
$$(2)^2 - 5 \stackrel{?}{=} 4(2) + 7 \quad \text{Substitute 2 for } x.$$
$$4 - 5 \stackrel{?}{=} 8 + 7 \quad \text{Simplify.}$$
$$-1 \neq 15 \quad \text{Solution does not check. } \times$$

So, $x = -2$ is a solution, but $x = 2$ is not a solution.

Exercises Within Reach®

Solutions in English & Spanish and tutorial videos at AlgebraWithinReach.com

Checking Solutions of an Equation In Exercises 1–10, **determine** whether each value of x is a solution of the equation.

Equation	Values		Equation	Values	
1. $2x - 18 = 0$	(a) $x = 0$	(b) $x = 9$	2. $3x - 3 = 0$	(a) $x = 4$	(b) $x = 1$
3. $6x + 1 = -11$	(a) $x = 2$	(b) $x = -2$	4. $2x + 5 = -15$	(b) $x = -10$	(b) $x = 5$
5. $x + 5 = 2x$	(a) $x = -1$	(b) $x = 5$	6. $15 - 2x = 3x$	(a) $x = 3$	(b) $x = 5$
7. $7x + 1 = 4(x - 2)$	(a) $x = 1$	(b) $x = 12$	8. $5x - 1 = 3(x + 5)$	(a) $x = 8$	(b) $x = -2$
9. $2x + 10 = 7(x + 1)$	(a) $x = \frac{3}{5}$	(b) $x = -\frac{2}{3}$	10. $3(3x + 2) = 9 - x$	(a) $x = -\frac{3}{4}$	(b) $x = \frac{3}{10}$

EXAMPLE 2 Comparing Equations and Expressions

Make a table that compares algebraic expressions and algebraic equations.

SOLUTION

Algebraic Expression	Algebraic Equation
• Example: $4(x - 1)$	• Example: $4(x - 1) = 12$
• Contains no equal sign	• Contains an equal sign and is true for only certain values of the variable
• Can be evaluated for any real number for which the expression is defined	• Solution is found by forming equivalent equations using the properties of equality:
• Can sometimes be simplified to an equivalent form: $4(x - 1)$ simplifies to $4x - 4$	$4(x - 1) = 12$ $4x - 4 = 12$ $4x = 16$ $x = 4$

Exercises Within Reach®

Solutions in English & Spanish and tutorial videos at AlgebraWithinReach.com

Comparing Equations and Expressions In Exercises 11–16, determine whether the statement describes an algebraic expression or an algebraic equation.

11. Contains an equal sign
12. Contains no equal sign
13. Can be evaluated for any real number for which it is defined
14. Is true for only certain values of the variable
15. The solution is found by using the properties of equality
16. A statement of equality

Comparing Equations and Expressions In Exercises 17–26, determine whether an algebraic expression or an algebraic equation is given.

17. $3x$
18. $\frac{1}{2}x$
19. $-7x - 8 = 0$
20. $7 = 9 - x$
21. $\frac{5}{6}x + 1$
22. $x - 4$
23. $2x + 1 = 4(x - 10)$
24. $\frac{1}{9}(4 + 8x) = -15$
25. $x^2 + 2x + 1$
26. $x^2 - 4$

Forming Equivalent Equations

Forming Equivalent Equations: Properties of Equality

An equation can be transformed into an **equivalent equation** using one or more of the following procedures.

	Original Equation	Equivalent Equation(s)
1. *Simplify either side:* Remove symbols of grouping, combine like terms, or simplify fractions on one or both sides of the equation.	$3x - x = 8$	$2x = 8$
2. *Apply the Addition Property of Equality:* Add (or subtract) the same quantity to (from) *each* side of the equation.	$x - 2 = 5$	$x - 2 + 2 = 5 + 2$ $x = 7$
3. *Apply the Multiplication Property of Equality:* Multiply (or divide) each side of the equation by the same *nonzero* quantity.	$3x = 9$	$\dfrac{3x}{3} = \dfrac{9}{3}$ $x = 3$
4. *Interchange the two sides of the equation.*	$7 = x$	$x = 7$

EXAMPLE 3 Forming Equivalent Equations

The second and third procedures in the above list can be used to eliminate terms or factors of an equation. For example, to solve the equation $x - 5 = 1$, you need to eliminate the term -5 on the left side. This is accomplished by adding its opposite, 5, to each side.

$x - 5 = 1$ Write original equation.

$x - 5 + 5 = 1 + 5$ Add 5 to each side.

$x + 0 = 6$ Combine like terms.

$x = 6$ Solution

These four equations are equivalent, and they are called the *steps* of the solution.

Exercises Within Reach®

Solutions in English & Spanish and tutorial videos at AlgebraWithinReach.com

Solving an Equation In Exercises 27–32, **solve** the equation.

27. $x - 8 = 5$

28. $4 = x - 11$

29. $x + 3 = 19$

30. $10 + x = 3$

31. $3x = 30$

32. $\dfrac{3}{2}x = 9$

EXAMPLE 4 Identifying Properties of Equality

Identify the property of equality used to solve each equation.

a. $x - 5 = 0$ Original equation
    $x - 5 + 5 = 0 + 5$ Add 5 to each side.
    $x = 5$ Solution

b. $\frac{x}{5} = -2$ Original equation
    $\frac{x}{5}(5) = -2(5)$ Multiply each side by 5.
    $x = -10$ Solution

c. $4x = 9$ Original equation
    $\frac{4x}{4} = \frac{9}{4}$ Divide each side by 4.
    $x = \frac{9}{4}$ Solution

SOLUTION

a. The Addition Property of Equality is used to add 5 to each side of the equation in the second step. Adding 5 eliminates the term -5 from the left side of the equation.

b. The Multiplication Property of Equality is used to multiply each side of the equation by 5 in the second step. Multiplying by 5 eliminates the denominator from the left side of the equation.

c. The Multiplication Property of Equality is used to divide each side of the equation by 4 (or multiply each side by 1/4) in the second step. Dividing by 4 eliminates the coefficient from the left side of the equation.

Study Tip

In Example 4(c), each side of the equation is divided by 4 to eliminate the coefficient 4 on the left side. You could just as easily *multiply* each side by 1/4. Both techniques are legitimate—which one you decide to use is a matter of personal preference.

Exercises Within Reach®

Solutions in English & Spanish and tutorial videos at AlgebraWithinReach.com

Identifying Properties of Equality In Exercises 33–38, **justify** each step of the equation. Then **identify** any properties of equality used to solve the equation.

33. $x - 8 = 3$
    $x - 8 + 8 = 3 + 8$
    $x = 11$

34. $x + 4 = 16$
    $x + 4 - 4 = 16 - 4$
    $x = 12$

35. $\frac{2}{3}x = 12$
    $\frac{3}{2}\left(\frac{2}{3}x\right) = \frac{3}{2}(12)$
    $x = 18$

36. $\frac{4}{5}x = -28$
    $\frac{5}{4}\left(\frac{4}{5}x\right) = \frac{5}{4}(-28)$
    $x = -35$

37. $5x + 12 = 22$
    $5x + 12 - 12 = 22 - 12$
    $5x = 10$
    $\frac{5x}{5} = \frac{10}{5}$
    $x = 2$

38. $14 - 3x = 5$
    $14 - 3x - 14 = 5 - 14$
    $14 - 14 - 3x = 5 - 14$
    $-3x = -9$
    $\frac{-3x}{-3} = \frac{-9}{-3}$
    $x = 3$

Writing Equations

It is helpful to use two phases in writing equations that model real-life situations, as shown below.

Value description → Verbal model → Assign labels → Algebraic equation

Phase 1 (Value description → Verbal model) Phase 2 (Assign labels → Algebraic equation)

In the first phase, you translate the verbal description into a verbal model. In the second phase, you assign labels and translate the verbal model into a mathematical model or an algebraic equation.

Study Tip

When you write an equation, be sure to check that both sides of the equation represent the *same* unit of measure. For instance, in Example 5, both sides of the equation $40{,}950 = 52x$ represent dollar amounts.

Application

EXAMPLE 5 Using a Verbal Model to Write an Equation

Write an algebraic equation for the following problem.

The total income that an employee received in a year was $40,950. How much was the employee paid each week? Assume that each weekly paycheck contained the same amount, and that the year consisted of 52 weeks.

SOLUTION

Verbal Model: Income for year = Number of weeks in a year · Weekly pay

Labels:
Income for year $= 40{,}950$ (dollars)
Weekly pay $= x$ (dollars per week)
Number of weeks $= 52$ (weeks)

Expression: $40{,}950 = 52x$

Exercises Within Reach®

Solutions in English & Spanish and tutorial videos at AlgebraWithinReach.com

Writing an Algebraic Equation **In Exercises 39–42, write an algebraic equation. Do *not* solve the equation.**

39. After your instructor added 6 points to each student's test score, your score is 94. What was your original score?

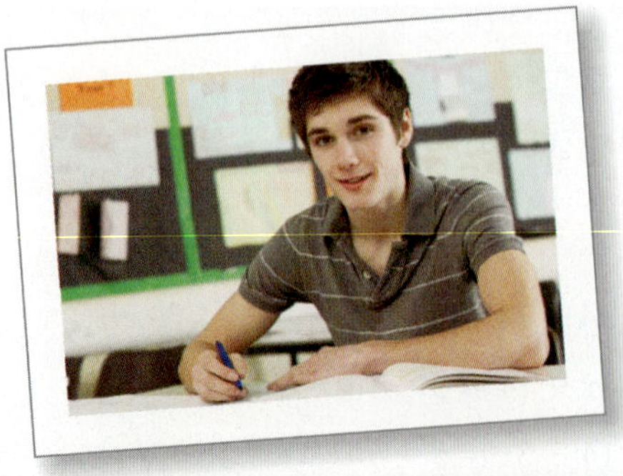

40. With the 1.2-inch rainfall today, the total for the month is 4.5 inches. How much had been recorded for the month before today's rainfall?

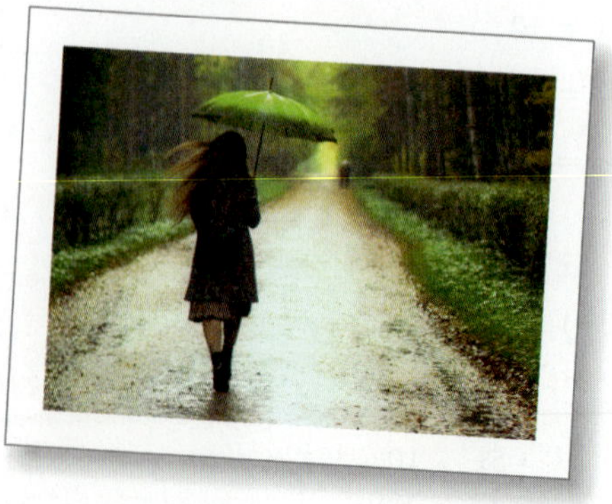

41. During a football game, a running back carried the ball 18 times and his average number of yards per carry was 4.5. How many yards did the running back gain for the game?

42. The total cost of admission for 6 adults at an aquarium is $132. What is the cost per adult?

Monkey Business Images/Shutterstock.com; Petr Malyshev/Shutterstock.com

Application

EXAMPLE 6 Using a Verbal Model to Write an Equation

Write an algebraic equation for the following problem.

Returning to college after spring break, you travel 3 hours and stop for lunch. You know that it takes 45 minutes to complete the last 36 miles of the 180-mile trip. What was the average speed during the first 3 hours of the trip?

SOLUTION

Verbal Model: Distance = Rate · Time

Labels:
Distance = 180 − 36 = 144 (miles)
Rate = r (miles per hour)
Time = 3 (hours)

Expression: $144 = 3r$

Study Tip

In Example 6, the information that it takes 45 minutes to complete the last part of the trip is unnecessary information. This type of unnecessary information in an applied problem is sometimes called a *red herring*.

Application

EXAMPLE 7 Using a Verbal Model to Write an Equation

Write an algebraic equation for the following problem.

Tickets for a concert cost $175 for each floor seat and $95 for each stadium seat. There were 2500 seats on the main floor, and these were sold out. The total revenue from ticket sales was $865,000. How many stadium seats were sold?

SOLUTION

Verbal Model: Total revenue = Revenue from floor seats + Revenue from stadium seats

Labels:
Total revenue = 865,000 (dollars)
Price per floor seat = 175 (dollars per seat)
Number of floor seats = 2500 (seats)
Price per stadium seat = 95 (dollars per seat)
Number of stadium seats = x (seats)

Expression: $865{,}000 = 175(2500) + 95x$

Exercises Within Reach®

Solutions in English & Spanish and tutorial videos at AlgebraWithinReach.com

Writing an Algebraic Equation In Exercises 43 and 44, **write** an algebraic equation. **Do *not* solve the equation.**

43. You want to volunteer at a soup kitchen for 150 hours over a 15-week period. After 8 weeks, you have volunteered for 72 hours. How many hours will you have to work per week over the remaining 7 weeks to reach your goal?

44. A textile corporation buys equipment with an initial purchase price of $750,000. It is estimated that its useful life will be 3 years and at that time its value will be $75,000. The total depreciation is divided equally among the three years. (Depreciation is the difference between the initial price of an item and its current value.) What is the total amount of depreciation declared each year?

Concept Summary: Writing and Solving Equations

What
You can use verbal models to write **equations** and algebra to **solve** equations.

EXAMPLE

The sale price of a coat is $250. The discount is $25. What is the original price?

How
To write an equation that models a real-life situation, first translate the verbal description into a verbal model. Then assign labels and translate the verbal model into an algebraic equation. You can solve the equation by creating **equivalent equations**.

EXAMPLE

Verbal Model:

$$\boxed{\text{Original price}} - \boxed{\text{Discount}} = \boxed{\text{Sale price}}$$

Labels: Original price = x
Discount = 25
Sale price = 250

Equation: $x - 25 = 250$

Why
You can use verbal models to write algebraic equations that model many real-life situations. Knowing how to use properties of equality will help you solve these equations.

Exercises Within Reach®
Worked-out solutions to odd-numbered exercises at AlgebraWithinReach.com

Concept Summary Check

45. Vocabulary Is there more than one way to write a verbal model? Explain.

46. Writing Describe the steps that can be used to transform an equation into an equivalent equation.

47. Number Sense When dividing each side of an equation by the same quantity, why must the quantity be nonzero?

48. Writing Describe how to solve $x - 25 = 250$.

Extra Practice

Identifying Properties of Equality In Exercises 49 and 50, *justify* each step of the equation. Then *identify* any properties of equality used to solve the equation.

49.
$$\frac{x}{3} = x + 1$$
$$3\left(\frac{x}{3}\right) = 3(x+1)$$
$$x = 3x + 3$$
$$x - 3x = 3x + 3 - 3x$$
$$x - 3x = 3x - 3x + 3$$
$$-2x = 3$$
$$\frac{-2x}{-2} = \frac{3}{-2}$$
$$x = -\frac{3}{2}$$

50.
$$\frac{4}{5}x = 4x - 16$$
$$\frac{5}{4}\left(\frac{4}{5}x\right) = \frac{5}{4}(4x - 16)$$
$$x = 5x - 20$$
$$x - 5x = 5x - 20 - 5x$$
$$x - 5x = 5x - 5x - 20$$
$$-4x = -20$$
$$\frac{-4x}{-4} = \frac{-20}{-4}$$
$$x = 5$$

Checking Solutions of an Equation In Exercises 51–54, determine whether each value of x is a solution of the equation.

	Equation	Values		Equation	Values
51.	$\frac{2}{x} - \frac{1}{x} = 1$	(a) $x = 0$ (b) $x = \frac{1}{3}$	52.	$\frac{4}{x} + \frac{2}{x} = 1$	(a) $x = 0$ (b) $x = 6$
53.	$\frac{5}{x-1} + \frac{1}{x} = 5$	(a) $x = 3$ (b) $x = \frac{1}{6}$	54.	$\frac{3}{x-2} = x$	(a) $x = -1$ (b) $x = 3$

Writing an Algebraic Equation In Exercises 55–58, write an algebraic equation. Do *not* solve the equation.

55. A student has n quarters and seven $1 bills totaling $8.75. How many quarters does the student have?

56. A school science club conducts a car wash to raise money. The club spends $12 on supplies and charges $5 per car. After the car wash, the club has a profit of $113. How many cars did the members of the science club wash?

57. A high school earned $986 in revenue for a play. Tickets for the play cost $10 for adults and $6 for students. The number of students attending the play was $\frac{3}{4}$ the number of adults attending the play. How many adults and student attended the play?

58. An ice show earns a revenue of $11,041 one night. Tickets for the ice show cost $18 for adults and $13 for children. The number of adults attending the ice show was 33 more than the number of children attending the show. How many adults and children attended the show?

Explaining Concepts

59. *Number Sense* Are there any equations of the form $ax = b$ ($a \neq 0$) that are true for more than one value of x? Explain.

60. *Structure* Determine which equations are equivalent to $14 = x + 8$.
(a) $x + 8 = 14$
(b) $8x = 14$
(c) $x - 8 = 14$
(d) $8 + x = 14$
(e) $2(x + 4) - x = 14$
(f) $3(x + 6) - 2x + 5 = 14$

61. *Modeling* Describe a real-life problem that uses the following verbal model.

$$\boxed{\text{Revenue of \$840}} = \boxed{\text{\$35 per case}} + \boxed{\text{Number of cases}}$$

62. *Writing* Explain the difference between simplifying an expression and solving an equation. Give an example of each.

Cumulative Review

In Exercises 63–68, simplify the expression.

63. $t^2 \cdot t^5$

64. $(-3y^3)y^2$

65. $6x + 9x$

66. $4 - 3t + t$

67. $-(-8b)$

68. $7(-10x)$

In Exercises 69–72, translate the phrase into an algebraic expression. Let x represent the real number.

69. 23 more than x

70. c divided by 6

71. Seven more than 4 times y

72. Nine times the difference of h and 3

2 Chapter Summary

	What did you learn?	Explanation and Examples	Review Exercises
2.1	Define and identify terms, variables, and coefficients of algebraic expressions *(p. 56)*.	In the expression $4x + 3y$, $4x$ and $3y$ are the terms of the expression, x and y are the variables, and 4 and 3 are the coefficients.	1–10
	Define exponential form and interpret exponential expressions *(p. 57)*.	Repeated multiplication can be expressed in exponential form using a base a and an exponent n, where a is a real number, variable, or algebraic expression, and n is a positive integer. $$a^n = \underbrace{a \cdot a \cdot a \cdot a \cdots a}_{n \text{ factors}}$$	11–16
	Evaluate algebraic expressions using real numbers *(p. 58)*.	To evaluate an algebraic expression, replace every occurrence of the specified variable in the expression with the appropriate real number, and perform the operation(s).	17–24
2.2	Use the properties of algebra *(p. 64)*.	Let a, b, and c represent real numbers, variables, or algebraic expressions. Commutative Property of Addition $\quad a + b = b + a$ Commutative Property of Multiplication $\quad ab = ba$ Associative Property of Addition $\quad (a + b) + c = a + (b + c)$ Associative Property of Multiplication $\quad (ab)c = a(bc)$ Distributive Property $\quad a(b + c) = ab + ac \qquad a(b - c) = ab - ac$ $\quad (a + b)c = ac + bc \qquad (a - b)c = ac - bc$ Additive Identity Property $\quad a + 0 = 0 + a = a$ Multiplicative Identity Property $\quad a \cdot 1 = 1 \cdot a = a$ Additive Inverse Property $\quad a + (-a) = 0$ Multiplicative Inverse Property $\quad a \cdot \dfrac{1}{a} = 1, a \neq 0$	25–30
	Combine like terms of an algebraic expression *(p. 67)*.	To combine like terms in an algebraic expression, add their respective coefficients and attach the common variable factor(s).	31–42
	Simplify an algebraic expression by rewriting the terms *(p. 68)*.	To simplify an algebraic expression, remove symbols of grouping and combine like terms.	43–54
	Use the Distributive Property to remove symbols of grouping *(p. 69)*.	The main tool for removing symbols of grouping is the Distributive Property.	55–66
2.3	Construct verbal mathematical models from written statements *(p. 76)*.	A person riding a scooter travels at a constant rate of 32 miles per hour. Write an expression showing how far the person can ride in t hours. Verbal Model: $\boxed{\text{Rate of travel}} \cdot \boxed{\text{Time traveled}}$ Expression: $32t$	67, 68

Chapter Summary 95

	What did you learn?	Explanation and Examples	Review Exercises
2.3	Translate verbal phases into algebraic expressions *(p. 78)*.	When translating verbal phrases into algebraic expressions, look for key words and phrases that indicate the four different operations of arithmetic. **Addition:** Sum, plus, greater than, increased by, more than, exceeds, total of **Subtraction:** Difference, minus, less than, subtracted from, reduced by **Multiplication:** Product, multiplied by, twice, times, percent of **Division:** Quotient, divided by, ratio, per	69–78
	Identify hidden operations when writing algebraic expressions *(p. 80)*.	Most real-life problems do not contain verbal expressions that clearly identify all the arithmetic operations involved. You need to rely on past experience and the physical nature of the problem in order to identify the operations hidden in the problem statement. Multiplication is the operation most commonly hidden in real-life applications.	79–82
	Use problem-solving strategies to solve application problems *(p. 82)*.	1. **Guess, Check, and Revise** Guess a reasonable solution based on the given data. Check the guess, and revise it, if necessary. Continue guessing, checking, and revising until a correct solution is found. 2. **Make a Table/Look for a Pattern** Make a table using the data in the problem. Look for a number pattern. Then use the pattern to complete the table or find a solution. 3. **Draw a Diagram** Draw a diagram that shows the facts of the problem. Use the diagram to visualize the action of the problem. Use algebra to find a solution. Then check the solution against the facts. 4. **Solve a Simpler Problem** Construct a simpler problem that is similar to the original problem. Solve the simpler problem. Then use the same procedure to solve the original problem.	83, 84
2.4	Check whether a given value is a solution of an equation *(p. 86)*.	To check whether a given value is a solution of an equation, substitute the value into the original equation. If the substitution results in a true statement, then the value is a solution of the equation. If the substitution results in a false statement, then the value is not a solution of the equation.	85–94
	Use properties of equality to form equivalent equations *(p. 88)*.	An equation can be transformed into an equivalent equation using one or more of the following procedures. 1. *Simplify either side:* Remove symbols of grouping, combine like terms, or simplify fractions on one or both sides of the equation. 2. *Apply the Addition Property of Equality:* Add (or subtract) the same quantity to (from) *each* side of the equation. 3. *Apply the Multiplication Property of Equality:* Multiply (or divide) each side of the equation by the same *nonzero* quantity. 4. *Interchange the two sides of the equation.*	95–98
	Use a verbal model to write an algebraic equation *(p. 90)*.	First, construct a verbal model. Then, assign labels to the known and unknown quantities and translate the verbal model into an algebraic equation.	99–102

Review Exercises

Worked-out solutions to odd-numbered exercises at AlgebraWithinReach.com

2.1

Identifying Variables In Exercises 1−4, identify the variable(s) in the expression.

1. $15 - x$
2. $t - 5^2$
3. $a - 3b$
4. $y + z$

Identifying Terms and Coefficients In Exercises 5−10, identify the terms of the expression and their coefficients.

5. $12y + y^2$
6. $4x - \frac{1}{2}x^3$
7. $5x^2 - 3xy + 10y^2$
8. $y^2 - 10yz + \frac{2}{3}z^2$
9. $\frac{2y}{3} - \frac{4x}{y}$
10. $-\frac{4b}{9} + \frac{11a}{b}$

Rewriting in Exponential Form In Exercises 11−16, rewrite the product in exponential form.

11. $5z \cdot 5z \cdot 5z$
12. $\frac{3}{8}y \cdot \frac{3}{8}y \cdot \frac{3}{8}y \cdot \frac{3}{8}y$
13. $(-3x) \cdot (-3x) \cdot (-3x) \cdot (-3x) \cdot (-3x)$
14. $\left(-\frac{2}{7}\right) \cdot \left(-\frac{2}{7}\right) \cdot \left(-\frac{2}{7}\right)$
15. $(b - c) \cdot (b - c) \cdot 6 \cdot 6$
16. $2 \cdot (a + b) \cdot 2 \cdot (a + b) \cdot 2$

Evaluating an Algebraic Expression In Exercises 17−24, evaluate the algebraic expression at the given values of the variable(s).

	Expression	Values
17.	$x^2 - 2x + 5$	(a) $x = 0$ (b) $x = 2$
18.	$x^3 - 8$	(a) $x = 2$ (b) $x = 4$
19.	$x^2 - x(y + 1)$	(a) $x = 2, y = -1$ (b) $x = 1, y = 2$
20.	$2r + r(t^2 - 3)$	(a) $r = 3, t = -2$ (b) $r = -2, t = 3$
21.	$\frac{x + 5}{y}$	(a) $x = -5, y = 3$ (b) $x = 2, y = -1$
22.	$\frac{a - 9}{2b}$	(a) $a = 7, b = -3$ (b) $a = -4, b = 5$
23.	$x^2 - 2y + z$	(a) $x = 1, y = 2, z = 0$ (b) $x = 2, y = -3, z = -4$
24.	$\frac{m + 2n}{-p}$	(a) $m = -1, n = 2, p = 3$ (b) $m = 4, n = -2, p = -7$

2.2

Identifying Properties In Exercises 25−30, identify the property of algebra illustrated by the statement.

25. $xy \cdot \frac{1}{xy} = 1$
26. $u(vw) = (uv)w$
27. $(x - y)(2) = 2(x - y)$
28. $(a + b) + 0 = a + b$
29. $2x + (3y - z) = (2x + 3y) - z$
30. $x(y + z) = xy + xz$

Combining Like Terms In Exercises 31−40, simplify the expression by combining like terms.

31. $3a - 5a$
32. $6c - 2c$
33. $3p - 4q + q + 8p$
34. $10x - 4y - 25x + 6y$
35. $\frac{1}{4}s - 6t + \frac{7}{2}s + t$

36. $\frac{2}{3}a + \frac{3}{5}a - \frac{1}{2}b + \frac{2}{3}b$

37. $x^2 + 3xy - xy + 4$

38. $uv^2 + 10 - 2uv^2 + 2$

39. $5x - 5y + 3xy - 2x + 2y$

40. $y^3 + 2y^2 + 2y^3 - 3y^2 + 1$

41. **Number Sense** Simplify the algebraic expression that represents the sum of three consecutive odd integers, $2n - 1$, $2n + 1$, and $2n + 3$.

42. **Number Sense** Simplify the algebraic expression that represents the sum of three consecutive even integers, $2n$, $2n + 2$, and $2n + 4$.

Simplifying an Algebraic Expression In Exercises 43–50, simplify the expression.

43. $12(4t)$

44. $8(7x)$

45. $-5(-9x^2)$

46. $-10(-3b^3)$

47. $(-6x)(2x^2)$

48. $(-3y^2)(15y)$

49. $\frac{12x}{5} \cdot \frac{10}{3}$

50. $\frac{4z}{15} \cdot \frac{9}{2}$

51. **Geometry** Write and simplify an expression for (a) the perimeter and (b) the area of the rectangle.

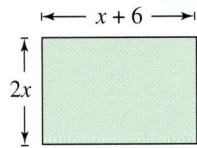

52. **Geometry** Write and simplify an expression for the area of the triangle.

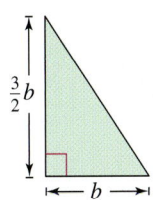

53. **Geometry** The face of a DVD player has the dimensions shown in the figure. Write an algebraic expression that represents the area of the face of the DVD player excluding the compartment holding the disc.

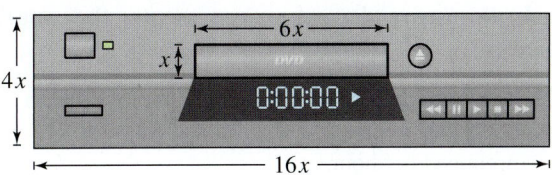

54. **Geometry** Write an expression for the perimeter of the figure. Use the rules of algebra to simplify the expression.

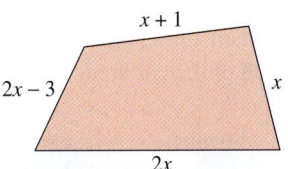

Removing Symbols of Grouping In Exercises 55–66, simplify the expression by removing symbols of grouping and combining like terms.

55. $5(u - 4) + 10$

56. $16 - 3(v + 2)$

57. $3s - (r - 2s)$

58. $50x - (30x + 100)$

59. $-3(1 - 10z) + 2(1 - 10z)$

60. $8(15 - 3y) - 5(15 - 3y)$

61. $\frac{1}{3}(42 - 18z) - 2(8 - 4z)$

62. $\frac{1}{4}(100 + 36s) - (15 - 4s)$

63. $10 - [8(5 - x) + 2]$

64. $3[2(4x - 5) + 4] - 3$

65. $2[x + 2(y - x)]$

66. $2t[4 - (3 - t)] + 5t$

2.3

Writing an Algebraic Expression In Exercises 67 and 68, **construct** a verbal model and then **write** an algebraic expression that represents the specified quantity.

67. The total hourly wage for an employee who earns $8.25 per hour and an additional $0.60 for each unit produced per hour.

68. The total cost for a family to stay one night at a campground when the charge is $18 for the parents plus $3 for each of the children.

Translating a Verbal Phrase In Exercises 69–78, **translate** the verbal phrase into an algebraic expression.

69. The sum of two-thirds of x and 5

70. One hundred decreased by the product of 5 and b

71. Ten less than twice y

72. The ratio of c and 10

73. Fifty increased by the product of 7 and z

74. Ten decreased by the quotient of a and 2

75. The sum of s and 10, all divided by 8

76. The product of 15 and d, all decreased by 2

77. The sum of the square of g and 64

78. The absolute value of the sum of t and -10

79. **Commission** A salesperson earns 5% commission on his total weekly sales x. Write an algebraic expression that represents the amount of commission that the salesperson earns in a week.

80. **Sale Price** A cordless phone is advertised for 20% off the list price of L dollars. Write an algebraic expression that represents the sale price of the phone.

81. **Rent** The monthly rent for your apartment is $625. Write an algebraic expression that represents the total rent for n months.

82. **Distance** A car travels for 10 hours at an average speed of s miles per hour. Write an algebraic expression that represents the total distance traveled by the car.

83. **Finding a Pattern** Describe the pattern, and use your description to find the value of the expression when $n = 20$.

n	0	1	2	3	4	5
Value of expression	4	7	10	13	16	19

84. **Finding a Pattern** Find values of a and b such that the expression $an + b$ yields the values in the table.

n	0	1	2	3	4	5
$an + b$	4	9	14	19	24	29

2.4

Checking Solutions of an Equation In Exercises 85–94, **determine** whether each value of x is a solution of the equation.

Equation   Values

85. $5x + 6 = 36$ (a) $x = 3$ (b) $x = 6$

86. $17 - 3x = 8$ (a) $x = 3$ (b) $x = -3$

87. $3x - 12 = x$ (a) $x = -1$ (b) $x = 6$

Equation	Values	
88. $8x + 24 = 2x$	(a) $x = 0$	(b) $x = -4$
89. $4(2 - x) = 3(2 + x)$	(a) $x = \frac{2}{7}$	(b) $x = -\frac{2}{3}$
90. $5x + 2 = 3(x + 10)$	(a) $x = 14$	(b) $x = -10$
91. $\frac{4}{x} - \frac{2}{x} = 5$	(a) $x = -1$	(b) $x = \frac{2}{5}$
92. $\frac{x}{3} + \frac{x}{6} = 1$	(a) $x = \frac{2}{9}$	(b) $x = -\frac{2}{9}$
93. $x(x - 7) = -12$	(a) $x = 3$	(b) $x = 4$
94. $x(x + 1) = 2$	(a) $x = 1$	(b) $x = -2$

Identifying Properties of Equality In Exercises 95–98, justify each step of the equation. Then identify any properties of equality used to solve the equation.

95.
$$-7x + 20 = -1$$
$$-7x + 20 - 20 = -1 - 20$$
$$-7x = -21$$
$$\frac{-7x}{-7} = \frac{-21}{-7}$$
$$x = 3$$

96.
$$3(x - 2) = x + 2$$
$$3x - 6 = x + 2$$
$$3x - 6 - x = x + 2 - x$$
$$3x - x - 6 = x - x + 2$$
$$2x - 6 = 2$$
$$2x - 6 + 6 = 2 + 6$$
$$2x = 8$$
$$\frac{2x}{2} = \frac{8}{2}$$
$$x = 4$$

97.
$$x = -(x - 14)$$
$$x = -x + 14$$
$$x + x = -x + 14 + x$$
$$x + x = -x + x + 14$$
$$2x = 14$$
$$\frac{2x}{2} = \frac{14}{2}$$
$$x = 7$$

98.
$$\frac{x}{4} = x - 2$$
$$4\left(\frac{x}{4}\right) = 4(x - 2)$$
$$x = 4x - 8$$
$$x - 4x = 4x - 8 - 4x$$
$$x - 4x = 4x - 4x - 8$$
$$-3x = -8$$
$$\frac{-3x}{-3} = \frac{-8}{-3}$$
$$x = \frac{8}{3}$$

Writing an Algebraic Equation In Exercises 99–102, write an algebraic equation. Do *not* solve the equation.

99. The sum of a number and its reciprocal is $\frac{37}{6}$. What is the number?

100. A car travels 135 miles in t hours with an average speed of 45 miles per hour (see figure). How many hours did the car travel?

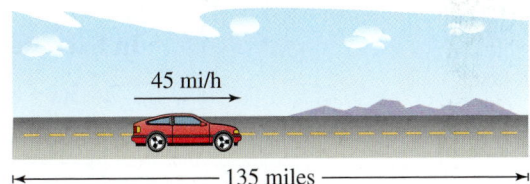

101. The area of the shaded region in the figure is 24 square inches. What is the length of the rectangle?

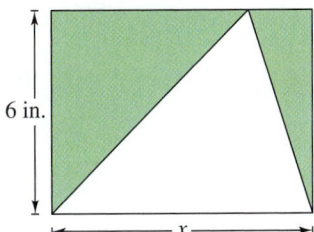

102. The perimeter of the face of a rectangular traffic light is 72 inches (see figure). What are the dimensions of the traffic light?

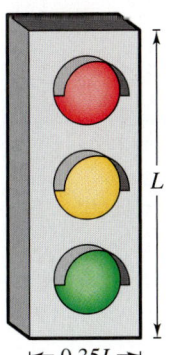

Chapter Test

Solutions in English & Spanish and tutorial videos at AlgebraWithinReach.com

Take this test as you would take a test in class. After you are done, check your work against the answers in the back of the book.

1. Identify the terms of the expression and their coefficients.
   $2x^2 - 7xy + 3y^3$

2. Rewrite the product in exponential form.
   $x \cdot (x + y) \cdot x \cdot (x + y) \cdot x$

In Exercises 3–6, identify the property of algebra illustrated by the statement.

3. $(5x)y = 5(xy)$
4. $2 + (x - y) = (x - y) + 2$
5. $7xy + 0 = 7xy$
6. $(x + 5) \cdot \dfrac{1}{(x + 5)} = 1$

In Exercises 7–10, use the Distributive Property to expand the expression.

7. $3(x + 8)$
8. $5(4r - s)$
9. $-y(3 - 2y)$
10. $-9(4 - 2x + x^2)$

In Exercises 11–14, simplify the expression.

11. $3b - 2a + a - 10b$
12. $15(u - v) - 7(u - v)$
13. $3z - (4 - z)$
14. $2[10 - (t + 1)]$

In Exercises 15 and 16, evaluate the expression for $x = 2$ and $y = -10$.

15. $x^3 - 2$
16. $x^2 + 4(y + 2)$

17. Explain why it is not possible to evaluate $\dfrac{a + 2b}{3a - b}$ when $a = 2$ and $b = 6$.

18. Translate the phrase "four less than one-third of n" into an algebraic expression.

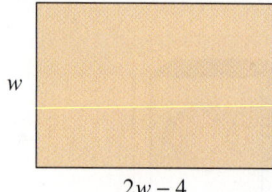

19. (a) Write expressions for the perimeter and area of the rectangle at the left. Simplify each expression.

    (b) Evaluate each expression for $w = 7$.

20. The prices of concert tickets for adults and children are $25 and $20, respectively.

    (a) Write an algebraic expression that represents the total income from the concert for m adults and n children.

    (b) How much will it cost two adults and three children to attend the concert?

21. Determine whether the values of x are solutions of $6(3 - x) - 5(2x - 1) = 7$.
    (a) $x = -2$
    (b) $x = 1$

3 Linear Equations and Problem Solving

- **3.1** Solving Linear Equations
- **3.2** Equations that Reduce to Linear Form
- **3.3** Problem Solving with Percents
- **3.4** Ratios and Proportions
- **3.5** Geometric and Scientific Applications
- **3.6** Linear Inequalities

MASTERY IS WITHIN REACH!

"When I study with a tutor, I always understand better and quicker when I do the problem on the board. I just see it better and it clicks. It turns out that I am a visual and kinesthetic learner when I study math. I spend a lot of time in the tutoring area where there are whiteboards."

Cynthia
Art

See page 127 for suggestions about using your preferred learning modality.

3.1 Solving Linear Equations

▶ Solve a linear equation in standard form.
▶ Solve a linear equation in nonstandard form.

Solving Linear Equations in Standard Form

A **linear equation in one variable** x is an equation that can be written in the standard form

$$ax + b = 0. \qquad a \text{ and } b \text{ are real numbers with } a \neq 0.$$

EXAMPLE 1 Solving Linear Equations in Standard Form

Solve $3x - 15 = 0$. Then check the solution.

SOLUTION

$3x - 15 = 0$	Write original equation.
$3x - 15 + 15 = 0 + 15$	Add 15 to each side.
$3x = 15$	Combine like terms.
$\dfrac{3x}{3} = \dfrac{15}{3}$	Divide each side by 3.
$x = 5$	Simplify.

The solution is $x = 5$. You can check this as shown at the right.

Check

$3x - 15 = 0$	Write original equation.
$3(5) - 15 \stackrel{?}{=} 0$	Substitute 5 for x.
$15 - 15 \stackrel{?}{=} 0$	Multiply.
$0 = 0$	Solution checks.

Study Tip

Remember that to *solve* an equation involving x means to find all values of x that make the equation true.

Exercises Within Reach ®

Solutions in English & Spanish and tutorial videos at AlgebraWithinReach.com

Mental Math In Exercises 1–4, **solve** the equation mentally.

1. $x - 9 = 0$
2. $u - 3 = 0$
3. $x + 6 = 0$
4. $a + 5 = 0$

Deciding How to Start In Exercises 5–12, **decide** which operation you would use first to solve the equation.

5. $x + 8 = 0$
6. $p - 1 = 0$
7. $\dfrac{y}{2} = 0$
8. $3z = 0$
9. $2x + 4 = 0$
10. $2d + 10 = 0$
11. $3x - 6 = 0$
12. $4x - 20 = 0$

Justifying Steps In Exercises 13 and 14, **justify** each step of the solution.

13.
$$5x + 15 = 0$$
$$5x + 15 - 15 = 0 - 15$$
$$5x = -15$$
$$\dfrac{5x}{5} = \dfrac{-15}{5}$$
$$x = -3$$

14.
$$-2x - 8 = 0$$
$$-2x - 8 + 8 = 0 + 8$$
$$-2x = 8$$
$$\dfrac{-2x}{-2} = \dfrac{8}{-2}$$
$$x = -4$$

EXAMPLE 2 Solving Linear Equations in Standard Form

a.
$2x + 18 = 0$ Original equation
$2x + 18 - 18 = 0 - 18$ Subtract 18 from each side.
$2x = -18$ Combine like terms.
$\dfrac{2x}{2} = -\dfrac{18}{2}$ Divide each side by 2.
$x = -9$ Simplify. Check in original equation.

b.
$5x - 12 = 0$ Original equation
$5x - 12 + 12 = 0 + 12$ Add 12 to each side.
$5x = 12$ Combine like terms.
$\dfrac{5x}{5} = \dfrac{12}{5}$ Divide each side by 5.
$x = \dfrac{12}{5}$ Simplify. Check in original equation.

c.
$\dfrac{x}{3} + 3 = 0$ Original equation
$\dfrac{x}{3} + 3 - 3 = 0 - 3$ Subtract 3 from each side.
$\dfrac{x}{3} = -3$ Combine like terms.
$3\left(\dfrac{x}{3}\right) = 3(-3)$ Multiply each side by 3.
$x = -9$ Simplify. Check in original equation.

Study Tip

As you gain experience, you will find that you can perform some of the solution steps in your head. For instance, you might solve part (c) by writing only the following steps.

$\dfrac{x}{3} + 3 = 0$

$\dfrac{x}{3} = -3$

$x = -9$

Exercises Within Reach®

Solutions in English & Spanish and tutorial videos at AlgebraWithinReach.com

Solving a Linear Equation In Exercises 15–32, solve the equation and check your solution.

15. $10x + 10 = 0$
16. $8f + 8 = 0$
17. $9g - 18 = 0$
18. $8x - 16 = 0$
19. $4x - 24 = 0$
20. $7x - 21 = 0$
21. $2x + 52 = 0$
22. $3x + 21 = 0$
23. $8x - 2 = 0$
24. $6x - 4 = 0$
25. $\dfrac{y}{4} + 7 = 0$
26. $\dfrac{x}{2} + 1 = 0$
27. $-x + 9 = 0$
28. $-q + 1 = 0$
29. $-5x - 15 = 0$
30. $-11y - 44 = 0$
31. $-3p + 1 = 0$
32. $-9w + 6 = 0$

Solving Linear Equations in Nonstandard Form

EXAMPLE 3 Solving a Linear Equation in Nonstandard Form

$3y + 8 - 5y = 4$	Original equation
$3y - 5y + 8 = 4$	Group like terms.
$-2y + 8 = 4$	Combine like terms.
$-2y + 8 - 8 = 4 - 8$	Subtract 8 from each side.
$-2y = -4$	Combine like terms.
$\dfrac{-2y}{-2} = \dfrac{-4}{-2}$	Divide each side by -2.
$y = 2$	Simplify. Check in original equation.

EXAMPLE 4 Solving Linear Equations: Special Cases

a.

$2x + 3 = 2(x + 4)$	Original equation
$2x + 3 = 2x + 8$	Apply Distributive Property.
$2x - 2x + 3 = 2x - 2x + 8$	Subtract $2x$ from each side.
$3 \ne 8$	Simplify.

Because 3 does not equal 8, the original equation has no solution.

b.

$4(x + 3) = 4x + 12$	Original equation
$4x + 12 = 4x + 12$	Apply Distributive Property.
$4x - 4x + 12 = 4x - 4x + 12$	Subtract $4x$ from each side.
$12 = 12$	Simplify.

Because the last equation is true for any value of x, the original equation has infinitely many solutions.

Study Tip

Equations like the one in Example 4(b) that are true for all values of x are called **identities**.

Exercises Within Reach®

Solutions in English & Spanish and tutorial videos at AlgebraWithinReach.com

Solving a Linear Equation in Nonstandard Form In Exercises 33–50, solve the equation and check your solution. (Some of the equations have no solution.)

33. $3y - 2 = 2y$

34. $2s - 13 = 28s$

35. $4 - 7x = 5x$

36. $24 - 5x = x$

37. $4 - 5t = 16 + t$

38. $3x + 4 = x + 10$

39. $-3t + 5 = -3t$

40. $4z + 2 = 4z$

41. $4x - 2 = 3x + 1$

42. $7x + 9 = 3x + 1$

43. $4x - 6 = 4x - 6$

44. $5 - 3x = 5 - 3x$

45. $2x + 4 = -3(x - 2)$

46. $4(y + 1) = -y + 5$

47. $5(3 - x) = x - 12$

48. $12 - w = -2(3w - 1)$

49. $2x = -3x$

50. $6t = 9t$

Application

EXAMPLE 5 *Geometry:* Finding Dimensions

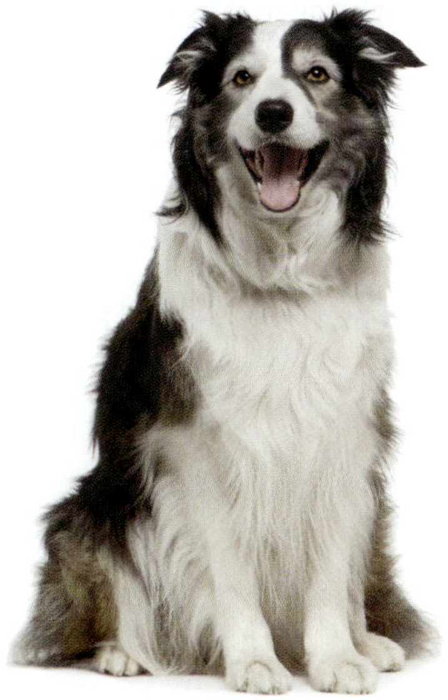

The Border Collie is a herding dog breed developed in the English-Scottish border region for herding livestock, especially sheep. Typically energetic and athletic, they often compete with success in dog sports.

You have 96 feet of fencing to enclose a rectangular exercise area for your Border Collie. The exercise area is to be three times as long as it is wide. Find the dimensions of the exercise area.

SOLUTION

Begin by drawing and labeling a diagram.

Labels & Diagram:

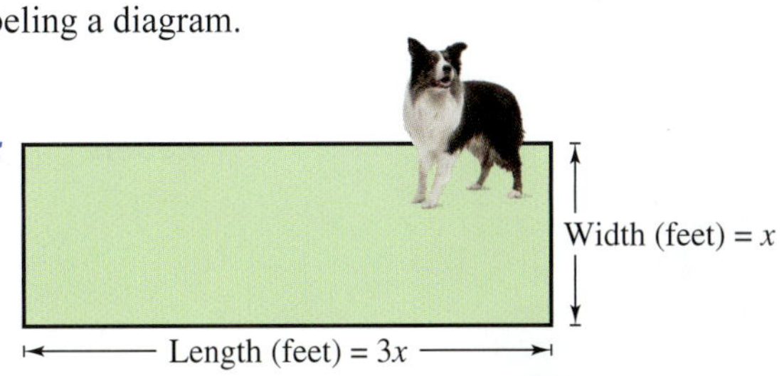

The perimeter of a rectangle is the sum of twice its length and twice its width.

Verbal Model: Perimeter = 2 • Length + 2 • Width

Equation: $96 = 2(3x) + 2x$

You can solve this equation as follows.

$96 = 6x + 2x$	Multiply.
$96 = 8x$	Combine like terms.
$\frac{96}{8} = \frac{8x}{8}$	Divide each side by 8.
$12 = x$	Simplify.

So, the width is 12 feet and the length is $3(12) = 36$ feet.

Exercises Within Reach®

Solutions in English & Spanish and tutorial videos at AlgebraWithinReach.com

51. ***Geometry*** The sides of a yield sign all have the same length (see figure). The perimeter of a roadway yield sign is 225 centimeters. Find the length of each side.

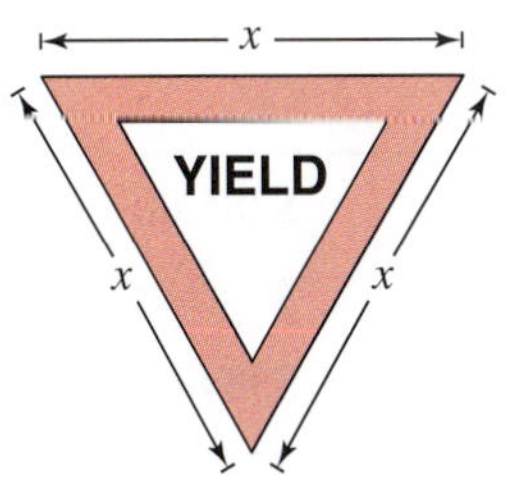

Figure for 51

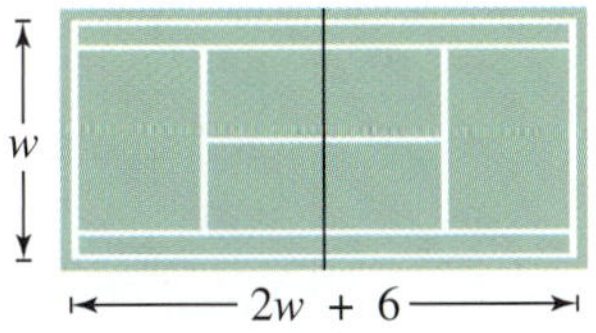

Figure for 52

52. ***Geometry*** The length of a tennis court is 6 feet more than twice the width (see figure). Find the width of the court when the length is 78 feet.

53. ***Geometry*** The perimeter of the Jamaican flag is 120 inches. Its length is twice its width. Find the dimensions of the flag.

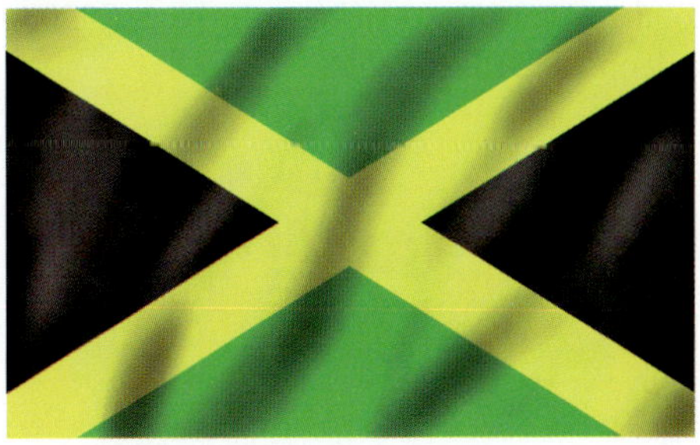

54. ***Geometry*** You are asked to cut a 12-foot board into 3 pieces. Two pieces are to have the same length and the third is to be twice as long as the others. How long are the pieces?

Application

EXAMPLE 6 Finding the Number of Stadium Seats Sold

Tickets for a concert cost $175 for each floor seat and $95 for each stadium seat. There were 2500 seats on the main floor, and these were sold out. The total revenue from ticket sales was $865,000. How many stadium seats were sold?

SOLUTION

Verbal Model: Total revenue = Revenue from floor seats + Revenue from stadium seats

Labels:
Total revenue = 865,000 (dollars)
Price per floor seat = 175 (dollars per seat)
Number of floor seats = 2500 (seats)
Price per stadium seat = 95 (dollars per seat)
Number of stadium seats = x (seats)

Equation: $865{,}000 = 175(2500) + 95x$ (See page 91.)

You can solve this equation as follows.

$865{,}000 = 175(2500) + 95x$	Write equation.
$865{,}000 = 437{,}500 + 95x$	Simplify.
$865{,}000 - 437{,}500 = 437{,}500 - 437{,}500 + 95x$	Subtract 437,500 from each side.
$427{,}500 = 95x$	Combine like terms.
$\dfrac{427{,}500}{95} = \dfrac{95x}{95}$	Divide each side by 95.
$4500 = x$	Simplify.

There were 4500 stadium seats sold. To check this, go back to the original statement of the problem.

Exercises Within Reach®

Solutions in English & Spanish and tutorial videos at AlgebraWithinReach.com

55. Ticket Sales Tickets for a community theater cost $10 for each main floor seat and $8 for each balcony seat. There are 400 seats on the main floor, and these seats were sold out for the evening performance. The total revenue from ticket sales was $5200. How many balcony seats were sold?

56. Ticket Sales Tickets for a drumline competition cost $5 at the gate and $3 in advance. Eight hundred tickets were sold at the gate. The total revenue from tickets sales was $5500. How many advance tickets were sold?

57. Car Repair The bill (including parts and labor) for the repair of your car is shown. Some of the bill is unreadable. From what is given, can you determine how many hours were spent on labor? Explain.

Parts	$285.00
Labor ($44 per hour)	$
Total	$384.00

58. Car Repair The bill for the repair of your car was $553. The cost for parts was $265. The cost for labor was $48 per hour. How many hours did the repair work take?

Section 3.1 Solving Linear Equations 107

Application   **EXAMPLE 7**   **Finding Your Gross Pay per Paycheck**

Write an algebraic equation that represents the following problem. Then solve the equation and answer the question.

> *You have accepted a job offer at an annual salary of $40,830. This salary includes a year-end bonus of $750. You are paid twice a month. What will your gross pay be for each paycheck?*

SOLUTION

You will receive 24 paychecks and 1 bonus check during the year.

Verbal Model: Income for year $= 24 \times$ Amount of each paycheck $+$ Bonus

Labels:
Income for year $= 40{,}830$ (dollars)
Amount of each paycheck $= x$ (dollars)
Bonus $= 750$ (dollars)

Equation:

$40{,}830 = 24x + 750$ Write equation.

$40{,}080 = 24x$ Subtract 750 from each side.

$\dfrac{40{,}080}{24} = \dfrac{24x}{24}$ Divide each side by 24.

$1670 = x$ Simplify.

Each paycheck will be $1670. Check this in the original statement of the problem.

Exercises Within Reach ®

Solutions in English & Spanish and tutorial videos at AlgebraWithinReach.com

59. **Job Offer** You have accepted a job offer at an annual salary of $37,120. This salary includes a year-end bonus of $2800. You are paid twice per month. What will your gross pay be for each paycheck?

60. **Assembly Line Production** You have a job on an assembly line for which you earn $10 per hour plus $0.75 for each unit you produce. Your earnings for an 8-hour day are $146. Find the number of units you produced.

61. **Internship** An internship pays $320 per week plus an additional $75 for a training session. The total pay for the internship and training is $2635. How many weeks long is the internship?

62. **Sales Position** You have a job as a salesperson for which you are paid $6 per hour plus $1.25 for each sale made. Your earnings for an 8-hour day are $88. Find the number of sales you made during the day.

Concept Summary: Solving Linear Equations

What
When given a **linear equation in one variable**, the goal is usually to solve the equation by isolating the variable on one side of the equation.

EXAMPLE
Solve the equation
$3x + 2 = 11$.

How
To isolate the variable, you "get rid" of terms and factors by using inverse operations.

EXAMPLE

$3x + 2 = 11$

$3x + 2 - 2 = 11 - 2$ Subtract.

$3x = 9$

$\dfrac{3x}{3} = \dfrac{9}{3}$ Divide.

$x = 3$

The solution is $x = 3$. ✓

Why
An equation is like a scale. To keep the equation balanced, you must do the same thing to each side of the equation. The resulting equation is said to be equivalent to the original equation.

Equivalent equations

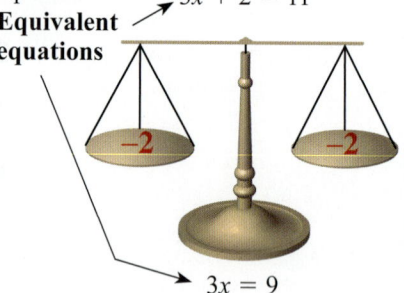

Exercises Within Reach®

Worked-out solutions to odd-numbered exercises at AlgebraWithinReach.com

Concept Summary Check

63. Checking a Solution Explain how to check a solution of an equation. Then illustrate your explanation using the equation $3x + 2 = 11$.

64. Isolating the Variable Is it okay to isolate the variable on the right side of the equation? Illustrate your answer using the equation $11 = 3x + 2$.

65. Inverse Operations In the solution above, explain how you know to subtract 2 from each side of the equation. What operation is "subtract 2" the inverse of?

66. Equivalent Equations Two equations are called *equivalent* if they have exactly the same solutions. Are each of the five equations in the solution steps above equivalent? Justify your answer.

Extra Practice

Solving a Linear Equation In Exercises 67–78, solve the equation and check your solution. (Some of the equations have no solution.)

67. $10x = 50$

68. $-3x = 21$

69. $\dfrac{x}{3} = 10$

70. $-\dfrac{x}{2} = 3$

71. $15x - 3 = 15 - 3x$

72. $2x - 5 = 7x + 10$

73. $2x - 5 + 10x = 3$

74. $-4x + 10 + 10x = 4$

75. $5t - 4 + 3t = 8t - 4$

76. $7z - 5z - 8 = 2z - 8$

77. $3(2 - 7x) = 3(4 - 7x)$

78. $2(5 + 6x) = 4(3x - 1)$

79. Geometry The perimeter of a rectangle is 260 meters. The length is 30 meters greater than the width. Find the dimensions of the rectangle.

80. Geometry A 10-foot board is cut so that 1 piece is 4 times as long as the other. Find the length of each piece.

81. Hourly Wage Your hourly wage is $8.30 per hour plus $0.60 for each unit you produce. How many units must you produce in 1 hour so that your hourly wage is $15.50?

82. Labor Cost The total cost for a new deck (including materials and labor) was $1830. The materials cost $1500 and the cost of labor was $55 per hour. How many hours did it take to build the deck?

83. Summer Jobs During the summer, you work 30 hours per week at a gas station and earn $8.75 per hour. You also work as a landscaper for $11.00 an hour and can work as many hours as you want. You want to earn a total of $400 per week. How many hours must you work at the landscaping job?

84. Summer Jobs During the summer, you work 40 hours per week at a coffee shop and earn $9.25 per hour. You also tutor for $10.00 per hour and can work as many hours as you want. You want to earn a total of $425 per week. How many hours must you tutor?

Explaining Concepts

85. Reasoning The scale below is balanced. Each blue box weighs 1 ounce. If you remove three blue boxes from each side, would the scale still balance? What property of equality does this illustrate? How much does each red box weigh?

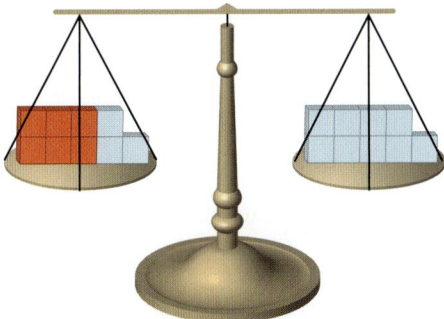

True or False? In Exercises 86–89, **determine** whether the statement is true or false. **Justify** your answer.

86. Subtracting 0 from each side of an equation yields an equivalent equation.

87. Multiplying each side of an equation by 0 yields an equivalent equation.

88. The sum of two odd integers is even.

89. The sum of an odd integer and an even integer is even.

90. Finding a Pattern The length of a rectangle is t times its width (see figure). The perimeter of the rectangle is 1200 meters, which implies that $2w + 2(tw) = 1200$, where w is the width of the rectangle.

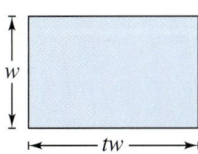

(a) Complete the table.

t	Width	Length	Area
1			
1.5			
2			
3			
4			
5			

(b) Use the completed table to write a conclusion about the area of a rectangle of fixed perimeter as its length increases relative to its width.

Cumulative Review

In Exercises 91–94, plot the numbers on the real number line.

91. 2, −3

92. −2.5, 0

93. $\frac{3}{2}$, 1, −1

94. 4, $-\frac{1}{2}$, 2.6

In Exercises 95–98, determine whether (a) $x = -1$ or (b) $x = 2$ is a solution of the equation.

95. $x - 8 = -9$

96. $x + 1.5 = 3.5$

97. $x + 4 = 2x$

98. $-2(x - 1) = 2 - 2x$

Chapter 3 Linear Equations and Problem Solving

3.2 Equations that Reduce to Linear Form

▶ Solve linear equations containing symbols of grouping.
▶ Solve linear equations involving fractions.
▶ Solve linear equations involving decimals.

Equations Containing Symbols of Grouping

To solve a linear equation that contains symbols of grouping, use the following guidelines.

1. Remove symbols of grouping from each side by using the Distributive Property.
2. Combine like terms.
3. Isolate the variable using properties of equality.
4. Check your solution in the original equation.

EXAMPLE 1   Solving Linear Equations Involving Parentheses

a.
$4(x - 3) = 8$	Original equation
$4 \cdot x - 4 \cdot 3 = 8$	Distributive Property
$4x - 12 = 8$	Simplify.
$4x - 12 + 12 = 8 + 12$	Add 12 to each side.
$4x = 20$	Combine like terms.
$\dfrac{4x}{4} = \dfrac{20}{4}$	Divide each side by 4.
$x = 5$	Simplify.

b.
$3(2x - 1) + x = 11$	Original equation
$6x - 3 + x = 11$	Distributive Property
$6x + x - 3 = 11$	Group like terms.
$7x - 3 = 11$	Combine like terms.
$7x - 3 + 3 = 11 + 3$	Add 3 to each side.
$7x = 14$	Combine like terms.
$x = 2$	Divide each side by 7.

Exercises Within Reach ®   Solutions in English & Spanish and tutorial videos at AlgebraWithinReach.com

Solving a Linear Equation Involving Parentheses In Exercises 1–14, **solve** the equation and check your solution.

1. $2(y - 4) = 0$
2. $9(y - 7) = 0$
3. $12(x - 3) = 0$
4. $4(z - 8) = 0$
5. $7(x + 5) = 49$
6. $4(x + 1) = 24$
7. $-5(t + 3) = 10$
8. $-3(x + 1) = 18$
9. $15(x + 1) - 8x = 29$
10. $7x - 2(x - 2) = 12$
11. $4 - (z + 6) = 8$
12. $25 - (y + 3) = 10$
13. $3 - (2x - 18) = 3$
14. $16 - (3x - 10) = 5$

EXAMPLE 2 Equations Involving Symbols of Grouping

a.
$5(x + 2) = 2(x - 1)$	Original equation
$5x + 10 = 2x - 2$	Distributive Property
$5x - 2x + 10 = 2x - 2x - 2$	Subtract $2x$ from each side.
$3x + 10 = -2$	Combine like terms.
$3x + 10 - 10 = -2 - 10$	Subtract 10 from each side.
$3x = -12$	Combine like terms.
$x = -4$	Divide each side by 3.

b.
$2(x - 7) - 3(x + 4) = 4 - (5x - 2)$	Original equation
$2x - 14 - 3x - 12 = 4 - 5x + 2$	Distributive Property
$-x - 26 = -5x + 6$	Combine like terms.
$-x + 5x - 26 = -5x + 5x + 6$	Add $5x$ to each side.
$4x - 26 = 6$	Combine like terms.
$4x - 26 + 26 = 6 + 26$	Add 26 to each side.
$4x = 32$	Combine like terms.
$x = 8$	Divide each side by 4.

c.
$5x - 2[4x + 3(x - 1)] = 8 - 3x$	Original equation
$5x - 2[4x + 3x - 3] = 8 - 3x$	Distributive Property
$5x - 2[7x - 3] = 8 - 3x$	Combine like terms inside brackets.
$5x - 14x + 6 = 8 - 3x$	Distributive Property
$-9x + 6 = 8 - 3x$	Combine like terms.
$-9x + 3x + 6 = 8 - 3x + 3x$	Add $3x$ to each side.
$-6x + 6 = 8$	Combine like terms.
$-6x + 6 - 6 = 8 - 6$	Subtract 6 from each side.
$-6x = 2$	Combine like terms.
$x = -\frac{2}{6} = -\frac{1}{3}$	Divide each side by -6.

Technology Tip
Try using your calculator to check the solution found in Example 2(b). This will give you practice working with parentheses on a calculator.

Exercises Within Reach®

Solutions in English & Spanish and tutorial videos at AlgebraWithinReach.com

Symbols of Grouping In Exercises 15–30, solve the equation and check your solution. (Some of the equations have no solution.)

15. $5(x - 4) = 2(2x + 5)$
16. $-(4x + 10) = 6(x + 2)$
17. $3(2x - 1) = 3(2x + 5)$
18. $4(z - 2) = 2(2z - 4)$
19. $-3(x + 4) = 4(x + 4)$
20. $-8(x - 6) = 3(x - 6)$
21. $-6(3 + x) + 2(3x + 5) = 0$
22. $-3(5x + 2) + 5(1 + 3x) = 0$
23. $7 = 3(x + 2) - 3(x - 5)$
24. $24 = 12(z + 1) - 3(4z - 2)$
25. $4 - (y - 3) = 3(y + 1) - 4(1 - y)$
26. $12 - 2(y + 3) = 4(y - 6) - (y - 1)$
27. $2[(3x + 5) - 7] = 3(4x - 3)$
28. $3[(5x + 1) - 4] = 4(2x - 3)$
29. $4x + 3[x - 2(2x - 1)] = 4 - 3x$
30. $16 + 4[5x - 4(x + 2)] = 7 - 2x$

Equations Involving Fractions

To solve a linear equation that contains one or more fractions, it is usually best to first clear the equation of fractions by multiplying each side by the least common multiple (LCM) of the denominators.

EXAMPLE 3 **Solving Linear Equations Involving Fractions**

a.
$$\frac{3x}{2} - \frac{1}{3} = 2$$ Original equation

$$6\left(\frac{3x}{2} - \frac{1}{3}\right) = 6 \cdot 2$$ Multiply each side by LCM 6.

$$6 \cdot \frac{3x}{2} - 6 \cdot \frac{1}{3} = 12$$ Distributive Property

$$9x - 2 = 12$$ Simplify.

$$9x = 14$$ Add 2 to each side.

$$x = \frac{14}{9}$$ Divide each side by 9.

b.
$$\frac{x}{5} + \frac{3x}{4} = 19$$ Original equation

$$20\left(\frac{x}{5}\right) + 20\left(\frac{3x}{4}\right) = 20(19)$$ Multiply each side by LCM 20.

$$4x + 15x = 380$$ Simplify.

$$19x = 380$$ Combine like terms.

$$x = 20$$ Divide each side by 19.

c.
$$\frac{2}{3}\left(x + \frac{1}{4}\right) = \frac{1}{2}$$ Original equation

$$\frac{2}{3}x + \frac{2}{12} = \frac{1}{2}$$ Distributive Property

$$12 \cdot \frac{2}{3}x + 12 \cdot \frac{2}{12} = 12 \cdot \frac{1}{2}$$ Multiply each side by LCM 12.

$$8x + 2 = 6$$ Simplify.

$$8x = 4$$ Subtract 2 from each side.

$$x = \frac{4}{8} = \frac{1}{2}$$ Divide each side by 8.

Study Tip

Notice in Example 3(c) that to clear all fractions in the equation, you multiply by 12, which is the LCM of 3, 4, and 2.

Solutions in English & Spanish and tutorial videos at AlgebraWithinReach.com

Solving a Linear Equation Involving Fractions In Exercises 31–42, solve the equation and check your solution.

31. $\dfrac{6x}{25} = \dfrac{3}{5}$

32. $\dfrac{8x}{9} = \dfrac{2}{3}$

33. $\dfrac{5x}{4} + \dfrac{1}{2} = 0$

34. $\dfrac{3z}{7} + \dfrac{5}{14} = 0$

35. $\dfrac{x}{5} - \dfrac{1}{2} = 3$

36. $\dfrac{y}{6} - \dfrac{5}{8} = 2$

37. $\dfrac{x}{5} - \dfrac{x}{2} = 1$

38. $\dfrac{x}{3} + \dfrac{x}{4} = 1$

39. $\dfrac{2}{3}\left(x - \dfrac{5}{4}\right) = -\dfrac{1}{3}$

40. $\dfrac{1}{2}\left(1 - \dfrac{4}{3}x\right) = \dfrac{1}{4}$

41. $3x + \dfrac{1}{4} = \dfrac{3}{4}$

42. $2x - \dfrac{3}{8} = \dfrac{5}{8}$

Section 3.2   Equations that Reduce to Linear Form   113

Application   **EXAMPLE 4**   **Finding a Test Score**

To get an A in a course, you must have an average of at least 90 points for 4 tests of 100 points each. For the first 3 tests, your scores are 87, 92, and 84. What must you score on the fourth test to earn a 90% average for the course?

SOLUTION

Verbal Model: $\dfrac{\text{Sum of 4 tests}}{4} = 90$

Labels: Score on 4th test $= x$ (points)

Scores on first 3 tests: 87, 92, 84 (points)

Equation: $\dfrac{87 + 92 + 84 + x}{4} = 90$

You can solve this equation by multiplying each side by 4.

$\dfrac{87 + 92 + 84 + x}{4} = 90$ Write equation.

$4\left(\dfrac{87 + 92 + 84 + x}{4}\right) = 4(90)$ Multiply each side by LCM 4.

$87 + 92 + 84 + x = 360$ Simplify.

$263 + x = 360$ Combine like terms.

$x = 97$ Subtract 263 from each side.

You need to score 97 on the fourth test to earn a 90% average.

Exercises Within Reach®

Solutions in English & Spanish and tutorial videos at AlgebraWithinReach.com

43. **Time to Complete a Task** Two people can complete 80% of a task in t hours, where t must satisfy the equation $\dfrac{t}{10} + \dfrac{t}{15} = 0.8$. How long will it take for the two people to complete 80% of the task?

44. **Time to Complete a Task** Two machines can complete a task in t hours, where t must satisfy the equation $\dfrac{t}{10} + \dfrac{t}{15} = 1$. How long will it take for the two machines to complete the task?

45. **Course Grade** To get a B in a course, you must have an average of at least 80 points for 4 tests of 100 points each. For the first 3 tests, your scores are 79, 83, and 81. What must you score on the fourth test to earn an 80% average for the course?

46. **Weight Loss** You want to lose an average of 4 pounds per month on a new weight loss program. In the first 3 months, you lost 3 pounds, 7 pounds, and 4 pounds. How much weight must you lose during the fourth month to maintain an average weight loss of 4 pounds per month?

Equations Involving Decimals

Many real-life applications of linear equations involve decimal coefficients. To solve such an equation, you can clear it of decimals in much the same way you clear an equation of fractions. Multiply each side by a power of 10 that converts all decimal coefficients to integers, as shown in the next example.

EXAMPLE 5 Solving a Linear Equation Involving Decimals

Solve $0.3x + 0.2(10 - x) = 0.15(30)$. Then check your solution.

SOLUTION

$0.3x + 0.2(10 - x) = 0.15(30)$	Write original equation.
$0.3x + 2 - 0.2x = 4.5$	Distributive Property
$0.1x + 2 = 4.5$	Combine like terms.
$10(0.1x + 2) = 10(4.5)$	Multiply each side by 10.
$x + 20 = 45$	Simplify.
$x = 25$	Subtract 20 from each side.

Check

$0.3(25) + 0.2(10 - 25) \stackrel{?}{=} 0.15(30)$	Substitute 25 for x in original equation.
$0.3(25) + 0.2(-15) \stackrel{?}{=} 0.15(30)$	Perform subtraction within parentheses.
$7.5 - 3.0 \stackrel{?}{=} 4.5$	Multiply.
$4.5 = 4.5$	Solution checks. ✓

The solution is $x = 25$.

Study Tip

There are other ways to solve the decimal equation in Example 5. You could first clear the equation of decimals by multiplying each side by 100. Or, you could keep the decimals and use a calculator to do the arithmetic operations. The method you choose is a matter of personal preference.

Exercises Within Reach®

Solutions in English & Spanish and tutorial videos at AlgebraWithinReach.com

Solving a Linear Equation Involving Decimals In Exercises 47–64, solve the equation. Round your answer to two decimal places.

47. $0.2x + 5 = 6$

48. $4 - 0.3x = 1$

49. $5.6 = 1.1x - 1.2$

50. $7.2x - 4.7 = 62.3$

51. $1.2x - 4.3 = 1.7$

52. $16 - 2.4x = -8$

53. $0.234x + 1 = 2.805$

54. $2.75x - 3.13 = 5.12$

55. $3 + 0.03x = 5$

56. $0.4x - 0.1 = 2$

57. $1.205x - 0.003 = 0.5$

58. $5.225 + 3.001x = 10.275$

59. $0.42x - 0.4(x + 2.4) = 0.3(5)$

60. $1.6x + 0.25(12 - x) = 0.43(-12)$

61. $\dfrac{x}{3.25} + 1 = 2.08$

62. $\dfrac{x}{4.08} + 7.2 = 5.14$

63. $\dfrac{x}{3.155} = 2.850$

64. $\dfrac{3x}{4.5} = \dfrac{1}{8}$

Application

EXAMPLE 6 Analyzing Postsecondary Enrollment

The enrollment y (in millions) at postsecondary schools from 2000 through 2009 can be approximated by the linear model $y = 0.35t + 9.1$, where t represents the year, with $t = 0$ corresponding to 2000. Use the model to predict the year in which the enrollment will be 14 million students. (*Source:* U.S. Department of Education)

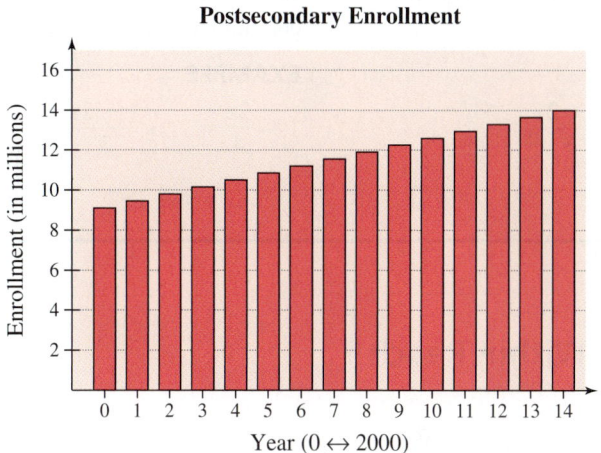

SOLUTION

To find the year in which the enrollment will be 14 million students, substitute 14 for y in the original equation and solve the equation for t.

$14 = 0.35t + 9.1$	Substitute 14 for y in original equation.
$4.9 = 0.35t$	Subtract 9.1 from each side.
$14 = t$	Divide each side by 0.35.

Because $t = 0$ corresponds to 2000, the enrollment at postsecondary schools will be 14 million during 2014. Check this in the original statement of the problem.

Exercises Within Reach®   Solutions in English & Spanish and tutorial videos at AlgebraWithinReach.com

65. *Data Analysis* The table shows the projected numbers N (in millions) of people living in the United States. (*Source:* U.S. Census Bureau)

Year	N
2020	341.4
2025	357.5
2030	373.5
2035	389.5

A model for the data is $N = 3.21t + 277.3$, where t represents time in years, with $t = 20$ corresponding to the year 2020. According to the model, in what year will the population exceed 450 million?

66. *Data Analysis* The table shows the sales S (in billions) of Coach for the years 2005 through 2010. (*Source:* Coach, Inc.)

Year	S
2005	1.71
2006	2.11
2007	2.61
2008	3.18
2009	3.23
2010	3.61

A model for the data is $S = 0.384t - 0.14$, where t represents time in years, with $t = 5$ corresponding to the year 2005. According to the model, in what year will the sales exceed 6 billion?

Concept Summary: Solving Equations

What
Some linear equations involve symbols of grouping, fractions, and decimals. To solve such equations, you must first reduce the equation to linear form.

EXAMPLE

Solve $\dfrac{2x}{4} = \dfrac{1-2x}{3}$.

How
To solve a linear equation involving fractions, first clear the equation of fractions. To do this, multiply each side of the equation by the least common multiple of the denominators.

EXAMPLE

$12 \cdot \dfrac{2x}{4} = 12 \cdot \dfrac{1-2x}{3}$ Multiply by LCM 12.

$6x = 4 - 8x$ Simplify.

Why
Many real-life applications are modeled by equations involving symbols of grouping, fractions, and decimals. Knowing how to reduce such equations to linear form will help you solve these equations.

Exercises Within Reach®

Worked-out solutions to odd-numbered exercises at AlgebraWithinReach.com

Concept Summary Check

67. Least Common Multiple What is the least common multiple of the denominators of two or more fractions?

68. Writing Discuss one method for finding the least common multiple of the denominators of two fractions.

69. Structure When solving an equation that contains fractions, explain what is accomplished by multiplying each side of the equation by the least common multiple of the denominators.

70. Number Sense What is the LCM of the denominators in the equation $\dfrac{2x}{7} = \dfrac{3x}{2}$?

Extra Practice

Solving a Linear Equation In Exercises 71–82, solve the equation and check your solution. (Some of the equations have no solution.)

71. $5z - 2 = 2(3z - 4)$

72. $3 - 4x = 5(x - 3)$

73. $7(y + 7) = 5y + 59$

74. $40 + 14k = 2(-4k - 13)$

75. $3.7y + 7 = 8.1y - 19.4$

76. $5(1.2x + 6) = 7.1x + 34.4$

77. $2s + \dfrac{3}{2} = 2s + 2$

78. $\dfrac{3}{4} + 5s = -2 + 5s$

79. $\dfrac{x}{4} = \dfrac{1 - 2x}{3}$

80. $\dfrac{x+1}{6} = \dfrac{3x}{10}$

81. $\dfrac{100 - 4u}{3} = \dfrac{5u+6}{4} + 6$

82. $\dfrac{8 - 3x}{2} - 4 = \dfrac{x}{6}$

Geometry In Exercises 83 and 84, the perimeter of the figure is 15. Find the value of x.

83.
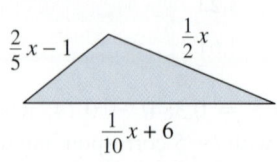

84.
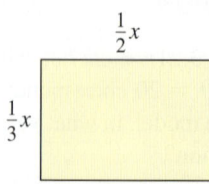

Solving a Mixture Problem In Exercises 85 and 86, use the equation and solve for x.

$$p_1 x + p_2(a - x) = p_3 a$$

85. *Mixture Problem* Determine the number of quarts of a 10% solution that must be mixed with a 30% solution to obtain 100 quarts of a 25% solution. ($p_1 = 0.1$, $p_2 = 0.3$, $p_3 = 0.25$, and $a = 100$.)

86. *Mixture Problem* Determine the number of gallons of a 25% solution that must be mixed with a 50% solution to obtain 5 gallons of a 30% solution. ($p_1 = 0.25$, $p_2 = 0.5$, $p_3 = 0.3$, and $a = 5$.)

Fireplace Construction In Exercises 87 and 88, use the following information. A fireplace is 93 inches wide. Each brick in the fireplace has a length of 8 inches, and there is $\frac{1}{2}$ inch of mortar between adjoining bricks (see figure). Let n be the number of bricks per row.

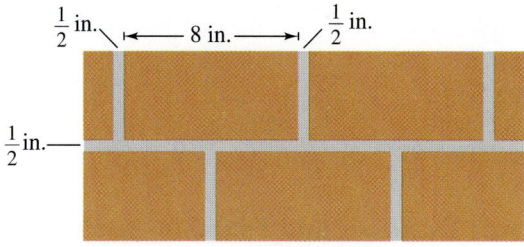

87. Explain why the number of bricks per row is the solution of the equation $8n + \frac{1}{2}(n - 1) = 93$.

88. Find the number of bricks per row in the fireplace.

Explaining Concepts

89. *Structure* You could solve $3(x - 7) = 15$ by applying the Distributive Property as the first step. However, there is another way to begin. What is it?

90. *Error Analysis* Describe and correct the error.

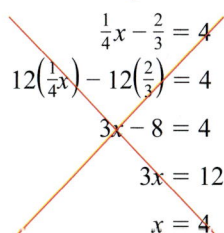

91. *Logic* Explain what happens when you divide each side of a linear equation by a variable factor.

92. *Writing* When simplifying an algebraic *expression* involving fractions, why can't you simplify the expression by multiplying by the least common multiple of the denominators?

Cumulative Review

In Exercises 93–100, simplify the expression.

93. $(-2x)^2 x^4$

94. $-y^2(-2y)^3$

95. $5z^3(z^2)$

96. $a^2 + 3a + 4 - 2a - 6$

97. $\dfrac{5x}{3} - \dfrac{2x}{3} - 4$

98. $2x^2 - 4 + 5 - 3x^2$

99. $-y^2(y^2 + 4) + 6y^2$

100. $5t(2 - t) + t^2$

In Exercises 101–104, solve the equation and check your solution.

101. $3x - 5 = 12$

102. $-5x + 9 = 9$

103. $4 - 2x = 22$

104. $x - 3 + 7x = 29$

3.3 Problem Solving with Percents

- Convert percents to decimals and fractions, and vice versa
- Solve linear equations involving percents.
- Solve problems involving markups and discounts.

Percents

"Cent" implies 100, as in the word *century*. A **percent** is the number of parts per one hundred.

$$25\% = 0.25 = \frac{25}{100}$$

Parts per one hundred

Study Tip

In Examples 1 and 2, there is a quick way to convert between percent form and decimal form.

- To convert from percent form to decimal form, move the decimal point two places to the left. For instance,
  $3.5\% = 0.035.$

- To convert from decimal form to percent form, move the decimal point two places to the right. For instance,
  $1.20 = 120\%$

EXAMPLE 1 Converting Decimals and Fractions to Percents

a. $1.20 = \frac{120}{100} = 120\%$

b. $\frac{3}{5} = \frac{3(20)}{5(20)} = \frac{60}{100} = 60\%$

EXAMPLE 2 Converting Percents to Decimals and Fractions

a. Convert 3.5% to a decimal.

b. Convert 55% to a fraction.

SOLUTION

a. $3.5\% = \frac{3.5}{100}$

$= \frac{3.5(10)}{100(10)}$

$= \frac{35}{1000}$

$= 0.035$

b. $55\% = \frac{55}{100}$

$= \frac{\cancel{5}(11)}{\cancel{5}(20)}$

$= \frac{11}{20}$

Exercises Within Reach ®   Solutions in English & Spanish and tutorial videos at AlgebraWithinReach.com

Converting to a Percent In Exercises 1−8, convert the decimal or fraction to a percent.

1. 0.62
2. 0.57
3. 0.075
4. 0.005
5. $\frac{4}{5}$
6. $\frac{1}{4}$
7. $\frac{5}{4}$
8. $\frac{6}{5}$

Converting to a Decimal In Exercises 9−16, convert the percent to a decimal.

9. 12%
10. 95%
11. 125%
12. 250%
13. 8.5%
14. 0.3%
15. $\frac{3}{4}\%$
16. $4\frac{4}{5}\%$

Converting to a Fraction In Exercises 17−24, convert the percent to a fraction.

17. 30%
18. 85%
19. 130%
20. 350%
21. 1.4%
22. 0.7%
23. $\frac{1}{2}\%$
24. $2\frac{3}{10}\%$

Section 3.3   Problem Solving with Percents   119

The Percent Equation

The primary use of percents is to compare two numbers. For example, 2 is 50% of 4, and 5 is 25% of 20. The following model is helpful.

Verbal Model: $a = p$ percent of b

Labels: b = base number
p = percent (in decimal form)
a = number being compared to b

Equation: $a = p \cdot b$

Application   EXAMPLE 3   Solving Percent Equations for a

a. What number is 30% of 70?

b. A union negotiates for a cost-of-living raise of 7%. What is the raise for a union member whose salary is $40,240? What is this person's new salary?

SOLUTION

a. **Verbal Model:**   What number = 30% of 70

   Labels: a = unknown number

   Equation: $a = (0.3)(70) = 21$

So, 21 is 30% of 70.

b. **Verbal Model:**   Raise = Percent (in decimal form) · Salary

   Labels: Raise = a (dollars)
   Percent = 7% = 0.07   (decimal form)
   Salary = 40,240   (dollars)

   Equation: $a = 0.07(40,240) = 2816.80$

So, the raise is $2816.80 and the new salary is 40,240.00 + 2816.80 = $43,056.80.

Exercises Within Reach®   Solutions in English & Spanish and tutorial videos at AlgebraWithinReach.com

Solving a Percent Equation In Exercises 25–28, **solve the percent equation.**

25. What number is 30% of 150?

26. What number is 60% of 820?

27. What number is 0.75% of 56?

28. What number is 325% of 450?

29. **Salary Raise** You accept a job with a salary of $35,600. After 6 months, you receive a 5% raise. What is your new salary?

30. **Salary Raise** A union negotiates for a cost-of-living raise of 4.5%. What is the raise for a union member whose salary is $37,380? What is this person's new salary?

EXAMPLE 4 Solving Percent Equations for *b*

a. 14 is 25% of what number?

b. You missed an A in your chemistry course by only 3 points. Your point total for the course was 402. How many points were possible in the course? (Assume that you needed 90% of the course total for an A.)

SOLUTION

Study Tip

It may help to think of *a* as a "new" amount and *b* as the "original" amount.

a. **Verbal Model:** $14 = 25\%$ of what number

Labels: $b =$ unknown number

Equation: $14 = 0.25b$

$$\frac{14}{0.25} = b$$

$$56 = b$$

So, 14 is 25% of 56.

b. **Verbal Model:** Your points + 3 points = Percent (in decimal form) · Total points

Labels: Your points = 402 (points)
Percent = 90% = 0.9 (decimal form)
Total points for course = b (points)

Equation: $402 + 3 = 0.9b$ Write equation.

$405 = 0.9b$ Add.

$450 = b$ Divide each side by 0.9

So, there were 450 possible points in the course.

Exercises Within Reach ®

Solutions in English & Spanish and tutorial videos at AlgebraWithinReach.com

Solving a Percent Equation In Exercises 31–34, **solve** the percent equation.

31. 903 is 43% of what number?

32. 275 is $12\frac{1}{2}\%$ of what number?

33. 594 is 450% of what number?

34. 51.2 is 0.08% of what number?

35. **Course Grade** You missed an A in your art course by only 5 points. Your point total for the course was 382. How many points were possible in the course? (Assume that you needed 90% of the course total for an A.)

36. **Course Grade** You were 6 points shy of a B in your mathematics course. Your point total for the course was 394. How many points were possible in the course? (Assume that you needed 80% of the course total for a B.)

Application

EXAMPLE 5 Solving Percent Equations for p

a. 135 is what percent of 27?

b. A real estate agency receives a commission of $8092.50 for the sale of a $124,500 house. What percent commission is this?

SOLUTION

a. *Verbal Model:* $\boxed{135} = \boxed{\text{What percent of 27}}$

Labels: p = unknown percent (in decimal form)

Equation:
$$135 = p(27)$$
$$\frac{135}{27} = p$$
$$5 = p$$

So, 135 is 500% of 27.

b. *Verbal Model:* $\boxed{\text{Commission}} = \boxed{\text{Percent (in decimal form)}} \cdot \boxed{\text{Sale price}}$

Labels:
Commission = 8092.50 (dollars)
Percent = p (decimal form)
Sale Price = 124,500 (dollars)

Equation:
$$8092.50 = p(124{,}500) \quad \text{Write equation.}$$
$$\frac{8092.50}{124{,}500} = p \quad \text{Divide each side by 124,500.}$$
$$0.065 = p \quad \text{Simplify.}$$

So, the real estate agency receives a commission of 6.5%.

In 2010, approximately 321,000 new houses were sold, with a median price of $221,900. (*Source:* U.S. Census Bureau)

Exercises Within Reach®

Solutions in English & Spanish and tutorial videos at AlgebraWithinReach.com

Solving a Percent Equation In Exercises 37–40, **solve** the percent equation.

37. 576 is what percent of 800?

38. 38 is what percent of 5700?

39. 22 is what percent of 800?

40. 148.8 is what percent of 960?

41. Commission A real estate agency receives a commission of $14,506.50 for the sale of a $152,700 house. What percent commission is this?

42. Commission A car salesman receives a commission of $1145 for the sale of a $45,800 car. What percent commission is this?

122 Chapter 3 Linear Equations and Problem Solving

Markups and Discounts

Retail stores generally sell items for more than what they paid for the items. The difference is called **markup**.

$$\boxed{\text{Selling price}} = \boxed{\text{Cost}} + \boxed{\text{Markup}}$$

In markup problems, the markup may be known or it may be expressed as a percent of the cost. This percent is called the **markup rate**.

$$\boxed{\text{Markup}} = \boxed{\text{Markup rate}} \cdot \boxed{\text{Cost}}$$

Study Tip
The cost of an item is sometimes called the wholesale cost.

EXAMPLE 6 Solving Markup Problems

a. The cost is $45. The markup rate is 55%. What is the selling price?
b. The selling price is $98. The markup rate is 60%. What is the cost?
c. The selling price is $60. The cost is $24. What is the markup rate?

SOLUTION

a. $\boxed{\text{Selling price}} = \boxed{\text{Cost}} + \boxed{\text{Markup}}$

$\qquad = 45 + (0.55)(45)$ Substitute known values.
$\qquad = 45 + 24.75$ Multiply.
$\qquad = \$69.75$ Add.

b. $\boxed{\text{Selling price}} = \boxed{\text{Cost}} + \boxed{\text{Markup}}$

$\qquad 98 = C + 0.6C$ Substitute known values.
$\qquad 98 = 1.6C$ Combine like terms.
$\qquad \dfrac{98}{1.6} = C$ Divide each side by 1.6.
$\qquad \$61.25 = C$ Simplify.

c. $\boxed{\text{Selling price}} = \boxed{\text{Cost}} + \boxed{\text{Markup rate}} \cdot \boxed{\text{Cost}}$

$\qquad 60 = 24 + p(24)$ Substitute known values.
$\qquad 36 = 24p$ Subtract 24 from each side.
$\qquad \dfrac{36}{24} = p$ Divide each side by 24.
$\qquad 1.5 = p$ Simplify.

So, the markup rate is 150%.

Exercises Within Reach ®

Solutions in English & Spanish and tutorial videos at AlgebraWithinReach.com

Solving a Markup Problem In Exercises 43–50, find the missing quantities.

	Cost	Selling Price	Markup	Markup Rate		Cost	Selling Price	Markup	Markup Rate
43.	$26.97	$49.95			44.	$71.97	$119.95		
45.		$74.38		81.5%	46.		$69.99		55.5%
47.		$125.98	$56.69		48.		$350.00	$80.77	
49.	$13,250.00			20%	50.	$107.97			85.2%

Section 3.3   Problem Solving with Percents   123

Retail stores also sometimes sell items at a **discount**.

$$\text{Sale price} = \text{Original price} - \text{Discount}$$

The discount is given in dollars, and the **discount rate** is given as a percent of the original price.

$$\text{Discount} = \text{Discount rate} \cdot \text{Original price}$$

Application   EXAMPLE 7   Solving Discount Problems

a. The original price of a lawn mower was $199.95. During a midsummer sale, the lawn mower is on sale for $139.95. What is the discount rate?

b. A drug store advertises 40% off the prices of all summer tanning products. The original price of a bottle of suntan oil is $3.49. What is the sale price?

SOLUTION

a. *Verbal Model:* $\text{Discount} = \text{Discount rate} \cdot \text{Original price}$

   Labels: Discount = 199.95 − 139.95 = 60   (dollars)
   Original price = 199.95   (dollars)
   Discount rate = p (decimal form)

   Equation: $60 = p(199.95)$ Write equation.
   $0.30 \approx p$ Divide each side by 199.95.

   So, the discount rate is 30%.

b. *Verbal Model:* $\text{Sale price} = \text{Original price} - \text{Discount}$

   Labels: Original price = 3.49   (dollars)
   Discount rate = 0.4   (decimal form)
   Discount = 0.4(3.49) = 1.396   (dollars)
   Sale Price = x (dollars)

   Equation: $x = 3.49 - 1.396 = \$2.09$

Exercises Within Reach ®   Solutions in English & Spanish and tutorial videos at AlgebraWithinReach.com

Solving a Discount Problem In Exercises 51−60, find the missing quantities.

	Original Price	Sale Price	Discount	Discount Rate		Original Price	Sale Price	Discount	Discount Rate
51.	$39.95	$29.95			52.	$50.99	$45.99		
53.		$18.95		20%	54.		$189.00		40%
55.	$189.99		$30.00		56.	$18.95		$8.00	
57.	$119.96			50%	58.	$84.95			65%
59.		$695.00	$300.00		60.		$259.97	$135.00	

Chas/Shutterstock.com; Kraska/Shutterstock.com

Concept Summary: Using the Percent Equation

What
The primary use of **percents** is to compare two numbers. You can use the percent equation to solve three basic types of percent problems.

EXAMPLE
1. What number is p percent of b?
2. a is p percent of what number?
3. a is what percent of b?

How
To solve these types of problems, substitute the known quantities into the percent equation and solve for the unknown quantity.

Percent Equation
Verbal Model:

$$a = p \text{ percent of } b$$

Equation: $a = p \cdot b$

Why
When you know how to use the percent equation, you can apply it in real-life situations. For instance, you can use the percent equation to find **markups** and **discounts**.

Exercises Within Reach®

Worked-out solutions to odd-numbered exercises at AlgebraWithinReach.com

Concept Summary Check

61. Writing Explain what is meant by the word *percent*.

62. Number Sense Can any positive terminating decimal be written as a percent? Explain.

63. Structure Write an equation that can be used to find the number x that is 25% of a number y.

64. Writing In your own words, explain what each variable in the percent equation represents.

Extra Practice

Finding Equivalent Forms of a Percent In Exercises 65–74, **complete** the table showing the equivalent forms of the percent.

	Percent	Parts out of 100	Decimal	Fraction		Percent	Parts out of 100	Decimal	Fraction
65.	40%				66.	16%			
67.	7.5%				68.	75%			
69.		63			70.		10.5		
71.			0.155		72.			0.80	
73.				$\frac{3}{5}$	74.				$\frac{3}{20}$

Finding a Percent In Exercises 75–78, **determine** the percent of the figure that is shaded. (There are 360° in a circle.)

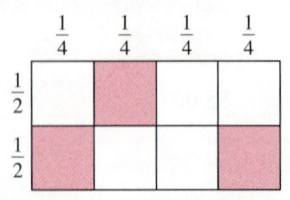 75.

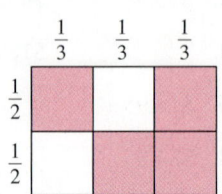 76.

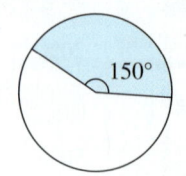 77.

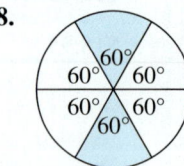 78.

79. *Lawn Tractor* You purchase a lawn tractor for $3750, and 1 year later you note that the price has increased to $3900. Find the percent increase in the price of the lawn tractor.

80. *Membership Drive* Because of a membership drive for a public television station, the current membership is 125% of what it was a year ago. The current number of members is 7815. How many members did the station have last year?

81. *Geometry* A rectangular plot of land measures 650 feet by 825 feet. A square garage with side lengths of 24 feet is built on the plot of land. What percent of the plot of land is occupied by the garage?

82. *Geometry* A circular target is attached to a rectangular board, as shown in the figure. The radius of the circle is $4\frac{1}{2}$ inches, and the measurements of the board are 12 inches by 15 inches. What percent of the board is covered by the target? (The area of a circle is $A = \pi r^2$, where r is the radius of the circle.)

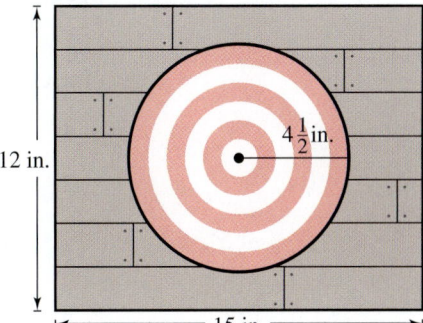

Explaining Concepts

83. *Number Sense* The fraction $\frac{a}{b}$ ($a > 0, b > 0$) is converted to a percent. For what values of a and b is the percent greater than 100%? Less than 100%? Equal to 100%? Explain.

84. *Reasoning* Would you rather receive a 3% raise followed by a 9% raise or a 9% raise followed by a 3% raise? Explain.

True or False? In Exercises 85–88, **decide** whether the statement is true or false. **Justify** your answer.

85. $1 = 1\%$

86. Every percent can be written as a fraction.

87. The question "What is 68% of 50?" can be answered by solving the equation $a = 68(50)$.

88. $\frac{1}{2}\% = 50\%$

Cumulative Review

In Exercises 89 and 90, evaluate the expression.

89. $8 - |-7 + 11| + (-4)$

90. $34 - [54 - (-16 + 4) + 6]$

In Exercises 91 and 92, evaluate the algebraic expression for the specified values of the variables.

91. $x^2 - y^2$
  (a) $x = 4, y = 3$ (b) $x = -5, y = 3$

92. $\dfrac{z^2 + 2}{x^2 - 1}$
  (a) $x = 1, z = 2$ (b) $x = 2, z = 1$

In Exercises 93 and 94, use the Distributive Property to expand the expression.

93. $4(2x - 5)$ **94.** $-z(xz - 2y^2)$

In Exercises 95–98, solve the equation and check your solution.

95. $4(x + 3) = 0$

96. $-3(y - 2) = 21$

97. $22 - (z + 1) = 33$

98. $\dfrac{w}{3} = \dfrac{8}{12}$

Mid-Chapter Quiz: Sections 3.1–3.3

Solutions in English & Spanish and tutorial videos at AlgebraWithinReach.com

Take this quiz as you would take a quiz in class. After you are done, check your work against the answers in the back of the book.

In Exercises 1–10, solve the equation.

1. $74 - 12x = 2$

2. $10(y - 8) = 0$

3. $3x + 1 = x + 20$

4. $6x + 8 = 8 - 2x$

5. $-10x + \dfrac{2}{3} = \dfrac{7}{3} - 5x$

6. $\dfrac{x}{5} + \dfrac{x}{7} = 1$

7. $\dfrac{9 + x}{3} = 15$

8. $3 - 5(4 - x) = -6$

9. $\dfrac{x + 3}{6} = \dfrac{4}{3}$

10. $\dfrac{x + 7}{5} = \dfrac{x + 9}{7}$

In Exercises 11 and 12, solve the equation. Round your answer to two decimal places.

11. $32.86 - 10.5x = 11.25$

12. $\dfrac{x}{5.45} + 3.2 = 12.6$

13. What number is 62% of 25?

14. What number is $\dfrac{1}{2}$% of 8400?

15. 300 is what percent of 150?

16. 145.6 is 32% of what number?

Applications

17. The perimeter of a rectangle is 60 meters. Its length is $1\dfrac{1}{2}$ times its width. Find the dimensions of the rectangle.

18. You work 40 hours per week at a candy store and earn $7.50 per hour. You also earn $7.00 per hour babysitting and can work as many hours as you want. You want to earn $370 a week. How many hours must you babysit?

19. A region has an area of 42 square meters. It must be divided into the three subregions so that the second has twice the area of the first, and the third has twice the area of the second. Find the area of each subregion.

20. To get an A in a psychology course, you must have an average of at least 90 points for 3 tests of 100 points each. For the first 2 tests, your scores are 84 and 93. What must you score on the third test to earn a 90% average for the course?

21. You budget 30% of your annual after-tax income for housing. Your after-tax income is $38,500. What amount can you spend on housing?

22. The price of a television set is approximately 108% of what it was 2 years ago. The current price is $1099. What was the approximate price 2 years ago?

23. Two people can paint a room in t hours, where t must satisfy the equation $\dfrac{t}{4} + \dfrac{t}{12} = 1$. How long will it take for the two people to paint the room?

Study Skills in Action

Knowing Your Preferred Learning Modality

Math is a specific system of rules, properties, and calculations used to solve problems. However, you can take different approaches to learning this specific system based on learning modalities. A learning modality is a preferred way of taking in information that is then transferred into the brain for processing. The three modalities are *visual, auditory,* and *kinesthetic*. The following are brief descriptions of these modalities.

- **Visual** You take in information more productively if you can see the information.
- **Auditory** You take in information more productively when you listen to an explanation and talk about it.
- **Kinesthetic** You take in information more productively if you can experience it or use physical activity in studying.

You may find that one approach, or even a combination of approaches, works best for you.

Smart Study Strategy

Use Your Preferred Learning Modality

Visual *Draw a picture of a word problem.*

- Draw a picture of a word problem before writing a verbal model. You do not have to be an artist.
- When making a review card for a word problem, include a picture. This will help recall the information while taking a test.
- Make sure your notes are visually neat for easy recall.

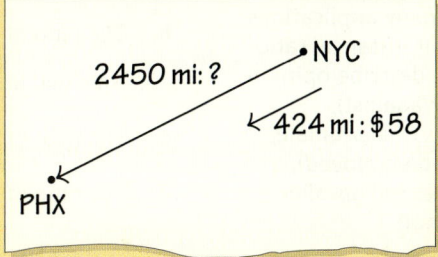

Auditory *Talk about a word problem.*

- Explain how to do a word problem to another student. This is a form of thinking out loud. Write the instructions down on a review card.
- Find several students as serious as you are about math and form a study group.
- Teach the material to an imaginary person when studying alone.

"It takes $58 worth of gas to travel 424 miles. To find the cost of traveling 2450 miles, I can set up and solve a proportion."

Kinesthetic *Incorporate a physical activity.*

- Act out a word problem as much as possible. Use props when you can.
- Solve a word problem on a large whiteboard—the physical action of writing is more kinesthetic when the writing is larger and you can move around while doing it.
- Make a review card.

3.4 Ratios and Proportions

- Compare relative sizes using ratios.
- Find the unit price of a consumer item.
- Solve proportions that equate two ratios.

Setting Up Ratios

A **ratio** is a comparison of one number with another by division. The ratio of the real number a to the real number b is given by

$$\frac{a}{b}.$$ Ratio of a to b

The ratio of a to b is sometimes written as $a:b$. Note the order implied by a ratio. The ratio of a to b means $a:b$, whereas the ratio of b to a means $b:a$.

EXAMPLE 1 Writing Ratios in Fractional Form

Study Tip

There are many applications of ratios. For instance, ratios are used to describe opinion surveys (for/against), populations (male/female, unemployed/employed), and mixtures (oil/gasoline, water/alcohol).

a. The ratio of 7 to 5 is given by $\frac{7}{5}$.

b. The ratio of 12 to 8 is given by $\frac{12}{8} = \frac{3}{2}$.

   Note that the fraction $\frac{12}{8}$ can be written in simplest form as $\frac{3}{2}$.

c. The ratio of 10 to 2 is given by $\frac{10}{2} = \frac{5}{1}$.

d. The ratio of $3\frac{1}{2}$ to $5\frac{1}{4}$ is given by

$$\frac{3\frac{1}{2}}{5\frac{1}{4}} = \frac{\frac{7}{2}}{\frac{21}{4}}$$ Rewrite mixed numbers as fractions.

$$= \frac{7}{2} \cdot \frac{4}{21}$$ Invert divisor and multiply.

$$= \frac{2}{3}$$ Simplify.

Exercises Within Reach®

Solutions in English & Spanish and tutorial videos at AlgebraWithinReach.com

Writing a Ratio in Fractional Form In Exercises 1–12, write the ratio as a fraction in simplest form.

1. 36 to 9
2. 45 to 15
3. 27 to 54
4. 27 to 63
5. $5\frac{2}{3}$ to $1\frac{1}{3}$
6. $2\frac{1}{4}$ to $3\frac{3}{8}$
7. $14:21$
8. $12:30$
9. $144:16$
10. $60:45$
11. $3\frac{1}{5}:5\frac{3}{10}$
12. $1\frac{2}{7}:\frac{1}{2}$

When comparing two measurements by a ratio, you should use the *same unit of measurement* in both the numerator and the denominator. For example, to find the ratio of 4 feet to 8 inches, you could write either of the following.

$$\frac{4 \text{ feet}}{8 \text{ inches}} = \frac{48 \text{ inches}}{8 \text{ inches}} = \frac{48}{8} = \frac{6}{1}$$ Convert feet to inches.

$$\frac{4 \text{ feet}}{8 \text{ inches}} = \frac{4 \text{ feet}}{\frac{8}{12} \text{ foot}} = 4 \cdot \frac{12}{8} = \frac{6}{1}$$ Convert inches to feet.

EXAMPLE 2   Comparing Measurements

Find ratios to compare the relative sizes of the following.

a. 5 gallons to 7 gallons   **b.** 3 meters to 40 centimeters

c. 200 cents to 3 dollars   **d.** 30 months to $1\frac{1}{2}$ years

SOLUTION

a. Because the units of measurement are the same, the ratio is $\frac{5}{7}$.

b. Because the units of measurement are different, begin by converting meters to centimeters *or* centimeters to meters. Here, it is easier to convert meters to centimeters by multiplying by 100.

$$\frac{3 \text{ meters}}{40 \text{ centimeters}} = \frac{3(100) \text{ centimeters}}{40 \text{ centimeters}}$$ Convert meters to centimeters.

$$= \frac{300}{40}$$ Multiply in numerator.

$$= \frac{15}{2}$$ Simplify.

c. Because 200 cents is the same as 2 dollars, the ratio is

$$\frac{200 \text{ cents}}{3 \text{ dollars}} = \frac{2 \text{ dollars}}{3 \text{ dollars}} = \frac{2}{3}.$$

d. Because $1\frac{1}{2}$ years = 18 months, the ratio is

$$\frac{30 \text{ months}}{1\frac{1}{2} \text{ years}} = \frac{30 \text{ months}}{18 \text{ months}} = \frac{30}{18} = \frac{5}{3}.$$

Exercises Within Reach ®
Solutions in English & Spanish and tutorial videos at AlgebraWithinReach.com

Comparing Measurements In Exercises 13−24, **find** a ratio that compares the relative sizes of the quantities. (Use the same units of measurement for both quantities.)

13. 42 inches to 21 inches   **14.** 81 feet to 27 feet

15. $40 to $60   **16.** 24 pounds to 30 pounds

17. 60 milliliters to 1 liter   **18.** 3 inches to 2 feet

19. 7 nickels to 3 quarters   **20.** 24 ounces to 3 pounds

21. 3 hours to 90 minutes   **22.** 21 feet to 35 yards

23. 75 centimeters to 2 meters   **24.** 2 weeks to 7 days

Unit Prices

The **unit price** of an item is given by the ratio of the total price to the total number of units.

$$\text{Unit price} = \frac{\text{Total price}}{\text{Total units}}$$

EXAMPLE 3 Finding a Unit Price

Find the unit price (in dollars per ounce) for a 5-pound, 4-ounce box of detergent that sells for $7.14.

SOLUTION

$$\text{Unit price} = \frac{\text{Total price}}{\text{Total units}} = \frac{\$7.14}{84 \text{ ounces}} = \$0.085 \text{ per ounce}$$

[5(16 oz) + 4 oz]

Study Tip

The word *per* is used to state unit prices. For instance, the unit price for a particular brand of coffee might be $4.69 per pound.

EXAMPLE 4 Comparing Unit Prices

Which has the lower unit price: a 12-ounce box of breakfast cereal for $2.79 or a 16-ounce box of the same cereal for $3.49?

SOLUTION

The unit price for the smaller box is

$$\text{Unit price} = \frac{\text{Total price}}{\text{Total units}} = \frac{\$2.79}{12 \text{ ounces}} = \$0.23 \text{ per ounce.}$$

The unit price for the larger box is

$$\text{Unit price} = \frac{\text{Total price}}{\text{Total units}} = \frac{\$3.49}{16 \text{ ounces}} = \$0.22 \text{ per ounce.}$$

So, the larger box has a slightly lower unit price.

16 oz: $3.49   12 oz: $2.79

Exercises Within Reach®

Solutions in English & Spanish and tutorial videos at AlgebraWithinReach.com

Finding a Unit Price In Exercises 25–28, find the unit price (in dollars per ounce).

25. A 20-ounce can of pineapple for $0.98
26. An 18-ounce box of cereal for $4.29
27. A 1-pound, 4-ounce loaf of bread for $1.46
28. A 1-pound package of cheese for $3.08

Comparing Unit Prices In Exercises 29–32, determine which product has the lower unit price.

29. (a) An 18-ounce jar of peanut butter for $1.92
    (b) A 28-ounce jar of peanut butter for $3.18
30. (a) A 16-ounce bag of chocolates for $1.99
    (b) An 18-ounce bag of chocolates for $2.29
31. (a) A 4-pound bag of sugar for $1.89
    (b) A 10-pound bag of sugar for $4.49
32. (a) A gallon of orange juice for $3.49
    (b) A half-gallon of orange juice for $1.70

Solving Proportions

A **proportion** is a statement that equates two ratios. For example, if the ratio of a to b is the same as the ratio of c to d, you can write the proportion as

$$\frac{a}{b} = \frac{c}{d}.$$

In typical applications, you know three of the values and are required to find the fourth. To solve a proportion, you can use *cross multiplication*. If $\frac{a}{b} = \frac{c}{d}$, then $ad = bc$.

EXAMPLE 5 Solving Proportions

a.
$$\frac{50}{x} = \frac{2}{28} \quad \text{Original proportion}$$

$$50(28) = 2x \quad \text{Cross-multiply.}$$

$$\frac{1400}{2} = x \quad \text{Divide each side by 2.}$$

$$700 = x \quad \text{Simplify.}$$

So, the ratio of 50 to 700 is the same as the ratio of 2 to 28.

b.
$$\frac{x-2}{5} = \frac{4}{3} \quad \text{Original proportion}$$

$$3(x-2) = 20 \quad \text{Cross-multiply.}$$

$$3x - 6 = 20 \quad \text{Distributive Property}$$

$$3x = 26 \quad \text{Add 6 to each side.}$$

$$x = \frac{26}{3} \quad \text{Divide each side by 3.}$$

Exercises Within Reach®

Solutions in English & Spanish and tutorial videos at AlgebraWithinReach.com

Solving a Proportion In Exercises 33–44, solve the proportion.

33. $\dfrac{5}{3} = \dfrac{20}{y}$

34. $\dfrac{9}{x} = \dfrac{18}{5}$

35. $\dfrac{5}{x} = \dfrac{3}{2}$

36. $\dfrac{4}{t} = \dfrac{2}{25}$

37. $\dfrac{z}{35} = \dfrac{5}{8}$

38. $\dfrac{y}{25} = \dfrac{12}{10}$

39. $\dfrac{0.5}{0.8} = \dfrac{n}{0.3}$

40. $\dfrac{2}{4.5} = \dfrac{t}{0.5}$

41. $\dfrac{x+1}{5} = \dfrac{3}{10}$

42. $\dfrac{z-3}{8} = \dfrac{3}{10}$

43. $\dfrac{x+6}{3} = \dfrac{x-5}{2}$

44. $\dfrac{x-2}{4} = \dfrac{x+10}{10}$

Application EXAMPLE 6 Geometry: Using Similar Triangles

A triangular lot has perpendicular sides with lengths of 100 feet and 210 feet. You are making a proportional sketch of this lot using 8 inches as the length of the shorter side. How long should you make the longer side?

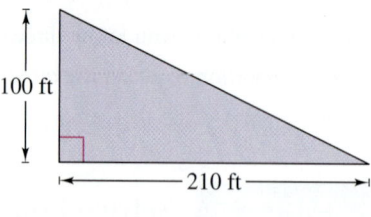
Triangular lot

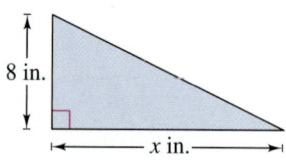
Sketch

SOLUTION

This is a case of similar triangles in which the ratios of the corresponding sides are equal.

$$\frac{\text{Shorter side of lot}}{\text{Longer side of lot}} = \frac{\text{Shorter side of sketch}}{\text{Longer side of sketch}}$$ Proportion for similar triangles

$$\frac{100}{210} = \frac{8}{x}$$ Substitute.

$$100x = 210(8)$$ Cross-multiply.

$$100x = 1680$$ Simplify.

$$x = \frac{1680}{100}$$ Divide each side by 100.

$$= 16.8$$ Simplify.

So, the length of the longer side of the sketch should be 16.8 inches.

Exercises Within Reach®

Solutions in English & Spanish and tutorial videos at AlgebraWithinReach.com

Geometry In Exercises 45–48, **find** the length x of the side of the larger triangle. (Assume that the two triangles are similar, and use the fact that corresponding sides of similar triangles are proportional.)

45.
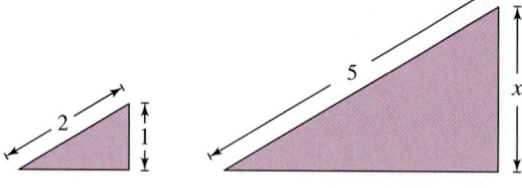

46.
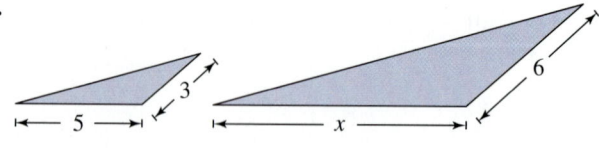

47.
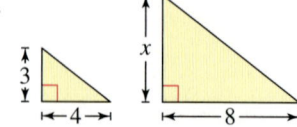

48.
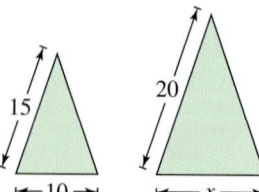

Application EXAMPLE 7 Finding Gasoline Cost

You are driving from New York to Phoenix, a trip of 2450 miles. You begin the trip with a full tank of gas, and after traveling 424 miles, you refill the tank for $58. Assuming gas prices will be the same for the duration of your trip, how much should you plan to spend on gasoline for the entire trip?

SOLUTION

Verbal Model: $\dfrac{\text{Cost for entire trip}}{\text{Cost for one tank}} = \dfrac{\text{Miles for entire trip}}{\text{Miles for one tank}}$

Labels:
Cost of gas for entire trip $= x$ (dollars)
Cost of gas for one tank $= 58$ (dollars)
Miles for entire trip $= 2450$ (miles)
Miles for one tank $= 424$ (miles)

Equation:
$\dfrac{x}{58} = \dfrac{2450}{424}$ Write proportion.

$x = 58\left(\dfrac{2450}{424}\right)$ Multiply each side by 58.

$x \approx 335.14$ Simplify.

You should plan to spend approximately $335 for gasoline on the trip. Check this in the original statement of the problem.

Exercises Within Reach®

Solutions in English & Spanish and tutorial videos at AlgebraWithinReach.com

49. Amount of Fuel A car uses 20 gallons of gasoline for a trip of 500 miles. How many gallons would be used on a trip of 400 miles?

50. Amount of Fuel A tractor requires 4 gallons of diesel fuel to plow for 90 minutes. How many gallons of fuel would be required to plow for 8 hours?

51. Polling Results In a poll, 624 people from a sample of 1100 indicated they would vote for the Republican candidate. How many votes can the candidate expect to receive from 40,000 votes cast?

52. Quality Control A quality control engineer found two defective units in a sample of 50. At this rate, what is the expected number of defective units in a shipment of 10,000 units?

53. Building Material One hundred cement blocks are required to build a 16-foot wall. How many blocks are needed to build a 40-foot wall?

54. Force on a Spring A force of 50 pounds stretches a spring 4 inches. How much force is required to stretch the spring 6 inches?

Concept Summary: Solving Proportions

What
A **proportion** is a statement that equates two **ratios**. When solving a proportion, you usually know three of the values and are asked to find the fourth.

EXAMPLE

Solve $\dfrac{6}{18} = \dfrac{x}{6}$.

How
One way to solve a proportion is to use cross-multiplication.

EXAMPLE

$\dfrac{6}{18} = \dfrac{x}{6}$ Write proportion.

$6(6) = 18x$ Cross-multiply.

$\dfrac{36}{18} = x$ Divide each side by 18.

$2 = x$ Simplify.

Why
There are many real-life applications involving ratios and proportions. For example, knowing how to use ratios will help you identify **unit prices**.

Exercises Within Reach®
Worked-out solutions to odd-numbered exercises at AlgebraWithinReach.com

Concept Summary Check

55. Creating an Example Give an example of a real-life problem that you can represent with a ratio.

56. Writing In your own words, explain what the following proportion represents.

$$\dfrac{1 \text{ gallon of milk}}{\$3.89} = \dfrac{3 \text{ gallons of milk}}{\$11.67}$$

57. Structure Explain how to solve a proportion.

58. Number Sense Determine whether the following statement is a proportion. Explain.

$$\dfrac{5}{25} \stackrel{?}{=} \dfrac{6}{36}$$

Extra Practice

Comparing Measurements In Exercises 59–64, find a ratio that compares the relative sizes of the quantities. (Use the same units of measurement for both quantities.)

59. 1 quart to 1 gallon
60. 3 miles to 2000 feet
61. 2 kilometers to 2500 meters
62. $5\frac{1}{2}$ pints to 2 quarts
63. 3000 pounds to 5 tons
64. 4 days to 30 hours

Comparing Unit Prices In Exercises 65 and 66, determine which product has the lower unit price.

65. (a) A 2 liter bottle (67.6 ounces) of soft drink for $1.09
    (b) Six 12-ounce cans of soft drink for $1.69

66. (a) A 1-quart container of oil for $2.12
    (b) A 2.5-gallon container of oil for $19.99

Writing a Ratio In Exercises 67 and 68, express the statement as a ratio in simplest form.

67. You study 4 hours per day and are in class 6 hours per day. Find the ratio of the number of study hours to class hours.

68. You have $22 of state tax withheld from your paycheck per week when your gross pay is $750. Find the ratio of tax to gross pay.

69. **Map Scale** On a map, $1\frac{1}{4}$ inches represents 80 miles. Estimate the distance between two cities that are 6 inches apart on the map.

70. **Map Scale** On a map, $1\frac{1}{2}$ inches represents 40 miles. Estimate the distance between two cities that are 4 inches apart on the map.

71. Geometry Find the length of the shadow of the man shown in the figure. (*Hint:* Use similar triangles to create a proportion.)

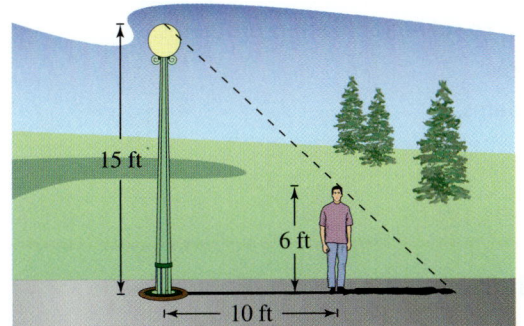

72. Geometry Find the height of a tree shown in the figure. (*Hint:* Use similar triangles to create a proportion.)

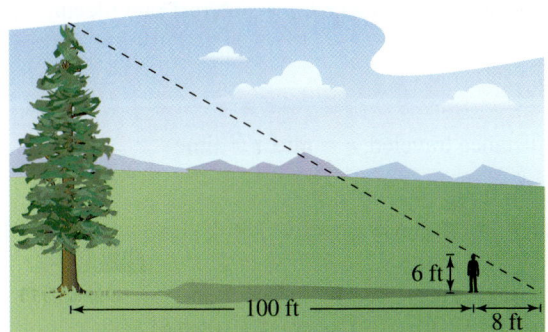

73. Pumping Time A pump can fill a 750-gallon tank in 35 minutes. How long will it take to fill a 1000-gallon tank with this pump?

74. Recipe Two cups of flour are required to make one batch of cookies. How many cups are required for $2\frac{1}{2}$ batches?

75. Salt Water The fresh water to salt ratio for a mixture is 25 to 1. How much fresh water is required to produce a mixture that contains one-half pound of salt?

76. Building Material The ratio of cement to sand in an 80-pound bag of dry mix is 1 to 4. Find the number of pounds of sand in the bag. (*Note:* Dry mix is composed of only cement and sand).

77. Resizing a Picture You have an 8-by-10-inch photo of a soccer player that must be reduced to a size of 1.6 inches by 2 inches for the school newsletter. What percent does the photo need to be reduced to in order for it to fit in the allotted space?

78. Resizing a Picture You have an 7-by-5-inch photo of the math club that must be reduced to a size of 5.6 inches by 4 inches for the school yearbook. What percent does the photo need to be reduced to in order for it to fit in the allotted space?

Explaining Concepts

79. Writing You are told that the ratio of men to women in a class is 2 to 1. Does this information tell you the total number of people in the class? Explain.

80. Writing Explain the following statement. "When setting up a ratio, be sure you are comparing apples to apples and not apples to oranges."

81. Creating a Problem Create a proportion problem. Exchange problems with another student and solve the problem you receive.

82. Writing Explain how to find the unit price of an item.

Cumulative Review

In Exercises 83–88, evaluate the expression.

83. $3^2 - (-4)$

84. $(-5)^3 + 3$

85. 9.3×10^6

86. $\dfrac{-|7 + 3^2|}{4}$

87. $(-4)^2 - (30 \div 50)$

88. $(8 \cdot 9) + (-4)^3$

In Exercises 89–92, solve the percent equation.

89. What number is 25% of 250?

90. What number is 45% of 90?

91. 150 is 250% of what number?

92. 465 is what percent of 500?

3.5 Geometric and Scientific Applications

- Use common formulas to solve application problems.
- Solve mixture problems.
- Solve work-rate problems.

Using Formulas

Miscellaneous Common Formulas

Temperature: F = degrees Fahrenheit, C = degrees Celsius

$$F = \frac{9}{5}C + 32$$

Simple Interest: I = interest, P = principal, r = interest rate (decimal form), t = time (years)

$$I = Prt$$

Distance: d = distance traveled, r = rate, t = time

$$d = rt$$

Application

EXAMPLE 1 Using the Simple Interest Formula

You deposit $5000 in an account paying simple interest. After 6 months, the account has earned $162.50 in interest. What is the annual interest rate?

SOLUTION

$I = Prt$	Simple interest formula
$162.50 = 5000(r)(0.5)$	Substitute for *I*, *P*, and *t*.
$162.50 = 2500r$	Simplify.
$\dfrac{162.50}{2500} = r$	Divide each side by 2500.
$0.065 = r$	Simplify.

The annual interest rate is $r = 0.065$ (or 6.5%). Check this solution in the original statement of the problem.

Exercises Within Reach®

Solutions in English & Spanish and tutorial videos at AlgebraWithinReach.com

Solving a Simple Interest Problem In Exercises 1–4, **find** the missing interest, principal, interest rate, or time.

1. $I =$
   $P = \$870$
   $r = 3.8\%$
   $t = 18$ months

2. $I = \$180$
   $P =$
   $r = 4.5\%$
   $t = 3$ years

3. $I = \$54$
   $P = \$450$
   $r =$
   $t = 2$ years

4. $I = \$97.30$
   $P = \$1200$
   $r = 4.5\%$
   $t =$

Study Tip

- Perimeter is always measured in linear units, such as inches, feet, miles, centimeters, and meters.
- Area is always measured in square units, such as square inches, square feet, and square meters.
- Volume is always measured in cubic units, such as cubic inches, cubic feet, and cubic meters.

Common Formulas for Area, Perimeter, and Volume

Square
$A = s^2$
$P = 4s$

Rectangle
$A = lw$
$P = 2l + 2w$

Circle
$A = \pi r^2$
$C = 2\pi r$

Triangle
$A = \frac{1}{2}bh$
$P = a + b + c$

Cube
$V = s^3$

Rectangular Solid
$V = lwh$

Circular Cylinder
$V = \pi r^2 h$

Sphere
$V = \frac{4}{3}\pi r^3$

Application

EXAMPLE 2 Using a Geometric Formula

You own a rectangular lot that is 500 feet deep and has an area of 100,000 square feet. To pay for a new sewer system, you are assessed $5.50 per foot of lot frontage. (a) Find the frontage of your lot. (b) How much will you be assessed for the new sewer system?

SOLUTION

a.
$A = lw$ Area of a rectangle
$100,000 = 500(w)$ Substitute 100,000 for A and 500 for l.
$200 = w$ Divide each side by 500.

The frontage of the rectangular lot is 200 feet.

b. If each foot of frontage costs $5.50, then your total assessment will be $200(5.50) = \$1100$.

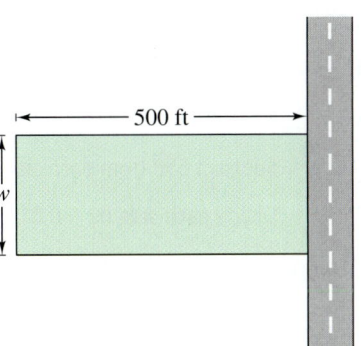

Exercises Within Reach®

Solutions in English & Spanish and tutorial videos at AlgebraWithinReach.com

5. *Geometry* A triangular piece of stained glass has an area of 6 square inches and a height of 3 inches. What is the length of the base?

6. *Geometry* A dime has a circumference of about 56.27 millimeters. What is the radius of a dime? Round your answer to two decimal places.

7. *Geometry* An Olympic-size swimming pool in the shape of a rectangular solid has a volume of 3125 cubic meters, a length of 50 meters, and a width of 25 meters. What is the depth of the pool?

8. *Geometry* A cylindrical bass drum has a volume of about 3054 cubic inches and a radius of 9 inches. What is the height of the drum? Round your answer to one decimal place.

Solving Mixture Problems

Many real-life problems involve combinations of two or more quantities that make up a new or different quantity. Such problems are called **mixture problems**.

$$\underbrace{\text{First rate} \cdot \text{Amount}}_{\text{First component}} + \underbrace{\text{Second rate} \cdot \text{Amount}}_{\text{Second component}} = \underbrace{\text{Final rate} \cdot \text{Final Amount}}_{\text{Final mixture}}$$

Application

EXAMPLE 3 Solving an Investment Mixture Problem

You invested a total of $10,000 in 2 funds earning $4\frac{1}{2}\%$ and $5\frac{1}{2}\%$ simple interest. During 1 year, the 2 funds earned a total of $508.75 in interest. How much did you invest in each fund?

SOLUTION

Verbal Model: $\boxed{\text{Interest earned at } 4\frac{1}{2}\%} + \boxed{\text{Interest earned at } 5\frac{1}{2}\%} = \boxed{\text{Total interest earned}}$

Labels:
- Amount invested at $4\frac{1}{2}\% = x$ (dollars)
- Amount invested at $5\frac{1}{2}\% = 10{,}000 - x$ (dollars)
- Interest earned at $4\frac{1}{2}\% = (x)(0.045)(1)$ (dollars)
- Interest earned at $5\frac{1}{2}\% = (10{,}000 - x)(0.055)(1)$ (dollars)
- Total interest earned $= 508.75$ (dollars)

Equation:

$0.045x + 0.055(10{,}000 - x) = 508.75$	Write equation.
$0.045x + 550 - 0.055x = 508.75$	Distributive Property
$550 - 0.01x = 508.75$	Simplify.
$-0.01x = -41.25$	Subtract 550 from each side.
$x = 4125$	Divide each side by -0.01.

So, you invested $4125 at $4\frac{1}{2}\%$ and $10{,}000 - x = 10{,}000 - 4125 = \$5875$ at $5\frac{1}{2}\%$. Check this in the original statement of the problem.

Exercises Within Reach®

Solutions in English & Spanish and tutorial videos at AlgebraWithinReach.com

9. *Investment Mixture* You invested a total of $6000 in 2 funds earning 7% and 9% simple interest. During 1 year, the 2 funds earned a total of $500 in interest. How much did you invest in each fund?

10. *Investment Mixture* You invested a total of $30,000 in 2 funds earning 8.5% and 10% simple interest. During 1 year, the 2 funds earned a total of $2700 in interest. How much did you invest in each fund?

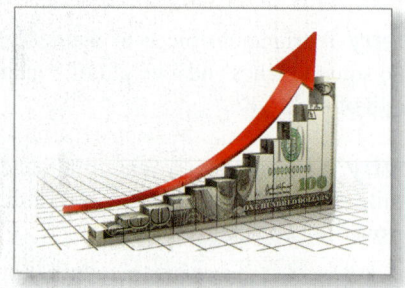

Section 3.5   Geometric and Scientific Applications   **139**

Application   EXAMPLE 4   Solving a Solution Mixture Problem

A pharmacist needs to strengthen a 15% alcohol solution with a pure alcohol solution to obtain a 32% solution. How much pure alcohol should be added to 100 milliliters of the 15% solution?

SOLUTION

In this problem, the rates are the alcohol *percents* of the solutions.

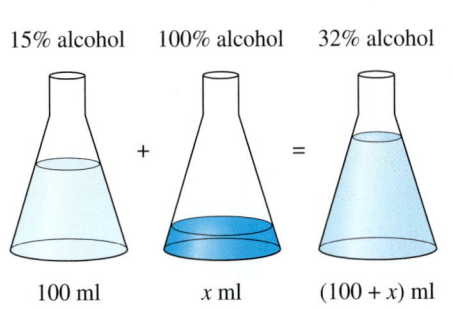

15% alcohol   100% alcohol   32% alcohol

100 ml x ml $(100 + x)$ ml

Verbal Model: Amount of alcohol in 15% alcohol solution + Amount of alcohol in 100% alcohol solution = Amount of alcohol in final alcohol solution

Labels:
15% solution: Percent alcohol = 0.15   (decimal form)
Amount of alcohol solution = 100   (milliliters)

100% solution: Percent alcohol = 1.00   (decimal form)
Amount of alcohol solution = x (milliliters)

32% solution: Percent alcohol = 0.32   (decimal form)
Amount of alcohol solution = $100 + x$ (milliliters)

Equation:
$0.15(100) + 1.00(x) = 0.32(100 + x)$ Write equation.
$15 + x = 32 + 0.32x$ Simplify.
$0.68x = 17$ Simplify.
$x = \dfrac{17}{0.68}$ Divide each side by 0.68.
$= 25$ Simplify.

So, the pharmacist should add 25 milliliters of pure alcohol to the 15% solution. This will result in $100 + x = 100 + 25 = 125$ milliliters of the 32% solution.

Exercises Within Reach®

Solutions in English & Spanish and tutorial videos at AlgebraWithinReach.com

11. Mixture You need to strengthen a 19% alcohol solution with a pure alcohol solution to obtain a 40% solution. How much pure alcohol should you add to 100 milliliters of the 19% solution?

12. Chemistry You need 175 milliliters of a 6% hydrochloric acid solution for an experiment. Your chemistry lab has a bottle of 3% hydrochloric acid solution and a bottle of 10% hydrochloric acid solution. How many milliliters of each solution should you mix together?

Solving Work-Rate Problems

Although not generally referred to as such, **work-rate problems** are actually mixture problems because they involve two or more rates.

$$\text{First rate} \cdot \text{Time} + \text{Second rate} \cdot \text{Time} = 1 \text{ (one whole job completed)}$$

Application

EXAMPLE 5 Solving a Work-Rate Problem

Consider two machines in a paper manufacturing plant. Machine 1 can complete one job in 3 hours. Machine 2 is newer and can complete one job in $2\frac{1}{2}$ hours. How long will it take the two machines working together to complete one job?

SOLUTION

Verbal Model: Portion done by Machine 1 + Portion done by Machine 2 = 1 (one whole job completed)

Labels:
- One whole job completed = 1 (job)
- Rate (Machine 1) = $\frac{1}{3}$ (job per hour)
- Time (Machine 1) = t (hours)
- Rate (Machine 2) = $\frac{2}{5}$ (job per hour)
- Time (Machine 2) = t (hours)

Equation:
$\left(\frac{1}{3}\right)t + \left(\frac{2}{5}\right)t = 1$ Write equation.
$\left(\frac{11}{15}\right)t = 1$ Combine like terms.
$t = \frac{15}{11}$ Multiply each side by $\frac{15}{11}$.

It will take $\frac{15}{11}$ hours (or about 1.36 hours) for the machines to complete the job working together. Check this solution in the original statement of the problem.

Exercises Within Reach®

Solutions in English & Spanish and tutorial videos at AlgebraWithinReach.com

13. **Mowing a Lawn** You can mow a lawn in 2 hours using a riding mower, and your friend can mow the same lawn in 3 hours using a push mower. Using both machines together, how long will it take you and your friend to mow the lawn?

14. **Typing Project** One person can complete a typing project in 6 hours, and another can complete the same project in 8 hours. How long will it take the two people working together to complete the project?

15. **Work Rate** One worker can complete a task in m minutes while a second can complete the task in $9m$ minutes. Show that by working together they can complete the task in $t = \frac{9}{10}m$ minutes.

16. **Work Rate** One worker can complete a task in h hours while a second can complete the task in $3h$ hours. Show that by working together they can complete the task in $t = \frac{3}{4}h$ hours.

Application

EXAMPLE 6 Solving a Fluid-Rate Problem

An above ground swimming pool has a capacity of 15,600 gallons. A drain pipe can empty the pool in $6\frac{1}{2}$ hours. At what rate (in gallons per minute) does the water flow through the drain pipe?

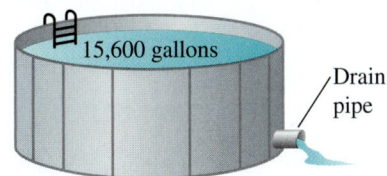

SOLUTION

To begin, change the time from hours to minutes by multiplying by 60. That is, $6\frac{1}{2}$ hours is equal to $(6.5)(60)$ or 390 minutes.

Verbal Model: $\boxed{\text{Volume of pool}} = \boxed{\text{Rate}} \cdot \boxed{\text{Time}}$

Labels:
Volume = 15,600 (gallons)
Rate = r (gallons per minute)
Time = 390 (minutes)

Equation:
$15{,}600 = r(390)$ Write equation.
$40 = r$ Divide each side by 390.

The water is flowing through the drain pipe at a rate of 40 gallons per minute. Check this solution in the original statement of the problem.

Exercises Within Reach®

Solutions in English & Spanish and tutorial videos at AlgebraWithinReach.com

17. Intravenous Bag A 1000-milliliter intravenous bag is attached to a patient with a tube and is empty after 8 hours. At what rate does the solution flow through the tube?

18. Swimming Pool A swimming pool has a capacity of 10,800 gallons. A drain pipe empties the pool at a rate of 12 gallons per minute. How long, in hours, will it take for the pool to empty?

19. Flower Order A floral shop receives a $384 order for roses and carnations. The order contains twice as many roses as carnations. The prices per dozen for the roses and carnations are $18 and $12, respectively. How many of each type of flower are in the order?

20. Ticket Sales Ticket sales for a play totaled $1700. The number of tickets sold to adults was three times the number sold to children. The prices of the tickets for adults and children were $5 and $2, respectively. How many of each type were sold?

Tracy Whiteside/Shutterstock.com

Concept Summary: Solving Geometric and Scientific Applications

What
Many real-life problems involve geometric applications such as perimeter, area, and volume. Other problems might involve distance, temperature, or interest.

EXAMPLE
You jog at an average rate of 8 kilometers per hour. How long will it take you to jog 14 kilometers?

How
To solve such problems, you can use formulas. For example, to solve a problem involving distance, rate, and time, you can use the distance formula.

EXAMPLE

Distance = Rate · Time

$d = rt$ Distance formula
$14 = 8(t)$ Substitute for d and r.
$\dfrac{14}{8} = t$ Divide each side by 8.
$1.75 = t$ Simplify.

It will take you 1.75 hours.

Why
You can use formulas to solve many real-life applications. Some formulas occur so frequently that it is to your benefit to memorize them.

Exercises Within Reach®

Worked-out solutions to odd-numbered exercises at AlgebraWithinReach.com

Concept Summary Check

21. **Formulas** What is the formula for the volume of a circular cylinder?

22. **Create an Example** Give an example in which you need to find the perimeter of a real-life object.

23. **Writing** In your own words, describe the units of measure used for perimeter, area, and volume. Give examples of each.

24. **Structure** Rewrite the formula for simple interest by solving for P.

Extra Practice

Using the Distance Formula In Exercises 25–28, find the missing distance, rate, or time.

	Distance, d	Rate, r	Time, t		Distance, d	Rate, r	Time, t
25.		4 m/min	12 min	26.		62 mi/hr	$2\frac{1}{2}$ hr
27.	210 mi	50 mi/hr		28.	2054 m		18 sec

Solving for a Variable In Exercises 29 and 30, solve for the specified variable.

29. Solve for h: $A = \frac{1}{2}bh$

30. Solve for r: $A = P + Prt$

Mixture Problem In Exercises 31–34, determine the numbers of units of solution 2 required to obtain the desired percent alcohol concentration of the final solution. Then find the amount of the final solution.

	Concentration Solution 1	Amount of Solution 1	Concentration Solution 2	Concentration Final Solution		Concentration Solution 1	Amount of Solution 1	Concentration Solution 2	Concentration Final Solution
31.	10%	25 gal	30%	25%	32.	25%	4 L	50%	30%
33.	15%	5 qt	45%	30%	34.	70%	18.75 gal	90%	75%

35. **Interest Rate** Find the annual interest rate on a certificate of deposit that earned $128.98 interest in 1 year on a principal of $1500.

36. **Interest** How long must $700 be invested at an annual interest rate of 6.25% to earn $460 interest?

37. *Geometry* Two sides of a triangle have the same length. The third side is 7 meters less than 4 times that length. The perimeter is 83 meters. What are the lengths of the three sides of the triangle?

38. *Geometry* The longest side of a triangle is 3 times the length of the shortest side. The third side of the triangle is 4 inches longer than the shortest side. The perimeter is 49 inches. What are the lengths of the three sides of the triangle?

39. *Distance* Two cars start at a given point and travel in the same direction at average speeds of 45 miles per hour and 52 miles per hour (see figure). How far apart will they be in 4 hours?

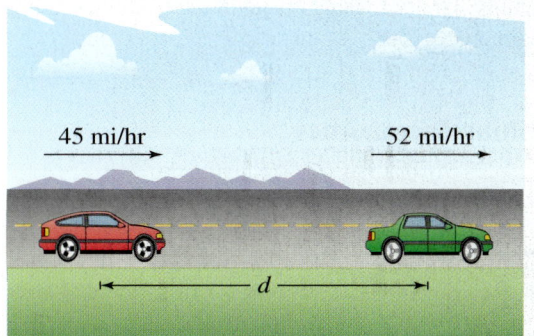

40. *Distance* Two planes leave Orlando International Airport approximately the same time and fly in opposite directions (see figure). Their speeds are 510 miles per hour and 600 miles per hour. How far apart will the planes be after $1\frac{1}{2}$ hours?

41. *Travel Time* On the first part of a 225-mile automobile trip, you averaged 55 miles per hour. On the last part of the trip, you averaged 48 miles per hour because of increased traffic congestion. The total trip took 4 hours and 15 minutes. Find the travel time for each part of the trip.

42. *Time* A jogger leaves a point on a fitness trail running at a rate of 4 miles per hour. Ten minutes later, a second jogger leaves from the same location running at 5 miles per hour. How long will it take the second jogger to overtake the first? How far will each have run at that point?

Explaining Concepts

43. *Mental Math* It takes you 4 hours to drive 180 miles. Explain how to use mental math to find your average speed. Then explain how your method is related to the formula $d = rt$.

44. *Error Analysis* A student solves the equation $S = 2lw + 2lh + 2wh$ for w and his answer is

$$w = \frac{S - 2lw - 2lh}{2h}.$$

Describe and correct the student's error.

45. *Structure* Write three equations that are equivalent to $A = \frac{1}{2}(x + y)h$ by solving for each variable, where A is the area, h is the height, and x and y are the bases of a trapezoid. Explain when you would use each equation.

46. *Think About It* When the height of a triangle doubles, does the area of the triangle double? Explain.

47. *Think About It* When the radius of a circle doubles, does its circumference double? Does its area double? Explain.

Cumulative Review

In Exercises 48 and 49, determine which of the numbers in the set are (a) natural numbers, (b) integers, (c) rational numbers, and (d) irrational numbers.

48. $\left\{-6, \frac{7}{4}, 2.1, \sqrt{49}, -8, \frac{4}{3}\right\}$

49. $\left\{1.8, \frac{1}{10}, 7, -2.75, 1, -3\right\}$

In Exercises 50–55, solve the proportion.

50. $\frac{x}{3} = \frac{28}{12}$

51. $\frac{1}{4} = \frac{y}{36}$

52. $\frac{z}{18} = \frac{8}{12}$

53. $\frac{3}{2} = \frac{9}{x}$

54. $\frac{5}{t} = \frac{75}{165}$

55. $\frac{34}{x} = \frac{102}{48}$

3.6 Linear Inequalities

▶ Graph the solution set of an inequality on the real number line, and determine whether a given value is a solution of an inequality.
▶ Solve a linear inequality and graph its solution set.
▶ Solve application problems involving inequalities.

Inequalities and Their Graphs

As with an equation, you can **solve an inequality** involving the variable x by finding all values of x for which the inequality is true. Such values are **solutions** and are said to **satisfy** the inequality. The **solution set** of an inequality is the set of all real numbers that are solutions of the inequality. To get a visual image of the solution set, it is helpful to sketch its **graph** on the real number line.

EXAMPLE 1 Graphing Inequalities

Study Tip

In the graph of an inequality, a parenthesis is used to exclude an endpoint from the solution interval. A square bracket is used to include an endpoint in the solution interval.

Inequality	Graph of Solution Set	Verbal Description
a. $x < 2$		x is less than 2.
b. $x > 2$		x is greater than 2.
c. $x \geq -2$		x is greater than or equal to -2.
d. $x \leq 0$		x is less than or equal to 0.
e. $-1 \leq x \leq 2$	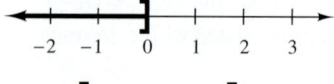	x is greater than or equal to -1 and less than or equal to 2.
f. $2 \leq x < 5$	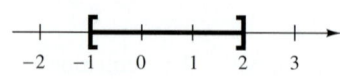	x is greater than or equal to 2 and less than 5.
g. $-3 < x \leq -1$	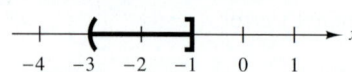	x is greater than -3 and less than or equal to -1.

Graphing an Inequality In Exercises 1–10, write a verbal description of the inequality and sketch its graph.

1. $x \geq 3$
2. $z > 8$
3. $x \leq 10$
4. $t < 0$
5. $y < -9$
6. $x \leq -2$
7. $5 \leq z \leq 10$
8. $-3 < x < 4$
9. $-\frac{3}{2} < y \leq 5$
10. $-3 \geq t > -3.8$

The procedure for checking solutions of inequalities is similar to the procedure for checking solutions of equations.

EXAMPLE 2 Checking Solutions of an Inequality

Determine whether (a) $x = 9$, (b) $x = -5$, or (c) $x = -1$ is a solution of $-3x + 8 \leq 11$.

SOLUTION

a.
$-3x + 8 \leq 11$ Write original inequality.
$-3(9) + 8 \stackrel{?}{\leq} 11$ Substitute 9 for x.
$-27 + 8 \stackrel{?}{\leq} 11$ Multiply.
$-19 \leq 11$ Solution checks. ✓

Because -19 is less than 11, you can conclude that $x = 9$ *is* a solution of the original inequality.

b.
$-3x + 8 \leq 11$ Write original inequality.
$-3(-5) + 8 \stackrel{?}{\leq} 11$ Substitute -5 for x.
$15 + 8 \stackrel{?}{\leq} 11$ Multiply.
$23 \leq 11$ Solution does not check. ✗

Because 23 is not less than or equal to 11, you can conclude that $x = -5$ *is not* a solution of the original inequality.

c.
$-3x + 8 \leq 11$ Write original inequality.
$-3(-1) + 8 \stackrel{?}{\leq} 11$ Substitute -1 for x.
$3 + 8 \stackrel{?}{\leq} 11$ Multiply.
$11 \leq 11$ Solution checks. ✓

Because 11 is equal to 11, you can conclude that $x = -1$ *is* a solution of the original inequality.

Exercises Within Reach®

Solutions in English & Spanish and tutorial videos at AlgebraWithinReach.com

Checking Solutions of an Inequality In Exercises 11–16, determine whether each value of x is a solution of the inequality.

Inequality   *Values*

11. $5x - 12 > 0$ (a) $x = 3$ (b) $x = -3$
                        (c) $x = \frac{5}{2}$ (d) $x = \frac{3}{2}$

12. $2x + 1 < 3$ (a) $x = 0$ (b) $x = 4$
                        (c) $x = -\frac{2}{5}$ (d) $x = \frac{1}{2}$

13. $3 - \frac{1}{2}x > 0$ (a) $x = 10$ (b) $x = 6$
                                  (c) $x = -\frac{3}{4}$ (d) $x = 0$

14. $\frac{2}{3}x + 4 < 6$ (a) $x = 7$ (b) $x = 0$
                                  (c) $x = -\frac{1}{2}$ (d) $x = 3$

15. $5x + 3 \leq x - 5$ (a) $x = 1$ (b) $x = -2$
                               (c) $x = -1$ (d) $x = 2$

16. $4 - 3x \geq x + 12$ (a) $x = 0$ (b) $x = 3$
                                (c) $x = -2$ (d) $x = -4$

Solving Linear Inequalities

The procedures for solving linear inequalities in one variable are much like those for solving linear equations. To isolate the variable, you can use the **properties of inequalities**. These properties are similar to the properties of equality, but there are two important exceptions. *When both sides of an inequality are multiplied or divided by a negative number, the direction of the inequality symbol must be reversed.* Here is an example.

$-2 < 5$ Original inequality

$(-3)(-2) > (-3)(5)$ Multiply each side by -3 and reverse inequality.

$6 > -15$ Simplify.

Two inequalities that have the same solution set are called **equivalent**.

Study Tip

Each of the properties below is true if the symbol < is replaced by ≤ and/or the symbol > is replaced by ≥.

Properties of Inequalities

Let *a*, *b*, and *c* be real numbers, variables, or algebraic expressions.

Property	Verbal Description	Algebraic Description
Addition	Add the same quantity to each side.	If $a < b$, then $a + c < b + c$.
Subtraction	Subtract the same quantity from each side.	If $a < b$, then $a - c < b - c$.
Multiplication	Multiply each side by a *positive* quantity.	If $a < b$ and c is positive, then $ac < bc$.
	Multiply each side by a *negative* quantity and reverse the inequality symbol.	If $a < b$ and c is negative, then $ac > bc$.
Division	Divide each side by a *positive* quantity.	If $a < b$ and c is positive, then $\dfrac{a}{c} < \dfrac{b}{c}$.
	Divide each side by a *negative* quantity and reverse the inequality symbol.	If $a < b$ and c is negative, then $\dfrac{a}{c} > \dfrac{b}{c}$.
Transitive	Consider three quantities of which the first is less than the second, and the second is less than the third. It follows that the first quantity must be less than the third quantity.	If $a < b$ and $b < c$, then $a < c$.

Exercises Within Reach®   Solutions in English & Spanish and tutorial videos at AlgebraWithinReach.com

Matching In Exercises 17−20, match the statement with the property it represents.

(a) Addition Property of Inequality   (b) Subtraction Property of Inequality

(c) Multiplication Property of Inequality   (d) Division Property of Inequality

17. $10 < 12$, so $\dfrac{10}{2} < \dfrac{12}{2}$.

18. $2 > -6$, so $2(7) > -6(7)$.

19. $3 < 9$, so $3 - 1 < 9 - 1$.

20. $-8 > -9$, so $-8 + 4 > -9 + 4$.

EXAMPLE 3 Solving a Linear Inequality

Solve and graph the inequality $x + 5 < 8$.

SOLUTION

$x + 5 < 8$	Write original inequality.
$x + 5 - 5 < 8 - 5$	Subtract 5 from each side.
$x < 3$	Solution set

The solution set is $x < 3$.

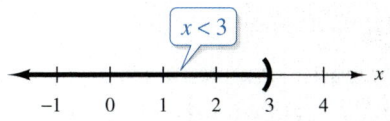

All real numbers less than 3

EXAMPLE 4 Solving a Linear Inequality

Solve and graph the inequality $3y - 1 \leq -13$.

SOLUTION

$3y - 1 \leq -13$	Write original inequality.
$3y - 1 + 1 \leq -13 + 1$	Add 1 to each side.
$3y \leq -12$	Combine like terms.
$\dfrac{3y}{3} \leq \dfrac{-12}{3}$	Divide each side by 3.
$y \leq -4$	Solution set

The solution set is $y \leq -4$.

All real numbers less than or equal to -4

Exercises Within Reach®

Solutions in English & Spanish and tutorial videos at AlgebraWithinReach.com

Solving a Linear Inequality In Exercises 21–38, solve and graph the inequality.

21. $t - 3 \geq 2$

22. $t + 1 < 6$

23. $x + 4 \leq 6$

24. $z - 2 > 0$

25. $4x < 12$

26. $2x > 3$

27. $-10x < 40$

28. $-6x > 18$

29. $\frac{2}{3}x \leq 12$

30. $\frac{1}{5}x > 3$

31. $3x + 2 \leq 14$

32. $8y + 4 > 0$

33. $4 - 2x < 3$

34. $14 - 3x > 5$

35. $12 - x > 4$

36. $18 - y \leq 5$

37. $3x + 9 < 2x$

38. $12 + 5x \geq 3x$

148 Chapter 3 Linear Equations and Problem Solving

EXAMPLE 5 Solving a Linear Inequality

Solve and graph the inequality $3(2x - 6) < 10x + 2$.

SOLUTION

$3(2x - 6) < 10x + 2$	Write original inequality.
$6x - 18 < 10x + 2$	Distributive Property
$6x - 10x - 18 < 10x - 10x + 2$	Subtract $10x$ from each side.
$-4x - 18 < 2$	Combine like terms.
$-4x - 18 + 18 < 2 + 18$	Add 18 to each side.
$-4x < 20$	Combine like terms.
$\dfrac{-4x}{-4} > \dfrac{20}{-4}$	Divide each side by -4 and reverse inequality.
$x > -5$	Solution set

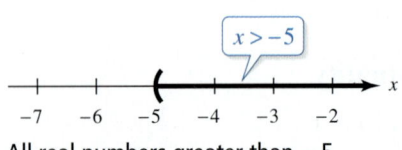

All real numbers greater than -5

EXAMPLE 6 Solving a Linear Inequality

Solve and graph the inequality $1 - \dfrac{3x}{2} \geq x - 4$.

SOLUTION

$1 - \dfrac{3x}{2} \geq x - 4$	Write original inequality
$2 - 3x \geq 2x - 8$	Multiply each side by 2.
$2 - 2 - 3x \geq 2x - 8 - 2$	Subtract 2 from each side.
$-3x \geq 2x - 10$	Combine like terms.
$-3x - 2x \geq 2x - 2x - 10$	Subtract $2x$ from each side.
$-5x \geq -10$	Combine like terms.
$\dfrac{-5x}{-5} \leq \dfrac{-10}{-5}$	Divide each side by -5 and reverse inequality.
$x \leq 2$	Solution set

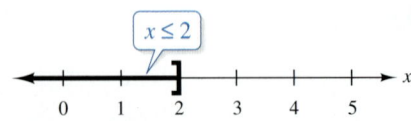

All real numbers less than or equal to 2

Exercises Within Reach®

Solutions in English & Spanish and tutorial videos at AlgebraWithinReach.com

Solving a Linear Inequality In Exercises 39–48, **solve** and **graph** the inequality.

39. $7t + 9 < 14 + 6t$

40. $25x + 4 \leq 10x + 19$

41. $2(x + 7) > 12$

42. $9(y - 4) \leq 36$

43. $6 + \dfrac{2x}{3} < x + 7$

44. $1 - \dfrac{5x}{4} \geq 2x + 13$

45. $-2(z + 1) \geq 3(z + 1)$

46. $8(t - 3) < 4(t - 3)$

47. $\dfrac{x}{4} + \dfrac{1}{2} > 0$

48. $\dfrac{y}{4} - \dfrac{5}{8} < 2$

Applications

Linear inequalities in real-life problems arise from statements that involve key phrases such as "at least," "no more than," "minimum value," and so on.

EXAMPLE 7 Translating Verbal Phrases

	Verbal Statement	Inequality		Verbal Statement	Inequality
a.	x is at most 2.	$x \leq 2$	b.	x is no more than 2.	$x \leq 2$
c.	x is at least 2.	$x \geq 2$	d.	x is more than 2.	$x > 2$
e.	x is less than 2.	$x < 2$	f.	x is a minimum of 2.	$x \geq 2$

Application EXAMPLE 8 Comparing Costs

A car can be rented from Company A for $240 per week with no extra charge for mileage. A similar car can be rented from Company B for $100 per week, plus 25 cents for each mile driven. How many miles must you drive per week so that the rental fee for Company B is more than that for Company A?

SOLUTION

Miles Driven	Company A	Company B
557	$240.00	$239.25
558	$240.00	$239.50
559	$240.00	$239.75
560	$240.00	$240.00
561	$240.00	$240.25
562	$240.00	$240.50
563	$240.00	$240.75

Verbal Model: Weekly cost for Company B $>$ Weekly cost for Company A

Labels: Number of miles driven per week $= m$ (miles)
Weekly cost for A $= 240$ (dollars)
Weekly cost for B $= 100 + 0.25m$ (dollars)

Inequality:
$100 + 0.25m > 240$ Write inequality.
$0.25m > 140$ Subtract 100 from each side.
$m > 560$ Divide each side by 0.25.

The car from Company B is more expensive if you drive more than 560 miles per week. The table at the left confirms this conclusion.

Exercises Within Reach®

Solutions in English & Spanish and tutorial videos at AlgebraWithinReach.com

Translating a Verbal Phrase In Exercises 49–52, translate the verbal statement into a linear inequality.

49. x is at most -1.

50. x is a minimum of 12.

51. x is greater than 0 and less than or equal to 6.

52. x is at least -3 and at most 5.

53. **Cargo Weight** The weight of a truck is 4350 pounds. The legal gross weight of the loaded truck is 6000 pounds. Find an interval for the number of bushels of grain that the truck can haul when each bushel weighs 48 pounds.

54. **Calling Card** An international calling card charges $0.25 to connect a call and $0.18 for each minute used. Your calling card has $7 on it. Find the interval of time that is available for a call.

Concept Summary: Solving Linear Inequalities

What
When you **solve an inequality**, you find all the values for which the inequality is true. The procedures for solving inequalities are similar to those for solving equations.

EXAMPLE

Solve the inequality $-2x + 4 > 8$.

How
To solve an inequality, isolate the variable by using the properties of inequalities.

EXAMPLE

$-2x + 4 > 8$ Write inequality.
$-2x + 4 - 4 > 8 - 4$ Subtract 4.
$-2x > 4$ Combine terms.
$\dfrac{-2x}{-2} < \dfrac{4}{-2}$ Divide by -2 and reverse inequality.
$x < -2$ Solution set

Why
Many real-life applications involve phrases like "at least" or "no more than." Using inequalities, you will be able to solve such problems.

Exercises Within Reach®

Worked-out solutions to odd-numbered exercises at AlgebraWithinReach.com

Concept Summary Check

55. *Writing* Does the graph of $x < -2$ contain a parenthesis or a square bracket? Explain.

56. *Solving an Inequality* Does dividing each side of an inequality by 5 yield the same result as multiplying each side by $\frac{1}{5}$? Give an example.

57. *Writing* Compare solving linear equations with solving linear inequalities.

58. *Solution Sets* How many real numbers are in the solution set of a linear inequality? Give an example.

Extra Practice

Checking Solutions of an Inequality In Exercises 59–62, determine whether each value of x is a solution of the inequality.

	Inequality	Values			Inequality	Values	
59.	$5(x - 2) + 1 < 12$	(a) $x = 4$	(b) $x = 1$	**60.** $3(x + 5) - 4 > 2$	(a) $x = 3$	(b) $x = 0$
		(c) $x = 5$	(d) $x = -3$		(c) $x = -4$	(d) $x = -10$
61.	$15 - (x + 8) \geq 13$	(a) $x = 0$	(b) $x = -6$	**62.** $9 - (x + 3) \leq 10$	(a) $x = -4$	(b) $x = 4$
		(c) $x = 2$	(d) $x = -10$		(c) $x = 0$	(d) $x = -6$

Solving an Inequality In Exercises 63–70, **solve** and **graph** the inequality.

63. $-3(x + 11) \leq 6$

64. $-7(z + 4) > 14$

65. $3(x + 1) \geq 2(x + 5)$

66. $8(z - 2) < 4(z + 1)$

67. $10(1 - y) < -4(y - 2)$

68. $6(3 - z) \geq 5(3 + z)$

69. $\dfrac{x}{5} - \dfrac{x}{2} \leq 1$

70. $\dfrac{x}{3} + \dfrac{x}{4} \geq 1$

71. Astronomy Mars is farther from the Sun than Venus, and Venus is farther from the Sun than Mercury. What can be said about the relationship between the distances of Mars and Mercury from the Sun? Identify the property of inequalities that is demonstrated.

72. Budgets The personnel department's budget is less than the finance department's budget, and the finance department's budget is less than the research and development department's budget. What can you say about the relationship between the budgets of personnel department and the research and development department? Identify the property of inequalities that is demonstrated.

73. Course Grade You are taking a biology course in which your grade is based on five 100-point tests. To earn an A in the course, you must have a total of at least 90% of the points. On the first 4 tests, your scores were 93, 88, 91, and 82. How many points do you have to obtain on the fifth test in order to earn an A in the course?

74. Quiz Grade In your statistics class, there are eight 20-point quizzes. To earn an A for the quizzes, you must have a total of at least 90% of the points. On the first 7 quizzes, your scores were 19, 20, 20, 15, 19, 14, and 18. How many points do you have to obtain on the eighth quiz in order to earn an A for the quizzes?

75. Comparing Costs You can rent a minivan from Company A for $270 per week with unlimited mileage. A similar minivan can be rented from Company B for $180 per week plus an additional 25 cents for each mile driven. How many miles must you drive per week so that the rental fee for Company B is more than that for Company A?

76. Comparing Costs A cell phone company offers two text messaging plans. With the first plan, you can send an unlimited number of text messages for $14 per month. With the second plan, you can send 200 text messages for $5 and then pay $0.15 for every additional message sent. How many text messages must be sent in a month for the first plan to be less expensive than the second plan?

Explaining Concepts

True or False? In Exercises 77–82, **determine** whether the statement is true or false. **Justify** your answer.

77. The inequality $x + 6 > 0$ is equivalent to $x > -6$.

78. The statement that z is nonnegative is equivalent to the inequality $z > 0$.

79. The statement that u is at least 10 is equivalent to the inequality $u \geq 10$.

80. The inequality $-\frac{1}{2}x + 6 > 0$ is equivalent to $x > 12$.

81. $x < x + 4$

82. $2x - 5 \geq 2x$

83. Error Analysis Describe and correct the error in graphing $x < 5$.

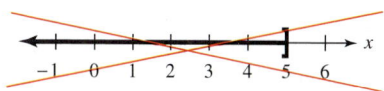

Cumulative Review

In Exercises 84–89, write a verbal description of the algebraic expression, without using a variable. (There is more than one correct answer.)

84. $x - 12$

85. $2x + 7$

86. $4 - 5x$

87. $-5(x + 7)$

88. $\dfrac{x - 5}{9}$

89. $\dfrac{1}{3} + \dfrac{x}{2}$

In Exercises 90–95, find the missing distance, rate, or time.

	Distance, d	Rate, r	Time, t
90.		35 mi/hr	$\frac{3}{4}$ hr
91.		48 ft/sec	38 sec
92.	200 mi	60 mi/hr	
93.	384 m	6 m/min	
94.	1028 ft		5 min
95.	240 km		20 hr

3 Chapter Summary

	What did you learn?	Explanation and Examples	Review Exercises
3.1	Solve a linear equation in standard form *(p. 102)*.	To solve a linear equation, use inverse operations to isolate the variable.	1–6
	Solve a linear equation in nonstandard form *(p. 104)*.	To solve a linear equation in nonstandard form, simplify each side of the equation before using inverse operations to isolate the variable.	7–20
3.2	Solve linear equations containing symbols of grouping *(p. 110)*.	To solve a linear equation that contains symbols of grouping, remove symbols of grouping from each side using the Distributive Property, combine like terms, isolate the variable using properties of equality, and check your solution in the original equation.	21–26
	Solve linear equations involving fractions *(p. 112)*.	To clear an equation of fractions, multiply each side by the least common multiple (LCM) of the denominators.	27–34, 39, 40
	Solve linear equations involving decimals *(p. 114)*.	To clear an equation of decimals, multiply each side by a power of 10 that converts all decimal coefficients to integers.	35–38
3.3	Convert percents to decimals and fractions, and vice versa *(p. 118)*.	To convert from percent form to decimal form, move the decimal point two places to the left. To convert from decimal form to percent form, move the decimal point two places to the right.	41–48
	Solve linear equations involving percents *(p. 119)*.	Use the percent equation. The percent equation $a = p \cdot b$ compares two numbers. b = base number p = percent (in decimal form) a = number being compared to b	49–54

	Solve problems involving markups and discounts *(p. 122)*.	1. Write a verbal model that describes the problem.
		  Selling price = Cost + Markup
		  Markup = Market rate · Cost
		  Sale price = Original price − Discount
		  Discount = Discount rate · Original price
		2. Assign labels to fixed quantities and variable quantities.
		3. Rewrite the verbal model as an algebraic equation using the assigned labels.
		4. Solve the resulting algebraic equation.
		5. Check to see that your solution satisfies the original problem as stated.
			55, 56

Chapter Summary

	What did you learn?	Explanation and Examples	Review Exercises
3.4	Compare relative sizes using ratios *(p. 128)*.	When comparing the relative sizes of two quantities, be sure to use the same units of measurement for both quantities.	57–62
	Find the unit price of a consumer item *(p. 130)*.	The unit price of an item is given by the ratio of the total price to the total number of units.	63, 64
	Solve proportions that equate two ratios *(p. 131)*.	A proportion equates two ratios. If $\frac{a}{b} = \frac{c}{d}$, then $ad = bc$.	65–72
3.5	Use common formulas to solve application problems *(p. 136)*.	Temperature: F = degrees Fahrenheit C = degrees Celsius $F = \frac{9}{5}C + 32$ Simple Interest: I = interest P = principal r = interest rate (decimal form) t = time (years) $I = Prt$ Distance: d = distance traveled r = rate t = time $d = rt$ Also see page 137 for common formulas for area, perimeter, and volume.	73–84
	Solve mixture problems *(p. 138)*.	Mixture problems involve combinations of two or more quantities that make up a new or different quantity.	85, 86
	Solve work-rate problems *(p. 140)*.	Work-rate problems are actually mixture problems because they involve two or more rates.	87, 88
3.6	Graph the solution set of an inequality on the real number line, and determine whether a given value is a solution of an inequality *(p. 144)*.	When graphing a solution set on a number line, a parenthesis is used to exclude an endpoint from the solution interval and a square bracket is used to include an endpoint in the solution interval. To determine whether a given value is a solution of an inequality, substitute the value into the original inequality and simplify each side.	89–98
	Solve a linear inequality and graph its solution set *(p. 146)*.	To solve a linear inequality, use the properties of inequality which are similar to the properties of equality, with two exceptions. When both sides of an inequality are multiplied or divided by a negative number, the direction of the inequality symbol must be reversed.	99–110
	Solve application problems involving inequalities *(p. 149)*.	When translating inequalities, remember that "at most" means "less than or equal to" and "at least" means "greater than or equal to."	111–116

Review Exercises

Worked-out solutions to odd-numbered exercises at AlgebraWithinReach.com

3.1

Solving a Linear Equation In Exercises 1–16, solve the equation and check your solution.

1. $2x - 10 = 0$
2. $12y + 72 = 0$
3. $-3y - 12 = 0$
4. $-7x + 21 = 0$
5. $5x - 3 = 0$
6. $-8x + 6 = 0$
7. $x + 10 = 13$
8. $x - 3 = 8$
9. $10x = 50$
10. $-3x = 21$
11. $8x + 7 = 39$
12. $12x - 5 = 43$
13. $15x - 4 = 16$
14. $3x - 8 = 2$
15. $\dfrac{x}{5} = 4$
16. $-\dfrac{x}{14} = \dfrac{1}{2}$

17. **Earnings** Your hourly wage is $9.75 per hour plus $0.80 for each unit you produce. How many units must you produce in 1 hour so that your hourly wage is $19.35?

18. **Carpet** The total cost for carpet installation (including materials and labor) was $990. The materials cost $710 and the cost of labor was $70 per hour. How many hours did it take to install the carpet?

19. **Geometry** The perimeter of a rectangle is 312 meters. The length is 50 meters greater than the width. Find the dimensions of the rectangle.

20. **Geometry** A 12-foot board is cut so that one piece is 5 times as long as the other. Find the length of each piece.

3.2

Solving a Linear Equation In Exercises 21–34, solve the equation and check your solution.

21. $3x - 2(x + 5) = 10$
22. $4x + 2(7 - x) = 5$
23. $2(x + 3) = 6(x - 3)$
24. $8(x - 2) = 3(x + 2)$
25. $7 - [2(3x + 4) - 5] = x - 3$
26. $14 + [3(6x - 15) + 4] = 5x - 1$
27. $\dfrac{2}{3}x - \dfrac{1}{6} = \dfrac{9}{2}$
28. $\dfrac{1}{8}x + \dfrac{3}{4} = \dfrac{5}{2}$
29. $\dfrac{u}{10} + \dfrac{u}{5} = 6$
30. $\dfrac{x}{3} + \dfrac{x}{5} = 1$
31. $\dfrac{2x}{9} = \dfrac{2}{3}$
32. $\dfrac{5y}{13} = \dfrac{2}{5}$
33. $\dfrac{x+3}{5} = \dfrac{x+7}{12}$
34. $\dfrac{y-2}{6} = \dfrac{y+1}{15}$

Solving a Linear Equation In Exercises 35–38, solve the equation. Round your answer to two decimal places.

35. $5.16x - 87.5 = 32.5$
36. $2.825x + 3.125 = 12.5$
37. $\dfrac{x}{4.625} = 48.5$
38. $5x + \dfrac{1}{4.5} = 18.125$

39. **Time to Complete a Task** Two people can complete 50% of a task in t hours, where t must satisfy the equation $\dfrac{t}{10} + \dfrac{t}{15} = 0.5$. How long will it take for the two people to complete 50% of the task?

40. **Course Grade** To get an A in a course, you must have an average of at least 90 points for 4 tests of 100 points each. For the first 3 tests, your scores are 85, 96, and 89. What must you score on the fourth test to earn a 90% average for the course?

3.3

Finding Equivalent Forms of a Percent In Exercises 41–48, complete the table showing the equivalent forms of the percent.

	Percent	Parts out of 100	Decimal	Fraction
41.	60%			
42.	35%			
43.				$\dfrac{4}{5}$
44.				$\dfrac{5}{8}$
45.			0.20	
46.			1.35	
47.		55		
48.		12.5		

Solving Percent Equations In Exercises 49−54, solve the percent equation.

49. What number is 125% of 16?
50. What number is 0.8% of 3250?
51. 150 is $37\frac{1}{2}$% of what number?
52. 323 is 95% of what number?
53. 150 is what percent of 250?
54. 130.6 is what percent of 3265?

55. **Selling Price** An electronics store uses a markup rate of 78% of all items. The cost of a CD player is $48. What is the selling price of the CD player?

56. **Sale Price** A sporting goods store advertises 30% off the original price of all golf equipment. A set of golf clubs has an original price of $229.99. What is the sale price?

3.4

Comparing Measurements In Exercises 57−60, find a ratio that compares the relative sizes of the quantities. (Use the same units of measurement for both quantities.)

57. Eighteen inches to 4 yards
58. One pint to 2 gallons
59. Two hours to 90 minutes
60. Four meters to 150 centimeters

61. **Hours** You are in school for 7.5 hours per day and you sleep for 6 hours per day. Find the ratio of the number of hours you sleep to the number of hours you are in school.

62. **Grandchildren** A grandmother has 2 grandsons and 8 granddaughters. Find the ratio of the number of grandsons to the number of granddaughters.

Comparing Unit Prices In Exercises 63 and 64, determine which product has the lower unit price.

63. (a) An 18-ounce container of cooking oil for $1.79
    (b) A 24-ounce container of cooking oil for $1.99

64. (a) A 17.4-ounce box of pasta noodles for $1.32
    (b) A 32-ounce box of pasta noodles for $2.62

Solving a Proportion In Exercises 65−70, solve the proportion.

65. $\dfrac{7}{16} = \dfrac{z}{8}$
66. $\dfrac{x}{12} = \dfrac{5}{4}$
67. $\dfrac{x+2}{4} = -\dfrac{1}{3}$
68. $\dfrac{x-4}{1} = \dfrac{9}{4}$
69. $\dfrac{x-3}{2} = \dfrac{x+6}{5}$
70. $\dfrac{x+1}{3} = \dfrac{x+2}{4}$

71. **Entertainment** A band charges $200 to play for three hours. How much would they charge to play for two hours?

72. **Resizing a Picture** You have a 4-by-6-inch photo of the student council that must be reduced to a size of 3.2 inches by 4.8 inches for the school yearbook. What percent does the photo need to be reduced to in order for it to fit in the allotted space?

3.5

Using the Distance Formula In Exercises 73−78, find the missing distance, rate, or time.

	Distance, d	Rate, r	Time, t
73.		65 mi/hr	8 hr
74.		45 mi/hr	2 hr
75.	855 m	5 m/min	
76.	205 mi	60 mi/hr	
77.	3000 mi		50 hr
78.	1000 km		25 hr

79. **Distance** An airplane has an average speed of 475 miles per hour. How far will it travel in $2\frac{1}{3}$ hours?

80. **Average Speed** You can walk 20 kilometers in 3 hours and 47 minutes. What is your average speed?

81. **Geometry** The width of a rectangular swimming pool is 4 feet less than its length. The perimeter of the pool is 112 feet. Find the dimensions of the pool.

82. **Geometry** The perimeter of an isosceles triangle is 65 centimeters. Find the length of the two equal sides if each is 10 centimeters longer than the third side. (An isosceles triangle has two sides of equal length.)

Simple Interest In Exercises 83 and 84, use the simple interest formula.

83. Find the total interest you will earn on a $1000 corporate bond that matures in 5 years and has an annual interest rate of 9.5%.

84. Find the annual interest rate on a certificate of deposit that pays $60 per year in interest on a principal of $750.

85. **Numbers of Coins** You have 30 coins in dimes and quarters with a combined value of $5.55. Determine the number of coins of each type.

86. **Bird Seed Mixture** A pet store owner mixes two types of bird seed that cost $1.25 per pound and $2.20 per pound to make 20 pounds of a mixture that costs $1.82 per pound. How many pounds of each kind of bird seed are in the mixture?

87. **Work Rate** One person can complete a task in 5 hours, and another can complete the same task in 6 hours. How long will it take both people working together to complete the task?

88. **Work Rate** The person in Exercise 87 who can complete the task in 5 hours has already worked 1 hour when the second person starts. How long will they work together to complete the task?

3.6

Graphing an Inequality In Exercises 89–92, write a verbal description of the inequality and sketch its graph.

89. $x < 3$

90. $x \geq 20$

91. $1 \leq x < 4$

92. $-3 < x < 0$

Checking Solutions of an Inequality In Exercises 93–98, determine whether each value of x is a solution of the inequality.

	Inequality	Values	
93.	$7x - 10 > 0$	(a) $x = 3$	(b) $x = \frac{1}{2}$
94.	$3x + 2 < 1$	(a) $x = 0$	(b) $x = -4$
95.	$\frac{1}{4}x - 2 < 1$	(a) $x = 4$	(b) $x = 32$
96.	$5 + \frac{2}{9}x > 4$	(a) $x = -9$	(b) $x = 18$
97.	$3 - (x - 4) < 0$	(a) $x = 2$	(b) $x = 9$
98.	$2 + x > 3x - 4$	(a) $x = -1$	(b) $x = 0$

Solving a Linear Inequality In Exercises 99–110, solve and graph the inequality.

99. $x + 9 \geq 7$

100. $x - 2 \leq 1$

101. $3x - 8 < 1$

102. $4x + 3 > 15$

103. $-11x \leq -22$

104. $-7x \geq 21$

105. $\frac{4}{5}x > 8$

106. $\frac{2}{3}n < -4$

107. $14 - \frac{1}{2}t < 12$

108. $32 + \frac{7}{8}k > 11$

109. $3 - 3y \geq 2(4 + y)$

110. $4 - 3y \leq 8(10 - y)$

Translating a Verbal Phrase In Exercises 111–114, translate the verbal statement into a linear inequality.

111. z is at least 10.

112. x is less than -3.

113. y is greater than 0 and no more than 100.

114. x is at least 2 and at most 12.

115. **Budget** You have budgeted $250 for a car rental on your vacation. The rental agency charges $184 per week plus an additional $0.75 per mile. To stay within your budget, how many miles can you drive the rental car?

116. **Fundraiser** A neighbor sponsors you in a bowl-a-thon for a school fundraiser. Your neighbor is going to give you $1.60 plus $0.18 per pin that you hit. How many pins do you need to hit to earn at least $25.00 from your neighbor?

Chapter Test

Solutions in English & Spanish and tutorial videos at AlgebraWithinReach.com

Take this test as you would take a test in class. After you are done, check your work against the answers in the back of the book.

In Exercises 1–6, solve the equation and check your solution.

1. $8x + 104 = 0$
2. $4x - 3 = 18$
3. $5 - 3x = -2x - 2$
4. $4 - (x - 3) = 5x + 1$
5. $\frac{2}{3}x = \frac{1}{9} + x$
6. $\frac{t + 2}{3} = \frac{2t}{9}$

7. Solve $4.08(x + 10) = 9.50(x - 2)$. Round your answer to two decimal places.

8. The bill (including parts and labor) for the repair of an oven was $142. The cost of parts was $62 and the cost of labor was $32 per hour. How many hours were spent repairing the oven?

9. Write the fraction $\frac{5}{16}$ as a percent and as a decimal.

10. 324 is 27% of what number?

11. 90 is what percent of 250?

12. What number is 25% of 24?

13. Write the ratio of 40 inches to 2 yards as a fraction in simplest form. Use the same units for both quantities, and explain how you made this conversion.

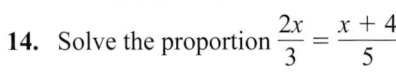

14. Solve the proportion $\frac{2x}{3} = \frac{x + 4}{5}$.

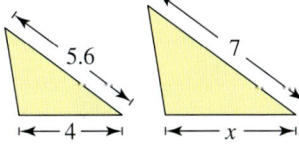

15. Find the length x of the side of the larger triangle shown in the figure at the left. (Assume that the two triangles are similar, and use the fact that corresponding sides of similar triangles are proportional.)

16. You traveled 264 miles in 4 hours. What was your average speed?

17. You can paint a building in 9 hours. Your friend can paint the same building in 12 hours. How long will it take the two of you working together to paint the building?

18. How much must you deposit in an account to earn $500 per year at 8% simple interest?

In Exercises 19–24, solve and graph the inequality.

19. $x + 3 \leq 7$
20. $-\frac{2x}{3} > 4$
21. $21 - 3x \leq 6$
22. $3(6 + 2x) > 8$
23. $-4(9 + 2x) \geq -40$
24. $-(3 + x) < 2(3x - 5)$

Cumulative Test: Chapters 1–3

Solutions in English & Spanish and tutorial videos at AlgebraWithinReach.com

Take this test as you would take a test in class. After you are done, check your work against the answers in the back of the book.

1. Place the correct inequality symbol (< or >) between the real numbers.

   $-\frac{3}{4}$ ☐ $\left|-\frac{7}{8}\right|$

In Exercises 2–7, evaluate the expression.

2. $(-200)(2)(-3)$
3. $\frac{3}{8} - \frac{5}{6}$
4. $-\frac{2}{9} \div \frac{8}{75}$
5. $-(-2)^3$
6. $3 + 2(6) - 1$
7. $24 + 12 \div 3$

In Exercises 8 and 9, evaluate the expression when $x = -2$ and $y = 3$.

8. $-3x - (2y)^2$
9. $\frac{5}{6}y + x^3$

10. Use exponential form to write the product $3 \cdot (x + y) \cdot (x + y) \cdot 3 \cdot 3$.

11. Use the Distributive Property to expand $-2x(x - 3)$.

12. Identify the property of real numbers illustrated by

    $2 + (3 + x) = (2 + 3) + x$.

In Exercises 13–15, simplify the expression.

13. $(3x^3)(5x^4)$
14. $2x^2 - 3x + 5x^2 - (2 + 3x)$
15. $4(x^2 + x) + 7(2x - x^2)$

In Exercises 16–18, solve the equation and check your solution.

16. $12x - 3 = 7x + 27$
17. $2x - \dfrac{5x}{4} = 13$
18. $5(x + 8) = -2x - 9$

19. Solve and graph the inequality.

    $-8(x + 5) \le 16$

20. The sticker on a new car gives the fuel efficiency as 28.3 miles per gallon. In your own words, explain how to estimate the annual fuel cost for the buyer when the car will be driven approximately 15,000 miles per year and the fuel cost $3.599 per gallon.

21. Write the ratio "24 ounces to 2 pounds" as a fraction in simplest form.

22. The original price of a digital camcorder is $1150. The camcorder is on sale for "20% off" the original price. Find the sale price.

23. The figure at the left shows two pieces of property. The assessed values of the properties are proportional to their areas. The value of the larger piece is $95,000. What is the value of the smaller piece?

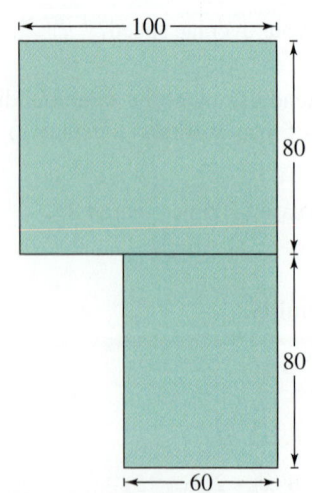

4
Equations and Inequalities in Two Variables

4.1 Ordered Pairs and Graphs
4.2 Graphs of Equations in Two Variables
4.3 Slope and Graphs of Linear Equations
4.4 Equations of Lines
4.5 Graphs of Linear Inequalities

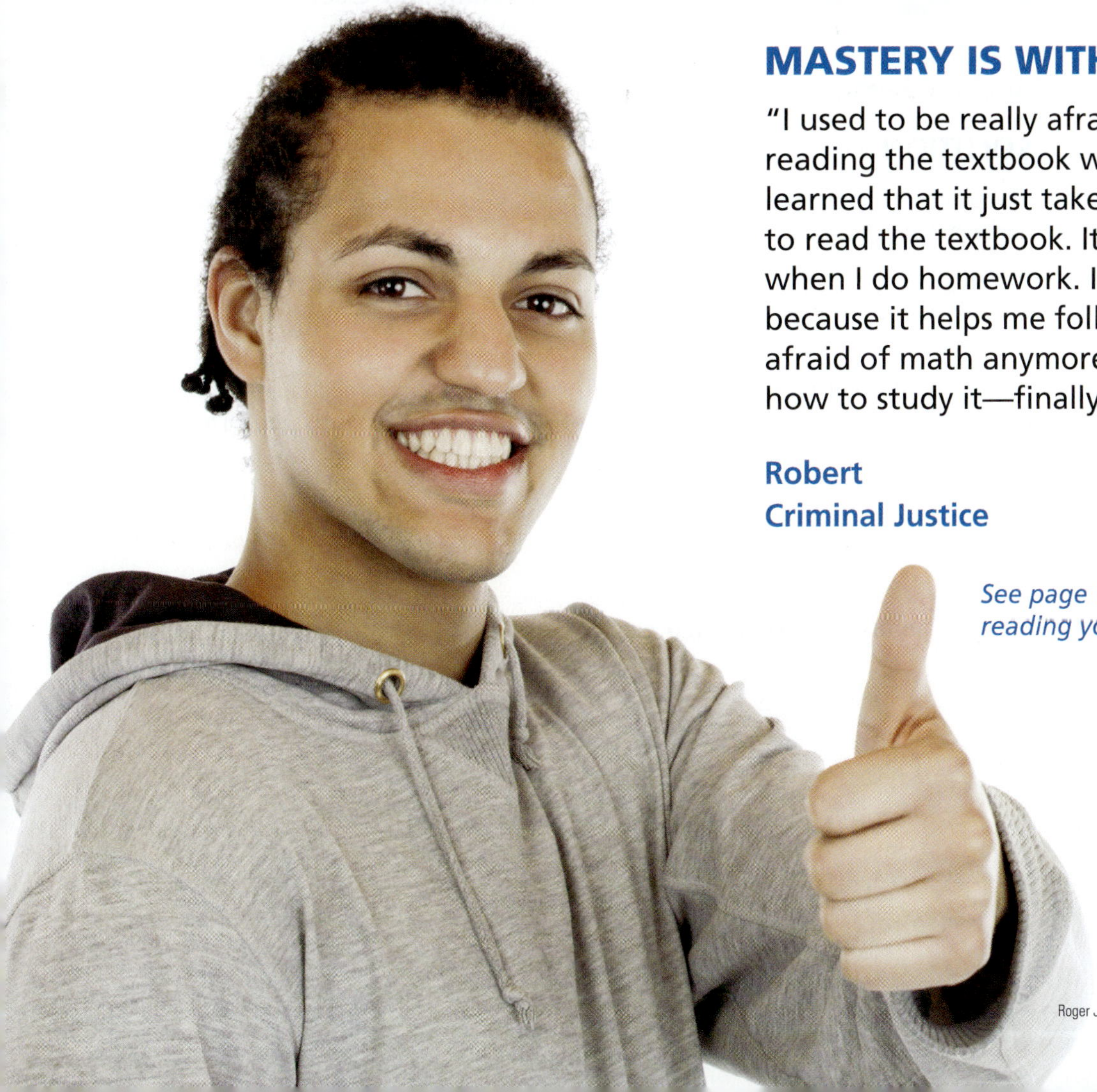

MASTERY IS WITHIN REACH!

"I used to be really afraid of math, so reading the textbook was difficult. I have learned that it just takes different strategies to read the textbook. It's my resource book when I do homework. I take it to class because it helps me follow along. I'm not afraid of math anymore because I know how to study it—finally."

Robert
Criminal Justice

See page 185 for suggestions about reading your textbook like a manual.

4.1 Ordered Pairs and Graphs

- Plot points on a rectangular coordinate system.
- Determine whether ordered pairs are solutions of equations.
- Use the verbal problem-solving method to plot points on a rectangular coordinate system.

The Rectangular Coordinate System

A **rectangular coordinate system** is formed by two real number lines intersecting at right angles. The horizontal number line is the ***x*-axis** and the vertical number line is the ***y*-axis**. (The plural of axis is *axes*.)

The point of intersection of the two axes is called the **origin**, and the axes separate the plane into four regions called **quadrants**. Each point in the plane corresponds to an **ordered pair** (x, y) of real numbers x and y called the **coordinates** of the point.

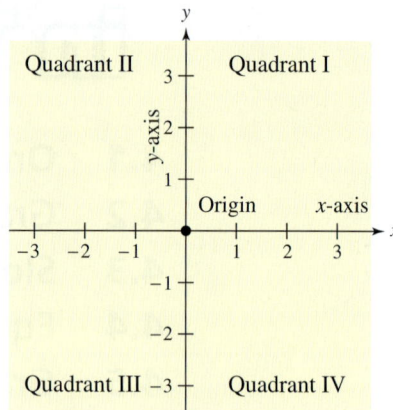

EXAMPLE 1 Plotting Points on a Rectangular Coordinate System

Plot the points $(-1, 2)$, $(3, 0)$, $(2, -1)$, $(3, 4)$, $(0, 0)$, and $(-2, -3)$ on a rectangular coordinate system.

SOLUTION

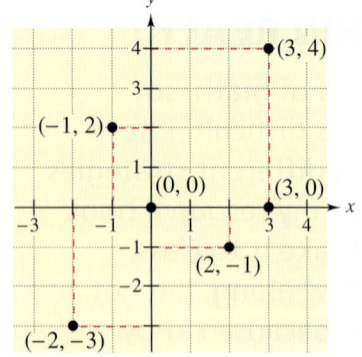

The point $(-1, 2)$ is one unit to the *left* of the vertical axis and two units *above* the horizontal axis.

One unit to the left of the vertical axis ⎫ Two units above the horizontal axis
$$(-1, 2)$$

Similarly, the point $(3, 0)$ is three units to the *right* of the vertical axis and *on* the horizontal axis. (It is on the horizontal axis because the *y*-coordinate is zero.) The other four points can be plotted in a similar way.

Exercises Within Reach®   Solutions in English & Spanish and tutorial videos at AlgebraWithinReach.com

Plotting Points In Exercises 1−8, **plot** the points on a rectangular coordinate system.

1. $(3, 2), (-4, 2), (2, -4)$
2. $(-1, 6), (-1, -6), (4, 6)$
3. $(-10, -4), (4, -4), (0, 0)$
4. $(-6, 4), (0, 0), (3, -2)$
5. $(-3, 4), (0, -1), (2, -2), (5, 0)$
6. $(-1, 3), (0, 2), (-4, -4), (-1, 0)$
7. $\left(\frac{3}{2}, -1\right), \left(-3, \frac{3}{4}\right), \left(\frac{1}{2}, -\frac{1}{2}\right)$
8. $\left(-\frac{2}{3}, 4\right), \left(\frac{1}{2}, -\frac{5}{2}\right), \left(-4, -\frac{5}{4}\right)$

Determining the Quadrant In Exercises 9−14, **determine** the quadrant in which the point is located without plotting it.

9. $(-3, 1)$
10. $(4, -3)$
11. $\left(-\frac{1}{8}, -\frac{2}{7}\right)$
12. $\left(\frac{3}{11}, \frac{7}{8}\right)$
13. $(-100, -365.6)$
14. $(-157.4, 305.6)$

Application

EXAMPLE 2 Graphing Super Bowl Scores

The scores of the Super Bowl games from 1992 through 2012 are shown in the table. Plot these points on a rectangular coordinate system. (*Source:* National Football League)

Year	Winning Score	Losing Score
1992	37	24
1993	52	17
1994	30	13
1995	49	26
1996	27	17
1997	35	21
1998	31	24
1999	34	19
2000	23	16
2001	34	7
2002	20	17
2003	48	21
2004	32	29
2005	24	21
2006	21	10
2007	29	17
2008	17	14
2009	27	23
2010	31	17
2011	31	25
2012	21	17

SOLUTION

The *x*-coordinates of the points represent the year, and the *y*-coordinates represent the winning and losing scores. The winning scores are shown as black dots, and the losing scores are shown as blue dots. Note that the break in the *x*-axis indicates that the numbers between 0 and 1992 have been omitted.

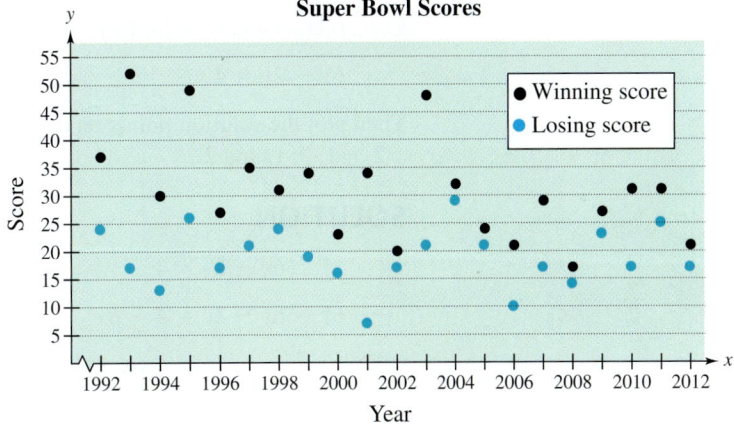

Exercises Within Reach®

Solutions in English & Spanish and tutorial videos at AlgebraWithinReach.com

15. *Organizing Data* The table shows the normal average temperatures *y* (in degrees Fahrenheit) in Anchorage, Alaska, for each month *x* of the year, with $x = 1$ corresponding to January. (*Source:* National Climatic Data Center)

x	1	2	3	4	5	6
y	16	19	26	36	47	55

x	7	8	9	10	11	12
y	58	56	48	34	22	18

(a) Plot the data in the table. Did you use the same scale on both axes? Explain.

(b) Using the graph, find the month for which the normal temperature changed the least from the previous month.

16. *Organizing Data* The table shows the annual net profits *y* (in billions of dollars) for Hewlett-Packard for the years 2002 through 2011, where *x* represents the year. (*Source:* Hewlett-Packard Company)

x	2002	2003	2004	2005	2006
y	2.4	3.6	4.1	4.7	5.8

x	2007	2008	2009	2010	2011
y	7.3	8.3	7.7	8.8	7.1

(a) Plot the data in the table.

(b) Use the graph to determine the two consecutive years between which the greatest decrease occurred in the annual profits for Hewlett-Packard.

Ordered Pairs as Solutions of Equations

> **Three Approaches to Problem Solving**
> 1. **Algebraic Approach** Use algebra to find several solutions.
> 2. **Numerical Approach** Construct a table that shows several solutions.
> 3. **Graphical Approach** Draw a graph that shows several solutions.

EXAMPLE 3 Constructing a Table of Values

Construct a table of values showing five solution points for the equation

$$6x - 2y = 4.$$

Then plot the solution points on a rectangular coordinate system. Choose x-values of $-2, -1, 0, 1,$ and 2.

SOLUTION

$6x - 2y = 4$	Write original equation.
$6x - 6x - 2y = 4 - 6x$	Subtract $6x$ from each side.
$-2y = -6x + 4$	Combine like terms.
$\dfrac{-2y}{-2} = \dfrac{-6x + 4}{-2}$	Divide each side by -2.
$y = 3x - 2$	Simplify.

Now, using the equation $y = 3x - 2$, you can construct a table of values.

x	−2	−1	0	1	2
y = 3x − 2	−8	−5	−2	1	4
Solution point	(−2, −8)	(−1, −5)	(0, −2)	(1, 1)	(2, 4)

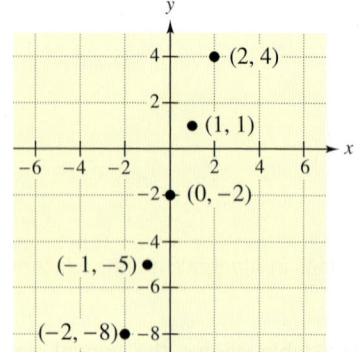

From the table, you can plot the solution points as shown at the left.

Exercises Within Reach®

Solutions in English & Spanish and tutorial videos at AlgebraWithinReach.com

Constructing a Table of Values In Exercises 17−22, **complete** the table of values. Then **plot** the solution points on a rectangular coordinate system.

17.
x	−2	0	2	4	6
y = 3x − 4					

18.
x	−2	0	2	4	6
y = 2x + 1					

19.
x	−2	0	4	6	8
$y = -\frac{3}{2}x + 5$					

20.
x	−4	−2	0	2	4
$y = -\frac{1}{2}x + 3$					

21.
x	−2	−1	0	1	2
y = −4x − 5					

22.
x	−2	0	$\frac{1}{2}$	2	4
$y = -\frac{7}{2}x + 3$					

Guidelines for Verifying Solutions

To verify that an ordered pair (x, y) is a solution of an equation with variables x and y, use the following steps.

1. Substitute the values of x and y into the equation.
2. Simplify each side of the equation.
3. If each side simplifies to the same number, then the ordered pair is a solution. If the two sides yield different numbers, then the ordered pair is not a solution.

EXAMPLE 4 Verifying Solutions of an Equation

Determine whether each ordered pair is a solution of $x + 3y = 6$.

a. $(1, 2)$ **b.** $(0, 2)$

SOLUTION

a. For the ordered pair $(1, 2)$, substitute $x = 1$ and $y = 2$ into the original equation.

$$x + 3y = 6 \quad \text{Write original equation.}$$
$$1 + 3(2) \stackrel{?}{=} 6 \quad \text{Substitute 1 for } x \text{ and 2 for } y.$$
$$7 \neq 6 \quad \text{Not a solution} \; \boldsymbol{X}$$

b. For the ordered pair $(0, 2)$, substitute $x = 0$ and $y = 2$ into the original equation.

$$x + 3y = 6 \quad \text{Write original equation.}$$
$$0 + 3(2) \stackrel{?}{=} 6 \quad \text{Substitute 0 for } x \text{ and 2 for } y.$$
$$6 = 6 \quad \text{Solution} \; \checkmark$$

Exercises Within Reach® Solutions in English & Spanish and tutorial videos at AlgebraWithinReach.com

---

Verifying Solutions of an Equation In Exercises 23–30, determine whether each ordered pair is a solution of the equation.

23. $y = 2x + 4$ (a) $(3, 10)$ (b) $(-1, 3)$ (c) $(0, 0)$ (d) $(-2, 0)$

24. $y = 5x - 2$ (a) $(2, 0)$ (b) $(-2, -12)$ (c) $(6, 28)$ (d) $(1, 1)$

25. $2y - 3x + 1 = 0$ (a) $(1, 1)$ (b) $(5, 7)$ (c) $(-3, -1)$ (d) $(-3, -5)$

26. $x - 8y + 10 = 0$ (a) $(-2, 1)$ (b) $(6, 2)$ (c) $(0, -1)$ (d) $(2, -2)$

27. $y = \frac{2}{3}x$ (a) $(6, 6)$ (b) $(-9, -6)$ (c) $(0, 0)$ (d) $\left(-1, \frac{2}{3}\right)$

28. $y = \frac{7}{8}x$ (a) $(-5, -2)$ (b) $(0, 0)$ (c) $(8, 8)$ (d) $\left(\frac{3}{5}, 1\right)$

29. $y = 3 - 4x$ (a) $\left(-\frac{1}{2}, 5\right)$ (b) $(1, 7)$ (c) $(0, 0)$ (d) $\left(-\frac{3}{4}, 0\right)$

30. $y = \frac{3}{2}x + 1$ (a) $\left(0, \frac{3}{2}\right)$ (b) $(4, 7)$ (c) $\left(\frac{2}{3}, 2\right)$ (d) $(-2, -2)$

Determining Solutions of an Equation In Exercises 31 and 32, complete each ordered pair so that it satisfies the equation.

31. $y = 3x + 4$ (a) $(\;\;\;, 0)$ (b) $(4, \;\;\;)$ (c) $(\;\;\;, -2)$

32. $y = -2x - 7$ (a) $(0, \;\;\;)$ (b) $(-1, \;\;\;)$ (c) $(\;\;\;, 1)$

Applications

Application — EXAMPLE 5 Finding the Total Cost

You set up a small business to assemble computer keyboards. Your initial cost is $120,000, and your unit cost of assembling each keyboard is $40. Write an equation that relates your total cost to the number of keyboards produced. Then plot the total costs of producing 1000, 2000, 3000, 4000, and 5000 keyboards.

SOLUTION

The total cost equation must represent both the unit cost and the initial cost. A verbal model for this problem is as follows.

Verbal Model: Total cost = Unit cost · Number of keyboards + Initial cost

Labels:
- Total cost = C (dollars)
- Unit cost = 40 (dollars per keyboard)
- Number of keyboards = x (keyboards)
- Initial cost = 120,000 (dollars)

Expression: $C = 40x + 120{,}000$

Using this equation, you can construct the following table of values.

x	1000	2000	3000	4000	5000
$C = 40x + 120{,}000$	160,000	200,000	240,0000	280,000	320,000

From the table, you can plot the ordered pairs.

Computer Keyboards (graph: Total cost (in dollars) vs. Number of keyboards)

Exercises Within Reach®

Solutions in English & Spanish and tutorial videos at AlgebraWithinReach.com

33. Spring Compression The distance y (in centimeters) a spring is compressed by a force x (in kilograms) is given by $y = 0.066x$. Complete a table of values for $x = 20, 40, 60, 80$, and 100 to determine the distance the spring is compressed for each of the specified forces. Plot the results on a rectangular coordinate system.

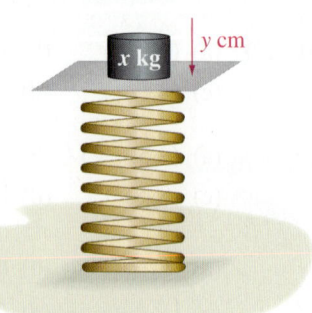

34. Copier Value A company buys a new copier for $9500. Its value y after x years is given by $y = -800x + 9500$. Complete a table of values for $x = 0, 2, 4, 6$, and 8 to determine the value of the copier at each specified time. Plot the results on a rectangular coordinate system.

35. Video Games With an initial cost of $5000, a company will produce x units of a video game at $25 per unit. Write an equation that relates the total cost of producing x units to the number of units produced. Plot the cost of producing 100, 150, 200, 250, and 300 units.

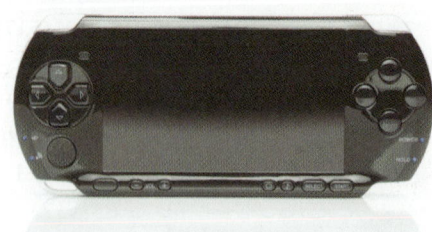

36. Assembly Line An employee earns $10 plus $0.50 for every x units produced per hour. Write an equation that relates the employee's total hourly wage to the number of units produced. Plot the hourly wages for producing 2, 5, 8, 10, and 20 units per hour.

Application

EXAMPLE 6 Identifying Misleading Graphs

The graphs shown below represent the yearly profits for a truck rental company. Which graph is misleading? Why?

a.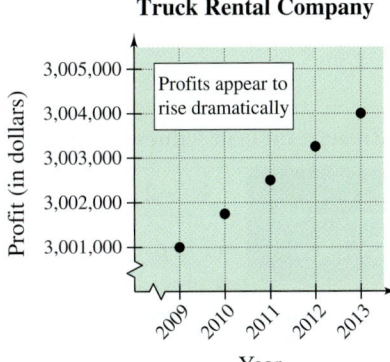

b.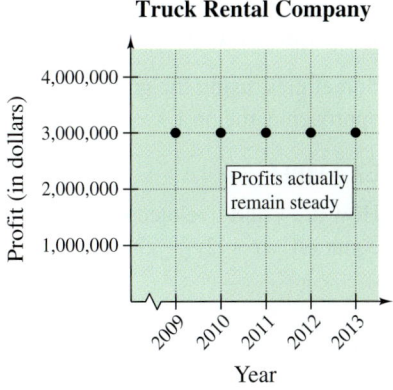

SOLUTION

a. This graph is misleading. The scale on the vertical axis makes it appear that the change in profits from 2009 to 2013 is dramatic, but the total change is only $3000, which is small in comparison with $3,000,000.

b. This graph is truthful. By showing the full scale on the y-axis, you can see that, relative to the overall size of the profit, there was almost no change from one year to the next.

Exercises Within Reach®

Solutions in English & Spanish and tutorial videos at AlgebraWithinReach.com

37. *Government Surplus* The graph below shows the estimated U.S. government surplus for the years 2012 through 2016. Is the graph misleading? Explain your reasoning. (*Source:* U.S. Office of Management and Budget)

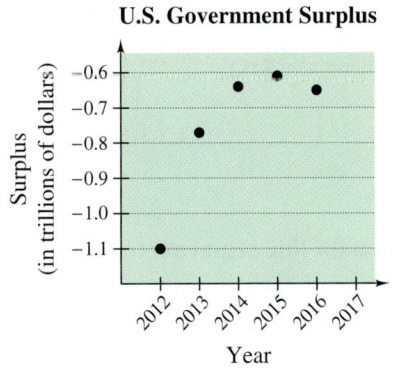

38. *Government Surplus* The graph below shows the estimated U.S. government surplus for the years 2012 through 2016. Is the graph misleading? Explain your reasoning. (*Source:* U.S. Office of Management and Budget)

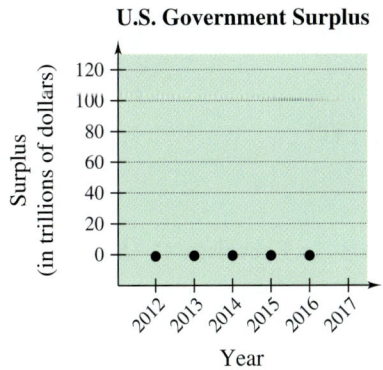

Concept Summary: Using a Graphical Approach to Solve Problems

What
Here are three ways to solve problems.
- **Algebraic Approach**
- **Numerical Approach**
- **Graphical Approach**

Sometimes, using the graphical approach is the best way.

EXAMPLE
How are x and y related in the equation $-x + 2y = 2$?

How
To use a graphical approach to solve a problem involving a linear equation:

1. Solve the equation for y.

$$-x + 2y = 2$$
$$2y = x + 2$$
$$y = \tfrac{1}{2}x + 1$$

2. Create a **table of values**.

x	-4	-2	0	2	4
$y = \tfrac{1}{2}x + 1$	-1	0	1	2	3

3. **Plot** the **ordered pairs** on a **rectangular coordinate system**.

Why
Using a graphical approach helps you see the relationship between the variables. For example, you can see from the graph that as x increases, y also increases.

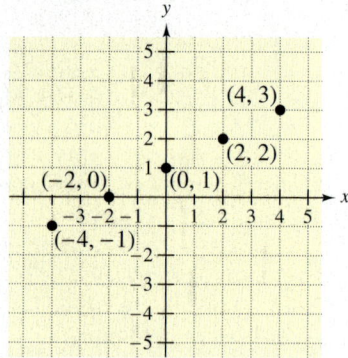

Exercises Within Reach®
Worked-out solutions to odd-numbered exercises at **AlgebraWithinReach.com**

Concept Summary Check

39. *Representing a Solution* Use the table of values above to write the coordinates of the point representing the solution of $-x + 2y = 2$ when $x = 2$.

40. *Creating a Table of Values* Explain how each y-value was determined in the table of values above.

41. *Identifying Points* Identify the point(s) in Quadrant III on the rectangular coordinate system above. Identify the point(s) on the y-axis.

42. *Describing a Graph* The phrase "x increases" describes a movement in what direction along the graph?

Extra Practice

Finding Coordinates In Exercises 43–46, **determine** the coordinates of the points.

43. 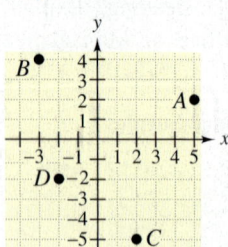   44. 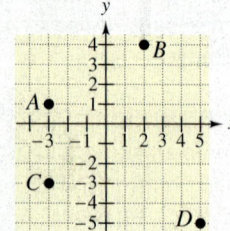   45. 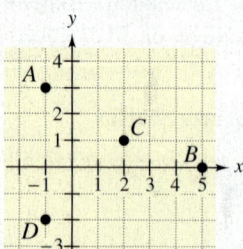   46.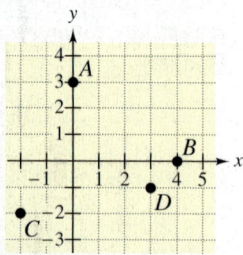

Plotting Points In Exercises 47–54, **plot** the points and **connect** them with line segments to form the figure.

47. Triangle: $(-1, 1), (2, -1), (3, 4)$

48. Triangle: $(0, 3), (-1, -2), (4, 8)$

49. Square: $(2, 4), (5, 1), (2, -2), (-1, 1)$

50. Rectangle: $(2, 1), (4, 2), (1, 8), (-1, 7)$

51. Parallelogram: $(5, 2), (7, 0), (1, -2), (-1, 0)$

52. Parallelogram: $(-1, 1), (0, 4), (5, 1), (4, -2)$

53. Rhombus: $(0, 0), (3, 2), (5, 5), (2, 3)$

54. Rhombus: $(0, 0), (1, 2), (3, 3), (2, 1)$

Solving for y In Exercises 55–58, solve the equation for *y*.

55. $7x + y = 8$ **56.** $2x + y = 1$ **57.** $10x - y = 2$ **58.** $12x - y = 7$

Determining Quadrants In Exercises 59 and 60, determine the quadrant(s) in which the point may be located without plotting it. Assume $x \neq 0$ and $y \neq 0$.

59. $(6, y)$, *y* is a real number.

60. $(x, -2)$, *x* is a real number.

61. *Graphical Interpretation* The table shows the numbers of hours *x* that a student studied for five different algebra exams, and the resulting scores *y*.

x	3.5	1	8	4.5	0.5
y	72	67	95	81	53

(a) Plot the data in the table.

(b) Use the graph to describe the relationship between the number of hours studied and the resulting exam score.

62. *Organizing Data* The table shows the speeds of a car *x* (in miles per hour) and the approximate stopping distance *y* (in feet).

x	20	30	40	50	60
y	63	109	164	229	303

(a) Plot the data in the table.

(b) The *x*-coordinates increase in equal increments of 10 miles per hour. Describe the pattern of the *y*-coordinates. What are the implications for the driver?

Explaining Concepts

63. (a) Plot the points $(3, 2)$, $(-5, 4)$, and $(6, -4)$ on a rectangular coordinate system.

(b) Change the sign of the *y*-coordinate of each point plotted in part (a). Plot the three new points on the same rectangular coordinate system used in part (a).

(c) What can you infer about the location of a point when the sign of its *y*-coordinate is changed?

64. (a) Plot the points $(3, 2)$, $(-5, 4)$, and $(6, -4)$ on a rectangular coordinate system.

(b) Change the sign of the *x*-coordinate of each point plotted in part (a). Plot the three new points on the same rectangular coordinate system used in part (a).

(c) What can you infer about the location of a point when the sign of its *x*-coordinate is changed?

65. *Think About It* The points $(6, -1)$, $(-2, -1)$, and $(-2, 4)$ are three vertices of a rectangle. Find the coordinates of the fourth vertex.

66. *Writing* Discuss the significance of the word "ordered" when referring to an ordered pair (x, y).

67. *Think About It* On a rectangular coordinate system, must the scales on the *x*-axis and *y*-axis be the same? If not, give an example in which the scales differ.

68. *Writing* Review the tables in Exercises 17–22 and observe that in some cases the *y*-coordinates of the solution points increase and in others the *y*-coordinates decrease. What factor in the equation causes this? Explain.

Cumulative Review

In Exercises 69–72, solve the equation.

69. $-y = 10$

70. $10 - t = 6$

71. $3x - 42 = 0$

72. $64 - 16x = 0$

In Exercises 73–76, solve the inequality.

73. $x + 3 > 2$

74. $y - 4 < -8$

75. $3x < 12$

76. $2(z - 4) > 10z$

4.2 Graphs of Equations in Two Variables

▶ Sketch graphs of equations using the point-plotting method.
▶ Find and use x- and y-intercepts as aids to sketching graphs.
▶ Use the verbal problem-solving method to write an equation and sketch its graph.

The Graph of an Equation in Two Variables

The solutions of an equation involving two variables can be represented by points on a rectangular coordinate system. The set of *all* such points is called the **graph** of the equation.

> **The Point-Plotting Method of Sketching a Graph**
> 1. If possible, rewrite the equation by isolating one of the variables.
> 2. Make a table of values showing several solution points.
> 3. Plot these points on a rectangular coordinate system.
> 4. Connect the points with a smooth curve or line.

EXAMPLE 1 Sketching the Graph of an Equation

Sketch the graph of $3x + y = 5$.

SOLUTION

Begin by solving the equation for y, so that y is isolated on the left.

$3x + y = 5$ Write original equation.
$y = -3x + 5$ Subtract $3x$ from each side.

Next, create a table of values, as shown below.

x	−2	−1	0	1	2	3
$y = -3x + 5$	11	8	5	2	−1	−4
Solution point	(−2, 11)	(−1, 8)	(0, 5)	(1, 2)	(2, −1)	(3, −4)

Now, plot the solution points. It appears that all six points lie on a line, so complete the sketch by drawing a line through the points.

Exercises Within Reach®

Solutions in English & Spanish and tutorial videos at AlgebraWithinReach.com

Sketching the Graph of an Equation In Exercises 1 and 2, complete the table and use the results to sketch the graph of the equation.

1. $y = 9 - x$

x	−2	−1	0	1	2
y					

2. $y = x - 1$

x	−2	−1	0	1	2
y					

Section 4.2 Graphs of Equations in Two Variables

EXAMPLE 2 Graphing a Nonlinear Equation

Sketch the graph of $x^2 + y = 4$.

SOLUTION

Begin by solving the equation for y, so that y is isolated on the left.

$$y = -x^2 + 4$$

Next, create a table of values, as shown below.

x	−3	−2	−1	0	1	2	3
$y = -x^2 + 4$	−5	0	3	4	3	0	−5
Solution point	(−3, −5)	(−2, 0)	(−1, 3)	(0, 4)	(1, 3)	(2, 0)	(3, −5)

Now, plot the solution points. Finally, connect the points with a smooth curve.

EXAMPLE 3 Graphing an Absolute Value Equation

Sketch the graph of $y = |x - 1|$.

SOLUTION

x	−2	−1	0	1	2	3	4
$y = |x - 1|$	3	2	1	0	1	2	3
Solution point	(−2, 3)	(−1, 2)	(0, 1)	(1, 0)	(2, 1)	(3, 2)	(4, 3)

Now, plot the solution points. It appears that the points lie in a "V-shaped" pattern, with the point (1, 0) lying at the bottom of the "V." Following this pattern, connect the points to form the graph shown at the left.

Exercises Within Reach®

Solutions in English & Spanish and tutorial videos at AlgebraWithinReach.com

Sketching the Graph of an Equation In Exercises 3−6, complete the table and use the results to sketch the graph of the equation.

3. $y = x^2 + 3$

x	−2	−1	0	1	2
y					

4. $x^2 + y = -1$

x	−2	−1	0	1	2
y					

5. $y = |x + 1|$

x	−3	−2	−1	0	1
y					

6. $y = |x| - 2$

x	−2	−1	0	1	2
y					

Sketching the Graph of an Equation In Exercises 7−12, sketch the graph of the equation and label the coordinates of at least three solution points.

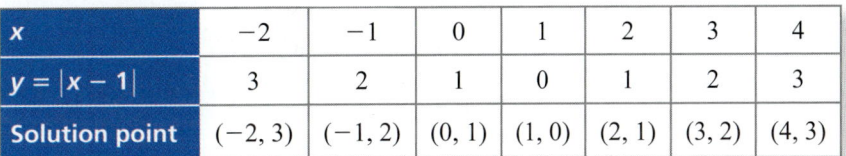

7. $4x + y = 6$

8. $2x - y = 5$

9. $y = -x^2 + 9$

10. $y = x^2 - 1$

11. $y = 5 - |x|$

12. $y = |x| + 3$

170 Chapter 4 Equations and Inequalities in Two Variables

Intercepts: Aids to Sketching Graphs

The point $(a, 0)$ is called an **x-intercept** of the graph of an equation if it is a solution point of the equation. To find the x-intercept(s), let $y = 0$ and solve the equation for x.

The point $(0, b)$ is called a **y-intercept** of the graph of an equation if it is a solution point of the equation. To find the y-intercept(s), let $x = 0$ and solve the equation for y.

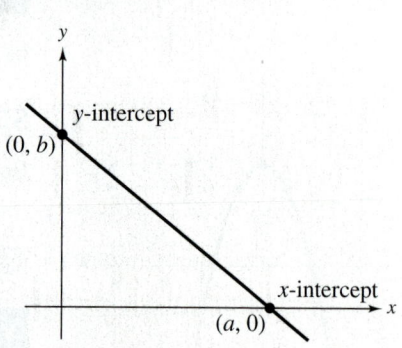

EXAMPLE 4 Identifying the Intercepts of Graphs

a.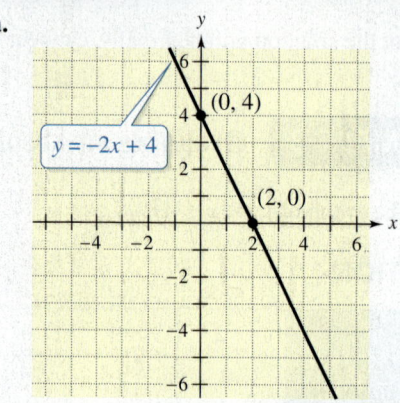

b.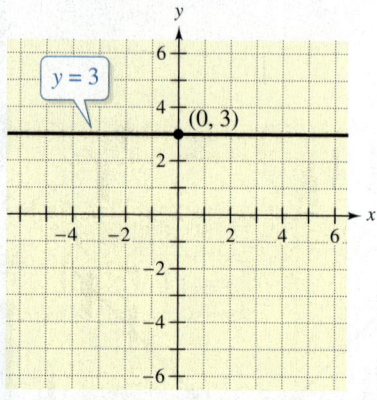

c.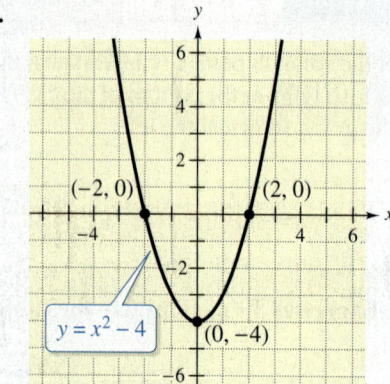

d.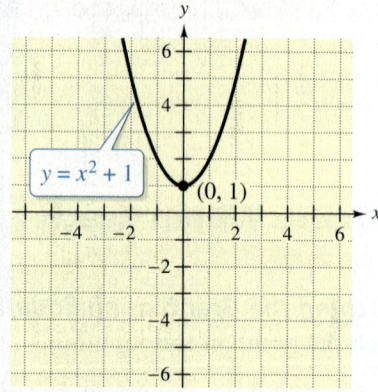

Exercises Within Reach ®
Solutions in English & Spanish and tutorial videos at AlgebraWithinReach.com

Identifying the Intercepts of a Graph In Exercises 13–16, **identify** the x- and y-intercepts of the graph visually.

13.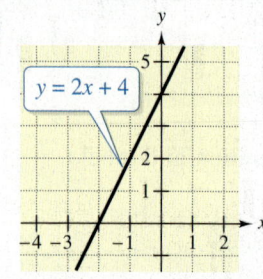

14.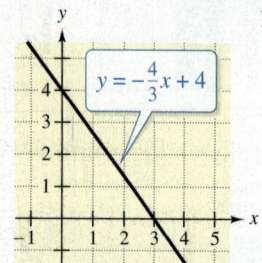

15.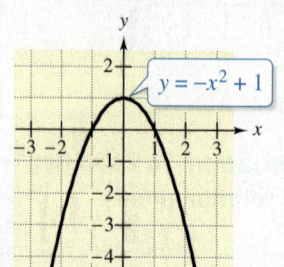

16.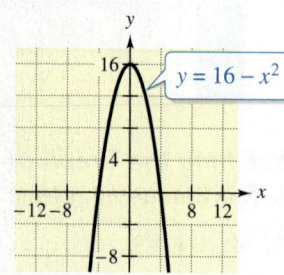

Section 4.2 Graphs of Equations in Two Variables 171

EXAMPLE 5 **Finding the Intercepts of a Graph**

Find the intercepts and sketch the graph of $y = 2x - 5$.

SOLUTION

To find any x-intercepts, let $y = 0$ and solve the resulting equation for x.

$y = 2x - 5$	Write original equation.
$0 = 2x - 5$	Let $y = 0$.
$\frac{5}{2} = x$	Solve equation for x.

To find any y-intercepts, let $x = 0$ and solve the resulting equation for y.

$y = 2x - 5$	Write original equation.
$y = 2(0) - 5$	Let $x = 0$.
$y = -5$	Solve equation for y.

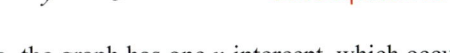

So, the graph has one x-intercept, which occurs at the point $\left(\frac{5}{2}, 0\right)$, and one y-intercept, which occurs at the point $(0, -5)$. To sketch the graph of the equation, create a table of values. (Include the intercepts in the table.) Then plot the points and connect the points with a line, as shown at the left.

x	-1	0	1	2	$\frac{5}{2}$	3	4
$y = 2x - 5$	-7	-5	-3	-1	0	1	3
Solution point	$(-1, -7)$	$(0, -5)$	$(1, -3)$	$(2, -1)$	$\left(\frac{5}{2}, 0\right)$	$(3, 1)$	$(4, 3)$

Exercises Within Reach® Solutions in English & Spanish and tutorial videos at AlgebraWithinReach.com

Finding the Intercepts of a Graph In Exercises 17–20, **find** the x- and y-intercepts and **sketch** the graph of the equations.

17. $y = \frac{1}{2}x - 1$
18. $y = -\frac{1}{2}x + 3$
19. $2x + y = -2$
20. $3x - 2y = 1$

21. **Car Rental** A car rental costs $50 per day plus an additional $0.50 for each mile driven. The daily cost y is given by the equation

    $y = 0.50x + 50$

    where x is the number of miles driven. Find the y-intercept of the graph of the equation.

22. **Hot-Air Balloon** A hot-air balloon descends 4 feet per second from a height of 400 feet. The height y of the balloon after x seconds is represented by the equation

    $y = 400 - 4x$.

    Find the x-intercept of the graph of the equation.

Kurhan/Shutterstock.com; goldsaint/Shutterstock.com

Applications

Application EXAMPLE 6 Depreciation

The value of a $35,500 sport utility vehicle (SUV) depreciates over 10 years (the depreciation is the same each year). At the end of the 10 years, the salvage value is expected to be $5500.

a. Write an equation that relates the value of the SUV and its age in years.
b. Sketch the graph of the equation.
c. What is the y-intercept of the graph, and what does it represent in the context of the problem?

SOLUTION

a. The depreciation over the 10 years is $35,500 - 5500 = \$30,000$. Because the same amount is depreciated each year, it follows that the annual depreciation is $30,000/10 = \$3000$.

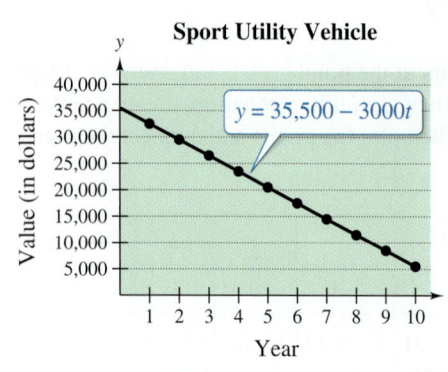

Verbal Model: Value after t years = Original value − Annual depreciation · Number of years

Labels:
Value after t years = y (dollars)
Original value = 35,500 (dollars)
Annual depreciation = 3000 (dollars per year)
Number of years = t (years)

Equation: $y = 35{,}500 - 3000t$

b. A sketch of the graph of the depreciation equation is shown at the left.

c. To find the y-intercept of the graph, let $t = 0$ and solve the equation for y.

$y = 35{,}500 - 3000t$ Write original equation.
$y = 35{,}500 - 3000(0)$ Substitute 0 for t.
$y = 35{,}500$ Simplify.

So, the y-intercept is $(0, 35{,}500)$, and it corresponds to the original value of the SUV.

Exercises Within Reach®

Solutions in English & Spanish and tutorial videos at AlgebraWithinReach.com

23. Hot-Air Balloon A hot-air balloon at 1120 feet descends at a rate of 80 feet per minute. Let y represent the height of the balloon and let x represent the number of minutes the balloon descends.

(a) Write an equation that relates the height of the hot-air balloon and the number of minutes it descends.

(b) Sketch the graph of the equation.

(c) What is the y-intercept of the graph, and what does it represent in the context of the problem?

24. Fitness You run and walk on a trail that is 6 miles long. You run 4 miles per hour and walk 3 miles per hour. Let y be the number of hours you walk and let x be the number of hours you run.

(a) Write an equation that relates the number of hours you run and the number of hours you walk to the total length of the trail.

(b) Sketch the graph of the equation.

(c) What is the y-intercept of the graph, and what does it represent in the context of the problem?

Section 4.2 Graphs of Equations in Two Variables 173

Application   EXAMPLE 7   Life Expectancy

The table shows the life expectancies y (in years) in the United States for a male child at birth for the years 2002 through 2007.

DATA Year	2002	2003	2004	2005	2006	2007
y	74.3	74.5	74.9	74.9	75.1	75.4

A model for this data is $y = 0.21t + 73.9$, where t is the year, with $t = 2$ corresponding to 2002. (*Source:* U.S. National Center for Health Statistics)

a. Plot the data and graph the model on the same set of coordinate axes.

b. Use the model to predict the life expectancy for a male child born in 2020.

SOLUTION

a. The points from the table are plotted with the graph of $y = 0.21t + 73.9$, as shown at the left.

b. To predict the life expectancy for a male child in 2020, let $t = 20$, and solve the equation for y.

$y = 0.21t + 73.9$ Write model.

$\quad = 0.21(20) + 73.9$ Substitute 20 for t.

$\quad = 78.1$ Simplify.

So, you can predict that the life expectancy in 2020 will be 78.1 years.

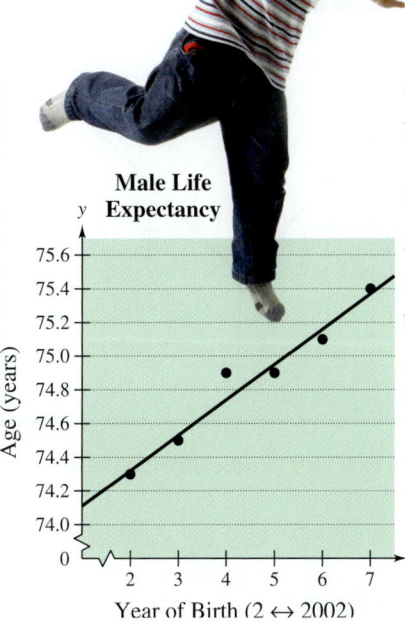

Male Life Expectancy

Exercises Within Reach®

Solutions in English & Spanish and tutorial videos at AlgebraWithinReach.com

25. *Life Expectancy* The table shows the life expectancies y (in years) in the United States for a female child at birth for the years 2002 through 2007.

DATA Year	y
2002	79.5
2003	79.6
2004	79.9
2005	79.9
2006	80.2
2007	80.4

A model for this data is $y = 0.18t + 79.1$, where t is the year, with $t = 2$ corresponding to 2002. (*Source:* U.S. National Center for Health Statistics)

(a) Plot the data and graph the model on the same set of coordinate axes.

(b) Use the model to predict the life expectancy for a female child born in 2020.

26. *Marital Status* The table shows the numbers y (in millions) of adults (over 18 years of age) never married in the United States for the years 2006 through 2011.

DATA Year	y
2006	55.3
2007	56.1
2008	58.3
2009	59.1
2010	61.5
2011	63.3

A model for this data is $y = 1.63t + 45.1$, where t is the year, with $t = 6$ corresponding to 2006. (*Source:* U.S. Census Bureau)

(a) Plot the data and graph the model on the same set of coordinate axes.

(b) Use the model to predict the number of adults over the age of 18 in 2020 who will never have married.

174 Chapter 4 Equations and Inequalities in Two Variables

Concept Summary: Graphing Equations in Two Variables

What
You can use a **graph** to represent all the solutions of an equation in two variables.

EXAMPLE
Sketch the graph of
$y = |x| - 1$.

How
Use these steps to sketch the graph of such an equation.

1. Make a table of values. Include any *x*- and *y*-intercepts.
2. Plot the points.
3. Connect the points.

x	−2	−1	0	1	2
$y = |x| - 1$	1	0	−1	0	1

Why
You can use the graph of an equation in two variables to see the relationship between the variables. For example, the graph shows that *y* decreases as *x* increases to zero, and *y* increases as *x* increases above zero.

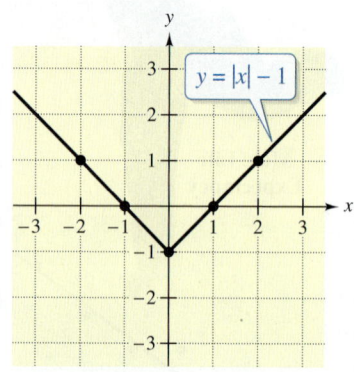

Exercises Within Reach®
Worked-out solutions to odd-numbered exercises at AlgebraWithinReach.com

Concept Summary Check

27. Using a Table of Values Use the table of values above to write the coordinates of the point representing the solution of $y = |x| - 1$ when $x = 2$.

28. Using a Table of Values Use the table of values above to determine the *x*- and *y*-intercepts of the graph of $y = |x| - 1$.

29. Finding an Intercept Explain how to find the *y*-intercept of the graph of an equation.

30. Interpreting a Table of Values Briefly describe the point-plotting method of sketching the graph of an equation.

Extra Practice

Sketching the Graph of an Equation In Exercises 31–38, **sketch** the graph of the equation and **label** the coordinates of at least three solution points.

31. $y = 3x$
32. $y = -2x$
33. $7x + 7y = 14$
34. $10x + 5y = 20$
35. $y = \frac{3}{8}x + 15$
36. $y = 14 - \frac{2}{3}x$
37. $y = |x - 5|$
38. $y = |x + 3|$

Finding Intercepts In Exercises 39–42, graphically **estimate** the *x*- and *y*-intercepts of the graph. Then check your results algebraically.

39. $x + 3y = 6$
40. $y = |x| - 3$
41. $y = 4 - |x|$
42. $y = x^2 - 4$

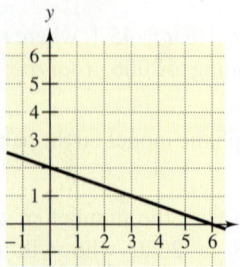
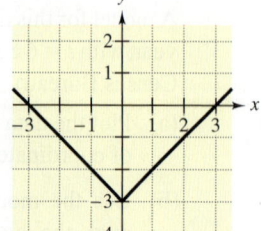
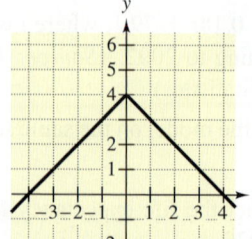
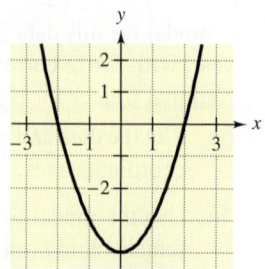

Comparing Graphs
In Exercises 43–46, sketch the graphs of both equations on the same rectangular coordinate system. Are the graphs identical? If so, what property of real numbers is being illustrated?

43. $y_1 = \frac{1}{3}x - 1$
    $y_2 = 1 - \frac{1}{3}x$

44. $y_1 = 3\left(\frac{1}{4}x\right)$
    $y_2 = \left(3 \cdot \frac{1}{4}\right)x$

45. $y_1 = 2(x - 2)$
    $y_2 = 2x - 4$

46. $y_1 = 2 + (x + 4)$
    $y_2 = (2 + x) + 4$

47. **Creating a Model** Let y represent the distance traveled by a car that is moving at a constant speed of 35 miles per hour. Let t represent the number of hours the car has traveled. Write an equation that relates y and t, and sketch its graph.

48. **Creating a Model** The cost of printing a book is $500, plus $5 per book. Let C represent the total cost and let x represent the number of books. Write an equation that relates C and x, and sketch its graph.

Explaining Concepts

49. **Writing** Do all graphs of linear equations in two variables have a y-intercept? Explain.

50. **Reasoning** When the graph of a linear equation in two variables has a negative x-intercept and a positive y-intercept, does the line rise or fall from left to right? Through which quadrant(s) does the line pass? Use a graph to illustrate your answer.

51. **Writing** You walk toward a tree at a constant speed. Let x represent the time (in seconds) and let y represent the distance (in feet) between you and the tree. Sketch a possible graph of this situation. Explain how x and y are related. What does the x-intercept mean?

52. **Writing** How many solution points does a linear equation in two variables have? Explain.

Cumulative Review

In Exercises 53–56, evaluate the expression.

53. $-4 + (-7) - 3 + 1$

54. $-6 + 3 - (-1) + 11$

55. $-(-3) + 5 - 4 + 9$

56. $-18 - (-6) + 2 - 8$

In Exercises 57–62, evaluate the expression and write the result in simplest form.

57. $\frac{3}{4}\left(-\frac{2}{9}\right)$

58. $\left(-\frac{1}{6}\right)\left(-\frac{8}{15}\right)$

59. $\left(-\frac{7}{12}\right)\left(-\frac{18}{35}\right)$

60. $-\frac{6}{7} \div \frac{5}{21}$

61. $\frac{12}{5} \div \left(-\frac{1}{3}\right)$

62. $\left(-\frac{16}{25}\right) \div \left(-\frac{4}{5}\right)$

In Exercises 63–66, determine whether each ordered pair is a solution of the equation.

63. $y = 3x - 5$
    (a) $(0, 5)$ (b) $(-1, -2)$
    (c) $(3, 4)$ (d) $(-2, -11)$

64. $y = 2x + 1$
    (a) $(-3, -5)$ (b) $(-1, 3)$
    (c) $(5, 8)$ (d) $(2, 5)$

65. $3y - 4x = 7$
    (a) $(1, 1)$ (b) $(-5, 9)$
    (c) $(4, -3)$ (d) $(7, 7)$

66. $x - 2y = -2$
    (a) $(-6, 2)$ (b) $(-2, 2)$
    (c) $(4, 3)$ (d) $(2, 0)$

4.3 Slope and Graphs of Linear Equations

- Determine the slope of a line through two points.
- Write linear equations in slope-intercept form and graph the equations.
- Use slopes to determine whether lines are parallel, perpendicular, or neither.

The Slope of a Line

The **slope** m of a nonvertical line that passes through the points (x_1, y_1) and (x_2, y_2) is

$$m = \frac{y_2 - y_1}{x_2 - x_1}$$

$$= \frac{\text{Change in } y}{\text{Change in } x}$$

$$= \frac{\text{Rise}}{\text{Run}}$$

where $x_1 \neq x_2$.

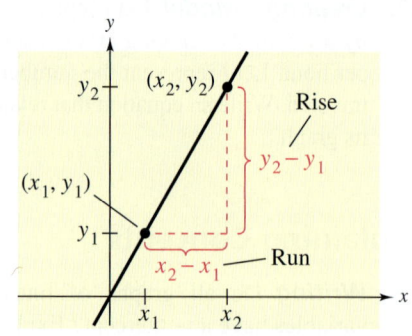

EXAMPLE 1 Finding the Slope of a Line Through Two Points

a. The slope of the line through $(x_1, y_1) = (-2, 0)$ and $(x_2, y_2) = (3, 1)$ is

$$m = \frac{y_2 - y_1}{x_2 - x_1}$$

$$= \frac{1 - 0}{3 - (-2)} \quad \Leftarrow \text{ Difference in } y\text{-values}$$
$$\quad\quad\quad\quad \Leftarrow \text{ Difference in } x\text{-values}$$

$$= \frac{1}{5}. \quad\quad \text{Simplify.}$$

The graph of the line is shown at the left.

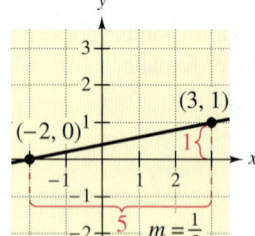

b. The slope of the line through $(0, 0)$ and $(1, -1)$ is

$$m = \frac{-1 - 0}{1 - 0} \quad \Leftarrow \text{ Difference in } y\text{-values}$$
$$\quad\quad\quad\quad \Leftarrow \text{ Difference in } x\text{-values}$$

$$= \frac{-1}{1} \quad\quad \text{Simplify.}$$

$$= -1. \quad\quad \text{Simplify.}$$

The graph of the line is shown at the left.

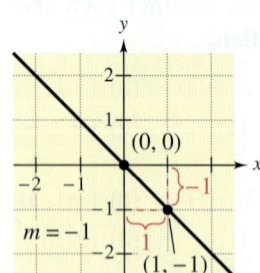

Exercises Within Reach®

Solutions in English & Spanish and tutorial videos at AlgebraWithinReach.com

Finding the Slope of a Line In Exercises 1–6, **plot** the points and **find** the slope of the line that passes through the points.

1. $(0, 0), (4, 5)$
2. $(0, 0), (3, 6)$
3. $(0, 0), (8, -4)$
4. $(0, 0), (-1, 3)$
5. $(6, 0), (0, 4)$
6. $(0, -3), (5, 0)$

Section 4.3   Slope and Graphs of Linear Equations   177

> **Slope of a Line**
> 1. A line with positive slope ($m > 0$) *rises* from left to right.
> 2. A line with negative slope ($m < 0$) *falls* from left to right.
> 3. A line with zero slope ($m = 0$) is *horizontal*.
> 4. A line with undefined slope is *vertical*.

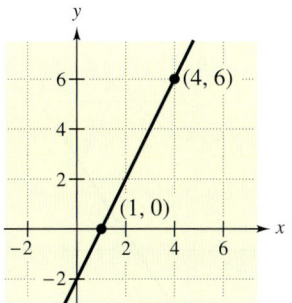
Line rises: positive slope

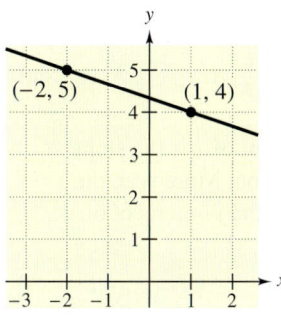
Line falls: negative slope

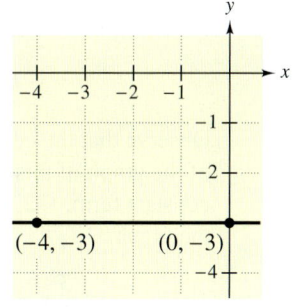
Horizontal line: zero slope

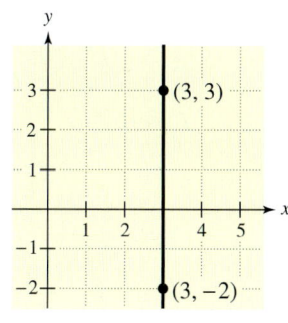
Vertical line: undefined slope

Application

EXAMPLE 2   Finding the Slope of a Ladder

Find the slope of the ladder leading up to the tree house.

SOLUTION

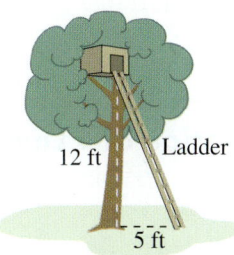

Consider the tree trunk as the *y*-axis and the level ground as the *x*-axis. The endpoints of the ladder are $(0, 12)$ and $(5, 0)$. So, the slope of the ladder is

$$m = \frac{y_2 - y_1}{x_2 - x_1} = \frac{0 - 12}{5 - 0} = -\frac{12}{5}.$$

Exercises Within Reach®

Solutions in English & Spanish and tutorial videos at AlgebraWithinReach.com

Using Slope to Describe a Line In Exercises 7–14, **find** the slope of the line that passes through the points. Use the slope to **state** whether the line rises, falls, is horizontal, or is vertical. Then **sketch** the line.

7. $(-4, -1), (2, 6)$

8. $(5, 3), (-3, 1)$

9. $(-6, -1), (-6, 4)$

10. $(-4, -10), (-4, 0)$

11. $(3, -4), (8, -4)$

12. $(1, -2), (-2, -2)$

13. $\left(\frac{1}{4}, \frac{3}{2}\right), \left(\frac{9}{2}, -3\right)$

14. $\left(\frac{5}{4}, \frac{1}{4}\right), \left(\frac{7}{8}, 2\right)$

15. ***Slide*** The ladder of a straight slide in a playground is 8 feet high. The distance along the ground from the ladder to the foot of the slide is 12 feet. Approximate the slope of the slide.

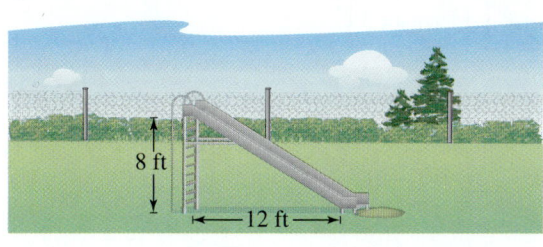

16. ***Ladder*** Find the slope of the ladder shown in the figure.

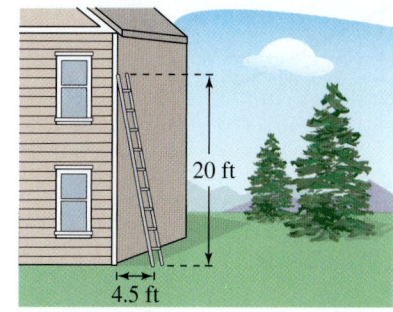

Slope as a Graphing Aid

You saw in Section 4.1 that before creating a table of values for an equation, it is helpful first to solve the equation for y. When you do this for a linear equation, you obtain some very useful information. Consider the following.

$$3x - 2y = 4 \qquad \text{Original equation}$$
$$3x - 3x - 2y = -3x + 4 \qquad \text{Subtract } 3x \text{ from each side.}$$
$$-2y = -3x + 4 \qquad \text{Combine like terms.}$$
$$\frac{-2y}{-2} = \frac{-3x + 4}{-2} \qquad \text{Divide each side by } -2.$$
$$y = \frac{3}{2}x - 2 \qquad \text{Simplify.}$$

Observe that the coefficient of x is the slope of the graph of this equation. Moreover, the constant term, -2, gives the y-intercept of the graph.

$$y = \underset{\text{slope}}{\frac{3}{2}}x + \underset{y\text{-intercept }(0,\,-2)}{-2}$$

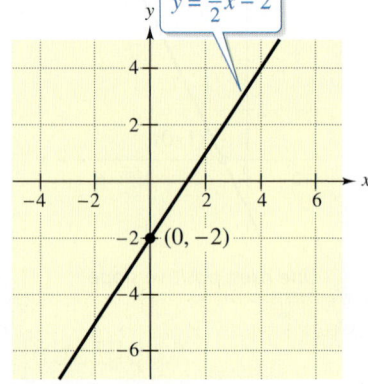

This form is called the **slope-intercept form** of the equation of the line.

Slope-Intercept Form of the Equation of a Line

The graph of the equation

$$y = mx + b$$

is a line whose slope is m and whose y-intercept is $(0, b)$.

Exercises Within Reach®

Solutions in English & Spanish and tutorial videos at AlgebraWithinReach.com

Slope-Intercept Form In Exercises 17–22, write the equation in slope-intercept form. Use the equation to identify the slope and y-intercept.

17. $\frac{1}{2}x + y = 2$

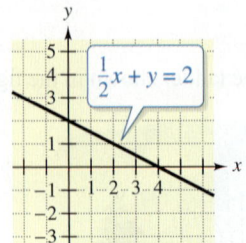

18. $\frac{3}{4}x - y = 3$

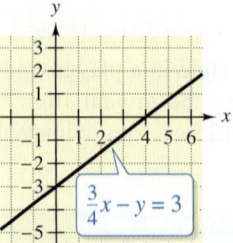

19. $2x + 3y = 6$

20. $4x + 2y = -8$

21. $3x - 4y + 2 = 0$

22. $2x - 3y + 1 = 0$

Section 4.3 Slope and Graphs of Linear Equations   179

EXAMPLE 3 **Using Slope-Intercept Form**

Use the slope and y-intercept to sketch the graph of

$$x - 3y = -6.$$

SOLUTION

First, write the equation in slope-intercept form.

$x - 3y = -6$	Write original equation.
$-3y = -x - 6$	Subtract x from each side.
$y = \dfrac{-x - 6}{-3}$	Divide each side by -3.
$y = \dfrac{1}{3}x + 2$	Simplify to slope-intercept form.

So, the slope of the line is $m = \frac{1}{3}$ and the y-intercept is $(0, b) = (0, 2)$. Now you can sketch the graph of the equation. First, plot the y-intercept. Then, using a slope of $\frac{1}{3}$,

$$m = \frac{1}{3} = \frac{\text{Change in } y}{\text{Change in } x}$$

locate a second point on the line by moving three units to the right and one unit up (or one unit up and 3 units to the right). Finally, obtain the graph by drawing a line through the two points.

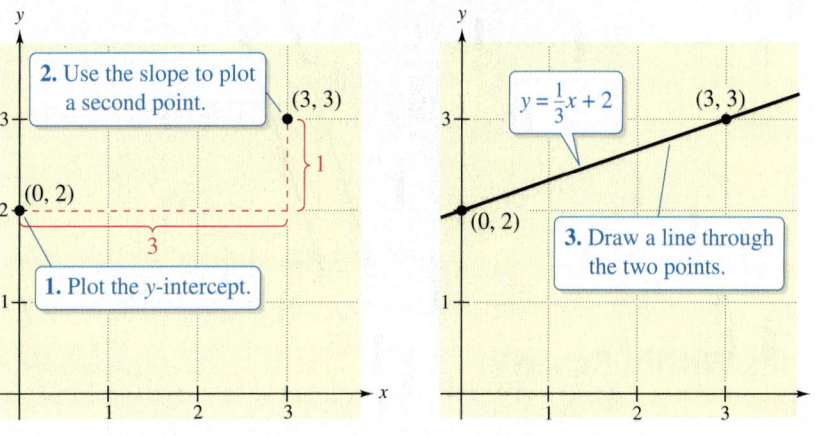

Exercises Within Reach®

Solutions in English & Spanish and tutorial videos at **AlgebraWithinReach.com**

Using Slope-Intercept Form In Exercises 23–28, **write** the equation in slope-intercept form. Use the slope and y-intercept to **sketch** the graph of the line.

23. $2x - y = 3$
24. $x - y = -2$
25. $x - 2y = -2$
26. $x + 3y = 12$
27. $2x - 6y - 15 = 0$
28. $10x + 6y - 3 = 0$

Using Slope and a Point on the Line In Exercises 29–34, **sketch** the graph of the line through the point $(3, 0)$ having the given slope.

29. $m = 0$
30. m is undefined.
31. $m = 2$
32. $m = -1$
33. $m = -\dfrac{2}{3}$
34. $m = \dfrac{3}{5}$

Parallel and Perpendicular Lines

You know from geometry that two lines in a plane are **parallel** if they do not intersect, and two lines in a plane are **perpendicular** if they intersect at right angles.

> ### Parallel Lines and Perpendicular Lines
>
> **Parallel Lines:** Two distinct nonvertical lines are parallel if and only if they have the same slope.
>
> **Perpendicular Lines:** Two lines are perpendicular if and only if their slopes are negative reciprocals of each other. That is,
> $$m_1 = -\frac{1}{m_2}.$$

Study Tip

The phrase "if and only if" is used in mathematics as a way to write two statements in one. In the rule for parallel lines, the first statement says that if two distinct nonvertical lines have the same slope, they must be parallel. The second (or reverse) statement says that if two distinct nonvertical lines are parallel, they must have the same slope.

EXAMPLE 4 Parallel and Perpendicular Lines

a. The lines $y = 3x$ and $y = 3x - 4$ each have a slope of 3. So, the lines are parallel.

b. The lines $y = 5x + 2$ and $y = -\frac{1}{5}x - 4$ have slopes that are negative reciprocals of each other. So, the lines are perpendicular.

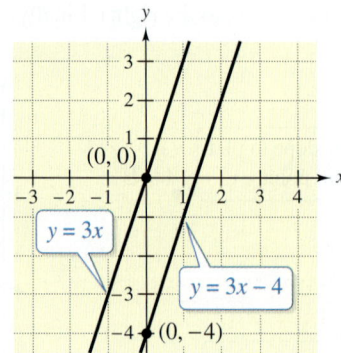
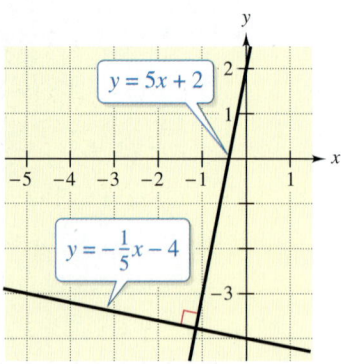

Exercises Within Reach®

Solutions in English & Spanish and tutorial videos at AlgebraWithinReach.com

Parallel and Perpendicular Lines In Exercises 35–38, *determine* whether the lines are parallel, perpendicular, or neither. *Explain* your reasoning.

35.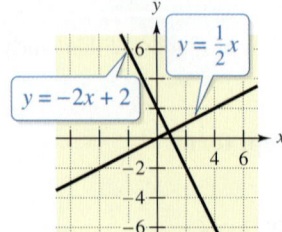

36.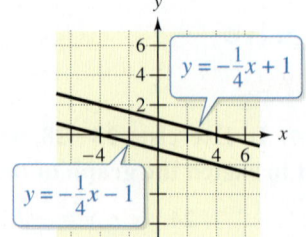

37.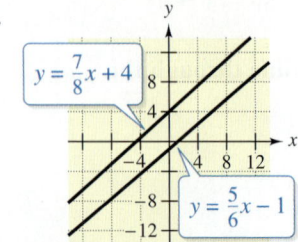

38.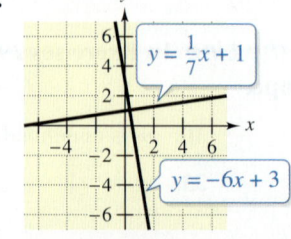

EXAMPLE 5 Parallel or Perpendicular?

Determine whether the pairs of lines are parallel, perpendicular, or neither.

a. $y = -3x - 2,\ y = \frac{1}{3}x + 1$ **b.** $y = \frac{1}{2}x + 1,\ y = \frac{1}{2}x - 1$

SOLUTION

a. This first line has a slope of $m_1 = -3$ and the second line has a slope of $m_2 = \frac{1}{3}$. Because these slopes are negative reciprocals of each other, the two lines must be perpendicular, as shown below on the left.

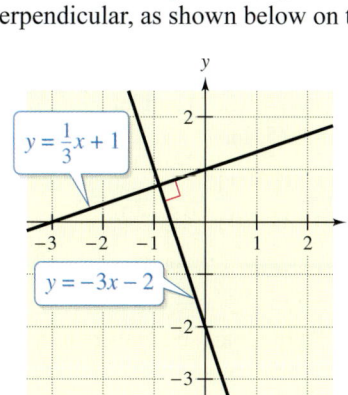
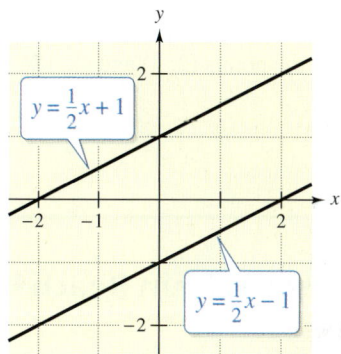

b. Both lines have a slope of $m = \frac{1}{2}$. So, the two lines must be parallel, as shown above on the right.

Parallel and perpendicular lines are common in architecture.

Exercises Within Reach®

Solutions in English & Spanish and tutorial videos at AlgebraWithinReach.com

Parallel or Perpendicular? In Exercises 39–44, use the equations of the lines to **determine** whether the lines are parallel, perpendicular, or neither. **Explain** your reasoning.

39. $y_1 = 2x - 3$
$y_2 = 2x + 1$

40. $y_1 = -\frac{1}{3}x - 3$
$y_2 = \frac{1}{3}x + 1$

41. $y_1 = 4x + 3$
$y_2 = 4x + 3$

42. $y_1 = 2x - 3$
$y_2 = -\frac{1}{2}x + 1$

43. $y_1 = -\frac{1}{3}x - 3$
$y_2 = 3x + 1$

44. $y_1 = \frac{3}{4}x + 5$
$y_2 = \frac{3}{4}x - 2$

Slopes of Parallel and Perpendicular Lines In Exercises 45–48, **determine** the slope of a line that is (a) parallel and (b) perpendicular to the given line.

45.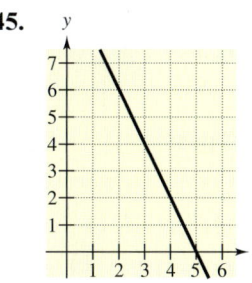

46.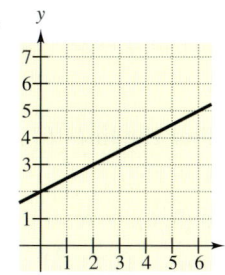

47.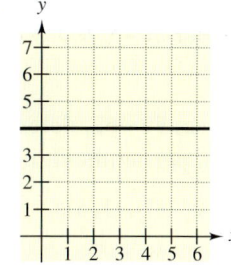

48.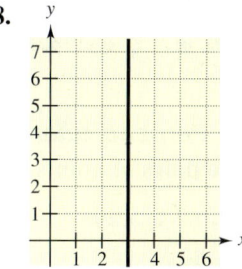

Concept Summary: Finding the Slope of a Line

What
The **slope** of a line is a measure of the "steepness" of the line. The slope also indicates whether the line *rises* or *falls* from left to right, is *horizontal*, or is *vertical*.

EXAMPLE
Find the slope of the line that passes through the points (2, 3) and (5, 1).

How
You can use a formula to find the slope of a nonvertical line that passes through any two points (x_1, y_1) and (x_2, y_2).

$$\text{Slope: } m = \frac{y_2 - y_1}{x_2 - x_1} = \frac{\text{Rise}}{\text{Run}}$$

EXAMPLE
Let $(x_1, y_1) = (2, 3)$ and let $(x_2, y_2) = (5, 1)$.

$$m = \frac{y_2 - y_1}{x_2 - x_1} = \frac{1 - 3}{5 - 2} = -\frac{2}{3}$$

Why
You can use slope to

- help sketch the graphs of linear equations.
- determine whether lines are **parallel**, **perpendicular**, or neither.

Exercises Within Reach® Worked-out solutions to odd-numbered exercises at AlgebraWithinReach.com

Concept Summary Check

49. Rise and Run In the solution above, what is the rise from (2, 3) to (5, 1)? What is the run?

50. Steepness Which slope is steeper, $-\frac{2}{3}$ or $\frac{1}{3}$? Explain.

51. The Sign of a Slope Does a line with a negative slope rise or fall from left to right?

52. Slope and Parallel Lines How can you use slope to determine whether two lines are parallel?

Extra Practice

Identifying the Slope of a Line In Exercises 53 and 54, **identify** the line in the figure that has each slope.

53. (a) $m = \frac{3}{2}$
    (b) $m = 0$
    (c) $m = -\frac{2}{3}$
    (d) $m = -2$

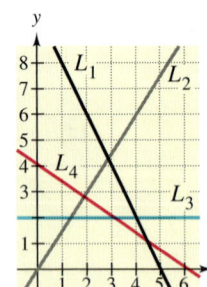

54. (a) $m = -\frac{3}{4}$
    (b) $m = \frac{1}{2}$
    (c) m is undefined.
    (d) $m = 3$

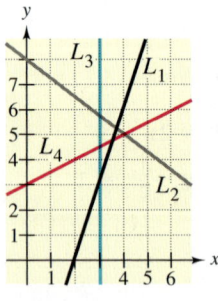

Using Slope and a Point on the Line In Exercises 55–58, **plot** the point and use the slope to **find** two additional points on the line. (There are many correct answers.)

55. $(-4, 0), m = \frac{2}{3}$

56. $(-1, -1), m = \frac{1}{4}$

57. $(0, 1), m = -2$

58. $(5, 6), m = -3$

Parallel or Perpendicular? In Exercises 59–62, **determine** whether the lines L_1 and L_2 that pass through the pairs of points are parallel, perpendicular, or neither.

59. L_1: $(0, -1), (5, 9)$
    L_2: $(0, 3), (4, 1)$

60. L_1: $(-2, -1), (1, 5)$
    L_2: $(1, 3), (5, 5)$

61. L_1: $(3, 6), (-6, 0)$
    L_2: $(0, -1), \left(5, \frac{7}{3}\right)$

62. L_1: $(4, 8), (-4, 2)$
    L_2: $(3, -5), \left(-1, \frac{1}{3}\right)$

63. Roof Pitch Determine the slope, or pitch, of the roof of the house shown in the figure.

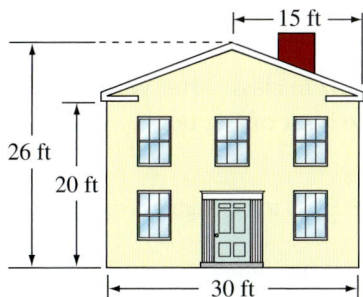

64. Skateboarding Ramp A wedge-shaped skateboarding ramp rises to a height of 12 inches over a 50-inch horizontal distance.

(a) Draw a diagram of the ramp and label the rise and run.

(b) Find the slope of the ramp.

65. Skateboarding Ramp Is a ramp that rises to a height of 12 inches over a 60-inch horizontal distance steeper than the ramp is Exercises 64? Explain.

66. Gold Prices The graph shows the average prices (in dollars) of a troy ounce of gold for the years 2006 through 2011.

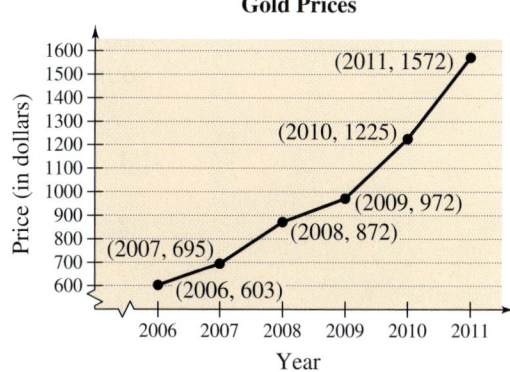

(a) Find the slopes of the five line segments.

(b) Find the slope of the line segment connecting the years 2006 and 2011. Interpret the meaning of this slope in the context of the problem.

Explaining Concepts

67. Think About It Can two perpendicular lines have positive slopes? Explain.

68. Think About It The slope of a line is $\frac{3}{2}$. How much will y change if x is increased by eight units? Explain.

69. Reasoning When a quantity y is increasing or decreasing at a constant rate over time t, the graph of y versus t is a line. What is another name for the rate of change?

70. Reasoning Explain how to use slopes to determine if the points $(-2, -3)$, $(1, 1)$, and $(3, 4)$ lie on the same line.

71. Writing When determining the slope of a line through two points, does the order of subtracting the coordinates of the points matter? Explain.

72. Structure The equations below give the heights h_1 and h_2 (in inches) of two bamboo plants over a period of 30 days, where t represents the time in days.

Plant 1: $h_1 = t + 30$, $0 \leq t \leq 30$

Plant 2: $h_2 = 2t + 10$, $0 \leq t \leq 30$

(a) Explain how you can use the equations to determine the heights of the plants on Day 0.

(b) Which plant grew at a faster rate? Explain.

(c) Were the two plants ever the same height? Explain.

Cumulative Review

In Exercises 73–80, simplify the expression.

73. $x^2 \cdot x^3$

74. $z^2 \cdot z^2$

75. $(-y^2)y$

76. $5x^2(x^5)$

77. $(25x^3)(2x^2)$

78. $(3yz)(6yz^3)$

79. $x^2 - 2x - x^2 + 3x + 2$

80. $x^2 - 5x - 2 + x$

In Exercises 81–84, find the x- and y-intercepts (if any) of the graph of the equation.

81. $y = 6x - 3$

82. $y = -\frac{4}{3}x - 8$

83. $2x + y = -3$

84. $3x - 5y = 15$

Mid-Chapter Quiz: Sections 4.1–4.3

Solutions in English & Spanish and tutorial videos at AlgebraWithinReach.com

Take this quiz as you would take a quiz in class. After you are done, check your work against the answers in the back of the book.

1. Plot the points $(4, -2)$ and $\left(-1, -\frac{5}{2}\right)$ on a rectangular coordinate system.

2. Determine the quadrant(s) in which the point $(3, y)$ may be located without plotting it. Assume $y \neq 0$. (y is a real number.)

3. Determine whether each ordered pair is a solution of the equation $y = 9 - |x|$.

    (a) $(2, 7)$ (b) $(-3, 12)$ (c) $(-9, 0)$ (d) $(0, -9)$

In Exercises 4 and 5, find the x- and y-intercepts of the graph of the equation.

4. $x - 3y = 12$

5. $y = -7x + 2$

In Exercises 6–11, sketch the graph of the equation.

6. $y = x - 1$
7. $y = 5 - 2x$
8. $y = 4 - x^2$
9. $y = (x + 2)^2$
10. $y = |x + 3|$
11. $y = 1 - |x|$

In Exercises 12–14, find the slope of the line that passes through the points. Use the slope to state whether the line rises, falls, is horizontal, or is vertical. Then sketch the line.

12. $(0, 0), (-3, 9)$
13. $(-4, 3), (-4, -5)$
14. $\left(4, \frac{1}{2}\right), \left(-1, -\frac{5}{2}\right)$

In Exercises 15 and 16, write the equation in slope-intercept form. Use the slope and y-intercept to sketch the graph of the line.

15. $3x - 3y + 9 = 0$
16. $-2x + 3y - 6 = 0$

Applications

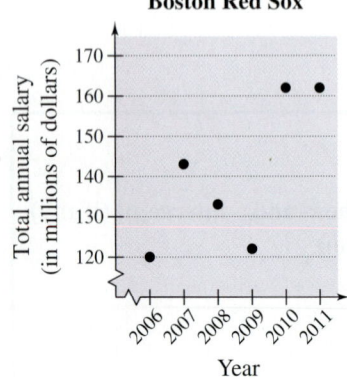

17. The scatter plot at the left shows the total annual salaries (in millions of dollars) of the Boston Red Sox in the years 2006 through 2011. Estimate the total salary each year. (*Source:* USA Today)

18. A new computer system sells for approximately $2000 and depreciates at the rate of $500 per year.

    (a) Write an equation that relates the value of the computer system and the number of years.

    (b) Sketch the graph of the equation.

    (c) What is the y-intercept of the graph, and what does it represent in the context of the problem?

Study Skills in Action

Reading Your Textbook Like a Manual

Many students avoid opening their textbooks for the same reason many people avoid opening their checkbooks—anxiety and frustration. The truth? Not opening your math textbook will cause more anxiety and frustration! Your textbook is a manual designed to help you master skills and understand and remember concepts. It contains many features and resources that can help you be successful in your course.

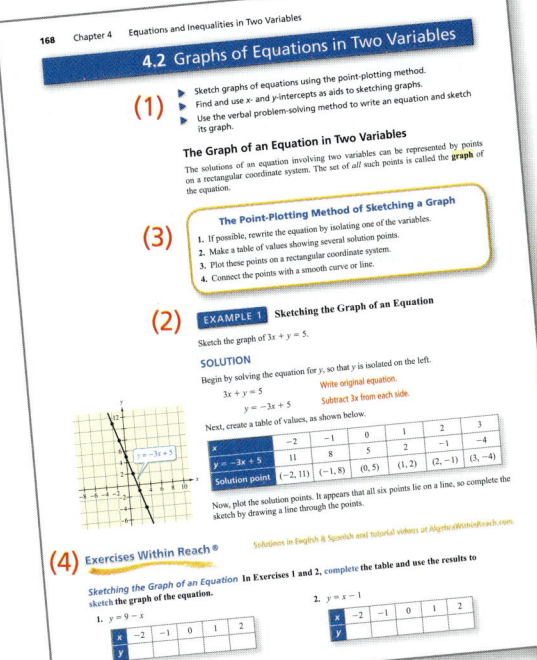

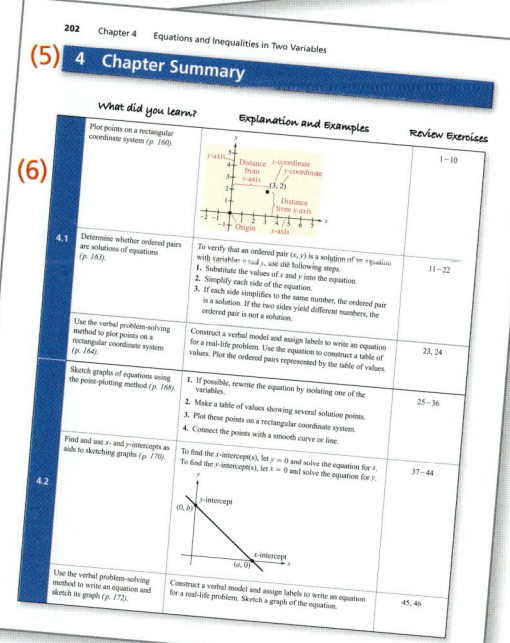

Smart Study Strategy

Use the Features of Your Textbook

To review what you learned in a previous class:

- Read the list of skills you should learn (1) at the beginning of this section. If you cannot remember how to perform a skill, review the appropriate example (2) in the section.

- Read and understand the contents of all tinted concept boxes (3)—these contain important definitions and rules.

To prepare for homework:

- Complete a few of the exercises (4) following each example. If you have difficulty with any of these, reread the example or seek help from a peer or instructor.

To review for quizzes and tests:

- Make use of the Chapter Summary (5). Check off the concepts (6) you know, and review those you do not know.

- Complete the Review Exercises. Then take the Mid-Chapter Quiz, Chapter Test, or Cumulative Test, as appropriate.

4.4 Equations of Lines

- Write equations of lines using the point-slope form.
- Write equations of horizontal and vertical lines.
- Use linear models to solve application problems.

Study Tip

In Example 1, it is concluded that $y = 3x - 5$ is "an" equation of the line rather than "the" equation of the line. The reason for this is that every equation can be written in many equivalent forms. For instance,

$y = 3x - 5$
    Slope-intercept form

and

$3x - y - 5 = 0$
    General Form

are both equations of the line in Example 1.

The Point-Slope Form of the Equation of a Line

There are two basic types of problems in analytic geometry.

1. Given an equation, sketch its graph.

2. Given a graph, write its equation.

In Section 4.3, you worked primarily with the first type of problem. In this section, you will study the second type.

> **Point-Slope Form of the Equation of a Line**
>
> The **point-slope form** of the equation of a line with slope m that passes through the point (x_1, y_1) is $y - y_1 = m(x - x_1)$.

EXAMPLE 1   Using Point-Slope Form

Write an equation of the line that passes through the point $(1, -2)$ and has slope $m = 3$.

SOLUTION

$y - y_1 = m(x - x_1)$ Use point-slope form.

$y - (-2) = 3(x - 1)$ Substitute -2 for y_1, 1 for x_1, and 3 for m.

$y + 2 = 3x - 3$ Simplify.

$y = 3x - 5$ Equation of line

So, an equation of the line is $y = 3x - 5$.

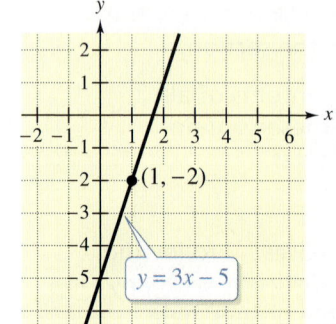

Exercises Within Reach ®

Solutions in English & Spanish and tutorial videos at AlgebraWithinReach.com

Using Point-Slope Form In Exercises 1–4, use the point-slope form to write an equation of the line. Write the equation in slope-intercept form.

1.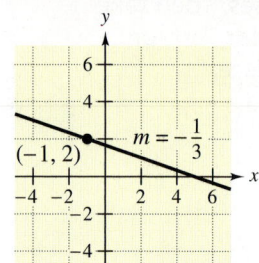
2.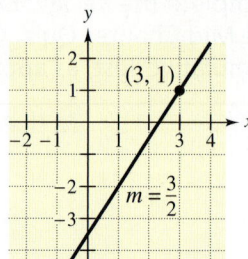
3.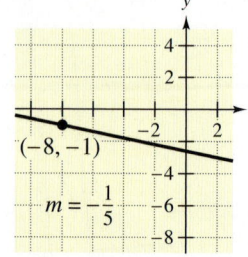
4.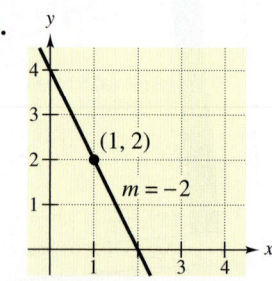

EXAMPLE 2 An Equation of a Line Through Two Points

Write an equation of the line that passes through the points $(3, 1)$ and $(-3, 4)$.

SOLUTION

Let $(x_1, y_1) = (3, 1)$ and $(x_2, y_2) = (-3, 4)$. The slope of the line passing through these points is

$$m = \frac{y_2 - y_1}{x_2 - x_1} \quad \text{Formula for slope}$$

$$= \frac{4 - 1}{-3 - 3} \quad \text{Substitute for } x_1, y_1, x_2, \text{ and } y_2.$$

$$= \frac{3}{-6} \quad \text{Subtract.}$$

$$= -\frac{1}{2}. \quad \text{Simplify.}$$

Now, use the point-slope form to find an equation of the line.

$$y - y_1 = m(x - x_1) \quad \text{Point-slope form}$$

$$y - 1 = -\frac{1}{2}(x - 3) \quad \text{Substitute 1 for } y_1, 3 \text{ for } x_1, \text{ and } -\frac{1}{2} \text{ for } m.$$

$$y - 1 = -\frac{1}{2}x + \frac{3}{2} \quad \text{Distributive Property}$$

$$y = -\frac{1}{2}x + \frac{5}{2} \quad \text{Equation of line}$$

The graph of this line is shown at the left.

Exercises Within Reach®   Solutions in English & Spanish and tutorial videos at AlgebraWithinReach.com

An Equation of a Line Through Two Points In Exercises 5–10, use the point-slope form to write an equation of the line. Write the equation in slope-intercept form.

5.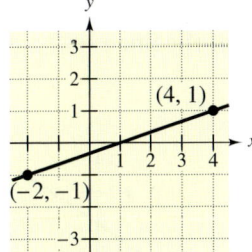

6.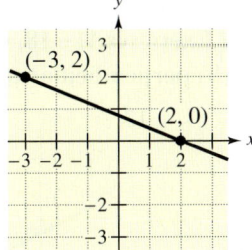

7.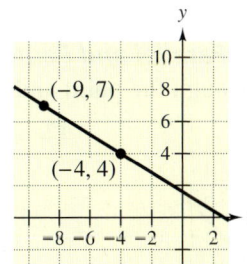

8.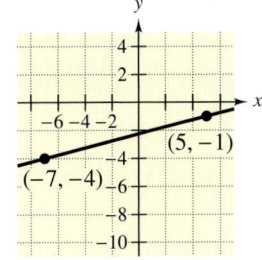

9.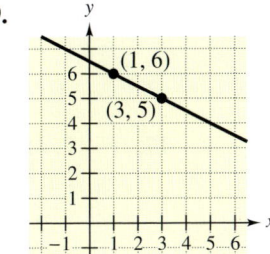

10.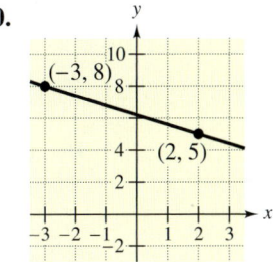

EXAMPLE 3 Equations of Parallel and Perpendicular Lines

Write an equation of the line that passes through the point $(2, -1)$ and is (a) parallel and (b) perpendicular to the line

$$y = \frac{2}{3}x - \frac{5}{3}.$$

SOLUTION

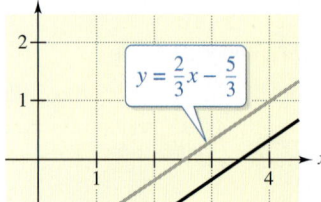

a. Because the line has a slope of $m = \frac{2}{3}$, it follows that any parallel line must have the same slope. So, an equation of the line through $(2, -1)$, parallel to the original line, is

$y - y_1 = m(x - x_1)$ Point-slope form

$y - (-1) = \frac{2}{3}(x - 2)$ Substitute -1 for y_1, 2 for x_1, and $\frac{2}{3}$ for m.

$y + 1 = \frac{2}{3}x - \frac{4}{3}$ Simplify.

$y = \frac{2}{3}x - \frac{7}{3}.$ Equation of parallel line

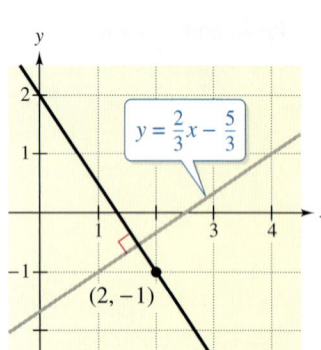

b. Because the line has a slope of $m = \frac{2}{3}$, it follows that any perpendicular line must have a slope of $-\frac{3}{2}$. So, an equation of the line through $(2, -1)$, perpendicular to the original line, is

$y - y_1 = m(x - x_1)$ Point-slope form

$y - (-1) = -\frac{3}{2}(x - 2)$ Substitute -1 for y_1, 2 for x_1, and $-\frac{3}{2}$ for m.

$y + 1 = -\frac{3}{2}x + 3$ Simplify.

$y = -\frac{3}{2}x + 2.$ Equation of perpendicular line

Exercises Within Reach®

Solutions in English & Spanish and tutorial videos at AlgebraWithinReach.com

Equations of Parallel and Perpendicular Lines In Exercises 11–18, write an equation of the line that passes through the point and is (a) parallel and (b) perpendicular to the given line.

11.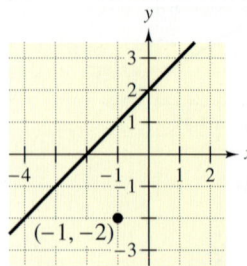

12.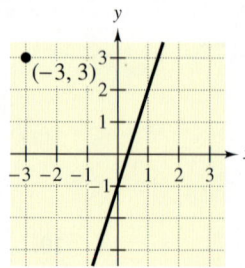

13.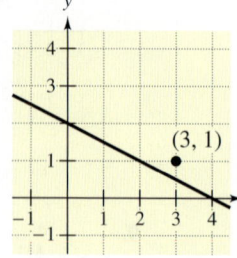

14.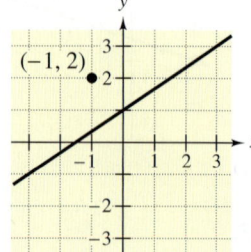

15. $(2, 1)$
    $x - y = 3$

16. $(-3, 2)$
    $x + y = 7$

17. $(-12, 4)$
    $3x + 4y = 7$

18. $(15, -2)$
    $5x - 3y = 0$

Equations of Horizontal and Vertical Lines

EXAMPLE 4 Equations of Horizontal and Vertical Lines

Write an equation for each line.

a. Vertical line through $(-3, 2)$
b. Line through $(-1, 2)$ and $(4, 2)$
c. Line through $(0, 2)$ and $(0, -2)$
d. Horizontal line through $(0, -4)$

SOLUTION

a. Because the line is vertical and passes through the point $(-3, 2)$, every point on the line has an *x*-coordinate of -3. So, an equation of the line is

   $x = -3$.   Vertical line

b. Because both points have the same *y*-coordinate, the line through $(-1, 2)$ and $(4, 2)$ is horizontal. So, its equation is

   $y = 2$.   Horizontal line

c. Because both points have the same *x*-coordinate, the line through $(0, 2)$ and $(0, -2)$ is vertical. So, its equation is

   $x = 0$.   Vertical line (*y*-axis)

d. Because the line is horizontal and passes through the point $(0, -4)$, every point on the line has a *y*-coordinate of -4. So, an equation of the line is

   $y = -4$.   Horizontal line

Exercises Within Reach®

Solutions in English & Spanish and tutorial videos at AlgebraWithinReach.com

An Equation of a Horizontal or Vertical Line In Exercises 19–22, write an equation of the line of the given type that passes through the point shown.

19. Vertical line
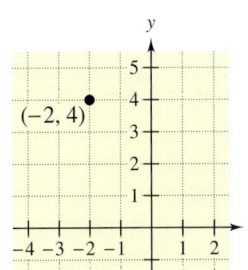

20. Horizontal line
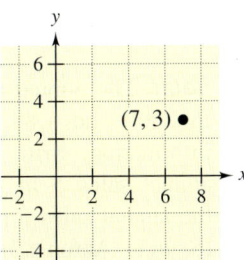

21. Horizontal line
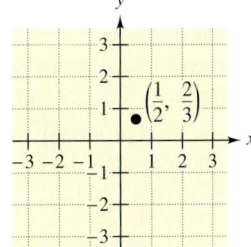

22. Vertical line
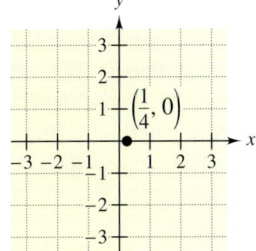

An Equation of a Horizontal or Vertical Line In Exercises 23–26, write an equation of the line that passes through the points shown.

23.
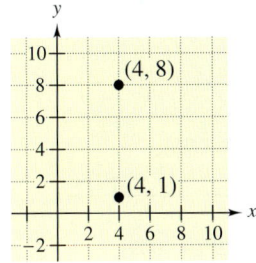

24.
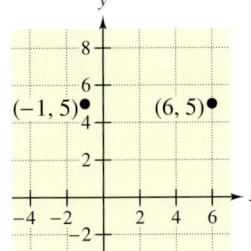

25.
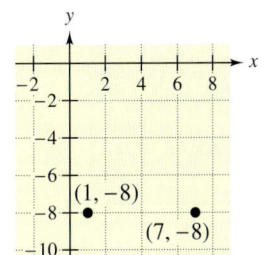

26.
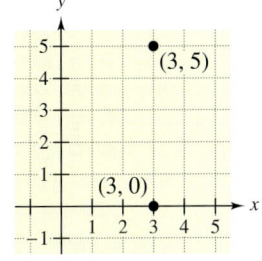

Applications

Application — EXAMPLE 5 Predicting Annual Sales

The annual sales of AutoZone were $6.82 billion in 2009 and $7.36 billion in 2010. Using only this information, write a linear equation that models the annual sales in terms of the year. Then predict the sales for 2011. (*Source:* AutoZone, Inc.)

SOLUTION

Let $t = 9$ represent 2009. Then the two given values are represented by the data points $(9, 6.82)$ and $(10, 7.36)$. The slope of the line through these points is

$$m = \frac{7.36 - 6.82}{10 - 9}$$

$$= 0.54.$$

Using the point-slope form, you can find an equation that relates the annual sales y and the year t to be

$y - y_1 = m(t - t_1)$	Point-slope form
$y - 7.36 = 0.54(t - 10)$	Substitute 7.36 for y_1, 10 for t_1, and 0.54 for m.
$y - 7.36 = 0.54t - 5.4$	Distributive Property
$y = 0.54t + 1.96$	Write in slope-intercept form.

Using this equation, a prediction of the annual sales in 2011 ($t = 11$) is

$$y = 0.54(11) + 1.96 = 7.9 \text{ billion.}$$

In this case, the prediction is fairly good—the actual annual sales in 2011 was $8.07 billion. The graph of this equation is shown at the left.

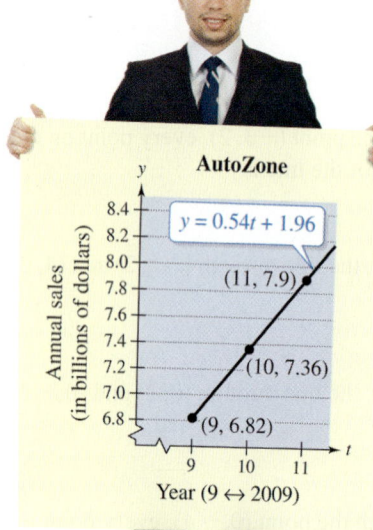

Exercises Within Reach®

Solutions in English & Spanish and tutorial videos at AlgebraWithinReach.com

27. *Net Profit* The net profit of Coach was $735 million in 2010 and $881 million in 2011. Using only this information, write a linear equation that models the net profit P in terms of the year t. Then predict the net profit for 2012 (Let $t = 0$ represent 2010.) (*Source:* Coach, Inc.)

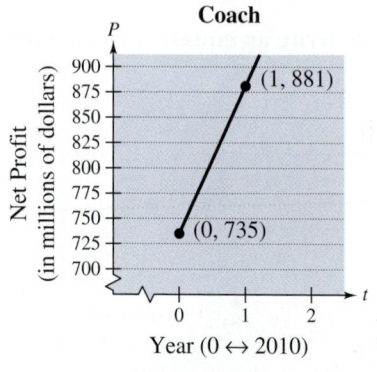

28. *Sales* The annual sales of Aaron's was $1.75 billion in 2009 and 1.88 billion in 2010. Using only this information, write a linear equation that models the annual sales S in terms of the year t. Then predict the annual sales for 2011. (Let $t = 9$ represent 2009.) (*Source:* Aaron's, Inc.)

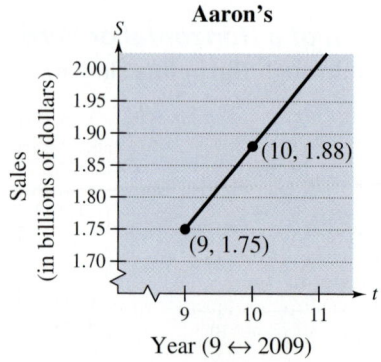

In the linear equation $y = mx + b$, you know that m represents the slope of the line. In applications, the slope of a line can often be interpreted as the *rate of change of y with respect to x*. Rates of change should always be described with appropriate units of measure.

Application

EXAMPLE 6 Using Slope as a Rate of Change

A rock climber is climbing up a 500-foot cliff. By 1 P.M., the rock climber has climbed 115 feet up the cliff. By 4 P.M., the climber has reached a height of 280 feet, as shown at the left. Find the average rate of change of the climber and use this rate of change to write a linear model that relates the height of the climber to the time.

SOLUTION

Let y represent the height of the climber and let t represent the time. Then the two points that represent the climber's two positions are $(t_1, y_1) = (1, 115)$ and $(t_2, y_2) = (4, 280)$. So, the average rate of change of the climber is

$$\text{Average rate of change} = \frac{y_2 - y_1}{t_2 - t_1} = \frac{280 - 115}{4 - 1} = 55 \text{ feet per hour.}$$

So, an equation that relates the height of the climber to the time is

$y - y_1 = m(t - t_1)$ Point-slope form

$y - 115 = 55(t - 1)$ Substitute 115 for y_1, 1 for t_1, and 55 for m.

$y = 55t + 60$.   Linear model

Exercises Within Reach®

Solutions in English & Spanish and tutorial videos at AlgebraWithinReach.com

29. **Depreciation** You purchase a boat for $25,000. After 1 year, its depreciated value is $22,700. The depreciation is linear.

    (a) Write a linear model that relates the value V of the boat to the time t in years.

    (b) Use the model to estimate the value of the boat after 3 years.

30. **Depreciation** A sub shop purchases a used pizza oven for $875. After 1 year, its depreciated value is $790. The depreciation is linear.

    (a) Write a linear model that relates the value V of the oven to the time t in years.

    (b) Use the model to estimate the value of the oven after 5 years.

31. **Bike Path** A city is paving a bike path. The same length of path is paved each day. After 4 days, 14 miles of the path remain to be paved. After 6 more days, 11 miles of the path remain to be paved. Find the average rate of change and use it to write a linear model that relates the distance remaining to be paved to the number of days.

32. **Swimming Pool** A swimming pool already contains a small amount of water when you start filling it at a constant rate. The pool contains 45 gallons of water after 5 minutes and 120 gallons after 30 minutes. Find the average rate of change and use it to write a linear model that relates the amount of water in the pool to the time.

Concept Summary: Writing Equations of Lines

What
You can use the **point-slope form** to write an equation of a line passing through any two points.

EXAMPLE
Write an equation of the line passing through the points (2, 3) and (3, 5).

How
Here are the steps to write an equation of the line.

1. Find the slope of the line.

$$m = \frac{y_2 - y_1}{x_2 - x_1} = \frac{5 - 3}{3 - 2} = 2$$

2. Use the point-slope form.

$$y - y_1 = m(x - x_1)$$
$$y - 3 = 2(x - 2)$$
$$y - 3 = 2x - 4$$
$$y = 2x - 1$$

Why
When you are given

- the slope of a line and a point on the line, or
- any two points on a line,

you can use point-slope form to write an equation of the line.

Exercises Within Reach®

Worked-out solutions to odd-numbered exercises at AlgebraWithinReach.com

Concept Summary Check

33. **Extra Steps** Why was finding the slope the first step of the solution above?

34. **Applying Point-Slope Form** What two pieces of information were used to write the equation in point-slope form in Step 2 above?

35. **The Form of an Equation** Describe the form of the final equation $y = 2x - 1$ in the solution above.

36. **Choosing a Form** When is it more convenient to apply the point-slope form than the slope-intercept form to write an equation of a line?

Extra Practice

Using Point-Slope Form In Exercises 37–40, use the point-slope form to write an equation of the line that passes through the point and has the specified slope. Write the equation in slope-intercept form.

37. $\left(0, \frac{3}{2}\right), m = \frac{2}{3}$

38. $\left(-\frac{5}{2}, 0\right), m = \frac{3}{4}$

39. $(2, 4), m = -0.8$

40. $(6, -3), m = 0.67$

Finding the Slope In Exercises 41–44, determine the slope of the line.

41. $y - 2 = 5(x + 3)$

42. $y + 3 = -2(x - 6)$

43. $3x - 2y + 10 = 0$

44. $5x + 4y - 8 = 0$

An Equation of a Line Through Two Points In Exercises 45–48, write an equation of the line that passes through the points. Write the equation in general form.

45. $\left(2, \frac{1}{2}\right), \left(\frac{1}{2}, \frac{5}{2}\right)$

46. $\left(\frac{1}{4}, 1\right), \left(-\frac{3}{4}, -\frac{2}{3}\right)$

47. $(1, 0.6), (2, -0.6)$

48. $(-8, 0.6), (2, -2.4)$

49. Wages A sales representative receives a monthly salary of $2000 plus a commission of 2% of the total monthly sales. Write a linear model that relates total monthly wages W to sales S.

50. Reimbursed Expenses A sales representative is reimbursed $250 per day for lodging and meals plus $0.30 per mile driven. Write a linear model that relates the daily cost C to the number of miles driven x.

51. Graphical Interpretation Match each of the situations labeled (a), (b), (c), and (d) with one of the graphs labeled (e), (f), (g), and (h). Then determine the slope of each line and interpret the slope in the context of the real-life situation.

(a) A friend is paying you $10 per week to repay a $100 loan.

(b) An employee is paid $12.50 per hour plus $1.50 for each unit produced per hour.

(c) A sales representative receives $40 per day for food plus $0.32 for each mile traveled.

(d) A television purchased for $600 depreciates $100 per year.

(e)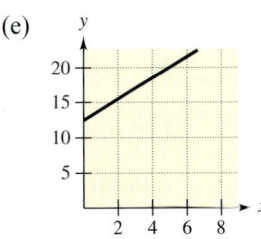

(f)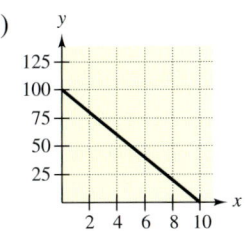

(g)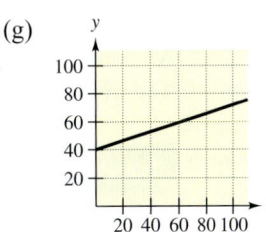

(h)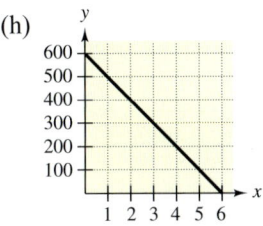

Explaining Concepts

52. Writing Can the equation of a vertical line be written in slope-intercept form? Explain.

53. Writing Explain how to find algebraically the x-intercept of the line given by $y = mx + b$.

54. Think About It Find the slope of the line given by $5x + 7y - 21 = 0$. Use the same process to find a formula for the slope of the line $ax + by + c = 0$, where $b \neq 0$.

55. Think About It What is implied about the graphs of the lines $a_1x + b_1y + c_1 = 0$ and $a_2x + b_2y + c_2 = 0$ if $\dfrac{a_1}{b_1} = \dfrac{a_2}{b_2}$?

56. Research Project Use a newspaper or weekly news magazine to find an example of data that are *increasing* linearly with time. Write a linear model that relates the data to time. Repeat the project for data that are *decreasing*.

Cumulative Review

In Exercises 57–60, simplify the expression.

57. $4(3 - 2x)$

58. $x^2(xy^3)$

59. $3x - 2(x - 5)$

60. $u - [3 + (u - 4)]$

In Exercises 61–64, solve for y in terms of x.

61. $3x + y = 4$

62. $4 - y + x = 0$

63. $4x - 5y = -2$

64. $3x + 4y - 5 = 0$

In Exercises 65–68, determine the slope of the line passing through the points.

65. $(3, 0)$ and $(4, 2)$

66. $(-1, 3)$ and $(2, -5)$

67. $(0, -2)$ and $(-7, -1)$

68. $(-2, 4)$ and $(-5, 10)$

4.5 Graphs of Linear Inequalities

- Determine whether an ordered pair is a solution of a linear inequality in two variables.
- Sketch graphs of linear inequalities in two variables.
- Use linear inequalities to model and solve real-life problems.

Linear Inequalities in Two Variables

A **linear inequality in two variables**, x and y, is an inequality that can be written in one of the forms below (where a and b are not both zero).

$$ax + by < c, \quad ax + by > c, \quad ax + by \leq c, \quad ax + by \geq c$$

An ordered pair (x_1, y_1) is a **solution** of a linear inequality in x and y if the inequality is true when x_1 and y_1 are substituted for x and y, respectively.

EXAMPLE 1 Verifying Solutions of a Linear Inequality

Determine whether each ordered pair is a solution of $3x - y \geq -1$.

a. $(0, 0)$ **b.** $(1, 4)$ **c.** $(-1, 2)$

SOLUTION

a.
$3x - y \geq -1$ Write original inequality.
$3(0) - 0 \stackrel{?}{\geq} -1$ Substitute 0 for x and 0 for y.
$0 \geq -1$ Inequality is satisfied. ✓

Because the inequality is satisfied, the point $(0, 0)$ *is* a solution.

b.
$3x - y \geq -1$ Write original inequality.
$3(1) - 4 \stackrel{?}{\geq} -1$ Substitute 1 for x and 4 for y.
$-1 \geq -1$ Inequality is satisfied. ✓

Because the inequality is satisfied, the point $(1, 4)$ *is* a solution.

c.
$3x - y \geq -1$ Write original inequality.
$3(-1) - 2 \stackrel{?}{\geq} -1$ Substitute -1 for x and 2 for y.
$-5 \not\geq -1$ Inequality is not satisfied. ✗

Because the inequality is not satisfied, the point $(-1, 2)$ *is not* a solution.

Exercises Within Reach®   Solutions in English & Spanish and tutorial videos at AlgebraWithinReach.com

Verifying Solutions of a Linear Inequality In Exercises 1–4, **determine** whether each ordered pair is a solution of the inequality.

1. $x + 4y > 10$ (a) $(0, 0)$ (b) $(3, 2)$
                   (c) $(1, 2)$ (d) $(-2, 4)$

2. $2x + 3y > 9$ (a) $(0, 0)$ (b) $(1, 1)$
                   (c) $(2, 2)$ (d) $(-2, 5)$

3. $-3x + 5y \leq 12$ (a) $(1, 2)$ (b) $(2, -3)$
                      (c) $(1, 3)$ (d) $(2, 8)$

4. $5x + 3y < 100$ (a) $(25, 10)$ (b) $(6, 10)$
                      (c) $(0, -12)$ (d) $(4, 5)$

Section 4.5 Graphs of Linear Inequalities

The Graph of a Linear Inequality in Two Variables

Sketching the Graph of a Linear Inequality in Two Variables

1. Replace the inequality sign by an equal sign and sketch the graph of the resulting equation. (Use a dashed line for < or > and a solid line for ≤ or ≥.)
2. Test one point in each of the half-planes formed by the graph in Step 1. If the point satisfies the inequality, then shade the entire half-plane to denote that every point in the region satisfies the inequality.

Study Tip

When the inequality is less than (<) or greater than (>), the graph of the corresponding equation is a dashed line because the points on the line are *not* solutions of the inequality. When the inequality is less than or equal to (≤) or greater than or equal to (≥), the graph of the corresponding equation is a solid line because the points on the line *are* solutions of the inequality.

EXAMPLE 2 Sketching the Graphs of Linear Inequalities

Sketch the graphs of (a) $x > -2$ and (b) $y \leq 3$.

SOLUTION

a. The graph of the corresponding equation is a vertical line. The points that satisfy the inequality are those lying to the right of the line.

b. The graph of the corresponding equation is a horizontal line. The points that satisfy the inequality are those lying on or below the line.

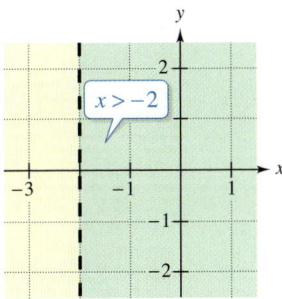
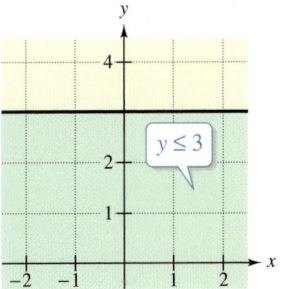

Exercises Within Reach®
Solutions in English & Spanish and tutorial videos at AlgebraWithinReach.com

Identifying the Type of Boundary In Exercises 5–8, state whether the boundary of the graph of the inequality should be dashed or solid.

5. $2x + 3y < 6$
6. $2x + 3y \leq 6$
7. $2x + 3y > 6$
8. $2x + 3y > 6$

Matching In Exercises 9–12, match the inequality with its graph.

(a)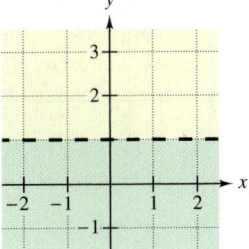
(b)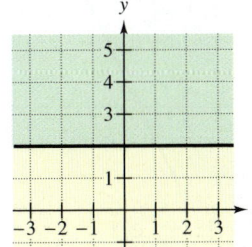
(c)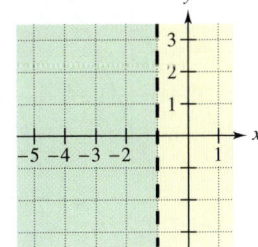
(d)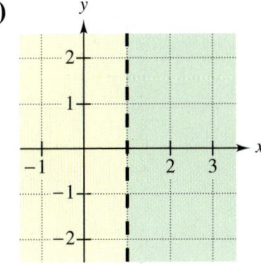

9. $x < -1$
10. $x > 1$
11. $y \geq 2$
12. $y < 1$

Sketching the Graph of a Linear Inequality In Exercises 13–16, sketch the graph of the linear inequality.

13. $y \geq 3$
14. $x > -4$
15. $x \leq 3$
16. $y > -5$

Study Tip

You can use any point that is not on the line as a test point. However, the origin is often the most convenient test point because it is easy to evaluate expressions in which 0 is substituted for each variable.

EXAMPLE 3 Sketching the Graph of a Linear Inequality

Sketch the graph of the linear inequality

$$x - y < 2.$$

SOLUTION

The graph of the corresponding equation

$x - y = 2$ Write corresponding equation.

is the line show below. Because the origin (0, 0) does not lie on the line, use it as the test point.

$x - y < 2$ Write original inequality

$0 - 0 \stackrel{?}{<} 2$ Substitute 0 for x and 0 for y.

$0 < 2$ Inequality is satisfied. ✓

Because (0, 0) satisfies the inequality, the graph consists of the half-plane lying above the line. Try checking a point below the line. Regardless of which point you choose, you will see that it does not satisfy the inequality.

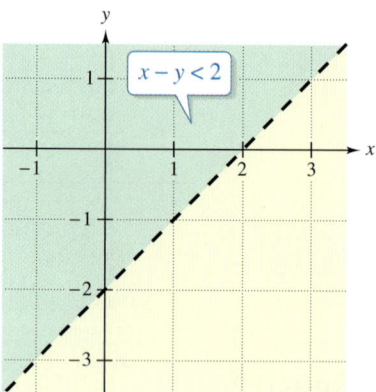

Exercises Within Reach®

Solutions in English & Spanish and tutorial videos at AlgebraWithinReach.com

Matching In Exercises 17–20, match the inequality with its graph.

(a) 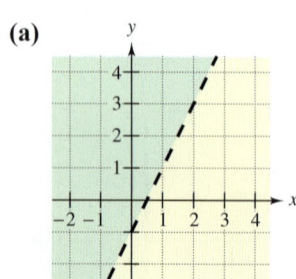 (b) 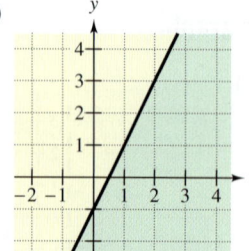 (c) 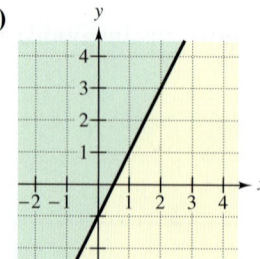 (d)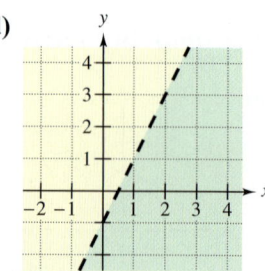

17. $2x - y \leq 1$ **18.** $2x - y < 1$ **19.** $2x - y \geq 1$ **20.** $2x - y > 1$

Sketching the Graph of a Linear Inequality In Exercises 21–28, sketch the graph of the linear inequality.

21. $y \leq 3x$ **22.** $y \geq 5x$ **23.** $y \leq 2x - 1$ **24.** $y \geq 2x - 1$

25. $x - y < 0$ **26.** $x + y > 0$ **27.** $y > -2x + 10$ **28.** $y < 3x + 1$

Section 4.5 Graphs of Linear Inequalities 197

EXAMPLE 4 Sketching the Graph of a Linear Inequality

Use the slope-intercept form of a linear equation as an aid in sketching the graph of the inequality $5x + 4y \leq 12$.

SOLUTION

To begin, rewrite the inequality in slope-intercept form.

$5x + 4y \leq 12$ Write original inequality.

$4y \leq -5x + 12$ Subtract 5x from each side.

$y \leq -\dfrac{5}{4}x + 3$ Divide each side by 4.

From this form, you can conclude that the solution is the half-plane lying *on* or *below* the line $y = -\dfrac{5}{4}x + 3$. The graph is shown below. You can verify this by testing the solution point (0, 0).

$5x + 4y \leq 12$ Write original inequality.

$5(0) + 4(0) \stackrel{?}{\leq} 12$ Substitute 0 for x and 0 for y.

$0 \leq 12$ Inequality is satisfied.

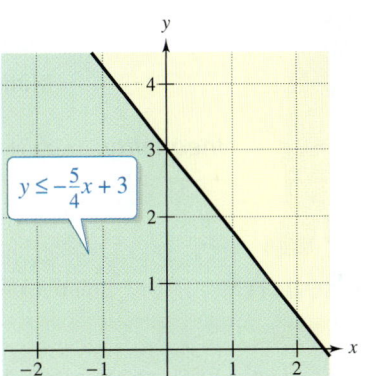

Exercises Within Reach®

Solutions in English & Spanish and tutorial videos at AlgebraWithinReach.com

Matching In Exercises 29–32, match the inequality with its graph.

(a) 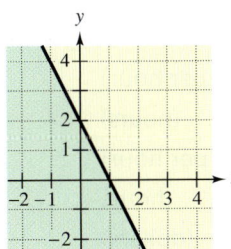 (b) 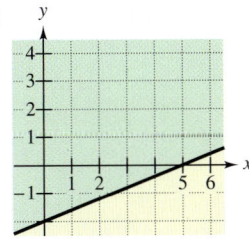 (c) 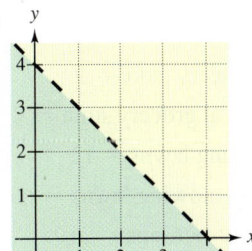 (d)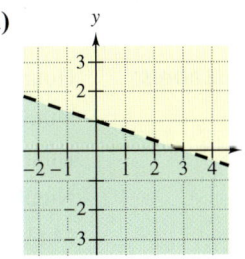

29. $y < -\dfrac{1}{3}x + 1$ **30.** $y \geq \dfrac{2}{5}x - 2$ **31.** $x + y < 4$ **32.** $y + 2x \leq 2$

Sketching the Graph of a Linear Inequality In Exercises 33–38, sketch the graph of the linear inequality.

33. $-3x + 2y < 6$ **34.** $x - 2y \leq -6$ **35.** $x \geq 3y - 5$

36. $x \leq -2y + 10$ **37.** $y - 3 > \dfrac{1}{2}(x - 4)$ **38.** $y + 1 < -2(x - 3)$

Applications

Application

EXAMPLE 5 Working to Meet a Budget

Your budget requires you to earn *at least* $250 per week. You work two part-time jobs. One is at a fast-food restaurant, which pays $8 per hour, and the other is tutoring for $10 per hour. Let x represent the number of hours you work at the restaurant and let y represent the number of hours you work as a tutor. Write a linear inequality that represents the different numbers of hours you can work at each job in order to meet your budget requirements.

SOLUTION

To write the inequality, use the problem-solving method.

Verbal Model: Hourly pay at fast-food restaurant · Number of hours at fast-food restaurant + Hourly pay tutoring · Number of hours tutoring ≥ Minimum weekly earnings

Labels:
Hourly pay at fast-food restaurant = 8 (dollars per hour)
Number of hours at fast-food restaurant = x (hours)
Hourly pay for tutoring = 10 (dollars per hour)
Number of hours tutoring = y (hours)
Minimum weekly earnings = 250 (dollars)

Inequality: $8x + 10y \geq 250$

Gelpi/Shutterstock.com

Exercises Within Reach ® Solutions in English & Spanish and tutorial videos at AlgebraWithinReach.com

---

39. *Part-Time Jobs* Your budget requires you to earn at least $210 per week. You work two part-time jobs. One is at a grocery store, which pays $9 per hour, and the other is mowing lawns, which pays $12 per hour. Use the verbal model and labels below to write a linear inequality that represents the different numbers of hours you can work at each job in order to meet your budget requirements.

Verbal Model: Hourly pay at grocery store · Number of hours at grocery store + Hourly pay mowing lawns · Number of hours mowing lawns ≥ Minimum weekly earnings

Labels:
Hourly pay at grocery store = 9 (dollars per hour)
Number of hours at grocery store = x (hours)
Hourly pay mowing lawns = 12 (dollars per hour)
Number of hours mowing lawns = y (hours)
Minimum weekly earnings = 210 (dollars)

40. *Inventory* A store sells two models of central air conditioning units. The costs to the store of the two models are $2000 and $3000, and the owner of the store does not want more than $30,000 invested in the inventory for these two models. Write a linear inequality that represents the different numbers of each model that can be held in inventory.

Section 4.5   Graphs of Linear Inequalities   199

Application

EXAMPLE 6   Working to Meet a Budget

Graph the inequality in Example 5 and find at least two ordered pairs (x, y) that identify the numbers of hours you can work at each job in order to meet your budget requirements.

SOLUTION

To sketch the graph, rewrite the inequality in slope-intercept form.

$8x + 10y \geq 250$ Write original inequality.

$10y \geq -8x + 250$ Subtract 8x from each side.

$y \geq -\dfrac{4}{5}x + 25$ Divide each side by 10.

Graph the corresponding equation

$y = -\dfrac{4}{5}x + 25$

and shade the half-plane lying above the line, as shown at the left. From the graph, you can see that two solutions that will yield the desired weekly earnings of at least $250 are (10, 17) and (20, 15). In other words, you can work 10 hours at the restaurant and 17 hours as a tutor, or 20 hours at the restaurant and 15 hours as a tutor, to meet your budget requirements. There are many other solutions.

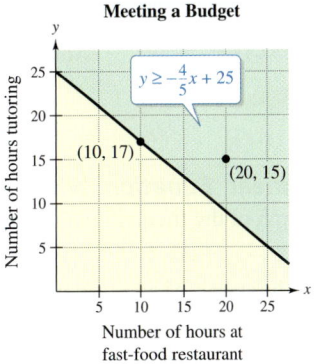

Exercises Within Reach®

Solutions in English & Spanish and tutorial videos at AlgebraWithinReach.com

41. **Part Time Jobs** Graph the inequality in Exercise 39 and find three ordered pairs that are solutions of the inequality.

42. **Inventory** Graph the inequality in Exercise 40 and find three ordered pairs that are solutions of the inequality.

Golden Pixels LLC/Shutterstock.com; GSPhotography/Shutterstock.com

Concept Summary: Graphing Linear Inequalities

What
You can use a graph to represent all the **solutions** of a **linear inequality in two variables**.

EXAMPLE
Sketch the **graph** of the linear inequality $y < x + 1$.

How
Use these steps to sketch the graph of such an inequality.

1. Replace the inequality sign by an equal sign.
2. Sketch the graph of the resulting equation.
   - Use a dashed line for $<$ and $>$.
   - Use a solid line for $\leq$ and $\geq$.
3. Shade the **half-plane** that contains points that satisfy the inequality.

Why
You can use linear inequalities to model and solve many real-life problems. The graphs of inequalities help you see all the possible solutions. For instance, one solution of the inequality $y < x + 1$ is $(1, 1)$.

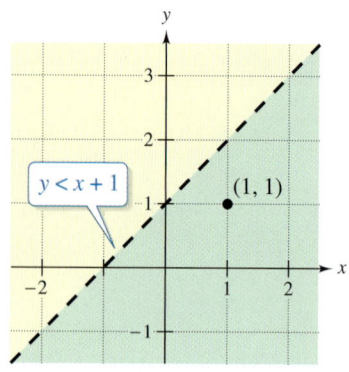

Exercises Within Reach®

Worked-out solutions to odd-numbered exercises at AlgebraWithinReach.com

Concept Summary Check

43. *The Corresponding Linear Equation* Why is a dashed line used in the graph of the inequality above?

44. *Using the Graph* Use the graph above to determine whether the point $(0, 2)$ represents a solution of the inequality $y < x + 1$. Explain.

45. *Using the Graph* Use the graph above to determine whether the point $(0, 1)$ represents a solution of the inequality $y < x + 1$. Explain.

46. *Testing Half-Planes* What point is often the most convenient test point to use when sketching the graph of a linear inequality?

Extra Practice

Writing a Linear Inequality In Exercises 47–52, **write** the statement as a linear inequality. Then **sketch** the graph of the inequality.

47. y is more than six times x.

48. x is at most three times y.

49. The sum of x and y is at least 9.

50. The difference of x and y is less than 20.

51. y is no more than the sum of x and 3.

52. The sum of x and 7 is more than three times y.

Writing a Linear Inequality In Exercises 53–56, **write** an inequality that is represented by the graph.

53.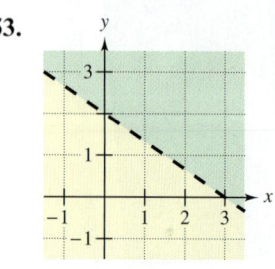

54.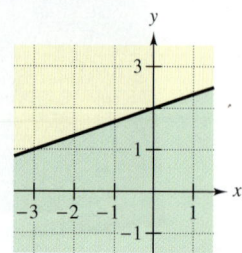

55.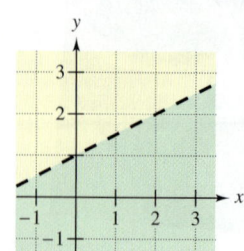

56.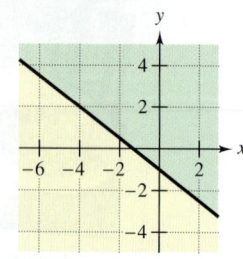

57. Nutrition A nutritionist recommends that the fat calories consumed per day should be at most 35% of the total calories consumed per day.

(a) Write a linear inequality that represents the different numbers of total calories and fat calories that are recommended for one day.

(b) Graph the inequality and find three ordered pairs that are solutions of the inequality.

58. Money A cash register must have at least $25 in change consisting of d dimes and q quarters.

(a) Write a linear inequality that represents the different numbers of dimes and quarters that can satisfy the requirement.

(b) Graph the inequality and find three ordered pairs that are solutions of the inequality.

59. Manufacturing Each table produced by a furniture company requires 1 hour in the assembly center. The matching chair requires $1\frac{1}{2}$ hours in the assembly center. A total of 12 hours per day is available in the assembly center.

(a) Write a linear inequality that represents the different numbers of hours that can be spent assembling tables and chairs.

(b) Graph the inequality and find three ordered pairs that are solutions of the inequality.

Explaining Concepts

60. Think About It Are there any points in the coordinate plane that are not solutions of either $y > x - 2$ or $y < x - 2$? Explain.

61. Think About It Write the inequality whose graph consists of all points above the x-axis.

62. Think About It Write an inequality whose graph has no points in the first quadrant.

63. Writing Does $2x < 2y$ have the same graph as $y > x$? Explain.

64. Writing Explain the difference between graphing the solution of the inequality $x \geq 1$ (a) on the real number line and (b) on a rectangular coordinate system.

Cumulative Review

In Exercises 65–70, find a ratio that compares the relative sizes of the quantities. (Use the same units of measurement for both quantities).

65. 36 feet to 8 feet

66. 8 dimes to 36 nickels

67. 4 hours to 45 minutes

68. 50 centimeters to 3 meters

69. 9 pounds to 8 ounces

70. 46 inches to 4 feet

In Exercises 71–76, write an equation of the line that passes through the points.

71. $(0, 0), (-3, 6)$

72. $(5, 8), (1, -4)$

73. $(5, 0), (0, -2)$

74. $(4, 3), (-4, 5)$

75. $(6, -1), (-3, -3)$

76. $\left(\frac{5}{6}, -2\right), \left(\frac{1}{6}, 4\right)$

4 Chapter Summary

	What did you learn?	Explanation and Examples	Review Exercises
4.1	Plot points on a rectangular coordinate system *(p. 160)*.	*(Diagram showing the point (3, 2) plotted on a rectangular coordinate system, with labels for y-axis, x-axis, Origin, x-coordinate, y-coordinate, Distance from y-axis, and Distance from x-axis.)*	1–10
	Determine whether ordered pairs are solutions of equations *(p. 163)*.	To verify that an ordered pair (x, y) is a solution of an equation with variables x and y, use the following steps. 1. Substitute the values of x and y into the equation. 2. Simplify each side of the equation. 3. If each side simplifies to the same number, then the ordered pair is a solution. If the two sides yield different numbers, then the ordered pair is not a solution.	11–22
	Use the verbal problem-solving method to plot points on a rectangular coordinate system *(p. 164)*.	Construct a verbal model and assign labels to write an equation for a real-life problem. Use the equation to construct a table of values. Plot the ordered pairs represented by the table of values.	23, 24
4.2	Sketch graphs of equations using the point-plotting method *(p. 168)*.	1. If possible, rewrite the equation by isolating one of the variables. 2. Make a table of values showing several solution points. 3. Plot these points on a rectangular coordinate system. 4. Connect the points with a smooth curve or line.	25–36
	Find and use x- and y-intercepts as aids to sketching graphs *(p. 170)*.	To find the x-intercept(s), let $y = 0$ and solve the equation for x. To find the y-intercept(s), let $x = 0$ and solve the equation for y. *(Diagram showing a line with y-intercept at $(0, b)$ and x-intercept at $(a, 0)$.)*	37–44
	Use the verbal problem-solving method to write an equation and sketch its graph *(p. 172)*.	Construct a verbal model and assign labels to write an equation for a real-life problem. Sketch a graph of the equation.	45, 46

What did you learn?	Explanation and Examples	Review Exercises
4.3 Determine the slope of a line through two points *(p. 176)*.	The slope m of a nonvertical line that passes through the points (x_1, y_1) and (x_2, y_2) is $$m = \frac{y_2 - y_1}{x_2 - x_1} = \frac{\text{Change in } y}{\text{Change in } x} = \frac{\text{Rise}}{\text{Run}}$$ where $x_1 \neq x_2$. 1. A line with positive slope ($m > 0$) *rises* from left to right. 2. A line with negative slope ($m < 0$) *falls* from left to right. 3. A line with zero slope ($m = 0$) is *horizontal*. 4. A line with undefined slope is *vertical*.	47–62
Write linear equations in slope-intercept form and graph the equations *(p. 178)*.	The slope-intercept form of the equation of a line is $y = mx + b$. The graph of the equation $y = mx + b$ is a line whose slope is m and whose y-intercept is $(0, b)$.	63–70
Use slopes to determine whether lines are parallel, perpendicular, or neither *(p. 180)*.	**Parallel Lines:** Two distinct nonvertical lines are parallel if and only if they have the same slope. **Perpendicular Lines:** Two lines are perpendicular if and only if their slopes are negative reciprocals of each other. That is, $$m_1 = -\frac{1}{m_2}.$$	71–74
4.4 Write equations of lines using the point-slope form *(p. 186)*.	The point-slope form of the equation of a line with slope m that passes through the point (x_1, y_1) is $y - y_1 = m(x - x_1)$.	75–94
Write equations of horizontal and vertical lines *(p. 189)*.	A horizontal line has a slope of zero and an equation of the form $y = b$. A vertical line has an undefined slope and an equation of the form $x = a$.	95–100
Use linear models to solve application problems *(p. 190)*.	The slope of a line can often be interpreted as the *rate of change of y with respect to x*. Use linear models to make predictions about real-life situations.	101, 102
4.5 Determine whether an ordered pair is a solution of a linear inequality in two variables *(p. 194)*.	An ordered pair (x_1, y_1) is a solution of a linear inequality in x and y if the inequality is true when x_1 and y_1 are substituted for x and y, respectively.	103–106
Sketch graphs of linear inequalities in two variables *(p. 195)*.	1. Replace the inequality sign by an equal sign and sketch the graph of the resulting equation. (Use a dashed line for $<$ or $>$ and a solid line for $\leq$ or $\geq$.) 2. Test one point in each of the half-planes formed by the graph in Step 1. If the point satisfies the inequality, then shade the entire half-plane to denote that every point in the region satisfies the inequality.	107–118
Use linear inequalities to model and solve real-life problems *(p. 198)*.	Construct a verbal model and assign labels to write a linear inequality for a real-life problem. Graph the inequality to find the solutions for the problem.	119, 120

Review Exercises

Worked-out solutions to odd-numbered exercises at AlgebraWithinReach.com

4.1

Plotting Points In Exercises 1–4, plot the points on a rectangular coordinate system.

1. $(-1, 6), (4, -3), (-2, 2), (3, 5)$
2. $(0, -1), (-4, 2), (5, 1), (3, -4)$
3. $(-2, 0), \left(\frac{3}{2}, 4\right), (-1, -3)$
4. $\left(3, -\frac{5}{2}\right), \left(-5, 2\frac{3}{4}\right), (4, 6)$

Finding Coordinates In Exercises 5 and 6, determine the coordinates of the points.

5.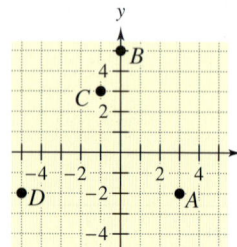

6.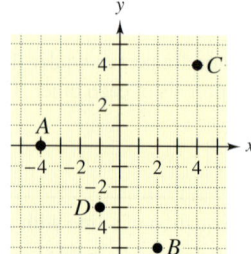

Determining the Quadrant In Exercises 7–10, determine the quadrant in which the point is located.

7. $(-5, 3)$
8. $(4, -6)$
9. $(-3, y), y < 0$
10. $(x, 5), x < 0$

Constructing a Table of Values In Exercises 11–14, complete the table of values. Then plot the solution points on a rectangular coordinate system.

11.
x	−1	0	1	2
y = 4x − 1				

12.
x	−1	0	1	2
y = 2x + 1				

13.
x	−1	0	1	2
y = −½x − 1				

14.
x	−1	0	1	2
y = 3/2 x + 5				

Solving for y In Exercises 15–18, solve the equation for y.

15. $3x + 4y = 12$
16. $2x + 3y = 6$
17. $9x - 3y = 12$
18. $-x - 3y = 9$

Verifying Solutions of an Equation In Exercises 19–22, determine whether each ordered pair is a solution of the equation.

19. $x - 3y = 4$
    (a) $(1, -1)$ (b) $(0, 0)$
    (c) $(2, 1)$ (d) $(5, -2)$

20. $y - 2x = -1$
    (a) $(3, 7)$ (b) $(0, -1)$
    (c) $(-2, -5)$ (d) $(-1, 0)$

21. $y = \frac{2}{3}x + 3$
    (a) $(3, 5)$ (b) $(-3, 1)$
    (c) $(-6, 0)$ (d) $(0, 3)$

22. $y = \frac{1}{4}x + 2$
    (a) $(-4, 1)$ (b) $(-8, 0)$
    (c) $(12, 5)$ (d) $(0, 2)$

23. **Organizing Data** The data from a study measuring the relationship between the wattage x of a standard 120-volt light bulb and the energy rate y (in lumens) is shown in the table.

x	25	40	60	100	150	200
y	235	495	840	1675	2650	3675

(a) Plot the data in the table.

(b) Use the graph to describe the relationship between the wattage and the energy rate.

24. **Organizing Data** An employee earns $12 plus $0.25 for every x units produced per hour. Write an equation that relates the employee's total hourly wage to the number of units produced. Plot the hourly wages for producing 10, 20, 30, 40, and 50 units per hour.

4.2

Sketching the Graph of an Equation In Exercises 25–28, complete the table and use the results to sketch the graph of the equation.

25. $y = x - 5$

x	−2	−1	0	1	2
y					

26. $3x + y = -4$

x	−2	−1	0	1	2
y					

27. $y = x^2 - 1$

x	−2	−1	0	1	2
y					

28. $y = |x - 2|$

x	0	1	2	3	4
y					

Sketching the Graph of an Equation In Exercises 29–36, sketch the graph of the equation using the point-plotting method.

29. $y = 7$
30. $x = -2$
31. $y = 3x$
32. $y = -2x$
33. $y = 4 - \frac{1}{2}x$
34. $y = \frac{3}{2}x - 3$
35. $y - 2x - 4 = 0$
36. $3x + 2y + 6 = 0$

Finding the Intercepts of a Graph In Exercises 37–44, find the x- and y-intercepts and sketch the graph of the equation.

37. $y = 6x + 2$
38. $y = -3x + 5$
39. $y = \frac{2}{5}x - 2$
40. $y = \frac{1}{3}x + 1$
41. $2x - y = 4$
42. $3x - y = 10$
43. $4x + 2y = 8$
44. $9x + 3y = 6$

45. **Creating a Model** The cost of producing a DVD is $125, plus $3 per DVD. Let C represent the total cost and let x represent the number of DVDs. Write an equation that relates C and x, and sketch its graph.

46. **Creating a Model** Let y represent the distance traveled by a train that is moving at a constant speed of 80 miles per hour. Let t represent the number of hours the train has traveled. Write an equation that relates y and t, and sketch its graph.

4.3

Finding the Slope of a Line In Exercises 47–50, find the slope of the line that passes through the points.

47.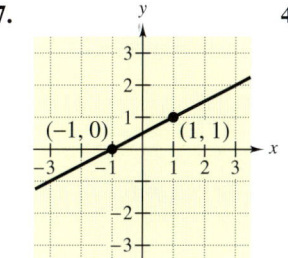

48.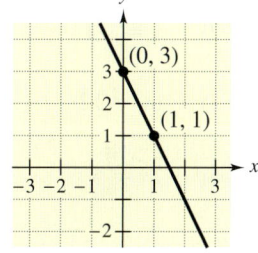

49.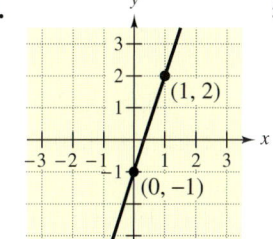

50.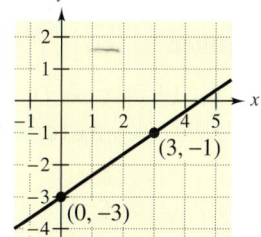

Finding the Slope of a Line In Exercises 51–60, plot the points and find the slope of the line that passes through the points. State whether the line rises, falls, is horizontal, or is vertical.

51. $(2, 1), (14, 6)$
52. $(-2, 2), (3, -10)$
53. $(-1, 0), (6, -2)$
54. $(1, 6), (4, 2)$
55. $(4, 0), (4, 6)$
56. $(1, 3), (4, 3)$
57. $(-1, -4), (-5, -10)$
58. $(-3, -3), (-8, -6)$
59. $\left(0, \frac{5}{2}\right), \left(\frac{5}{6}, 0\right)$
60. $(0, 0), \left(3, \frac{4}{5}\right)$

61. Loading Ramp The bed of a truck is 4 feet above ground level. The end of the ramp used in loading the truck rests on the ground 6 feet behind the truck. Determine the slope of the ramp.

62. Flight Path An aircraft is on its approach to an airport. As it flies over a town, its altitude is 15,000 feet. The town is about 10 miles from the airport. Approximate the slope of the linear path followed by the aircraft during landing.

Using Slope-Intercept Form In Exercises 63–70, write the equation in slope-intercept form. Use the slope and the y-intercept to sketch the line.

63. $x + y = 6$
64. $x - y = -3$
65. $2x - y = -1$
66. $-4x + y = -2$
67. $3x + 6y = 12$
68. $7x + 21y = -14$
69. $5y - 2x = 5$
70. $3y - x = 6$

Parallel or Perpendicular? In Exercises 71–74, determine whether the lines L_1 and L_2 that pass through the pairs of points are parallel, perpendicular, or neither.

71. L_1: $(0, 3), (-2, 1)$
    L_2: $(-8, -3), (4, 9)$
72. L_1: $(-3, -1), (2, 5)$
    L_2: $(2, 11), (8, 6)$
73. L_1: $(3, 6), (-1, -5)$
    L_2: $(-2, 3), (4, 7)$
74. L_1: $(-1, 2), (-1, 4)$
    L_2: $(7, 3), (4, 7)$

4.4

Using Point-Slope Form In Exercises 75–82, use the point-slope form to write an equation of the line that passes through the point and has the specified slope. Write the equation in slope-intercept form.

75. $(4, -1)$, $m = 2$
76. $(-5, 2)$, $m = 3$
77. $(1, 2)$, $m = -4$
78. $(7, -3)$, $m = -1$
79. $(-1, 3)$, $m = -\frac{8}{3}$
80. $(4, -2)$, $m = \frac{8}{5}$
81. $(3, 8)$, m is undefined.
82. $(-4, 6)$, $m = 0$

An Equation of a Line Through Two Points In Exercises 83–90, write an equation of the line that passes through the points. Write the equation in general form.

83. $(-4, 0), (0, -2)$
84. $(-4, -2), (4, 6)$
85. $(0, 8), (6, 8)$
86. $(2, -6), (2, 5)$
87. $(-1, -2), (-4, -7)$
88. $\left(0, \frac{4}{3}\right), (3, 0)$
89. $(2.4, 3.3), (6, 7.8)$
90. $(-1.4, 0), (3.2, 9.2)$

Equations of Parallel and Perpendicular Lines In Exercises 91–94, write an equation of the line that passes through the point and is (a) parallel and (b) perpendicular to the given line.

91. $(-6, 3)$
    $x - y = -2$
92. $\left(\frac{1}{5}, -\frac{4}{5}\right)$
    $5x + y = 2$
93. $\left(\frac{3}{8}, 4\right)$
    $y - 9 = 0$
94. $(-2, 1)$
    $5x = 2$

An Equation of a Horizontal or Vertical Line In Exercises 95–100, write an equation of the line.

95. Horizontal line through $(-4, 5)$
96. Horizontal line through $(3, -7)$
97. Vertical line through $(5, -1)$
98. Vertical line through $(-10, 4)$

99.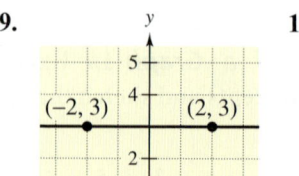
100.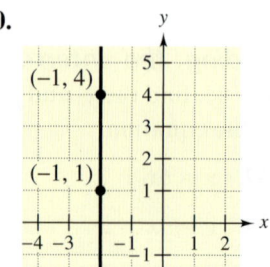

101. Wages A pharmaceutical salesperson receives a monthly salary of $5500 plus a commission of 7% of the total monthly sales. Write a linear model that relates the total monthly wages W to the sales S.

102. *Rental Demand* An apartment complex has 50 units. When the rent per unit is $425 per month, all 50 units are occupied. When the rent is $480 per month, the average number of occupied units drops to 47. Assume that the relationship between the monthly rent p and the demand x is linear.

(a) Write a linear model that relates the monthly rent p to the demand x.

(b) Use the model to estimate the number of units occupied when the rent is $475.

4.5

Verifying Solutions of a Linear Inequality In Exercises 103–106, determine whether each ordered pair is a solution of the inequality.

103. $x - y > 4$

(a) $(-1, -5)$ (b) $(0, 0)$
(c) $(3, -2)$ (d) $(8, 1)$

104. $y - 2x \leq -1$

(a) $(0, 0)$ (b) $(-2, 1)$
(c) $(-3, 4)$ (d) $(-1, -6)$

105. $3x - 2y < -1$

(a) $(3, 4)$ (b) $(-1, 2)$
(c) $(1, 8)$ (d) $(0, 0)$

106. $-4y + 5x > 3$

(a) $(1, 2)$ (b) $(-3, 6)$
(c) $(-1, -3)$ (d) $(4, 4)$

Matching In Exercises 107–110, match the inequality with its graph.

107. $x \geq -1$ **108.** $y \leq x + 1$

109. $y > -\frac{1}{3}x$ **110.** $y < 2$

(a)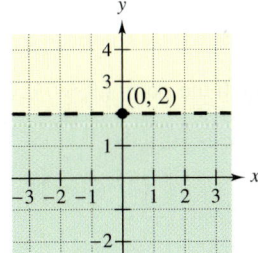

(b)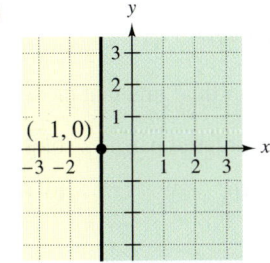

(c)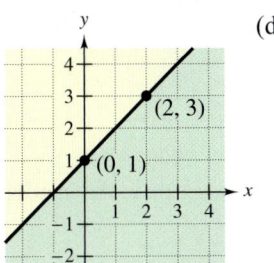

(d)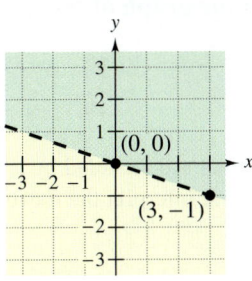

Sketching the Graph of a Linear Inequality In Exercises 111–118, sketch the graph of the linear inequality.

111. $x - 2 \geq 0$

112. $y + 3 < 0$

113. $2x + y < 1$

114. $3x - 4y > 2$

115. $x \leq 4y - 2$

116. $x \geq 3 - 2y$

117. $3x - y + 4 < -3$

118. $-2y - 2x - 7 > -2$

119. *Manufacturing* Each DVD player produced by an electronics manufacturer requires 2 hours in the assembly center. Each camcorder produced by the same manufacturer requires 3 hours in the assembly center. A total of 120 hours per week is available in the assembly center.

(a) Write a linear inequality that represents the different numbers of hours that can be spent assembling DVD players and camcorders.

(b) Graph the inequality and find three ordered pairs that are solutions of the inequality.

120. *Manufacturing* A company produces two types of lawn mowers, Economy and Deluxe. The Deluxe model requires 3 hours in the assembly center and the Economy model requires $1\frac{1}{2}$ hours in the assembly center. A total of 24 hours per day is available in the assembly center.

(a) Write a linear inequality that represents the different numbers of hours that can be spent assembling the two models.

(b) Graph the inequality and find three ordered pairs that are solutions of the inequality.

Chapter Test

Solutions in English & Spanish and tutorial videos at AlgebraWithinReach.com

Take this test as you would take a test in class. After you are done, check your work against the answers in the back of the book.

1. Plot the points $(-1, 2)$, $(1, 4)$, and $(2, -1)$ on a rectangular coordinate system. Connect the points with line segments to form a right triangle.

2. Determine whether each ordered pair is a solution of $y + 2x = 2$.
   (a) $(0, -2)$ (b) $(0, 2)$ (c) $(-4, 10)$ (d) $(-2, -2)$

3. What is the y-coordinate of any point on the x-axis?

4. Find the x- and y-intercepts of the graph of $8x - 2y = -16$.

5. Complete the table and use the results to sketch the graph of the equation.
   $3x + y = -4$

x	−2	−1	0	1	2
y					

In Exercises 6−9, sketch the graph of the equation.

6. $x + 2y = 6$
7. $y = \frac{1}{4}x - 1$
8. $y = |x + 2|$
9. $y = (x - 3)^2$

10. Find the slope of the line that passes through the points $(-5, 0)$ and $\left(2, \frac{3}{2}\right)$. Then write an equation of the line in slope-intercept form.

11. A line with slope $m = -2$ passes through the point $(-3, 4)$. Plot the point and use the slope to find two additional points on the line. (There are many correct answers.)

12. Find the slope of a line *perpendicular* to the line $7x - 8y + 5 = 0$.

13. Write the equation in slope-intercept form of the line that passes through the point $(0, 6)$ with slope $m = -\frac{3}{8}$.

14. Write an equation of the vertical line that passes through the point $(3, -7)$.

15. Determine whether each ordered pair is a solution of $3x + 5y \leq 16$.
    (a) $(2, 2)$ (b) $(6, -1)$ (c) $(-2, 4)$ (d) $(7, -1)$

In Exercises 16−19, sketch the graph of the linear inequality.

16. $y \geq -2$
17. $y < 5 - 2x$
18. $x \geq 2$
19. $-y + 4x > 3$

20. The unit sales y of a product are modeled by $y = 230x + 5000$, where x represents the time in years. Interpret the meaning of the slope in this model.

5 Exponents and Polynomials

5.1 Negative Exponents and Scientific Notation
5.2 Adding and Subtracting Polynomials
5.3 Multiplying Polynomials: Special Products
5.4 Dividing Polynomials

MASTERY IS WITHIN REACH!

"I used to waste time reviewing for a test by just reading through my notes over and over again. Then I learned how to make note cards and keep working through them, saving the important ones for my mental cheat sheets. Now a couple of my friends and I go through our notes and make mental cheat sheets together. We are all doing well in our math class."

Stephanie
Art

See page 227 for suggestions about managing test anxiety.

5.1 Negative Exponents and Scientific Notation

▶ Apply the rules of exponents to rewrite exponential expressions.
▶ Use the negative exponent rule to rewrite exponential expressions.
▶ Write very large and very small numbers in scientific notation.

Rules of Exponents

Study Tip

Rules 1 and 3 can be extended to three or more factors, such as $a^m \cdot a^n \cdot a^k = a^{m+n+k}$ and $(abc)^m = a^m \cdot b^m \cdot c^m$.

Product and Power Rules of Exponents

Let m and n be positive integers, and let a and b be real numbers, variables, or algebraic expressions.

1. Product rule: $a^m \cdot a^n = a^{m+n}$
2. Power-to-power rule: $(a^m)^n = a^{m \cdot n}$
3. Product-to-power rule: $(ab)^m = a^m \cdot b^m$

EXAMPLE 1 Applying the Product and Power Rules of Exponents

Using the Product Rule

a. $5^2 \cdot 5^3 = 5^{2+3} = 5^5 = 3125$
b. $a \cdot a^9 = a^{1+9} = a^{10}$
c. $b^4 b^2 b = b^{4+2+1} = b^7$
d. $3^2 x^3 \cdot x = (3^2)(x^{3+1}) = 9x^4$

Using the Power-to-Power Rule

e. $(2^3)^4 = 2^{3 \cdot 4} = 2^{12} = 4096$
f. $(y^2)^3 = y^{2 \cdot 3} = y^6$

Using the Product-to-Power Rule

g. $(3x)^3 = 3^3 \cdot x^3 = 27x^3$
h. $(-x)^4 = (-1)^4 x^4 = x^4$

EXAMPLE 2 Applying the Product and Power Rules of Exponents

a. $(2x^2 y)(-xy^4) = (2)(-1)(x^2 \cdot x)(y \cdot y^4) = -2x^{2+1}y^{1+4} = -2x^3 y^5$
b. $x(x^3 y^2)^3 = x(x^{3 \cdot 3} y^{2 \cdot 3}) = x(x^9 y^6) = x^{1+9} y^6 = x^{10} y^6$
c. $(-9x^2)^2(-3x^5) = (-9)^2(x^2)^2(-3x^5)$
$= 81 \cdot (-3) \cdot x^{2 \cdot 2} \cdot x^5$
$= -243 \cdot x^{4+5}$
$= -243 x^9$

Exercises Within Reach®   Solutions in English & Spanish and tutorial videos at AlgebraWithinReach.com

Applying the Product and Power Rules of Exponents In Exercises 1–16, simplify the expression.

1. $3 \cdot 3^4$
2. $2^3 \cdot 2^5$
3. $u^2 \cdot u^3 \cdot u$
4. $z^2 \cdot z^2 \cdot z^3$
5. $5x(x^6)$
6. $6x^2(x^4)$
7. $(-t)^6$
8. $(-y)^5$
9. $(t^2)^4$
10. $(v^3)^2$
11. $(2v)^3$
12. $(7u)^2$
13. $(-2s)^3$
14. $(-3z)^2$
15. $(a^2 b)^3(ab^2)^4$
16. $(st)^5(s^2 t)^4$

Quotient Rules of Exponents

Let m and n be positive integers and let a represent a real number, a variable, or an algebraic expression.

1. $\dfrac{a^m}{a^n} = a^{m-n},\ m > n,\ a \neq 0$

2. $\dfrac{a^n}{a^n} = 1 = a^0,\ a \neq 0$

Note the special definition for raising a *nonzero* quantity to the zero power. That is, if $a \neq 0$, then $a^0 = 1$.

Study Tip

The quotient rules of exponents work only for expressions with the *same variable* for a base. For instance, the quotient rule does not apply to x^5/y^3 because you cannot divide out any variables. So, for the expression

$$\dfrac{x^5}{y^3} = \dfrac{x \cdot x \cdot x \cdot x \cdot x}{y \cdot y \cdot y}$$

no simplifying is possible.

EXAMPLE 3 Applying the Quotient Rules of Exponents

a. $\dfrac{y^8}{y^5} = y^{8-5} = y^3$

b. $\dfrac{6x^4}{x^2} = 6 \cdot \dfrac{x^4}{x^2} = 6 \cdot x^{4-2} = 6x^2$

c. $\dfrac{16x^4}{3x^4} = \dfrac{16}{3} \cdot \dfrac{x^4}{x^4} = \dfrac{16}{3} \cdot x^{4-4} = \dfrac{16}{3}(1) = \dfrac{16}{3}$

d. $\dfrac{12x^8 y^2}{4x^4 y} = \dfrac{12}{4} \cdot \dfrac{x^8}{x^4} \cdot \dfrac{y^2}{y} = \dfrac{12}{4} \cdot x^{8-4} \cdot y^{2-1} = 3x^4 y$

e. $\dfrac{3x^3 y^6}{4xy^3} = \dfrac{3}{4} \cdot \dfrac{x^3}{x} \cdot \dfrac{y^6}{y^3} = \dfrac{3}{4} \cdot x^{3-1} \cdot y^{6-3} = \dfrac{3}{4} x^2 y^3 = \dfrac{3x^2 y^3}{4}$

Exercises Within Reach®

Solutions in English & Spanish and tutorial videos at AlgebraWithinReach.com

Applying the Quotient Rules of Exponents In Exercises 17–30, simplify the expression. (Assume that no denominator is zero.)

17. $\dfrac{x^5}{x^2}$

18. $\dfrac{y^7}{y^3}$

19. $\dfrac{x^2}{x}$

20. $\dfrac{y^3}{y}$

21. $\dfrac{z^9}{z^3}$

22. $\dfrac{y^8}{y^3}$

23. $\dfrac{3u^4}{u^3}$

24. $\dfrac{5z^6}{z^4}$

25. $\dfrac{-12z^3}{3z}$

26. $-\dfrac{16y^5}{8y^3}$

27. $\dfrac{32b^4}{-12b^3}$

28. $\dfrac{54x^4}{-24x^2}$

29. $\dfrac{24u^2 v^6}{18u^2 v^4}$

30. $\dfrac{15x^3 y^8}{27x^3 y^6}$

Negative Exponents

The rules of exponents can be extended to include **negative exponents**.

Technology Tip

The keystrokes used to evaluate expressions with negative exponents vary. For instance, to evaluate 13^{-2} on a calculator, you can try one of the following keystroke sequences.

Keystrokes

13 [y^x] 2 [+/−] [=]   Scientific

13 [^] [(−)] 2 [ENTER]   Graphing

With either of these sequences, your calculator should display .00591716. If it does not, consult the user's guide for your calculator to find the correct keystrokes.

Negative Exponent Rule

Let n be an integer and let a be a real number, variable, or algebraic expression such that $a \neq 0$.

$$a^{-n} = \frac{1}{a^n} \quad \text{and} \quad \frac{1}{a^{-n}} = a^n$$

EXAMPLE 4 Negative Exponents

a. $6^{-2} = \dfrac{1}{6^2} = \dfrac{1}{36}$
b. $x^{-7} = \dfrac{1}{x^7}$
c. $5x^{-4} = \dfrac{5}{x^4}$
d. $\dfrac{1}{2x^{-3}} = \dfrac{x^3}{2}$
e. $x^{-2}y^3 = \dfrac{y^3}{x^2}$
f. $(2x)^{-1} = \dfrac{1}{2x}$

Summary of Rules of Exponents

Let m and n be integers, and let a and b be a real numbers, variables, or algebraic expressions, such that $a \neq 0$ and $b \neq 0$.

Rule	Example
1. $a^m a^n = a^{m+n}$	$y^2 \cdot y^4 = y^{2+4} = y^6$
2. $\dfrac{a^m}{a^n} = a^{m-n}$	$\dfrac{x^7}{x^4} = x^{7-4} = x^3$
3. $(ab)^m = a^m b^m$	$(5x)^4 = 5^4 x^4 = 625 x^4$
4. $\left(\dfrac{a}{b}\right)^m = \dfrac{a^m}{b^m}$	$\left(\dfrac{2}{x}\right)^3 = \dfrac{2^3}{x^3} = \dfrac{8}{x^3}$
5. $(a^m)^n = a^{mn}$	$(y^3)^{-4} = y^{3(-4)} = y^{-12}$
6. $a^{-n} = \dfrac{1}{a^n}$	$y^{-4} = \dfrac{1}{y^4}$
7. $a^0 = 1$	$(x^2 + 1)^0 = 1$

Exercises Within Reach®

Solutions in English & Spanish and tutorial videos at AlgebraWithinReach.com

Negative Exponents In Exercises 31–46, rewrite the expression using only positive exponents.

31. 2^{-3}
32. 4^{-2}
33. y^{-5}
34. z^{-6}
35. $8x^{-7}$
36. $6x^{-2}$
37. $\dfrac{1}{2z^{-4}}$
38. $\dfrac{7}{y^{-3}}$
39. $7x^4 y^{-1}$
40. $9u^{-5} v^2$
41. $3(4x)^{-1}$
42. $5(3x)^{-1}$
43. $(-2)^{-5}$
44. $(-5)^{-4}$
45. $\dfrac{2^{-4}}{3}$
46. $\dfrac{4^{-3}}{2}$

Section 5.1 Negative Exponents and Scientific Notation   213

EXAMPLE 5 Using the Rules of Exponents

a. $5^0 x^3 (2x^{-4}) = 5^0(2)(x^3)(x^{-4})$ Regroup factors.

$\qquad = 1(2)x^{3+(-4)}$ Apply rules of exponents.

$\qquad = 2x^{-1}$ Simplify.

$\qquad = \dfrac{2}{x}$ Negative exponent rule

b. $(-3ab^4)(4ab^{-3}) = (-3)(4)(a)(a)(b^4)(b^{-3})$ Regroup factors.

$\qquad = (-3)(4)(a^{1+1})(b^{4+(-3)})$ Product rule of exponents

$\qquad = -12a^2 b$ Simplify.

c. $\dfrac{y^{-2}}{3y^{-5}} = \dfrac{1}{3}y^{-2-(-5)}$ Quotient rule of exponents

$\qquad = \dfrac{1}{3}y^3$ Simplify.

$\qquad = \dfrac{y^3}{3}$ Simplify.

d. $\left(\dfrac{a^2}{3}\right)^{-2} = \dfrac{a^{-4}}{3^{-2}}$ Apply rules of exponents.

$\qquad = \dfrac{3^2}{a^4}$ Negative exponent rule

$\qquad = \dfrac{9}{a^4}$ Simplify.

Study Tip

There is more than one way to solve problems such as the one in Example 5(c). For instance, you might prefer to write Example 5(c) as

$\dfrac{y^{-2}}{3y^{-5}} = \dfrac{y^5}{3y^2} = \dfrac{y^{5-2}}{3} = \dfrac{y^3}{3}.$

Exercises Within Reach®   Solutions in English & Spanish and tutorial videos at AlgebraWithinReach.com

Using the Rules of Exponents In Exercises 47–62, use the rules of exponents to simplify the expression using only positive exponents. (Assume that no variable is zero.)

47. $4^{-2} \cdot 4^3$

48. $5^{-3} \cdot 5^2$

49. $9^0 x^{-4} \cdot x^6$

50. $8^0 a^{-5} \cdot a^2$

51. $(-2x^2)(4x^{-3})$

52. $(4y^{-2})(3y^4)$

53. $xy^{-3} \cdot y^2 \cdot x^0$

54. $u^{-2}v \cdot u^2 \cdot v^0$

55. $(3x^2 y)^{-2}$

56. $(4x^{-3} y^2)^{-5}$

57. $\dfrac{x^2}{x^{-3}}$

58. $\dfrac{z^4}{z^{-2}}$

59. $\dfrac{x^{-4}}{x^{-2}}$

60. $\dfrac{t^{-5}}{t^{-1}}$

61. $\left(\dfrac{3z^2}{x}\right)^{-5}$

62. $\left(\dfrac{x^{-3} y^4}{5}\right)^{-4}$

63. **Volume** The volume of the cube in terms of its side lengths is $(2x)^3$ cubic units. Simplify this expression.

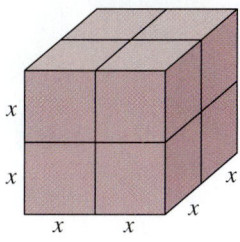

64. **Volume** The volume of the cube in terms of the smaller cubes is $3^3 \cdot 3^3$ cubic units. The volume of the cube in terms of its side lengths is 9^3 cubic units. Use the rules of exponents to show that the two expressions are equal.

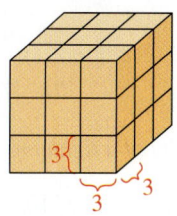

Technology Tip

Most scientific calculators automatically switch to scientific notation when they are displaying large (or small) numbers that exceed the display range. Try multiplying 98,900.000 × 5000. If your calculator follows standard conventions, its display should show

$$\boxed{4.945 \ \ 11}$$

or

$$\boxed{4.945 \ E \ 11}.$$

This means that $c = 4.945$ and the exponent of 10 is $n = 11$, which implies that the number is 4.945×10^{11}.

Scientific Notation

It is common to write very large or very small numbers in **scientific notation**. This notation has the form $c \times 10^n$, where $1 \leq c < 10$ and n is an integer.

EXAMPLE 6 Converting from Decimal to Scientific Notation

a. The speed of light is

$$299{,}792{,}458 = 2.99792458 \times 10^8 \text{ meters per second.}$$

Eight places

Large number yields positive exponent.

b. The relative density of hydrogen is

$$0.08988 = 8.988 \times 10^{-2} \text{ gram per cubic centimeter.}$$

Two places

Small number yields negative exponent.

EXAMPLE 7 Converting from Scientific to Decimal Notation

a. The number of air sacs in human lungs is

$$3.5 \times 10^8 = 350{,}000{,}000.$$

Eight places

Positive exponent yields large number.

b. The width of a human hair is

$$8.0 \times 10^{-5} = 0.00008 \text{ meter.}$$

Five places

Negative exponent yields small number.

Exercises Within Reach® Solutions in English & Spanish and tutorial videos at AlgebraWithinReach.com

Converting from Decimal to Scientific Notation
In Exercises 65–70, write the number in scientific notation.

65. 93,000,000
66. 900,000,000
67. 0.0000212
68. 0.000008736
69. *Length of a leaf beetle:* 0.0042 meter
70. *Volume of Earth:* 1,080,000,000,000 cubic kilometers

Converting from Scientific to Decimal Notation
In Exercises 71–76, write the number in decimal notation.

71. 8.67×10^2
72. 9.4675×10^4
73. 8.52×10^{-3}
74. 7.021×10^{-5}
75. *Potential electricity of a lightning bolt:* 1.0×10^9 volts
76. *Weight of a mosquito:* 2.5×10^{-6} kilogram

EXAMPLE 8 Using Scientific Notation

Use a calculator to evaluate 78,000 × 2,400,000,000.

SOLUTION

Because $78,000 = 7.8 \times 10^4$ and $2,400,000,000 = 2.4 \times 10^9$, you can evaluate the product as follows.

7.8 [EE] 4 [×] 2.4 [EE] 9 [=]   Calculator with EE key

7.8 [EXP] 4 [×] 2.4 [EXP] 9 [ENTER]   Calculator with EXP key

After these keystrokes have been entered, the calculator display should show [1.872 14] or [1.872 E 14]. So, the product of the two numbers is

$$(7.8 \times 10^4)(2.4 \times 10^9) = 1.872 \times 10^{14}$$
$$= 187,200,000,000,000.$$

Application

EXAMPLE 9 Federal Debt

In 2010, the population of the United States was about 310 million and the federal debt was about $13.5 trillion. Estimate the amount each person would have to pay to eliminate the federal debt.

SOLUTION

$$\frac{\text{Amount per}}{\text{person}} = \frac{\text{Federal debt}}{\text{Population}}$$ Write verbal model.

$$= \frac{13.5 \text{ trillion}}{310 \text{ million}}$$ Substitute.

$$= \frac{1.35 \times 10^{13}}{3.1 \times 10^8}$$ Write in scientific notation.

$$= \frac{1.35}{3.1} \times 10^5$$ Quotient rule of exponents

$$\approx 43,548.39$$ Simplify.

So, each person would have to pay about $43,548.39 to eliminate the federal debt.

Exercises Within Reach®

Solutions in English & Spanish and tutorial videos at AlgebraWithinReach.com

Using Scientific Notation In Exercises 77–84, use a calculator to **evaluate** the expression. Write your answer in scientific notation. Round your answer to four decimal places.

77. $5,000,000 \times 654,000$ **78.** $187,000 \times 0.000058$ **79.** $3,200,000^5$ **80.** $75,000,000^6$

81. $\dfrac{8.6 \times 10^4}{3.9 \times 10^7}$ **82.** $\dfrac{4.1 \times 10^5}{9.6 \times 10^9}$ **83.** $\dfrac{848,000,000}{1,620,000}$ **84.** $\dfrac{67,000,000}{0.0052}$

85. *Land Area* In 2012, the United States had a population of about 3.16×10^8 people and a land area of about 3.537×10^6 square miles. Find the amount of land per person in the United States. Write your answer in scientific notation.

86. *Hydraulic Compression* A hydraulic cylinder in a large press contains 2 gallons of oil. When the cylinder is under full pressure, the actual volume of the oil is decreased by $2(150)(20 \times 10^{-6})$ gallons. Write the decreased volume in decimal notation.

Concept Summary: Using Scientific Notation

What
You can use exponents and **scientific notation** to write very large or very small numbers.

EXAMPLE
Write each number in scientific notation.

a. 240,000,000,000
b. 0.000000012

How
A number written in scientific notation has the form $c \times 10^n$, where $1 \leq c < 10$ and n is an integer.

A positive exponent indicates a large number.

a. $240{,}000{,}000{,}000 = 2.4 \times 10^{11}$

11 places

A **negative exponent** indicates a small number.

b. $0.000000012 = 1.2 \times 10^{-8}$

8 places

Why
When multiplying or dividing very large or very small numbers, you can use scientific notation and the rules of exponents to find the products or quotients more efficiently.

Exercises Within Reach®

Worked-out solutions to odd-numbered exercises at AlgebraWithinReach.com

Concept Summary Check

87. Scientific Notation In the solution above, 240,000,000,000 is rewritten in the form $c \times 10^n$. What is the value of c?

88. Scientific Notation In the solution above, 0.000000012 is rewritten in the form $c \times 10^n$. What is the value of n?

89. Writing Explain how to write a small number in scientific notation.

90. Writing Explain how to write a large number in scientific notation.

Extra Practice

Using the Rules of Exponents In Exercises 91–98, use the rules of exponents to **simplify** the expression. (Assume that no variable is zero.)

91. $\dfrac{4^{-2}}{3^{-4}}$

92. $\dfrac{5^{-3}}{3^{-1}}$

93. $\left(\dfrac{2}{3}\right)^{-2}$

94. $\left(\dfrac{5}{4}\right)^{-3}$

95. $(uv^3)(-2uv^2)^5$

96. $(-3y^2z)^2(2yz^2)^3$

97. $(-3x^2)(5x^{-4})$

98. $(2y^{-2})(6y^3)$

Identifying Exponents In Exercises 99–106, **determine** the exponent that makes the statement true.

99. $5^{\square} = \dfrac{1}{25}$

100. $2^{\square} = \dfrac{1}{16}$

101. $\dfrac{1}{3^{\square}} = 27$

102. $\dfrac{1}{4^{\square}} = 16$

103. $(x^{\square} y^3)^{-2} = \dfrac{x^4}{y^6}$

104. $(x^{\square} y^{-2})^3 = \dfrac{1}{x^9 y^6}$

105. $(x^4 y^{-3})^{\square} = \dfrac{y^9}{x^{12}}$

106. $(x^{-2} y^5)^{\square} = \dfrac{x^2}{y^5}$

107. Numerical and Graphical Analysis

(a) Complete the table by evaluating the indicated powers of 2.

x	−1	−2	−3	−4	−5
y = 2ˣ					

(b) Plot the data in the table.

(c) Use the table or the graph to describe the value of 2^{-n} when n is very large. Will the value of 2^{-n} ever be negative?

108. *Astronomy* The star Vega is approximately 25 light-years from Earth (see figure). (A light-year is the distance light can travel in 1 year.) A light-year is approximately 5.8657×10^{12} miles. Approximate the distance from Earth to Vega and write the answer in scientific notation. (*Source:* NASA)

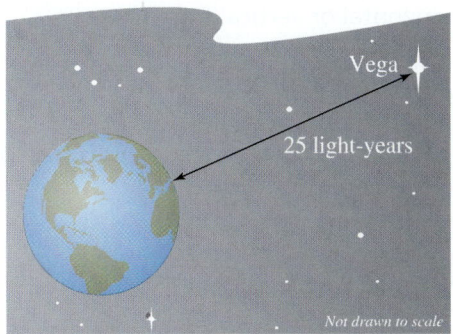

109. *Astronomy* One astronomical unit AU is the average distance between the Sun and Earth (approximately 149,605,000 kilometers). The table shows the average distances between selected planets and the Sun in astronomical units. Approximate each distance in kilometers and write the answer in scientific notation.

Planet	Venus	Mars	Jupiter	Uranus
AU	0.72	1.52	5.20	19.18

110. *Boltzmann's Constant* The study of the kinetic energy of an ideal gas uses Boltzmann's constant

$$k = \frac{8.31}{6.02 \times 10^{23}}.$$

Evaluate the expression and write the result in scientific notation.

Explaining Concepts

True or False? In Exercises 111–114, state whether the equation is true or false. If it is false, find values of x and y that show it to be false.

111. $x^3 y^3 = xy^3$

112. $\dfrac{x^{-4}}{x^{-3}} = x$

113. $(x \times 10^3)^4 = x^4 \times 10^{12}$

114. $\dfrac{2x \times 10^{-5}}{x \times 10^{-3}} = 2 \times 10^{-2}$

115. *Think About It* Does $a^{-1}b^{-1} = \dfrac{1}{ab}$? Explain your answer.

116. *Think About It* Does $a^{-1} + b^{-1} = \dfrac{1}{a+b}$? Explain your answer.

117. *Finding a Reciprocal* Find the reciprocal of 4×10^{-3}.

118. (a) Find as many equivalent pairs as possible among the following expressions.

$$\dfrac{2}{x^{-3}}, \dfrac{1}{2x^3}, 2x^{-3}, \dfrac{1}{(2x)^{-3}}, 2x^3, \dfrac{x^3}{8}, \dfrac{1}{8x^3}, \dfrac{x^{-3}}{2}, 8x^3, (2x)^{-3}$$

(b) Use a table to illustrate the equivalence of each pair. Organize the table with four columns—an expression, the equivalent expression, and each expression evaluated at $x = 3$.

119. *Reasoning* Justify each step.

$$(3 \times 10^5)(4 \times 10^6) = (3 \times 10^5)(10^6 \times 4)$$
$$= 3(10^5 \times 10^6)(4)$$
$$= 4(10^{5+6})(4)$$
$$= (3 \cdot 4)10^{11}$$
$$= 12 \times 10^{11}$$
$$= 1.2 \times 10^{12}$$

Cumulative Review

In Exercises 120–123, sketch the graph of the equation, Identify any intercepts.

120. $3x + y = 4$

121. $y = |2x + 1|$

122. $y = x^2 - 2x + 1$

123. $y = \sqrt{x + 4}$

In Exercises 124–127, Sketch the graph of the inequality.

124. $y > -8$

125. $x + y < 4$

126. $y \leq \tfrac{1}{3}x + 4$

127. $x \geq \tfrac{3}{5}(y - 10)$

5.2 Adding and Subtracting Polynomials

- Identify the degrees and leading coefficients of polynomials.
- Add polynomials using a horizontal or vertical format.
- Subtract polynomials using a horizontal or vertical format.

Basic Definitions

Let $a_n, a_{n-1}, \ldots, a_2, a_1, a_0$ be real numbers and let n be a nonnegative integer. A **polynomial in x** is an expression of the form

$$a_n x^n + a_{n-1} x^{n-1} + \cdots + a_2 x^2 + a_1 x + a_0$$

where $a_n \neq 0$. The polynomial is of **degree n**, and the number a_n is called the **leading coefficient**. The number a_0 is called the **constant term**.

Study Tip

A polynomial with only one term is called a **monomial**. A polynomial with two unlike terms is called a **binomial**, and a polynomial with three unlike terms is called a **trinomial**. For example, $3x^2$ is a monomial, $-3x + 1$ is a binomial, and $4x^3 - 5x + 6$ is a trinomial.

EXAMPLE 1 Identifying Polynomials

a. $3x^4 - 8x + x^{-1}$ is *not* a polynomial because the third term, x^{-1}, has a negative exponent.

b. $x^2 - 3x + 1$ *is* a polynomial of degree 2 with integer coefficients.

c. $x^3 + 3x^{1/2}$ is *not* a polynomial because the exponent in the second term, $3x^{1/2}$, is not an integer.

d. $-\dfrac{1}{3}x + \dfrac{x^3}{4}$ *is* a polynomial of degree 3 with rational coefficients.

EXAMPLE 2 Determining Degrees and Leading Coefficients

	Polynomial	Standard Form	Degree	Leading Coefficient
a.	$4x^2 - 5x^7 - 2 + 3x$	$-5x^7 + 4x^2 + 3x - 2$	7	-5
b.	$4 - 9x^2$	$-9x^2 + 4$	2	-9
c.	8	8	0	8
d.	$2 + x^3 - 5x^2$	$x^3 - 5x^2 + 2$	3	1

In part (c), note that a polynomial with only a constant term has a degree of zero.

Exercises Within Reach®

Solutions in English & Spanish and tutorial videos at AlgebraWithinReach.com

Identifying a Polynomial In Exercises 1–6, determine whether the expression is a polynomial. If it is not, explain why.

1. $9 - z$
2. $t^2 - 4$
3. $p^{3/4} - 16$
4. $9 - z^{1/2}$
5. $6x^{-1}$
6. $4 - 9x^4$

Determining the Degree and Leading Coefficient In Exercises 7–10, write the polynomial in standard form. Then identify its degree and leading coefficient.

7. $7x - 5x^2 + 10$
8. $1 - 4z + 12z^3$
9. $6m - 3m^5 - m^2 + 12$
10. $5x^3 - 3x^2 + 10$

Adding Polynomials

EXAMPLE 3 Adding Polynomials Horizontally

a. $(2x^2 + 4x - 1) + (x^2 - 3)$ Original polynomials
$= (2x^2 + x^2) + (4x) + (-1 - 3)$ Group like terms.
$= 3x^2 + 4x - 4$ Combine like terms.

b. $(x^3 + 2x^2 + 4) + (3x^2 - x + 5)$ Original polynomials
$= (x^3) + (2x^2 + 3x^2) + (-x) + (4 + 5)$ Group like terms.
$= x^3 + 5x^2 - x + 9$ Combine like terms.

c. $(2x^2 - x + 3) + (4x^2 - 7x + 2) + (-x^2 + x - 2)$ Original polynomials
$= (2x^2 + 4x^2 - x^2) + (-x - 7x + x) + (3 + 2 - 2)$ Group like terms.
$= 5x^2 - 7x + 3$ Combine like terms.

Study Tip

When you use a vertical format to add polynomials, be sure that you line up the like terms.

EXAMPLE 4 Adding Polynomials Vertically

a. $-4x^3 - 2x^2 + x - 5$
$ 2x^3 + 3x + 4$
$\overline{-2x^3 - 2x^2 + 4x - 1}$

b. $5x^3 + 2x^2 - x + 7$
$ 3x^2 - 4x + 7$
$-x^3 + 4x^2 - 2x - 8$
$\overline{4x^3 + 9x^2 - 7x + 6}$

Exercises Within Reach®

Solutions in English & Spanish and tutorial videos at AlgebraWithinReach.com

Adding Polynomials In Exercises 11–16, use a horizontal format to find the sum.

11. $(4w + 5) + (16w - 9)$

12. $(-2x + 4) + (x - 6)$

13. $(3z^2 - z + 2) + (z^2 - 4)$

14. $(6x^4 + 8x) + (4x - 6)$

15. $b^2 + (b^3 - 2b^2 + 3) + (b^3 - 3)$

16. $(3x^2 - x) + 5x^3 + (-4x^3 + x^2 - 8)$

Adding Polynomials In Exercises 17–22, use a vertical format to find the sum.

17. $2x + 5$
$\underline{3x + 8}$

18. $11x + 5$
$\underline{7x - 6}$

19. $-2x + 10$
$\underline{x - 38}$

20. $4x^2 + 13$
$\underline{3x^2 - 11}$

21. $(x^2 - 2x + 2) + (x^2 + 4x) + 2x^2$

22. $(5y + 10) + (y^2 - 3y - 2) + (2y^2 + 4y - 3)$

Geometry In Exercises 23 and 24, find an expression for the perimeter of the figure.

23.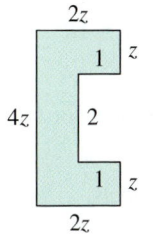

24.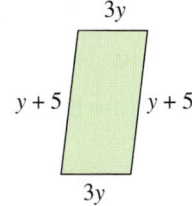

Application

EXAMPLE 5 Modeling School Enrollments

The projected enrollments (in millions) at public colleges and private colleges for the years 2015 through 2020 can be modeled by the following, where t represents the year, with $t = 15$ corresponding to 2015. (*Source:* U.S. National Center for Education Statistics)

$P = 0.20t + 12.7, \quad 15 \le t \le 20$ — Public college enrollment

$R = -0.004t^2 + 0.21t + 3.7, \quad 15 \le t \le 20$ — Private college enrollment

a. Add the polynomials to find a model for the projected total enrollment at public *and* private colleges.

b. Make a bar graph for all three models.

SOLUTION

a. $T = 0.20t + 12.7 + (-0.004t^2 + 0.21t + 3.7)$

$= -0.t^2 + 0.41t + 16.4$ — Total school enrollment

b. A spreadsheet is useful for making a bar graph.

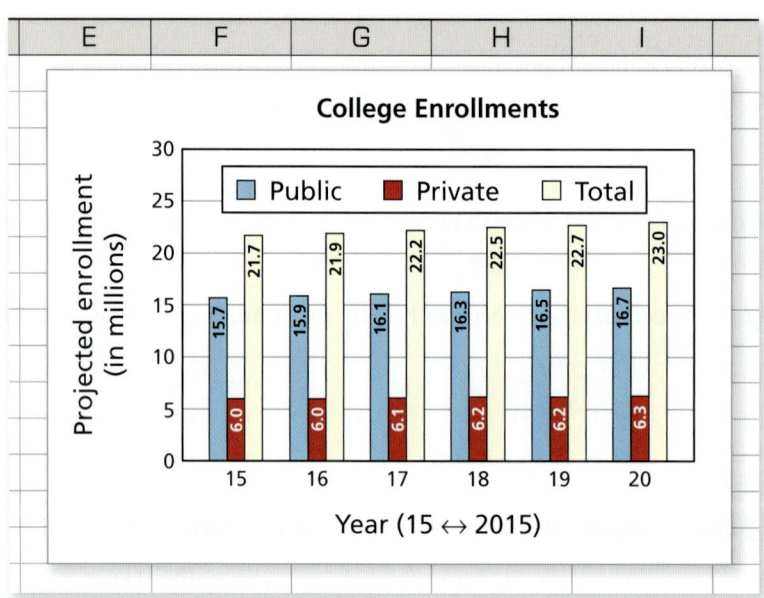

Exercises Within Reach®

Solutions in English & Spanish and tutorial videos at AlgebraWithinReach.com

25. The Civilian Labor Force The numbers of men M and women W in the civilian labor force that were 65 years old or older in the years 2006 through 2010 can be modeled by the following, where t represents the year, with $t = 6$ corresponding to 2006. (*Source:* U.S. Bureau of Labor Statistics)

$M = 0.16t + 2.1, \quad 6 \le t \le 10$ — Men
$W = -0.026t^2 + 0.57t - 0.1, \quad 6 \le t \le 10$ — Women

(a) Add the polynomials to find a model for the total number of people.

(b) Make a bar graph for all three models.

26. The Civilian Labor Force The numbers of men M and women W in the civilian labor force that were 35 to 44 years old in the years 2006 through 2010 can be modeled by the following, where t represents the year, with $t = 6$ corresponding to 2006. (*Source:* U.S. Bureau of Labor Statistics)

$M = -0.051t^2 + 0.47t + 18.4, \quad 6 \le t \le 10$ — Men
$W = -0.054t^2 + 0.57t + 14.9, \quad 6 \le t \le 10$ — Women

(a) Add the polynomials to find a model for the total number of people.

(b) Make a bar graph for all three models.

Subtracting Polynomials

EXAMPLE 6 Subtracting Polynomials Horizontally

a. $(2x^2 + 3) - (3x^2 - 4) = 2x^2 + 3 - 3x^2 + 4$ Distributive Property
$= (2x^2 - 3x^2) + (3 + 4)$ Group like terms.
$= -x^2 + 7$ Combine like terms.

b. $(3x^3 - 4x^2 + 3) - (x^3 + 3x^2 - x - 4)$ Original polynomials
$= 3x^3 - 4x^2 + 3 - x^3 - 3x^2 + x + 4$ Distributive Property
$= (3x^3 - x^3) + (-4x^2 - 3x^2) + (x) + (3 + 4)$ Group like terms.
$= 2x^3 - 7x^2 + x + 7$ Combine like terms.

EXAMPLE 7 Subtracting Polynomials Vertically

a.
$$\begin{array}{r} 3x^2 + 7x - 6 \\ -3x^2 - 7x \phantom{{}-6} \\ \hline -6 \end{array}$$
Change signs and add.

b.
$$\begin{array}{r} 5x^3 - 2x^2 + x \phantom{{}-2} \\ -4x^2 + 3x - 2 \\ \hline 5x^3 - 6x^2 + 4x - 2 \end{array}$$
Change signs and add.

c.
$$\begin{array}{r} 4x^4 - 2x^3 + 5x^2 - x + 8 \\ -3x^4 + 2x^3 \phantom{{}+5x^2} - 3x + 4 \\ \hline x^4 \phantom{{}-2x^3} + 5x^2 - 4x + 12 \end{array}$$

Study Tip
When using a vertical format, write the polynomial being subtracted underneath the one from which it is being subtracted. Be sure to line up like terms in vertical columns.

Exercises Within Reach®

Solutions in English & Spanish and tutorial videos at AlgebraWithinReach.com

Subtracting Polynomials In Exercises 27–32, use a horizontal format to find the difference.

27. $(11x - 8) - (2x + 3)$
28. $(5x + 1) - (18x - 7)$
29. $(x^2 - x) - (x - 2)$
30. $(x^2 - 4) - (x^2 - 4x)$
31. $(4z^3 - 6) - (-z^3 + z - 2)$
32. $(4t^3 - 3t + 5) - (3t^2 - 3t - 10)$

Subtracting Polynomials In Exercises 33–40, use a vertical format to find the difference.

33.
$$\begin{array}{r} 2x - 2 \\ -(x - 1) \\ \hline \end{array}$$

34.
$$\begin{array}{r} 9x + 7 \\ -(3x + 9) \\ \hline \end{array}$$

35.
$$\begin{array}{r} 2x^2 - x + 2 \\ -(3x^2 + x - 1) \\ \hline \end{array}$$

36.
$$\begin{array}{r} y^4 - 2x - 3 \\ -(y^4 \phantom{{}-2x} + 2) \\ \hline \end{array}$$

37. $(7x^2 - x) - (x^3 - 2x^2 + 10)$
38. $(y^2 - 3y + 8) - (y^3 + y^2 - 3)$
39. $(-3x^3 - 4x^2 + 2x - 5) - (2x^4 + 2x^3 - 4x + 5)$
40. $(12x^3 + 25x^2 - 15) - (-2x^3 + 18x^2 - 3x)$

EXAMPLE 8 Combining Polynomials Horizontally

a. $(x^2 - 2x + 1) - [(x^2 + x - 3) + (-2x^2 - 4x)]$ Original polynomials

$= (x^2 - 2x + 1) - [(x^2 - 2x^2) + (x - 4x) + (-3)]$ Group like terms.

$= (x^2 - 2x + 1) - [-x^2 - 3x - 3]$ Combine like terms.

$= x^2 - 2x + 1 + x^2 + 3x + 3$ Distributive Property

$= (x^2 + x^2) + (-2x + 3x) + (1 + 3)$ Group like terms.

$= 2x^2 + x + 4$ Combine like terms.

b. $(3x^2 - 7x + 2) - (4x^2 + 6x - 1) + (-x^2 + 4x + 5)$

$= 3x^2 - 7x + 2 - 4x^2 - 6x + 1 - x^2 + 4x + 5$

$= (3x^2 - 4x^2 - x^2) + (-7x - 6x + 4x) + (2 + 1 + 5)$

$= -2x^2 - 9x + 8$

c. $(-2x^2 + 4x - 3) - [(4x^2 - 5x + 8) - 2(-x^2 + x + 3)]$

$= (-2x^2 + 4x - 3) - [4x^2 - 5x + 8 + 2x^2 - 2x - 6]$

$= (-2x^2 + 4x - 3) - [(4x^2 + 2x^2) + (-5x - 2x) + (8 - 6)]$

$= (-2x^2 + 4x - 3) - [6x^2 - 7x + 2]$

$= -2x^2 + 4x - 3 - 6x^2 + 7x - 2$

$= (-2x^2 - 6x^2) + (4x + 7x) + (-3 - 2)$

$= -8x^2 + 11x - 5$

Exercises Within Reach®

Solutions in English & Spanish and tutorial videos at AlgebraWithinReach.com

Combining Polynomials In Exercises 41–54, **perform** the indicated operations and simplify.

41. $(6x - 5) - (8x + 15)$

42. $(2x^2 + 1) + (x^2 - 2x + 1)$

43. $-(x^3 - 2) + (4x^3 - 2x)$

44. $-(5x^2 - 1) - (-3x^2 + 5)$

45. $2(x^4 + 2x) + (5x + 2)$

46. $(z^4 - 2z^2) + 3(z^4 + 4)$

47. $5z - [3z - (10z + 8)]$

48. $9w^2 - [2w - (w^2 + 3w)]$

49. $(y^3 + 1) - [(y^2 + 1) + (3y - 7)]$

50. $(a^2 - a) - [(2a^2 + 3a) - (5a^2 - 12)]$

51. $2(t^2 + 5) - 3(t^2 + 5) + 5(t^2 + 5)$

52. $-10(u + 1) + 8(u - 1) - 3(u + 6)$

53. $8v - 6(3v - v^2) + 10(10v + 3)$

54. $3(x^2 - 2x + 3) - 4(4x + 1) - (3x^2 - 2x)$

Section 5.2 Adding and Subtracting Polynomials   223

Application   **EXAMPLE 9**   Geometry: Area of a Region

Find an expression for the area of the shaded region.

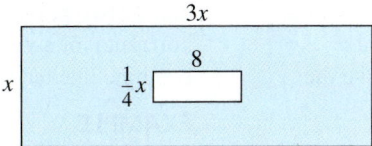

SOLUTION

To find a polynomial that represents the area of the shaded region, subtract the area of the inner rectangle from the area of the outer rectangle, as follows.

Area of shaded region = Area of outer rectangle − Area of inner rectangle

$$= 3x(x) - 8\left(\frac{1}{4}x\right)$$

$$= 3x^2 - 2x$$

Exercises Within Reach®   Solutions in English & Spanish and tutorial videos at AlgebraWithinReach.com

Geometry In Exercises 55−60, find an expression for the area of the shaded region of the figure.

55.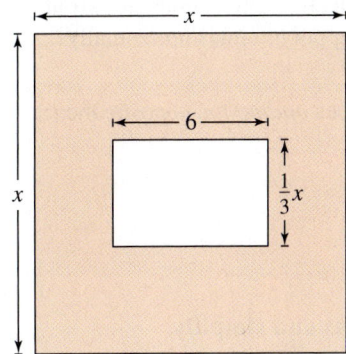

56.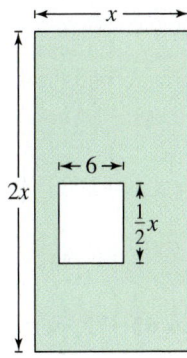

57.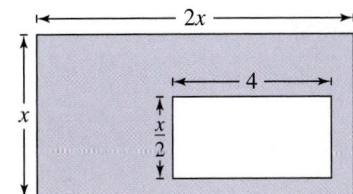

58.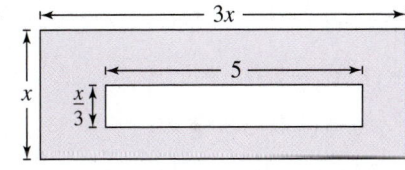

59.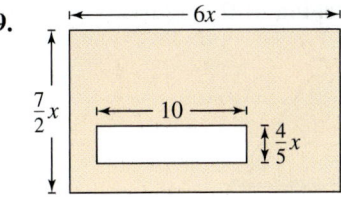

60.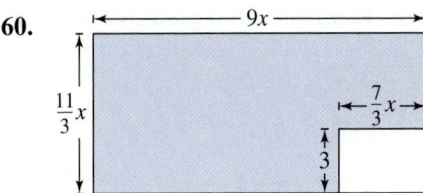

Concept Summary: Adding and Subtracting Polynomials

What
The key to adding or subtracting **polynomials** is to recognize like terms. Like terms have the same degree.

EXAMPLE

Find the sum $(3x^2 + 2x) + (x^2 - 4x)$.

How
To add two polynomials, use a horizontal or a vertical format to combine like terms.

EXAMPLE

Horizontal format:

$(3x^2 + 2x) + (x^2 - 4x)$
$= (3x^2 + x^2) + (2x - 4x)$
$= 4x^2 - 2x$

Vertical format: $\begin{array}{r} 3x^2 + 2x \\ x^2 - 4x \\ \hline 4x^2 - 2x \end{array}$

Why
When you know how to add any two polynomials, you can subtract any two polynomials by *adding the opposite*. Use the Distributive Property to change all the signs in the polynomial that is being subtracted.

Exercises Within Reach ®
Worked-out solutions to odd-numbered exercises at AlgebraWithinReach.com

Concept Summary Check

61. *Identifying Like Terms* What are the like terms in the expression $(3x^2 + 2x) + (x^2 - 4x)$?

62. *Combining Like Terms* Explain how to find the sum $3x^2 + x^2$.

63. *Adding Polynomials* What step in adding polynomials vertically corresponds to grouping like terms when adding polynomials horizontally?

64. *Writing* What does *adding the opposite* mean?

Extra Practice

Combining Polynomials In Exercises 65–72, **perform** the indicated operations and simplify.

65. $(x^5 - 3x^4 + x^3 - 5x + 1) - (4x^5 - x^3 + x - 5)$

66. $(t^4 + 5t^3 - t^2 + 8t - 10) - (t^4 + t^3 + 2t^2 + 4t - 7)$

67. $\left(\frac{2}{3}y^2 - \frac{3}{4}\right) + \left(\frac{5}{6}y^2 + 2\right)$

68. $\left(\frac{3}{4}x^3 - \frac{1}{2}\right) + \left(\frac{1}{8}x^3 + 3\right)$

69. $(0.1t^3 - 3.4t^2) + (1.5t^3 - 7.3)$

70. $(0.7x^2 - 0.2x + 2.5) + (7.4x - 3.9)$

71. $(2ab - 3) + (a^2 - 2ab) + (4b^2 - a^2)$

72. $(uv - 3) + (4uv - v^2) + (u^2 - 8uv)$

Geometry In Exercises 73 and 74, **find** an expression for the area of the shaded region of the figure.

73.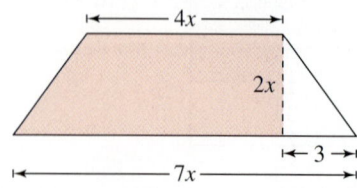

74.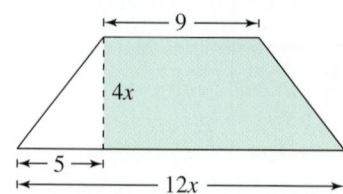

75. **Cost, Revenue, and Profit** The cost C (in dollars) of producing x dome tents is $C = 200 + 45x$. The revenue R (in dollars) for selling x dome tents is $R = 120x - x^2$, where $0 \leq x \leq 60$. The profit P is the difference between revenue and cost.

   (a) Perform the subtraction required to find the polynomial representing profit P.

   (b) Determine the profit when 40 tents are produced and sold.

   (c) Find the change in profit when the number of tents produced and sold increases from 40 to 50.

76. **Cost, Revenue, and Profit** The cost C (in dollars) of producing x multimedia projectors is $C = 1000 + 150x$. The revenue R (in dollars) for selling x multimedia projectors is $R = 400x - \frac{1}{2}x^2$, where $0 \leq x \leq 600$. The profit P is the difference between revenue and cost.

   (a) Perform the subtraction required to find the polynomial representing profit P.

   (b) Determine the profit when 200 projectors are produced and sold.

   (c) Find the change in profit when the number of projectors produced and sold decreases from 200 to 100.

Explaining Concepts

77. Is the sum of two binomials always a binomial? Explain.

78. Determine which of the two statements is always true. Is the other statement always false? Explain.

   (a) A polynomial is a trinomial.

   (b) A trinomial is a polynomial.

79. **Writing** In your own words, define "like terms." What are the only factors of like terms that can differ?

80. **Precision** Describe how to combine like terms. What operations are used?

81. **Structure** Is a polynomial an algebraic expression? Explain.

82. **Writing** Write a paragraph that explains how the adage "You can't add apples and oranges" might relate to adding two polynomials. Include several examples to illustrate the applicability of this statement.

Cumulative Review

In Exercises 83–86, solve the equation and check your solution.

83. $\dfrac{4x}{27} = \dfrac{8}{9}$

84. $\dfrac{x}{6} - \dfrac{x}{18} = 3$

85. $\dfrac{x+3}{6} = \dfrac{2}{5}$

86. $\dfrac{x-5}{2} = \dfrac{4x}{3}$

In Exercises 87 and 88, graph the equation.

87. $y = 2 - \dfrac{3}{2}x$

88. $y = |x - 1|$

In Exercises 89 and 90, determine the exponent that makes the statement true.

89. $2^{\square} = \dfrac{1}{32}$

90. $(x^{\square} y^2)^{-3} = \dfrac{1}{x^{12} y^6}$

In Exercises 91 and 92, use a calculator to evaluate the expression. Write your answer in scientific notation. Round your answer to four decimal places.

91. $(4.15 \times 10^3)^{-4}$

92. $\dfrac{1.5 \times 10^8}{2.3 \times 10^5}$

Mid-Chapter Quiz: Sections 5.1–5.2

Solutions in English & Spanish and tutorial videos at AlgebraWithinReach.com

Take this quiz as you would take a quiz in class. After you are done, check your work against the answers in the back of the book.

In Exercises 1–4, simplify the expression. (Assume that no denominator is zero.)

1. $(9m^3)^2$
2. $(-3xy)(2x^2)^3$
3. $\dfrac{-12x^3y}{9x^5y^2}$
4. $\dfrac{3t^3}{(-6t)^2}$

In Exercises 5 and 6, rewrite the expression using only positive exponents.

5. $5x^{-2}y^{-3}$
6. $\dfrac{3x^{-2}y}{5z^{-1}}$

In Exercises 7 and 8, use the rules of exponents to simplify the expression using only positive exponents. (Assume that no variable is zero.)

7. $(3a^{-3}b^2)^{-2}$
8. $(4t^{-3})^0$

9. Write the number 8,168,000,000,000 in scientific notation.

10. Write the number 5.021×10^{-3} in decimal notation.

11. Explain why $x^2 + 2x - 3x^{-1}$ is not a polynomial.

12. Determine the degree and the leading coefficient of the polynomial $10 + x^2 - 4x^3$.

13. Write the polynomial $5x - 3 + x^2$ in standard form.

In Exercises 14–17, perform the indicated operations and simplify.

14. $(y^2 + 3y - 1) + (4 + 3y)$
15. $(3v^2 - 5) - (v^3 + 2v^2 - 6v)$
16. $9s - [6 - (s - 5)]$
17. $-3(4 - x) + 4(x^2 + 2) - (x^2 - 2x)$

In Exercises 18 and 19, use a vertical format to find the sum.

18. $\quad 2x^2 + x - 3$
    $\quad 3x^3 - 2x^2 - 3x + 5$

19. $\quad x^3 - 3x^2 - 15$
    $\quad 2x^2 + 5x - 4$

In Exercises 20 and 21, use a vertical format to find the difference.

20. $\quad x^2 - x + 2$
    $\quad - (x - 4)$

21. $\quad 6x^4 + 3x^3 + 8$
    $\quad -(x^4 + 4x^2 + 2)$

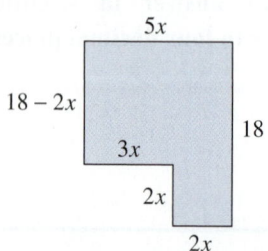

22. Find an expression for the perimeter of the figure at the left.

Study Skills in Action

Managing Test Anxiety

Test anxiety is different from the typical nervousness that usually occurs during tests. It interferes with the thinking process. After leaving the classroom, have you suddenly been able to recall what you could not remember during the test? It is likely that this was a result of test anxiety. Test anxiety is a learned reaction or response—no one is born with it. The good news is that most students can learn to manage test anxiety.

It is important to get as much information as you can into your long-term memory and to practice retrieving the information before you take a test. The more you practice retrieving information, the easier it will be during the test.

Smart Study Strategy

Make Mental Cheat Sheets

No, we are not asking you to cheat! Just prepare as if you were going to and then memorize the information you've gathered.

1. ▶ Write down important information on note cards. This can include:
   - formulas
   - examples of problems you find difficult
   - concepts that always trip you up

2. ▶ Memorize the information on the note cards. Flash through the cards, placing the ones containing information you know in one stack and the ones containing information you do not know in another stack. Keep working on the information you do not know.

3. ▶ As soon as you receive your test, turn it over and write down all the information you remember, starting with things you have the greatest difficulty remembering. Having this information available should boost your confidence and free up mental energy for focusing on the test.

Do not wait until the night before the test to make note cards. Make them after you study each section. Then review them two or three times a week.

The FOIL Method

To multiply two binomials, you can combine the products of the First, Outer, Inner, and Last terms.

$(2x + 1)(x - 5)$

$$= \underset{F}{2x(x)} + \underset{O}{2x(-5)} + \underset{I}{1(x)} + \underset{L}{1(-5)}$$
$$= 2x^2 - 10x + x - 5$$
$$= 2x^2 - 9x - 5$$

Special Products:

$(a + b)(a - b) = a^2 - b^2$

$(a + b)^2 = a^2 + 2ab + b^2$

$(a - b)^2 = a^2 - 2ab + b^2$

5.3 Multiplying Polynomials: Special Products

▶ Find products with monomial multipliers.
▶ Multiply binomials using the Distributive Property and the FOIL Method.
▶ Multiply polynomials using a horizontal or vertical format.
▶ Identify and use special binomial products.

Monomial Multipliers

To multiply polynomials, you use many of the rules for simplifying algebraic expressions. You may want to review these rules in Section 2.2 and Section 5.1.

1. The Distributive Property
2. Combining like terms
3. Removing symbols of grouping
4. Rules of exponents

EXAMPLE 1 Finding Products with Monomial Multipliers

Find each product.

a. $(3x - 7)(-2x)$
b. $3x^2(5x - x^3 + 2)$
c. $(-x)(2x^2 - 3x)$

SOLUTION

a. $(3x - 7)(-2x) = 3x(-2x) - 7(-2x)$ Distributive Property
$\qquad = -6x^2 + 14x$ Write in standard form.

b. $3x^2(5x - x^3 + 2)$
$\qquad = (3x^2)(5x) - (3x^2)(x^3) + (3x^2)(2)$ Distributive Property
$\qquad = 15x^3 - 3x^5 + 6x^2$ Rules of exponents
$\qquad = -3x^5 + 15x^3 + 6x^2$ Write in standard form.

c. $(-x)(2x^2 - 3x) = (-x)(2x^2) - (-x)(3x)$ Distributive Property
$\qquad = -2x^3 + 3x^2$ Write in standard form.

Exercises Within Reach ® Solutions in English & Spanish and tutorial videos at AlgebraWithinReach.com

Finding a Product with a Monomial Multiplier In Exercises 1–12, find the product.

1. $-x(x^2 - 4)$
2. $-t(10 - 9t^2)$
3. $-3x(2x^2 + 5)$
4. $-5u(u^2 + 4)$
5. $-4x(3 + 3x^2 - 6x^3)$
6. $-5v(5 - 4v + 5v^2)$
7. $2x(x^2 - 2x + 8)$
8. $7y(y^2 - y + 5)$
9. $x^2(4x^2 - 3x + 1)$
10. $y^2(2y^2 + y - 5)$
11. $4t^3(t - 3)$
12. $-2t^4(t + 6)$

Section 5.3 Multiplying Polynomials: Special Products 229

Multiplying Binomials

EXAMPLE 2 Multiplying Binomials with the Distributive Property

a. $(x - 1)(x + 5) = x(x + 5) - 1(x + 5)$ Distributive Property
$= x^2 + 5x - x - 5$ Distributive Property
$= x^2 + (5x - x) - 5$ Group like terms.
$= x^2 + 4x - 5$ Combine like terms.

b. $(2x + 3)(x - 2) = 2x(x - 2) + 3(x - 2)$ Distributive Property
$= 2x^2 - 4x + 3x - 6$ Distributive Property
$= 2x^2 + (-4x + 3x) - 6$ Group like terms.
$= 2x^2 - x - 6$ Combine like terms.

Study Tip

You can write the product of two binomials in just one step. This is called the **FOIL Method**. Note that the words first, outer, inner, and last refer to the positions of the terms in the original product.

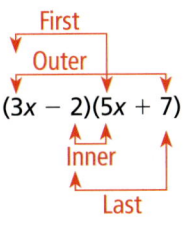

EXAMPLE 3 Multiplying Binomials using the FOIL Method

 F O I L

a. $(x + 4)(x - 4) = x^2 - 4x + 4x - 16$
$= x^2 - 16$ Combine like terms.

 F O I L

b. $(3x + 5)(2x + 1) = 6x^2 + 3x + 10x + 5$
$= 6x^2 + 13x + 5$ Combine like terms.

Exercises Within Reach®

Solutions in English & Spanish and tutorial videos at AlgebraWithinReach.com

Multiplying with the Distributive Property In Exercises 13–16, use the Distributive Property to find the product.

13. $(x + 3)(x + 4)$ 14. $(x - 5)(x + 10)$ 15. $(3x - 5)(x + 1)$ 16. $(7x - 2)(x - 3)$

Using the FOIL Method In Exercises 17 and 18, use the FOIL Method to complete the expression.

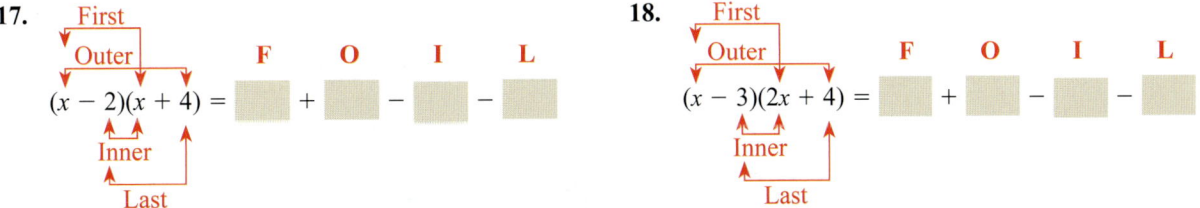

Multiplying Binomials using the FOIL Method In Exercises 19–26, use the FOIL Method to find the product.

19. $(2x - y)(x - 2y)$ 20. $(x + y)(x + 2y)$ 21. $(5x + 6)(3x + 1)$ 22. $(4x + 3)(2x - 1)$

23. $(6 - 2x)(4x + 3)$ 24. $(8x - 6)(5 - 4x)$ 25. $(3x - 2y)(x - y)$ 26. $(7x + 5y)(x + y)$

EXAMPLE 4   A Geometric Model of a Polynomial Product

Use the geometric model to show that

$$x^2 + 3x + 2 = (x + 1)(x + 2).$$

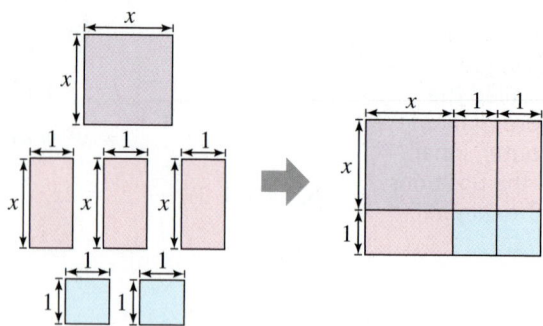

SOLUTION

The left part of the model shows that the sum of the areas of the six rectangles is

$$x^2 + (x + x + x) + (1 + 1) = x^2 + 3x + 2.$$

The right part of the model shows that the area of the rectangle is

$$(x + 1)(x + 2) = x^2 + 2x + x + 2$$
$$= x^2 + 3x + 2.$$

So, $x^2 + 3x + 2 = (x + 1)(x + 2)$.

Exercises Within Reach®

Solutions in English & Spanish and tutorial videos at AlgebraWithinReach.com

A Geometric Model of a Polynomial Product In Exercises 27–30, write the polynomial product represented by the geometric model.

27.
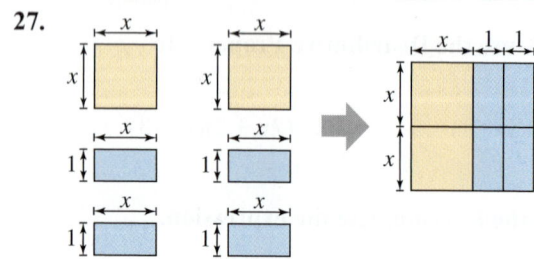

28.
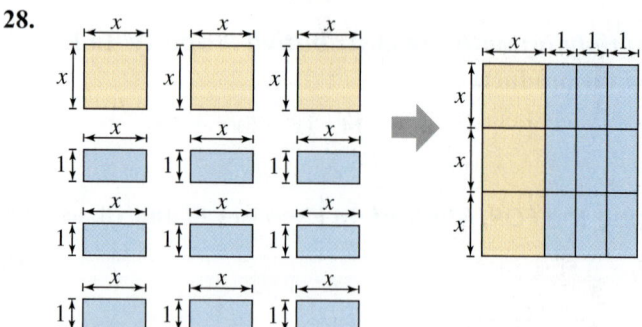

29.
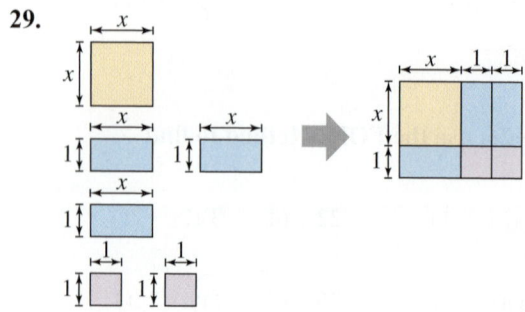

30.
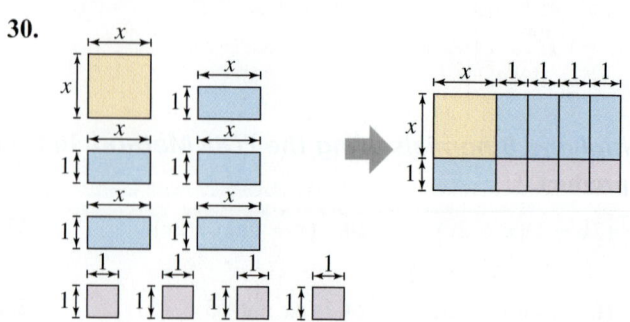

Section 5.3   Multiplying Polynomials: Special Products   231

EXAMPLE 5   Simplifying a Polynomial Expression

Simplify the expression and write the result in standard form.

$$(4x + 5)^2$$

SOLUTION

$(4x + 5)^2 = (4x + 5)(4x + 5)$ Repeated multiplication

$\qquad\qquad = 16x^2 + 20x + 20x + 25$ Use FOIL Method.

$\qquad\qquad = 16x^2 + 40x + 25$ Combine like terms.

EXAMPLE 6   Simplifying a Polynomial Expression

Simplify the expression and write the result in standard form.

$$(3x^2 - 2)(4x + 7) - (4x)^2$$

SOLUTION

$(3x^2 - 2)(4x + 7) - (4x)^2$

$\qquad = 12x^3 + 21x^2 - 8x - 14 - (4x)^2$ Use FOIL Method.

$\qquad = 12x^3 + 21x^2 - 8x - 14 - 16x^2$ Square monomial.

$\qquad = 12x^3 + 5x^2 - 8x - 14$ Combine like terms.

Exercises Within Reach ®   Solutions in English & Spanish and tutorial videos at AlgebraWithinReach.com

Simplifying a Polynomial Expression In Exercises 31–40, simplify the expression and write the result in standard form.

31. $(2x + 4)^2$
32. $(7x - 3)^2$
33. $(8x + 2)^2$
34. $(5x - 1)^2$
35. $(3x^2 - 4)(x + 2)$
36. $(5x^2 - 2)(x - 1)$
37. $(3s + 1)(3s + 4) - (3s)^2$
38. $(2t + 5)(4t - 2) - (2t)^2$
39. $(4x^2 - 1)(2x + 8) + (-x^2)^3$
40. $(3 - 3x^2)(4 - 5x^2) - (-x^4)^2$

41. *Geometry* Add the areas of the four rectangular regions shown in the figure. What product does the geometric model represent?

42. *Geometry* Use the geometric model to write an equation that demonstrates the FOIL Method.

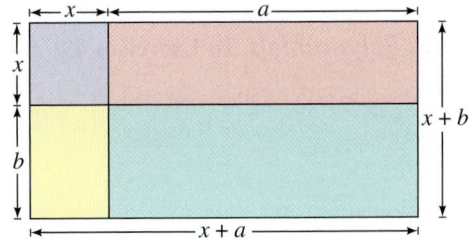

Multiplying Polynomials

EXAMPLE 7 **Multiplying Polynomials (Horizontal Format)**

a. $(x - 4)(x^2 - 4x + 2)$

$= x(x^2 - 4x + 2) - 4(x^2 - 4x + 2)$ Distributive Property

$= x^3 - 4x^2 + 2x - 4x^2 + 16x - 8$ Distributive Property

$= x^3 - 8x^2 + 18x - 8$ Combine like terms.

b. $(2x^2 - 7x + 1)(4x + 3)$

$= (2x^2 - 7x + 1)(4x) + (2x^2 - 7x + 1)(3)$ Distributive Property

$= 8x^3 - 28x^2 + 4x + 6x^2 - 21x + 3$ Distributive Property

$= 8x^3 - 22x^2 - 17x + 3$ Combine like terms.

EXAMPLE 8 **Multiplying Polynomials (Vertical Format)**

a.
$$
\begin{array}{r}
3x^2 + x - 5 \\
\times \quad 2x - 1 \\
\hline
-3x^2 - x + 5 \\
6x^3 + 2x^2 - 10x \\
\hline
6x^3 - x^2 - 11x + 5
\end{array}
$$

Place polynomial with most terms on top. Line up like terms.

$-1(3x^2 + x - 5)$

$2x(3x^2 + x - 5)$

Combine like terms in columns.

b.
$$
\begin{array}{r}
4x^3 + 8x - 1 \\
\times \quad 2x^2 + 3 \\
\hline
12x^3 + 24x - 3 \\
8x^5 + 16x^3 - 2x^2 \\
\hline
8x^5 + 28x^3 - 2x^2 + 24x - 3
\end{array}
$$

Place polynomial with most terms on top. Line up like terms.

$3(4x^3 + 8x - 1)$

$2x^2(4x^3 + 8x - 1)$

Combine like terms in columns.

Exercises Within Reach®

Solutions in English & Spanish and tutorial videos at AlgebraWithinReach.com

Multiplying Polynomials In Exercises 43–48, use a horizontal format to find the product.

43. $(x + 1)(x^2 + 2x - 1)$

44. $(x - 3)(x^2 - 3x + 4)$

45. $(x^2 - 2x + 1)(x - 5)$

46. $(x^2 - x + 1)(x + 4)$

47. $(x - 2)(5x^2 + 2x + 4)$

48. $(x + 9)(2x^2 - x - 4)$

Multiplying Polynomials In Exercises 49–54, use a vertical format to find the product.

49. $\begin{array}{r} x + 3 \\ \times \ x - 2 \end{array}$

50. $\begin{array}{r} 2x - 1 \\ \times \ 5x + 1 \end{array}$

51. $\begin{array}{r} 4x^2 - 6x + 9 \\ \times \quad 2x + 3 \end{array}$

52. $\begin{array}{r} x^2 - 3x + 9 \\ \times \quad x + 3 \end{array}$

53. $(3x^3 + x + 7)(x^2 + 1)$

54. $(5x^4 - 3x + 2)(2x^2 - 4)$

Section 5.3 Multiplying Polynomials: Special Products 233

EXAMPLE 9 Multiplying Polynomials (Vertical Format)

Write the polynomials in standard form and use a vertical format to find the product of $(x + 3x^2 - 4)$ and $(5 + 3x - x^2)$.

SOLUTION

Study Tip
When multiplying two polynomials, it is best to write each in standard form before using either the horizontal or the vertical format.

$$
\begin{array}{r}
3x^2 + x - 4 \\
\times \quad -x^2 + 3x + 5 \\
\hline
15x^2 + 5x - 20 \\
9x^3 + 3x^2 - 12x \\
-3x^4 - x^3 + 4x^2 \\
\hline
-3x^4 + 8x^3 + 22x^2 - 7x - 20
\end{array}
$$

Write in standard form.
Write in standard form.
⟵ $5(3x^2 + x - 4)$
⟵ $3x(3x^2 + x - 4)$
⟵ $-x^2(3x^2 + x - 4)$
Combine like terms.

EXAMPLE 10 Raising a Polynomial to a Power

Use two steps to expand $(x - 3)^3$.

SOLUTION

Step 1: $(x - 3)^2 = (x - 3)(x - 3)$ Repeated multiplication
$ = x^2 - 3x - 3x + 9$ Use FOIL Method.
$ = x^2 - 6x + 9$ Combine like terms.

Step 2: $(x^2 - 6x + 9)(x - 3) = (x^2 - 6x + 9)(x) - (x^2 - 6x + 9)(3)$
$ = x^3 - 6x^2 + 9x - 3x^2 + 18x - 27$
$ = x^3 - 9x^2 + 27x - 27$

So, $(x - 3)^3 = x^3 - 9x^2 + 27x - 27$.

Exercises Within Reach® Solutions in English & Spanish and tutorial videos at AlgebraWithinReach.com

Multiplying Polynomials In Exercises 55–58, use a vertical format to **find** the product.

55. $(x^2 - x + 2)(x^2 + x - 2)$

56. $(x^2 + 2x + 5)(2x^2 - x - 1)$

57. $(x + 3 - 2x^2)(5x + x^2 - 4)$

58. $(1 - x - x^2)(x + 1 - x^3)$

Raising a Polynomial to a Power In Exercises 59 and 60, **expand** the power.

59. $(x - 2)^3$

60. $(x + 3)^3$

Special Products

Special Products

Let a and b be real numbers, variables, or algebraic expressions.

Special Product	Example
Sum and Difference of Two Terms:	
$(a + b)(a - b) = a^2 - b^2$	$(2x - 5)(2x + 5) = 4x^2 - 25$
Square of a Binomial:	
$(a + b)^2 = a^2 + 2ab + b^2$	$(3x + 4)^2 = 9x^2 + 2(3x)(4) + 16$
	$= 9x^2 + 24x + 16$
$(a - b)^2 = a^2 - 2ab + b^2$	$(x - 7)^2 = x^2 - 2(x)(7) + 49$
	$= x^2 - 14x + 49$

EXAMPLE 11   Finding Special Products

a. $(5x - 6)(5x + 6) = (5x)^2 - (6)^2 = 25x^2 - 36$
      Difference   Sum   (1st term)²   (2nd term)²

b. $(3x + 7)^2 = (3x)^2 + 2(3x)(7) + (7)^2 = 9x^2 + 42x + 49$

c. $(4x - 9)^2 = (4x)^2 - 2(4x)(9) + (9)^2 = 16x^2 - 72x + 81$

d. $(6 - 5x^2)^2 = (6)^2 - 2(6)(5x^2) + (5x^2)^2$
   $= 36 - 60x^2 + (5)^2(x^2)^2 = 36 - 60x^2 + 25x^4$

Exercises Within Reach®

Solutions in English & Spanish and tutorial videos at AlgebraWithinReach.com

Finding a Special Product In Exercises 61–76, use a special product pattern to find the product.

61. $(x + 3)(x - 3)$
62. $(x - 5)(x + 5)$
63. $(4t - 6)(4t + 6)$
64. $(3u + 7)(3u - 7)$
65. $(4x + y)(4x - y)$
66. $(5u + 12v)(5u - 12v)$
67. $(x + 6)^2$
68. $(a - 2)^2$
69. $(t - 3)^2$
70. $(x + 10)^2$
71. $(8 - 3z)^2$
72. $(1 - 5t)^2$
73. $(4 + 7s^2)^2$
74. $(3 + 8v^2)^2$
75. $(2x - 5y)^2$
76. $(4s + 3t)^2$

Section 5.3 Multiplying Polynomials: Special Products

Application **EXAMPLE 12** **Finding the Dimensions of a Golf Tee**

A landscaper wants to reshape a square tee area for the ninth hole of a golf course. The new tee area will have one side 2 feet longer and the adjacent side 6 feet longer than the original tee. The area of the new tee will be 204 square feet greater than the area of the original tee. What are the dimensions of the original tee?

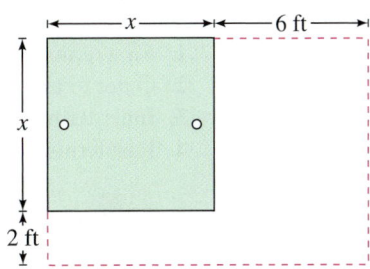

SOLUTION

Verbal Model: New area = Original area + 204

Labels:
Original length = original width = x (feet)
Original area = x^2 (square feet)
New length = $x + 6$ (feet)
New width = $x + 2$ (feet)

Equation:
$(x + 6)(x + 2) = x^2 + 204$ Write equation.
$x^2 + 8x + 12 = x^2 + 204$ Multiply factors.
$8x + 12 = 204$ Subtract x^2 from each side.
$8x = 192$ Subtract 12 from each side.
$x = 24$ Divide each side by 8.

The original tee measured 24 feet by 24 feet.

Exercises Within Reach®

Solutions in English & Spanish and tutorial videos at AlgebraWithinReach.com

77. Geometry A park recreation manager wants to reshape a square sandbox. The new sandbox will have one side 2 feet longer and the adjacent side 3 feet longer than the original sandbox. The area of the new sandbox will be 26 square feet greater than the area of the original sandbox. What are the dimensions of the original sandbox?

78. Geometry A carpenter wants to expand a square room. The new room will have one side 4 feet longer and the adjacent side 6 feet longer than the original room. The area of the new room will be 144 square feet greater than the area of the original room. What are the dimensions of the original room?

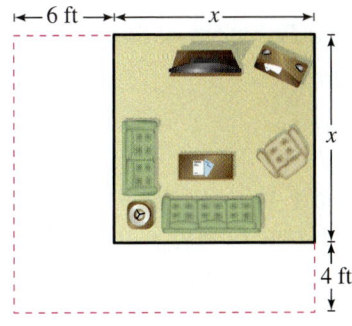

Paul Yates/Shutterstock.com

Concept Summary: Multiplying Two Binomials

What
Here are two methods you can use to multiply binomials.

1. The Distributive Property.
2. The FOIL Method.

Each method applies many of the rules for simplifying algebraic expressions.

EXAMPLE

Find the product $(x - y)(x + 2)$.

How
To use the FOIL Method, add the products of the following pairs of terms.

1. First terms
2. Outer terms
3. Inner terms
4. Last terms

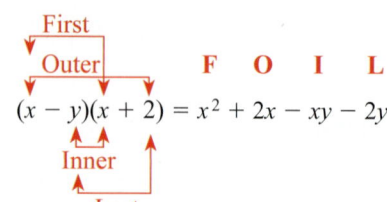

$(x - y)(x + 2) = x^2 + 2x - xy - 2y$

Why
When multiplying two binomials, you can use the FOIL Method. It is a useful tool to help you remember all the products.

To multiply any two polynomials, use the Distributive Property.

Exercises Within Reach®
Worked-out solutions to odd-numbered exercises at AlgebraWithinReach.com

Concept Summary Check

79. *Identifying Binomials* Identify the binomials in the example above.

80. *Signs of Terms* Identify the positive terms and the negative terms in the two binomials in the example above.

81. *Identifying Terms* Identify the first terms, outer terms, inner terms, and last terms in the example above.

82. *Alternate Methods* Describe a method that you can use to multiply any two polynomials.

Extra Practice

Geometry In Exercises 83 and 84, find a polynomial product that represents the area of the region. Then simplify the product.

83.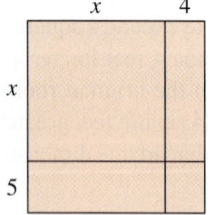

84.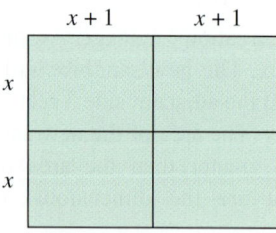

Simplifying a Polynomial Expression In Exercises 85–90, perform the multiplication and simplify.

85. $(x + 2)^2 - (x - 2)^2$

86. $(u + 5)^2 + (u - 5)^2$

87. $(x + 2)^2(x - 4)$

88. $(x - 4)^2(x - 1)$

89. $[(x + 1) + y]^2$

90. $[(x - 3) - y]^2$

Think About It In Exercises 91 and 92, decide whether the equation is an identity. Explain. (An identity is true for any values of the variables.)

91. $(x + y)^3 = x^3 + 3x^2y + 3xy^2 + y^3$

92. $(x - y)^3 = x^3 - 3x^2y + 3xy^2 - y^3$

Using a Result In Exercises 93 and 94, use the result of Exercise 91 to find the product.

93. $(x + 2)^3$

94. $(x + 1)^3$

95. *Geometry* The base of a triangular sail is $2x$ feet and its height is $(x + 10)$ feet (see figure). Find an expression for the area of the sail.

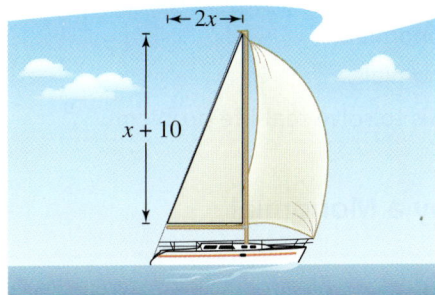

96. *Compound Interest* After 2 years, an investment of $500 compounded annually at interest rate r (in decimal form) will yield an amount $500(1 + r)^2$. Find this product.

Explaining Concepts

True or False? **In Exercises 97 and 98, determine whether the statement is true or false. Justify your answer.**

97. The expressions $(3x)^2$ and $3x^2$ represent the same quantity.

98. Because the product of two monomials is a monomial, it follows that the product of two binomials is a binomial.

99. *Writing* Explain why an understanding of the Distributive Property is essential in multiplying polynomials. Illustrate your explanation with an example.

100. *Reasoning* What is the degree of the product of two polynomials of degrees m and n? Explain.

101. *Reasoning* A polynomial with m terms is multiplied by a polynomial with n terms. How many monomial-by-monomial products must be found? Explain.

102. *Repeated Reasoning* Perform each multiplication.
(a) $(x - 1)(x + 1)$
(b) $(x - 1)(x^2 + x + 1)$
(c) $(x - 1)(x^3 + x^2 + x + 1)$
(d) From the pattern formed in the first three products, can you predict the product of
$(x - 1)(x^4 + x^3 + x^2 + x + 1)?$
Verify your prediction by multiplying.

Cumulative Review

In Exercises 103–106, perform the indicated operations and simplify.

103. $(12x - 3) + (3x - 4)$

104. $(9x - 5) - (x + 7)$

105. $(-8x + 11) - (-4x - 6)$

106. $-(5x - 10) + (-13x + 40)$

In Exercises 107–110, solve the percent equation.

107. What number is 25% of 45?

108. 78 is 10% of what number?

109. 20 is what percent of 60?

110. What number is 55% of 62?

In Exercises 111–114, solve the proportion.

111. $\dfrac{2}{5} = \dfrac{x}{10}$

112. $\dfrac{3}{2} = \dfrac{15}{y}$

113. $\dfrac{z}{6} = \dfrac{5}{8}$

114. $\dfrac{9}{w} = \dfrac{6}{7}$

5.4 Dividing Polynomials

- Divide polynomials by monomials.
- Divide polynomials by binomials.
- Use division of polynomials to solve real-life problems.

Dividing a Polynomial by a Monomial

Dividing a Polynomial by a Monomial

Let a, b, and c be real numbers, variables, or algebraic expressions, such that $c \neq 0$.

1. $\dfrac{a+b}{c} = \dfrac{a}{c} + \dfrac{b}{c}$
2. $\dfrac{a-b}{c} = \dfrac{a}{c} - \dfrac{b}{c}$

EXAMPLE 1 Dividing Polynomials by Monomials

a. $\dfrac{6x+5}{3} = \dfrac{6x}{3} + \dfrac{5}{3}$ Divide each term separately.

$= 2x + \dfrac{5}{3}$ Simplify.

b. $\dfrac{4x^2 - 3x}{3x} = \dfrac{4x^2}{3x} - \dfrac{3x}{3x}$ Divide each term separately.

$= \dfrac{4x}{3} - 1, \quad x \neq 0$ Simplify.

c. $\dfrac{8x^3 - 6x^2 + 10x}{2x} = \dfrac{8x^3}{2x} - \dfrac{6x^2}{2x} + \dfrac{10x}{2x}$ Divide each term separately.

$= 4x^2 - 3x + 5, \quad x \neq 0$ Simplify.

d. $\dfrac{5x^3 - 4x^2 - x + 6}{2x} = \dfrac{5x^3}{2x} - \dfrac{4x^2}{2x} - \dfrac{x}{2x} + \dfrac{6}{2x}$ Divide each term separately.

$= \dfrac{5x^2}{2} - 2x - \dfrac{1}{2} + \dfrac{3}{x}$ Simplify.

Exercises Within Reach®

Solutions in English & Spanish and tutorial videos at AlgebraWithinReach.com

Dividing a Polynomial by a Monomial In Exercises 1–12, perform the division and simplify. (Assume that no denominator is zero.)

1. $\dfrac{3z+3}{3}$

2. $\dfrac{7x+7}{7}$

3. $\dfrac{9x-5}{3}$

4. $\dfrac{8a+5}{-4}$

5. $\dfrac{b^2-2b}{b}$

6. $\dfrac{3x+2x^3}{x}$

7. $\dfrac{25z^3+10z^2}{-5z}$

8. $\dfrac{12c^4-36c}{-6c}$

9. $\dfrac{8z^3+3z^2-2z}{2z}$

10. $\dfrac{12l^5-6l^2+2l}{2l}$

11. $\dfrac{6x^4-2x^3+3x^2-x+4}{2x^3}$

12. $\dfrac{12x^4+4x^3-3x^2+8x-7}{4x^3}$

Dividing a Polynomial by a Binomial

Long Division of Polynomials

1. Write the dividend and divisor in descending powers of the variable.
2. Insert placeholders with zero coefficients for missing powers of the variable. (See Example 4.)
3. Perform the long division of the polynomials as you would with integers.
4. Continue the process until the degree of the remainder is less than the degree of the divisor.

Study Tip

In the illustration below, the number 6982 is the **dividend**, 27 is the **divisor**, 258 is the **quotient**, and 16 is the **remainder**.

EXAMPLE 2 Long Division Algorithm for Polynomials

Use the long division algorithm to divide $x^2 + 3x + 5$ by $x + 1$.

SOLUTION

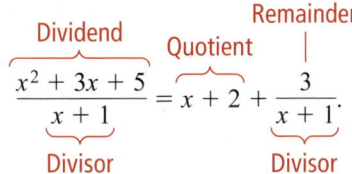

$$\begin{array}{r} x + 2 \\ x+1\overline{)x^2 + 3x + 5} \\ \underline{x^2 + x} \\ 2x + 5 \\ \underline{2x + 2} \\ 3 \end{array}$$

- Multiply x by $(x + 1)$.
- Subtract and bring down 5.
- Multiply 2 by $(x + 1)$.
- Remainder

Considering the remainder as a fractional part of the divisor, the result is

$$\underbrace{\frac{\overbrace{x^2 + 3x + 5}^{\text{Dividend}}}{\underbrace{x + 1}_{\text{Divisor}}}} = \overbrace{x + 2}^{\text{Quotient}} + \frac{\overbrace{3}^{\text{Remainder}}}{\underbrace{x + 1}_{\text{Divisor}}}.$$

Exercises Within Reach®

Solutions in English & Spanish and tutorial videos at AlgebraWithinReach.com

Dividing Polynomials In Exercises 13–22, use the long division algorithm to **perform** the division. (Assume that no denominator is zero.)

13. $\dfrac{x^2 - x - 2}{x + 1}$

14. $\dfrac{x^2 - 5x + 6}{x - 2}$

15. $\dfrac{x^2 + 9x + 20}{x + 4}$

16. $\dfrac{x^2 - 7x - 30}{x - 10}$

17. $\dfrac{x^2 + 8x + 7}{x + 4}$

18. $\dfrac{x^2 - 7x + 12}{x - 1}$

19. $\dfrac{2z^2 - 5z - 3}{2z + 1}$

20. $\dfrac{5m^2 + 23m + 12}{5m + 3}$

21. $\dfrac{3y^2 + 4y - 4}{3y + 1}$

22. $\dfrac{7t^2 - 10t - 8}{7t - 3}$

EXAMPLE 3 Long Division Algorithm for Polynomials

Study Tip

When using long division to divide polynomials, remember to write the divisor and dividend in standard form before using the algorithm.

a.

$$
\begin{array}{r}
6x^2 - 7x + 2 \\
x - 2 \overline{) 6x^3 - 19x^2 + 16x - 4} \\
\underline{6x^3 - 12x^2} \\
-7x^2 + 16x \\
\underline{-7x^2 + 14x} \\
2x - 4 \\
\underline{2x - 4} \\
0
\end{array}
$$

Think $\frac{6x^3}{x} = 6x^2$.

Think $-\frac{7x^2}{x} = -7x$.

Think $\frac{2x}{x} = 2$.

Multiply $6x^2$ by $(x - 2)$.
Subtract and bring down $16x$.
Multiply $-7x$ by $(x - 2)$.
Subtract and bring down -4.
Multiply 2 by $(x - 2)$.
Remainder

So, $\overbrace{(6x^3 - 19x^2 + 16x - 4)}^{\text{Dividend}} \div \overbrace{(x - 2)}^{\text{Divisor}} = \overbrace{6x^2 - 7x + 2}^{\text{Quotient}}, \quad x \neq 2.$

Check this result by multiplying the quotient $6x^2 - 7x + 2$ by the divisor $x - 2$.

b.

$$
\begin{array}{r}
-5x^3 - x^2 + 2x - 1 \\
-2x + 3 \overline{) 10x^4 - 13x^3 - 7x^2 + 8x + 4} \\
\underline{10x^4 - 15x^3} \\
2x^3 - 7x^2 \\
\underline{2x^3 - 3x^2} \\
-4x^2 + 8x \\
\underline{-4x^2 + 6x} \\
2x + 4 \\
\underline{2x - 3} \\
7
\end{array}
$$

Multiply $-5x^3$ by $(-2x + 3)$.
Subtract and bring down $-7x^2$.
Multiply $-x^2$ by $(-2x + 3)$.
Subtract and bring down $8x$.
Multiply $2x$ by $(-2x + 3)$.
Subtract and bring down 4.
Multiply -1 by $(-2x + 3)$.
Remainder

This shows that

$$\underbrace{\frac{10x^4 - 13x^3 - 7x^2 + 8x + 4}{-2x + 3}}_{\substack{\text{Dividend} \\ \text{Divisor}}} = \underbrace{-5x^3 - x^2 + 2x - 1}_{\text{Quotient}} + \underbrace{\frac{7}{-2x + 3}}_{\substack{\text{Remainder} \\ \text{Divisor}}}.$$

Exercises Within Reach® Solutions in English & Spanish and tutorial videos at AlgebraWithinReach.com

Dividing Polynomials In Exercises 23–28, use the long division algorithm to **perform** the division. (Assume that no denominator is zero.)

23. $(18t^2 - 21t - 4) \div (3t - 4)$

24. $(20t^2 + 32t - 16) \div (2t + 4)$

25. $(x^3 - 4x^2 + 9x - 7) \div (x - 2)$

26. $(2x^3 - 2x^2 + 3x + 9) \div (x + 1)$

27. $(9x^4 + 7x^2 - 12x^3 + 9 + 10x) \div (-3x + 5)$

28. $(-16x^3 + 16x^4 + 12x^2 - 4x + 5) \div (-2 + 4x)$

Section 5.4 Dividing Polynomials 241

EXAMPLE 4 **Accounting for Missing Powers of *x***

Use the long division algorithm to perform the division.

a. $\dfrac{x^3 - 1}{x - 1}$

b. $\dfrac{2x^4 - 3x^2 + 8x - 1}{x + 2}$

SOLUTION

a. Because there are no x^2- or x-terms in the dividend, you can line up the subtractions by using *zero* coefficients for the missing terms.

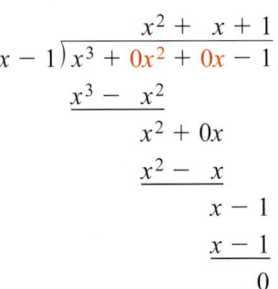

	Insert $0x^2$ and $0x$.
	Multiply x^2 by $(x - 1)$.
	Subtract and bring down $0x$.
	Multiply x by $(x - 1)$.
	Subtract and bring down -1.
	Multiply 1 by $(x - 1)$.
	Remainder

Study Tip

When the dividend is missing one or more powers of *x* the long division algorithm requires that you account for the missing powers.

So, $x - 1$ divides evenly into $x^3 - 1$ and you can write

$$\dfrac{x^3 - 1}{x - 1} = x^2 + x + 1, \quad x \neq 1.$$

b.
$$\begin{array}{r} 2x^3 - 4x^2 + 5x - 2 \\ x+2\overline{\smash{\big)}\,2x^4 + 0x^3 - 3x^2 + 8x - 1} \end{array}$$

Insert $0x^3$.
Multiply $2x^3$ by $(x + 2)$.
Subtract and bring down $-3x^2$.
Multiply $-4x^2$ by $(x + 2)$.
Subtract and bring down $8x$.
Multiply $5x$ by $(x + 2)$.
Subtract and bring down -1.
Multiply -2 by $(x + 2)$.
Remainder

This shows that $\dfrac{2x^4 - 3x^2 + 8x - 1}{x + 2} = 2x^3 - 4x^2 + 5x - 2 + \dfrac{3}{x + 2}.$

Exercises Within Reach® Solutions in English & Spanish and tutorial videos at AlgebraWithinReach.com

Dividing Polynomials In Exercises 29–36, use the long division algorithm to **perform** the division. (Assume that no denominator is zero.)

29. $\dfrac{x^3 - 8}{x - 2}$

30. $\dfrac{y^3 - 27}{y - 3}$

31. $\dfrac{3x^4 - 40x^2 + 28x - 18}{x + 4}$

32. $\dfrac{5x^4 + 6x^3 + 9x - 2}{x - 1}$

33. $(9x^2 - 1) \div (3x + 1)$

34. $(25y^2 - 4) \div (5y - 2)$

35. $(x^4 - 1) \div (x - 1)$

36. $x^4 \div (x - 1)$

Applications

Application — **EXAMPLE 5** — Geometry: **Finding the Width of a Rectangle**

The area of a rectangle is $(x^2 + 2x - 15)$ square feet and its length is $(x + 5)$ feet. Find the width of the rectangle.

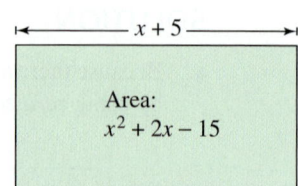

SOLUTION

Verbal Model: Area = Length × Width

Labels: Area $= x^2 + 2x - 15$ (square feet)
Length $= x + 5$ (feet)

Equation: Width $= \dfrac{x^2 + 2x - 15}{x + 5}$

$$
\begin{array}{r}
x - 3 \\
x + 5 \overline{) x^2 + 2x - 15} \\
\underline{x^2 + 5x} \\
-3x - 15 \\
\underline{-3x - 15} \\
0
\end{array}
$$

Multiply x by $(x + 5)$.
Subtract and bring down -15.
Multiply -3 by $(x + 5)$.
Remainder

The width of the rectangle is $(x - 3)$ feet.

Exercises Within Reach® — Solutions in English & Spanish and tutorial videos at AlgebraWithinReach.com

37. *Geometry* The area of a rectangle is $(x^2 + 5x - 6)$ square meters and its width is $(x - 1)$ meters (see figure).

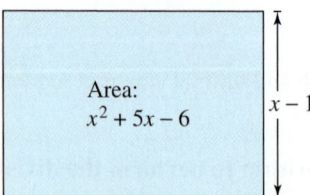

(a) Use the formula for the area of a rectangle to complete the verbal model.

Verbal Model: Area =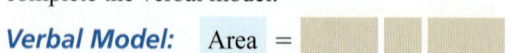

(b) Use the figure to write labels for the area and width.

(c) Use the labels and the verbal model to write and solve an equation for the length of the rectangle.

38. *Geometry* The area of a rectangle is $(3x^2 + x - 2)$ square inches and its length is $(3x - 2)$ inches (see figure).

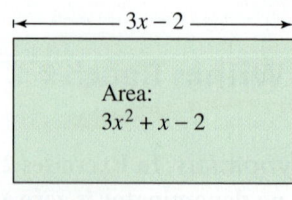

(a) Use the formula for the area of a rectangle to complete the verbal model.

Verbal Model: Area =

(b) Use the figure to write labels for the area and length.

(c) Use the labels and the verbal model to write and solve an equation for the width of the rectangle.

Application   EXAMPLE 6   Comparing Salaries

At a manufacturing company, Employee A and Employee B earn $40,000 and $50,000, respectively. After a company-wide raise of m dollars, their salaries will be $(40,000 + m)$ dollars and $(50,000 + m)$ dollars.

a. Write the ratio of Employee B's salary to Employee A's salary. Use long division to rewrite the ratio.

b. Complete the table.

m	0	1000	2000	3000	4000
$\dfrac{50,000 + m}{40,000 + m}$					

c. What happens to the value of the ratio as m increases? Use the result of part (a) to explain your conclusion.

SOLUTION

a. $\dfrac{m + 50,000}{m + 40,000} = 1 + \dfrac{10,000}{m + 40,000}$

b.

m	0	1000	2000	3000	4000
$\dfrac{50,000 + m}{40,000 + m}$	1.250	1.244	1.238	1.233	1.227

c. As m increases, the value of the expression $\dfrac{10,000}{m + 40,000}$ approaches 0.

So, the value of the ratio approaches $1 + 0$, or 1.

Exercises Within Reach ®

Solutions in English & Spanish and tutorial videos at AlgebraWithinReach.com

39. Comparing Ages Your neighbor has two children: one is 12 years old and the other is 8 years old. In t years, their ages will be $t + 12$ and $t + 8$.

(a) Write the ratio of the older child's age to the younger child's age. Use long division to rewrite the ratio.

(b) Complete the table.

t	0	10	20	30	40	50	60
$\dfrac{t + 12}{t + 8}$	1.5						

(c) What happens to the value of the ratio as t increases? Use the result of part (a) to explain your conclusion.

40. Brine Solution A brine solution contains 20 kilograms of salt and 80 kilograms of water. When w kilograms of water are added, the amount of water will be $80 + w$ and the total amount of brine solution will be $100 + w$.

(a) Write the ratio of the amount of water to the total amount of brine solution. Use long division to rewrite the ratio.

(b) Complete the table.

w	0	50	100	150	200
$\dfrac{80 + w}{100 + w}$					

(c) What happens to the value of the ratio as w increases? Use the result of part (a) to explain your conclusion.

Concept Summary: Dividing Polynomials

What
You can use long division to divide polynomials just as you do for integers.

EXAMPLE
Divide $(-4 + x^2)$ by $(2 + x)$.

How
Use these steps to perform long division.

1. Write the **dividend** and **divisor** in standard form.
2. Insert placeholders with zero coefficients for missing powers of the variable.
3. Use the long division algorithm.

EXAMPLE
$$\begin{array}{r} x - 2 \\ x + 2 \overline{\smash{)}x^2 + 0x - 4} \\ \underline{x^2 + 2x} \\ -2x - 4 \\ \underline{-2x - 4} \\ 0 \end{array}$$

Why
You can use long division of polynomials to check the product of two polynomials.

Exercises Within Reach®
Worked-out solutions to odd-numbered exercises at AlgebraWithinReach.com

Concept Summary Check

41. Rewriting the Dividend Write the polynomials $-4 + x^2$ and $2 + x$ in standard form.

42. Inserting Placeholders What is the missing power of x in the dividend in the example above? Explain.

43. Analyzing the Solution In the example above, does the divisor divide evenly into the dividend? Explain.

44. Checking the Solution How can you check the solution of the example above algebraically?

Extra Practice

Operations with Polynomials In Exercises 45−56, **simplify** the expression. (Assume that no denominator is zero.)

45. $\dfrac{4x^3}{x^2} - \dfrac{8x}{4}$

46. $\dfrac{9z^3}{z^2} - \dfrac{6z}{3}$

47. $\dfrac{25x^2}{10x} + \dfrac{3x}{2}$

48. $\dfrac{21y^2}{14y} + \dfrac{5y}{2}$

49. $\dfrac{8u^2v}{2u} + \dfrac{(uv)^2}{uv}$

50. $\dfrac{18xy^2}{6y} + \dfrac{(xy)^2}{xy}$

51. $\dfrac{9x^5y}{3x^4} - \dfrac{(x^2y)^3}{x^5y^2}$

52. $\dfrac{35a^3b^2}{5a^2b} - \dfrac{(a^2b^3)^2}{a^3b^5}$

53. $\dfrac{x^2 + 2x + 1}{x + 1} - (3x - 4)$

54. $\dfrac{x^2 - 2x - 3}{x + 1} - (4x - 1)$

55. $\dfrac{x^2 - 3x + 2}{x - 1} + (4x - 3)$

56. $\dfrac{x^2 + x - 6}{x - 2} + (2x - 7)$

57. *Exploration* Consider the equation

$(x + 3)(x^2 + 2x - 1) = x^3 + 5x^2 + 5x - 3$.

(a) Verify that the equation is an identity by multiplying the polynomials on the left side of the equation.

(b) Verify that the equation is an identity by performing the long division

$$\frac{x^3 + 5x^2 + 5x - 3}{x + 3}.$$

58. *Exploration* Consider the equation

$2x^3 - 5x^2 + 2x - 5 = (2x - 5)(x^2 + 1)$.

(a) Verify that the equation is an identity by multiplying the polynomials on the right side of the equation.

(b) Verify that the equation is an identity by performing the long division

$$\frac{2x^3 - 5x^2 + 2x - 5}{2x - 5}.$$

59. *Geometry* The volume of a rectangular box is $(2h^3 + 3h^2 + h)$ cubic inches and the height is $(h + 1)$ inches (see figure). Find the area of the bottom of the box.

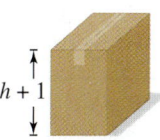

60. *Floor Space* The volume of a rectangular building is $(x^3 + 55x^2 + 650x + 2000)$ cubic feet. The height of the building is $(x + 5)$ feet (see figure). Find the area of the ground floor of the building.

Explaining Concepts

61. *Reasoning* A polynomial of degree 4 is divided by a monomial of degree 2. What is the degree of the quotient?

62. *Structure* Label each part of the equation with its name:

$$\frac{x^2 + 2}{x - 3} = x + 3 + \frac{11}{x - 3}.$$

(a) Dividend (b) Divisor

(c) Quotient (d) Remainder

63. *Writing* When dividing a polynomial by a monomial, explain how you can use the rule for adding two fractions with like denominators.

64. *Creating Problems* Create two division problems using polynomials: one that divides evenly and one that has a remainder.

Cumulative Review

In Exercises 65−68, simplify the fraction.

65. $\frac{-8}{12}$

66. $\frac{18}{-144}$

67. $\frac{-60}{-150}$

68. $\frac{-175}{-42}$

In Exercises 69−72, write an equation of the line that passes through the points. Write the equation in slope-intercept form.

69. $(4, 5), (8, 9)$

70. $(-3, 2), (-1, 7)$

71. $(-5, -8), (1, 5)$

72. $(0, 4), (-5, 2)$

In Exercises 73−76, find the product and simplify.

73. $-2x^2(5x^3)$

74. $(2z + 1)(2z - 1)$

75. $(x + 7)^2$

76. $(x + 4)(2x - 5)$

5 Chapter Summary

	What did you learn?	Explanation and Examples	Review Exercises
5.1	Apply the rules of exponents to rewrite exponential expressions (p. 210).	Let m and n be integers, and let a and b be real numbers, variables, or algebraic expressions, such that $a \neq 0$ and $b \neq 0$. *Rule* *Example* 1. $a^m a^n = a^{m+n}$ $y^2 \cdot y^4 = y^{2+4} = y^6$ 2. $\dfrac{a^m}{a^n} = a^{m-n}$ $\dfrac{x^7}{x^4} = x^{7-4} = x^3$ 3. $(ab)^m = a^m b^m$ $(5x)^4 = 5^4 x^4 = 625x^4$ 4. $\left(\dfrac{a}{b}\right)^m = \dfrac{a^m}{b^m}$ $\left(\dfrac{2}{x}\right)^3 = \dfrac{2^3}{x^3} = \dfrac{8}{x^3}$ 5. $(a^m)^n = a^{mn}$ $(y^3)^{-4} = y^{3(-4)} = y^{-12}$	1–12
	Use the negative exponent rule to rewrite exponential expressions (p. 212).	Let n be an integer and let a be a real number, variable, or algebraic expression such that $a \neq 0$. $a^{-n} = \dfrac{1}{a^n}$ and $\dfrac{1}{a^{-n}} = a^n$	13–36
	Write very large and very small numbers in scientific notation (p. 214).	Scientific notation has the form $c \times 10^n$, where $1 \leq c < 10$ and n is an integer. If $n > 0$, the number is large (10 or more). If $n < 0$, the number is small (less than 1).	37–40
5.2	Identify the degrees and leading coefficients of polynomials (p. 218).	Let $a_n, a_{n-1}, \ldots, a_2, a_1, a_0$ be real numbers and let n be a nonnegative integer. A polynomial in x is an expression of the form $a_n x^n + a_{n-1} x^{n-1} + \cdots + a_2 x^2 + a_1 x + a_0$ where $a_n \neq 0$. The polynomial is of degree n, and the number a_n is called the leading coefficient. The number a_0 is called the constant term.	41–46
	Add polynomials using a horizontal or vertical format (p. 219).	The key to adding two polynomials is to recognize *like* terms—those having the same *degree*. To use a horizontal format, group like terms and add. To use a vertical format, line up like terms and add. *Horizontal format:* *Vertical format:* $(2x^2 + 3x) + (x^2 - 2x)$ $2x^2 + 3x$ $= (2x^2 + x^2) + (3x - 2x)$ $\underline{x^2 - 2x}$ $= 3x^2 + x$ $3x^2 + x$	47–58
	Subtract polynomials using a horizontal or vertical format (p. 221).	The subtract one polynomial from another, *add the opposite* by changing the sign of each term being subtracted. Then add like terms. *Horizontal format:* *Vertical format:* $(x^2 + 3) - (x^2 - 2)$ $(x^2 + 3) \Rightarrow x^2 + 3$ $= x^2 + 3 - x^2 + 2$ $-(x^2 - 2) \Rightarrow \underline{-x^2 + 2}$ $= (x^2 - x^2) + (3 + 2)$ 5 $= 5$	59–68

Chapter Summary

	What did you learn?	Explanation and Examples	Review Exercises
5.3	Find products with monomial multipliers *(p. 228)*.	To multiply a polynomial by a monomial, use the Distributive Property. $(3x - 7)(-2x) = 3x(-2x) - 7(-2x)$ $= -6x^2 + 14x$	69–72
	Multiply binomials using the Distributive Property and the FOIL Method *(p. 229)*.	*Distributive Property:* $(3x - 2)(5x + 7) = 3x(5x + 7) - 2(5x + 7)$ $= 15x^2 + 11x - 14$ *FOIL Method:* Product of First terms, Product of Outer terms, Product of Inner terms, Product of Last terms $(3x - 2)(5x + 7) = 15x^2 + 21x - 10x - 14$ $= 15x^2 + 11x - 14$	73–84
	Multiply polynomials using a horizontal or vertical format *(p. 232)*.	When multiplying two polynomials, it is best to write each in standard form before using either the horizontal or vertical format. To multiply using a horizontal format, use the Distributive Property. To multiply using a vertical format, line up the like terms to help with combining like terms after multiplying.	85–96
	Identify and use special binomial products *(p. 234)*.	Let a and b be real numbers, variables, or algebraic expressions. *Sum and Difference of Two Terms:* $(a + b)(a - b) = a^2 - b^2$ *Square of a Binomial:* $(a + b)^2 = a^2 + 2ab + b^2$ $(a - b)^2 = a^2 - 2ab + b^2$	97–106
5.4	Divide polynomials by monomials *(p. 238)*.	Let a, b, and c be real numbers, variables, or algebraic expressions, such that $c \neq 0$. 1. $\dfrac{a + b}{c} = \dfrac{a}{c} + \dfrac{b}{c}$ 2. $\dfrac{a - b}{c} = \dfrac{a}{c} - \dfrac{b}{c}$	107–110
	Divide polynomials by binomials *(p. 239)*.	*Long Division of Polynomials:* 1. Write the dividend and divisor in descending powers of the variable. 2. Insert placeholders with zero coefficients for missing powers of the variable. 3. Perform the long division of the polynomials as you would with integers. 4. Continue the process until the degree of the remainder is less than the degree of the divisor.	111–116
	Use division of polynomials to solve real-life problems *(p. 242)*.	Long division of polynomials can be used to find the width of a rectangle when the area and the length of the rectangle are given as polynomials. (See Example 5.)	117, 118

Review Exercises

Worked-out solutions to odd-numbered exercises at AlgebraWithinReach.com

5.1

Applying the Rules of Exponents In Exercises 1–12, simplify the expression. (Assume that no denominator is zero.)

1. $x^2 \cdot x^2 \cdot x^4$
2. $y^5 \cdot y^8 \cdot y$
3. $(x^9)^2$
4. $(t^4)^3$
5. $t^5(-2t^7)$
6. $u^2(3u^2)$
7. $(xy)^4(-5x^2y^3)$
8. $(3uv)(-2uv^2)^3$
9. $\dfrac{24x^3}{12x^2}$
10. $\dfrac{-36v^6}{4v^3}$
11. $\dfrac{64u^3v^2}{32uv}$
12. $\dfrac{21x^3y^7}{-7x^3y}$

Negative Exponents In Exercises 13–20, rewrite the expressions using only positive exponents.

13. 2^{-2}
14. 3^{-5}
15. $6x^{-3}$
16. $5x^{-5}$
17. $\dfrac{4}{t^{-8}}$
18. $\dfrac{7}{y^{-1}}$
19. $7(2x)^{-1}$
20. $6(5x)^{-1}$

Using the Rules of Exponents In Exercises 21–36, use the rules of exponents to simplify the expression using only positive exponents. (Assume that no variable is zero.)

21. $t^{-2} \cdot t$
22. $-u^{-3} \cdot u^5$
23. $\dfrac{a^3}{a^{-4}}$
24. $\dfrac{b^5}{b^{-5}}$
25. $4x^{-6}y^2 \cdot x^6$
26. $-2u^5v^{-4} \cdot v^4$
27. $(-3a^2)^{-2}(a^2)^0$
28. $(3u^{-3})(9u^6)^0$
29. $(2x^2y^{-3})^2$
30. $(5x^{-1}y^3)^{-2}$
31. $\left(\dfrac{y}{5}\right)^{-2}$
32. $\left(\dfrac{7}{x^4}\right)^{-1}$
33. $\dfrac{(-3y)^{-4}}{-9y^{-6}}$
34. $\dfrac{5z^{-3}}{(-10z^2)^{-2}}$
35. $(2u^{-2}v)^3(4u^{-5}v^4)^{-1}$
36. $(3x^2y^4)^3(3x^2y^4)^{-3}$

Converting from Decimal to Scientific Notation In Exercises 37 and 38, write the number in scientific notation.

37. *Length of a dust mite:* 0.0004 meter

Sebastian Kaulitzki/Shutterstock.com

38. *Circumference of the Sun* 4,370,000,000 meters

Converting from Scientific to Decimal Notation In Exercises 39 and 40, write the number in decimal notation.

39. 1.809×10^8

40. 4×10^{-6}

5.2

Determining the Degree and Leading Coefficient In Exercises 41–46, write the polynomial in standard form. Then identify its degree and leading coefficient.

41. $10x - 4 - 5x^3$

42. $2x^2 + 9$

43. $4x^3 - 2x + 5x^4 - 7x^2$

44. $6 - 3x + 6x^2 - x^3$

45. $7x^4 - 1 + 11x^2$

46. $12x^2 + 2x - 8x^5 + 1$

Adding Polynomials In Exercises 47–52, use a horizontal format to find the sum.

47. $(8x + 4) + (x - 4)$

48. $\left(\frac{1}{2}x + \frac{2}{3}\right) + \left(4x + \frac{1}{3}\right)$

49. $(3y^3 + 5y^2 - 9y) + (2y^3 - 3y + 10)$

50. $(6 - x + x^2) + (3x^2 + x)$

51. $(3u + 4u^2) + 5(u + 1) + 3u^2$

52. $6(u^2 + 2) + 12u + (u^2 - 5u + 2)$

Adding Polynomials In Exercises 53–56, use a vertical format to find the sum.

53. $x^3 + 2x - 3$
       $\phantom{x^3 + {}} 4x + 5$

54. $-x^3 + 3$
       $3x^3 + 2x^2 + 5$

55. $-x^4 - 2x^2 + 3$
       $3x^4 - 5x^2$

56. $5z^3 - 4z - 7$
       $-z^2 - 4z$

57. *Geometry* The length of a rectangular wall is x units, and its height is $(x - 3)$ units (see figure). Find an expression for the perimeter of the wall.

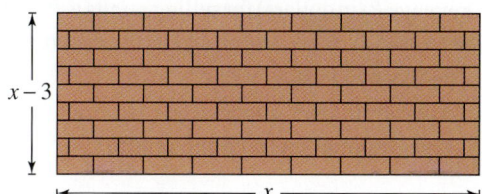

58. *Geometry* A rectangular garden has length $(t + 5)$ feet and width $2t$ feet (see figure). Find an expression for the perimeter of the garden.

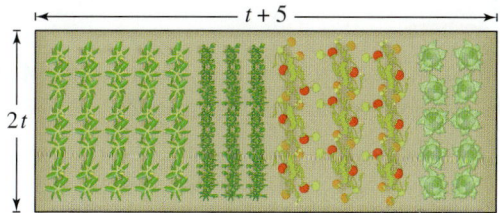

Subtracting Polynomials In Exercises 59–64, use a horizontal format to find the difference.

59. $(3t - 5) - (3t - 9)$

60. $\left(2x - \frac{1}{5}\right) - \left(\frac{1}{4}x + \frac{1}{4}\right)$

61. $(6x^2 - 9x - 5) - 3(4x^2 - 6x + 1)$

62. $(5t^2 + 2) - 2(4t^2 + 1)$

63. $4y^2 - [y - 3(y^2 + 2)]$

64. $(6a^3 + 3a) - 2[a - (a^3 + 2)]$

Subtracting Polynomials In Exercises 65 and 66, use a vertical format to find the difference.

65. $5x^2 + 2x - 27$
    $-(2x^2 - 2x - 13)$

66. $12y^4 - 15y^2 + 7$
    $-(18y^4 - 9)$

67. **Geometry** Find an expression for the area of the shaded region of the figure.

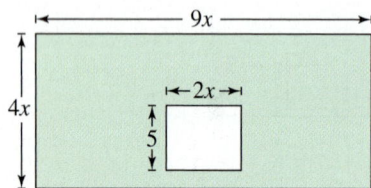

68. **Cost, Revenue, and Profit** The cost C (in dollars) of producing x units of a product is $C = 15 + 26x$. The revenue R (in dollars) for selling x units is $R = 40x - \frac{1}{2}x^2$, where $0 \le x \le 20$. The profit P is the difference between revenue and cost.

   (a) Perform the subtraction required to find the polynomial representing profit P.

   (b) Determine the profit when 14 units are produced and sold.

5.3

Finding A Product with a Monomial Multiplier In Exercises 69–72, find the product.

69. $2x(x + 4)$

70. $3y(y - 1)$

71. $(4x + 2)(-3x^2)$

72. $(5 - 7y)(-6y^2)$

Multiplying with the Distributive Property In Exercises 73–76, use the Distributive Property to find the product.

73. $(x + 10)(x + 2)$

74. $(x - 1)(x + 3)$

75. $(2x - 5)(7x + 2)$

76. $(3x - 2)(2x - 3)$

Multiplying Binomials using the FOIL Method In Exercises 77–84, use the FOIL Method to find the product.

77. $(x + 4)(x + 6)$

78. $(u + 5)(u - 2)$

79. $(4x - 3)(3x + 4)$

80. $(6x - 7)(2x + 5)$

81. $(3 - 4x)(7x - 6)$

82. $(6 - 2x)(7x - 10)$

83. $(x + y)(x + 3y)$

84. $(x - y)(x + 4y)$

Multiplying Polynomials In Exercises 85–88, use a horizontal format to find the product.

85. $(x^2 + 5x + 2)(x - 6)$

86. $(s^2 + 4s - 3)(s - 3)$

87. $(2t - 1)(t^2 - 3t + 3)$

88. $(4x + 2)(x^2 + 6x - 5)$

Multiplying Polynomials In Exercises 89–92, use a vertical format to find the product.

89. $3x^2 + x - 2$
    $\times 4x - 5$

90. $5y^2 - 2y + 9$
    $\times 3y + 4$

91. $y^2 - 4y + 5$
    $\times y^2 + 2y - 3$

92. $x^2 + 8x - 12$
    $\times x^2 - 9x + 2$

Raising a Polynomial to a Power In Exercises 93 and 94, **expand** the power.

93. $(2x + 1)^3$

94. $(3y - 2)^3$

95. *Geometry* The width of a rectangular window is $(2x + 6)$ inches, and its height is $(3x + 10)$ inches (see figure). Find an expression for the area of the window.

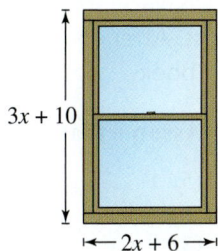

96. *Geometry* The width of a rectangular parking lot is $(x + 25)$ meters, and its length is $(x + 30)$ meters (see figure). Find an expression for the area of the parking lot.

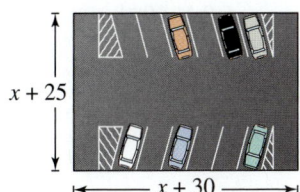

Finding a Special Product In Exercises 97–106, use a special product pattern to **find** the product.

97. $(x + 3)^2$

98. $(x - 5)^2$

99. $\left(\frac{1}{2}x - 4\right)^2$

100. $(4 + 3b)^2$

101. $(u - 6)(u + 6)$

102. $(r + 7)(r - 7)$

103. $(2r - 5t)^2$

104. $(3a + b)^2$

105. $(2x - 4y)(2x + 4y)$

106. $(4u + 5v)(4u - 5v)$

5.4

Dividing a Polynomial by a Monomial In Exercises 107–110, **perform** the division and simplify. (Assume that no denominator is zero.)

107. $\dfrac{24x^3 - 12x}{3x^2}$

108. $\dfrac{8u^3 + 4u^2}{2u}$

109. $(18 - 3x + 9x^2) \div (12x^2)$

110. $(5x^2 + 15x - 25) \div (-5x)$

Dividing Polynomials In Exercises 111–116, use the long division algorithm to **perform** the division. (Assume that no denominator is zero.)

111. $\dfrac{x^2 - x - 6}{x - 3}$

112. $\dfrac{x^2 + x - 20}{x + 5}$

113. $\dfrac{24x^2 - 2x - 8}{4x + 3}$

114. $\dfrac{16x^2 - 5}{2x + 1}$

115. $\dfrac{2x^3 + 2x^2 - x + 2}{x - 1}$

116. $\dfrac{6x^4 - 4x^3 - 27x^2 + 18x}{3x - 2}$

117. *Geometry* The area of a rectangle is $2x^2 - 5x - 12$, and its width is $x - 4$. Find the length of the rectangle.

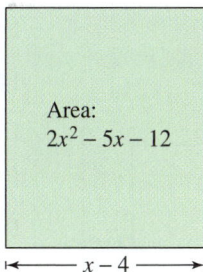

118. *Geometry* The area of a rectangle is $3x^2 + 5x - 3$, and its length is $x + 3$. Find the width of the rectangle.

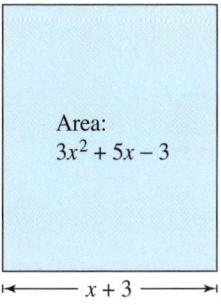

Chapter Test

Solutions in English & Spanish and tutorial videos at AlgebraWithinReach.com

Take this test as you would take a test in class. After you are done, check your work against the answers in the back of the book.

In Exercises 1–6, simplify the expression. (Assume that no variable is zero.)

1. $x^2 \cdot x^7$
2. $(5y^7)^2$
3. $\dfrac{-6a^2b}{-9ab}$
4. $(3x^{-2}y^3)^2$
5. $(4u^{-3})^{-2}(8u^{-1}v^{-2})^0$
6. $\dfrac{12x^{-3}y^5}{4x^{-2}y^{-1}}$

7. Evaluate each expression *without* using a calculator.

   (a) $\dfrac{2^{-3}}{3^{-1}}$ (b) $(2.0 \times 10^5)^2$ (c) $\dfrac{6.0 \times 10^{-3}}{2.0 \times 10^2}$

8. Write 0.00015 in scientific notation.

9. Write 8×10^7 in decimal notation.

10. Determine the degree and the leading coefficient of $-3x^4 - 5x^2 + 2x - 10$.

In Exercises 11–22, simplify the expression. (Assume that no variable or denominator is zero.)

11. $(3z^2 - 3z + 7) + (8 - z^2)$
12. $(8u^3 + 3u^2 - 2u - 1) - (u^3 + 3u^2 - 2u)$
13. $6y + [2y - (3 - y^2)]$
14. $-5(x^2 - 1) + 3(4x + 7) - (x^2 + 26)$
15. $(x - 7)^2$
16. $(2x - 3)(2x + 3)$
17. $(z + 2)(2z^2 - 3z + 5)$
18. $(y + 3)(y^2 - 4)$
19. $\dfrac{4z^3 + z}{2z}$
20. $\dfrac{16x^2 - 12}{-8}$
21. $\dfrac{x^3 - x - 6}{x - 2}$
22. $\dfrac{4x^3 + 10x^2 - 2x - 5}{2x - 4}$

23. Find an expression for the area of the shaded region shown in the figure.

24. Find an expression for the area of the triangle shown in the figure.

25. After 2 years, an investment of $1500 compounded annually at interest rate r (in decimal form) will yield an amount of $1500(1 + r)^2$. Find this product.

26. The area of a rectangle is $x^2 - 2x - 3$, and its length is $x + 1$. Find the width of the rectangle.

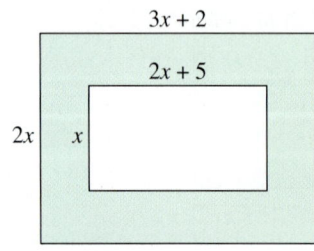

Figure for 23

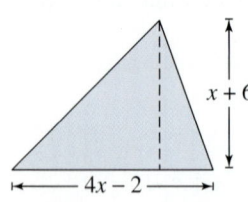

Figure for 24

6 Factoring and Solving Equations

- **6.1** Factoring Polynomials with Common Factors
- **6.2** Factoring Trinomials
- **6.3** More About Factoring Trinomials
- **6.4** Factoring Polynomials with Special Forms
- **6.5** Solving Quadratic Equations by Factoring

MASTERY IS WITHIN REACH!

"I failed my first college math class because, for some reason, I thought just showing up and listening would be enough. I was wrong. Now, I get to class a little early to review my notes, sit where I can see the instructor, and ask questions. I try to learn and remember as much as possible in class because I am so busy juggling work and college."

Julianna
Business

See page 279 for suggestions about making the most of class time.

6.1 Factoring Polynomials with Common Factors

- Find the greatest common factor of two or more expressions.
- Factor out the greatest common monomial factor from polynomials.
- Factor polynomials by grouping.

Greatest Common Factor

In Chapter 5, you used the Distributive Property to multiply polynomials. In this chapter, you will study the *reverse* process, which is **factoring**.

Multiplying Polynomials

$2x(7 - 3x) \Rightarrow 14x - 6x^2$
Factor Factor — Product

Factoring Polynomials

$14x - 6x^2 \Rightarrow 2x(7 - 3x)$
Product — Factor Factor

EXAMPLE 1 Finding the Greatest Common Factor

To find the greatest common factor of $5x^2y^2$ and $30x^3y$, first factor each term.

$$5x^2y^2 = 5 \cdot x \cdot x \cdot y \cdot y = (5x^2y)(y)$$
$$30x^3y = 2 \cdot 3 \cdot 5 \cdot x \cdot x \cdot x \cdot y = (5x^2y)(6x)$$

So, you can conclude that the greatest common factor is $5x^2y$.

EXAMPLE 2 Finding the Greatest Common Factor

To find the greatest common factor of $8x^5$, $20x^3$, and $16x^4$, first factor each term.

$$8x^5 = 2 \cdot 2 \cdot 2 \cdot x \cdot x \cdot x \cdot x \cdot x = (4x^3)(2x^2)$$
$$20x^3 = 2 \cdot 2 \cdot 5 \cdot x \cdot x \cdot x = (4x^3)(5)$$
$$16x^4 = 2 \cdot 2 \cdot 2 \cdot 2 \cdot x \cdot x \cdot x \cdot x = (4x^3)(4x)$$

So, you can conclude that the greatest common factor is $4x^3$.

Exercises Within Reach®

Solutions in English & Spanish and tutorial videos at AlgebraWithinReach.com

Finding the Greatest Common Factor In Exercises 1–12, find the greatest common factor of the expressions.

1. $z^2, -z^6$
2. t^4, t^7
3. $2x^2, 12x$
4. $36x^4, 18x^3$
5. u^2v, u^3v^2
6. $r^6s^4, -rs$
7. $9y^8z^4, -12y^5z^4$
8. $-15x^6y^3, 45xy^3$
9. $14x^2, 1, 7x^4$
10. $5y^4, 10x^2y^2, 1$
11. $28a^4b^2, 14a^3, 42a^2b^5$
12. $16x^2y, 12xy^2, 36x^2$

Section 6.1 Factoring Polynomials with Common Factors

Common Monomial Factors

Study Tip
To find the greatest common monomial factor of a polynomial, answer these two questions.
1. What is the greatest integer factor common to each coefficient of the polynomial?
2. What is the highest-power variable factor common to each term of the polynomial?

EXAMPLE 3 Greatest Common Monomial Factor

Factor out the greatest common monomial factor from $6x - 18$.

SOLUTION

The greatest common integer factor of $6x$ and 18 is 6. There is no common variable factor.

$6x - 18 = 6(x) - 6(3)$ Greatest common monomial factor is 6.
$ = 6(x - 3)$ Factor 6 out of each term.

EXAMPLE 4 Greatest Common Monomial Factor

Factor out the greatest common monomial factor from $10y^3 - 25y^2$.

SOLUTION

For the terms $10y^2$ and $25y^2$, 5 is the greatest common integer factor and y^2 is the highest-power common variable factor.

$10y^3 - 25y^2 = 5y^2(2y) - 5y^2(5)$ Greatest common factor is $5y^2$.
$ = 5y^2(2y - 5)$ Factor $5y^2$ out of each term.

EXAMPLE 5 Greatest Common Monomial Factor

Factor out the greatest common monomial factor from $45x^3 - 15x^2 - 15$.

SOLUTION

The greatest common integer factor of $45x^3$, $15x^2$, and 15 is 15. There is no common variable factor.

$45x^3 - 15x^2 - 15 = 15(3x^3) - 15(x^2) - 15(1)$
$ = 15(3x^3 - x^2 - 1)$

Exercises Within Reach®   Solutions in English & Spanish and tutorial videos at AlgebraWithinReach.com

Greatest Common Monomial Factor In Exercises 13–30, **factor** out the greatest common monomial factor from the polynomial. (*Note:* Some of the polynomials have no common monomial factor.)

13. $3x + 3$
14. $5y - 5$
15. $8t - 16$
16. $4u + 12$
17. $24y^2 - 18$
18. $8z^3 + 12$
19. $x^2 + x$
20. $s^3 - s$
21. $25u^2 - 14u$
22. $36t^4 + 24t^2$
23. $2x^4 + 6x^3$
24. $9z^6 + 27z^4$
25. $7s^2 + 9t^2$
26. $12x^2 - 5y^3$
27. $12x^2 + 16x - 8$
28. $9 - 3y + 15y^2$
29. $100 + 75z + 50z^2$
30. $42t^3 - 21t^2 + 7$

EXAMPLE 6 Greatest Common Monomial Factor

Factor out the greatest common monomial factor from $3xy^2 - 15x^2y + 12xy$.

SOLUTION

$3xy^2 - 15x^2y + 12xy = 3xy(y) - 3xy(5x) + 3xy(4)$ Greatest common factor is $3xy$.

$\qquad\qquad\qquad\qquad\quad = 3xy(y - 5x + 4)$ Factor $3xy$ out of each term.

EXAMPLE 7 Greatest Common Monomial Factor

Factor out the greatest common monomial factor from $35y^3 - 7y^2 - 14y$.

SOLUTION

$35y^3 - 7y^2 - 14y = 7y(5y^2) - 7y(y) - 7y(2)$ Greatest common factor is $7y$.

$\qquad\qquad\qquad\quad = 7y(5y^2 - y - 2)$ Factor $7y$ out of each term.

EXAMPLE 8 A Negative Common Monomial Factor

Factor the polynomial $-2x^2 + 8x - 12$ in two ways.

a. Factor out a common monomial factor of 2.
b. Factor out a common monomial factor of -2.

SOLUTION

a. To factor out the common monomial factor of 2, write the following.

$-2x^2 + 8x - 12 = 2(-x^2) + 2(4x) + 2(-6)$ Factor each term.

$\qquad\qquad\qquad\quad = 2(-x^2 + 4x - 6)$ Factored form

b. To factor -2 out the polynomial, write the following.

$-2x^2 + 8x - 12 = -2(x^2) + (-2)(-4x) + (-2)(6)$ Factor each term.

$\qquad\qquad\qquad\quad = -2(x^2 - 4x + 6)$ Factored form

Check this result by multiplying $(x^2 - 4x + 6)$ by -2. When you do, you will obtain the original polynomial.

Study Tip

The greatest common monomial factor of the terms of a polynomial is usually considered to have a positive coefficient. However, sometimes it is convenient to factor a negative number out of a polynomial, as shown in Example 8.

Greatest Common Monomial Factor In Exercises 31–38, factor out the greatest common monomial factor from the polynomial. (*Note:* Some of the polynomials have no common monomial factor.)

31. $9x^4 + 6x^3 + 18x^2$

32. $32a^5 - 2a^3 + 6a$

33. $5u^2 + 5u^2 + 5u$

34. $11y^3 - 22y^2 + 11y^2$

35. $10ab + 10a^2b$

36. $21x^2z^5 + 35x^6z$

37. $4xy - 8x^2y + 24x^4y^5$

38. $15m^4n^3 - 25m^7n + 30m^4n^8$

A Negative Common Monomial Factor In Exercises 39–44, factor out a negative real number from the polynomial and then write the polynomial factor in standard form.

39. $5 - 10x$

40. $3 - 6x$

41. $-15x^2 + 5x + 10$

42. $-4x^2 - 8x + 20$

43. $4 + 12x - 2x^2$

44. $8 - 4x - 12x^2$

Section 6.1 Factoring Polynomials with Common Factors

Factoring by Grouping

EXAMPLE 9 **Common Binomial Factors**

Factor each expression.

a. $5x^2(7x - 1) - 3(7x - 1)$

b. $2x(3x - 4) + (3x - 4)$

c. $3y^2(y - 3) + 4(3 - y)$

SOLUTION

a. Each of the terms of this expression has a binomial factor of $(7x - 1)$.

$$5x^2(7x - 1) - 3(7x - 1) = (7x - 1)(5x^2 - 3)$$

b. Each of the terms of this expression has a binomial factor of $(3x - 4)$.

$$2x(3x - 4) + (3x - 4) = (3x - 4)(2x + 1)$$

Be sure you see that when $(3x - 4)$ is factored out of itself, you are left with the factor 1. This follows from the fact that $(3x - 4)(1) = (3x - 4)$.

c. $3y^2(y - 3) + 4(3 - y) = 3y^2(y - 3) - 4(y - 3)$ Write $4(3 - y)$ as $-4(y - 3)$.

$\qquad\qquad\qquad\qquad\quad = (y - 3)(3y^2 - 4)$ Common factor is $(y - 3)$.

In Example 9, the polynomials were already grouped so that it was easy to determine the common binomial factors. In practice, you will have to do the grouping as well as the factoring. To see how this works, consider the expression

$$x^3 + 2x^2 + 3x + 6$$

and try to factor it. Note first that there is no common monomial factor to take out of all four terms. But suppose you *group* the first two terms together and the last two terms together.

$x^3 + 2x^2 + 3x + 6 = (x^3 + 2x^2) + (3x + 6)$ Group terms.

$\qquad\qquad\qquad\quad = x^2(x + 2) + 3(x + 2)$ Factor out common monomial factor in each group.

$\qquad\qquad\qquad\quad = (x + 2)(x^2 + 3)$ Factored form

When factoring by grouping, be sure to group terms that have a common monomial factor. For example, in the polynomial above, you should not group the first term x^3 with the fourth term 6.

Exercises Within Reach®   Solutions in English & Spanish and tutorial videos at AlgebraWithinReach.com

A Common Binomial Factor In Exercises 45–50, **factor** the polynomial by grouping.

45. $x(x - 3) + 5(x - 3)$

46. $x(x + 6) + 3(x + 6)$

47. $y(q - 5) - (q - 5)$

48. $a^2(b + 2) - (b + 2)$

49. $x^3(x - 4) + 2(4 - x)$

50. $x^3(x - 2) + 6(2 - x)$

Chapter 6 Factoring and Solving Equations

EXAMPLE 10 **Factoring by Grouping**

Factor $x^3 + 2x^2 + x + 2$.

SOLUTION

$$x^3 + 2x^2 + x + 2 = (x^3 + 2x^2) + (x + 2) \quad \text{Group terms.}$$

$$= x^2(x + 2) + (x + 2) \quad \text{Factor out common monomial factor in each group.}$$

$$= (x + 2)(x^2 + 1) \quad \text{Factored form}$$

Note that in Example 10 the polynomial is factored by grouping the first and second terms and the third and fourth terms. You could just as easily have grouped the first and third terms and the second and fourth terms, as follows.

$$x^3 + 2x^2 + x + 2 = (x^3 + x) + (2x^2 + 2) \quad \text{Group terms.}$$

$$= x(x^2 + 1) + 2(x^2 + 1) \quad \text{Factor out common monomial factor in each group.}$$

$$= (x^2 + 1)(x + 2) \quad \text{Factored form}$$

EXAMPLE 11 **Factoring by Grouping**

Factor $3x^2 - 12x - 5x + 20$.

SOLUTION

$$3x^2 - 12x - 5x + 20 = (3x^2 - 12x) + (-5x + 20) \quad \text{Group terms.}$$

$$= 3x(x - 4) - 5(x - 4) \quad \text{Factor out common monomial factor in each group.}$$

$$= (x - 4)(3x - 5) \quad \text{Factored form}$$

Note how a -5 is factored out so that the common binomial factor $x - 4$ appears.

Exercises Within Reach ®   Solutions in English & Spanish and tutorial videos at AlgebraWithinReach.com

Factoring by Grouping In Exercises 51–62, factor the polynomial by grouping.

51. $x^2 + 10x + x + 10$

52. $x^2 - 5x + x - 5$

53. $x^2 + 3x + 4x + 12$

54. $x^2 - 6x + 5x - 30$

55. $x^2 + 3x - 5x - 15$

56. $x^2 + 4x + 2x + 8$

57. $4x^2 - 14x + 14x - 49$

58. $4x^2 - 6x + 6x - 9$

59. $6x^2 + 3x - 2x - 1$

60. $4x^2 + 20x - x - 5$

61. $8x^2 + 32x + x + 4$

62. $8x^2 - 4x - 2x + 1$

Section 6.1 Factoring Polynomials with Common Factors

Application

EXAMPLE 12 Geometry: Area of a Rectangle

The area of a rectangle of width $(2x - 1)$ feet is $(2x^3 + 4x - x^2 - 2)$ square feet, as shown below. Factor this expression to determine the length of the rectangle.

Length

$2x - 1$ | Area $= 2x^3 + 4x - x^2 - 2$

Study Tip

Notice in Example 12 that the polynomial is not written in standard form. You could have rewritten the polynomial before factoring and still obtained the same result.

$2x^3 + 4x - x^2 - 2$
$= 2x^3 - x^2 + 4x - 2$
$= (2x^3 - x^2) + (4x - 2)$
$= x^2(2x - 1) + 2(2x - 1)$
$= (2x - 1)(x^2 + 2)$

SOLUTION

Verbal Model: Area $=$ Length $\times$ Width

Labels: Area $= 2x^3 + 4x - x^2 - 2$ (square feet)
Width $= 2x - 1$ (feet)

Expression: $2x^3 + 4x - x^2 - 2 = $ (Length)$(2x - 1)$

$2x^3 + 4x - x^2 - 2 = (2x^3 + 4x) + (-x^2 - 2)$ Group terms.

$= 2x(x^2 + 2) - (x^2 + 2)$ Factor out common monomial factor in each group.

$= (x^2 + 2)(2x - 1)$ Factored form

You can see that the length of the rectangle is $(x^2 + 2)$ feet.

Exercises Within Reach®   Solutions in English & Spanish and tutorial videos at AlgebraWithinReach.com

Geometry In Exercises 63–66, factor the polynomial to **find** an expression for the length of the rectangle.

63. Area $= 2x^2 + 2x$

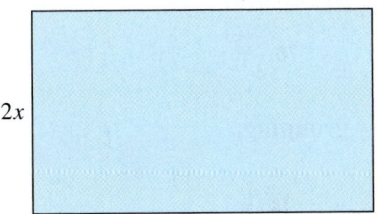

$2x$

64. Area $= 12x^3 - 9x^2$

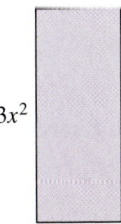

$3x^2$

65. Area $= 3x^2(4x - 1) - 2(4x - 1)$

$4x - 1$

66. Area $= x^2 + 2x + 10x + 20$

$x + 2$

260 Chapter 6 Factoring and Solving Equations

Concept Summary: Factoring Polynomials by Grouping

What
You can use **factoring by grouping** to **factor** some polynomials.

EXAMPLE

Factor $x^3 + x^2 + 2x + 2$.

How
To factor polynomials by grouping, do the following.

1. Group terms that have a common monomial factor.
2. **Factor out** the **greatest common monomial factor** from each group.
3. Use the Distributive Property to write the factored form.

EXAMPLE

$x^3 + x^2 + 2x + 2$
$= (x^3 + x^2) + (2x + 2)$
$= x^2(x + 1) + 2(x + 1)$
$= (x + 1)(x^2 + 2)$

Why
You can use factoring by grouping to find an expression for the length of a rectangle given polynomial expressions for its area and its width.

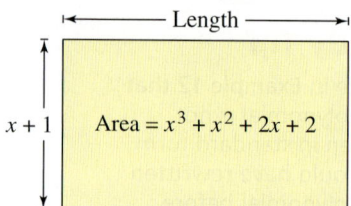

Exercises Within Reach®

Worked-out solutions to odd-numbered exercises at AlgebraWithinReach.com

Concept Summary Check

67. *Grouping Terms* Explain why the polynomial in the example above was rewritten as $(x^3 + x^2) + (2x + 2)$.

68. *Finding the GCF* Explain how to find the greatest common monomial factor of a binomial.

69. *The Common Binomial Factor* In the solution above, what is the common binomial factor of the two groups of terms?

70. *Checking the Result* Explain how you can check the factored form in the solution above.

Extra Practice

Factoring Out a Fraction In Exercises 71–76, determine the missing factor.

71. $\frac{1}{4}x + \frac{3}{4} = \frac{1}{4}(\quad)$

72. $\frac{5}{6}x - \frac{1}{6} = \frac{1}{6}(\quad)$

73. $2y - \frac{1}{5} = \frac{1}{5}(\quad)$

74. $3z + \frac{5}{4} = \frac{1}{4}(\quad)$

75. $\frac{7}{8}x + \frac{5}{16}y = \frac{1}{16}(\quad)$

76. $\frac{5}{12}u - \frac{5}{8}v = \frac{1}{24}(\quad)$

Factoring by Grouping In Exercises 77–80, **factor** the polynomial by grouping.

77. $(a + b)(a - b) + a(a + b)$

78. $(x + y)(x - y) - x(x - y)$

79. $ky^2 - 4ky + 2y - 8$

80. $ay^2 + 3ay + 3y + 9$

Geometry In Exercises 81 and 82, write an expression for the area of the shaded region and factor the expression if possible.

81.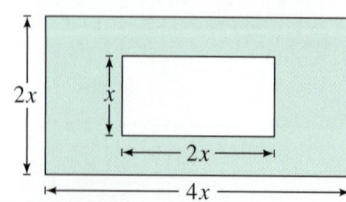

82.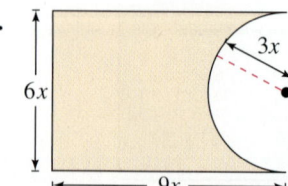

83. Geometry The surface area of a right circular cylinder is given by

$$2\pi r^2 + 2\pi rh$$

where r is the radius of the base of the cylinder and h is the height of the cylinder. Factor this expression.

84. Simple Interest A principal of P dollars earns simple interest at a rate r (in decimal form). The amount after t years is given by

$$P + Prt.$$

Factor this expression.

85. Chemical Reaction The rate of change in a chemical reaction is

$$kQx - kx^2$$

where Q is the original amount, x is the new amount, and k is a constant of proportionality. Factor this expression.

86. Unit Price The revenue R from selling x units of a product at a price of p dollars per unit is given by $R = xp$. For a pool table, the revenue is

$$R = 900x - 0.1x^2.$$

Factor the revenue model and determine an expression that represents the price p in terms of x.

Explaining Concepts

87. Reasoning Explain why $x^2(2x + 1)$ is in factored form.

88. Reasoning Explain why $3x$ is the greatest common monomial factor of $3x^3 + 3x^2 + 3x$.

89. Creating an Example Give an example of the use of the Distributive Property to factor out the greatest common monomial factor from a polynomial.

90. Creating an Example Give an example of the use of the Distributive Property to factor by grouping.

91. Structure Give an example of a trinomial with no common monomial factors.

92. Structure Give an example of a polynomial with four terms that can be factored by grouping.

Cumulative Review

In Exercises 93–96, determine whether each value of x is a solution of the equation.

93. $x + 2 - 3$ (a) $x = 5$
                             (b) $x = 1$

94. $x - 5 = 10$ (a) $x = 0$
                             (b) $x = 15$

95. $2x - 4 = 0$ (a) $x = 2$
                             (b) $x = -2$

96. $-7x - 8 = 6$ (a) $x = 1$
                             (b) $x = -2$

In Exercises 97–100, simplify the expression.

97. $\left(\dfrac{3y}{2x^3}\right)^2$ **98.** $\left(\dfrac{b^{-4}}{4a^{-5}}\right)^2$

99. $z^2 \cdot z^{-6}$

100. $(x^2 y)^5 (x^7 y^6)^2$

In Exercises 101–104, perform the division and simplify. (Assume that no denominator is zero.)

101. $\dfrac{3m^2 n}{m}$ **102.** $\dfrac{12x^3 y^5}{3x^2 y^5}$

103. $\dfrac{x^2 + 9x + 8}{x + 8}$

104. $\dfrac{2x^2 + 5x - 12}{2x - 3}$

6.2 Factoring Trinomials

▶ Factor trinomials of the form $x^2 + bx + c$.
▶ Factoring trinomials in two variables.
▶ Factor trinomials completely.

Factoring Trinomials of the Form $x^2 + bx + c$

Guidelines for Factoring $x^2 + bx + c$

To factor $x^2 + bx + c$, you need to find two numbers m and n whose product is c and whose sum is b.

$$x^2 + bx + c = (x + m)(x + n)$$

1. If c is positive, then m and n have like signs that match the sign of b.
2. If c is negative, then m and n have unlike signs.
3. If $|b|$ is small relative to $|c|$, first try those factors of c that are closest to each other in absolute value.

EXAMPLE 1 **Factoring a Trinomial**

Factor the trinomial $x^2 + 5x - 6$.

SOLUTION

You need to find two numbers whose product is -6 and whose sum is 5.

The product of -1 and 6 is -6.

$$x^2 + 5x - 6 = (x - 1)(x + 6)$$

The sum of -1 and 6 is 5.

EXAMPLE 2 **Factoring a Trinomial**

Factor the trinomial $x^2 - x - 6$.

SOLUTION

The product of -3 and 2 is -6.

$$x^2 - x - 6 = (x - 3)(x + 2)$$

The sum of -3 and 2 is -1.

Study Tip

Use a list to help you find the two numbers with the required product and sum. For Example 2:

Factors of -6	Sum
1, -6	-5
-1, 6	5
2, -3	-1
-2, 3	1

Because -1 is the required sum, the correct factorization is
$x^2 - x - 6 = (x - 3)(x + 2)$.

Exercises Within Reach®

Solutions in English & Spanish and tutorial videos at AlgebraWithinReach.com

Finding a Missing Factor In Exercises 1 and 2, determine the missing factor. Then check your answer by multiplying the factors.

1. $x^2 + 8x + 7 = (x + 7)()$
2. $x^2 + 2x - 3 = (x + 3)()$

Factoring a Trinomial In Exercises 3–6, factor the trinomial.

3. $x^2 + 6x + 8$
4. $x^2 + 12x + 35$
5. $x^2 + 2x - 15$
6. $x^2 + 4x - 21$

Section 6.2 Factoring Trinomials

Study Tip

Not all trinomials are factorable using integer factors. For instance, $x^2 - 2x - 6$ is not factorable using integer factors because there is no pair of factors of -6 whose sum is -2. Such nonfactorable trinomials are called **prime polynomials**.

EXAMPLE 3 Factoring Trinomials

Factor the trinomial.

a. $x^2 - 5x + 6$ b. $14 + 5x - x^2$

c. $x^2 - 8x - 48$ d. $x^2 + 7x - 30$

SOLUTION

a. You need to find two numbers whose product is 6 and whose sum is -5.

The product of -2 and -3 is 6.

$$x^2 - 5x + 6 = (x - 2)(x - 3)$$

The sum of -2 and -3 is -5.

b. First, factor out -1 and write the polynomial factor in standard form. Then find two numbers whose product is -14 and whose sum is -5.

The product of -7 and 2 is -14.

$$14 + 5x - x^2 = -(x^2 - 5x - 14) = -(x - 7)(x + 2).$$

The sum of -7 and 2 is -5.

c. You need to find two numbers whose product is -48 and whose sum is -8.

The product of -12 and 4 is -48.

$$x^2 - 8x - 48 = (x - 12)(x + 4)$$

The sum of -12 and 4 is -8.

d. You need to find two numbers whose product is -30 and whose sum is 7.

The product of -3 and 10 is -30.

$$x^2 + 7x - 30 = (x - 3)(x + 10)$$

The sum of -3 and 10 is 7.

Exercises Within Reach®

Solutions in English & Spanish and tutorial videos at AlgebraWithinReach.com

Finding the Possible Binomial Factors In Exercises 7–10, **find** all possible products of the form $(x + m)(x + n)$ where $m \cdot n$ is the specified product. (Assume that m and n are integers.)

7. $m \cdot n = 11$
8. $m \cdot n = 5$
9. $m \cdot n = 14$
10. $m \cdot n = -10$

Factoring a Trinomial In Exercises 11–18, **factor** the trinomial.

11. $x^2 - 9x - 22$
12. $x^2 - 9x - 10$
13. $x^2 - 9x + 14$
14. $x^2 + 10x + 24$
15. $2x + 15 - x^2$
16. $3x + 18 - x^2$
17. $x^2 + 3x - 70$
18. $x^2 - 13x + 40$

Application

EXAMPLE 4 Geometry: Area of a Rectangle

The area of a rectangle shown in the figure is $(x^2 + 30x + 200)$ square feet. What is the area of the shaded region?

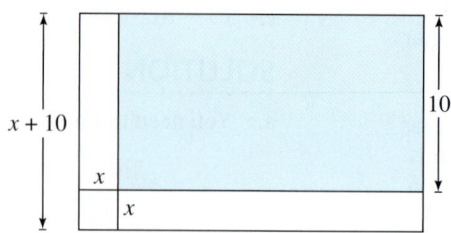

SOLUTION

Verbal Model: Area = Length × Width

Labels: Area = $x^2 + 30x + 200$ (square feet)
Width = $x + 10$ (feet)

Equation: $x^2 + 30x + 200 = $ (Length)$(x + 10)$
$= (x + 20)(x + 10)$

The length of the rectangle is $(x + 20)$ feet. So, the length of the shaded region is 20 feet, which means that the area of the shaded region is 200 square feet.

Exercises Within Reach®

Solutions in English & Spanish and tutorial videos at AlgebraWithinReach.com

19. *Geometry* The area of the rectangle shown in the figure is $x^2 + 10x + 21$. What is the area of the shaded region?

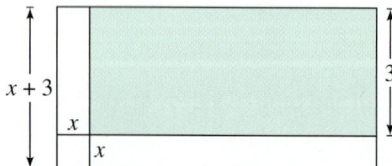

21. *Geometry* The area of the rectangle shown in the figure is $x^2 + 13x + 36$. What is the length of the shaded region?

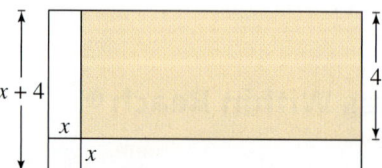

20. *Geometry* The area of the rectangle shown in the figure is $x^2 + 17x + 70$. What is the area of the shaded region?

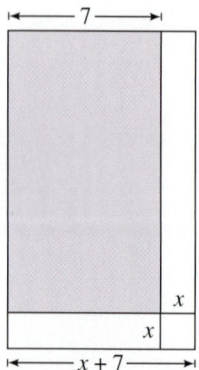

22. *Geometry* The area of the rectangle shown in the figure is $x^2 + 24x + 135$. What is the length of the shaded region?

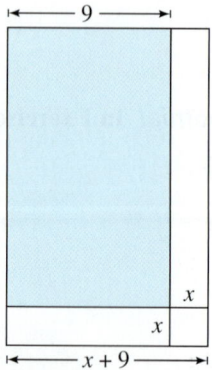

Factoring Trinomials in Two Variables

Study Tip

With *any* factoring problem, remember that you can check your result by multiplying. For instance, in Example 5, you can check the result by multiplying $(x - 4y)$ by $(x + 3y)$ to obtain $x^2 - xy - 12y^2$.

EXAMPLE 5 Factoring a Trinomial in Two Variables

Factor the trinomial $x^2 - xy - 12y^2$.

SOLUTION

You need to find two numbers whose product is -12 and whose sum is -1.

The product of -4 and 3 is -12.
$$x^2 - xy - 12y^2 = (x - 4y)(x + 3y)$$
The sum of -4 and 3 is -1.

EXAMPLE 6 Factoring a Trinomial in Two Variables

Factor the trinomial $x^2 + 11xy + 10y^2$.

SOLUTION

You need to find two numbers whose product is 10 and whose sum is 11.

The product of 1 and 10 is 10.
$$x^2 + 11xy + 10y^2 = (x + y)(x + 10y)$$
The sum of 1 and 10 is 11.

EXAMPLE 7 Factoring a Trinomial in Two Variables

Factor the trinomial $y^2 - 6xy + 8x^2$.

SOLUTION

You need to find two numbers whose product is 8 and whose sum is -6.

The product of -2 and -4 is 8.
$$y^2 - 6xy + 8x^2 = (y - 2x)(y - 4x)$$
The sum of -2 and -4 is -6.

Exercises Within Reach ®

Solutions in English & Spanish and tutorial videos at AlgebraWithinReach.com

Factoring a Trinomial In Exercises 23–30, **factor** the trinomial.

23. $x^2 - 7xz - 18z^2$
24. $u^2 - 4uv - 5v^2$
25. $x^2 - 5xy + 6y^2$
26. $x^2 + xy - 2y^2$
27. $x^2 + 8xy + 15y^2$
28. $x^2 + 15xy + 50y^2$
29. $a^2 + 2ab - 15b^2$
30. $y^2 + 4yz - 60z^2$

Factoring Completely

Some trinomials have a common monomial factor. In such cases, you should first factor out the common monomial factor. Then you can try to factor the resulting trinomial by the methods of this section. This "multiple-stage factoring process" is called **factoring completely**. The trinomial below is completely factored.

$$2x^2 - 4x - 6 = 2(x^2 - 2x - 3)$$ Factor out common monomial factor 2.
$$= 2(x - 3)(x + 1)$$ Factor trinomial.

EXAMPLE 8 Factoring Completely

Factor the trinomial $2x^2 - 12x + 10$ completely.

SOLUTION

$$2x^2 - 12x + 10 = 2(x^2 - 6x + 5)$$ Factor out common monomial factor 2.
$$= 2(x - 5)(x - 1)$$ Factor trinomial.

EXAMPLE 9 Factoring Completely

Factor the trinomial $3x^3 - 27x^2 + 54x$ completely.

SOLUTION

$$3x^3 - 27x^2 + 54x = 3x(x^2 - 9x + 18)$$ Factor out common monomial factor $3x$.
$$= 3x(x - 3)(x - 6)$$ Factor trinomial.

EXAMPLE 10 Factoring Completely

Factor the trinomial $4y^4 + 32y^3 + 28y^2$ completely.

SOLUTION

$$4y^4 + 32y^3 + 28y^2 = 4y^2(y^2 + 8y + 7)$$ Factor out common monomial factor $4y^2$.
$$= 4y^2(y + 1)(y + 7)$$ Factor trinomial.

Exercises Within Reach®   Solutions in English & Spanish and tutorial videos at AlgebraWithinReach.com

Factoring Completely In Exercises 31–42, **factor** the trinomial completely.

31. $4x^2 - 32x + 60$
32. $4y^2 - 8y - 12$
33. $9x^2 + 18x - 18$
34. $6x^2 - 24x - 6$
35. $x^3 - 13x^2 + 30x$
36. $x^3 + x^2 - 2x$
37. $3x^3 + 18x^2 + 24x$
38. $4x^3 + 8x^2 - 12x$
39. $x^4 - 5x^3 + 6x^2$
40. $x^4 + 3x^3 - 10x^2$
41. $2x^4 - 20x^3 + 42x^2$
42. $5x^4 - 10x^3 - 240x^2$

Application

EXAMPLE 11 Geometry: Volume of an Open Box

An open box is to be made from a four-foot-by-six-foot sheet of metal by cutting equal squares from the corners and turning up the sides. The volume of the box can be modeled by $V = 4x^3 - 20x^2 + 24x$, $0 < x < 2$.

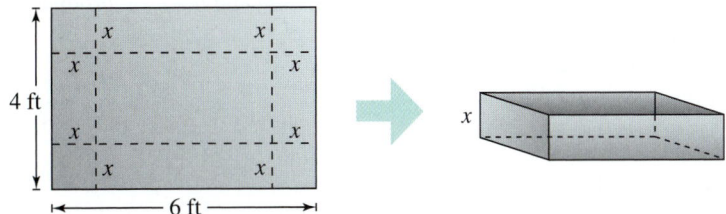

A	B
x	Volume
0.0	0.00
0.1	2.20
0.2	4.03
0.3	5.51
0.4	6.66
0.5	7.50
0.6	8.06
0.7	8.37
0.8	8.45
0.9	8.32
1.0	8.00
1.1	7.52
1.2	6.91
1.3	6.19
1.4	5.38
1.5	4.50
1.6	3.58
1.7	2.65
1.8	1.73
1.9	0.84
2.0	0.00

a. Factor the trinomial that models the volume of the box. Use the factored form to explain how the model was found.

b. Use a spreadsheet to approximate the size of the squares to be cut from the corners so that the box has the maximum volume.

SOLUTION

a. $4x^3 - 20x^2 + 24x = 4x(x^2 - 5x + 6)$ Factor out common monomial factor $4x$.

$\qquad = 4x(x - 3)(x - 2)$ Factored form

Because $4 = (-2)(-2)$, you can rewrite the factored form as

$4x(x - 3)(x - 2) = x[(-2)(x - 3)][(-2)(x - 2)]$

$\qquad = x(6 - 2x)(4 - 2x)$

$\qquad = (6 - 2x)(4 - 2x)(x)$.

The model was found by multiplying the length, width, and height of the box.

$$\text{Volume} = \underset{\text{Length}}{(6 - 2x)}\underset{\text{Width}}{(4 - 2x)}\underset{\text{Height}}{(x)}$$

b. From the spreadsheet at the left, you can see that the maximum volume of the box is about 8.45 cubic feet. This occurs when the value of x is about 0.8 foot.

Exercises Within Reach®

Solutions in English & Spanish and tutorial videos at AlgebraWithinReach.com

43. Geometry The box in Example 11 is to be made from a six-foot-by-eight-foot sheet of metal. The volume of the box is modeled by

$V = 4x^3 - 28x^2 + 48x$, $0 < x < 3$.

(a) Factor the trinomial that models the volume of the box. Use the factored form to explain how the model was found.

(b) Use a spreadsheet to approximate the size of the squares to be cut from the corners so that the box has the maximum volume.

44. Geometry The box in Example 11 is to be made from an eight-foot-by-ten-foot sheet of metal. The volume of the box is modeled by

$V = 4x^3 - 36x^2 + 80x$, $0 < x < 4$.

(a) Factor the trinomial that models the volume of the box. Use the factored form to explain how the model was found.

(b) Use a spreadsheet to approximate the size of the squares to be cut from the corners so that the box has the maximum volume.

Concept Summary: Factoring Trinomials

What
The most common technique for factoring trinomials is guess, check, and revise. But the guidelines for factoring $x^2 + bx + c$ can help make the process more efficient.

EXAMPLE

Factor the trinomial $x^2 + 2x - 35$.

How
To factor $x^2 + bx + c$, find two numbers m and n whose product is c and whose sum is b.

EXAMPLE

The product of 7 and -5 is -35.

$x^2 + 2x - 35 = (x + 7)(x - 5)$

The sum of 7 and -5 is 2.

Why
Knowing how to factor trinomials can help you factor trinomials completely. Also, the techniques for factoring trinomials will help you when you learn to solve quadratic equations.

Exercises Within Reach ®

Worked-out solutions to odd-numbered exercises at AlgebraWithinReach.com

Concept Summary Check

45. Unlike Signs For the trinomial $x^2 + 2x - 35$, how do you know that m and n will have unlike signs in the factored form $(x + m)(x + n)$?

46. Possible Factors When factoring $x^2 + 2x - 35$, why should you first try factors of 35 that are close to each other in absolute value?

47. Factoring a Trinomial To factor $x^2 + 2x - 35$ into the form $(x + m)(x + n)$, you need to find two numbers m and n with what product and what sum?

48. Determining Signs What are the signs of m and n for $x^2 + bx + c = (x + m)(x + n)$ when b is negative and c is positive?

Extra Practice

Factoring Completely In Exercises 49–56, factor the trinomial completely. (*Note:* some of the trinomials may be prime.)

49. $y^2 + 5y + 11$
50. $x^2 - x - 36$
51. $x^3 + 5x^2y + 6xy^2$
52. $x^2y - 6xy^2 + y^3$
53. $3z^2 + 5z + 6$
54. $7x^2 + 5x + 10$
55. $2x^3y + 4x^2y^2 - 6xy^3$
56. $2x^3y - 10x^2y^2 + 6xy^3$

Finding Coefficients In Exercises 57 and 58, find all integers b such that the trinomial can be factored.

57. $x^2 + bx + 18$
58. $x^2 + bx + 10$

Finding Constant Terms In Exercises 59 and 60, find two integers c such that the trinomial can be factored. (There are many correct answers.)

59. $x^2 + 3x + c$
60. $x^2 + 5x + c$

Geometric Model of Factoring In Exercises 61 and 62, factor the trinomial and draw a geometric model of the result. [The sample shows a geometric model for factoring $x^2 + 3x + 2 = (x + 1)(x + 2)$.]

61. $x^2 + 4x + 3$
62. $x^2 + 5x + 6$

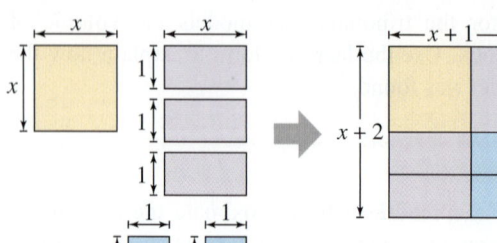

63. **Geometry** An open box is to be made from a sheet of metal that is y feet long and z feet wide by cutting equal squares from the corners and turning up the sides. The volume of the box can be modeled by

$V = 4x^3 - 24x^2 + 32x$, $0 < x < 2$.

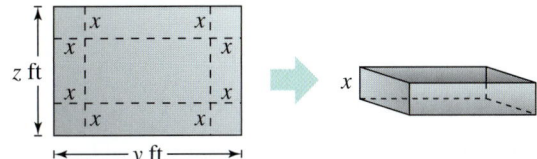

(a) Use x, y, and z to write an expression for the volume V of the box in the form

$V =$ Length $\times$ Width $\times$ Height. (Do not simplify.)

(b) Factor the trinomial that models the volume of the box.

(c) Use the expressions for the volume of the box in parts (a) and (b) to find the dimensions y and z.

64. **Geometry** An open box is to be made from a sheet of metal that is y feet long and z feet wide by cutting equal squares from the corners and turning up the sides. The volume of the box can be modeled by

$V = 4x^3 - 44x^2 + 120x$, $0 < x < 5$.

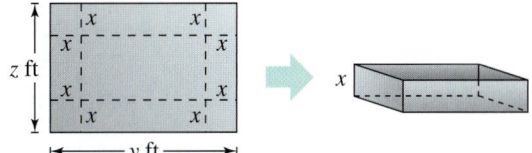

(a) Use x, y, and z to write an expression for the volume V of the box in the form

$V =$ Length $\times$ Width $\times$ Height. (Do not simplify.)

(b) Factor the trinomial that models the volume of the box.

(c) Use the expressions for the volume of the box in parts (a) and (b) to find the dimensions y and z.

Explaining Concepts

65. **Precision** State which of the following are factorizations of $2x^2 + 6x - 20$. For each correct factorization, state whether or not it is completely factored.

   (a) $(2x - 4)(x + 5)$
   (b) $(2x - 4)(2x + 10)$
   (c) $(x - 2)(x + 5)$
   (d) $2(x - 2)(x + 5)$

66. **Vocabulary** What is a prime trinomial?

67. **Structure** In factoring $x^2 - 4x + 3$, why is it unnecessary to test $(x - 1)(x + 3)$ and $(x + 1)(x - 3)$?

68. **Writing** In factoring the trinomial $x^2 + bx + c$, is the process easier if c is a prime number such as 5 or if c is a composite number such as 120? Explain.

Cumulative Review

In Exercises 69–76, solve the equation and check your solution.

69. $5x - 9 = 26$
70. $3x - 5 = 16$
71. $5 - 3x = 6$
72. $7 - 2x = 9$
73. $7x - 12 = 3x$
74. $10x + 24 = 2x$
75. $5x - 16 = 7x - 9$
76. $3x - 8 = 9x + 4$

In Exercises 77–84, find the greatest common factor of the expressions.

77. $6x^3$, $52x^6$
78. $35t^2$, $7t^8$
79. a^5b^4, a^3b^7
80. xy^3, x^2y^2
81. $18r^3s^3$, $-54r^5s^3$
82. $12xy^3$, $28x^3y^4$
83. $16u^2v^5$, $8u^4v^3$, $2u^3v^7$
84. $21xy^4$, $42x^2y^2$, $9x^4y$

6.3 More About Factoring Trinomials

- Factor trinomials of the form $ax^2 + bx + c$.
- Factor trinomials completely.
- Factor trinomials by grouping.

Factoring Trinomials of the Form $ax^2 + bx + c$

To see how to factor a trinomial whose leading coefficient is not 1, consider the following.

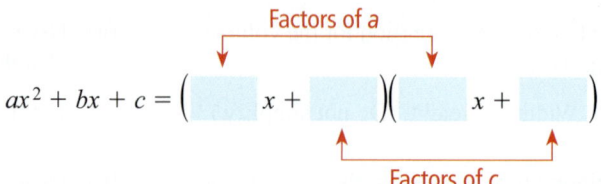

$$ax^2 + bx + c = (\boxed{}\,x + \boxed{})(\boxed{}\,x + \boxed{})$$

Factors of a (top) • Factors of c (bottom)

The goal is to find a combination of factors of a and c such that the outer and inner products add up to the middle term bx.

EXAMPLE 1 Factoring a Trinomial

Factor the trinomial $4x^2 - 4x - 3$.

SOLUTION

First, observe that $4x^2 - 4x - 3$ has no common monomial factor. For this trinomial, $a = 4$ and $c = -3$. You need to find a combination of the factors of 4 and -3 such that the outer and inner products add up to $-4x$. The possible combinations are as follows.

Factors	$O + I$	
$(x + 1)(4x - 3)$	$-3x + 4x = x$	x does not equal $-4x$.
$(x - 1)(4x + 3)$	$3x - 4x = -x$	$-x$ does not equal $-4x$.
$(x + 3)(4x - 1)$	$-x + 12x = 11x$	$11x$ does not equal $-4x$.
$(x - 3)(4x + 1)$	$x - 12x = -11x$	$-11x$ does not equal $-4x$.
$(2x + 1)(2x - 3)$	$-6x + 2x = -4x$	$-4x$ equals $-4x$.
$(2x - 1)(2x + 3)$	$6x - 2x = 4x$	$4x$ does not equal $-4x$.

In the first row: Inner product $= 4x$, Outer product $= -3x$.

So, the correct factorization is $4x^2 - 4x - 3 = (2x + 1)(2x - 3)$.

Exercises Within Reach® Solutions in English & Spanish and tutorial videos at AlgebraWithinReach.com

Factoring a Trinomial In Exercises 1–4, determine the missing factor.

1. $2x^2 + 7x - 4 = (2x - 1)(\boxed{})$
2. $3x^2 + x - 4 = (3x + 4)(\boxed{})$
3. $3t^2 + 4t - 15 = (3t - 5)(\boxed{})$
4. $5t^2 + t - 18 = (5t - 9)(\boxed{})$

Factoring a Trinomial In Exercises 5–8, factor the trinomial.

5. $2x^2 + 5x + 3$
6. $3x^2 + 7x + 2$
7. $4y^2 + 5y + 1$
8. $3x^2 + 5x - 2$

Section 6.3   More About Factoring Trinomials   271

> **Guidelines for Factoring $ax^2 + bx + c$ ($a > 0$)**
>
> 1. When a trinomial has a common monomial factor, you should factor out the common factor before trying to find the binomial factors.
> 2. Because the resulting trinomial has no common monomial factors, you do not have to test any binomial factors that have a common monomial factor.
> 3. Switch the signs of the factors of c when the middle term (O + I) is correct except in sign.

EXAMPLE 2   Factoring a Trinomial

Factor the trinomial $2x^2 + x - 15$.

SOLUTION

First, observe that $2x^2 + x - 15$ has no common monomial factor. For this trinomial, $a = 2$, which factors as $(1)(2)$, and $c = -15$, which factors as $(1)(-15)$, $(-1)(15)$, $(3)(-5)$, and $(-3)(5)$.

$(2x + 1)(x - 15) = 2x^2 - 29x - 15$

$(2x + 15)(x - 1) = 2x^2 + 13x - 15$

$(2x + 3)(x - 5) = 2x^2 - 7x - 15$

$(2x + 5)(x - 3) = 2x^2 - x - 15$ Middle term has opposite sign.

$(2x - 5)(x + 3) = 2x^2 + x - 15$ Correct factorization

So, the correct factorization is $2x^2 + x - 15 = (2x - 5)(x + 3)$.

EXAMPLE 3   Factoring a Trinomial

Factor the trinomial $6x^2 + 5x - 4$.

SOLUTION

$(x + 4)(6x - 1) = 6x^2 + 23x - 4$ 23x does not equal 5x.

$(2x + 1)(3x - 4) = 6x^2 - 5x - 4$ Opposite sign

$(2x - 1)(3x + 4) = 6x^2 + 5x - 4$ Correct factorization

So, the correct factorization is $6x^2 + 5x - 4 = (2x - 1)(3x + 4)$.

Exercises Within Reach®

Solutions in English & Spanish and tutorial videos at AlgebraWithinReach.com

Factoring a Trinomial In Exercises 9–16, **factor** the trinomial. (*Note:* Some of the trinomials may be prime.)

9. $6y^2 - 7y + 1$
10. $7a^2 - 9a + 2$
11. $5x^2 - 2x + 1$
12. $4z^2 - 8z + 1$
13. $4x^2 + 13x - 12$
14. $16y^2 + 24y - 27$
15. $9x^2 - 18x + 8$
16. $25a^2 - 40a + 12$

Factoring Completely

EXAMPLE 4 **Factoring Completely**

Factor $4x^3 - 30x^2 + 14x$ completely.

SOLUTION

Begin by factoring out the common monomial factor.

$$4x^3 - 30x^2 + 14x = 2x(2x^2 - 15x + 7)$$

Now, for the new trinomial $2x^2 - 15x + 7$, $a = 2$, and $c = 7$. The possible factorizations of the trinomial are as follows.

$$(2x - 7)(x - 1) = 2x^2 - 9x + 7$$

$$(2x - 1)(x - 7) = 2x^2 - 15x + 7 \quad \Longleftarrow \quad \text{Correct factorization}$$

So, the complete factorization of the original trinomial is

$$4x^3 - 30x^2 + 14x = 2x(2x^2 - 15x + 7)$$
$$= 2x(2x - 1)(x - 7).$$

EXAMPLE 5 **A Negative Leading Coefficient**

Factor the trinomial $-5x^2 + 7x + 6$.

SOLUTION

This trinomial has a negative leading coefficient, so you should begin by factoring -1 out of the trinomial.

$$-5x^2 + 7x + 6 = (-1)(5x^2 - 7x - 6)$$

Now, for the new trinomial $5x^2 - 7x - 6$, $a = 5$, and $c = -6$. After testing the possible factorizations, you can conclude that

$$(x - 2)(5x + 3) = 5x^2 - 7x - 6. \quad \Longleftarrow \quad \text{Correct factorization}$$

So, the correct factorization is

$$-5x^2 + 7x + 6 = (-1)(x - 2)(5x + 3)$$
$$= (-x + 2)(5x + 3). \quad \text{Distributive Property}$$

Another correct factorization is $(x - 2)(-5x - 3)$.

Exercises Within Reach ®

Solutions in English & Spanish and tutorial videos at AlgebraWithinReach.com

Factoring Completely In Exercises 17−22, factor the trinomial completely.

17. $2v^2 + 8v - 42$
18. $4z^2 - 12z - 40$
19. $9z^2 - 24z + 15$
20. $6x^2 + 8x - 8$
21. $16s^3 - 28s^2 + 6s$
22. $18v^3 + 3v^2 - 6v$

A Negative Leading Coefficient In Exercises 23−26, factor the trinomial.

23. $-2x^2 + 7x + 9$
24. $-5x^2 + x + 4$
25. $-6x^2 + 7x + 10$
26. $3 + 2x - 8x^2$

Section 6.3 More About Factoring Trinomials 273

Application   **EXAMPLE 6**   **Geometry: The Dimensions of a Sandbox**

The sandbox shown in the figure below has a height of x feet and a width of $(x + 2)$ feet. The volume of the sandbox is $(2x^3 + 7x^2 + 6x)$ cubic feet.

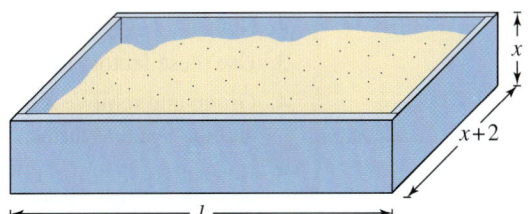

a. Find the length of the sandbox.

b. How many cubic yards of sand are needed to fill the sandbox when $x = 2$?

SOLUTION

a. **Verbal Model:** Volume = Length × Width × Height

   Labels: Volume = $2x^3 + 7x^2 + 6x$ (cubic feet)
   Width = $x + 2$ (feet)
   Height = x (feet)

   Equation: $2x^3 + 7x^2 + 6x = x(2x^2 + 7x + 6)$
   $= x(x + 2)(2x + 3)$

   So, the length of the sandbox is $(2x + 3)$ feet.

b. When $x = 2$, the volume of the sandbox is

   Volume $= 2(2^3) + 7(2^2) + 6(2)$
   $= 2(8) + 7(4) + 12$
   $= 56$ ft^3.

   One cubic yard contains 27 cubic feet. So, you need 56/27 or about 2 cubic yards of sand to fill the sandbox.

Exercises Within Reach®   Solutions in English & Spanish and tutorial videos at AlgebraWithinReach.com

27. **Geometry** The shower stall has a width of x feet and a depth of $(x + 2)$ feet. The volume of the shower stall is $(2x^3 + 3x^2 - 2x)$ cubic feet.

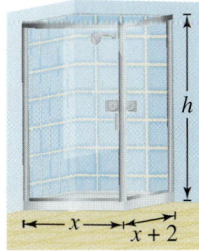

(a) Find the height of the shower stall.

(b) What is the volume of the shower stall when $x = 4$?

28. **Geometry** The box has a width of x centimeters and a height of $(5x - 1)$ centimeters. The volume of the box is $(15x^3 + 7x^2 - 2x)$ cubic centimeters.

(a) Find the length of the box.

(b) What is the volume of the box when $x = 5$?

Chapter 6 Factoring and Solving Equations

Factoring Trinomials by Grouping

> **Guidelines for Factoring $ax^2 + bx + c$ by Grouping**
>
> 1. If necessary, write the trinomial in standard form.
> 2. Choose factors of the product ac that add up to b.
> 3. Use these factors to rewrite the middle term as a sum or difference.
> 4. Group and remove any common monomial factors from the first two terms and the last two terms.
> 5. If possible, factor out the common binomial factor.

EXAMPLE 7 Factoring a Trinomial by Grouping

Use factoring by grouping to factor the trinomial $2x^2 + 5x - 3$.

SOLUTION

$2x^2 + 5x - 3 = 2x^2 + 6x - x - 3$ Rewrite middle term.

$\qquad = (2x^2 + 6x) + (-x - 3)$ Group terms.

$\qquad = 2x(x + 3) - (x + 3)$ Factor out common monomial factor in each group.

$\qquad = (x + 3)(2x - 1)$ Factor out common binomial factor.

So, the trinomial factors as $2x^2 + 5x - 3 = (x + 3)(2x - 1)$.

EXAMPLE 8 Factoring a Trinomial by Grouping

Use factoring by grouping to factor the trinomial $6x^2 - 11x - 10$.

SOLUTION

$6x^2 - 11x - 10 = 6x^2 - 15x + 4x - 10$ Rewrite middle term.

$\qquad = (6x^2 - 15x) + (4x - 10)$ Group terms.

$\qquad = 3x(2x - 5) + 2(2x - 5)$ Factor out common monomial factor in each group.

$\qquad = (2x - 5)(3x + 2)$ Factor out common binomial factor.

So, the trinomial factors as $6x^2 - 11x - 10 = (2x - 5)(3x + 2)$.

Exercises Within Reach®   Solutions in English & Spanish and tutorial videos at AlgebraWithinReach.com

Factoring a Trinomial by Grouping In Exercises 29–36, **factor** the trinomial by grouping.

29. $3x^2 + 4x + 1$

30. $2x^2 + 5x + 2$

31. $7x^2 + 20x - 3$

32. $5x^2 - 14x - 3$

33. $6x^2 + 5x - 4$

34. $12y^2 + 11y + 2$

35. $15x^2 - 11x + 2$

36. $12x^2 - 13x + 1$

Application

EXAMPLE 9 Geometry: The Dimensions of a Swimming Pool

The swimming pool shown in the figure has a depth of d feet and a length of $(5d + 2)$ feet. The volume of the swimming pool is $(15d^3 - 14d^2 - 8d)$ cubic feet.

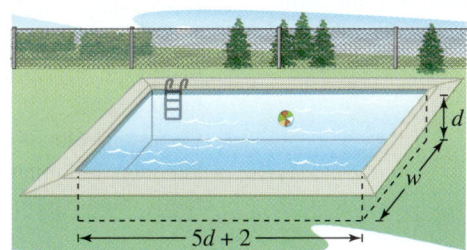

a. Find the width of the swimming pool.

b. How many cubic feet of water are needed to fill the swimming pool when $d = 6$?

SOLUTION

a. **Verbal Model:** $\boxed{\text{Volume}} = \boxed{\text{Length}} \times \boxed{\text{Width}} \times \boxed{\text{Depth}}$

Labels:
Volume $= 15d^3 - 14d^2 - 8d$ (cubic feet)
Width $= 5d + 2$ (feet)
Depth $= d$ (feet)

Equation:
$$15d^3 - 14d^2 - 8d = d(15d^2 - 14d - 8)$$
$$= d(15d^2 + 6d - 20d - 8)$$
$$= d[3d(5d + 2) - 4(5d + 2)]$$
$$= d(5d + 2)(3d - 4)$$

So, the width of the swimming pool is $(3d - 4)$ feet.

b. When $d = 6$, the dimensions of the swimming pool are 6 feet by 32 feet by 14 feet. So, the volume is $6(32)(14) = 2688$ cubic feet. There are about 7.5 gallons in 1 cubic foot. So, the swimming pool holds about 20,160 gallons of water.

Exercises Within Reach®

Solutions in English & Spanish and tutorial videos at AlgebraWithinReach.com

37. Geometry The fire box of a wood stove is x inches deep and $(2x - 7)$ inches wide. The volume of the fire box is $(2x^3 - 3x^2 - 14x)$ cubic inches.

(a) Find the height of the fire box.

(b) What is the volume of the fire box when $x = 15$?

38. Geometry The block of ice has a width of x inches and a length of $(2x + 5)$ inches. The volume of the block is $(6x^3 + 17x^2 + 5x)$ cubic inches.

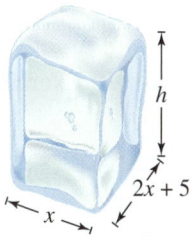

(a) Find the height of the block of ice.

(b) What is the volume of the block of ice when $x = 10$?

Chapter 6 Factoring and Solving Equations

> **Concept Summary:** Factoring Trinomials of the Form $ax^2 + bx + c$
>
> ### What
> Using some simple guidelines can help you factor trinomials of the form $ax^2 + bx + c$, where $a > 0$ and $a \neq 1$.
>
> **EXAMPLE**
> Factor the trinomial
> $3x^2 + 13x - 16$.
>
> ### How
> Guidelines for factoring $ax^2 + bx + c$
> 1. If necessary, write the trinomial in standard form.
> 2. Factor out any common monomial factors.
> 3. Choose factors of a and c such that the outer and inner products add up to the middle term bx.
>
> $$ax^2 + bx + c = (\quad x + \quad)(\quad x + \quad)$$
> Factors of a / Factors of c
>
> **EXAMPLE**
> $(3x - 1)(x + 16) = 3x^2 + 47x - 16$
> $(3x - 16)(x + 1) = 3x^2 - 13x - 16$
> $(3x + 16)(x - 1) = 3x^2 + 13x - 16$
> So, $3x^2 + 13x - 16 = (3x + 16)(x - 1)$.
>
> ### Why
> Trinomials are used in many geometric applications. Knowing how to factor trinomials will help you solve these applications.

Exercises Within Reach® Worked-out solutions to odd-numbered exercises at AlgebraWithinReach.com

Concept Summary Check

39. Common Monomial Factor Does the trinomial $3x^2 + 13x - 16$ have a common monomial factor?

40. Testing Factors How do you know that $(3x - 1)(x + 16)$ is not the correct factorization of $3x^2 + 13x - 16$?

41. Reasoning Use the first step in the solution above to explain why the factorization $(3x + 1)(x - 16)$ was not tested.

42. Reasoning Use the second step in the solution above to explain why the factorization $(3x + 16)(x - 1)$ was tested.

Extra Practice

Factoring Completely In Exercises 43–48, factor the trinomial completely.

43. $-15x^4 - 2x^3 + 8x^2$

44. $15y^2 - 7y^3 - 2y^4$

45. $6x^3 + 24x^2 - 192x$

46. $35x + 28x^2 - 7x^3$

47. $18u^4 + 18u^3 - 27u^2$

48. $12x^5 - 16x^4 + 8x^3$

Finding Coefficients In Exercises 49–54, find all integers b such that the trinomial can be factored.

49. $3x^2 + bx + 10$

50. $4x^2 + bx + 3$

51. $2x^2 + bx - 6$

52. $5x^2 + bx - 6$

53. $6x^2 + bx + 20$

54. $8x^2 + bx - 18$

Finding Constant Terms In Exercises 55–60, find two integers c such that the trinomial can be factored. (There are many correct answers.)

55. $4x^2 + 3x + c$

56. $2x^2 + 5x + c$

57. $3x^2 - 10x + c$

58. $8x^2 - 3x + c$

59. $6x^2 - 5x + c$

60. $4x^2 - 9x + c$

61. Geometry The area of the rectangle shown in the figure is $2x^2 + 9x + 10$. What is the area of the shaded region?

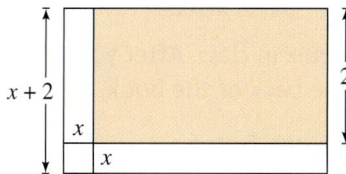

62. Geometry The area of the rectangle shown in the figure is $3x^2 + 10x + 3$. What is the area of the shaded region?

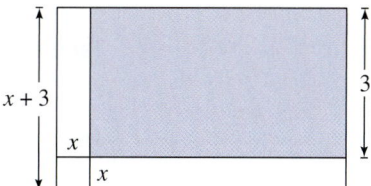

Explaining Concepts

63. Writing What is the first step in factoring $ax^2 + bx + c$ by grouping?

64. Error Analysis Describe and correct the error.
$$9x^2 - 9x - 54 = (3x + 6)(3x - 9)$$
$$= 3(x + 2)(x - 3)$$

65. Reasoning Without multiplying the factors, explain why $(2x + 3)(x + 5)$ is not a factorization of $2x^2 + 7x - 15$.

66. Structure Give an example of a prime trinomial of the form $ax^2 + bx + c$.

67. Structure Give an example of a trinomial of the form $ax^3 + bx^2 + cx$ that has a common monomial factor of $2x$.

68. Reasoning Can a trinomial with a leading coefficient not equal to 1 have two identical factors? If so, give an example.

69. Reasoning How many possible factorizations are there to consider for a trinomial of the form $ax^2 + bx + c$, when a and c are prime? Explain your reasoning.

70. Explanation Many people think the technique of factoring a trinomial by grouping is more efficient than the *guess, check, and revise* strategy, especially when the coefficients a and c have many factors. Try factoring $6x^2 - 13x + 6$, $2x^2 + 5x - 12$, and $3x^2 + 11x - 4$ using both methods. Which method do you prefer? Explain the advantages and disadvantages of each method.

Cumulative Review

In Exercises 71–74, write the prime factorization of the number.

71. 500

72. 315

73. 792

74. 2275

In Exercises 75–78, perform the multiplication and simplify.

75. $(2x - 5)(x + 7)$

76. $(3x - 2)^2$

77. $(7y + 2)(7y - 2)$

78. $(3y + 8)(9y + 3)$

In Exercises 79–82, factor the trinomial.

79. $x^2 - 4x - 45$

80. $y^2 + 2y - 15$

81. $z^2 + 22z + 40$

82. $x^2 - 12x + 20$

Mid-Chapter Quiz: Sections 6.1–6.3

Solutions in English & Spanish and tutorial videos at AlgebraWithinReach.com

Take this quiz as you would take a quiz in class. After you are done, check your work against the answers in the back of the book.

In Exercises 1–4, determine the missing factor.

1. $\frac{2}{3}x - \frac{1}{3} = \frac{1}{3}(\underline{})$
2. $x^2y - xy^2 = xy(\underline{})$
3. $y^2 + y - 42 = (y + 7)(\underline{})$
4. $3y^2 - y - 30 = (3y - 10)(\underline{})$

In Exercises 5–16, factor the polynomial completely.

5. $9x^2 + 21$
6. $5a^3 - 25a^2$
7. $x(x + 7) - 6(x + 7)$
8. $t^3 - 3t^2 + t - 3$
9. $y^2 + 11y + 30$
10. $u^2 + u - 56$
11. $x^3 - x^2 - 30x$
12. $2x^2y + 8xy - 64y$
13. $2y^2 - 3y - 27$
14. $6 - 13z - 5z^2$
15. $12x^2 - 5x - 2$
16. $10s^4 - 14s^3 + 2s^2$

17. Find all integers b such that the trinomial
   $x^2 + bx + 12$
   can be factored. Describe the method you used.

18. Find two integers c such that the trinomial
   $x^2 - 10x + c$
   can be factored. Describe the method you used. (There are many correct answers.)

19. Find all possible products of the form
   $(3x + m)(x + n)$
   such that $m \cdot n = 6$.

Application

20. The area of the rectangle shown in the figure is $3x^2 + 38x + 80$. What is the area of the shaded region?

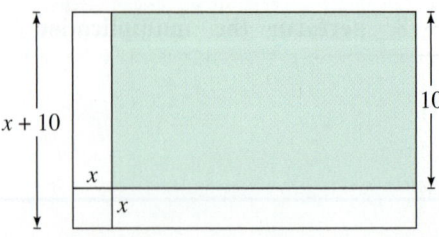

Study Skills in Action

Making the Most of Class Time

Have you ever slumped at your desk while in class and thought, "I'll just get the notes down and study later—I'm too tired"? Learning math in college is a team effort, between instructor and student. The more you understand in class, the more you will be able to learn while studying outside of class.

Approach math class with the intensity of a navy pilot during a mission briefing. The pilot has strategic plans to learn during the briefing. He or she listens intensely, takes notes, and memorizes important information. The goal is for the pilot to leave the briefing with a clear picture of the mission. It is the same with a student in a math class.

These students are sitting in the front row, where they are more likely to pay attention.

Smart Study Strategy

Take Control of Your Class Time

1. ▶ **Sit where you can easily see and hear the instructor, and the instructor can see you.** The instructor may be able to tell when you are confused just by the look on your face, and may adjust the lesson accordingly. In addition, sitting in this strategic place will keep your mind from wandering.

2. ▶ **Pay attention to what the instructor says about the math, not just what is written on the board.** Write problems on the left side of your notes and what the instructor says about the problems on the right side.

3. ▶ **If the instructor is moving through the material too fast, ask a question.** Questions help to slow the pace for a few minutes and also to clarify what is confusing to you.

4. ▶ **Try to memorize new information while learning it.** Repeat in your head what you are writing in your notes. That way you are reviewing the information twice.

5. ▶ **Ask for clarification.** If you don't understand something at all and don't even know how to phrase a question, just ask for clarification. You might say something like, "Could you please explain the step in this problem one more time?"

6. ▶ **Think as intensely as if you were going to take a quiz on the material at the end of class.** This kind of mindset will help you process new information.

7. ▶ **If the instructor asks for someone to go up to the board, volunteer.** The student at the board often receives additional attention and instruction to complete the problem.

8. ▶ **At the end of class, identify concepts or problems on which you still need clarification.** Make sure you see the instructor or a tutor as soon as possible.

6.4 Factoring Polynomials with Special Forms

▶ Factor the difference of two squares.
▶ Factor a polynomial completely.
▶ Identify and factor perfect square trinomials.
▶ Factor the sum or difference of two cubes.

Difference of Two Squares

Difference of Two Squares

Let *a* and *b* be real numbers, variables, or algebraic expressions.

$$a^2 - b^2 = (a + b)(a - b)$$

↑   ↑   ↑
Difference   Opposite signs

To recognize perfect square terms, look for coefficients that are squares of integers and for variables raised to *even* powers. Here are some examples.

Original Polynomials	*Difference of Squares*	*Factored Form*
$x^2 - 1$	$(x)^2 - (1)^2$	$(x + 1)(x - 1)$
$4x^2 - 9$	$(2x)^2 - (3)^2$	$(2x + 3)(2x - 3)$

EXAMPLE 1   Factoring the Difference of Two Squares

Study Tip

When factoring a polynomial, remember that you can check your result by multiplying the factors. For instance, you can check the factorization in Example 1(a) as follows.

$(x + 6)(x - 6)$
$= x^2 - 6x + 6x - 36$
$= x^2 - 36$

a. $x^2 - 36 = x^2 - 6^2$ Write as difference of two squares.

           $= (x + 6)(x - 6)$ Factored form

b. $x^2 - \frac{4}{25} = x^2 - \left(\frac{2}{5}\right)^2$ Write as difference of two squares.

           $= \left(x + \frac{2}{5}\right)\left(x - \frac{2}{5}\right)$ Factored form

c. $81x^2 - 49 = (9x)^2 - 7^2$ Write as difference of two squares.

           $= (9x + 7)(9x - 7)$ Factored form

d. $(x + 1)^2 - 4 = (x + 1)^2 - 2^2$ Write as difference of two squares.

           $= [(x + 1) + 2][(x + 1) - 2]$ Factored form

           $= (x + 3)(x - 1)$ Simplify.

Exercises Within Reach®

Solutions in English & Spanish and tutorial videos at **AlgebraWithinReach.com**

Factoring the Difference to Two Squares   In Exercises 1–8, **factor** the difference of two squares.

1. $x^2 - 9$
2. $y^2 - 49$
3. $u^2 - \frac{1}{4}$
4. $v^2 - \frac{4}{9}$
5. $16y^2 - 9$
6. $36z^2 - 121$
7. $(x - 1)^2 - 4$
8. $(t + 2)^2 - 9$

Application

EXAMPLE 2 Removing a Common Monomial Factor First

A hammer is dropped from the roof of a building. The height of the hammer is given by the expression $-16t^2 + 64$, where t is the time in seconds.

A	B
t	Height
0	64
0.1	63.84
0.2	63.36
0.3	62.56
0.4	61.44
0.5	60
0.6	58.24
0.7	56.16
0.8	53.76
0.9	51.04
1	48
1.1	44.64
1.2	40.96
1.3	36.96
1.4	32.64

a. Factor the expression.

b. How many seconds does it take the hammer to fall to a height of 41 feet?

SOLUTION

a. $-16t^2 + 64 = -16(t^2 - 4)$ Factor out common monomial factor.

$\qquad = -16(t^2 - 2^2)$ Write a difference of two squares.

$\qquad = -16(t + 2)(t - 2)$ Factored form

b. Use a spreadsheet to find the heights of the hammer at 0.1-second intervals of the time t. From the spreadsheet at the left, you can see that the hammer falls to a height of 41 feet in about 1.2 seconds.

Exercises Within Reach®

Solutions in English & Spanish and tutorial videos at AlgebraWithinReach.com

9. *Free-Falling Object* The height of an object that is dropped from the top of the U.S. Steel Tower in Pittsburgh is given by the expression $-16t^2 + 841$, where t is the time in seconds.

(a) Factor this expression.

(b) Use a spreadsheet to determine how many seconds it takes the object to fall to a height of 805 feet. (Use 0.5-second intervals for t.

10. *Geometry* An *annulus* is the region between two concentric circles. The area of the annulus shown in the figure is $\pi R^2 - \pi r^2$.

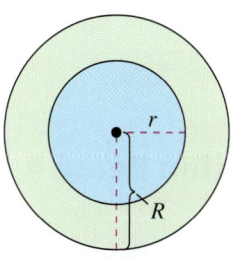

(a) Factor this expression.

(b) Use a spreadsheet to determine the value of R for which the annulus has an area of about 8 square feet when $r = 3$ feet. (Use 3.14 for π and increments of 0.1 for R.)

Factoring Completely

EXAMPLE 3 Removing a Common Monomial Factor First

Factor the polynomial $20x^3 - 5x$.

SOLUTION

$20x^3 - 5x = 5x(4x^2 - 1)$ Factor out common monomial factor $5x$.

$ = 5x[(2x)^2 - 1^2]$ Write as difference of two squares.

$ = 5x(2x + 1)(2x - 1)$ Factored form

EXAMPLE 4 Factoring Completely

Factor the polynomial completely.

a. $x^4 - 16$ b. $48x^4 - 3$

SOLUTION

a. Recognizing $x^4 - 16$ as a difference of two squares, you can write

$x^4 - 16 = (x^2)^2 - 4^2$ Write as difference of two squares.

$ = (x^2 + 4)(x^2 - 4).$ Factored form

Note that the second factor, $(x^2 - 4)$, is itself a difference of two squares, and so

$x^4 - 16 = (x^2 + 4)(x^2 - 4)$ Factor as difference of two squares.

$ = (x^2 + 4)(x + 2)(x - 2).$ Factor completely.

b. Start by removing the common monomial factor.

$48x^4 - 3 = 3(16x^4 - 1)$ Remove common monomial factor 3.

Recognizing $16x^4 - 1$ as the difference of two squares, you can write

$48x^4 - 3 = 3(16x^4 - 1)$ Factor out common monomial.

$ = 3[(4x^2)^2 - 1^2]$ Write as difference of two squares.

$ = 3(4x^2 + 1)(4x^2 - 1)$ Recognize $4x^2 - 1$ as a difference of two squares.

$ = 3(4x^2 + 1)[(2x)^2 - 1^2]$ Write as difference of two squares.

$ = 3(4x^2 + 1)(2x + 1)(2x - 1).$ Factor completely.

Study Tip

Note in Example 4 that no attempt is made to factor the *sum of two squares*. A second-degree polynomial that is the sum of two squares cannot be factored as the product of binomials (using integer coefficients). In general, *the sum of two squares is not factorable*.

Exercises Within Reach®

Solutions in English & Spanish and tutorial videos at AlgebraWithinReach.com

Removing a Common Monomial Factor First In Exercises 11–16, factor the polynomial.

11. $2x^2 - 72$ 12. $3x^2 - 27$ 13. $4x - 25x^3$

14. $a^3 - 16a$ 15. $8y^3 - 50y$ 16. $20x^3 - 180x$

Factoring Completely In Exercises 17–22, factor the polynomial completely.

17. $y^4 - 81$ 18. $z^4 - 625$ 19. $1 - x^4$

20. $256 - u^4$ 21. $2x^4 - 162$ 22. $5x^4 - 80$

Perfect Square Trinomials

A **perfect square trinomial** is the square of a binomial. For instance,

$$x^2 + 4x + 4 = (x + 2)(x + 2)$$
$$= (x + 2)^2$$

is the square of the binomial $(x + 2)$.

Perfect Square Trinomials

Let a and b be real numbers, variables, or algebraic expressions.

1. $a^2 + 2ab + b^2 = (a + b)^2$
   Same sign

2. $a^2 - 2ab + b^2 = (a - b)^2$
   Same sign

Study Tip

To recognize a perfect square trinomial, remember that the first and last terms must be perfect squares and positive, and the middle term must be twice the product of a and b. (The middle term can be positive or negative.) Watch for squares of fractions.

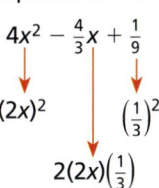

EXAMPLE 5 Identifying Perfect Square Trinomials

Which of the following are prefect square trinomials?

a. $m^2 - 4m + 4$ b. $4x^2 - 2x + 1$

c. $y^2 + 6y - 9$ d. $x^2 + x + \frac{1}{4}$

SOLUTION

a. This polynomial *is* a perfect square trinomial. It factors as $(m - 2)^2$.

b. This polynomial *is not* a perfect square trinomial because the middle term is not twice the product of $2x$ and 1.

c. This polynomial *is not* a perfect square trinomial because the last term, -9, is not positive.

d. This polynomial *is* a perfect square trinomial. It factors as $\left(x + \frac{1}{2}\right)^2$.

Exercises Within Reach®

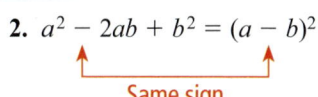

Solutions in English & Spanish and tutorial videos at AlgebraWithinReach.com

Identifying a Perfect Square Trinomial   In Exercises 23–28, **determine** whether the polynomial is a perfect square trinomial.

23. $9b^2 + 24b + 16$

24. $y^2 - 2y + 6$

25. $m^2 - 2m - 1$

26. $16n^2 + 2n + 1$

27. $4k^2 - 20k + 25$

28. $x^2 + 20x + 100$

Geometric Model of Factoring   In Exercises 29 and 30, **write** the factoring problem represented by the geometric factoring model.

29.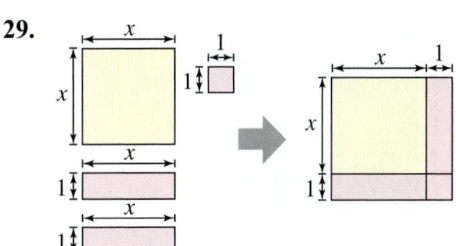

30.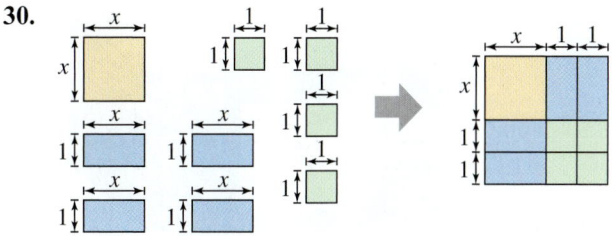

EXAMPLE 6 Factoring a Perfect Square Trinomial

Factor the trinomial $y^2 - 6y + 9$.

SOLUTION

$y^2 - 6y + 9 = y^2 - 2(3y) + 3^2$ Recognize the pattern.

$ = (y - 3)^2$ Write in factored form.

EXAMPLE 7 Factoring a Perfect Square Trinomial

Factor the trinomial $16x^2 + 40x + 25$.

SOLUTION

$16x^2 + 40x + 25 = (4x)^2 + 2(4x)(5) + 5^2$ Recognize the pattern.

$ = (4x + 5)^2$ Write in factored form.

EXAMPLE 8 Factoring a Perfect Square Trinomial

Factor the trinomial $9x^2 - 24xy + 16y^2$.

SOLUTION

$9x^2 - 24xy + 16y^2 = (3x)^2 - 2(3x)(4y) + (4y)^2$ Recognize the pattern.

$ = (3x - 4y)^2$ Write in factored form.

Exercises Within Reach®

Solutions in English & Spanish and tutorial videos at AlgebraWithinReach.com

Factoring a Perfect Square Trinomial In Exercises 31–42, **factor** the perfect square trinomial.

31. $x^2 - 8x + 16$
32. $x^2 + 10x + 25$
33. $x^2 + 14x + 49$
34. $a^2 - 12a + 36$
35. $4t^2 + 4t + 1$
36. $9x^2 - 12x + 4$
37. $25y^2 - 10y + 1$
38. $16z^2 + 24z + 9$
39. $x^2 - 6xy + 9y^2$
40. $16x^2 - 8xy + y^2$
41. $4y^2 + 20yz + 25z^2$
42. $36u^2 + 84uv + 49v^2$

43. **Geometry** The vaccine cooler shown in the figure has a square base and a height of x. The volume of the cooler is $(x^3 + 16x^2 + 64x)$ cubic centimeters. Find the dimensions of the cooler.

44. **Geometry** The building shown in the figure has a square base and a height of x. The volume of the building is $(x^3 - 80x^2 + 1600x)$ cubic feet. Find the dimensions of the building.

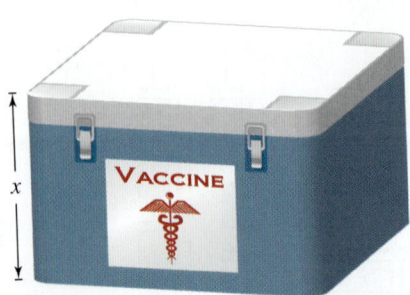

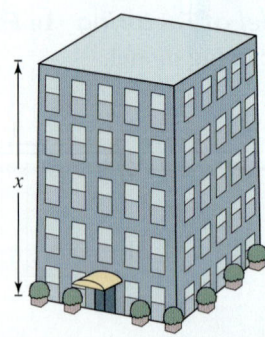

Sum or Difference of Two Cubes

Study Tip
When using either of the factoring patterns at the right, pay special attention to the signs. Remembering the "like" and "unlike" patterns for the signs is helpful.

Sum or Difference of Two Cubes

Let a and b be real numbers, variables, or algebraic expressions.

1. $a^3 + b^3 = (a + b)(a^2 - ab + b^2)$
   (Like signs on outer terms; Unlike signs on middle term)

2. $a^3 - b^3 = (a - b)(a^2 + ab + b^2)$
   (Like signs on outer terms; Unlike signs on middle term)

EXAMPLE 9 Factoring the Sum or Difference of Two Cubes

a. $y^3 + 27 = y^3 + 3^3$ Write as sum of two cubes.

$\quad\quad\quad\quad = (y + 3)[y^2 - (y)(3) + 3^2]$ Factored form

$\quad\quad\quad\quad = (y + 3)(y^2 - 3y + 9)$ Simplify.

b. $64 - x^3 = 4^3 - x^3$ Write as difference of two cubes.

$\quad\quad\quad\quad = (4 - x)[4^2 + (4)(x) + x^2]$ Factored form

$\quad\quad\quad\quad = (4 - x)(16 + 4x + x^2)$ Simplify.

c. $2x^3 - 16 = 2(x^3 - 8)$ Factor out common monomial factor 2.

$\quad\quad\quad\quad = 2(x^3 - 2^3)$ Write as difference of two cubes.

$\quad\quad\quad\quad = 2(x - 2)[x^2 + (x)(2) + 2^2]$ Factored form

$\quad\quad\quad\quad = 2(x - 2)(x^2 + 2x + 4)$ Simplify.

Exercises Within Reach®
Solutions in English & Spanish and tutorial videos at AlgebraWithinReach.com

Factoring the Sum or Difference of Two Cubes In Exercises 45–54, *factor* the sum or difference of two cubes.

45. $x^3 - 8$

46. $x^3 - 27$

47. $y^3 + 64$

48. $z^3 + 125$

49. $1 + 8t^3$

50. $1 + 27s^3$

51. $27u^3 - 8$

52. $64v^3 - 125$

53. $27x^3 + 64y^3$

54. $27y^3 + 125z^3$

Chapter 6 Factoring and Solving Equations

Concept Summary: Factoring Polynomials with Special Forms

What

When you are asked to factor a polynomial, you should identify whether the polynomial involves a special form.

EXAMPLE

Factor the polynomial.

a. $3x^2 - 12$

b. $16 - 2t^3$

c. $x^3 + 4x^2 + 4x$

How

Factor out any common monomial factor and then compare the remaining polynomial to the special polynomial forms.

EXAMPLE

a. $3x^2 - 12 = 3(\overbrace{x^2 - 4}^{\text{Difference of two squares}})$

b. $16 - 2t^3 = 2(\overbrace{8 - t^3}^{\text{Difference of two cubes}})$

c. $x^3 + 4x^2 + 4x = x(\overbrace{x^2 + 4x + 4}^{\text{Perfect square trinomial}})$

Why

Recognizing polynomials involving special forms can help you factor such polynomials *completely*.

EXAMPLE

a. $3x^2 - 12 = 3(x + 2)(x - 2)$

b. $16 - 2t^2 = 2(2 - t)(4 + 2t + t^2)$

c. $x^3 + 4x^2 + 4x = x(x + 2)^2$

Exercises Within Reach®   Worked-out solutions to odd-numbered exercises at AlgebraWithinReach.com

Concept Summary Check

55. Special Polynomial Forms Are any of the polynomials $3x^2 - 12$, $16 - 2t^3$, and $x^3 + 4x^2 + 4x$ in one of the special polynomial forms?

56. Finding Special Polynomial Forms What was done to each polynomial in the example above to reveal a special polynomial form?

57. Analyzing a Perfect Square Trinomial The polynomial $x^2 + 4x + 4$ is a perfect square trinomial of the form $a^2 + 2ab + b^2$. Identify a and b.

58. Factoring a Polynomial What is the first thing you should do when you factor a polynomial?

Extra Practice

Finding Coefficients In Exercises 59–62, find two real numbers b such that the expression is a perfect square trinomial.

59. $x^2 + bx + 1$

60. $x^2 + bx + 100$

61. $4x^2 + bx + 81$

62. $4x^2 + bx + 9$

Finding a Constant Term In Exercises 63–66, find a real number c such that the expression is a perfect square trinomial.

63. $x^2 + 6x + c$

64. $x^2 + 10x + c$

65. $y^2 - 4y + c$

66. $z^2 - 14z + c$

Factoring a Polynomial In Exercises 67–78, factor the polynomial completely. (*Note:* Some of the polynomials may be prime.)

67. $y^4 - 25y^2$

68. $x^6 - 49x^4$

69. $x^2 - 2x + 1$

70. $81 + 18x + x^2$

71. $(t - 1)^2 - 121$

72. $(x - 3)^2 - 100$

73. $x^2 + 81$

74. $x^2 + 16$

75. $x^4 - 81$

76. $2x^4 - 32$

77. $x^3 + 4x^2 - x + 4$

78. $y^3 + 3y^2 - 4y - 12$

Mental Math In Exercises 79–82, evaluate the quantity mentally using the two samples as models.

$29^2 = (30 - 1)^2 = 30^2 - 2 \cdot 30 \cdot 1 + 1^2$

$ = 900 - 60 + 1 = 841$

$48 \cdot 52 = (50 - 2)(50 + 2)$

$ = 50^2 - 2^2 = 2496$

79. 21^2

80. 49^2

81. $59 \cdot 61$

82. $28 \cdot 32$

Rewriting a Trinomial In Exercise 83 and 84, write the polynomial as the difference of two squares. Use the result to factor the polynomial completely.

83. $x^2 + 6x + 8 = (x^2 + 6x + 9) - 1$

    $= \underline{}^2 - \underline{}^2$

84. $x^2 + 8x + 12 = (x^2 + 8x + 16) - 4$

    $= \underline{}^2 - \underline{}^2$

85. *Modeling* The figure below shows two cubes: a large cube whose volume is a^3 and a smaller cube whose volume is b^3. If the smaller cube is removed from the larger, the remaining solid has a volume of $a^3 - b^3$ and is composed of three rectangular boxes, labeled Box 1, Box 2, and Box 3. Find the volume of each box and describe how these results are related to the following special product pattern.

    $a^3 - b^3 = (a - b)(a^2 + ab + b^2)$

    $ = (a - b)a^2 + (a - b)ab + (a - b)b^2$

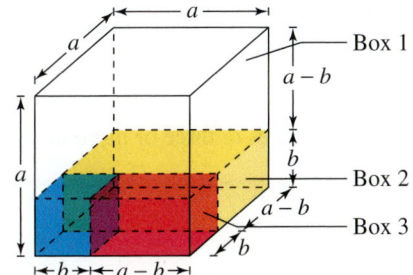

86. *Geometry* From the eight vertices of a cube of dimension x, cubes of dimension y are removed (see figure).

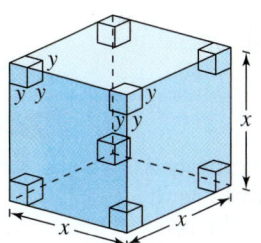

(a) Write an expression for the volume of the solid that remains after the eight cubes at the vertices are removed.

(b) Factor the expression for the volume in part (a).

(c) In the context of this problem, y must be less than what multiple of x? Explain your answer geometrically and from the result of part (b).

Explaining Concepts

True or False? In Exercise 87–90, determine whether the statement is true or false. Justify your answer.

87. The expression $x(x + 2) - 2(x + 2)$ is completely factored.

88. $x^2 + 4 = (x + 2)^2$

89. $x^3 - 27 = (x - 3)^3$

90. Because the sum of two squares cannot be factored, it follows that the sum of two cubes cannot be factored.

91. *Writing* Explain how to identify and factor the difference of two squares.

92. *Writing* Explain how to identify and factor a perfect square trinomial.

Cumulative Review

In Exercises 93–96, solve the equation and check your solution.

93. $7 + 5x = 7x - 1$

94. $2 - 5(x - 1) = 2[x + 10(x - 1)]$

95. $2(x + 1) = 0$

96. $\frac{3}{4}(12x - 8) = 10$

In Exercises 97–100, factor the trinomial.

97. $2x^2 + 7x + 3$

98. $3y^2 - 5y - 12$

99. $6m^2 + 7m - 20$

100. $15x^2 - 28x + 12$

6.5 Solving Quadratic Equations by Factoring

- Use the Zero-Factor Property to solve equations.
- Use factoring to solve quadratic equations.
- Solve application problems by factoring.

Zero-Factor Property

You have spent nearly two chapters developing skills for rewriting (simplifying and factoring) polynomials. You are now ready to use these skills, together with the **Zero-Factor Property**, to solve **second-degree polynomial equations** (quadratic equations).

Study Tip
The Zero-Factor Property is just another way of saying that the only way the product of two or more factors can equal zero is if one or more of the factors equals zero.

Zero-Factor Property

Let a and b be real numbers, variables, or algebraic expressions. If a and b are factors such that

$$ab = 0$$

then $a = 0$ or $b = 0$. This property also applies to three or more factors.

EXAMPLE 1 Using the Zero-Factor Property

Find the solutions of each equation.

a. $(x - 1)(x + 2) = 0$
b. $(x + 3)(2x - 5) = 0$

SOLUTION

a. You can use the Zero-Factor Property to conclude that either $(x - 1)$ or $(x + 2)$ equals 0. Setting the first factor equal to 0 implies that $x = 1$ is a solution.

$x - 1 = 0 \implies x = 1$ First solution

Similarly, setting the second factor equal to 0 implies that $x = -2$ is a solution.

$x + 2 = 0 \implies x = -2$ Second solution

b. Setting the first factor equal to 0 implies that $x = -3$ is a solution.

$x + 3 = 0 \implies x = -3$ First solution

Setting the second factor equal to 0 implies that $x = \frac{5}{2}$ is a solution.

$2x - 5 = 0 \implies x = \frac{5}{2}$ Second solution

Exercises Within Reach® Solutions in English & Spanish and tutorial videos at AlgebraWithinReach.com

Using the Zero-Factor Property In Exercises 1–6, use the Zero-Factor Property to **solve** the equation.

1. $x(x - 5) = 0$
2. $z(z - 3) = 0$
3. $(y - 2)(y - 3) = 0$
4. $(t - 3)(t + 8) = 0$
5. $(2t - 5)(3t + 1) = 0$
6. $(2 - 3x)(5 - 2x) = 0$

Solving Quadratic Equations by Factoring

A **quadratic equation** is an equation that can be written in the general form

$$ax^2 + bx + c = 0 \quad \text{Quadratic equation}$$

where a, b, and c are real numbers with $a \neq 0$.

Guidelines for Solving Quadratic Equations

1. Write the quadratic equation in general form.
2. Factor the left side of the equation.
3. Set each factor with a variable equal to zero.
4. Solve each linear equation.
5. Check each solution in the original equation.

EXAMPLE 2   Solving a Quadratic Equation by Factoring

Solve $x^2 - x - 6 = 0$.

SOLUTION

$x^2 - x - 6 = 0$ Write original equation.

$(x + 2)(x - 3) = 0$ Factor left side of equation.

$x + 2 = 0 \Rightarrow x = -2$ Set 1st factor equal to 0 and solve for x.

$x - 3 = 0 \Rightarrow x = 3$ Set 2nd factor equal to 0 and solve for x.

The solutions are $x = -2$ and $x = 3$.

Check

$(-2)^2 - (-2) - 6 \stackrel{?}{=} 0$ Substitute -2 for x in original equation.

$4 + 2 - 6 \stackrel{?}{=} 0$ Simplify.

$0 = 0$ Solution checks. ✓

$(3)^2 - (3) - 6 \stackrel{?}{=} 0$ Substitute 3 for x in original equation.

$9 - 3 - 6 \stackrel{?}{=} 0$ Simplify.

$0 = 0$ Solution checks. ✓

Exercises Within Reach®

Solutions in English & Spanish and tutorial videos at AlgebraWithinReach.com

Solving a Quadratic Equation by Factoring In Exercises 7–14, **solve** the equation.

7. $x^2 - 16 = 0$
8. $x^2 - 144 = 0$
9. $3y^2 - 27 = 0$
10. $25z^2 - 100 = 0$
11. $x^2 - 2x - 8 = 0$
12. $x^2 - 8x - 9 = 0$
13. $3 + 5x - 2x^2 = 0$
14. $33 + 5y - 2y^2 = 0$

EXAMPLE 3   **Solving a Quadratic Equation by Factoring**

Solve $x^2 - 2x + 16 = 6x$.

SOLUTION

$x^2 - 2x + 16 = 6x$	Write original equation.
$x^2 - 8x + 16 = 0$	Write in general form.
$(x - 4)^2 = 0$	Factor.
$x - 4 = 0$ or $x - 4 = 0$	Set factors equal to 0.
$x = 4$	Solve for x.

Note that even though the left side of this equation has two factors, the factors are the same. So, the only solution of the equation is $x = 4$. This solution is called a **repeated solution**.

Check

$x^2 - 2x + 16 = 6x$	Write original equation.
$(4)^2 - 2(4) + 16 \stackrel{?}{=} 6(4)$	Substitute 4 for x.
$16 - 8 + 16 \stackrel{?}{=} 24$	Simplify.
$24 = 24$	Solution checks. ✓

EXAMPLE 4   **Solving a Quadratic Equation by Factoring**

Solve $2x^2 + 5x = 12$.

SOLUTION

$2x^2 + 5x = 12$	Write original equation.
$2x^2 + 5x - 12 = 0$	Write in general form.
$(2x - 3)(x + 4) = 0$	Factor left side of equation.
$2x - 3 = 0$	Set 1st factor equal to 0.
$x = \frac{3}{2}$	Solve for x.
$x + 4 = 0$	Set 2nd factor equal to 0.
$x = -4$	Solve for x.

The solution are $x = \frac{3}{2}$ and $x = -4$. Check these solutions in the original equation.

Exercises Within Reach ®   Solutions in English & Spanish and tutorial videos at AlgebraWithinReach.com

Solving a Quadratic Equation by Factoring In Exercises 15–22, **solve** the equation.

15. $x^2 + 18x + 49 = 4x$
16. $x^2 - 2x + 25 = 8x$
17. $4t^2 + 2t + 9 = 14t$
18. $16x^2 + 24x + 49 = -32x$
19. $t^2 + 6t = 16$
20. $x^2 - 8x = 65$
21. $2y^2 - y = 3$
22. $3s^2 + 8s = 35$

EXAMPLE 5 Solving a Quadratic Equation by Factoring

Solve $(x + 3)(x + 6) = 4$.

SOLUTION

Begin by multiplying the factors on the left side.

$(x + 3)(x + 6) = 4$	Write original equation.
$x^2 + 9x + 18 = 4$	Multiply factors.
$x^2 + 9x + 14 = 0$	Write in general form.
$(x + 2)(x + 7) = 0$	Factor.
$x + 2 = 0 \implies x = -2$	Set 1st factor equal to 0 and solve for x.
$x + 7 = 0 \implies x = -7$	Set 2nd factor equal to 0 and solve for x.

The solutions are $x = -2$ and $x = -7$. Check these in the original equation.

Application

EXAMPLE 6 Solving a Consecutive Integers Problem

The product of two consecutive positive integers is 56. What are the integers?

SOLUTION

Verbal Model: First integer · Second integer = 56

Labels: First integer $= n$
Second integer $= n + 1$

Expression:

$n(n + 1) = 56$	Write equation.
$n^2 + n - 56 = 0$	Write in general form.
$(n + 8)(n - 7) = 0$	Factor.
$n = -8$ or $n = 7$	Solutions using Zero-Factor Property

Because the problem states that the integers are positive, discard -8 as a solution and choose $n = 7$. So, the two integers are $n = 7$ and $n + 1 = 7 + 1 = 8$. Check these in the original statement of the problem.

Exercises Within Reach®

Solutions in English & Spanish and tutorial videos at AlgebraWithinReach.com

Solving a Quadratic Equation by Factoring In Exercises 23–30, solve the equation.

23. $x(x + 10) = 24$
24. $x(x - 14) = -40$
25. $x(3x + 1) = 2$
26. $x(2x - 3) = 5$
27. $(x + 1)(x - 3) = 12$
28. $(x - 5)(x - 1) = 21$
29. $(x - 7)(x + 9) = 17$
30. $(x + 10)(x + 1) = 36$

31. **Number Problem** Find two consecutive positive integers whose product is 72.

32. **Number Problem** Find two consecutive positive integers whose product is 240.

33. **Number Problem** Find two consecutive positive even integers whose product is 440.

34. **Number Problem** Find two consecutive positive odd integers whose product is 323.

Applications

Application — EXAMPLE 7 Free-Falling Object

A rock is dropped from the top of a 256-foot river gorge, as shown in the figure. The height h (in feet) of the rock is modeled by the position equation

$$h = -16t^2 + 256$$

where t is the time measured in seconds. How long will it take for the rock to reach the bottom of the gorge?

SOLUTION

In the figure, note that the bottom of the gorge corresponds to a height of 0 feet. So, substitute a height of 0 for h into the equation and solve for t.

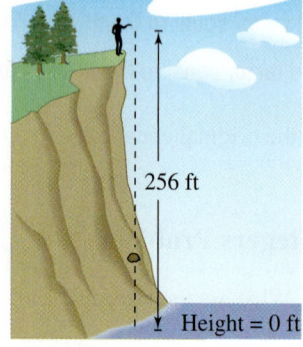

256 ft

Height = 0 ft

$0 = -16t^2 + 256$	Substitute 0 for h.
$16t^2 - 256 = 0$	Write in general form.
$16(t^2 - 16) = 0$	Factor out common factor.
$16(t + 4)(t - 4) = 0$	Factor.
$t + 4 = 0 \implies t = -4$	Set 1st factor equal to 0 and solve for t.
$t - 4 = 0 \implies t = 4$	Set 2nd factor equal to 0 and solve for t.

Because a time of -4 seconds does not make sense in this problem, choose the positive solution $t = 4$ and conclude that the rock reaches the bottom of the gorge 4 seconds after it is dropped. Check this solution in the original statement of the problem.

Exercises Within Reach® Solutions in English & Spanish and tutorial videos at AlgebraWithinReach.com

35. Free-Falling Object An object is dropped from a weather balloon 1600 feet above the ground. The height h (in feet) of the object is modeled by the position equation

$$h = -16t^2 + 1600$$

where t is the time measured in seconds. How long will it take for the object to reach the ground?

36. Height of a Diver A diver jumps from a diving board that is 32 feet above the water with an initial velocity of 16 feet per second (see figure). The height h (in feet) of the diver is modeled by the position equation

$$h = -16t^2 + 16t + 32$$

where t is the time measured in seconds. How long will it take for the diver to reach the water?

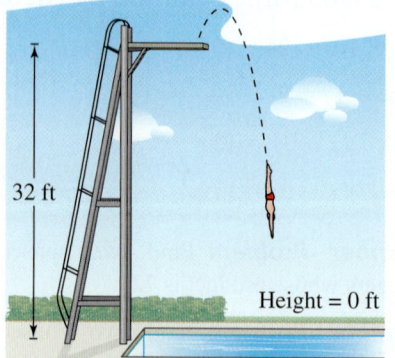

32 ft

Height = 0 ft

Application

EXAMPLE 8 Geometry: Dimensions of a Room

A rectangular family room has an area of 160 square feet. The length of the room is 6 feet greater than its width. Find the dimensions of the room.

SOLUTION

To begin, make a sketch of the room, as shown at the left. Label the width of the room as x and the length of the room as $x + 6$ because the length is 6 feet greater than the width.

Verbal Model: Length · Width = Area

Labels:
Width = x (feet)
Length = $x + 6$ (feet)
Area = 160 (square feet)

Expression:
$x(x + 6) = 160$ Write equation.
$x^2 + 6x = 160$ Distributive Property
$x^2 + 6x - 160 = 0$ Write in general form.
$(x + 16)(x - 10) = 0$ Factor.
$x + 16 = 0 \implies x = -16$ Set 1st factor equal to 0 and solve for x.
$x - 10 = 0 \implies x = 10$ Set 2nd factor equal to 0 and solve for x.

Because the dimensions must be positive, discard $x = -16$ as a solution and use the positive solution $x = 10$. So, the width of the room is 10 feet and the length of the room is

Length = $x + 6$
= $10 + 6$ Substitute 10 for x.
= 16 feet.   Simplify.

Exercises Within Reach®

Solutions in English & Spanish and tutorial videos at AlgebraWithinReach.com

37. *Geometry* The length of a rectangular painting is 3 inches greater than its width (see figure). The area of the painting is 108 square inches. Find the dimensions of the painting.

38. *Geometry* The width of a rectangular garden is 4 feet less than its length. The area of the garden is 320 square feet. Find the dimensions of the garden.

39. *Geometry* The length of a rectangular poster is 2 times its width (see figure). The area of the poster is 648 square inches. Find the dimensions of the poster.

40. *Geometry* The length of a rectangular room is $1\frac{1}{2}$ times its width. The area of the rectangular room is 216 square feet. Find the dimensions of the room.

294 Chapter 6 Factoring and Solving Equations

Concept Summary: Solving Quadratic Equations by Factoring

What
To solve a **quadratic equation** by factoring, use the guidelines for solving quadratic equations.

EXAMPLE
Solve $2x^2 + 18x = 0$.

How
- Write the quadratic equation in **general form**.
- Factor the left side of the equation.
- Set each factor with a variable equal to zero.
- Solve each linear equation.

EXAMPLE
$2x^2 + 18x = 0$
$2x(x + 9) = 0$
$2x = 0 \Rightarrow x = 0$
$x + 9 = 0 \Rightarrow x = -9$

Why
You can solve quadratic equations to answer many real-life problems, such as finding the length of time it will take a falling object to reach the ground.

Exercises Within Reach®

Worked-out solutions to odd-numbered exercises at AlgebraWithinReach.com

Concept Summary Check

41. General Form Is the equation $2x^2 + 18x = 0$ in general form? Explain.

42. Factoring the Left Side What are the factors of the left side of the equation $2x^2 + 18x = 0$?

43. Using a Property How is the Zero-Factor Property used in the example above?

44. Verifying Solutions How can you verify the solutions in the example above?

Extra Practice

Solving a Quadratic Equation by Factoring In Exercises 45–52, **solve** the equation.

45. $2x^2 + 4x = 0$
46. $6x^2 + 3x = 0$
47. $5x^2 - 24 = 37x$
48. $4x^2 - 40 = -27x$
49. $3x^2 - 2x = 9 - 8x$
50. $6x^2 - 8x = 10 - 4x$
51. $(x - 6)(x - 3) = -2$
52. $(x + 4)(x - 8) = -20$

Graphical Estimation In Exercises 53–56, **estimate** the x-intercepts of the graph of the equation. Set the quadratic equation equal to zero and solve. What do you notice?

53. $y = x^2 + 2x - 3$
54. $y = 2 + x - x^2$
55. $y = 12 + x - x^2$
56. $y = 2x^2 + x - 3$

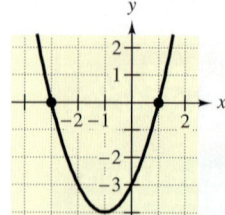
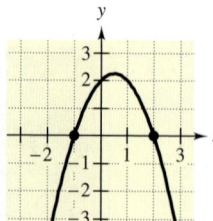
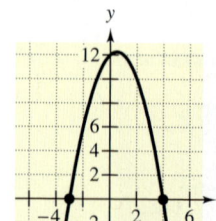
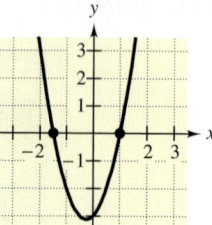

57. Sum of Natural Numbers A formula for the sum of the first n natural numbers is $1 + 2 + 3 + \cdots + n = \frac{1}{2}n(n + 1)$.

(a) Use the formula to find the sum of the first 15 natural numbers $(1 + 2 + 3 + \cdots + 15)$.

(b) Use the formula to find n when the sum of the first n natural numbers is 210.

58. *Geometry* An open box with a square base is to be made from a square piece of material. The side length of the base is x inches and the height of the box is 3 inches. The surface area of the box is given by $S = x^2 + 4xh$.

(a) Complete the table.

x	1	2	3	4
S				

(b) Find the dimensions of the base of the box when $S = 108$ square inches.

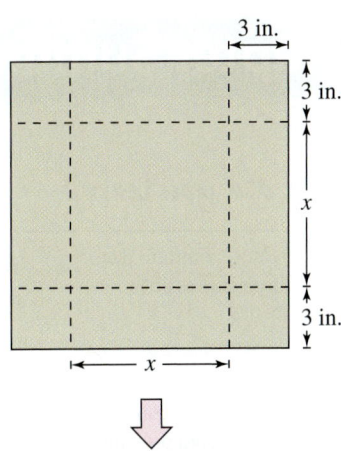

Explaining Concepts

True or False? In Exercises 59–62, determine whether the statement is true or false. **Justify** your answer.

59. The only solution of $x^2 = 4x$ is $x = 4$.

60. The only equation with solutions $x = 2$ and $x = -5$ is $(x - 2)(x + 5) = 0$.

61. If a and b are nonzero real numbers, then the solutions of the equations $ax^2 + bx = 0$ are $x = 0$ and $x = -\dfrac{b}{a}$.

62. If a is a nonzero real number, then the solutions of the equation $ax^2 - ax = 0$ are $x = 0$ and $x = a$.

63. *Error Analysis* A student submits the following steps in solving the equation $x^2 + 4x = 12$.

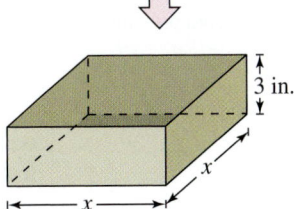

Write an explanation of why the method does not work. Solve the equation correctly and check your solutions.

64. *Reasoning* When does a quadratic equation have one repeated solution?

Cumulative Review

In Exercises 65 and 66, determine whether each ordered pair is a solution of the inequality.

65. $3x + 4y \leq 12$

(a) $(0, 0)$ (b) $(-1, 3)$

(c) $(7, -2)$ (d) $(-1, -1)$

66. $7x - 2y > -5$

(a) $(-2, -6)$ (b) $(1, 8)$

(c) $(0, 0)$ (d) $(3, 12)$

In Exercises 67–74, factor the polynomial completely.

67. $x^2 - 81$

68. $\dfrac{9}{16} - x^2$

69. $3x^2 - 48$

70. $9x^2 - 324$

71. $x^2 + 4x + 4$

72. $x^2 - 22x + 121$

73. $4x^2 - 20x + 25$

74. $9x^2 + 42x + 49$

6 Chapter Summary

	What did you learn?	Explanation and Examples	Review Exercises
6.1	Find the greatest common factor of two or more expressions *(p. 254)*.	$60x^4 = 2 \cdot 2 \cdot 3 \cdot 5 \cdot x \cdot x \cdot x \cdot x = (20x^3)(3x)$ $40x^3 = 2 \cdot 2 \cdot 2 \cdot 5 \cdot x \cdot x \cdot x = (20x^3)(2)$ The greatest common factor of $60x^4$ and $40x^3$ is $20x^3$.	1–8
	Factor out the greatest common monomial factor from polynomials *(p. 255)*.	Use the Distributive Property to factor out the greatest common monomial factor from each term of a polynomial.	9–22
	Factor polynomials by grouping *(p. 257)*.	For polynomials with four terms, group terms that have a common monomial factor. Factor the two groupings and then look for a common binomial factor.	23–32
6.2	Factor trinomials of the form $x^2 + bx + c$ *(p. 262)*.	To factor $x^2 + bx + c$, you need to find two numbers m and n whose product is c and whose sum is b. $$x^2 + bx + c = (x + m)(x + n)$$ 1. If c is positive, then m and n have like signs that match the sign of b. 2. If c is negative, then m and n have unlike signs. 3. If $\|b\|$ is small relative to $\|c\|$, first try those factors of c that are closest to each other in absolute value.	33–48
	Factor trinomials in two variables *(p. 265)*.	To factor $x^2 + bxy + cy^2$, find two factors of c whose sum is b. The product of -4 and 3 is -12. $$x^2 - xy - 12y^2 = (x - 4y)(x + 3y)$$ The sum of -4 and 3 is -1.	49–54
	Factor trinomials completely *(p. 266)*.	When a trinomial has a common monomial factor, you should factor out the common factor first. Then factor the resulting trinomial. Be sure to include the common monomial factor in the final factored form.	55–62
6.3	Factor trinomials of the form $ax^2 + bx + c$ *(p. 270)*.	1. When a trinomial has a common factor, you should factor out the common factor before trying to find the binomial factors. 2. You do not have to test any binomial factors that have a common monomial factor. 3. Switch the signs of the factors of c when the middle term (O + I) is correct except in sign.	63–80
	Factor trinomials completely *(p. 272)*.	Remember that when a trinomial has a common monomial factor, you should factor out the common factor first. The complete factorization will show all monomial and binomial factors.	81–92

Chapter Summary

	What did you learn?	Explanation and Examples	Review Exercises
6.3	Factor trinomials by grouping *(p. 274)*.	1. If necessary, write the trinomial in standard form. 2. Choose factors of the product ac that add up to b. 3. Use these factors to rewrite the middle term as a sum or difference. 4. Group and remove any common monomial factors from the first two terms and the last two terms. 5. If possible, factor out the common binomial factor.	93–100
	Factor the difference of two squares *(p. 280)*.	$a^2 - b^2 = (a + b)(a - b)$ Difference — Opposite signs	101–110
	Factor a polynomial completely *(p. 282)*.	1. Factor out any common factors. 2. Factor according to one of the special polynomial forms: difference of two squares, sum or difference of two cubes, or perfect square trinomials. 3. Factor trinomials, $ax^2 + bx + c$, with $a = 1$ or $a \neq 1$. 4. Factor by grouping—for polynomials with four terms. 5. Check to see whether the factors themselves can be factored.	111–118
6.4	Identify and factor perfect square trinomials *(p. 283)*.	$a^2 + 2ab + b^2 = (a + b)^2$ Same sign $a^2 - 2ab + b^2 = (a - b)^2$ Same sign	119–130
	Factor the sum or difference of two cubes *(p. 285)*.	Like signs $a^3 + b^3 = (a + b)(a^2 - ab + b^2)$ Unlike signs Like signs $a^3 - b^3 = (a - b)(a^2 + ab + b^2)$ Unlike signs	131–136
6.5	Use the Zero-Factor Property to solve equations *(p. 288)*.	Let a and b be real numbers, variables, or algebraic expressions. If a and b are factors such that $ab = 0$, then $a = 0$ or $b = 0$. This property also applies to three or more factors.	137–142
	Use factoring to solve quadratic equations *(p. 289)*.	1. Write the quadratic equation in general form. 2. Factor the left side of the equation. 3. Set each factor with a variable equal to zero. 4. Solve each linear equation. 5. Check each solution in the original equation.	143–156
	Solve application problems by factoring *(p. 292)*.	When you solve application problems by factoring, check your answers. Eliminate answers that are not appropriate in the context of the problem.	157–159

Review Exercises

Worked-out solutions to odd-numbered exercises at AlgebraWithinReach.com

6.1

Finding the Greatest Common Factor In Exercises 1–8, find the greatest common factor of the expressions.

1. t^2, t^5
2. $-y^3, y^8$
3. $3x^4, 21x^2$
4. $14z^2, 21z$
5. $14x^2y^3, -21x^3y^5$
6. $-15y^2z^2, 5y^2z$
7. $8x^2y, 24xy^2, 4xy$
8. $27ab^5, 9ab^6, 18a^2b^3$

Greatest Common Monomial Factor In Exercises 9–20, factor out the greatest common monomial factor from the polynomial.

9. $3x - 6$
10. $7 + 21x$
11. $3t - t^2$
12. $u^2 - 6u$
13. $5x^2 + 10x^3$
14. $7y - 21y^4$
15. $8a^2 - 12a^3$
16. $14x - 26x^4$
17. $5x^3 + 5x^2 - 5x$
18. $6u - 9u^2 + 15u^3$
19. $8y^2 + 4y + 12$
20. $3z^4 - 21z^3 + 10z$

Geometry In Exercises 21 and 22, write an expression for the area of the shaded region and factor the expression.

21.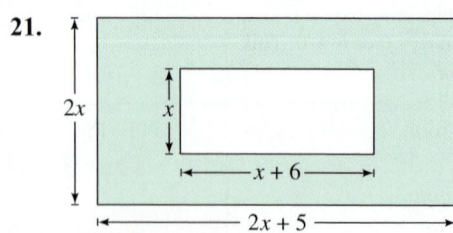

22.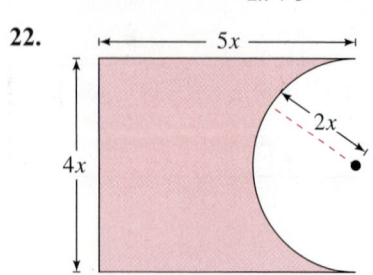

Factoring by Grouping In Exercises 23–32, factor the polynomial by grouping.

23. $x(x + 1) - 3(x + 1)$
24. $5(y - 3) - y(y - 3)$
25. $2u(u - 2) + 5(u - 2)$
26. $7(x + 8) + 3x(x + 8)$
27. $y^3 + 3y^2 + 2y + 6$
28. $z^3 - 5z^2 + z - 5$
29. $x^3 + 2x^2 + x + 2$
30. $x^3 - 5x^2 + 5x - 25$
31. $x^2 - 4x + 3x - 12$
32. $2x^2 + 6x - 5x - 15$

6.2

Factoring a Trinomial In Exercises 33–44, factor the trinomial.

33. $x^2 - 3x - 28$
34. $x^2 - 3x - 40$
35. $u^2 + 5u - 36$
36. $y^2 + 15y + 56$
37. $x^2 - 2x - 24$
38. $x^2 + 8x + 15$
39. $y^2 + 10y + 21$
40. $a^2 - 7a + 12$
41. $b^2 + 13b - 30$
42. $z^2 - 9z + 18$
43. $w^2 + 3w - 40$
44. $x^2 - 7x - 8$

Finding Coefficients In Exercises 45–48, find all integers b such that the trinomial can be factored.

45. $x^2 + bx + 9$
46. $y^2 + by + 25$
47. $z^2 + bz + 11$
48. $x^2 + bx + 14$

Factoring a Trinomial In Exercises 49–54, factor the trinomial.

49. $x^2 + 9xy - 10y^2$
50. $u^2 + 3uv - 4v^2$
51. $y^2 - 6xy - 27x^2$
52. $v^2 + 18uv + 32u^2$
53. $x^2 - 2xy - 8y^2$
54. $a^2 - ab - 30b^2$

Factoring Completely In Exercises 55–62, factor the trinomial completely.

55. $4x^2 - 24x + 32$
56. $3u^2 - 6u - 72$
57. $x^3 + 9x^2 + 18x$
58. $y^3 - 8y^2 + 15y$
59. $3x^2 + 18x - 81$
60. $8x^2 - 48x + 64$
61. $4x^3 + 36x^2 + 56x$
62. $2y^3 - 4y^2 - 30y$

6.3

Factoring a Trinomial In Exercises 63–76, factor the trinomial.

63. $3x^2 + 2x - 5$
64. $8x^2 - 18x + 9$
65. $2x^2 - 5x - 7$
66. $2x^2 - 3x - 35$
67. $6x^2 + 7x + 2$
68. $16x^2 + 13x - 3$
69. $4y^2 - 3y - 1$
70. $5x^2 - 12x + 7$
71. $3x^2 + 7x - 6$
72. $45y^2 - 8y - 4$
73. $3x^2 + 5x - 2$
74. $7x^2 - 4x - 3$
75. $2x^2 - 3x + 1$
76. $3x^2 + 8x + 4$

Finding Coefficients In Exercises 77 and 78, find all integers b such that the trinomial can be factored.

77. $x^2 + bx - 24$
78. $2x^2 + bx - 16$

Finding Constant Terms In Exercises 79 and 80, find two integers c such that the trinomial can be factored. (There are many correct answers.)

79. $2x^2 - 4x + c$
80. $5x^2 + 6x + c$

Factoring Completely In Exercises 81–90, factor the trinomial completely.

81. $3x^2 + 33x + 90$
82. $4x^2 + 12x - 16$
83. $6y^2 + 39y - 21$
84. $10b^2 - 38b + 24$
85. $6u^3 + 3u^2 - 30u$
86. $8x^3 - 8x^2 - 30x$
87. $8y^3 - 20y^2 + 12y$
88. $14x^3 + 26x^2 - 4x$
89. $6x^3 + 14x^2 - 12x$
90. $12y^3 + 36y^2 + 15y$

91. *Geometry* The pastry box shown in the figure has a height of x inches and a width of $(x + 1)$ inches. The volume of the box is $(3x^3 + 4x^2 + x)$. find the length of the box.

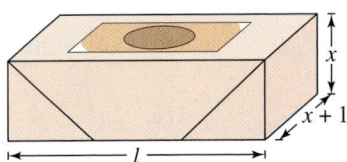

92. *Geometry* The area of the rectangle shown in the figure is $2x^2 + 5x + 3$. What is the area of the shaded region?

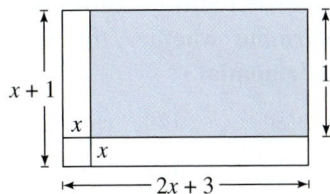

Factoring a Trinomial by Grouping In Exercises 93–100, factor the trinomial by grouping.

93. $2x^2 - 13x + 21$
94. $3a^2 - 13a - 10$
95. $4y^2 + y - 3$
96. $6z^2 - 43z + 7$
97. $6x^2 + 11x - 10$
98. $21x^2 - 25x - 4$
99. $14x^2 + 17x + 5$
100. $5t^2 + 27t - 18$

6.4

Factoring the Difference of Two Squares In Exercises 101–110, factor the difference of two squares.

101. $a^2 - 100$
102. $36 - b^2$
103. $25 - 4y^2$
104. $16b^2 - 1$
105. $12x^2 - 27$
106. $100x^2 - 64$
107. $(u + 1)^2 - 4$
108. $(v - 2)^2 - 9$
109. $16 - (z - 5)^2$
110. $81 - (x + 9)^2$

Factoring Completely In Exercises 111–118, factor the polynomial completely.

111. $3y^3 - 75y$
112. $16b^3 - 36b$
113. $s^3t - st^3$
114. $5x^3 - 20xy^2$
115. $x^4 - 81$
116. $2a^4 - 32$
117. $x^3 - 2x^2 + 4x - 8$
118. $b^3 - 3b^2 + 9b - 27$

Identifying a Perfect Square Trinomial In Exercises 119–122, determine whether the polynomial is a perfect square trinomial.

119. $x^2 + 25x + 10$
120. $x^2 + 12x + 36$
121. $4y^2 - 8y + 4$
122. $9b^2 - 18b - 9$

Factoring a Perfect Square Trinomial In Exercises 123–130, factor the perfect square trinomial.

123. $x^2 - 8x + 16$
124. $y^2 + 24y + 144$
125. $9s^2 + 12s + 4$
126. $16x^2 - 40x + 25$
127. $y^2 + 4yz + 4z^2$
128. $u^2 - 2uv + v^2$
129. $x^2 + \frac{2}{3}x + \frac{1}{9}$
130. $y^2 - \frac{4}{3}y + \frac{4}{9}$

Factoring the Sum or Difference of Two Cubes In Exercises 131–136, factor the sum or difference of two cubes.

131. $a^3 + 1$
132. $z^3 + 8$
133. $27 - 8t^3$
134. $z^3 - 125$
135. $8x^3 + y^3$
136. $125a^3 - 27b^3$

6.5

Using the Zero-Factor Property In Exercises 137–142, use the Zero-Factor Property to solve the equation.

137. $x(2x - 3) = 0$
138. $3x(5x + 1) = 0$
139. $(x + 3)(x - 2) = 0$
140. $(a - 5)(a + 8) = 0$
141. $(3x - 7)(2x + 1) = 0$
142. $(2y + 3)(4y - 5) = 0$

Solving a Quadratic Equation by Factoring In Exercises 143–154, solve the equation.

143. $x^2 - 81 = 0$
144. $121 - y^2 = 0$
145. $x^2 - 12x + 36 = 0$
146. $y^2 - y - 6 = 0$
147. $2t^2 - 3t - 2 = 0$
148. $4s^2 + s - 3 = 0$
149. $(z - 2)^2 - 4 = 0$
150. $(x + 1)^2 - 16 = 0$
151. $x(7 - x) = 12$
152. $x(x + 5) = 24$
153. $(x - 1)(x + 2) = 10$
154. $(x + 4)(x - 3) = 18$

155. **Number Problem** Find two consecutive positive even integers whose product is 168.

156. **Number Problem** Find two consecutive positive odd integers whose product is 399.

157. **Geometry** The height of a rectangular window is $1\frac{1}{2}$ times its width (see figure). The area of the window is 2400 square inches. Find the dimensions of the window.

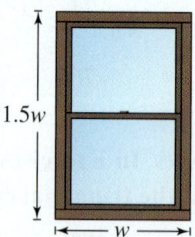

158. **Geometry** A box with a square base (see figure) has a surface area of 400 square inches. The side length of the base is x inches and the height of the box is 5 inches. Find the dimensions of the box. (*Hint:* The surface area of the box is given by $S = 2x^2 + 4xh$.)

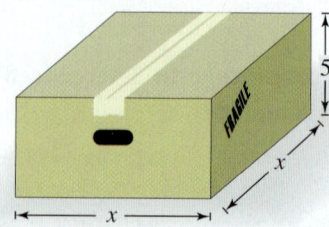

159. **Free-Falling Object** A rock is thrown vertically upward from a height of 48 feet with an initial velocity of 32 feet per second. The height h (in feet) of the rock is modeled by the position equation $h = -16t^2 + 32t + 48$, where t is the time measured in seconds. How long will it take for the rock to reach the ground?

Chapter Test

Solutions in English & Spanish and tutorial videos at AlgebraWithinReach.com

Take this test as you would take a test in class. After you are done, check your work against the answers in the back of the book.

In Exercises 1–10, factor the polynomial completely.

1. $9x^2 - 63x^5$
2. $z(z + 17) - 10(z + 17)$
3. $t^2 - 2t - 80$
4. $6x^2 - 11x + 4$
5. $3y^3 + 72y^2 - 75y$
6. $4 - 25v^2$
7. $x^3 + 8$
8. $100 - (z + 11)^2$
9. $x^3 + 2x^2 - 9x - 18$
10. $16 - z^4$

11. Determine the missing factor: $\frac{2}{5}x - \frac{3}{5} = \frac{1}{5}()$.

12. Find all integers b such that $x^2 + bx + 5$ can be factored.

13. Find a real number c such that $x^2 + 12x + c$ is a perfect square trinomial.

14. Explain why $(x + 1)(3x - 6)$ is not a complete factorization of $3x^2 - 3x - 6$.

In Exercises 15–18, solve the equation.

15. $(x + 4)(2x - 3) = 0$
16. $3x^2 + 7x - 6 = 0$
17. $y(2y - 1) = 6$
18. $2x^2 - 3x = 8 + 3x$

19. The suitcase shown below has a height of x inches and a width of $(x + 2)$ inches. The volume of the suitcase is $(x^3 + 6x^2 + 8x)$ cubic inches. Find the length of the suitcase.

20. The width of a rectangle is 5 inches less than its length. The area of the rectangle is 84 square inches. Find the dimensions of the rectangle.

21. An object is thrown upward from the top of the Aon Center in Chicago at a height of 1136 feet, with an initial velocity of 14 feet per second. The height h (in feet) of the object is modeled by the position equation

$$h = -16t^2 + 14t + 1136$$

where t is the time measured in seconds. How long will it take for the object to reach the ground?

22. Find two consecutive positive even integers whose product is 624.

Cumulative Test: Chapters 4–6

Solutions in English & Spanish and tutorial videos at AlgebraWithinReach.com

Take this test as you would take a test in class. After you are done, check your work against the answers in the back of the book.

1. Determine the quadrants in which the point $(-2, y)$ may be located. Assume $y \neq 0$. (y is a real number.)

2. Determine whether each ordered pair is a solution of the equation $9x - 4y + 36 = 0$.
   (a) $(-1, -1)$ (b) $(8, 27)$ (c) $(-4, 0)$ (d) $(3, -2)$

In Exercises 3 and 4, sketch the graph of the equation and determine any intercepts of the graph.

3. $y = 3 + |x|$
4. $x + 2y = 6$

5. The slope of a line is $-\frac{1}{4}$ and a point on the line is $(2, 1)$. Find the coordinates of a second point on the line. Explain why there are many correct answers.

6. Write the slope-intercept form of the equation of the line that passes through the point $\left(0, -\frac{3}{2}\right)$ and has slope $m = \frac{2}{5}$.

In Exercises 7 and 8, sketch the lines and determine whether they are parallel, perpendicular, or neither.

7. $y_1 = \frac{2}{3}x - 3$, $y_2 = -\frac{3}{2}x + 1$
8. $y_1 = 2 - 0.4x$, $y_2 = -\frac{2}{5}x$

9. Subtract: $(x^3 - 3x^2) - (x^3 + 2x^2 - 5)$.
10. Multiply: $(6z)(-7z)(z^2)$.
11. Multiply: $(3x + 5)(x - 4)$.
12. Multiply: $(5x - 3)(5x + 3)$.
13. Simplify: $(5x + 6)^2$.
14. Divide: $(6x^2 + 72x) \div 6x$.
15. Divide: $\dfrac{x^2 - 3x - 2}{x - 4}$.
16. Simplify: $\dfrac{(3xy^2)^{-2}}{6x^{-3}}$.
17. Factor: $2u^2 - 6u$.
18. Factor and simplify: $(x - 4)^2 - 36$.
19. Factor completely: $x^3 + 8x^2 + 16x$.
20. Factor completely: $x^3 + 2x^2 - 4x - 8$.
21. Solve: $u(u - 12) = 0$.
22. Solve: $5x^2 - 12x - 9 = 0$.

23. Rewrite the expression $\left(\dfrac{x}{2}\right)^{-2}$ using only positive exponents, and simplify.

24. A sales representative is reimbursed $150 per day for lodging and meals, plus $0.45 per mile driven. Write a linear equation giving the daily cost C to the company in terms of x, the number of miles driven. Find the cost for a day when the representative drives 70 miles.

25. You must perform two songs that are no longer than 10 minutes combined for a jazz band audition. Write a linear inequality that represents the numbers of minutes that can be spent performing each song. Then find the possible lengths of the second song when the first song is 6 minutes long.

7

Rational Expressions and Equations

- **7.1** Simplifying Rational Expressions
- **7.2** Multiplying and Dividing Rational Expressions
- **7.3** Adding and Subtracting Rational Expressions
- **7.4** Complex Fractions
- **7.5** Rational Equations and Applications

MASTERY IS WITHIN REACH!

"No matter what I did, I kept getting really worked up about tests in my math class. Finally, I took a math study skills course the college offered. One of the most helpful strategies I learned was the ten steps to taking a math test. I went in to my math tests with a plan and felt more confident. I still get nervous, but it doesn't keep me from doing well on my tests anymore."

Benjamin
Biology

See page 329 for suggestions about using a test-taking strategy.

7.1 Simplifying Rational Expressions

▶ Find the domain of a rational expression.
▶ Simplify rational expressions.

Rational Expressions and Their Domains

A **rational expression** is a fraction whose numerator and denominator are polynomials. The **domain** of a rational expression is the set of all real numbers for which it is defined. To find the values that are excluded from the domain, set the denominator equal to zero and find the solution(s) of that equation.

EXAMPLE 1 Finding the Domain of a Rational Expression

Find the domain of each rational expression.

a. $\dfrac{7}{x+3}$ b. $\dfrac{x-5}{9}$ c. $\dfrac{x}{x^2-16}$ d. $\dfrac{x^2+1}{x^2+4x-5}$

SOLUTION

a. The denominator is 0 when $x + 3 = 0$ or $x = -3$. So, the domain is all real values of x such that $x \neq -3$.

b. The denominator, 9, is never 0, and so the domain is the set of *all* real numbers.

c. For this rational expression, the denominator is 0 when $x^2 - 16 = 0$.

$$x^2 - 16 = 0 \quad \Longrightarrow \quad (x+4)(x-4) = 0 \qquad \text{Set denominator equal to 0 and factor.}$$

So, the domain is all real values of x such that $x \neq -4$ and $x \neq 4$.

d. For this rational expression, the denominator is 0 when $x^2 + 4x - 5 = 0$.

$$x^2 + 4x - 5 = 0 \quad \Longrightarrow \quad (x+5)(x-1) = 0 \qquad \text{Set denominator equal to 0 and factor.}$$

So, the domain is all real values of x such that $x \neq -5$ and $x \neq 1$.

Exercises Within Reach®

Solutions in English & Spanish and tutorial videos at AlgebraWithinReach.com

Finding the Domain of a Rational Expression In Exercises 1–10, find the domain of the rational expression.

1. $\dfrac{5}{x-4}$

2. $\dfrac{10}{x-6}$

3. $\dfrac{x}{x+2}$

4. $\dfrac{2z}{z+8}$

5. $\dfrac{x^2-4}{3}$

6. $\dfrac{y^2-1}{5}$

7. $\dfrac{3}{x^2+4}$

8. $\dfrac{4t}{t^2-25}$

9. $\dfrac{x^2}{x^2-x-2}$

10. $\dfrac{x^2}{x^2+3x-10}$

Section 7.1   Simplifying Rational Expressions   305

Application

EXAMPLE 2   **An Application Involving a Restricted Domain**

A publisher asks a printing company to print copies of a book. The printing company charges $5000 as a setup fee, plus $4 per book. The total cost (in dollars) of printing x books is

Total cost = $5000 + 4x$.   Total cost of printing x books.

The average cost per book depends on the number of books printed. For instance, the average cost per book for printing 100 books is

$$\text{Average cost} = \frac{\text{Total cost}}{\text{Number of books}}$$

$$= \frac{5000 + 4(100)}{100}$$

$= \$54$ per book.   Average cost per book for 100 books

On the other hand, the average cost per book for printing 10,000 books is

$$\frac{5000 + 4(10{,}000)}{10{,}000} = \$4.50 \text{ per book.}$$ Average cost per book for 10,000 books

In general, the average cost of printing x books is

$$\frac{5000 + 4x}{x}.$$ Average cost per book for x books

What is the domain of this rational expression?

SOLUTION

If you were simply considering the rational expression $(5000 + 4x)/x$ as a mathematical quantity, you would say that the domain is all real values of x such that $x \neq 0$. However, because this fraction is a mathematical model representing a real-life situation, you must decide which values of x make sense in real life. For this model, the variable x represents the number of books printed. Because the number of books printed cannot be negative and cannot be a fraction, you can conclude that the domain is the set of positive integers. That is,

Domain = $\{1, 2, 3, 4, \ldots\}$.

Study Tip

In applications involving rational expressions, it is often necessary to restrict the domain further. To indicate such a restriction, you should write the domain to the right of the fraction.

Exercises Within Reach®

Solutions in English & Spanish and tutorial videos at AlgebraWithinReach.com

11. *Average Cost* A manufacturer has a setup cost of $3000 for the production of a new alarm clock. The cost of labor and materials for producing each unit is $8.50.

  (a) Write a rational expression that models the average cost per unit when x units are produced.

  (b) Find the domain of the expression in part (a).

  (c) Find the average cost per unit when 100 units are produced.

12. *Average Cost* A manufacturer has a setup cost of $5000 for the production of a new tennis racquet. The cost of labor and materials for producing each unit is $12.75.

  (a) Write a rational expression that models the average cost per unit when x units are produced.

  (b) Find the domain of the expression in part (a).

  (c) Find the average cost per unit when 200 units are produced.

archidea/Shutterstock.com

Simplifying Rational Expressions

Simplifying Rational Expressions

Let a, b, and c represent real numbers, variables, or algebraic expressions such that $b \neq 0$ and $c \neq 0$. Then the following is valid.

$$\frac{ac}{bc} = \frac{a\cancel{c}}{b\cancel{c}} = \frac{a}{b}$$

Study Tip

Simplifying a rational expression requires two steps.

1. Completely factor the numerator and denominator.
2. Divide out any factors that are common to both the numerator and denominator.

Your success in simplifying rational expressions actually lies in your ability to factor completely the polynomials in both the numerator and denominator.

EXAMPLE 3 Simplifying Rational Expressions by Factoring

a. $\dfrac{15(x^2 - 2x)}{3x^2} = \dfrac{3 \cdot 5 \cdot x(x - 2)}{3 \cdot x \cdot x}$ Factor completely.

$\phantom{\dfrac{15(x^2 - 2x)}{3x^2}} = \dfrac{\cancel{3} \cdot 5 \cdot \cancel{x}(x - 2)}{\cancel{3} \cdot \cancel{x} \cdot x}$ Divide out common factors.

$\phantom{\dfrac{15(x^2 - 2x)}{3x^2}} = \dfrac{5(x - 2)}{x}$ Simplified form

b. $\dfrac{2x^2 - 4x}{(x - 2)^2} = \dfrac{2x(x - 2)}{(x - 2)(x - 2)}$ Factor completely.

$\phantom{\dfrac{2x^2 - 4x}{(x - 2)^2}} = \dfrac{2x\cancel{(x - 2)}}{(x - 2)\cancel{(x - 2)}}$ Divide out common factors.

$\phantom{\dfrac{2x^2 - 4x}{(x - 2)^2}} = \dfrac{2x}{x - 2}$ Simplified form

c. $\dfrac{2x - 6}{9 - 6x + x^2} = \dfrac{2(x - 3)}{(3 - x)(3 - x)}$ Factor completely.

$\phantom{\dfrac{2x - 6}{9 - 6x + x^2}} = \dfrac{2(x - 3)}{(-1)(x - 3)(3 - x)}$ $(3 - x) = (-1)(x - 3)$

$\phantom{\dfrac{2x - 6}{9 - 6x + x^2}} = \dfrac{2\cancel{(x - 3)}}{(-1)\cancel{(x - 3)}(3 - x)}$ Divide out common factor $(x - 3)$.

$\phantom{\dfrac{2x - 6}{9 - 6x + x^2}} = -\dfrac{2}{3 - x}$ Simplified form

Exercises Within Reach®

Solutions in English & Spanish and tutorial videos at AlgebraWithinReach.com

Simplifying a Rational Expression In Exercises 13–20, simplify the rational expression.

13. $\dfrac{14(x^2 + 3x)}{2x^2}$

14. $\dfrac{3x^2 - 9x}{12x^2}$

15. $\dfrac{5x^2 - 20x}{(x - 4)^2}$

16. $\dfrac{7x - x^2}{(x - 7)^2}$

17. $\dfrac{a + 2}{a^2 + 4a + 4}$

18. $\dfrac{a + 3}{a^2 + 6a + 9}$

19. $\dfrac{1 - y^3}{1 + y + y^2}$

20. $\dfrac{x^2 + x - 12}{x^2 - 6x + 9}$

Section 7.1 Simplifying Rational Expressions 307

EXAMPLE 4 **Adjusting the Domain after Simplifying**

a. $\dfrac{4x^3 - 8x^2}{4x^2} = \dfrac{4x^2(x - 2)}{4x^2}$ Factor numerator.

$= \dfrac{4x^2(x - 2)}{4x^2}$ Divide out common factor $4x^2$.

$= x - 2, \; x \neq 0$ Simplified form

b. $\dfrac{x^2 + 4x - 12}{3x - 6} = \dfrac{(x + 6)(x - 2)}{3(x - 2)}$ Factor numerator and denominator.

$= \dfrac{(x + 6)(x - 2)}{3(x - 2)}$ Divide out common factor $(x - 2)$.

$= \dfrac{x + 6}{3}, \; x \neq 2$ Simplified form

Study Tip
In Example 4, be sure you see that simplifying a rational expression can change its domain. For instance, in part (a), the domain of the original expression is all real values of *x* such that $x \neq 0$. So, to equate the original expression with the simplified expression, you must restrict the domain of the simplified expression to exclude 0. Similarly, 2 must be excluded in part (b).

EXAMPLE 5 **A Rational Expression Involving Two Variables**

Simplify the rational expression $\dfrac{x^2 - 2xy + y^2}{5x - 5y}$.

SOLUTION

$\dfrac{x^2 - 2xy + y^2}{5x - 5y} = \dfrac{(x - y)(x - y)}{5(x - y)}$ Factor numerator and denominator.

$= \dfrac{(x - y)(x - y)}{5(x - y)}$ Divide out common factor $(x - y)$.

$= \dfrac{x - y}{5}, \; x \neq y$ Simplified form

In simplified form, the domain of the expression is all real values of *x* such that $x \neq y$.

Exercises Within Reach®
Solutions in English & Spanish and tutorial videos at AlgebraWithinReach.com

Simplifying a Rational Expression In Exercises 21–32, simplify the rational expression.

21. $\dfrac{3(y^3 - 4y^2)}{27y^2}$

22. $\dfrac{x - 5}{2x - 10}$

23. $\dfrac{x^2 + 10x + 16}{4x + 8}$

24. $\dfrac{x^2 - x - 12}{4x^2 - 16x}$

25. $\dfrac{x^3 + 5x^2 + 6x}{x^2 - 4}$

26. $\dfrac{t^3 - t}{t^3 + 5t^2 - 6t}$

27. $\dfrac{x^3 - 2x^2 + x - 2}{x - 2}$

28. $\dfrac{x^2 - 9}{x^3 + x^2 - 9x - 9}$

29. $\dfrac{5r + 5s}{r^2 + 7rs + 6s^2}$

30. $\dfrac{6x + 12y}{x^2 - xy - 6y^2}$

31. $\dfrac{x^2 + 4xy - 21y^2}{x^2 + 2xy - 15y^2}$

32. $\dfrac{a^2 + 10ab + 21b^2}{a^2 + 11ab + 28b^2}$

Application EXAMPLE 6 An Application in Probability

A parachutist plans to land in the middle of a football stadium during a halftime show. The parachutist is certain of landing on the stadium field, but cannot guarantee a landing in a square area in the center of the field, as shown below.

The probability that the parachutist will land in the center square is equal to the ratio of the area of the center square to the area of the entire field. Find the probability.

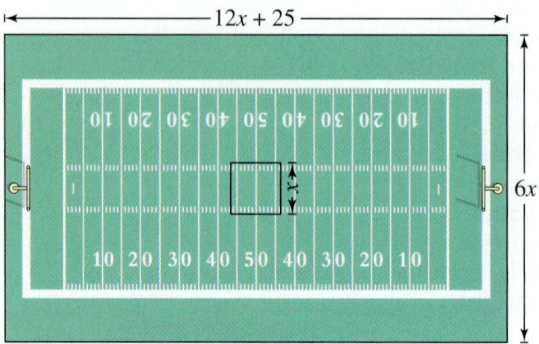

SOLUTION

The area of the center square is given by

$$\text{Area of center square} = (\text{Side})^2 = x^2.$$

The area of the entire field is given by

$$\text{Area of field} = (\text{Length})(\text{Width}) = (6x)(12x + 25).$$

So, the probability of the parachutist landing in the center square is

$$\frac{x^2}{(6x)(12x + 25)} = \frac{x}{6(12x + 25)}.$$

Exercises Within Reach ®   Solutions in English & Spanish and tutorial videos at AlgebraWithinReach.com

Probability In Exercises 33–36, the probability of hitting the shaded portion of the region with a dart is equal to the ratio of the shaded area to the total area of the figure. **Find** the probability.

33.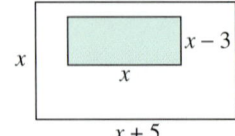

34.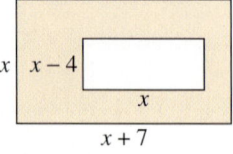

35.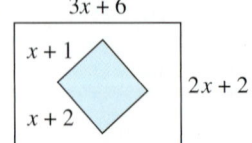

36.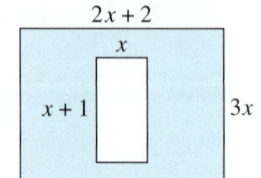

Application

EXAMPLE 7 An Environmental Application

A utility company burns coal to produce electricity. The cost C (in dollars) of removing p percent of the air pollutants in the stack emission of the utility company is given by

$$C = \frac{80{,}000p}{100 - p}.$$

a. Determine the domain of the expression.

b. Create a table showing the costs of removing 20%, 40%, 60%, and 80% of the pollutants in the stack emission.

c. Use the table in part (b) to describe the relationship between the cost and the percent. According to this model, can you remove 100% of the pollutants?

SOLUTION

a. The domain of the expression

$$C = \frac{80{,}000p}{100 - p}$$

is $0 \le p < 100$.

b.

Percent, p	Cost, C
20%	$20,000
40%	$53,333
60%	$120,000
80%	$320,000

c. From the table, you can see that as the company seeks to remove higher and higher percents of the pollutants, the cost increases dramatically. From this model, it would not be possible to remove 100% of the pollutants.

Exercises Within Reach®

Solutions in English & Spanish and tutorial videos at AlgebraWithinReach.com

37. *Boiling Temperature* As air pressure increases, the temperature at which water boils also increases. A model that relates air pressure x (in pounds per square inch) to boiling temperature B (in degrees Fahrenheit) is

$$B = \frac{156.89x + 7.34x^2}{x + 0.017x^2}, \quad 10 \le x \le 100.$$

(a) Simplify the rational expression.

(b) Use the model to estimate the boiling temperature of water when $x = 14.7$ pounds per square inch (approximate air pressure at sea level). Round your answer to three decimal places.

38. *Depreciation* The value V of an automobile t years after it is purchased is given by

$$V = \frac{1{,}500{,}000t}{54t + 24t^2}, \quad 1 < t < 5$$

(a) Simplify the rational expression.

(b) Use the model to estimate the value of the automobile 3 years after it is purchased.

Chapter 7 Rational Expressions and Equations

Concept Summary: Simplifying Rational Expressions

What

A **rational expression** is in **simplified form** when its numerator and denominator have no common factors (other than ± 1).

EXAMPLE

Simplify $\dfrac{3x}{3x+12}$.

How

Be sure to completely factor the numerator and denominator of a rational expression before concluding that there are no common factors.

EXAMPLE

$\dfrac{3x}{3x+12}$ Original expression

$= \dfrac{3 \cdot x}{3(x+4)}$ Factor completely.

$= \dfrac{\cancel{3} \cdot x}{\cancel{3}(x+4)}$ Divide out factors.

$= \dfrac{x}{x+4}$ Simplified form

Why

Simplifying a rational expression into a more usable form is a skill frequently used in algebra.

Exercises Within Reach®

Worked-out solutions to odd-numbered exercises at AlgebraWithinReach.com

Concept Summary Check

39. Vocabulary What is a rational expression?

40. Writing How can you determine whether a rational expression is in simplified form?

41. Precision While simplifying a rational expression, do you divide out factors or terms? Explain.

42. Reasoning Is the following expression in simplified form? Explain your reasoning.

$$\dfrac{3 \cdot x}{3(x+4)}$$

Extra Practice

Finding the Domain of a Rational Expression In Exercises 43 and 44, find the domain of the rational expression.

43. $\dfrac{z-3}{3z^2 - z - 2}$

44. $\dfrac{y+5}{4y^2 + y - 3}$

Evaluating a Rational Expression In Exercises 45 and 46, evaluate the rational expression for each specified value. If not possible, state the reason.

Expression	Values		Expression	Values	
45. $\dfrac{x}{x-3}$	(a) $x = 0$ (b) $x = 3$ (c) $x = 10$ (d) $x = -3$		**46.** $\dfrac{x+1}{x^2 - 4}$	(a) $x = 2$ (b) $x = 1$ (c) $x = -5$ (d) $x = -2$	

Simplifying a Rational Expression In Exercises 47–54, simplify the rational expression.

47. $\dfrac{x^2 - 36z^2}{x - 6z}$

48. $\dfrac{a - 9b}{a^2 - 81b^2}$

49. $\dfrac{7s^2 - 28t^2}{28t^2 - 7s^2}$

50. $\dfrac{9m^2 - 4n^2}{4n^2 - 9m^2}$

51. $\dfrac{x^2 - 2x + 1}{1 - x^2}$

52. $\dfrac{u^2 - 3u - 4}{16 - u^2}$

53. $\dfrac{x^4 - y^4}{(y - x)^4}$

54. $\dfrac{16y^4 - x^4}{(x^2 + 4y^2)(x - 2y)}$

Exploration In Exercises 55 and 56, complete the table. Explain why the values of the expressions agree for all values of x except one.

55.

x	2	2.5	3	3.5	4
$\dfrac{x^3 - 3x^2}{x - 3}$					
x^2					

56.

x	0	0.5	1	1.5	2
$\dfrac{x - 1}{x^2 + 2x - 3}$					
$\dfrac{1}{x + 3}$					

Explaining Concepts

57. *True or False?* To simplify a rational expression, you can divide out common factors of the numerator and denominator. Justify your answer.

58. *Error Analysis* Describe and correct the error.

$$\dfrac{2x^2 + 3x}{x^2 + 4x} \cdot \dfrac{2 + 3}{1 + 4} = 1$$

59. *Reasoning* Is $(4x)/(2x)$ equivalent to 2? Explain.

60. *Creating an Example* Construct a rational expression that cannot be simplified. (There are many correct answers.)

61. *Creating an Example* Construct a rational expression that can be simplified. (There are many correct answers.)

62. *Writing* Two algebra students hand in the following work.

Student A
$$\dfrac{5x - 25}{x - 5} = 5 - 5 = 0$$

Student B
$$\dfrac{3(x + 2)}{3x + 6} = \dfrac{x + 2}{x + 6}$$

Find all errors or misconceptions in each student's work. In a written, detailed explanation, show the students how to work each problem correctly.

Cumulative Review

In Exercises 63–66, solve the equation and check your solution.

63. $14 - 2x = x + 2$

64. $7 - 3(1 - 2p) = 2p + 4$

65. $\dfrac{x}{3} + 5 = 8$

66. $\dfrac{x}{3} + \dfrac{x}{2} = \dfrac{1}{3}$

In Exercises 67–70, use the Zero-Factor Property to solve the equation.

67. $x(x - 8) = 0$

68. $(y + 12)(y - 3) = 0$

69. $(5z + 6)(z + 14) = 0$

70. $7w(2w - 9)(w + 3) = 0$

7.2 Multiplying and Dividing Rational Expressions

▶ Multiply rational expressions and simplify.
▶ Divide rational expressions and simplify.

Multiplying Rational Expressions

> **Multiplying Rational Expressions**
>
> Let a, b, c, and d represent real numbers, variables, or algebraic expressions such that $b \neq 0$ and $d \neq 0$. Then the product of a/b and c/d is
>
> $$\frac{a}{b} \cdot \frac{c}{d} = \frac{ac}{bd}.$$

Study Tip

To recognize common factors when simplifying a product, use factoring in the numerator and denominator, as demonstrated in Example 1.

EXAMPLE 1 Multiplying Rational Expressions

Multiply and simplify: $\dfrac{3x^2y}{2xy^2} \cdot \dfrac{-10xy^3}{6x^3}$.

SOLUTION

$\dfrac{3x^2y}{2xy^2} \cdot \dfrac{-10xy^3}{6x^3} = \dfrac{(3x^2y)(-10xy^3)}{(2xy^2)(6x^3)}$ Multiply numerators and denominators.

$= \dfrac{-3(10)x^3y^4}{2(6)x^4y^2}$ Simplify.

$= \dfrac{-3(2)(5)(x^3)(y^2)(y^2)}{2(3)(2)(x^3)(x)(y^2)}$ Factor and divide out common factors.

$= -\dfrac{5y^2}{2x}, \; y \neq 0$ Simplified form

Exercises Within Reach® Solutions in English & Spanish and tutorial videos at AlgebraWithinReach.com

Multiplying Rational Expressions In Exercises 1–12, multiply and simplify.

1. $\dfrac{8x^2}{3} \cdot \dfrac{9}{16x}$

2. $\dfrac{6x}{5} \cdot \dfrac{1}{x}$

3. $\dfrac{12x^2}{6x} \cdot \dfrac{12x}{8x^2}$

4. $\dfrac{25x^2}{8x} \cdot \dfrac{8x}{5x}$

5. $\left(-\dfrac{5a^4}{6a}\right)\left(-\dfrac{2}{a}\right)$

6. $\left(-\dfrac{10}{t^7}\right)\left(-\dfrac{3t^2}{25t}\right)$

7. $\dfrac{20y^3}{6xy} \cdot \dfrac{4x^2y}{5y^2}$

8. $\dfrac{14x^2y^2}{3x^4} \cdot \dfrac{2x^4}{7y^3}$

9. $\dfrac{y-1}{5} \cdot \dfrac{5}{y-1}$

10. $\dfrac{x+1}{2} \cdot \dfrac{2}{x+1}$

11. $\dfrac{x+1}{2} \cdot \dfrac{4x}{x+1}$

12. $\dfrac{x-3}{6x} \cdot \dfrac{4}{x-3}$

Section 7.2 Multiplying and Dividing Rational Expressions

EXAMPLE 2 **Multiplying Rational Expressions**

a. $\dfrac{4x}{x^2 - 9} \cdot \dfrac{x - 3}{8x^2 + 12x}$

$= \dfrac{4x(x - 3)}{(x^2 - 9)(8x^2 + 12x)}$ Multiply numerators and denominators.

$= \dfrac{4x(x - 3)}{(x + 3)(x - 3)(4x)(2x + 3)}$ Factor.

$= \dfrac{4x(x - 3)}{(x + 3)(x - 3)(4x)(2x + 3)}$ Divide out common factors. Factor of 1 remains in numerator.

$= \dfrac{1}{(x + 3)(2x + 3)},\ x \ne 0,\ x \ne 3$ Simplified form

b. $\dfrac{x + 2}{x + 4} \cdot (3x) = \dfrac{x + 2}{x + 4} \cdot \dfrac{3x}{1}$ Rewrite in fractional form.

$= \dfrac{(x + 2)(3x)}{x + 4}$ Multiply numerators and denominators.

$= \dfrac{3x(x + 2)}{x + 4}$ Simplified form

c. $\dfrac{x}{2x^2 - x - 3} \cdot (2x^2 + 11x - 21)$

$= \dfrac{x}{2x^2 - x - 3} \cdot \dfrac{2x^2 + 11x - 21}{1}$ Rewrite in fractional form.

$= \dfrac{x(2x^2 + 11x - 21)}{2x^2 - x - 3}$ Multiply numerators and denominators.

$= \dfrac{x(2x - 3)(x + 7)}{(2x - 3)(x + 1)}$ Factor.

$= \dfrac{x(2x - 3)(x + 7)}{(2x - 3)(x + 1)}$ Divide out common factor.

$= \dfrac{x(x + 7)}{x + 1},\ x \ne \dfrac{3}{2}$ Simplified form

Exercises Within Reach ® Solutions in English & Spanish and tutorial videos at AlgebraWithinReach.com

Multiplying Rational Expressions In Exercises 13−20, multiply and simplify.

13. $\dfrac{x - 3}{x^2 - 16} \cdot \dfrac{x + 4}{2x^2 - 6x}$

14. $\dfrac{x + 7}{3x^2 - 15x} \cdot \dfrac{x - 5}{x^2 - 49}$

15. $\dfrac{y + 5}{y - 2} \cdot (2y)$

16. $\dfrac{z - 4}{z - 1} \cdot (-2z)$

17. $\dfrac{3x}{5x - 15} \cdot (3 - x)$

18. $\dfrac{7y}{24 - 6y} \cdot (y - 4)$

19. $\dfrac{(x - 5)^2}{x + 5} \cdot \dfrac{x + 5}{x - 5}$

20. $\dfrac{y + 2}{y - 2} \cdot \dfrac{(y - 2)^2}{y + 2}$

EXAMPLE 3 Multiplying Rational Expressions

$$\frac{x-y}{6x+4y} \cdot \frac{3x+2y}{y^2-x^2}$$

$$= \frac{(x-y)(3x+2y)}{(6x+4y)(y^2-x^2)} \quad \text{Multiply numerators and denominators.}$$

$$= \frac{(x-y)(3x+2y)}{2(3x+2y)(y+x)(y-x)} \quad \text{Factor.}$$

$$= \frac{(x-y)(3x+2y)}{2(3x+2y)(y+x)(-1)(x-y)} \quad (y-x) = (-1)(x-y)$$

$$= \frac{\cancel{(x-y)}\cancel{(3x+2y)}}{2\cancel{(3x+2y)}(y+x)(-1)\cancel{(x-y)}} \quad \text{Divide out common factors.}$$

$$= -\frac{1}{2(y+x)}, \; x \neq y, \; x \neq -\frac{2}{3}y \quad \text{Simplified form}$$

Study Tip
In Example 3, the factor $(y-x)$ in the denominator is factored as $(-1)(x-y)$. Because the factor $(x-y)$ appears in both the numerator and the denominator, it can be divided out.

EXAMPLE 4 Multiplying Three Rational Expressions

$$\frac{3}{x} \cdot \frac{x+1}{x+2} \cdot \frac{x^2-2x-8}{x+1}$$

$$= \frac{3(x+1)(x+2)(x-4)}{x(x+2)(x+1)} \quad \text{Multiply and factor.}$$

$$= \frac{3\cancel{(x+1)}\cancel{(x+2)}(x-4)}{x\cancel{(x+2)}\cancel{(x+1)}} \quad \text{Divide out common factors.}$$

$$= \frac{3(x-4)}{x}, \; x \neq -2, \; x \neq -1 \quad \text{Simplified form}$$

Exercises Within Reach® Solutions in English & Spanish and tutorial videos at AlgebraWithinReach.com

Multiplying Rational Expressions In Exercises 21–30, multiply and simplify.

21. $\dfrac{1-r}{3} \cdot \dfrac{3}{r-1}$

22. $\dfrac{t-6}{7} \cdot \dfrac{7}{6-t}$

23. $\dfrac{x^2+x-2}{x^3+x^2} \cdot \dfrac{x}{x^2+3x+2}$

24. $\dfrac{x^2-25}{x^2-3x-10} \cdot \dfrac{x+2}{x}$

25. $\dfrac{9y-15z}{7y+14z} \cdot \dfrac{2y+4z}{3y-5z}$

26. $\dfrac{4y-16x}{5y+15x} \cdot \dfrac{2y+6x}{y-4x}$

27. $\dfrac{4}{x} \cdot \dfrac{x+2}{x+6} \cdot \dfrac{x}{x+2}$

28. $\dfrac{x}{7} \cdot \dfrac{x-7}{x} \cdot \dfrac{x+1}{x-7}$

29. $\dfrac{a+1}{a-1} \cdot \dfrac{a^2-2a+1}{a} \cdot (3a^2+3a)$

30. $\dfrac{z^2-z-2}{z} \cdot \dfrac{2z^2+3z}{2z+3} \cdot \dfrac{z}{z-2}$

Dividing Rational Expressions

> **Dividing Rational Expressions**
>
> Let a, b, c, and d represent real numbers, variables, or algebraic expressions such that $b \neq 0$, $c \neq 0$, and $d \neq 0$. Then the quotient of a/b and c/d is
>
> $$\frac{a}{b} \div \frac{c}{d} = \frac{a}{b} \cdot \frac{d}{c} = \frac{ad}{bc}.$$

EXAMPLE 5 Dividing Rational Expressions

$\dfrac{x}{x+4} \div \dfrac{x+3}{x+4} = \dfrac{x}{x+4} \cdot \dfrac{x+4}{x+3}$ Invert divisor and multiply.

$= \dfrac{(x)(x+4)}{(x+4)(x+3)}$ Multiply numerators and denominators.

$= \dfrac{(x)\cancel{(x+4)}}{\cancel{(x+4)}(x+3)}$ Divide out common factor.

$= \dfrac{x}{x+3},\ x \neq -4$ Simplified form

EXAMPLE 6 Dividing Rational Expressions

$\dfrac{x^2 - 2x}{x^2 - 6x + 8} \div \dfrac{2x}{3x - 12}$

$= \dfrac{x^2 - 2x}{x^2 - 6x + 8} \cdot \dfrac{3x - 12}{2x}$ Invert divisor and multiply.

$= \dfrac{(x^2 - 2x)(3x - 12)}{(x^2 - 6x + 8)(2x)}$ Multiply numerators and denominators.

$= \dfrac{(x)(x-2)(3)(x-4)}{(x-2)(x-4)(2x)}$ Factor.

$= \dfrac{\cancel{(x)}\cancel{(x-2)}(3)\cancel{(x-4)}}{\cancel{(x-2)}\cancel{(x-4)}(2\cancel{x})}$ Divide out common factors.

$= \dfrac{3}{2},\ x \neq 0,\ x \neq 2,\ x \neq 4$ Simplified form

Exercises Within Reach®

Solutions in English & Spanish and tutorial videos at AlgebraWithinReach.com

Dividing Rational Expressions In Exercises 31–36, **divide** and simplify.

31. $\dfrac{y}{y+8} \div \dfrac{y-6}{y+8}$

32. $\dfrac{z-3}{z^2} \div \dfrac{z-3}{z+5}$

33. $\dfrac{3(x+4)}{8} \div \dfrac{x+4}{2}$

34. $\dfrac{10(x-3)}{7} \div \dfrac{x-3}{21}$

35. $\dfrac{(2x)^2}{(x+2)^2} \div \dfrac{4x}{(x+2)^3}$

36. $\dfrac{3(x+1)^2}{5x} \div \dfrac{9(x+1)}{10x^2}$

Chapter 7 Rational Expressions and Equations

EXAMPLE 7 **Dividing Rational Expressions**

$$\frac{x^2 + 3x - 10}{x^2} \div (x - 2)^2$$

$$= \frac{x^2 + 3x - 10}{x^2} \div \frac{(x - 2)^2}{1} \qquad \text{Rewrite in fractional form.}$$

$$= \frac{x^2 + 3x - 10}{x^2} \cdot \frac{1}{(x - 2)^2} \qquad \text{Invert divisor and multiply.}$$

$$= \frac{(x + 5)(x - 2)}{(x^2)(x - 2)(x - 2)} \qquad \text{Factor.}$$

$$= \frac{(x + 5)\cancel{(x - 2)}}{(x^2)(x - 2)\cancel{(x - 2)}} \qquad \text{Divide out common factor.}$$

$$= \frac{x + 5}{x^2(x - 2)} \qquad \text{Simplified form}$$

EXAMPLE 8 **Dividing Rational Expressions**

$$\frac{16 - x^2}{2x + 8} \div \frac{2x^2 - 9x + 4}{6x + 2}$$

$$= \frac{16 - x^2}{2x + 8} \cdot \frac{6x + 2}{2x^2 - 9x + 4} \qquad \text{Invert divisor and multiply.}$$

$$= \frac{(-1)(x^2 - 16)(6x + 2)}{(2x + 8)(2x^2 - 9x + 4)} \qquad 16 - x^2 = (-1)(x^2 - 16)$$

$$= \frac{(-1)(x + 4)(x - 4)(2)(3x + 1)}{(2)(x + 4)(x - 4)(2x - 1)} \qquad \text{Factor.}$$

$$= \frac{(-1)\cancel{(x + 4)}\cancel{(x - 4)}\cancel{(2)}(3x + 1)}{\cancel{(2)}\cancel{(x + 4)}\cancel{(x - 4)}(2x - 1)} \qquad \text{Divide out common factors.}$$

$$= -\frac{3x + 1}{2x - 1}, \; x \ne -4, x \ne -\frac{1}{3}, x \ne 4 \qquad \text{Simplified form}$$

Exercises Within Reach ®

Solutions in English & Spanish and tutorial videos at AlgebraWithinReach.com

Dividing Rational Expressions In Exercises 37–44, **divide** and simplify.

37. $\dfrac{x^2 - 7x + 12}{x + 4} \div (x - 3)^2$

38. $\dfrac{x^2 - x - 2}{x + 2} \div (x - 2)^2$

39. $(x - 3) \div \dfrac{x^2 + 3x - 18}{x}$

40. $(x + 3) \div \dfrac{3x^2 + 18x + 27}{x^2 + 1}$

41. $\dfrac{x + 2}{7 - x} \div \dfrac{x^2 - 5x + 6}{x^2 - 9x + 14}$

42. $\dfrac{x + 6}{4 - x} \div \dfrac{x^2 + 9x + 18}{x^2 - 3x - 4}$

43. $\dfrac{25 - y^2}{4y + 20} \div \dfrac{2y^2 - 7y - 15}{24y + 4}$

44. $\dfrac{3y - 6}{y^2 - 4} \div \dfrac{6y + 12}{3y^2 + 2y - 8}$

Application

EXAMPLE 9 Geometry: Analyzing Dimensions

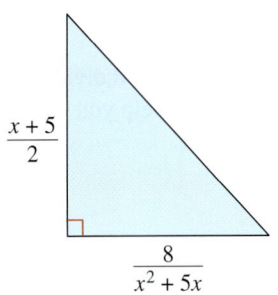

The base and height of a triangle are given by

$$\frac{8}{x^2 + 5x} \quad \text{and} \quad \frac{x+5}{2}$$

respectively. (Assume $x > 0$.)

a. Write an expression for the area of the triangle in terms of x. Simplify the expression.

b. Use the expression in part (a) to complete the table.

DATA x	2	4	6	8	10	12
Area						

c. As x increases, determine whether (i) the base, (ii) the height, and (iii) the area increase or decrease.

SOLUTION

a. $\text{Area} = \frac{1}{2}(\text{base})(\text{height})$ — Write formula for area.

$= \frac{1}{2} \cdot \frac{8}{x^2 + 5x} \cdot \frac{x+5}{2}$ — Substitute.

$= \frac{2(4)(x+5)}{4(x)(x+5)}$ — Factor.

$= \frac{2(4)(x+5)}{4(x)(x+5)}$ — Divide out common factors.

$= \frac{2}{x}$ — Simplify.

b.

x	2	4	6	8	10	12
Area	1	$\frac{1}{2}$	$\frac{1}{3}$	$\frac{1}{4}$	$\frac{1}{5}$	$\frac{1}{6}$

c. As x increases, (i) the base decreases, (ii) the height increases, and (iii) the area decreases.

Exercises Within Reach®

Solutions in English & Spanish and tutorial videos at AlgebraWithinReach.com

Finding the Area of a Triangle In Exercises 45 and 46, write an expression for the area of the triangle. (Assume $x > 0$.) Does the area increase or decrease as x increases?

45.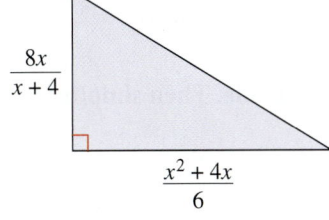

46.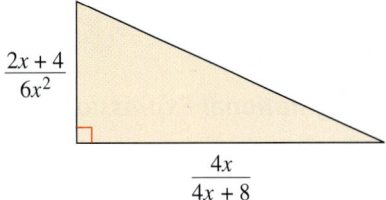

Concept Summary: Multiplying and Dividing Rational Expressions

What
The rules for multiplying and dividing rational expressions are the same as the rules for multiplying and dividing numerical fractions.

EXAMPLE
Divide:
$$\frac{x^2 - 2x - 24}{x - 6} \div \frac{x^2 + 6x + 8}{x + 3}$$

How
To divide two rational expressions, multiply the first expression by the reciprocal of the second.

EXAMPLE
$$\frac{x^2 - 2x - 24}{x - 6} \div \frac{x^2 + 6x + 8}{x + 3} \quad \text{Original}$$

$$= \frac{x^2 - 2x - 24}{x - 6} \cdot \frac{x + 3}{x^2 + 6x + 8} \quad \text{Invert and multiply.}$$

$$= \frac{(x - 6)(x + 4)(x + 3)}{(x - 6)(x + 2)(x + 4)} \quad \text{Factor.}$$

$$= \frac{\cancel{(x - 6)}\cancel{(x + 4)}(x + 3)}{\cancel{(x - 6)}(x + 2)\cancel{(x + 4)}} \quad \text{Divide out common factors.}$$

$$= \frac{x + 3}{x + 2}, \ x \neq -4, x \neq -3, x \neq 6 \quad \text{Simplified form}$$

Why
Knowing how to multiply and divide rational expressions will help you solve rational equations.

Exercises Within Reach®

Worked-out solutions to odd-numbered exercises at AlgebraWithinReach.com

Concept Summary Check

47. Writing Explain how to multiply two rational expressions.

48. Common Factors Describe how to divide out common factors.

49. Think About It Explain how dividing two rational expressions is different from multiplying two rational expressions.

50. Writing Define the reciprocal of a rational expression.

Extra Practice

Reasoning In Exercises 51−56, determine the missing factor.

51. $\dfrac{x - 7}{x + 10} \cdot \dfrac{\boxed{}}{x - 7} = \dfrac{x - 3}{x + 10}, \ x \neq 7$

52. $\dfrac{2x + 1}{x - 11} \cdot \dfrac{x - 11}{\boxed{}} = \dfrac{2x + 1}{x - 1}, \ x \neq 11$

53. $\dfrac{x^2 - 2x - 8}{x - 5} \cdot \dfrac{x - 5}{\boxed{}} = x + 2, \ x \neq 4, x \neq 5$

54. $\dfrac{x - 1}{x^2 + 8x + 12} \cdot \dfrac{\boxed{}}{x - 1} = x + 6, \ x \neq -6, x \neq -2, x \neq 1$

55. $\dfrac{x + 3}{x - 4} \cdot \dfrac{\boxed{}}{x + 3} = x + 2, \ x \neq -3, x \neq 4$

56. $\dfrac{x - 5}{x + 2} \cdot \dfrac{\boxed{}}{x - 5} = 2x - 3, \ x \neq -2, x \neq 5$

Multiplying or Dividing Rational Expressions In Exercises 57−60, multiply or divide. Then simplify.

57. $\dfrac{2}{z + 3} \cdot \dfrac{z^2 + 6z + 9}{z - 3} \cdot \dfrac{4}{z^2 - 9}$

58. $(x + 5)^2 \cdot \dfrac{x}{x^2 - 25} \cdot \dfrac{x^2 - x - 20}{x^2}$

59. $\dfrac{a^2 - 10a + 25}{a^2 + 7a + 12} \div \dfrac{a^2 - a - 20}{a^2 + 6a + 9}$

60. $\dfrac{a^2 + 5a + 4}{a^2 - 2a + 1} \div \dfrac{a^2 + 8a + 16}{a^2 - 5a - 6}$

Section 7.2 Multiplying and Dividing Rational Expressions

Using Order of Operations In Exercises 61–64, perform the indicated operations and simplify.

61. $\left[\left(\dfrac{x+2}{3}\right)^2 \cdot \left(\dfrac{x+1}{2}\right)^2\right] \div \dfrac{(x+1)(x+2)}{36}$

62. $\left[\left(\dfrac{4}{x-1}\right)^2 \cdot \left(\dfrac{x+1}{3}\right)^3\right] \div \dfrac{(x+1)^2}{27(x-1)}$

63. $\left[\dfrac{(t+2)^3}{(t+1)^3} \div \dfrac{t^2+4t+4}{t^2+2t+1}\right] \cdot \dfrac{t+1}{t+2}$

64. $\left(\dfrac{3y^3+6y^2}{y^2-y-12} \div \dfrac{y^2-y}{y^2-2y-8}\right) \cdot \dfrac{y^2+5y+6}{y^2}$

65. *Pump Rate* A gasoline pump can pump 5 gallons per minute. Determine the time required to pump (a) 1 gallon, (b) x gallons, and (c) 120 gallons.

66. *Printing Rate* A photo printer produces 12 photos per minute. Find the time required to print (a) 1 photo, (b) x photos, and (c) 32 photos.

Explaining Concepts

67. *Creating an Example* Give an example of three rational expressions whose product is 1. (There are many correct answers.)

68. *True or False?* $10 \div x = \dfrac{1}{10} \cdot x$. Justify your answer.

69. *Error Analysis* Describe and correct the error.

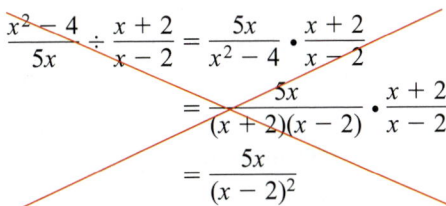

70. *Reasoning* The following expressions have the same value when $x = 4$.

(i) $\dfrac{3x^2+4x-4}{2x^2+5x+2}$

(ii) $\dfrac{x+6}{2x+1}$

(iii) $\dfrac{3x-2}{2x+1}$

Are they all equivalent? Is there an equivalent pair? Are there any excluded values?

Cumulative Review

In Exercises 71–74, factor the polynomial.

71. $3x^2 + 7x$

72. $16 - (x-11)^2$

73. $x^2 + 7x - 18$

74. $10x^2 + 13x - 3$

In Exercises 75–82, find the domain of the rational expression.

75. $\dfrac{3x}{48}$

76. $\dfrac{15x^3}{5x}$

77. $\dfrac{3(x+4)}{4(x+4)^2}$

78. $\dfrac{2x+6}{x+9}$

79. $\dfrac{x-4}{8-2x}$

80. $\dfrac{x^2+3x+2}{x^2+1}$

81. $\dfrac{x-2}{x^2-4}$

82. $\dfrac{x^2+5x+6}{x^2+10x+16}$

7.3 Adding and Subtracting Rational Expressions

▶ Add or subtract rational expressions with like denominators.
▶ Find the least common multiple of two or more polynomials.
▶ Add or subtract rational expressions with unlike denominators.

Rational Expressions with Like Denominators

Combining Rational Expressions with Like Denominators

Let a, b, and c represent real numbers, variables, or algebraic expressions such that $c \neq 0$.

1. $\dfrac{a}{c} + \dfrac{b}{c} = \dfrac{a+b}{c}$ — Add fractions with like denominators.

2. $\dfrac{a}{c} - \dfrac{b}{c} = \dfrac{a-b}{c}$ — Subtract fractions with like denominators.

EXAMPLE 1 Adding and Subtracting with Like Denominators

a. $\dfrac{x}{3} + \dfrac{2-x}{3} = \dfrac{x + (2-x)}{3} = \dfrac{2}{3}$ — Add numerators.

b. $\dfrac{5}{x+4} - \dfrac{2x}{x+4} = \dfrac{5-2x}{x+4}$ — Subtract numerators.

EXAMPLE 2 Adding Rational Expressions and Simplifying

$\dfrac{x}{x^2-4} + \dfrac{2}{x^2-4} = \dfrac{x+2}{x^2-4}$ — Add numerators.

$= \dfrac{(x+2) \cdot 1}{(x+2)(x-2)}$ — Factor and divide out common factor.

$= \dfrac{1}{x-2}, \quad x \neq -2$ — Simplified form

Exercises Within Reach®

Solutions in English & Spanish and tutorial videos at AlgebraWithinReach.com

Adding and Subtracting with Like Denominators In Exercises 1–12, **combine** and simplify.

1. $\dfrac{y}{4} + \dfrac{3y}{4}$

2. $\dfrac{2x}{5} - \dfrac{7x}{5}$

3. $\dfrac{5}{3a} + \dfrac{9}{3a}$

4. $\dfrac{16}{5z} - \dfrac{11}{5z}$

5. $\dfrac{x}{3} + \dfrac{1-x}{3}$

6. $\dfrac{-6t}{9} - \dfrac{12-8t}{9}$

7. $\dfrac{3}{x+2} - \dfrac{7x}{x+2}$

8. $\dfrac{4y}{y-3} - \dfrac{6}{y-3}$

9. $\dfrac{x}{x^2-1} + \dfrac{1}{x^2-1}$

10. $\dfrac{x}{x^2-9} - \dfrac{3}{x^2-9}$

11. $\dfrac{3x-2}{(x+1)^2} + \dfrac{4x+5}{(x+1)^2}$

12. $\dfrac{2x+1}{(x-5)^2} + \dfrac{3x-6}{(x-5)^2}$

Section 7.3 Adding and Subtracting Rational Expressions

EXAMPLE 3 **Subtracting Rational Expressions and Simplifying**

$$\frac{x}{x^2 - 2x} - \frac{5x - x^2}{x^2 - 2x} = \frac{x - (5x - x^2)}{x^2 - 2x} \quad \text{Subtract numerators.}$$

$$= \frac{x - 5x + x^2}{x^2 - 2x} \quad \text{Distributive Property}$$

$$= \frac{x^2 - 4x}{x^2 - 2x} \quad \text{Combine like terms.}$$

$$= \frac{x(x - 4)}{x(x - 2)} \quad \text{Factor and divide out common factor.}$$

$$= \frac{x - 4}{x - 2}, \quad x \neq 0 \quad \text{Simplified form}$$

Study Tip

To add or subtract rational expressions with opposite denominators (denominators that differ only in sign), multiply the numerator and denominator of either expression by -1. Then both expressions will have the same denominator.

EXAMPLE 4 **Adding Expressions with Opposite Denominators**

$$\frac{2x}{x - 4} + \frac{5x - 7}{4 - x} = \frac{2x}{x - 4} + \frac{(-1)(5x - 7)}{(-1)(4 - x)} \quad \text{Multiply numerator and denominator by } -1.$$

$$= \frac{2x}{x - 4} + \frac{(-5x + 7)}{(-4 + x)} \quad \text{Distributive Property}$$

$$= \frac{2x}{x - 4} + \frac{(-5x + 7)}{x - 4} \quad \text{Rewrite } -4 + x \text{ as } x - 4.$$

$$= \frac{2x - 5x + 7}{x - 4} \quad \text{Add numerators.}$$

$$= \frac{7 - 3x}{x - 4} \quad \text{Combine like terms.}$$

Exercises Within Reach®

Solutions in English & Spanish and tutorial videos at AlgebraWithinReach.com

Adding and Subtracting Rational Expressions In Exercises 13–22, combine and simplify.

13. $\dfrac{6x}{x^2 - 7x} - \dfrac{2x - x^2}{x^2 - 7x}$

14. $\dfrac{3x}{x^2 - 4x} - \dfrac{5x - x^2}{x^2 - 4x}$

15. $\dfrac{3}{x - 5} + \dfrac{9}{5 - x}$

16. $\dfrac{6}{x - 3} + \dfrac{2}{3 - x}$

17. $\dfrac{5x}{2x - 3} - \dfrac{9}{3 - 2x}$

18. $\dfrac{14}{6x - 1} - \dfrac{8x}{1 - 6x}$

19. $\dfrac{4x - 1}{x - 11} + \dfrac{2x}{11 - x}$

20. $\dfrac{10x - 3}{x - 7} - \dfrac{5x}{7 - x}$

21. $\dfrac{3x + 1}{x - 15} + \dfrac{x - 2}{15 - x}$

22. $\dfrac{5x - 1}{12 - x} + \dfrac{x + 2}{x - 12}$

Least Common Multiple

Guidelines for Finding the Least Common Multiple

1. Factor each polynomial completely.
2. The least common multiple must contain all the *different* factors of the polynomials, and each of these factors must be repeated the maximum number of times it occurs in any one of the factorizations.

EXAMPLE 5 Finding Least Common Multiples

Find the least common multiple of each set of polynomials.

a. $5x, x^3$ b. $x^2 - x, x - 1, x$ c. $3x^2 + 6x, -x^2 - 4x - 4$

SOLUTION

a. These polynomials factor as

$$5x = 5 \cdot x \quad \text{and} \quad x^3 = x \cdot x \cdot x$$

The different factors are 5 and x. The maximum number of times 5 occurs in any one factorization is 1. The maximum number of times x occurs in any one factorization is 3. So, the least common multiple is $5 \cdot x \cdot x \cdot x = 5x^3$.

b. These polynomials factor as

$$x^2 - x = x \cdot (x - 1), x - 1, \text{ and } x.$$

The different factors are x and $x - 1$. The maximum number of times x and $x - 1$ occur in any one factorization is 1. So, the least common multiple is $x(x - 1)$.

c. These polynomials factor as

$$3x^2 + 6x = 3 \cdot x \cdot (x + 2) \text{ and}$$

$$-x^2 - 4x - 4 = (-1) \cdot (x + 2) \cdot (x + 2).$$

The different factors are 3, x, -1, and $x + 2$. The maximum number of times 3, x, and -1 occur in any one factorization is 1. The maximum number of times $x + 2$ occurs in any one factorization is 2. So, the least common multiple is $-3x(x + 2)^2$.

Study Tip

Remember that the exponent on a number or variable factor indicates the number of repetitions of that factor. For example, in the term $25x^3$, the factorization $5^2 \cdot x^3$ shows that 5 is repeated twice and x is repeated three times.

Exercises Within Reach ®

Solutions in English & Spanish and tutorial videos at AlgebraWithinReach.com

Finding the Least Common Multiple In Exercises 23–34, find the least common multiple of the polynomials.

23. $9y^2, 12y$

24. $6x^3, 16x$

25. $2(y - 3), 6(y - 3)$

26. $4(x - 1), 8(x - 1)$

27. $16x, 12x(x + 2)$

28. $18y^2, 27y(y - 3)$

29. $x - 7, x^2, x(x + 7)$

30. $x - 1, x, 2(x + 3)$

31. $x + 2, x^2 - 4, x$

32. $x^2 - 1, x + 1, x$

33. $t^3 + 4t^2 + 4t, t^2 - 4t$

34. $y^3 - y^2, y^4 - y^2$

Rational Expressions with Unlike Denominators

EXAMPLE 6 Adding Expressions with Unlike Denominators

Add the rational expressions: $\dfrac{5}{3x} + \dfrac{7}{4x}$.

SOLUTION

By factoring the denominators, $3x = 3 \cdot x$ and $4x = 2^2 \cdot x$, you can conclude that the least common denominator of the fractions is $2^2 \cdot 3 \cdot x = 12x$.

$$\frac{5}{3x} + \frac{7}{4x} = \frac{5(4)}{3x(4)} + \frac{7(3)}{4x(3)} \qquad \text{Rewrite expressions using LCD of } 12x.$$

$$= \frac{20}{12x} + \frac{21}{12x} \qquad \text{Simplify.}$$

$$= \frac{20 + 21}{12x} = \frac{41}{12x} \qquad \text{Add numerators and simplify.}$$

EXAMPLE 7 Subtracting Expressions with Unlike Denominators

Subtract the rational expressions: $\dfrac{4}{x-2} - \dfrac{2}{x+1}$.

SOLUTION

The only factors of the denominators are $x - 2$ and $x + 1$. So, the least common denominator is $(x-2)(x+1)$.

$$\frac{4}{x-2} - \frac{2}{x+1}$$

$$= \frac{4(x+1)}{(x-2)(x+1)} - \frac{2(x-2)}{(x-2)(x+1)} \qquad \text{Rewrite expressions using LCD of } (x-2)(x+1).$$

$$= \frac{4(x+1) - 2(x-2)}{(x-2)(x+1)} \qquad \text{Subtract numerators.}$$

$$= \frac{4x + 4 - 2x + 4}{(x-2)(x+1)} \qquad \text{Distributive Property}$$

$$= \frac{2x + 8}{(x-2)(x+1)} \qquad \text{Combine like terms.}$$

Study Tip

Notice that in the subtraction of numerators in Example 7, the minus sign has to be distributed over the quantity $2(x - 2)$, or $(2x - 4)$. That is,

$$-(2x - 4) = -2x + 4.$$

Exercises Within Reach® Solutions in English & Spanish and tutorial videos at AlgebraWithinReach.com

Adding and Subtracting Rational Expressions In Exercises 35–46, **combine** and simplify.

35. $\dfrac{3}{2s} + \dfrac{1}{5s}$

36. $\dfrac{5}{6z} + \dfrac{3}{8z}$

37. $\dfrac{1}{5x} - \dfrac{3}{5}$

38. $\dfrac{2}{3} + \dfrac{1}{2x}$

39. $\dfrac{5}{u} + \dfrac{2}{u^2}$

40. $\dfrac{5}{z} - \dfrac{6}{z^2}$

41. $\dfrac{3}{2b} + \dfrac{5}{2b^2}$

42. $\dfrac{8}{6u^2} - \dfrac{2}{9u}$

43. $\dfrac{3}{x-5} + \dfrac{2}{x+3}$

44. $\dfrac{6}{x+4} + \dfrac{3}{x-1}$

45. $\dfrac{1}{x-1} - \dfrac{1}{x+2}$

46. $\dfrac{3}{x-4} - \dfrac{2}{x+1}$

Chapter 7 Rational Expressions and Equations

EXAMPLE 8 Adding Expressions with Unlike Denominators

$$\frac{2x}{x^2 - 5x + 6} + \frac{1}{3 - x}$$

$$= \frac{2x}{x^2 - 5x + 6} + \frac{1}{(-1)(x - 3)} \quad \text{$3 - x = -1(x - 3)$}$$

$$= \frac{2x}{(x - 2)(x - 3)} - \frac{1}{x - 3} \quad \text{Factor denominator.}$$

$$= \frac{2x}{(x - 2)(x - 3)} - \frac{x - 2}{(x - 2)(x - 3)} \quad \text{Rewrite expressions using LCD of $(x - 2)(x - 3)$.}$$

$$= \frac{2x - (x - 2)}{(x - 2)(x - 3)} \quad \text{Subtract numerators.}$$

$$= \frac{2x - x + 2}{(x - 2)(x - 3)} \quad \text{Distributive Property}$$

$$= \frac{x + 2}{(x - 2)(x - 3)} \quad \text{Combine like terms.}$$

Study Tip

In Example 8, notice that the denominator $3 - x$ is rewritten as $(-1)(x - 3)$ and then the problem is changed from addition to subtraction.

EXAMPLE 9 Combining Three Rational Expressions

$$\frac{2x - 5}{6x + 9} - \frac{4}{2x^2 + 3x} + \frac{1}{x}$$

$$= \frac{2x - 5}{3(2x + 3)} - \frac{4}{x(2x + 3)} + \frac{1}{x} \quad \text{Factor denominators.}$$

$$= \frac{(2x - 5)(x)}{3(2x + 3)(x)} - \frac{4(3)}{x(2x + 3)(3)} + \frac{3(2x + 3)}{x(3)(2x + 3)} \quad \text{Rewrite expressions using LCD of $3(2x + 3)(x)$.}$$

$$= \frac{(2x - 5)(x) - 4(3) + 3(2x + 3)}{3x(2x + 3)} \quad \text{Combine numerators.}$$

$$= \frac{2x^2 - 5x - 12 + 6x + 9}{3x(2x + 3)} \quad \text{Distributive Property}$$

$$= \frac{2x^2 + x - 3}{3x(2x + 3)} \quad \text{Combine like terms.}$$

$$= \frac{(x - 1)(2x + 3)}{3x(2x + 3)} \quad \text{Factor and divide out common factor.}$$

$$= \frac{x - 1}{3x}, \quad x \neq -\frac{3}{2} \quad \text{Simplified form}$$

Exercises Within Reach ® Solutions in English & Spanish and tutorial videos at AlgebraWithinReach.com

Adding and Subtracting Rational Expressions In Exercises 47−50, **combine** and simplify.

47. $\dfrac{x + 2}{x^2 - 5x + 6} + \dfrac{3}{2 - x}$

48. $\dfrac{2}{x + 1} + \dfrac{1 - x}{x^2 - 2x + 3}$

49. $\dfrac{x + 2}{3(x - 2)^2} + \dfrac{4}{3(x - 2)} + \dfrac{1}{2x}$

50. $\dfrac{5}{2(x + 1)} - \dfrac{1}{2x} - \dfrac{3}{2(x + 1)^2}$

Section 7.3 Adding and Subtracting Rational Expressions 325

EXAMPLE 10 **Combining Three Rational Expressions**

$$\frac{4x}{x^2-16} + \frac{x}{x+4} - 2$$

$$= \frac{4x}{(x+4)(x-4)} + \frac{x}{x+4} - \frac{2}{1} \qquad \text{Factor denominator.}$$

$$= \frac{4x}{(x+4)(x-4)} + \frac{x(x-4)}{(x+4)(x-4)} - \frac{2(x+4)(x-4)}{(x+4)(x-4)}$$

$$= \frac{4x + x(x-4) - 2(x^2-16)}{(x+4)(x-4)} \qquad \text{Combine numerators.}$$

$$= \frac{4x + x^2 - 4x - 2x^2 + 32}{(x+4)(x-4)} \qquad \text{Distributive Property}$$

$$= \frac{32 - x^2}{(x+4)(x-4)} \qquad \text{Combine like terms.}$$

EXAMPLE 11 **Adding Rational Expressions**

$$\frac{x}{x-3} + \frac{2}{3x+4} = \frac{x(3x+4) + 2(x-3)}{(x-3)(3x+4)} \qquad \text{Basic definition}$$

$$= \frac{3x^2 + 4x + 2x - 6}{(x-3)(3x+4)} \qquad \text{Distributive Property}$$

$$= \frac{3x^2 + 6x - 6}{(x-3)(3x+4)} \qquad \text{Combine like terms.}$$

Exercises Within Reach®

Solutions in English & Spanish and tutorial videos at AlgebraWithinReach.com

Adding and Subtracting Rational Expressions In Exercises 51–62, combine and simplify.

51. $\dfrac{3}{x-1} - 5$

52. $\dfrac{3}{2x-5} + 2$

53. $\dfrac{x}{x-3} + \dfrac{3}{2x+5}$

54. $\dfrac{y}{y-4} - \dfrac{7}{2y-6}$

55. $\dfrac{x}{x^2-9} + \dfrac{3}{x+3}$

56. $\dfrac{6}{z+2} - \dfrac{z-3}{z^2-4}$

57. $\dfrac{5v}{v(v+4)} + \dfrac{2v}{v^2}$

58. $\dfrac{2t}{t^2} - \dfrac{t}{t(t+1)}$

59. $\dfrac{x}{x^2-4} + \dfrac{3x}{x+2} + \dfrac{3x^2-5x}{4-x^2}$

60. $\dfrac{x+6}{4-x^2} - \dfrac{x+3}{x+2} + \dfrac{x-3}{2-x}$

61. $\dfrac{6x}{x^2+x-6} + \dfrac{x}{x+3} - 4$

62. $\dfrac{2}{x^2+3x+2} - \dfrac{x}{x+1} + 3$

Concept Summary: Adding and Subtracting Rational Expressions

What
The rules for adding and subtracting rational expressions are the same as the rules for adding and subtracting numerical fractions.

EXAMPLE
Add: $\dfrac{7}{4x^2} + \dfrac{5}{2x}$.

How
To add (or subtract) rational expressions with unlike denominators, do the following.

1. Use the **least common denominator** (LCD) to rewrite the fractions so they have like denominators.
2. Add (or subtract) the numerators.
3. Simplify the result.

EXAMPLE
$$\dfrac{7}{4x^2} + \dfrac{5}{2x}$$
$$= \dfrac{7}{4x^2} + \dfrac{5(2x)}{2x(2x)} \quad \text{Rewrite.}$$
$$= \dfrac{7}{4x^2} + \dfrac{10x}{4x^2} \quad \text{Simplify.}$$
$$= \dfrac{10x + 7}{4x^2} \quad \text{Add.}$$

Why
Knowing how to add and subtract rational expressions will help you solve rational equations.

Exercises Within Reach®
Worked-out solutions to odd-numbered exercises at AlgebraWithinReach.com

Concept Summary Check

63. Writing In your own words, explain how to add or subtract rational expressions with like denominators.

64. Least Common Multiple Explain how to find the least common multiple of two or more polynomials. Give an example.

65. Writing Describe how to add or subtract rational expressions with unlike denominators.

66. Opposite Denominators Explain how to add or subtract rational expressions with opposite denominators (denominators that differ only in sign).

Extra Practice

Finding the Least Common Denominator In Exercises 67–72, **find** the least common denominator of the expressions and **rewrite** one or both of the fractions using the least common denominator.

67. $\dfrac{x+5}{3x-6}, \dfrac{10}{x-2}$

68. $\dfrac{8x}{(x+2)}, \dfrac{3}{4x+8}$

69. $\dfrac{2}{(x+3)^2}, \dfrac{5}{x(x+3)}$

70. $\dfrac{5t}{(t-3)^2}, \dfrac{4}{t(t-3)}$

71. $\dfrac{x-8}{x^2-16}, \dfrac{9x}{x^2-8x+16}$

72. $\dfrac{3y}{y^2-y-6}, \dfrac{y+2}{y^2-3y}$

73. Geometry Find an expression for the perimeter of the rectangle (see figure).

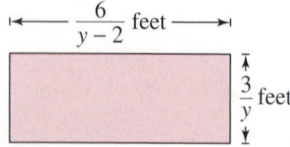

74. Geometry Find an expression for the perimeter of the triangle (see figure).

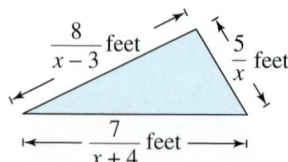

75. Kayaking A kayaker travels 8 miles upstream (against the current) and 8 miles downstream (with the current). The speed of the current is 2 miles per hour. An expression for the time (in hours) it takes the kayaker to travel upstream is

$$\frac{8}{r-2}.$$

An expression for the time (in hours) it takes the kayaker to travel downstream is

$$\frac{8}{r+2}.$$

Write and simplify an expression for the total travel time (in hours) of the kayaker.

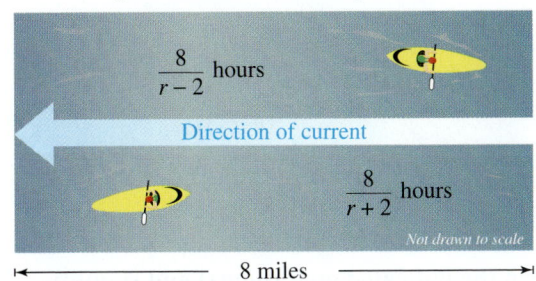

76. Boat Travel A boater travels 25 miles upstream (against the current) and 25 miles downstream (with the current). The speed of the current is 4 miles per hour. An expression for the time (in hours) it takes the boater to travel upstream is

$$\frac{25}{r-4}.$$

An expression for the time (in hours) it takes the boater to travel downstream is

$$\frac{25}{r+4}.$$

Write and simplify an expression for the total travel time (in hours) of the boater.

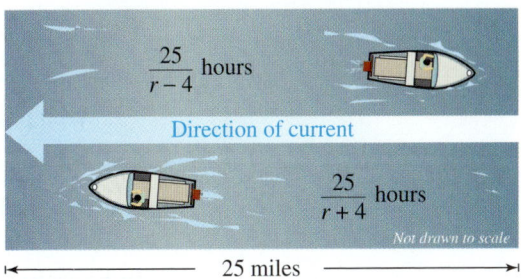

Explaining Concepts

77. Reasoning Is it possible for the least common multiple of two polynomials to be the same as one of the polynomials? If so, give an example.

78. Error Analysis Describe and correct the error.

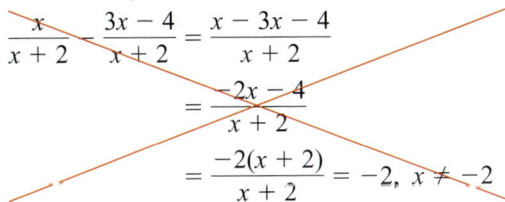

79. Writing When the least common multiple of a binomial and a trinomial is the trinomial, what can be said about the binomial? Explain by using an example.

80. Writing What is the advantage of using the least common denominator instead of another common denominator?

Cumulative Review

In Exercises 81 and 82, determine whether each value of x is a solution of the inequality.

81. $3x - 18 > 0$ (a) $x = 0$ (b) $x = 6$
  (c) $x = 4$ (d) $x = 11$

82. $4x - 9 \leq 6$ (a) $x = 3$ (b) $x = -1$
  (c) $x = -\frac{1}{2}$ (d) $x = 8$

In Exercises 83–86, perform the indicated operation and simplify.

83. $\dfrac{12x}{7x - 35} \cdot (5 - x)$

84. $\dfrac{4}{y + 1} \cdot \dfrac{y + 1}{24(y - 2)}$

85. $\dfrac{x^2 - 9}{x^3} \div \dfrac{x - 3}{9x}$

86. $\dfrac{x^2 - 6x + 8}{x + 4} \div (4 - x)$

Mid-Chapter Quiz: Sections 7.1–7.3

Solutions in English & Spanish and tutorial videos at AlgebraWithinReach.com

Take this quiz as you would take a quiz in class. After you are done, check your work against the answers in the back of the book.

1. In your own words, explain the meaning of *domain*. Find the domain of

   (a) $\dfrac{x^2}{x^2 + 4}$ and (b) $\dfrac{x^2}{x^2 - 4}$.

2. Evaluate $\dfrac{y - 3}{y + 2}$ for (a) $y = 10$, (b) $y = 3$, and (c) $y = -2$.

In Exercises 3–8, simplify the rational expression.

3. $\dfrac{14z^4}{35z}$

4. $\dfrac{60a^2 b}{45ab^3}$

5. $\dfrac{y^2 - 4}{8 - 4y}$

6. $\dfrac{15u(u - 3)^2}{25u^2(u - 3)}$

7. $\dfrac{b^2 + 3b}{b^3 + 2b^2 - 3b}$

8. $\dfrac{4x^2 - 12x + 9}{2x^2 - x - 3}$

In Exercises 9–18, perform the indicated operation(s) and simplify.

9. $\dfrac{3y^3}{5} \cdot \dfrac{25}{9y}$

10. $\dfrac{s - 5}{15} \cdot \dfrac{12s}{25 - s^2}$

11. $(x^3 + 4x^2) \cdot \dfrac{5x}{x^2 + 2x - 8}$

12. $\dfrac{r^2 - 16}{r} \div \dfrac{r + 4}{r^2}$

13. $\dfrac{x}{25} \div \dfrac{x^2 + 2x}{10} \cdot \dfrac{1}{x + 2}$

14. $\dfrac{10x^2}{3y} \div \left(\dfrac{y}{x} \cdot \dfrac{x^3 y}{6} \right)$

15. $\dfrac{x}{x^2 - 36} + \dfrac{4}{x - 6}$

16. $\dfrac{x + 5}{x - 5} - \dfrac{x - 5}{x + 5}$

17. $\dfrac{x - 5}{x^2 + 7x + 12} + \dfrac{x - 3}{x + 4}$

18. $\dfrac{5x + 10}{x^2 - 5x + 14} - \dfrac{4}{x - 7}$

Applications

19. A small business has a setup cost of $10,000 for the production of downhill skis. The cost of labor and materials for producing each unit is $225.

    (a) Write a rational expression that models the average cost per unit when x units are produced.

    (b) Complete the table and describe any trends you find.

x	2000	3000	4000	5000
Average cost				

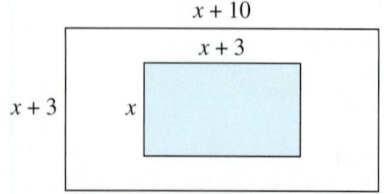

20. Find the ratio of the area of the shaded region to the total area of the figure. (Assume $x > 0$.)

Study Skills in Action

Using a Test-Taking Strategy

What do runners do before a race? They design a strategy for running their best. They make sure they get enough rest, eat sensibly, and get to the track early to warm up. In the same way, it is important for students to get a good night's sleep, eat a healthy meal, and get to class early to allow time to focus before a test.

The biggest difference between a runner's race and a math test is that a math student does not have to reach the finish line first! In fact, many students would increase their scores if they used all the test time instead of worrying about being the last student left in the class. This is why it is important to have a strategy for taking the test.

These runners are focusing on their techniques, not on whether other runners are ahead of or behind them.

Smart Study Strategy

Use Ten Steps for Test Taking

1. **Do a memory data dump.** As soon as you get the test, turn it over and write down anything that you still have trouble remembering sometimes (formulas, calculations, rules).

2. **Preview the test.** Look over the test and mark the questions you know how to do easily. These are the problems you should do first.

3. **Do a second memory data dump.** As you previewed the test, you may have remembered other information. Write this information on the back of the test.

4. **Develop a test progress schedule.** Based on how many points each question is worth, decide on a progress schedule. You should always have more than half the test done before half the time has elapsed.

5. **Answer the easiest problems first.** Solve the problems you marked while previewing the test.

6. **Skip difficult problems.** Skip the problems that you suspect will give you trouble.

7. **Review the skipped problems.** After solving all the problems that you know how to do easily, go back and reread the problems you skipped.

8. **Try your best at the remaining problems that confuse you.** Even if you cannot completely solve a problem, you may be able to get partial credit for a few correct steps.

9. **Review the test.** Look for any careless errors you may have made.

10. **Use all the allowed test time.** The test is not a race against the other students.

7.4 Complex Fractions

▶ Simplify complex fractions using rules for dividing rational expressions.
▶ Simplify complex fractions involving a sum or difference.
▶ Use complex fractions to solve application problems.

Complex Fractions

A **complex fraction** is a fraction that has a fraction in its numerator or denominator, or both. The rules for dividing rational expressions still apply.

EXAMPLE 1   Simplifying a Complex Fraction

a. $\dfrac{\left(\dfrac{5}{14}\right)}{\left(\dfrac{25}{8}\right)} = \dfrac{5}{14} \cdot \dfrac{8}{25}$ Invert divisor and multiply.

$= \dfrac{5 \cdot 2 \cdot 2 \cdot 2}{2 \cdot 7 \cdot 5 \cdot 5}$ Multiply and factor.

$= \dfrac{\cancel{5} \cdot \cancel{2} \cdot 2 \cdot 2}{\cancel{2} \cdot 7 \cdot \cancel{5} \cdot 5}$ Divide out common factors.

$= \dfrac{4}{35}$ Simplified form

b. $\dfrac{\left[\dfrac{4y^3}{(5x)^2}\right]}{\left[\dfrac{(2y)^2}{10x^3}\right]} = \dfrac{4y^3}{25x^2} \cdot \dfrac{10x^3}{4y^2}$ Invert divisor and multiply.

$= \dfrac{4y^2 \cdot y \cdot 2 \cdot 5x^2 \cdot x}{5 \cdot 5x^2 \cdot 4y^2}$ Multiply and factor.

$= \dfrac{\cancel{4y^2} \cdot y \cdot 2 \cdot \cancel{5x^2} \cdot x}{5 \cdot \cancel{5x^2} \cdot \cancel{4y^2}}$ Divide out common factors.

$= \dfrac{2xy}{5}, \; x \neq 0, \; y \neq 0$ Simplified form

Exercises Within Reach®   Solutions in English & Spanish and tutorial videos at AlgebraWithinReach.com

Simplifying a Complex Fraction   In Exercises 1−6, **simplify** the complex fraction.

1. $\dfrac{\left(\dfrac{3}{2}\right)}{\left(\dfrac{9}{10}\right)}$

2. $\dfrac{\left(\dfrac{7}{16}\right)}{\left(-\dfrac{4}{21}\right)}$

3. $\dfrac{\left(\dfrac{3}{x}\right)}{\left(\dfrac{6}{x^2}\right)}$

4. $\dfrac{\left(\dfrac{2}{3}\right)}{\left(\dfrac{u}{v}\right)}$

5. $\dfrac{\left[\dfrac{6x^3}{(5y)^2}\right]}{\left[\dfrac{(3x)^2}{15y^4}\right]}$

6. $\dfrac{\left[\dfrac{(3r)^3}{10t^4}\right]}{\left[\dfrac{9r}{(2t)^2}\right]}$

Section 7.4 Complex Fractions

Study Tip

In Example 2, the domain of the complex fraction is restricted by the denominators in the original expression and by the denominators in the expression after the divisor has been inverted. So, the domain of the original expression is all real values of x except $x = -2$, $x = -5$, and $x = -1$.

EXAMPLE 2 Simplifying a Complex Fraction

$$\frac{\left(\dfrac{x+1}{x+2}\right)}{\left(\dfrac{x+1}{x+5}\right)} = \frac{x+1}{x+2} \cdot \frac{x+5}{x+1} \qquad \text{Invert divisor and multiply.}$$

$$= \frac{(x+1)(x+5)}{(x+2)(x+1)} \qquad \text{Multiply.}$$

$$= \frac{\cancel{(x+1)}(x+5)}{(x+2)\cancel{(x+1)}} \qquad \text{Divide out common factors.}$$

$$= \frac{x+5}{x+2}, \; x \neq -1, \; x \neq -5 \qquad \text{Simplified form}$$

EXAMPLE 3 Simplifying a Complex Fraction

$$\frac{\left(\dfrac{x^2+4x+3}{x-2}\right)}{2x+6} = \frac{\left(\dfrac{x^2+4x+3}{x-2}\right)}{\left(\dfrac{2x+6}{1}\right)} \qquad \text{Rewrite denominator.}$$

$$= \frac{x^2+4x+3}{x-2} \cdot \frac{1}{2x+6} \qquad \text{Invert divisor and multiply.}$$

$$= \frac{(x+1)(x+3)}{(x-2)(2)(x+3)} \qquad \text{Multiply and factor.}$$

$$= \frac{(x+1)\cancel{(x+3)}}{(x-2)(2)\cancel{(x+3)}} \qquad \text{Divide out common factor.}$$

$$= \frac{x+1}{2(x-2)}, \; x \neq -3 \qquad \text{Simplified form}$$

Exercises Within Reach®

Solutions in English & Spanish and tutorial videos at AlgebraWithinReach.com

Simplifying a Complex Fraction In Exercises 7–12, simplify the complex fraction.

7. $\dfrac{\left(\dfrac{x+4}{x+7}\right)}{\left(\dfrac{x+4}{x+14}\right)}$

8. $\dfrac{\left(\dfrac{x-3}{x-6}\right)}{\left(\dfrac{x+3}{x-6}\right)}$

9. $\dfrac{\left(\dfrac{2x-10}{x+1}\right)}{\left(\dfrac{x-5}{x+1}\right)}$

10. $\dfrac{\left(\dfrac{a+5}{8a-20}\right)}{\left(\dfrac{a+5}{2a-5}\right)}$

11. $\dfrac{\left(\dfrac{x^2+3x-10}{x+4}\right)}{3x+15}$

12. $\dfrac{\left(\dfrac{x^2-2x-8}{x-1}\right)}{8x-32}$

Complex Fractions with Sums or Differences

EXAMPLE 4 Simplifying a Complex Fraction

Method 1: Invert and Multiply

$$\frac{\left(\frac{x}{3} + \frac{2}{3}\right)}{\left(1 - \frac{2}{x}\right)} = \frac{\left(\frac{x}{3} + \frac{2}{3}\right)}{\left(\frac{x}{x} - \frac{2}{x}\right)}$$ Rewrite with least common denominators.

$$= \frac{\left(\frac{x+2}{3}\right)}{\left(\frac{x-2}{x}\right)}$$ Add fractions.

$$= \frac{x+2}{3} \cdot \frac{x}{x-2}$$ Invert divisor and multiply.

$$= \frac{x(x+2)}{3(x-2)}, \; x \neq 0$$ Simplified form

Method 2: Multiply by Least Common Denominator

$$\frac{\left(\frac{x}{3} + \frac{2}{3}\right)}{\left(1 - \frac{2}{x}\right)} = \frac{\left(\frac{x}{3} + \frac{2}{3}\right)}{\left(1 - \frac{2}{x}\right)} \cdot \frac{3x}{3x}$$ $3x$ is the least common denominator.

$$= \frac{\frac{x}{3}(3x) + \frac{2}{3}(3x)}{(1)(3x) - \frac{2}{x}(3x)}$$ Distributive Property

$$= \frac{x^2 + 2x}{3x - 6}$$ Simplify.

$$= \frac{x(x+2)}{3(x-2)}, \; x \neq 0$$ Simplified form

Exercises Within Reach®

Solutions in English & Spanish and tutorial videos at AlgebraWithinReach.com

Simplifying a Complex Fraction In Exercises 13–18, simplify the complex fraction.

13. $\dfrac{\left(5 + \frac{3}{4}\right)}{\left(1 + \frac{1}{4}\right)}$

14. $\dfrac{\left(3 + \frac{4}{5}\right)}{\left(1 - \frac{9}{16}\right)}$

15. $\dfrac{\left(\frac{x}{2}\right)}{\left(2 + \frac{3}{x}\right)}$

16. $\dfrac{\left(1 - \frac{2}{x}\right)}{\left(\frac{x}{4}\right)}$

17. $\dfrac{\left(\frac{x}{3} - \frac{4}{3}\right)}{\left(5 + \frac{1}{x}\right)}$

18. $\dfrac{\left(\frac{x}{4} + \frac{3}{4}\right)}{\left(6 - \frac{2}{x}\right)}$

EXAMPLE 5 Simplifying a Complex Fraction

Simplify the complex fraction.

$$\frac{\left(\dfrac{2}{x+2}\right)}{\left(\dfrac{3}{x+2}+\dfrac{2}{x}\right)}$$

SOLUTION

$$\frac{\left(\dfrac{2}{x+2}\right)}{\left(\dfrac{3}{x+2}+\dfrac{2}{x}\right)} = \frac{\left(\dfrac{2}{x+2}\right)(x)(x+2)}{\left(\dfrac{3}{x+2}+\dfrac{2}{x}\right)(x)(x+2)} \qquad x(x+2) \text{ is the least common denominator.}$$

$$= \frac{\left(\dfrac{2}{x+2}\right)(x)(x+2)}{\left(\dfrac{3}{x+2}\right)(x)(x+2)+\left(\dfrac{2}{x}\right)(x)(x+2)} \qquad \text{Distributive Property}$$

$$= \frac{2x}{3x+2(x+2)} \qquad \text{Simplify.}$$

$$= \frac{2x}{3x+2x+4} \qquad \text{Distributive Property}$$

$$= \frac{2x}{5x+4}, \; x \neq -2, x \neq 0 \qquad \text{Simplify.}$$

Notice that the numerator and denominator of the complex fraction were multiplied by $(x)(x+2)$, which is the least common denominator of the fractions in the original complex fraction.

Exercises Within Reach®   Solutions in English & Spanish and tutorial videos at AlgebraWithinReach.com

Simplifying a Complex Fraction In Exercises 19–24, simplify the complex fraction.

19. $\dfrac{\left(\dfrac{x}{4}-\dfrac{4}{x}\right)}{(x-4)}$

20. $\dfrac{\left(\dfrac{4}{x^2}+\dfrac{2}{x}\right)}{\left(\dfrac{4}{x}+\dfrac{2}{x^2}\right)}$

21. $\dfrac{\left(\dfrac{10}{x+1}\right)}{\left(\dfrac{1}{2x+2}+\dfrac{3}{x+1}\right)}$

22. $\dfrac{\left(\dfrac{2}{x+5}\right)}{\left(\dfrac{2}{x+5}+\dfrac{1}{4x+20}\right)}$

23. $\dfrac{\left(\dfrac{1}{x}-\dfrac{1}{x+1}\right)}{\left(\dfrac{1}{x+1}\right)}$

24. $\dfrac{\left(\dfrac{5}{y}-\dfrac{6}{2y+1}\right)}{\left(\dfrac{5}{2y+1}\right)}$

Applications

Application **EXAMPLE 6** Finding an Average of Two Real Numbers

Find the average of the two real numbers $\frac{x}{5}$ and $\frac{x}{6}$.

Method 1: Invert and Multiply

$$\frac{\left(\frac{x}{5} + \frac{x}{6}\right)}{2} = \frac{\left(\frac{6x}{30} + \frac{5x}{30}\right)}{2} \qquad \text{Rewrite with least common denominator.}$$

$$= \frac{\left(\frac{11x}{30}\right)}{2} \qquad \text{Add fractions.}$$

$$= \frac{\left(\frac{11x}{30}\right)}{\left(\frac{2}{1}\right)} \qquad \text{Rewrite denominator.}$$

$$= \frac{11x}{30} \cdot \frac{1}{2} \qquad \text{Invert divisor and multiply.}$$

$$= \frac{11x}{60} \qquad \text{Simplified form}$$

Method 2: Multiply by Least Common Denominator

$$\frac{\left(\frac{x}{5} + \frac{x}{6}\right)}{2} = \frac{\left(\frac{x}{5} + \frac{x}{6}\right) \cdot 30}{2 \cdot 30} \qquad \text{30 is the least common denominator.}$$

$$= \frac{\left(\frac{x}{5}\right)(30) + \left(\frac{x}{6}\right)(30)}{2 \cdot 30} \qquad \text{Distributive Property}$$

$$= \frac{6x + 5x}{60} \qquad \text{Simplify.}$$

$$= \frac{11x}{60} \qquad \text{Simplified form}$$

Exercises Within Reach®

Solutions in English & Spanish and tutorial videos at AlgebraWithinReach.com

Average of Two Numbers In Exercises 25–28, **find** the average of the two real numbers.

25. $\frac{x}{4}$ and $\frac{x}{5}$

26. $\frac{x}{2}$ and $\frac{x}{7}$

27. $\frac{2x}{3}$ and $\frac{3x}{5}$

28. $\frac{4x}{7}$ and $\frac{5x}{8}$

Application

EXAMPLE 7 An Application in Electrical Resistance

When two resistors of resistances R_1 and R_2 (both in ohms) are connected in parallel, the total resistance (in ohms) is modeled by

$$\frac{1}{\left(\dfrac{1}{R_1} + \dfrac{1}{R_2}\right)}.$$

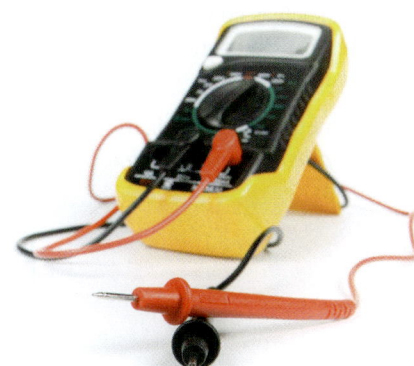

a. Simplify this complex fraction.

b. Find the total resistance when $R_1 = 10$ ohms and $R_2 = 20$ ohms.

SOLUTION

a.
$$\frac{1}{\left(\dfrac{1}{R_1} + \dfrac{1}{R_2}\right)} = \frac{1}{\dfrac{R_1 + R_2}{R_1 R_2}} \qquad \text{Add fractions in denominator.}$$

$$= 1 \cdot \frac{R_1 R_2}{R_1 + R_2} \qquad \text{Invert divisor and multiply.}$$

$$= \frac{R_1 R_2}{R_1 + R_2} \qquad \text{Simplify.}$$

b. When $R_1 = 10$ ohms and $R_2 = 20$ ohms, the total resistance is

$$\frac{R_1 R_2}{R_1 + R_2} = \frac{(10)(20)}{10 + 20} \qquad \text{Substitute.}$$

$$= \frac{200}{30} \qquad \text{Simplify.}$$

$$\approx 6.7 \text{ ohms.} \qquad \text{Simplify.}$$

Exercises Within Reach®

Solutions in English & Spanish and tutorial videos at AlgebraWithinReach.com

29. *Electronics* When three resistors of resistance R_1, R_2, and R_3 (all in ohms) are connected in parallel, the total resistance (in ohms) is modeled by

$$\frac{1}{\left(\dfrac{1}{R_1} + \dfrac{1}{R_2} + \dfrac{1}{R_3}\right)}.$$

Simplify this complex fraction.

30. *Using Results* Use the simplified fraction in Exercise 29 to find the total resistance when $R_1 = 6$ ohms, $R_2 = 12$ ohms, and $R_3 = 24$ ohms.

Concept Summary: Simplifying Complex Fractions

What
When simplifying a **complex fraction**, the rules for dividing rational expressions still apply.

EXAMPLE

Simplify $\dfrac{\left(\dfrac{x+2}{3}\right)}{\left(\dfrac{x+2}{x}\right)}$.

How
To simplify a complex fraction, invert the denominator fraction and multiply.

EXAMPLE

$$\dfrac{\left(\dfrac{x+2}{3}\right)}{\left(\dfrac{x+2}{x}\right)} = \dfrac{x+2}{3} \cdot \dfrac{x}{x+2}$$

$$= \dfrac{x(x+2)}{3(x+2)}$$

$$= \dfrac{x}{3}, \quad x \neq 0 \quad x \neq -2$$

Why
Knowing how to simplify "messy" expressions, such as complex fractions, will help you as you continue your study of algebra.

A useful strategy in algebra is to *rewrite complicated problems into simpler forms*.

Exercises Within Reach®
Worked-out solutions to odd-numbered exercises at AlgebraWithinReach.com

Concept Summary Check

31. Vocabulary Define the term *complex fraction*. Give an example.

32. Structure What operation does the fraction bar imply?

33. Precision Describe how to rewrite a complex fraction as a product.

34. Alternate Methods How can you simplify a complex fraction without inverting the divisor?

Extra Practice

Simplifying a Complex Fraction In Exercises 35–40, **simplify** the complex fraction.

35. $\dfrac{\left(\dfrac{y}{3-y}\right)}{\left(\dfrac{y^2}{y-3}\right)}$

36. $\dfrac{\left(\dfrac{x}{x-4}\right)}{\left(\dfrac{x}{4-x}\right)}$

37. $\dfrac{\left(\dfrac{6x^2 - 17x + 5}{3x^2 + 3x}\right)}{\left(\dfrac{3x-1}{3x+1}\right)}$

38. $\dfrac{\left(\dfrac{6x^2 - 13x - 5}{5x^2 + 5x}\right)}{\left(\dfrac{2x-5}{5x+1}\right)}$

39. $\dfrac{\left(1 + \dfrac{3}{y}\right)}{y}$

40. $\dfrac{x}{\left(\dfrac{5}{x} + 2\right)}$

Average of Two Numbers In Exercises 41 and 42, **find** the average of two real numbers.

41. $\dfrac{x+5}{4}$ and $\dfrac{2}{x}$

42. $\dfrac{5}{2x}$ and $\dfrac{x+1}{5}$

Section 7.4 Complex Fractions

43. *Equal Lengths* Find expressions for three real numbers that divide the real number line between $x/9$ and $x/6$ into four parts of equal length (see figure).

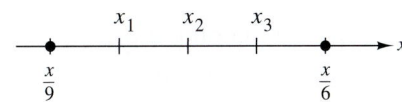

44. *Equal Lengths* Find expressions for two real numbers that divide the real number line between $x/3$ and $5x/4$ into three parts of equal length (see figure).

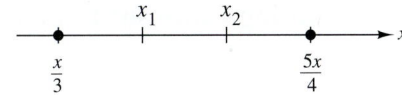

45. *Monthly Payment* The approximate annual percent rate r (in decimal form) of a monthly installment loan is

$$r = \frac{\left[\dfrac{24(MN - P)}{N}\right]}{\left(P + \dfrac{MN}{12}\right)}$$

where N is the total number of payments, M is the monthly payment, and P is the amount financed. Simplify the expression.

46. *Using Results* Use the simplified fraction in Exercise 45 to find the annual percent rate for a 4-year car loan of $15,000 with monthly payments of $400.

Explaining Concepts

47. *Writing* Of the two methods discussed in this section for simplifying complex fractions, select the method you prefer and explain the method in your own words.

48. *Reasoning* What are the numerator and denominator of each complex fraction?

(a) $\dfrac{\left(\dfrac{x-1}{5}\right)}{\left(\dfrac{2}{x^2 + 2x - 35}\right)}$ (b) $\dfrac{\left(\dfrac{1}{2y} + x\right)}{\left(\dfrac{3}{y} + x\right)}$

Cumulative Review

In Exercises 49–54, simplify the expression.

49. $(12x^2 + 4x - 8) + (-x^2 - 6x + 5)$

50. $(-2x^2 + 8x - 7) - (3x^2 + 12)$

51. $-4x(-6x^2 + 4x - 2)$

52. $(3x + 6)(-4x + 1)$

53. $\dfrac{8x^2 + 4x}{x}$

54. $(x^2 + x - 12) \div (x - 3)$

In Exercises 55–60, combine and simplify.

55. $\dfrac{x}{5} + \dfrac{3x}{5}$

56. $\dfrac{x}{7} - \dfrac{x-8}{7}$

57. $\dfrac{6}{5x} - \dfrac{2}{3x}$

58. $\dfrac{-9}{2x} + \dfrac{9 - 2x}{4x}$

59. $\dfrac{8 - 3x}{x + 2} + \dfrac{4 + 2x}{x + 2}$

60. $\dfrac{10x}{(x-3)^2} - \dfrac{7x}{(x-3)}$

7.5 Rational Equations and Applications

- Solve rational equations with constant denominators.
- Solve rational equations with variable denominators.
- Solve application problems using rational equations with variable denominators.

Equations Containing Constant Denominators

EXAMPLE 1 Equations Containing Constant Denominators

Study Tip

In Example 1, note that the key to solving an equation that involves fractions is to multiply each side of the equation by the least common denominator (LCD) of the fractions.

a.
$$\frac{x}{5} = 6 - \frac{x}{10}$$ Original equation

$$10\left(\frac{x}{5}\right) = 10\left(6 - \frac{x}{10}\right)$$ Multiply each side by LCD of 10.

$$2x = 60 - x$$ Distribute and simplify.

$$3x = 60$$ Add x to each side.

$$x = 20$$ Divide each side by 3.

The solution is $x = 20$. Check this in the original equation.

b.
$$\frac{x+6}{9} - \frac{x-2}{5} = \frac{4}{15}$$ Original equation

$$45\left(\frac{x+6}{9} - \frac{x-2}{5}\right) = 45\left(\frac{4}{15}\right)$$ Multiply each side by LCD of 45.

$$5(x+6) - 9(x-2) = 12$$ Distribute and simplify.

$$5x + 30 - 9x + 18 = 12$$ Distributive Property

$$-4x + 48 = 12$$ Combine like terms.

$$-4x = -36$$ Subtract 48 from each side.

$$x = 9$$ Divide each side by -4.

The solution is $x = 9$. Check this in the original equation.

Exercises Within Reach®   Solutions in English & Spanish and tutorial videos at AlgebraWithinReach.com

Solving a Rational Equation In Exercises 1–10, **solve** the rational equation.

1. $\dfrac{z}{3} - \dfrac{2z}{8} = 1$

2. $3 + \dfrac{y}{5} = \dfrac{y}{2}$

3. $\dfrac{t}{3} = 25 - \dfrac{t}{6}$

4. $\dfrac{x}{10} + \dfrac{x}{5} = 20$

5. $\dfrac{5x}{7} - \dfrac{2x}{3} = \dfrac{1}{2}$

6. $\dfrac{2}{3} - \dfrac{3x}{6} = -\dfrac{4x}{9}$

7. $\dfrac{a+3}{4} - \dfrac{a-1}{6} = \dfrac{4}{3}$

8. $\dfrac{u-5}{10} + \dfrac{u+8}{15} = \dfrac{7}{10}$

9. $\dfrac{x-4}{3} + \dfrac{2x+1}{4} = \dfrac{5}{6}$

10. $\dfrac{y+6}{5} + \dfrac{3y-2}{20} = \dfrac{3}{4}$

Equations Containing Variable Denominators

EXAMPLE 2 An Equation Containing Variable Denominators

$\dfrac{1}{x} - \dfrac{2}{3} = \dfrac{3}{x}$ — Original equation

$3x\left(\dfrac{1}{x} - \dfrac{2}{3}\right) = 3x\left(\dfrac{3}{x}\right)$ — Multiply each side by LCD of $3x$.

$\dfrac{3x}{x} - \dfrac{6x}{3} = \dfrac{9x}{x}$ — Distributive Property

$3 - 2x = 9,\ x \neq 0$ — Simplify.

$-2x = 6$ — Subtract 3 from each side.

$x = -3$ — Divide each side by -2.

The solution is $x = -3$. Check this in the original equation.

EXAMPLE 3 An Equation with No Solution

$\dfrac{2x}{x+3} = 1 - \dfrac{6}{x+3}$ — Original equation

$(x+3)\left(\dfrac{2x}{x+3}\right) = (x+3)\left(1 - \dfrac{6}{x+3}\right)$ — Multiply each side by LCD of $x+3$.

$\dfrac{(x+3)(2x)}{x+3} = (x+3) - \dfrac{(x+3)6}{x+3}$ — Distributive Property

$2x = (x+3) - 6,\ x \neq -3$ — Simplify.

$2x = x - 3$ — Combine like terms.

$x = -3$ — Subtract x from each side.

At this point, the solution appears to be $x = -3$. However, by performing a check, you can see that this "trial solution" is extraneous. So, the original equation has no solution.

Study Tip

A "trial solution" may not satisfy the original equation. This type of solution is called an **extraneous solution**. An extraneous solution of an equation does not, by definition, satisfy the original equation and so must not be listed as an actual solution.

Exercises Within Reach®

Solutions in English & Spanish and tutorial videos at AlgebraWithinReach.com

Solving a Rational Equation In Exercises 11–20, solve the rational equation. (Check for extraneous solutions.)

11. $\dfrac{3}{x} + \dfrac{1}{4} = \dfrac{2}{x}$

12. $\dfrac{3}{5} - \dfrac{7}{x} = -\dfrac{4}{x}$

13. $\dfrac{6}{12x} + \dfrac{3}{4} = \dfrac{2}{3x}$

14. $\dfrac{5}{2x} + \dfrac{5}{8} = -\dfrac{5}{8x}$

15. $\dfrac{10}{y+3} + \dfrac{10}{3} = 6$

16. $\dfrac{5}{2} - \dfrac{12}{x-4} = 6$

17. $\dfrac{1}{x-4} + 2 = \dfrac{2x}{x-4}$

18. $\dfrac{-2x}{x-1} + 2 = \dfrac{10}{x-1}$

19. $\dfrac{6x}{x-11} + 1 = \dfrac{3}{x-11}$

20. $\dfrac{z}{z+7} - 3 = \dfrac{-1}{z+7}$

EXAMPLE 4 An Equation with One Solution

$$\frac{6}{x-1} + \frac{2x}{x-2} = 2$$

$$(x-1)(x-2)\left(\frac{6}{x-1} + \frac{2x}{x-2}\right) = 2(x-1)(x-2)$$

$$6(x-2) + 2x(x-1) = 2(x^2 - 3x + 2), \quad x \neq 1, \, x \neq 2$$

$$6x - 12 + 2x^2 - 2x = 2x^2 - 6x + 4$$

$$10x = 16$$

$$x = \frac{8}{5}$$

The solution is $x = \frac{8}{5}$. Check this in the original equation.

EXAMPLE 5 An Equation with Two Solutions

$$\frac{2x}{x+2} = \frac{1}{x^2-4} + 1 \qquad \text{Original equation}$$

$$(x^2-4)\left(\frac{2x}{x+2}\right) = (x^2-4)\left(\frac{1}{x^2-4} + 1\right) \qquad \text{Multiply each side by LCD of } x^2 - 4.$$

$$(x-2)(2x) = 1 + (x^2-4), \quad x \neq \pm 2 \qquad \text{Distribute and simplify.}$$

$$2x^2 - 4x = 1 + x^2 - 4 \qquad \text{Distributive Property}$$

$$x^2 - 4x + 3 = 0 \qquad \text{Write in standard form.}$$

$$(x-3)(x-1) = 0 \qquad \text{Factor.}$$

$$x - 3 = 0 \implies x = 3 \qquad \text{Set 1st factor equal to 0.}$$

$$x - 1 = 0 \implies x = 1 \qquad \text{Set 2nd factor equal to 0.}$$

The solutions are $x = 1$ and $x = 3$. Check these in the original equation.

Study Tip

The symbol $\pm$ in Example 5 is read as "plus or minus." For instance, $x \neq \pm 2$ means that x is not equal to 2 or -2.

Exercises Within Reach®

Solutions in English & Spanish and tutorial videos at AlgebraWithinReach.com

Solving a Rational Equation In Exercises 21–34, solve the rational equation. (Check for extraneous solutions.)

21. $\dfrac{7x}{x+1} = \dfrac{5}{x-3} + 7$

22. $\dfrac{-3x}{x+4} = \dfrac{2}{x+1} - 3$

23. $\dfrac{4}{x+2} - \dfrac{1}{x} = \dfrac{1}{x}$

24. $\dfrac{3}{x+5} + \dfrac{5}{x} = \dfrac{10}{x}$

25. $10 - \dfrac{13}{x} = 4 + \dfrac{5}{x}$

26. $\dfrac{15}{x} - 4 = \dfrac{6}{x} + 3$

27. $\dfrac{2}{x-3} + \dfrac{1}{x} = \dfrac{x-1}{x-3}$

28. $\dfrac{x}{x-4} - \dfrac{9}{x} = \dfrac{3}{x-4}$

29. $\dfrac{20-x}{x} = x$

30. $\dfrac{x+30}{x} = x$

31. $2y = \dfrac{y+6}{y+1}$

32. $\dfrac{3x}{x+1} = \dfrac{2}{x-1}$

33. $x + \dfrac{1}{x} = \dfrac{5}{2}$

34. $\dfrac{4}{x} - \dfrac{x}{6} = \dfrac{5}{3}$

Applications

Application — EXAMPLE 6 Average Speeds

You and your friend travel to separate colleges in the same amount of time. You drive 380 miles and your friend drives 400 miles. Your friend's average speed is 3 miles per hour greater than your average speed. What is your average speed and what is your friend's average speed?

SOLUTION

Begin by setting your time equal to your friend's time. Then use the formula $t = \dfrac{d}{r}$ that gives the time t in terms of the distance d and the rate r.

Verbal Model: Your time = Your friend's time

$$\dfrac{\text{Your distance}}{\text{Your rate}} = \dfrac{\text{Friend's distance}}{\text{Friend's rate}}$$

Labels:
Your distance = 380 (miles)
Your rate = r (miles per hour)
Friend's distance = 400 (miles)
Friend's rate = $r + 3$ (miles per hour)

Equation:

$\dfrac{380}{r} = \dfrac{400}{r+3}$ Original equation

$380(r+3) = 400(r), \; r \neq 0, \; r \neq -3$ Cross-multiply.

$380r + 1140 = 400r$ Distribute and simplify.

$1140 = 20r$ Subtract $380r$.

$57 = r$ Divide by 20.

Your average speed is 57 miles per hour and your friend's average speed is $57 + 3 = 60$ miles per hour.

Exercises Within Reach®

Solutions in English & Spanish and tutorial videos at AlgebraWithinReach.com

35. *Average Speed* One car makes a trip of 440 miles in the same amount of time that it takes a second car to make a trip of 416 miles. The average speed of the second car is 3 miles per hour less than the average speed of the first car. What is the average speed of each car?

36. *Average Speed* One car makes a trip of 308 miles in the same amount of time that it takes a second car to make a trip of 352 miles. The average speed of the second car is 6 miles per hour greater than the average speed of the first car. What is the average speed of each car?

37. *Average Speed* One speed skater travels 3192 meters in the same amount of time that it takes a second skater to travel 2880 meters. The average speed of the second skater is 1.3 meters per second less than the average speed of the first skater. What is the average speed of each skater?

38. *Average Speed* One marathon runner travels 18 miles in the same amount of time that it takes a second marathon runner to travel 13.5 miles. The average speed of the first runner is 1.5 miles per hour greater than the average speed of the second runner. What is the average speed of each runner?

Application — EXAMPLE 7 Work Rate

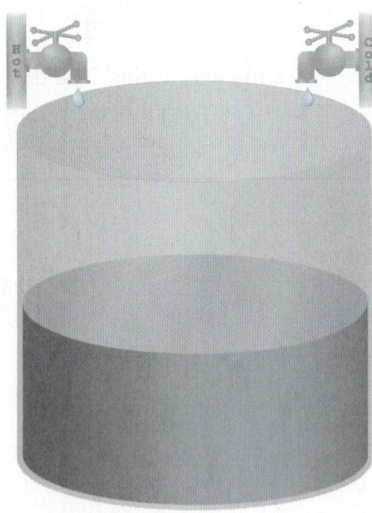

With only the cold water valve open, it takes 7 minutes to fill the tub of a washing machine. With both the hot and cold water valves open, it takes only 4 minutes. How long will it take with only the hot water valve open?

SOLUTION

Verbal Model: $\boxed{\text{Rate for cold water}} + \boxed{\text{Rate for hot water}} = \boxed{\text{Rate for warm water}}$

Labels:
Rate for warm water $= \dfrac{1}{4}$ (tub per minute)

Rate for cold water $= \dfrac{1}{7}$ (tub per minute)

Rate for hot water $= \dfrac{1}{t}$ (tub per minute)

Equation:

$\dfrac{1}{7} + \dfrac{1}{t} = \dfrac{1}{4}$ Original equation

$28t\left(\dfrac{1}{7} + \dfrac{1}{t}\right) = 28t\left(\dfrac{1}{4}\right)$ Multiply each side by LCD of $28t$.

$4t + 28 = 7t,\ t \neq 0$ Distribute and simplify.

$28 = 3t$ Subtract $4t$ from each side.

$\dfrac{28}{3} = t$ Divide each side by 3.

So, it will take $9\tfrac{1}{3}$ minutes to fill the tub with hot water alone.

Exercises Within Reach ®

Solutions in English & Spanish and tutorial videos at AlgebraWithinReach.com

39. *Work Rate* One person can paint a wall in 4 hours. The same person working with a friend can paint a similar wall in 1 hour. Working alone, how long would it take the second person to paint the wall?

40. *Work Rate* A pump empties a storage tank in 50 minutes. When a new pump is added to the system, the time to empty the tank using both pumps is 20 minutes. How long would it take to empty the tank using only the new pump?

41. *Work Rate* One person takes $1\tfrac{1}{2}$ times as long to rake a front lawn as a second person. The two people can rake the front lawn together in 30 minutes. Find the time it takes the first person to rake the front lawn.

42. *Flow Rate* The flow rate of one pipe is $1\tfrac{1}{4}$ times that of another pipe. A swimming pool can be filled in 5 hours using both pipes. Find the time required to fill the swimming pool using only the pipe with the lower flow rate.

Section 7.5 Rational Equations and Applications 343

Application EXAMPLE 8 Batting Average

A baseball player bats 280 times and hits the ball safely 70 times. The batting average for the player is 70/280 = .250. How many additional *consecutive* times must the player hit the ball safely to obtain a batting average of .300?

SOLUTION

Verbal Model: Batting average = Total hits ÷ Total times at bat

Labels: Current times at bat = 280
Current hits = 70
Additional consecutive hits = x

Equation:

$.300 = \dfrac{x + 70}{x + 280}$ Original equation.

$.300(x + 280) = x + 70, \ x \neq -280$ Multiply each side by LCD of $(x + 280)$.

$0.3x + 84 = x + 70$ Distributive Property.

$14 = 0.7x$ Simplify.

$20 = x$ Divide each side by 0.7.

The player must hit safely the next 20 times at bat. After doing so, the player's batting average will be 90/300 = .300.

Exercises Within Reach ®

Solutions in English & Spanish and tutorial videos at AlgebraWithinReach.com

43. **Batting Average** A softball player bats 35 times and hits the ball safely 6 times. How many additional consecutive times must the player hit the ball safely to obtain a batting average of .275?

44. **Batting Average** After 50 times at bat, a baseball player has a batting average of .160. How many additional consecutive times must the player hit the ball safely to obtain a batting average of .250?

45. **Serve Percentage** A volleyball player successfully serves the ball 26 times out of 32 attempts. How many additional consecutive times must the player serve the ball successfully to obtain a serve percentage (in decimal form) of 0.900?

46. **Free Throw Percentage** A basketball player makes 37 out of 46 attempted free throws. How many additional consecutive times must the player make a free throw to obtain a free throw percentage (in decimal form) of 0.850?

Richard Paul Kane/Shutterstock.com; Yuri Arcurs/Shutterstock.com

Chapter 7 Rational Expressions and Equations

> ### Concept Summary: Solving Rational Equations
>
> **What**
> The goal when solving rational equations is the same as when solving linear equations: You want to isolate the variable.
>
> **EXAMPLE**
> Solve $\dfrac{2}{x} - \dfrac{2}{5} = \dfrac{4}{x}$.
>
> **How**
> Use the following steps to solve rational equations.
> 1. Find the LCD of all fractions in the equation.
> 2. Multiply each side of the equation by the LCD.
> 3. Simplify each term.
> 4. Use the properties of equality to solve the resulting equation.
>
> **EXAMPLE**
>
> $\dfrac{2}{x} - \dfrac{2}{5} = \dfrac{4}{x}$ Original equation
>
> $5x\left(\dfrac{2}{x} - \dfrac{2}{5}\right) = 5x\left(\dfrac{4}{x}\right)$ Multiply by LCD.
>
> $\dfrac{10x}{x} - \dfrac{10x}{5} = \dfrac{20x}{x}$ Distributive Property
>
> $10 - 2x = 20$ Simplify.
>
> $-2x = 10$ Subtract 10.
>
> $x = -5, x \neq 0$ Divide by -2.
>
> **Why**
> You can use rational equations to model and solve many real-life problems. For example, knowing how to solve a rational equation can help you determine how long it might take a team to complete a job.

Exercises Within Reach ®

Worked-out solutions to odd-numbered exercises at AlgebraWithinReach.com

Concept Summary Check

47. *Making a Comparison* Compare the following.

$\dfrac{2}{x} - \dfrac{2}{5} = \dfrac{4}{x}$ $\dfrac{2}{x} - \dfrac{2}{5} + \dfrac{4}{x}$

48. *Writing* Describe the steps used to solve a rational equation.

49. *Vocabulary* What is an extraneous solution? How do you identify an extraneous solution?

50. *Cross-Multiplication* Can you use cross-multiplication to solve the rational equation in the example above?

Extra Practice

Solving a Rational Equation In Exercises 51–56, **solve** the rational equation. (Check for extraneous solutions.)

51. $\dfrac{3}{x-1} + \dfrac{5}{x+1} = \dfrac{8x+5}{x^2-1}$

52. $\dfrac{3}{x-3} + \dfrac{2}{x-5} = \dfrac{4}{x^2-8x+15}$

53. $\dfrac{x+3}{x^2-9} + \dfrac{4}{x-3} = -2$

54. $-1 - \dfrac{6}{x-4} = \dfrac{x+2}{x^2-16}$

55. $\dfrac{x}{2} = \dfrac{1 + \dfrac{3}{x}}{1 + \dfrac{1}{x}}$

56. $\dfrac{2x}{3} = \dfrac{1 + \dfrac{1}{x}}{1 + \dfrac{2}{x}}$

57. *Average Speed* A car leaves a town 20 minutes after a truck leaves. The speed of the car is approximately 10 miles per hour greater than that of the truck. After traveling 100 miles, the car overtakes the truck. Find the average speed of each vehicle.

58. *Average Speed* A car leaves a town 30 minutes after a bus leaves. The speed of the bus is 15 miles per hour less than that of the car. After traveling 150 miles, the car overtakes the bus. Find the average speed of each vehicle.

Explaining Concepts

59. Error Analysis Describe and correct the error.

$$\frac{12}{x+5} + \frac{1}{2} = 2$$

$$2(x+5)\left(\frac{12}{x+5} + \frac{1}{2}\right) = 2$$

$$2(12) + x + 5 = 2$$

$$x + 29 = 2$$

$$x = -27$$

True or False? **In Exercises 60–63, determine whether the statement is true or false. Justify your answer.**

60. The equation $1 - \frac{8}{x-5} = \frac{3}{x}$ has an extraneous solution of 5.

61. The equation $\frac{6}{x-3} + \frac{4x}{x+3} = \frac{8x^2}{x^2-9}$ has an extraneous solution of -3.

62. The equations $\frac{3}{x-7} + 1 = \frac{8}{x^2 - 9x + 14}$ and $x^2 - 6x = 0$ are equivalent.

63. The equations $\frac{3+x}{x} = \frac{7+2x}{2x}$ and $\frac{3}{x} = \frac{7}{2x}$ are equivalent.

64. Writing Can you assume that two people will complete a task twice as quickly as one of the two people working alone? Explain.

65. Writing A student submits the following argument to prove that 1 is equal to 0. Discuss what is wrong with the argument.

$x = 1$	Original equation
$x^2 = x$	Multiply each side by x.
$x^2 - x = 0$	Subtract x from each side.
$x(x-1) = 0$	Factor.
$\frac{x(x-1)}{x-1} = \frac{0}{x-1}$	Divide each side by $x - 1$.
$x = 0$	Simplify.

Cumulative Review

In Exercises 66–71, factor the polynomial by grouping.

66. $x^3 - x^2 + 4x - 4$

67. $2t^3 + 5t^2 - 6t - 15$

68. $3x^3 + 6x^2 + 2x + 4$

69. $6y^3 + 3y^2 - 2y - 1$

70. $15s^3 + 10s^2 - 12s - 8$

71. $4x^2 - x^3 - 8 + 2x$

In Exercises 72–77, simplify the complex fraction.

72. $\dfrac{\left(\frac{4}{9}\right)}{\left(\frac{2}{39}\right)}$

73. $\dfrac{\left(\frac{x^5}{12}\right)}{\left(\frac{x^2}{54}\right)}$

74. $\dfrac{\left(\frac{7xy^2}{2z^3}\right)}{\left(\frac{16x^3y^2}{21z^6}\right)}$

75. $\dfrac{\left[\frac{(2y)^3}{15x}\right]}{\left[\frac{22y^2}{(3x)^2}\right]}$

76. $\dfrac{\left(\frac{x+8}{x-4}\right)}{\left(\frac{3x+24}{x-4}\right)}$

77. $\dfrac{\left(\frac{x^2 + 3x - 18}{x+5}\right)}{2x - 6}$

7 Chapter Summary

	What did you learn?	Explanation and Examples	Review Exercises
7.1	Find the domain of a rational expression *(p. 304)*.	To find the values to exclude from the domain of a rational expression, follow these steps. 1. Set the denominator equal to zero. 2. Find the solution(s) of the equation in Step 1. Simplifying a rational expression can change its domain. To equate the original expressions with the simplified expression, you may have to restrict the domain of the simplified expression.	1–4
	Simplify rational expressions *(p. 306)*.	Let a, b, and c represent real numbers, variables, or algebraic expressions such that $b \neq 0$ and $c \neq 0$. Then the following is valid. $$\frac{ac}{bc} = \frac{a\cancel{c}}{b\cancel{c}} = \frac{a}{b}$$	5–24
7.2	Multiply rational expressions and simplify *(p. 312)*.	Let a, b, c, and d represent real numbers, variables, or algebraic expressions such that $b \neq 0$ and $d \neq 0$. Then the product of a/b and c/d is $$\frac{a}{b} \cdot \frac{c}{d} = \frac{ac}{bd}.$$	25–36, 49, 50
	Divide rational expressions and simplify *(p. 315)*.	Let a, b, c, and d represent real numbers, variables, or algebraic expressions such that $b \neq 0$, $c \neq 0$, and $d \neq 0$. Then the quotient of a/b and c/d is $$\frac{a}{b} \div \frac{c}{d} = \frac{a}{b} \cdot \frac{d}{c} = \frac{ad}{bc}.$$	37–50
7.3	Add or subtract rational expressions with like denominators *(p. 320)*.	Let a, b, and c represent real numbers, variables, or algebraic expressions such that $c \neq 0$. 1. $\dfrac{a}{c} + \dfrac{b}{c} = \dfrac{a+b}{c}$ 2. $\dfrac{a}{b} - \dfrac{b}{c} = \dfrac{a-b}{c}$	51–58
	Find the least common multiple of two or more polynomials *(p. 322)*.	1. Factor each polynomial completely. 2. The least common multiple must contain all the *different* factors of the polynomials, and each of these factors must be repeated the maximum number of times it occurs in any one of the factorizations.	59–64

Chapter Summary

	What did you learn?	Explanation and Examples	Review Exercises
7.3	Add or subtract rational expressions with unlike denominators *(p. 323)*.	1. Find the least common denominator (LCD) of the rational expressions. 2. Rewrite each rational expression so that it has the LCD in its denominator. 3. Combine these rational expressions with like denominators.	65–82
7.4	Simplify complex fractions using rules for dividing rational expressions *(p. 330)*.	1. Combine the numerator and denominator into single fractions, if necessary. 2. Divide by inverting the denominator and multiplying.	83–88
	Simplify complex fractions involving a sum or difference *(p. 332)*.	For complex fractions with numerators and/or denominators that are sums or differences of fractions, combine the numerator and/or denominator into single fractions. Then divide.	89–94
	Use complex fractions to solve application problems *(p. 334)*.	A common type of real-life problem modeled by complex fractions is electrical circuits.	95, 96
7.5	Solve rational equations with constant denominators *(p. 338)*.	To solve a rational equation with constant denominators, multiply each side of the equation by the LCD (a constant) of the fractions. Then solve the resulting equation. $$\frac{x}{3} = 2 - \frac{x}{6}$$ $$6\left(\frac{x}{3}\right) = 6\left(2 - \frac{x}{6}\right)$$ $$2x = 12 - x$$ $$3x = 12$$ $$x = 4$$	97–100
	Solve rational equations with variable denominators *(p. 339)*.	To solve a rational equation with variable denominators, multiply each side of the equation by the LCD (a variable term) of the fractions. Then solve the resulting equation. $$1 - \frac{2}{x} = \frac{1}{5}$$ $$5x\left(1 - \frac{2}{x}\right) = 5x\left(\frac{1}{5}\right)$$ $$5x - 10 = x$$ $$4x = 10$$ $$x = \frac{5}{2}$$	101–110
	Solve application problems using rational equations with variable denominators *(p. 341)*.	Many applications problems require the use of rational equations. (See Example 7.)	111–116

Review Exercises

Worked-out solutions to odd-numbered exercises at AlgebraWithinReach.com

7.1

Finding the Domain of a Rational Expression In Exercises 1–4, find the domain of the rational expression.

1. $\dfrac{8x}{x-5}$

2. $\dfrac{y+1}{y+3}$

3. $\dfrac{t}{t^2-3t+2}$

4. $\dfrac{x-10}{x(x^2-4)}$

Simplifying a Rational Expression In Exercises 5–22, simplify the rational expression.

5. $\dfrac{6t}{18}$

6. $\dfrac{45x}{15}$

7. $\dfrac{4x^5}{x^2}$

8. $\dfrac{88z^2}{33z}$

9. $\dfrac{7x^2y}{21xy^2}$

10. $\dfrac{2(yz)^2}{6yz^4}$

11. $\dfrac{3b-6}{4b-8}$

12. $\dfrac{2a+5}{10a+25}$

13. $\dfrac{4x-4y}{y-x}$

14. $\dfrac{x-y}{3y-3x}$

15. $\dfrac{x^2-9}{x^2-x-6}$

16. $\dfrac{x^2-4}{x^2+x-6}$

17. $\dfrac{1-x^3}{x^2-1}$

18. $\dfrac{x^2-4}{x^3+8}$

19. $\dfrac{x^2-3xy-18y^2}{x^2+4xy+3y^2}$

20. $\dfrac{x^2+2xy+y^2}{x^2-xy-2y^2}$

21. $\dfrac{x(x-4)+7(x-4)}{x^2+7x}$

22. $\dfrac{x^3-5x^2+2x-10}{x^3+2x}$

Probability In Exercises 23 and 24, the probability of hitting the shaded portion of the figure with a dart is equal to the ratio of the shaded area to the total area of the figure. Find the probability.

23.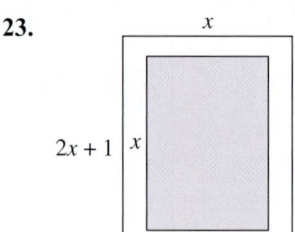

24.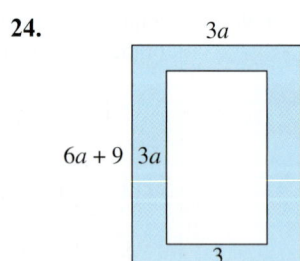

7.2

Multiplying Rational Expressions In Exercises 25–36, multiply and simplify.

25. $\dfrac{x^2}{6} \cdot \dfrac{2x}{x^3}$

26. $\dfrac{5y^4}{8y} \cdot \dfrac{12}{-15y^2}$

27. $\dfrac{5x^2y}{4} \cdot \dfrac{6x}{10y^3}$

28. $\dfrac{x^2y^3}{6} \cdot \dfrac{3}{(xy)^3}$

29. $\dfrac{x-5}{6} \cdot \dfrac{3}{x-5}$

30. $\dfrac{2-y}{9} \cdot \dfrac{9}{y-2}$

31. $\dfrac{2-x}{x+3} \cdot \dfrac{4x+12}{x^2-4}$

32. $\dfrac{8x-10}{7-x} \cdot \dfrac{x^2-49}{4x-5}$

33. $\dfrac{x+6}{x-8} \cdot (4x)$

34. $\dfrac{10x}{x+7} \cdot (x-4)$

35. $\dfrac{x^2-36}{6} \cdot \dfrac{3}{x^2-12x+36}$

36. $\dfrac{x^2-5x+4}{9} \cdot \dfrac{-18x}{x^2-8x+16}$

Dividing Rational Expressions In Exercises 37–48, divide and simplify.

37. $\dfrac{5}{8} \div \dfrac{u}{v}$

38. $\dfrac{x}{9} \div \dfrac{x^2}{3}$

39. $10y^2 \div \dfrac{y}{5}$

40. $\dfrac{5}{z^2} \div 3z^2$

41. $\dfrac{3(x+7)}{21x^2} \div \dfrac{12(x+7)^2}{9x}$

42. $\dfrac{u^2}{u^2-9} \div \dfrac{u}{u+3}$

43. $\dfrac{9-x^2}{2x+6} \div \dfrac{2x^2-5x-3}{6x-4}$

44. $\dfrac{25-x^2}{6x-30} \div \dfrac{7x^2+36x+5}{36x+6}$

45. $\dfrac{x^2-16}{x+2} \div \dfrac{x+4}{6x+12}$

46. $\dfrac{x^2-9x+20}{x+3} \div (4-x)$

47. $\dfrac{x^2-8x}{x-1} \div \dfrac{x^2-16x+64}{x^2-1}$

48. $\dfrac{x^2-x}{x+1} \div \dfrac{5x-5}{x^2+6x+5}$

Using Order of Operations In Exercises 49 and 50, perform the indicated operations and simplify.

49. $\left(\dfrac{3x^2}{8} \cdot \dfrac{x-7}{x}\right) \div \dfrac{x}{32}$

50. $\left[\left(\dfrac{x-4}{2}\right)^2 \cdot \left(\dfrac{x+9}{5}\right)^2\right] \div \dfrac{(x-4)(x+9)}{25}$

7.3

Adding and Subtracting Rational Expressions
In Exercises 51–58, combine and simplify.

51. $\dfrac{5x}{8} - \dfrac{3x}{8}$

52. $\dfrac{4t}{9} + \dfrac{11t}{9}$

53. $\dfrac{4x-5}{x+2} + \dfrac{2x+1}{x+2}$

54. $\dfrac{3y+4}{2y+1} - \dfrac{y+3}{2y+1}$

55. $\dfrac{5t+1}{2t-3} - \dfrac{2t-7}{2t-3}$

56. $\dfrac{3x+11}{x-5} + \dfrac{x-2}{x-5}$

57. $\dfrac{x-5}{x^2+2x-3} + \dfrac{3x+14}{x^2+2x-3}$

58. $\dfrac{x-9}{x^2-5x-12} - \dfrac{x+8}{x^2-5x-12}$

Finding the Least Common Multiple In Exercises 59–64, find the least common multiple of the polynomials.

59. $20x^2, 24, 30x^3$

60. $4y^2, 6y, 18z$

61. $2(x-2), 12(x-2)$

62. $x-5, 2x^2, x(x+5)$

63. $x^2-16, x(x+4)$

64. $5(x-1), 2(x^2+x+1)$

Adding and Subtracting Rational Expressions
In Exercises 65–82, **combine** and simplify.

65. $\dfrac{5t}{16} - \dfrac{5t}{24}$

66. $\dfrac{x}{8} + \dfrac{5x}{6}$

67. $\dfrac{5}{2x} + \dfrac{7}{3x}$

68. $\dfrac{1}{9x} - \dfrac{3}{5x}$

69. $\dfrac{1}{x+2} - \dfrac{1}{x+1}$

70. $\dfrac{4}{x-3} + \dfrac{1}{x+4}$

71. $\dfrac{4x}{x-1} - \dfrac{3}{1-x}$

72. $\dfrac{8}{x-10} - \dfrac{4x}{10-x}$

73. $\dfrac{2x-1}{5-3x} - \dfrac{3x+2}{3x-5}$

74. $\dfrac{5x+2}{2x-1} - \dfrac{x-4}{1-2x}$

75. $\dfrac{1}{x+4} - \dfrac{x-1}{x^2+4x+4}$

76. $\dfrac{1}{x-1} + \dfrac{1-x}{x^2+2x+1}$

77. $\dfrac{x-7}{x^2-16} - \dfrac{3}{4-x}$

78. $\dfrac{2x-7}{x^2-36} - \dfrac{1}{6-x}$

79. $x - 1 + \dfrac{1}{x+2} + \dfrac{1}{x-1}$

80. $\dfrac{2}{x} - \dfrac{3}{x-1} + \dfrac{4}{x+1}$

81. $2x + \dfrac{3}{2(x-4)} - \dfrac{1}{2(x+2)}$

82. $\dfrac{1}{x-2} + \dfrac{1}{(x-2)^2} + \dfrac{1}{x+2}$

7.4

Simplifying a Complex Fraction
In Exercises 83–88, **simplify** the complex fraction.

83. $\dfrac{\left(\dfrac{7}{18}\right)}{\left(\dfrac{3}{28}\right)}$

84. $\dfrac{\left(\dfrac{4}{x}\right)}{\left(\dfrac{1}{x^2}\right)}$

85. $\dfrac{5x}{\left(\dfrac{x}{y}\right)}$

86. $\dfrac{\left(\dfrac{6x-21}{x+3}\right)}{\left(\dfrac{2x-7}{x^2-9}\right)}$

87. $\dfrac{\left(\dfrac{x^2+x}{x^2+x-12}\right)}{\left(\dfrac{4x+4}{x^2-6x+9}\right)}$

88. $\dfrac{\left(\dfrac{3x^2+7x+4}{x^2+x}\right)}{\left(\dfrac{2x+8}{3x+1}\right)}$

Simplifying a Complex Fraction
In Exercises 89–94, **simplify** the complex fraction.

89. $\dfrac{\left(\dfrac{2}{x}+2\right)}{\left(1-\dfrac{1}{x}\right)}$

90. $\dfrac{\left(1+\dfrac{2}{x}\right)}{\left(\dfrac{2}{x}-1\right)}$

91. $\dfrac{\left(\dfrac{1}{x+1}-\dfrac{1}{4}\right)}{x-3}$

92. $\dfrac{\left(\dfrac{x}{x+1}-\dfrac{4}{5}\right)}{x-4}$

93. $\dfrac{\left(\dfrac{1}{x+3}\right)}{\left(\dfrac{4}{6x+18}+\dfrac{2}{x+3}\right)}$

94. $\dfrac{\left(\dfrac{5}{3x+6}\right)}{\left(\dfrac{9}{x+2}-\dfrac{7}{5x+10}\right)}$

95. **Average of Two Numbers** Find the average of the two real numbers $x/3$ and $5x/12$.

96. *Equal Lengths* Find the expressions for three real numbers that divide the real number line between $x/9$ and $x/5$ into four parts of equal length (see figure).

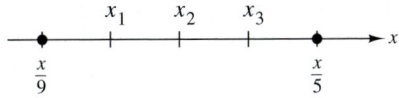

7.5

Solving a Rational Equation In Exercises 97–100, solve the rational equation.

97. $\dfrac{x}{6} + \dfrac{5x}{8} = 19$

98. $\dfrac{4x}{15} - 1 = \dfrac{x}{5}$

99. $\dfrac{t+1}{9} = \dfrac{2}{3} - t$

100. $\dfrac{2t-5}{14} = \dfrac{5}{7} - 2t$

Solving a Rational Equation In Exercises 101–110, solve the rational equation. (Check for extraneous solutions.)

101. $\dfrac{2}{x} + \dfrac{3}{8} = \dfrac{5}{x}$

102. $\dfrac{1}{6} - \dfrac{7}{x} = -\dfrac{5}{x}$

103. $\dfrac{7}{x} - 2 = \dfrac{3}{x} + 6$

104. $\dfrac{2}{x} + \dfrac{5}{3} = 1 + \dfrac{4}{x}$

105. $\dfrac{x}{x+3} - \dfrac{3x}{x^2-9} = 1$

106. $\dfrac{2x}{x-1} + \dfrac{4}{x+4} = 2$

107. $\dfrac{t}{t-4} + \dfrac{3}{t-2} = 0$

108. $\dfrac{2x}{x-3} - \dfrac{4}{x-1} = 4$

109. $\dfrac{2}{x-5} + \dfrac{x}{x-8} = \dfrac{-6}{x^2 - 13x + 40}$

110. $\dfrac{6}{x} - \dfrac{9}{x(x-3)} = \dfrac{-x}{x-3}$

111. *Number Problem* Find a number such that the sum of the number and its reciprocal is $\dfrac{41}{20}$.

112. *Number Problem* Find a number such that the sum of the number and its reciprocal is $\dfrac{34}{15}$.

113. *Average Speed* One car makes a trip of 159 miles in the same amount of time that it takes a second car to make a trip of 174 miles. The average speed of the second car is 5 miles per hour greater than the average speed of the first car. What is the average speed of each car?

114. *Average Speed* One car makes a trip of 455 miles in the same amount of time that it takes a second car to make a trip of 399 miles. The average speed of the second car is 8 miles per hour less than the average speed of the first car. What is the average speed of each car?

115. *Work Rate* One person can complete a landscaping job in 3 hours. The same person working with a coworker can complete the job in 1 hour. Working alone, how long would it take the second person to complete the job?

116. *Work Rate* One bricklayer lays $1\tfrac{1}{4}$ times as many bricks as a second bricklayer in the same amount of time. Find their individual times to complete a task if it takes them 12 hours working together.

Chapter Test

Solutions in English & Spanish and tutorial videos at AlgebraWithinReach.com

Take this test as you would take a test in class. After you are done, check your work against the answers in the back of the book.

1. Find the domain of the rational expression $\dfrac{x}{x^2 - 81}$.

2. Determine the missing factor: $\dfrac{2x^2}{x+1} = \dfrac{2x^2(\quad)}{x(x+1)^2}$.

3. Simplify: $\dfrac{3y(2y-1)}{6y^3(2y-1)^2}$.

4. Simplify: $\dfrac{x^2 - 64}{x^2 - 3x - 40}$.

In Exercises 5–10, perform the indicated operations and simplify.

5. $\dfrac{18x}{5} \cdot \dfrac{15}{3x^3}$

6. $(x+2)^2 \cdot \dfrac{x-2}{x^3 + 2x^2}$

7. $\dfrac{3x^2}{4} \div \dfrac{9x^3}{10}$

8. $\left[\left(\dfrac{x}{x-3} \right)^2 \cdot \dfrac{x^2}{x^2 - 3x} \right] \div (x-3)^5$

9. $\dfrac{3}{x+2} + 6$

10. $\dfrac{2}{x+1} - \dfrac{2x}{x^2 + 2x + 1}$

In Exercises 11–14, simplify the complex fraction.

11. $\dfrac{16}{\left(\dfrac{2}{x} + 8 \right)}$

12. $\dfrac{\left(\dfrac{3}{x+1} - \dfrac{3}{x-1} \right)}{\left(\dfrac{5}{x^2 - 1} \right)}$

13. $\dfrac{\left(\dfrac{t}{t-5} \right)}{\left(\dfrac{t^2}{5-t} \right)}$

14. $\dfrac{\left(9x - \dfrac{1}{x} \right)}{\left(\dfrac{1}{x} - 3 \right)}$

15. Determine whether each value of x is a solution of $\dfrac{x}{4} + \dfrac{2}{x} = \dfrac{3}{2}$.

    (a) $x = 1$ (b) $x = 2$ (c) $x = -\dfrac{1}{2}$ (d) $x = 4$

In Exercises 16–18, solve the rational equation.

16. $5 + \dfrac{t}{3} = t + 2$

17. $\dfrac{5}{x+1} - \dfrac{1}{x} = \dfrac{6}{x}$

18. $\dfrac{x-3}{x-2} + \dfrac{x+1}{x+3} = \dfrac{2x^2 - 15}{x^2 + x - 6}$

19. A van makes a trip of 54 miles in the same amount of time that it takes a car to make a trip of 63 miles. The average speed of the van is 6 miles per hour less than the average speed of the car. What is the average speed of each vehicle?

8
Systems of Linear Equations and Inequalities

- **8.1** Solving Systems of Equations by Graphing
- **8.2** Solving Systems of Equations by Substitution
- **8.3** Solving Systems of Equations by Elimination
- **8.4** Applications of Systems of Linear Equations
- **8.5** Systems of Linear Inequalities

MASTERY IS WITHIN REACH!

"I always get to know the instructor because then he or she is more willing to help me. I usually talk to the instructor before class starts, just to visit. I used to avoid going to instructors' offices, but now I realize that they are used to helping students, and it really is no big deal. I'm able to ask more questions about what is confusing me. I feel more connected in class too."

Manuel
Business

See page 379 for suggestions about being confident.

8.1 Solving Systems of Equations by Graphing

▶ Determine whether an ordered pair is a solution of a system of equations.
▶ Use a coordinate system to solve systems of linear equations graphically.
▶ Use systems of equations to model real-life problems.

Systems of Linear Equations

Consider the following problem, which will be solved in Section 8.2.

A total of $12,000 is invested in two funds paying 9% and 11% simple interest. (There is more risk in the 11% fund.) The combined annual interest for the two funds is $1180. How much of the $12,000 is invested at each rate?

Letting x and y denote the amounts (in dollars) in the two funds, you can translate this problem into the following pair of linear equations in two variables.

$$\begin{cases} x + y = 12{,}000 \\ 0.09x + 0.11y = 1{,}180 \end{cases}$$

Taken together, these two equations form a **system of linear equations**. A **solution** of a system of linear equations in two variables x and y is an ordered pair (x, y) that satisfies both equations.

EXAMPLE 1   Checking a Solution

Show that $(2, -1)$ is a solution of the system of linear equations.

$$\begin{cases} 3x + 2y = 4 & \text{Equation 1} \\ -x + 3y = -5 & \text{Equation 2} \end{cases}$$

SOLUTION

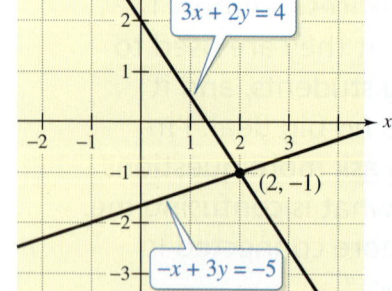

To check that a point is a solution, substitute the coordinates of the point into each equation. In the first equation, substitute 2 for x and -1 for y.

$$3(2) + 2(-1) \stackrel{?}{=} 4 \quad \Rightarrow \quad 6 - 2 = 4 \quad \text{Solution checks in 1st equation.} \checkmark$$

In the second equation, substitute 2 for x and -1 for y.

$$-(2) + 3(-1) \stackrel{?}{=} -5 \quad \Rightarrow \quad -2 - 3 = -5 \quad \text{Solution checks in 2nd equation.} \checkmark$$

Because the solution $(2, -1)$ checks in *both* equations, you can conclude that it is a solution of the original system of linear equations. You can also check by graphing both equations. In the graph, $(2, -1)$ is the point of intersection.

Exercises Within Reach®   Solutions in English & Spanish and tutorial videos at AlgebraWithinReach.com

Checking Solutions In Exercises 1–4, determine whether each ordered pair is a solution of the system.

	System	Ordered Pairs		System	Ordered Pairs	
1.	$\begin{cases} x + 3y = 11 \\ -x + 3y = 7 \end{cases}$	(a) $(2, 3)$	(b) $(5, 4)$	2. $\begin{cases} 3x - y = -2 \\ x - 3y = 2 \end{cases}$	(a) $(0, 2)$	(b) $(-1, -1)$
3.	$\begin{cases} 2x - 3y = -8 \\ x + y = 1 \end{cases}$	(a) $(5, -3)$	(b) $(-1, 2)$	4. $\begin{cases} 5x - 3y = -12 \\ x - 4y = 1 \end{cases}$	(a) $(-3, -1)$	(b) $(3, 1)$

Section 8.1 Solving Systems of Equations by Graphing

Solving a System of Linear Equations by Graphing

EXAMPLE 2 Solving a System of Linear Equations

Solve the system of linear equations by graphing each equation and locating the point of intersection.

$$\begin{cases} x + y = -2 & \text{Equation 1} \\ 2x - 3y = -9 & \text{Equation 2} \end{cases}$$

SOLUTION

One way to begin is to write each equation in slope-intercept form.

Equation 1
$x + y = -2$
$y = -x - 2$

Equation 2
$2x - 3y = -9$
$-3y = -2x - 9$
$y = \frac{2}{3}x + 3$

Then use a numerical approach by creating two tables of values.

Table of Values for Equation 1

x	−4	−3	−2	−1	0	1	2
$y = -x - 2$	2	1	0	−1	−2	−3	−4

Table of Values for Equation 2

x	−4	−3	−2	−1	0	1	2
$y = \frac{2}{3}x + 3$	$\frac{1}{3}$	1	$\frac{5}{3}$	$\frac{7}{3}$	3	$\frac{11}{3}$	$\frac{13}{3}$

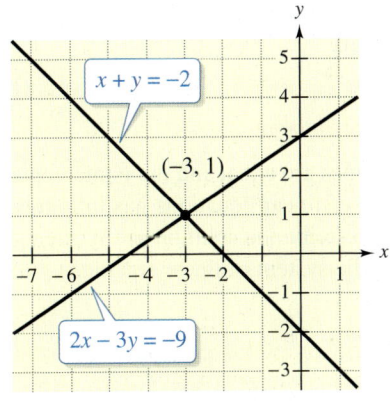

The tables show that the point $(-3, 1)$ is a common solution point. Another way to solve the system is to sketch the graphs of both equations, as shown at the left. From the graph, it appears that the lines intersect at the point $(-3, 1)$. To verify this, substitute the coordinates of the point into each of the two original equations.

Substitute into Equation 1
$x + y = -2$
$-3 + 1 \stackrel{?}{=} -2$
$-2 = -2$ ✓

Substitute into Equation 2
$2x - 3y = -9$
$2(-3) - 3(1) \stackrel{?}{=} -9$
$-9 = -9$ ✓

Because both equations are satisfied, the point $(-3, 1)$ is the solution of the system.

Exercises Within Reach®

Solutions in English & Spanish and tutorial videos at AlgebraWithinReach.com

Solving a System of Linear Equations In Exercises 5–10, solve the system by graphing.

5. $\begin{cases} y = -x + 3 \\ y = x + 1 \end{cases}$

6. $\begin{cases} y = 2x - 1 \\ y = x + 1 \end{cases}$

7. $\begin{cases} y = 2x - 4 \\ y = -\frac{1}{2}x + 1 \end{cases}$

8. $\begin{cases} y = \frac{1}{2}x + 2 \\ y = -x + 8 \end{cases}$

9. $\begin{cases} x - y = 3 \\ x + y = 3 \end{cases}$

10. $\begin{cases} x - y = 0 \\ x + y = 4 \end{cases}$

356 Chapter 8 Systems of Linear Equations and Inequalities

A system of linear equations can have exactly one solution, infinitely many solutions, or no solution.

Graphs			
Graphical interpretation	The two lines intersect.	The two lines coincide (are identical).	The two lines are parallel.
Intersection	Single point of intersection	Infinitely many points of intersection	No point of intersection
Slopes of lines	Slopes are not equal.	Slopes are equal.	Slopes are equal.
Number of solutions	Exactly one solution	Infinitely many solutions	No solution
Type of system	**Consistent system**	**Dependent (consistent) system**	**Inconsistent system**

EXAMPLE 3 A System with Infinitely Many Solutions

Solve the system of linear equations.

$$\begin{cases} x - y = 2 & \text{Equation 1} \\ -3x + 3y = -6 & \text{Equation 2} \end{cases}$$

SOLUTION

Begin by writing each equation in slope-intercept form.

$$\begin{cases} y = x - 2 & \text{Slope-intercept form of Equation 1} \\ y = x - 2 & \text{Slope-intercept form of Equation 2} \end{cases}$$

From these forms, you can see that the slopes of the lines are equal and the y-intercepts are the same (see the graph). So, the original system of linear equations has infinitely many solutions and is a dependent system. You can describe the solution set by saying that each point on the line $y = x - 2$ is a solution of the system of linear equations.

Exercises Within Reach®  Solutions in English & Spanish and tutorial videos at AlgebraWithinReach.com

Determining the Number of Solutions In Exercises 11–14, use the graphs of the equations to determine the number of solutions of the system of linear equations.

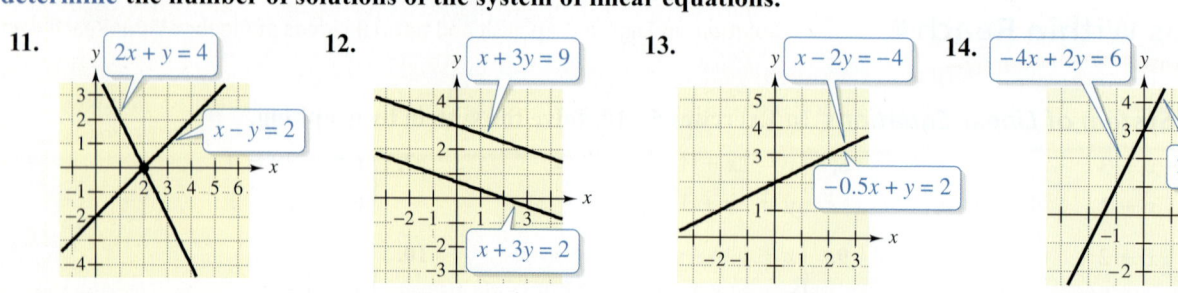

Section 8.1 Solving Systems of Equations by Graphing

EXAMPLE 4 A System with No Solution

Solve the system of linear equations.

$$\begin{cases} x - y = 2 & \text{Equation 1} \\ -3x + 3y = 6 & \text{Equation 2} \end{cases}$$

SOLUTION

Begin by writing each equation in slope-intercept form.

Equation 1

$x - y = 2$	Write original Equation 1.
$-y = -x + 2$	Subtract x from each side.
$y = x - 2$	Divide each side by -1.

Equation 2

$-3x + 3y = 6$	Write original Equation 2.
$3y = 3x + 6$	Add $3x$ to each side.
$y = x + 2$	Divide each side by 3.

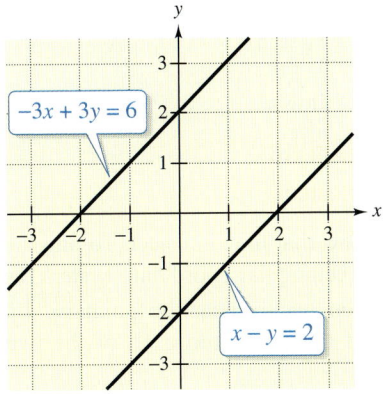

From these forms, you can see that the slopes of the lines are equal and the y-intercepts are different (see the graph). So, the original system of linear equations has no solution and is an inconsistent system.

Exercises Within Reach®

Solutions in English & Spanish and tutorial videos at AlgebraWithinReach.com

Solving a System of Linear Equations In Exercises 15–18, use the graphs of the equations to determine the solution (if any) of the system of linear equations. Check your solution.

15.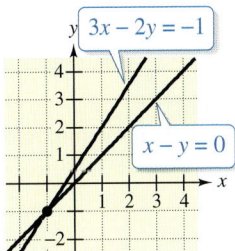
16.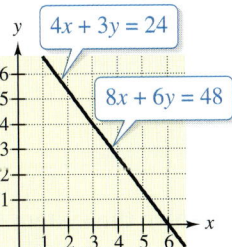
17.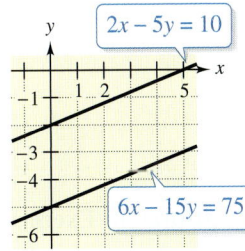
18.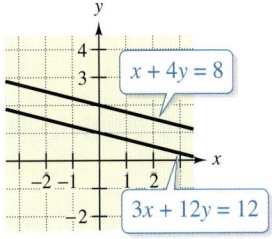

Solving a System of Linear Equations In Exercises 19–26, solve the system by graphing.

19. $\begin{cases} x + y = 4 \\ 4x + 4y = 2 \end{cases}$

20. $\begin{cases} 3x + 3y = 1 \\ x + y = 3 \end{cases}$

21. $\begin{cases} x + 2y = 3 \\ x - 3y = 13 \end{cases}$

22. $\begin{cases} -x + 10y = 30 \\ x + 10y = 10 \end{cases}$

23. $\begin{cases} -2x + y = -1 \\ x - 2y = -1 \end{cases}$

24. $\begin{cases} 2x + y = -4 \\ 4x - 2y = 8 \end{cases}$

25. $\begin{cases} x - 2y = 4 \\ 2x - 4y = 8 \end{cases}$

26. $\begin{cases} 2x + 3y = 6 \\ 4x + 6y = 12 \end{cases}$

EXAMPLE 5 A System with a Single Solution

Solve the system of linear equations.

$$\begin{cases} 2x + y = 4 & \text{Equation 1} \\ 4x + 3y = 9 & \text{Equation 2} \end{cases}$$

SOLUTION

Begin by writing each equation in slope-intercept form.

Equation 1
$2x + y = 4$
$y = -2x + 4$

Equation 2
$4x + 3y = 9$
$3y = -4x + 9$
$y = -\frac{4}{3}x + 3$

The slope-intercept forms of the two equations are as follows.

$$\begin{cases} y = -2x + 4 & \text{Slope-intercept form of Equation 1} \\ y = -\frac{4}{3}x + 3 & \text{Slope-intercept form of Equation 2} \end{cases}$$

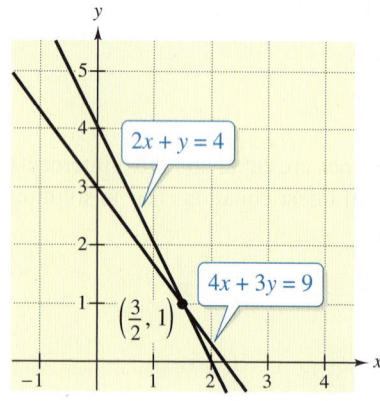

Because the lines do not have the same slope, you know that they intersect. To find the point of intersection, sketch both lines on the same rectangular coordinate system, as shown. From this sketch, it appears that the solution occurs near the point $\left(\frac{3}{2}, 1\right)$. To check this solution, substitute the coordinates of the point into each of the two original equations.

Substitute into Equation 1
$2x + y = 4$
$2\left(\frac{3}{2}\right) + 1 \stackrel{?}{=} 4$
$3 + 1 \stackrel{?}{=} 4$
$4 = 4$ ✓

Substitute into Equation 2
$4x + 3y = 9$
$4\left(\frac{3}{2}\right) + 3(1) \stackrel{?}{=} 9$
$6 + 3 \stackrel{?}{=} 9$
$9 = 9$ ✓

Because both equations are satisfied, the point $\left(\frac{3}{2}, 1\right)$ is the solution of the system.

Exercises Within Reach®

Solutions in English & Spanish and tutorial videos at AlgebraWithinReach.com

Solving a System of Linear Equations In Exercises 27–36, solve the system by graphing.

27. $\begin{cases} x + 7y = -5 \\ 3x - 2y = 8 \end{cases}$

28. $\begin{cases} x + 2y = 4 \\ 2x - 2y = -1 \end{cases}$

29. $\begin{cases} -4x + 2y = 12 \\ 2x + y = 6 \end{cases}$

30. $\begin{cases} 3x + 2y = -6 \\ 3x - 2y = 6 \end{cases}$

31. $\begin{cases} 4x - 5y = 0 \\ 6x - 5y = 10 \end{cases}$

32. $\begin{cases} \frac{1}{2}x + 2y = -4 \\ -3x + y = 11 \end{cases}$

33. $\begin{cases} x + \frac{5}{4}y = 5 \\ 4x + 5y = 20 \end{cases}$

34. $\begin{cases} 7x + 3y = 21 \\ \frac{7}{3}x + y = 7 \end{cases}$

35. $\begin{cases} 8x - 6y = -12 \\ x - \frac{3}{4}y = -2 \end{cases}$

36. $\begin{cases} -x + \frac{2}{3}y = 5 \\ 9x - 6y = 6 \end{cases}$

Section 8.1 Solving Systems of Equations by Graphing 359

Application

EXAMPLE 6 Break-Even Analysis

A small business invests $14,000 to produce a new, soy-free energy bar. Each bar costs $0.80 to produce and is sold for $1.50. The total cost C of producing x units of the bar is given by

$$C = 0.80x + 14,000. \quad \text{Cost equation}$$

The revenue R from selling x units of the bar is given by

$$R = 1.50x. \quad \text{Revenue equation}$$

How many energy bars must be sold before the business breaks even?

SOLUTION

The system of equations that represents this situation is as follows.

$$\begin{cases} C = 0.80x + 14,000 & \text{Equation 1} \\ R = 1.50x & \text{Equation 2} \end{cases}$$

The two equations are in slope-intercept form, and because the lines do not have the same slope, you know that they intersect. So, to find the break-even point, graph both equations and determine the point of intersection of the two graphs, as shown. From this graph, it appears that the break-even point occurs near the point (20,000, 30,000). To check this solution, substitute the coordinates of the point into each of the two original equations.

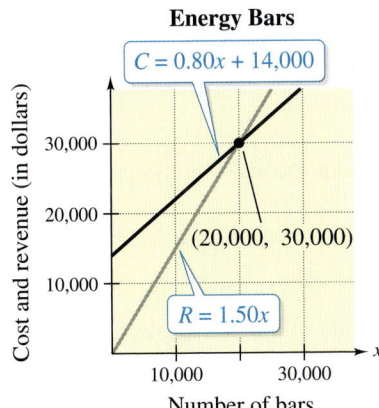

Energy Bars

Substitute into Equation 1

$$C = 0.80x + 14,000$$
$$30,000 \stackrel{?}{=} 0.80(20,000) + 14,000$$
$$30,000 \stackrel{?}{=} 16,000 + 14,000$$
$$30,000 = 30,000 \checkmark$$

Substitute into Equation 2

$$R = 1.50x$$
$$30,000 \stackrel{?}{=} 1.50(20,000)$$
$$30,000 = 30,000 \checkmark$$

Because both equations are satisfied, the business must sell 20,000 energy bars before it breaks even.

Exercises Within Reach®

Solutions in English & Spanish and tutorial videos at AlgebraWithinReach.com

37. Break-Even Analysis A small company produces bird feeders that sell for $23 per unit. The cost of producing each unit is $16.75, and the company has fixed costs of $400.

(a) Use a verbal model to show that the cost C of producing x units is $C = 16.75x + 400$ and the revenue R from selling x units is $R = 23x$.

(b) Use a graphing calculator to graph the cost and revenue functions in the same viewing window. Approximate the point of intersection of the graphs and interpret the result.

38. Break-Even Analysis A company produces hockey sticks that sell for $79 per unit. The cost of producing each unit is $53.25, and the company has fixed costs of $1000.

(a) Use a verbal model to show that the cost C of producing x units is $C = 53.25x + 1000$ and the revenue R from selling x units is $R = 79x$.

(b) Use a graphing calculator to graph the cost and revenue functions in the same viewing window. Approximate the point of intersection of the graphs and interpret the result.

Concept Summary: Solving Systems of Linear Equations by Graphing

What

A **system of linear equations** can have one solution, infinitely many solutions, or no solution.

One way to solve a system of linear equations is by graphing.

EXAMPLE

Solve the system by graphing.

$$\begin{cases} x + y = 4 \\ -2x + y = -2 \end{cases}$$

How

Use these steps to solve a linear system by graphing.

1. Write each equation in slope-intercept form.

$$\begin{cases} y = -x + 4 \\ y = 2x - 2 \end{cases}$$

2. Sketch the lines representing the equations.

3. Estimate the point of intersection.

4. Check that the point of intersection is the solution.

Why

By solving a system by graphing, you can easily determine whether the system is **consistent** (one solution), **dependent** (infinitely many solutions), or **inconsistent** (no solution). In this case, the system is consistent and the solution appears to be the point (2, 2).

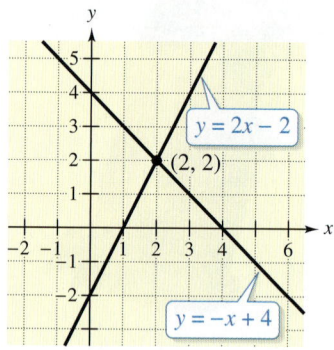

Exercises Within Reach®

Worked-out solutions to odd-numbered exercises at AlgebraWithinReach.com

Concept Summary Check

39. *Writing* Give graphical descriptions of the three types of systems of two linear equations in two variables based on the number of points of intersection.

40. *Vocabulary* What is a dependent system of linear equations?

41. *Vocabulary* What is an inconsistent system of linear equations?

42. *Reasoning* What is one of the drawbacks to solving a system of linear equations by graphing?

Extra Practice

Determining the Number of Solutions In Exercises 43–48, *write* the equations of the lines in slope-intercept form. What can you *conclude* about the number of solutions of the system?

43. $\begin{cases} 2x - 3y = -12 \\ -8x + 12y = -12 \end{cases}$

44. $\begin{cases} -5x + 8y = 8 \\ -5x + 8y = -28 \end{cases}$

45. $\begin{cases} -2x + 3y = 4 \\ 2x + 3y = 8 \end{cases}$

46. $\begin{cases} 2x + 5y = 15 \\ 2x - 5y = 5 \end{cases}$

47. $\begin{cases} -x + 4y = 7 \\ 3x - 12y = -21 \end{cases}$

48. $\begin{cases} 7x + 6y = -4 \\ 3.5x + 3y = -2 \end{cases}$

49. Number Problem The sum of two numbers x and y is 20 and the difference of the two numbers is 2. The system of equations that represents this situation is
$$\begin{cases} x + y = 20 \\ x - y = 2 \end{cases}.$$
Solve the system graphically to find the two numbers.

50. Number Problem The sum of two numbers x and y is 35 and the difference of the two numbers is 11. The system of equations that represents this situation is
$$\begin{cases} x + y = 35 \\ x - y = 11 \end{cases}.$$
Solve the system graphically to find the two numbers.

Think About It In Exercises 51 and 52, the graphs of the two equations appear to be parallel. **Determine whether the two lines are actually parallel. Does the system have a solution? If so, find the solution.**

51. $\begin{cases} x - 200y = -200 \\ x - 199y = 198 \end{cases}$

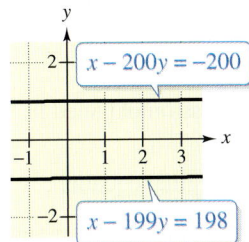

52. $\begin{cases} 25x - 24y = 0 \\ 13x - 12y = 24 \end{cases}$

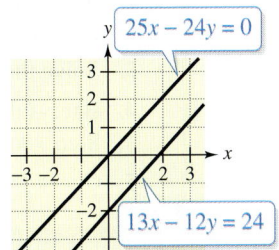

Explaining Concepts

53. *True or False?* It is possible for a consistent system of linear equations to have exactly two solutions. Justify your answer.

54. *Creating a System* Write a system of linear equations with integer coefficients that has the unique solution $(3, -4)$. (There are many correct answers.)

55. *Creating a System* Write a system of linear equations that has no solution. (There are many correct answers.)

56. *Creating a System* Write a system of linear equations that has infinitely many solutions. (There are many correct answers.)

Cumulative Review

In Exercises 57–60, evaluate the expression.

57. $\frac{2}{3} + \frac{1}{3}$

58. $\frac{5}{8} + \frac{3}{2}$

59. $\frac{2}{7} - \frac{1}{3}$

60. $\frac{3}{5} - \frac{7}{6}$

In Exercises 61–64, solve the equation and check your solution.

61. $x - 6 = 5x$

62. $2 - 3x = 14 + x$

63. $y - 3(4y - 2) = 1$

64. $y + 6(3 - 2y) = 4$

In Exercises 65–68, solve the rational equation.

65. $\dfrac{x}{5} + \dfrac{2x}{5} = 3$

66. $\dfrac{3x}{5} + \dfrac{4x}{8} = \dfrac{11}{10}$

67. $\dfrac{x - 3}{x + 1} = \dfrac{4}{3}$

68. $\dfrac{3}{x} = \dfrac{9}{2(x + 2)}$

8.2 Solving Systems of Equations by Substitution

▶ Use the method of substitution to solve systems of equations algebraically.
▶ Solve systems with no solution or infinitely many solutions.
▶ Use the method of substitution to solve application problems.

The Method of Substitution

Study Tip

The term **back-substitute** implies that you work backwards. After solving for one of the variables, substitute that value back into one of the equations in the original (or revised) system to find the value of the other variable.

The Method of Substitution

1. Solve one of the equations for one variable in terms of the other.
2. Substitute the expression obtained in Step 1 into the other equation to obtain an equation in one variable.
3. Solve the equation obtained in Step 2.
4. Back-substitute the solution from Step 3 into the expression obtained in Step 1 to find the value of the other variable.
5. Check the solution to see that it satisfies *both* of the original equations.

EXAMPLE 1   The Method of Substitution

Solve the system of linear equations.

$$\begin{cases} -x + y = 1 & \text{Equation 1} \\ 2x + y = -2 & \text{Equation 2} \end{cases}$$

SOLUTION

Begin by solving for y in Equation 1.

$y = x + 1$ Revised Equation 1

Next, substitute this expression for y in Equation 2 and solve for x.

$2x + y = -2$ Equation 2
$2x + (x + 1) = -2$ Substitute $x + 1$ for y.
$3x + 1 = -2$ Combine like terms.
$x = -1$ Solve for x.

Finally, *back-substitute* this x-value into the revised Equation 1.

$y = x + 1$ Revised Equation 1
$y = -1 + 1 = 0$ Substitute -1 for x.

So, the solution is $(-1, 0)$. Check this solution by substituting $x = -1$ and $y = 0$ into both of the original equations.

Exercises Within Reach®   Solutions in English & Spanish and tutorial videos at AlgebraWithinReach.com

Solving a System of Linear Equations   In Exercises 1–4, **solve** the system by the method of substitution.

1. $\begin{cases} y = 2x - 1 \\ y = -x + 5 \end{cases}$
2. $\begin{cases} y = -2x + 9 \\ y = 3x - 1 \end{cases}$
3. $\begin{cases} x - y = 0 \\ 2x + y = 9 \end{cases}$
4. $\begin{cases} x - y = 0 \\ 5x - 3y = 10 \end{cases}$

Section 8.2 Solving Systems of Equations by Substitution 363

EXAMPLE 2 The Method of Substitution

Solve the system of linear equations.

$$\begin{cases} 5x + 7y = 1 & \text{Equation 1} \\ x + 4y = -5 & \text{Equation 2} \end{cases}$$

SOLUTION

For this system, it is convenient to begin by solving for x in Equation 2.

$x + 4y = -5$	Original Equation 2
$x = -4y - 5$	Revised Equation 2

Next, substitute this expression for x in Equation 1 and solve for y.

$5x + 7y = 1$	Equation 1
$5(-4y - 5) + 7y = 1$	Substitute $-4y - 5$ for x.
$-20y - 25 + 7y = 1$	Distributive Property
$-13y - 25 = 1$	Combine like terms.
$-13y = 26$	Add 25 to each side.
$y = -2$	Divide each side by -13.

Finally, back-substitute this y-value into the revised Equation 2.

$x = -4(-2) - 5 = 3$ Substitute -2 for y in revised Equation 2.

So, the solution is $(3, -2)$. Check this by substituting $x = 3$ and $y = -2$ into both of the original equations, as follows.

Substitute into Equation 1

$5x + 7y = 1$
$5(3) + 7(-2) \stackrel{?}{=} 1$
$15 - 14 = 1$ ✓

Substitute into Equation 2

$x + 4y = -5$
$(3) + 4(-2) \stackrel{?}{=} -5$
$3 - 8 = -5$ ✓

Study Tip

When you use the method of substitution, it does not matter which variable you choose to solve for first. You should choose the variable and equation that are easier to work with. For instance, in the system

$$\begin{cases} 3x - 2y = 1 & \text{Equation 1} \\ x + 4y = 3 & \text{Equation 2} \end{cases}$$

it is easier first to solve for x in Equation 2. On the other hand, in the system

$$\begin{cases} 2x + y = 5 & \text{Equation 1} \\ 3x - 2y = 11 & \text{Equation 2} \end{cases}$$

it is easier first to solve for y in Equation 1.

Exercises Within Reach®   Solutions in English & Spanish and tutorial videos at AlgebraWithinReach.com

Solving a System of Linear Equations In Exercises 5–14, solve the system by the method of substitution.

5. $\begin{cases} x = 4y - 5 \\ x = 3y \end{cases}$

6. $\begin{cases} x = -5y - 2 \\ x = 2y - 23 \end{cases}$

7. $\begin{cases} 2x = 8 \\ x - 2y = 12 \end{cases}$

8. $\begin{cases} 2x - y = 0 \\ 3y = 6 \end{cases}$

9. $\begin{cases} x - 2y = -10 \\ 3x - y = 0 \end{cases}$

10. $\begin{cases} x - 2y = 5 \\ 3x - y = 0 \end{cases}$

11. $\begin{cases} 2x - y = -2 \\ 4x + y = 5 \end{cases}$

12. $\begin{cases} x + 6y = 7 \\ -x + 4y = -2 \end{cases}$

13. $\begin{cases} x + 2y = 1 \\ 5x - 4y = -23 \end{cases}$

14. $\begin{cases} -2x + y = -18 \\ 3x + 3y = 9 \end{cases}$

EXAMPLE 3 The Method of Substitution

Solve the system of linear equations.

$$\begin{cases} 5x + 3y = 18 & \text{Equation 1} \\ 2x - 7y = -1 & \text{Equation 2} \end{cases}$$

SOLUTION

Because neither variable has a coefficient of 1, you can choose to solve for either variable first. For instance, you can begin by solving for x in Equation 1.

$5x + 3y = 18$ Original Equation 1

$5x = -3y + 18$ Subtract $3y$ from each side.

$x = -\dfrac{3}{5}y + \dfrac{18}{5}$ Revised Equation 1

Next, substitute this expression for x in Equation 2 and solve for y.

$2x - 7y = -1$ Equation 2

$2\left(-\dfrac{3}{5}y + \dfrac{18}{5}\right) - 7y = -1$ Substitute $-\dfrac{3}{5}y + \dfrac{18}{5}$ for x.

$-\dfrac{6}{5}y + \dfrac{36}{5} - 7y = -1$ Distributive Property

$-6y + 36 - 35y = -5$ Multiply each side by 5.

$36 - 41y = -5$ Combine like terms.

$-41y = -41$ Subtract 36 from each side.

$y = 1$ Divide each side by -41.

Finally, back-substitute this y-value into the revised Equation 1.

$x = -\dfrac{3}{5}y + \dfrac{18}{5}$ Revised Equation 1

$x = -\dfrac{3}{5}(1) + \dfrac{18}{5}$ Substitute 1 for y.

$x = 3$ Simplify.

So, the solution is $(3, 1)$. Check this in the original system.

Exercises Within Reach®

Solutions in English & Spanish and tutorial videos at AlgebraWithinReach.com

Solving a System of Linear Equations In Exercises 15–20, solve the system by the method of substitution.

15. $\begin{cases} 8x + 4y = -2 \\ -12x + 5y = -8 \end{cases}$

16. $\begin{cases} 5x - 2y = 10 \\ 3x + 2y = 6 \end{cases}$

17. $\begin{cases} 3x + 7y = 2 \\ 5x - 3y = -26 \end{cases}$

18. $\begin{cases} 4x + 3y = 15 \\ 2x - 5y = 1 \end{cases}$

19. $\begin{cases} 2x - 3y = 16 \\ 3x + 4y = 7 \end{cases}$

20. $\begin{cases} -2x + 7y = 9 \\ 3x + 2y = -1 \end{cases}$

The No-Solution and Many-Solutions Cases

EXAMPLE 4 The Method of Substitution: No-Solution Case

Solve the system of linear equations.

$$\begin{cases} x - 3y = 2 & \text{Equation 1} \\ -2x + 6y = 2 & \text{Equation 2} \end{cases}$$

SOLUTION

Begin by solving for x in Equation 1 to obtain $x = 3y + 2$. Then substitute this expression for x in Equation 2.

$-2x + 6y = 2$	Equation 2
$-2(3y + 2) + 6y = 2$	Substitute $3y + 2$ for x.
$-6y - 4 + 6y = 2$	Distributive Property
$-4 = 2$	False statement

Because $-4 = 2$ is a false statement, you can conclude that the original system is inconsistent and has no solution. The graph confirms this result.

EXAMPLE 5 The Method of Substitution: Many-Solutions Case

Solve the system of linear equations.

$$\begin{cases} 9x + 3y = 15 & \text{Equation 1} \\ 3x + y = 5 & \text{Equation 2} \end{cases}$$

SOLUTION

Begin by solving for y in Equation 2 to obtain $y = -3x + 5$. Then substitute this expression for y in Equation 1.

$9x + 3y = 15$	Equation 1
$9x + 3(-3x + 5) = 15$	Substitute $-3x + 5$ for y.
$9x - 9x + 15 = 15$	Distributive Property
$15 = 15$	Simplify.

The equation $15 = 15$ is true for any value of x. This implies that any solution of Equation 2 is also a solution of Equation 1. In other words, the original system of linear equations is *dependent* and has infinitely many solutions. The solutions consist of all ordered pairs (x, y) lying on the line $3x + y = 5$, such as $(-1, 8)$, $(0, 5)$, and $(1, 2)$.

Study Tip

By writing both equations in Example 5 in slope-intercept form, you will get identical equations. This means that the lines coincide and the system has infinitely many solutions.

Exercises Within Reach®

Solutions in English & Spanish and tutorial videos at AlgebraWithinReach.com

Solving a System of Linear Equations In Exercises 21–26, solve the system by the method of substitution.

21. $\begin{cases} 4x - y = 2 \\ 2x - \frac{1}{2}y = 1 \end{cases}$

22. $\begin{cases} x - 5y = 5 \\ 3x - 15y = 15 \end{cases}$

23. $\begin{cases} -5x + 4y = 14 \\ 5x - 4y = 4 \end{cases}$

24. $\begin{cases} 3x - 2y = 3 \\ -6x + 4y = -6 \end{cases}$

25. $\begin{cases} -6x + 1.5y = 6 \\ 8x - 2y = -8 \end{cases}$

26. $\begin{cases} 0.3x - 0.3y = 0 \\ x - y = 4 \end{cases}$

Applications

Application

EXAMPLE 6 An Interest Rate Problem

A total of $12,000 is invested in two funds paying 9% and 11% simple interest. (There is more risk in the 11% fund.) The combined annual interest for the two funds is $1180. The system of equations that represents this situation is

$$\begin{cases} x + y = 12{,}000 & \text{Equation 1} \\ 0.09x + 0.11y = 1{,}180 & \text{Equation 2} \end{cases}$$

where x represents the amount invested in the 9% fund and y represents the amount invested in the 11% fund. Solve this system to determine how much of the $12,000 is invested at each rate.

SOLUTION

To begin, it is convenient to multiply each side of Equation 2 by 100. This eliminates the need to work with decimals.

$$9x + 11y = 118{,}000 \quad \text{Revised Equation 2}$$

Then solve for x in Equation 1.

$$x + y = 12{,}000 \quad \text{Equation 1}$$
$$x = 12{,}000 - y \quad \text{Revised Equation 1}$$

Next, substitute this expression for x in the revised Equation 2 and solve for y.

$$9x + 11y = 118{,}000 \quad \text{Revised Equation 2}$$
$$9(12{,}000 - y) + 11y = 118{,}000 \quad \text{Substitute } 12{,}000 - y \text{ for } x.$$
$$108{,}000 - 9y + 11y = 118{,}000 \quad \text{Distributive Property}$$
$$2y = 10{,}000 \quad \text{Simplify.}$$
$$y = 5000 \quad \text{Divide each side by 2.}$$

Finally, back-substitute this y-value into the revised Equation 1 to obtain $x = 12{,}000 - 5000 = 7000$. The solution is $(7000, 5000)$. So, $7000 is invested at 9% simple interest and $5000 is invested at 11% simple interest. Check this in the original statement of the problem.

Exercises Within Reach®

Solutions in English & Spanish and tutorial videos at AlgebraWithinReach.com

27. *Investment* A total of $15,000 is invested in two funds paying 5% and 8% simple interest. (There is more risk in the 8% fund.) The combined annual interest for the two funds is $900. The system of equations that represents this situation is

$$\begin{cases} x + y = 15{,}000 \\ 0.05x + 0.08y = 900 \end{cases}$$

where x represents the amount invested in the 5% fund and y represents the amount invested in the 8% fund. Solve this system to determine how much of the $15,000 is invested at each rate.

28. *Investment* A total of $10,000 is invested in two funds paying 7% and 10% simple interest. (There is more risk in the 10% fund.) The combined annual interest for the two funds is $775. The system of equations that represents this situation is

$$\begin{cases} x + y = 10{,}000 \\ 0.07x + 0.10y = 775 \end{cases}$$

where x represents the amount invested in the 7% fund and y represents the amount invested in the 10% fund. Solve this system to determine how much of the $10,000 is invested at each rate.

Application — EXAMPLE 7 Comparing Costs

Car model ES costs $18,000 and costs an average of $0.26 per mile to maintain and fuel. Car model LS costs $22,000 and costs an average of $0.22 per mile to maintain and fuel. The system of equations that represents this situation is

$$\begin{cases} y = 18{,}000 + 0.26x & \text{Equation 1} \\ y = 22{,}000 + 0.22x & \text{Equation 2} \end{cases}$$

where y represents the total cost of the car and x represents the number of miles driven. Solve this system to determine after how many miles the total costs of the two models will be the same. What will be the cost?

SOLUTION

Substitute the expression for y from Equation 1 into Equation 2.

$18{,}000 + 0.26x = 22{,}000 + 0.22x$	Substitute.
$0.04x = 4000$	Simplify.
$x = 100{,}000$	Divide each side by 0.04.

Substitute $x = 100{,}000$ into either of the original equations to find the total cost.

$y = 18{,}000 + 0.26x$	Equation 1
$\;\;= 18{,}000 + 0.26(100{,}000)$	Substitute 100,000 for x.
$\;\;= 18{,}000 + 26{,}000$	Multiply.
$\;\;= 44{,}000$	Simplify.

So, after 100,000 miles the total cost of the two models will be $44,000.

Exercises Within Reach®

Solutions in English & Spanish and tutorial videos at AlgebraWithinReach.com

29. Comparing Costs A solar heating system for a three-bedroom home costs $28,500 for installation and $125 per year to operate. An electric heating system for the same home costs $5750 for installation and $1000 per year to operate. The system of equations that represents this situation is

$$\begin{cases} y = 28{,}500 + 125x & \text{Solar heating} \\ y = \;\;5{,}750 + 1000x & \text{Electric heating} \end{cases}$$

where y represents the total cost of heating the home and x represents the number of years. Solve this system to determine after how many years the total costs for solar heating and electric heating will be the same. What will be the cost at that time?

30. Ticket Sales You are selling tickets for a football game. Student tickets cost $3 each and general admission tickets cost $5 each. You sell 1957 tickets and collect $8113. The system of equations that represents this situation is

$$\begin{cases} x + y = 1957 \\ 3x + 5y = 8113 \end{cases}$$

where x represents the number of students tickets sold and y represents the number of general admission tickets sold. Solve this system to determine how many of each type of ticket are sold.

Concept Summary: Solving Systems of Linear Equations by Substitution

What
You can solve a system of linear equations using an algebraic method called the **method of substitution**.

EXAMPLE
Solve the system of linear equations.
$$\begin{cases} y = x + 2 & \text{Equation 1} \\ x + y = 6 & \text{Equation 2} \end{cases}$$

How
The goal is to reduce the system of two linear equations in two variables to a single equation in one variable. Then solve the equation.

EXAMPLE
Substitute $x + 2$ for y in Equation 2. Then solve for x.
$$x + (x + 2) = 6$$
$$x = 2$$
Now substitute 2 for x in Equation 1 to find the value of y.
$$y = 2 + 2$$
$$y = 4$$

Why
Solving a system of equations using an algebraic method like substitution is more accurate than solving a system by graphing.

Exercises Within Reach®   Worked-out solutions to odd-numbered exercises at AlgebraWithinReach.com

Concept Summary Check

31. Precision In your own words, explain the basic steps in solving a system of linear equations by the method of substitution.

32. Reasoning When solving a system of linear equations by the method of substitution, how do you recognize that it has no solution?

33. Reasoning When solving a system of linear equations by the method of substitution, how do you recognize that it has infinitely many solutions?

34. Writing Explain how you can check the solution of a system of linear equations algebraically.

Extra Practice

Solving a System of Linear Equations In Exercises 35–46, solve the system by the method of substitution.

35. $\begin{cases} y = \frac{1}{4}x + \frac{19}{4} \\ y = \frac{8}{5}x - 2 \end{cases}$

36. $\begin{cases} y = \frac{5}{4}x + 3 \\ y = \frac{1}{2}x + 6 \end{cases}$

37. $\begin{cases} 3x + 2y = 12 \\ x - y = 3 \end{cases}$

38. $\begin{cases} 16x - 8y = 5 \\ 32x + 8y = 19 \end{cases}$

39. $\begin{cases} 5x + 3y = 15 \\ 2x - 3y = 6 \end{cases}$

40. $\begin{cases} 4x - 5y = 0 \\ 2x - 5y = -10 \end{cases}$

41. $\begin{cases} \frac{x}{3} - \frac{y}{4} = 2 \\ \frac{x}{2} + \frac{y}{6} = 3 \end{cases}$

42. $\begin{cases} -\frac{x}{5} + \frac{y}{2} = -3 \\ \frac{x}{4} - \frac{y}{4} = 0 \end{cases}$

43. $\begin{cases} \frac{x}{4} + \frac{y}{2} = 1 \\ \frac{x}{2} - \frac{y}{3} = 1 \end{cases}$

44. $\begin{cases} -\frac{x}{6} + \frac{y}{12} = 1 \\ \frac{x}{2} + \frac{y}{8} = 1 \end{cases}$

45. $\begin{cases} 2(x - 5) = y + 2 \\ 3x = 4(y + 2) \end{cases}$

46. $\begin{cases} 3(x - 2) + 5 = 4(y + 3) - 2 \\ 2x + 7 = 2y + 8 \end{cases}$

Think About It In Exercises 47 and 48, find a system of linear equations that has the given solution. (There are many correct answers.)

47. $(2, 1)$

48. $(-4, -3)$

49. **Number Problem** The sum of two numbers x and y is 40 and the difference of the two numbers is 10. The system of equations that represents this situation is
$$\begin{cases} x + y = 40 \\ x - y = 10 \end{cases}.$$
Solve this system to find the two numbers.

50. **Number Problem** The sum of two numbers x and y is 50 and the difference of the two numbers is 20. The system of equations that represents this situation is
$$\begin{cases} x + y = 50 \\ x - y = 20 \end{cases}.$$
Solve this system to find the two numbers.

51. **Geometry** Find an equation of the line with slope $m = 2$ passing through the intersection of the lines $x - 2y = 3$ and $3x + y = 16$.

52. **Geometry** Find an equation of the line with slope $m = -3$ passing through the intersection of the lines $4x + 6y = 26$ and $5x - 2y = -15$.

Explaining Concepts

53. **Writing** Describe any advantages of the method of substitution over the graphical method of solving a system of linear equations.

54. **Reasoning** Your instructor says, "An equation (not in standard form) such as $2x - 3 = 5x - 9$ can be considered a system of equations." Create the system, and find the solution point. How many solution points does the "system" $x^2 - 1 = 2x - 1$ have? Illustrate your results with a graph.

Think About It In Exercises 55–58, find the value of a or b such that the system of linear equations is inconsistent.

55. $\begin{cases} x + by = 1 \\ x + 2y = 2 \end{cases}$

56. $\begin{cases} ax + 3y = 6 \\ 5x - 5y = 2 \end{cases}$

57. $\begin{cases} -6x + y = 4 \\ 2x + by = 3 \end{cases}$

58. $\begin{cases} 6x - 3y = 4 \\ ax - y = -2 \end{cases}$

Cumulative Review

In Exercises 59–62, factor the expression.

59. $x(3 - x) - 2(3 - x)$

60. $4t^2 - 9$

61. $4y^2 - 20y + 25$

62. $6u^2 - 5u - 21$

In Exercises 63–66, solve the equation.

63. $14 - 2x = x + 2$

64. $\dfrac{9}{x + 3} = 15$

65. $z^2 - 4z - 12 = 0$

66. $t^3 + t^2 - 4t - 4 = 0$

In Exercises 67–70, solve the system by graphing.

67. $\begin{cases} 3x + 4y = 10 \\ 3x + 4y = -1 \end{cases}$

68. $\begin{cases} x - 2y = 6 \\ x + 2y = 2 \end{cases}$

69. $\begin{cases} 5x + y = -3 \\ x + 2y = -6 \end{cases}$

70. $\begin{cases} -x + 9y = 6 \\ \frac{1}{3}x - 3y = -2 \end{cases}$

8.3 Solving Systems of Equations by Elimination

▶ Solve systems of linear equations algebraically using the method of elimination.
▶ Solve systems with no solution or infinitely many solutions.
▶ Use the method of elimination to solve application problems.

The Method of Elimination

> **The Method of Elimination**
>
> 1. Obtain opposite coefficients of x (or y) by multiplying all terms of one or both equations by suitable constants.
> 2. Add the equations to eliminate one variable and solve the resulting equation.
> 3. Back-substitute the value obtained in Step 2 into either of the original equations and solve for the other variable.
> 4. Check your solution in *both* of the original equations.

EXAMPLE 1 The Method of Elimination

Solve the system of linear equations.

$$\begin{cases} 4x + 3y = 1 & \text{Equation 1} \\ 2x - 3y = 5 & \text{Equation 2} \end{cases}$$

SOLUTION

Begin by noting that the coefficients of y are opposites. So, by adding the two equations, you can eliminate y.

$$\begin{array}{rl} 4x + 3y = 1 & \text{Equation 1} \\ \underline{2x - 3y = 5} & \text{Equation 2} \\ 6x = 6 & \text{Add equations.} \end{array}$$

So, $x = 1$. Back-substitute this x-value into Equation 1 and solve for y.

$$\begin{aligned} 4(1) + 3y &= 1 & &\text{Substitute 1 for } x \text{ in Equation 1.} \\ 3y &= -3 & &\text{Subtract 4 from each side.} \\ y &= -1 & &\text{Divide each side by 3.} \end{aligned}$$

The solution is $(1, -1)$. Check this in both of the original equations.

Study Tip

Try solving the system in Example 1 by substitution. Notice that the method of elimination is more efficient for this system.

Exercises Within Reach ®   Solutions in English & Spanish and tutorial videos at AlgebraWithinReach.com

Solving a System of Linear Equations In Exercises 1–6, **solve** the system by the method of elimination.

1. $\begin{cases} x - y = 4 \\ x + y = 12 \end{cases}$

2. $\begin{cases} x + y = 7 \\ x - y = 3 \end{cases}$

3. $\begin{cases} -x + 2y = 12 \\ x + 6y = 20 \end{cases}$

4. $\begin{cases} x + 2y = 14 \\ x - 2y = 10 \end{cases}$

5. $\begin{cases} 3x - 5y = 1 \\ 2x + 5y = 9 \end{cases}$

6. $\begin{cases} -2x + 3y = -4 \\ 2x - 4y = 6 \end{cases}$

Section 8.3 Solving Systems of Equations by Elimination

To obtain opposite coefficients for one of the variables, you often need to multiply one or both of the equations by a suitable constant. This is demonstrated in the following example.

EXAMPLE 2 **The Method of Elimination**

Solve the system of linear equations.

$$\begin{cases} 2x - 3y = -7 & \text{Equation 1} \\ 3x + y = -5 & \text{Equation 2} \end{cases}$$

SOLUTION

For this system, you can obtain opposite coefficients of y by multiplying Equation 2 by 3.

$$\begin{cases} 2x - 3y = -7 \\ 3x + y = -5 \end{cases} \implies \begin{array}{l} 2x - 3y = -7 \quad \text{Equation 1} \\ 9x + 3y = -15 \quad \text{Multiply Equation 2 by 3.} \\ \hline 11x = -22 \quad \text{Add equations.} \end{array}$$

So, $x = -2$. Back-substitute this x-value into Equation 2 and solve for y.

$$\begin{aligned} 3x + y &= -5 & &\text{Equation 2} \\ 3(-2) + y &= -5 & &\text{Substitute } -2 \text{ for } x. \\ -6 + y &= -5 & &\text{Simplify.} \\ y &= 1 & &\text{Add 6 to each side.} \end{aligned}$$

The solution is $(-2, 1)$. Check this in both of the original equations, as follows.

Substitute into Equation 1
$$2x - 3y = -7$$
$$2(-2) - 3(1) \stackrel{?}{=} -7$$
$$-4 - 3 \stackrel{?}{=} -7$$
$$-7 = -7 \checkmark$$

Substitute into Equation 2
$$3x + y = -5$$
$$3(-2) + 1 \stackrel{?}{=} -5$$
$$-6 + 1 \stackrel{?}{=} -5$$
$$-5 = -5 \checkmark$$

Exercises Within Reach®

Solutions in English & Spanish and tutorial videos at AlgebraWithinReach.com

Solving a System of Linear Equations In Exercises 7–12, solve the system by the method of elimination.

7. $\begin{cases} 2a + 5b = 3 \\ 2a + b = 9 \end{cases}$

8. $\begin{cases} 4a + 5b = 9 \\ 2a + 5b = 7 \end{cases}$

9. $\begin{cases} -x + 2y = 6 \\ 2x + 5y = 6 \end{cases}$

10. $\begin{cases} -4x + 8y = 0 \\ 3x - 2y = 2 \end{cases}$

11. $\begin{cases} 2x - 5y = -1 \\ 2x - y = 1 \end{cases}$

12. $\begin{cases} 7x + 8y = 6 \\ 3x - 4y = 10 \end{cases}$

EXAMPLE 3 The Method of Elimination

Solve the system of linear equations.

$$\begin{cases} 5x + 3y = 6 & \text{Equation 1}\\ 2x - 4y = 5 & \text{Equation 2}\end{cases}$$

SOLUTION

You can obtain opposite coefficients of y by multiplying Equation 1 by 4 and Equation 2 by 3.

$$\begin{cases} 5x + 3y = 6 \\ 2x - 4y = 5 \end{cases} \Rightarrow \begin{array}{r} 20x + 12y = 24 \\ 6x - 12y = 15 \\ \hline 26x = 39 \end{array}$$

Multiply Equation 1 by 4.
Multiply Equation 2 by 3.
Add equations.

From this equation, you can see that $x = \frac{3}{2}$. Back-substitute this x-value into Equation 2 and solve for y.

$2x - 4y = 5$	Equation 2
$2\left(\frac{3}{2}\right) - 4y = 5$	Substitute $\frac{3}{2}$ for x.
$3 - 4y = 5$	Simplify.
$-4y = 2$	Subtract 3 from each side.
$y = -\frac{1}{2}$	Divide each side by -4.

The solution is $\left(\frac{3}{2}, -\frac{1}{2}\right)$. You can check this as follows.

Substitute into Equation 1
$5x + 3y = 6$
$5\left(\frac{3}{2}\right) + 3\left(-\frac{1}{2}\right) \stackrel{?}{=} 6$
$\frac{15}{2} - \frac{3}{2} = 6$ ✓

Substitute into Equation 2
$2x - 4y = 5$
$2\left(\frac{3}{2}\right) - 4\left(-\frac{1}{2}\right) \stackrel{?}{=} 5$
$3 + 2 = 5$ ✓

The graph of this system is shown at the left. From the graph, it appears that the solution $\left(\frac{3}{2}, -\frac{1}{2}\right)$ is reasonable.

Exercises Within Reach®

Solutions in English & Spanish and tutorial videos at AlgebraWithinReach.com

Solving a System of Linear Equations In Exercises 13–18, **solve** the system by the method of elimination.

13. $\begin{cases} 4x + 5y = 7 \\ 6x - 2y = -18 \end{cases}$

14. $\begin{cases} 5x + 3y = 18 \\ 2x - 7y = -1 \end{cases}$

15. $\begin{cases} 3x + 2y = 10 \\ 2x + 5y = 3 \end{cases}$

16. $\begin{cases} 5u + 6v = 14 \\ 3u + 5v = 7 \end{cases}$

17. $\begin{cases} 2x + 3y = 16 \\ 5x - 10y = 30 \end{cases}$

18. $\begin{cases} 3x - 4y = 1 \\ 4x + 3y = 1 \end{cases}$

Section 8.3   Solving Systems of Equations by Elimination   373

The next example shows how the method of elimination works with a system of linear equations with decimal coefficients.

EXAMPLE 4   The Method of Elimination

Solve the system of linear equations.

$$\begin{cases} 0.02x - 0.05y = -0.38 & \text{Equation 1} \\ 0.03x + 0.04y = 1.04 & \text{Equation 2} \end{cases}$$

SOLUTION

Because the coefficients in this system have two decimal places, begin by multiplying each equation by 100. This produces a system in which the coefficients are all integers.

$$\begin{cases} 2x - 5y = -38 & \text{Revised Equation 1} \\ 3x + 4y = 104 & \text{Revised Equation 2} \end{cases}$$

Now you can obtain opposite coefficients of x by multiplying Equation 1 by 3 and Equation 2 by -2.

$$\begin{cases} 2x - 5y = -38 \\ 3x + 4y = 104 \end{cases} \Rightarrow \begin{aligned} 6x - 15y &= -114 & \text{Multiply Equation 1 by 3.} \\ -6x - 8y &= -208 & \text{Multiply Equation 2 by } -2. \\ \hline -23y &= -322 & \text{Add equations.} \end{aligned}$$

So, the y-coordinate of the solution is

$$y = \frac{-322}{-23} = 14.$$

Back-substitute this y-value into revised Equation 2 and solve for x.

$3x + 4(14) = 104$ Substitute 14 for y in revised Equation 2.

$3x + 56 = 104$ Simplify.

$3x = 48$ Subtract 56 from each side.

$x = 16$ Divide each side by 3.

So, the solution is (16, 14). Check this in both of the original equations.

> **Study Tip**
> When multiplying an equation by a negative number, be sure to distribute the negative sign to each term of the equation.

Exercises Within Reach®

Solutions in English & Spanish and tutorial videos at AlgebraWithinReach.com

Solving a System of Linear Equations In Exercises 19–24, solve the system by the method of elimination.

19. $\begin{cases} 0.02x - 0.05y = -0.19 \\ 0.03x + 0.04y = 0.52 \end{cases}$

20. $\begin{cases} 0.05x - 0.03y = 0.21 \\ 0.01x + 0.01y = 0.09 \end{cases}$

21. $\begin{cases} 0.1x - 0.1y = 0 \\ 0.8x + 0.3y = 1.5 \end{cases}$

22. $\begin{cases} x - 2y = 0 \\ 0.2x + 0.8y = 2.4 \end{cases}$

23. $\begin{cases} 6r + 5s = 3 \\ \frac{3}{2}r - \frac{5}{4}s = \frac{3}{4} \end{cases}$

24. $\begin{cases} \frac{1}{4}x - y = \frac{1}{2} \\ 4x + 4y = 3 \end{cases}$

The No-Solution and Many-Solutions Cases

EXAMPLE 5 The Method of Elimination: No-Solution Case

Solve the system of linear equations.

$$\begin{cases} 2x - 6y = 5 & \text{Equation 1} \\ 3x - 9y = 2 & \text{Equation 2} \end{cases}$$

SOLUTION

You can obtain opposite coefficients by multiplying Equation 1 by 3 and Equation 2 by -2.

$$\begin{cases} 2x - 6y = 5 \\ 3x - 9y = 2 \end{cases} \Rightarrow \begin{array}{l} 6x - 18y = 15 \\ -6x + 18y = -4 \\ \hline 0 = 11 \end{array}$$

Multiply Equation 1 by 3.
Multiply Equation 2 by -2.
Add equations.

Because $0 = 11$ is a false statement, you can conclude that the system is inconsistent and has no solution. The lines corresponding to the two equations of this system are shown at the left. Note that the two lines are parallel and have no point of intersection.

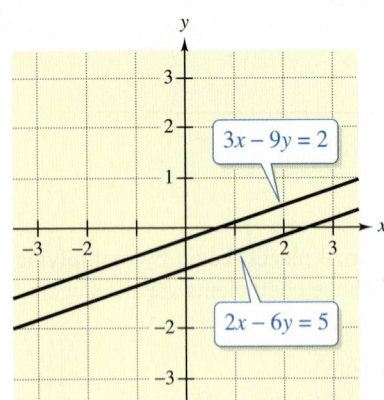

Study Tip

By writing both equations in Example 6 in slope-intercept form, you will obtain identical equations. This shows that the system has infinitely many solutions.

EXAMPLE 6 The Method of Elimination: Many-Solutions Case

Solve the system of linear equations.

$$\begin{cases} 2x - 6y = -5 & \text{Equation 1} \\ -4x + 12y = 10 & \text{Equation 2} \end{cases}$$

SOLUTION

You can obtain opposite coefficients of x by multiplying Equation 1 by 2.

$$\begin{cases} 2x - 6y = -5 \\ -4x + 12y = 10 \end{cases} \Rightarrow \begin{array}{l} 4x - 12y = -10 \\ -4x + 12y = 10 \\ \hline 0 = 0 \end{array}$$

Multiply Equation 1 by 2.
Equation 2
Add equations.

Because $0 = 0$ is a true statement, you can conclude that the system is dependent and has infinitely many solutions. The solutions consist of all ordered pairs (x, y) lying on the line $2x - 6y = -5$.

Exercises Within Reach®
Solutions in English & Spanish and tutorial videos at AlgebraWithinReach.com

Solving a System of Linear Equations In Exercises 25–28, **solve** the system by the method of elimination.

25. $\begin{cases} -3x - 12y = 3 \\ 5x + 20y = -5 \end{cases}$

26. $\begin{cases} 7x + 10y = 0 \\ 21x + 30y = 0 \end{cases}$

27. $\begin{cases} 0.4a + 0.7b = 3 \\ 0.8a + 1.4b = 7 \end{cases}$

28. $\begin{cases} 0.2u - 0.1v = 1 \\ -0.8u + 0.4v = 3 \end{cases}$

Application

EXAMPLE 7 Ticket Sales

A fundraising dinner was held on two consecutive nights. On the first night, 100 adult tickets and 175 children's tickets were sold, for a total of $1225. On the second night, 200 adult tickets and 316 children's tickets were sold, for a total of $2348. The system of linear equations that represents this situation is

$$\begin{cases} 100x + 175y = 1225 & \text{Equation 1} \\ 200x + 316y = 2348 & \text{Equation 2} \end{cases}$$

where x represents the price of an adult ticket and y represents the price of a child's ticket. Solve this system to find the price of each type of ticket.

SOLUTION

You can obtain opposite coefficients of x by multiplying Equation 1 by -2.

$$\begin{cases} 100x + 175y = 1225 \\ 200x + 316y = 2348 \end{cases} \implies \begin{array}{r} -200x - 350y = -2450 \\ 200x + 316y = 2348 \\ \hline -34y = -102 \end{array}$$

Multiply Equation 1 by -2.
Equation 2
Add equations.

So, the y-coordinate of the solution is

$$y = \frac{-102}{-34}$$

$$= 3.$$

Back-substitute this y-value into Equation 2 and solve for x.

$$200x + 316(3) = 2348 \quad \text{Substitute 3 for } y \text{ in Equation 2.}$$
$$200x = 1400 \quad \text{Simplify.}$$
$$x = 7 \quad \text{Divide each side by 200.}$$

The solution is $(7, 3)$. So the price of an adult ticket was $7 and the price of a child's ticket was $3. Check this solution in both of the original equations.

Exercises Within Reach ®

Solutions in English & Spanish and tutorial videos at AlgebraWithinReach.com

29. Ticket Sales Ticket sales for a play were $3799 on the first night and $4905 on the second night. On the first night, 213 student tickets and 632 general admission tickets were sold. On the second night, 275 student tickets and 816 general admission tickets were sold. The system of equations that represents this situation is

$$\begin{cases} 213x + 632y = 3799 \\ 275x + 816y = 4905 \end{cases}$$

where x represents the price of a student ticket and y represents the price of a general admission ticket. Solve this system to determine the price of each type of ticket.

30. Ticket Sales Ticket sales for an annual variety show were $540 on the first night and $850 on the second night. On the first night, 150 student tickets and 80 general admission tickets were sold. On the second night, 200 student tickets and 150 general admission tickets were sold. The system of equations that represents this situation is

$$\begin{cases} 150x + 80y = 540 \\ 200x + 150y = 850 \end{cases}$$

where x represents the price of a student ticket and y represents the price of a general admission ticket. Solve this system to determine the price of each type of ticket.

Concept Summary: Solving Systems of Linear Equations by Elimination

What
You can solve a system of linear equations using an algebraic method called the **method of elimination**.

EXAMPLE
Solve the system of linear equations.
$$\begin{cases} 3x + 5y = 7 & \text{Equation 1} \\ -3x - 2y = -1 & \text{Equation 2} \end{cases}$$

How
The key step in this method is to obtain opposite coefficients for one of the variables so that adding the two equations eliminates this variable.

EXAMPLE
$$\begin{array}{rl} 3x + 5y = 7 & \text{Equation 1} \\ -3x - 2y = -1 & \text{Equation 2} \\ \hline 3y = 6 & \text{Add equations.} \end{array}$$

After eliminating the variable, solve for the other variable. Then use back-substitution to find the value of the eliminated variable.

Why
When solving a system of linear equations, choose the method that is most efficient.

1. Use graphing to approximate the solution.
2. To find exact solutions, use substitution or elimination.
   - When one of the variables has a coefficient of 1, use substitution.
   - When the coefficients of one of the variables are opposites, use elimination.
3. When you are not sure, use elimination. It is usually more efficient.

Exercises Within Reach®
Worked-out solutions to odd-numbered exercises at AlgebraWithinReach.com

Concept Summary Check

31. *Solving a System* What is the solution of the system of linear equations in the example above?

32. *Precision* Explain how to solve a system of linear equations by elimination.

33. *Reasoning* When solving a system by the method of elimination, how do you recognize that it has no solution?

34. *Reasoning* When solving a system by the method of elimination, how do you recognize that it has infinitely many solutions?

Extra Practice

Solving a System of Linear Equations In Exercises 35–38, **solve** the system by the method of elimination.

35. $\begin{cases} -\dfrac{x}{4} + y = 1 \\ \dfrac{x}{4} + \dfrac{y}{2} = 1 \end{cases}$

36. $\begin{cases} \dfrac{x}{3} - \dfrac{y}{5} = 1 \\ \dfrac{x}{12} + \dfrac{y}{40} = 1 \end{cases}$

37. $\begin{cases} 3(x+5) - 7 = 2(3-2y) \\ 2x + 1 = 4(y+2) \end{cases}$

38. $\begin{cases} \dfrac{1}{2}(x-4) + 9 = y - 10 \\ -5(x+3) = 8 - 2(y-3) \end{cases}$

39. *Investment* You invest a total of $10,000 in two funds earning 7.5% and 10% simple interest. (There is more risk in the 10% fund.) Your goal is to have a total annual interest income of $850. The system of equations that represents this situation is

$$\begin{cases} x + y = 10{,}000 \\ 0.075x + 0.10y = 850 \end{cases}$$

where x is the amount invested in the 7.5% fund and y is the amount invested in the 10% fund. Solve this system to determine the smallest amount that you can invest at 10% in order to meet your objective.

40. *Investment* You invest a total of $12,000 in two funds earning 8% and 11.5% simple interest. (There is more risk in the 11.5% fund.) Your goal is to have a total annual interest income of $1065. The system of equations that represents this situation is

$$\begin{cases} x + y = 12{,}000 \\ 0.08x + 0.115y = 1{,}065 \end{cases}$$

where x is the amount invested in the 8% fund and y is the amount invested in the 11.5% fund. Solve this system to determine the smallest amount that you can invest at 11.5% in order to meet your objective.

Number Problem In Exercises 41 and 42, solve the system to find the two numbers.

41. The sum of two numbers x and y is 82 and the difference of the numbers is 14. The systems of equations that represents this problem is
$$\begin{cases} x + y = 82 \\ x - y = 14 \end{cases}.$$

42. The sum of two numbers x and y is 154 and the difference of the numbers is 38. The system of equations that represents this problem is
$$\begin{cases} x + y = 154 \\ x - y = 38 \end{cases}.$$

43. **Geometry** Find an equation of the line of slope $m = \frac{1}{3}$ passing through the intersection of the lines
$3x + 4y = 7$ and $5x - 4y = 1$.

44. **Geometry** Find an equation of the line of slope $m = -2$ passing through the intersection of the lines
$2x + 5y = 11$ and $4x - y = 11$.

Explaining Concepts

45. *Creating an Example* Explain how to "clear" a system of decimals. Give an example to justify your answer. (There are many correct answers.)

46. *Creating a System* Write a system of linear equations that is more efficiently solved by the method of elimination than by the method of substitution. (There are many correct answers.)

47. *Creating a System* Write a system of linear equations that is more efficiently solved by the method of substitution than by the method of elimination. (There are many correct answers.)

48. *Reasoning* Consider the system of linear equations.
$$\begin{cases} x + y = 8 \\ 2x + 2y = k \end{cases}$$

(a) Find the value(s) of k for which the system has an infinite number of solutions.

(b) Find one value of k for which the system has no solution. (There are many correct answers.)

(c) Can the system have a single solution for some value of k? Why or why not?

Cumulative Review

In Exercises 49–54, plot the points and find the slope (if possible) of the line that passes through the points. If not possible, state why.

49. $(-6, 4), (-3, -4)$

50. $(4, 6), (8, -2)$

51. $\left(\frac{7}{2}, \frac{9}{2}\right), \left(\frac{4}{3}, -3\right)$

52. $\left(-\frac{3}{4}, -\frac{7}{4}\right), \left(-1, \frac{5}{2}\right)$

53. $(-3, 6), (-3, 2)$

54. $(6, 2), (10, 2)$

In Exercises 55–58, solve and graph the inequality.

55. $x \leq 3$

56. $x > -4$

57. $x + 5 < 6$

58. $3x - 7 \geq 2x + 9$

In Exercises 59–62, solve the system by the method of substitution.

59. $\begin{cases} y = x \\ x + 3y = 20 \end{cases}$

60. $\begin{cases} x + y = 9 \\ 2x + 2y = 18 \end{cases}$

61. $\begin{cases} 2x + y = 5 \\ 5x + 3y = 12 \end{cases}$

62. $\begin{cases} 5x + 6y = 21 \\ 25x + 30y = 10 \end{cases}$

Mid-Chapter Quiz: Sections 8.1–8.3

Solutions in English & Spanish and tutorial videos at AlgebraWithinReach.com

Take this quiz as you would take a quiz in class. After you are done, check your work against the answers in the back of the book.

1. Is $(4, 2)$ a solution of $3x + 4y = 4$ and $5x - 3y = 14$? Explain.

2. Is $(2, -1)$ a solution of $2x - 3y = 7$ and $3x + 5y = 1$? Explain.

In Exercises 3–5, use the given graphs to solve the system of linear equations.

3. $\begin{cases} x + y = 5 \\ x - 3y = -3 \end{cases}$
4. $\begin{cases} x + 2y = 6 \\ 3x - 4y = 8 \end{cases}$
5. $\begin{cases} x + 2y = 2 \\ x - 2y = 6 \end{cases}$

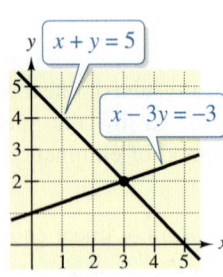

Figure for 3

In Exercises 6–8, solve the system by graphing.

6. $\begin{cases} x = -3 \\ x + y = 8 \end{cases}$
7. $\begin{cases} y = \frac{3}{2}x - 1 \\ y = -x + 4 \end{cases}$
8. $\begin{cases} 4x + y = 0 \\ -x + y = 5 \end{cases}$

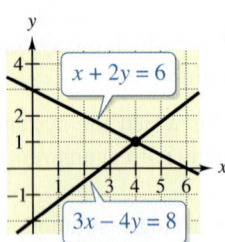

Figure for 4

In Exercises 9–11, solve the system by the method of substitution.

9. $\begin{cases} x - y = 4 \\ y = 2 \end{cases}$
10. $\begin{cases} y = -\frac{2}{3}x + 5 \\ y = 2x - 3 \end{cases}$
11. $\begin{cases} 2x - y = -7 \\ 4x + 3y = 16 \end{cases}$

In Exercises 12–14, solve the system by the method of elimination.

12. $\begin{cases} x + y = -2 \\ x - y = 4 \end{cases}$
13. $\begin{cases} 2x + y = 1 \\ 6x + 5y = 13 \end{cases}$
14. $\begin{cases} -x + 3y = 10 \\ 9x - 4y = 5 \end{cases}$

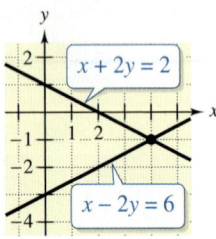

Figure for 5

In Exercises 15 and 16, find a system of linear equations that has the ordered pair as its only solution. (There are many correct answers.)

15. $(-1, 9)$
16. $(3, 0)$

In Exercises 17 and 18, find the value of k such that the system of linear equations is inconsistent.

17. $\begin{cases} 5x + ky = 3 \\ 10x - 4y = 1 \end{cases}$
18. $\begin{cases} 8x - 5y = 16 \\ kx - 0.5y = 3 \end{cases}$

19. The sum of two numbers x and y is 50 and their difference is 22. Solve the following system of equations to find the two numbers.

$$\begin{cases} x + y = 50 \\ x - y = 22 \end{cases}$$

Application

$\begin{cases} x + y = 32 \\ x = 4y + 2 \end{cases}$

System for 20

20. A student spent a total of $32 for a book and a calendar. The price of the book was $2 more than 4 times the price of the calendar. The system of equations that represents this situation is shown at the left, where x is the price of the book and y is the price of the calendar. Solve this system to determine the price of each item.

Study Skills in Action

Being Confident

How does someone "get" confidence? Confidence is linked to another attribute called self-efficacy. Self-efficacy is the belief that one has the ability to accomplish a specific task. It is possible for a student to have high self-efficacy when it comes to writing a personal essay, but to have low self-efficacy when it comes to learning math.

A good way to foster self-efficacy is by building a support system. A support system should include faculty and staff who can encourage and guide you, and other students who can help you study and stay focused.

Collegial friends, who share the same desire to do well, can be the best type of support.

Smart Study Strategy

Build a Support System

1. ▶ **Surround yourself with positive collegial friends.** Find another student in class with whom to study. Make sure this person is not anxious about math because you do not want another student's anxiety to increase your own. Arrange to meet on campus and compare notes, homework, and so on at least two times per week. Collegial friends can encourage each other.

2. ▶ **Find a place on campus to study where other students are also studying.** Libraries, learning centers, and tutoring centers are great places to study. While studying in such places, you will be able to ask for assistance when you have questions. You do not want to study alone if you typically get down on yourself with lots of negative self-talk.

3. ▶ **Establish a relationship with a learning assistant.** Get to know someone who can help you find assistance for any type of academic issue. Learning assistants, tutors, and instructors are excellent resources.

4. ▶ **Seek out assistance before you are overwhelmed.** Visit your instructor when you need help. Instructors are more than willing to help their students, particularly during office hours. Go with a friend if you are nervous about visiting your instructor.

5. ▶ **Be your own support.** Listen to what you tell yourself when frustrated with studying math. Replace any negative self-talk dialog with more positive statements. Here are some examples of positive statements:

"I may not have done well in the past, but I'm learning how to study math, and will get better."

"It does not matter what others believe—I know that I can get through this course."

"Wow, I messed up on a quiz. I need to talk to someone and figure out what I need to do differently."

Monkey Business Images/Shutterstock.com

8.4 Applications of Systems of Linear Equations

- Construct a system of linear equations from an application problem.
- Solve real-life applications using systems of linear equations.

Constructing Systems of Linear Equations

Application

EXAMPLE 1 Setting Up a Mixture Problem

A total of $12,000 is invested in two funds paying 9% and 11% simple interest. (There is more risk in the 11% fund.) The combined annual interest for the two funds is $1180. How much of the $12,000 is invested at each rate?

SOLUTION

Notice that this problem has two unknowns: the amount (in dollars) invested at 9% and the amount (in dollars) invested at 11%. You can construct a model for this problem as follows.

Verbal Model:

$$\boxed{\text{Amount in 9\% fund}} + \boxed{\text{Amount in 11\% fund}} = \boxed{\text{Total amount}}$$

$$\boxed{\text{Interest from 9\% fund}} + \boxed{\text{Interest from 11\% fund}} = \boxed{\text{Total interest}}$$

Labels:
Amount in 9% fund $= x$ (dollars)
Amount in 11% fund $= y$ (dollars)
Total amount $= 12,000$ (dollars)
Interest from 9% fund $= 0.09x$ (dollars)
Interest from 11% fund $= 0.11y$ (dollars)
Total interest $= 1180$ (dollars)

System:
$$\begin{cases} x + y = 12,000 & \text{Equation 1} \\ 0.09x + 0.11y = 1,180 & \text{Equation 2} \end{cases}$$

This system was solved in Example 6 on page 366. There you found that $x = 7000$ and $y = 5000$, which means that $7000 was invested at 9% simple interest and $5000 was invested at 11% simple interest.

Exercises Within Reach®

Solutions in English & Spanish and tutorial videos at AlgebraWithinReach.com

1. **Gasoline** The total cost of 15 gallons of regular gasoline and 10 gallons of premium gasoline is $97.15. Premium gasoline costs $0.24 more per gallon than regular gasoline. What is the cost per gallon of each type of gasoline?

   (a) Write a verbal model for this problem.
   (b) Assign labels to the verbal model.
   (c) Use the labels to write a linear system.
   (d) Solve the system and answer the question.

2. **Investment** A total of $12,000 is invested in two bonds that pay 10.5% and 12% simple interest. (There is more risk in the 12% bond.) The combined annual interest is $1380. How much is invested in each bond?

   (a) Write a verbal model for this problem.
   (b) Assign labels to the verbal model.
   (c) Use the labels to write a linear system.
   (d) Solve the system and answer the question.

Applications

Application

EXAMPLE 2 Solving a Mixture Problem

A company with two stores buys six large delivery vans and five small delivery vans. The first store receives 4 large vans and 2 small vans for a total cost of $212,000. The second store receives 2 large vans and 3 small vans for a total cost of $166,000. What is the cost of each type of van?

SOLUTION

The two unknowns in this problem are the costs of the two types of vans.

Verbal Model:

$$4\left(\begin{array}{c}\text{Cost of}\\\text{large van}\end{array}\right) + 2\left(\begin{array}{c}\text{Cost of}\\\text{small van}\end{array}\right) = \begin{array}{c}\text{Total cost}\\\text{for first store}\end{array}$$

$$2\left(\begin{array}{c}\text{Cost of}\\\text{large van}\end{array}\right) + 3\left(\begin{array}{c}\text{Cost of}\\\text{small van}\end{array}\right) = \begin{array}{c}\text{Total cost for}\\\text{second store}\end{array}$$

Labels:
Cost of large van = x (dollars)
Cost of small van = y (dollars)
Total cost for first store = 212,000 (dollars)
Total cost for second store = 166,000 (dollars)

System:
$$\begin{cases} 4x + 2y = 212{,}000 & \text{Equation 1} \\ 2x + 3y = 166{,}000 & \text{Equation 2} \end{cases}$$

To solve this system of linear equations, use the method of elimination.

$$\begin{cases} 4x + 2y = 212{,}000 \\ 2x + 3y = 166{,}000 \end{cases} \Rightarrow \begin{array}{r} 4x + 2y = 212{,}000 \\ -4x - 6y = -332{,}000 \\ \hline -4y = -120{,}000 \end{array}$$

So, $y = -120{,}000/-4 = 30{,}000$. To solve for x, back-substitute this value into Equation 1.

$4x + 2(30{,}000) = 212{,}000$	Substitute 30,000 for y in Equation 1.
$4x + 60{,}000 = 212{,}000$	Multiply.
$4x = 152{,}000$	Subtract 60,000 from each side.
$x = 38{,}000$	Divide each side by 4.

The cost of each large van is $x = \$38{,}000$, and the cost of each small van is $y = \$30{,}000$. Check this solution in the original system.

Exercises Within Reach®

Solutions in English & Spanish and tutorial videos at AlgebraWithinReach.com

3. Truck Cost A bakery with two stores buys three large delivery trucks and six small delivery trucks. One store receives one large delivery truck and four small delivery trucks for a total cost of $118,000. The second store receives two large delivery trucks and two small delivery trucks for a total cost of $107,000. What is the cost of each type of delivery truck?

4. Truck Cost A furniture company with two stores buys three large delivery trucks and four small delivery trucks. One store receives one large delivery truck and three small delivery trucks for a total cost of $157,000. The second store receives two large delivery trucks and one small delivery truck for a total cost of $139,000. What is the cost of each type of delivery truck?

Application

EXAMPLE 3 Solving a Mixture Problem

A chemist has a 50% alcohol solution and a 75% alcohol solution. How many liters of each type of solution should be mixed to obtain 8 liters of a 60% alcohol solution?

SOLUTION

The two unknowns in this problem are the amounts of the two types of solutions.

Verbal Model: Amount of 50% solution + Amount of 75% solution = Amount of 60% solution

Amount of alcohol in 50% solution + Amount of alcohol in 75% solution = Amount of alcohol in 60% solution

Labels:
Amount of 50% solution = x (liters)
Amount of 75% solution = y (liters)
Amount of 60% solution = 8 (liters)
Amount of alcohol in 50% solution = $0.50x$ (liters)
Amount of alcohol in 75% solution = $0.75y$ (liters)
Amount of alcohol in 60% solution = $0.60(8) = 4.8$ (liters)

System:
$$\begin{cases} x + y = 8 & \text{Equation 1} \\ 0.50x + 0.75y = 4.8 & \text{Equation 2} \end{cases}$$

To solve this system of linear equations, use the method of elimination.

$$\begin{cases} x + y = 8 \\ 0.50x + 0.75y = 4.8 \end{cases} \Rightarrow \begin{array}{r} -50x - 50y = -400 \\ 50x + 75y = 480 \\ \hline 25y = 80 \end{array}$$

So, $y = 3.2$. To solve for x, back-substitute this value into Equation 1.

$x + y = 8$ Equation 1
$x + 3.2 = 8$ Substitute 3.2 for y.
$x = 4.8$ Subtract 3.2 from each side.

So, the chemist should use 4.8 liters of the 50% alcohol solution and 3.2 liters of the 75% alcohol solution. Check this solution in the original system.

Exercises Within Reach®

Solutions in English & Spanish and tutorial videos at AlgebraWithinReach.com

5. **Mixture Problem** How many liters of a 35% alcohol solution and a 60% alcohol solution must be mixed to obtain 10 liters of a 50% alcohol solution?

6. **Mixture Problem** Ten gallons of a 30% acid solution is obtained by mixing a 20% acid solution with a 50% acid solution. How many gallons of each solution must be used to obtain the desired mixture?

Application

EXAMPLE 4 Selling Price and Wholesale Cost

The selling price of a pair of ski boots is $136.95. The markup rate is 65% of the wholesale cost. What is the wholesale cost?

SOLUTION

You could solve this problem using only one unknown (the wholesale cost). However, for the sake of illustration, use two unknowns: the wholesale cost and the markup (in dollars).

Verbal Model: Wholesale cost + Markup = Selling price

Markup = Markup rate · Wholesale cost

Labels:
Wholesale cost = C (dollars)
Markup = M (dollars)
Selling price = 136.95 (dollars)
Markup rate = 0.65 (percent in decimal form)

System:
$$\begin{cases} C + M = 136.95 & \text{Equation 1} \\ M = 0.65C & \text{Equation 2} \end{cases}$$

To solve this system of linear equations, use the method of substitution.

$C + M = 136.95$	Equation 1
$C + 0.65C = 136.95$	Substitute 0.65C for M.
$1.65C = 136.95$	Combine like terms.
$C = 83$	Divide each side by 1.65.

The wholesale cost of the pair of boots is $83. (You were not asked to find the markup, so it is unnecessary to back-substitute to find the value of M.)

Exercises Within Reach®

Solutions in English & Spanish and tutorial videos at AlgebraWithinReach.com

Using a System of Linear Equations In Exercises 7–10, use a system of linear equations to solve the problem.

7. The selling price of a watch is $108.75. The markup rate is 45% of the wholesale cost. Find the wholesale cost.

8. The selling price of a cellular phone is $149.92. The markup rate is 60% of the wholesale cost. Find the wholesale cost.

9. The sale price of a microwave oven is $110. The discount is 20% of the original price. Find the original price.

10. The sale price of a treadmill is $280. The discount is 30% of the original price. Find the original price.

Application

EXAMPLE 5 Geometry: Dimensions of a Rectangle

A rectangle is twice as long as it is wide, and its perimeter is 132 inches. Find the dimensions of the rectangle using a system of linear equations.

SOLUTION

The two unknowns in this problem are the length and width of the rectangle, as shown at the left.

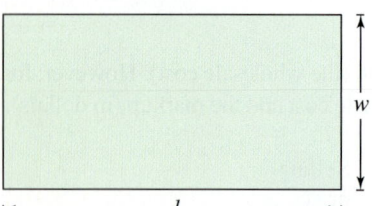

Verbal Model: $2\left(\text{Width of rectangle}\right) + 2\left(\text{Length of rectangle}\right) = \text{Perimeter of rectangle}$

$\left(\text{Length of rectangle}\right) = 2 \cdot \text{Width of rectangle}$

Labels:
Width of rectangle = w (inches)
Length of rectangle = l (inches)
Perimeter of rectangle = 132 (inches)

System: $\begin{cases} 2w + 2l = 132 & \text{Equation 1} \\ l = 2w & \text{Equation 2} \end{cases}$

To solve this system of linear equations, use the method of substitution.

$2w + 2l = 132$ Equation 1
$2w + 2(2w) = 132$ Substitute $2w$ for l.
$6w = 132$ Combine like terms.

So, $w = 22$. To solve for l, back-substitute this value into Equation 2.

$l = 2(22)$ Substitute 22 for w in Equation 2.
$l = 44$ Simplify.

This implies that the rectangle has a width of 22 inches and a length of 44 inches. Check this in the original statement of the problem as follows.

Check

The rectangle is twice as long as it is wide.
Length = 2(width) = 2(22) = 44 inches   Solution checks. ✓
The rectangle's perimeter is 132 inches.
2(width) + 2(length) = 2(22) + 2(44) = 132 inches.   Solution checks. ✓

Exercises Within Reach®

Solutions in English & Spanish and tutorial videos at AlgebraWithinReach.com

Geometry In Exercises 11–16, use a system of linear equations to find the dimensions of the rectangle that meet the specified conditions.

	Perimeter	Relationship Between Length and Width		Perimeter	Relationship Between Length and Width
11.	40 feet	The length is 4 feet greater than the width.	12.	220 inches	The width is 10 inches less than the length.
13.	16 yards	The width is one-third of the length.	14.	48 meters	The length is twice the width.
15.	35.2 meters	The length is 120% of the width.	16.	35 feet	The width is 75% of the length.

Application

EXAMPLE 6 An Application Involving Two Speeds

You are taking a motorboat trip on a river—18 miles upstream and 18 miles back downstream. You run the motor at the same speed going up and down the river, but because of the current of the river, the trip upstream takes longer than the trip downstream. You do not know the speed of the river's current, but you know that the trip upstream takes $1\frac{1}{2}$ hours and the trip downstream takes only 1 hour. From this information, determine the speed of the current.

SOLUTION

Verbal Model:

Boat speed (still water) − Speed of current = Upstream speed

Boat speed (still water) + Speed of current = Downstream speed

Labels:

Boat speed in still water = x (miles per hour)
Speed of current = y (miles per hour)
Upstream speed = $\dfrac{18}{1.5} = 12$ (miles per hour)
Downstream speed = $\dfrac{18}{1} = 18$ (miles per hour)

System: $\begin{cases} x - y = 12 & \text{Equation 1} \\ x + y = 18 & \text{Equation 2} \end{cases}$

To solve this system of linear equations, use the method of elimination.

$$\begin{array}{ll} x - y = 12 & \text{Equation 1} \\ x + y = 18 & \text{Equation 2} \\ \hline 2x = 30 & \text{Add equations.} \end{array}$$

So, the speed of the boat in still water is $x = 15$ miles per hour. To find the speed of the current, back-substitute this value into Equation 2.

$15 + y = 18$ Substitute 15 for x in Equation 2.

$y = 3$ Subtract 15 from each side.

So, the speed of the current is 3 miles per hour. Check this solution in the original statement of the problem.

Exercises Within Reach®

Solutions in English & Spanish and tutorial videos at AlgebraWithinReach.com

17. Current Speed You travel 10 miles upstream and 10 miles downstream on a motorboat trip. You run the motor at the same speed going up and down the river, but because of the speed of the current, the trip upstream takes $\frac{1}{2}$ hour and the trip downstream takes $\frac{1}{3}$ hour. Determine the speed of the current.

18. Boat Speed You travel 16 miles upstream and 16 miles downstream on a motorboat trip. You run the motor at the same speed going up and down the river, but because of the speed of the current, the trip upstream takes 2 hours and the trip downstream take $1\frac{1}{3}$ hours. Determine the speed of the boat in still water.

Concept Summary: Constructing Systems of Linear Equations

What
You can solve some application problems by constructing and solving a system of linear equations.

How
You already know how to construct a verbal model and write a linear equation for an application problem with one unknown.

In this section, you use the same process to create two or more verbal models and equations for application problems with two or more unknowns.

Why
Many of the problems solved in Chapters 3 and 4 using one variable can now be solved using systems of linear equations in two variables. Now you have another way to solve application problems.

Exercises Within Reach®
Worked-out solutions to odd-numbered exercises at AlgebraWithinReach.com

Concept Summary Check

19. **Identifying Methods** List the three methods for solving systems of linear equations.

20. **Writing** How can you determine whether a real-life problem may be modeled with a system of linear equations?

21. **Writing** What is a verbal model of a real-life problem?

22. **Precision** Compare the one-variable method with the two-variable method for modeling real-life problems.

Extra Practice

Coin Problems In Exercises 23–28, use a system of linear equations to determine the number of each type of coin.

	Total number of coins	Types of coins	Value		Total number of coins	Types of coins	Value
23.	21	Dimes and quarters	$4.05	24.	17	Dimes and quarters	$2.90
25.	35	Nickels and quarters	$5.75	26.	31	Nickels and quarters	$6.55
27.	44	Nickels and dimes	$3.00	28.	28	Nickels and dimes	$2.40

Using a System of Linear Equations In Exercises 29–34, use a system of linear equations to solve the problem.

29. The selling price of an air conditioner is $359. The markup rate is 30% of the wholesale cost. Find the wholesale cost.

30. The sale price of a surround sound system is $716. The discount is 20% of the original price. Find the original price.

31. The total cost of 8 gallons of regular gasoline and 12 gallons of premium gasoline is $75.60. Premium gasoline costs $0.15 more per gallon than regular gasoline. Find the price per gallon for each type of gasoline.

32. The total cost of 6 gallons of regular gasoline and 11 gallons of premium gasoline is $68.33. Premium gasoline costs $0.20 more per gallon than regular gasoline. Find the price per gallon for each type of gasoline.

33. A van travels for 2 hours at an average speed of 40 miles per hour. How much longer must the van travel at an average speed of 55 miles per hour so that the average speed for the entire trip will be 45 miles per hour?

34. A van travels for 3 hours at an average speed of 40 miles per hour. How much longer must the van travel at an average speed of 55 miles per hour so that the average speed for the entire trip will be 50 miles per hour?

Break-Even Analysis In Exercises 35–38, use a graphing calculator to **graph** the cost and revenue equations in the same viewing window. **Find** the sales x necessary to break even ($R = C$) and the corresponding revenue R obtained by selling x units. (Round x to the nearest whole unit.)

	Cost	Revenue		Cost	Revenue
35.	$C = 7650x + 125{,}000$	$R = 8950x$	36.	$C = 2175x + 85{,}000$	$R = 3525x$
37.	$C = 0.55x + 40{,}000$	$R = 0.85x$	38.	$C = 0.25x + 25{,}000$	$R = 0.45x$

Generating a Linear System In Exercises 39–42, find m and b such that $y = mx + b$ is the equation of the line through the points. (*Hint:* **Generate** a linear system in m and b by substituting the coordinates of the specified points into $y = mx + b$.)

39. $(2, -1), (6, 1)$

40. $(1, 3), (4, 9)$

41. $(-3, 6), (5, 2)$

42. $(0, 2), (4, -8)$

Explaining Concepts

43. **Writing** Consider the one-variable method and the two-variable method for solving Example 4. Which do you prefer? Explain.

44. **Error Analysis** Describe and correct the error in writing the system of linear equations for the problem below. Do not solve the system.

    A total of $9000 is invested in two funds paying 5% and 8% simple interest. (There is more risk in the 8% fund.) The combined annual interest for the two funds is $645. How much of the $9000 is invested at each rate?

    $$\begin{cases} x + y = 9000 \\ 5x + 8y = 645 \end{cases}$$

45. **Error Analysis** Describe and correct the error in writing the system of linear equations for the problem below. Do not solve the system.

    The sum of two numbers is 42, and the larger number is 3 less than twice the smaller number.

    $x + y = 42$
    $2x - y = -3$

Cumulative Review

In Exercises 46–49, determine whether the lines L_1 and L_2 that pass through the pairs of points are parallel, perpendicular, or neither.

46. L_1: $(8, 7), (4, 4)$
    L_2: $(2, 1), (-1, 5)$

47. L_1: $(-2, 3), (-4, 1)$
    L_2: $(-6, 1), (-8, -1)$

48. L_1: $(12, 0), (7, -2)$
    L_2: $(0, 7), (-5, 9)$

49. L_1: $(-10, 1), (-7, 2)$
    L_2: $(5, -2), (6, -5)$

In Exercises 50–53, solve the system by the method of elimination.

50. $\begin{cases} 3x + 3y = 7 \\ 3x + 5y = 3 \end{cases}$

51. $\begin{cases} -4x + 3y = 18 \\ -6x + y = -8 \end{cases}$

52. $\begin{cases} 6x + 5y = 19 \\ 2x + 3y = 5 \end{cases}$

53. $\begin{cases} 4x + 5y = 35 \\ -3x + 2y = -9 \end{cases}$

8.5 Systems of Linear Inequalities

- Solve systems of linear inequalities in two variables.
- Find the boundaries of a region.
- Use systems of linear inequalities to model and solve real-life problems.

Systems of Linear Inequalities in Two Variables

Study Tip

In Example 1, note that the two border lines of the region intersect at a point. Such a point is called a **vertex** of the region. The region shown in the figure has only one vertex. Some regions, however, have several vertices. When you are sketching the graph of a system of linear inequalities, you should find and label any vertices of the region.

Graphing a System of Linear Inequalities

1. Sketch the line that corresponds to each inequality. (Use dashed lines for inequalities with $<$ or $>$ and use solid lines for inequalities with $\leq$ or $\geq$.)
2. Lightly shade the half-plane that is the graph of each linear inequality. (Colored pencils may help distinguish different half-planes.)
3. The graph of the system is the intersection of the half-planes. (If you use colored pencils, it is the region that is shaded with *every* color.)

EXAMPLE 1 Graphing a System of Linear Inequalities

Sketch the graph of the system of linear inequalities: $\begin{cases} 2x - y \leq 5 \\ x + 2y \geq 2 \end{cases}$.

SOLUTION

Begin by rewriting each inequality in slope-intercept form. Then sketch the line for each corresponding equation.

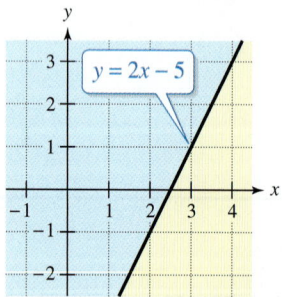

Graph of $2x - y \leq 5$ is all points on and above $y = 2x - 5$.

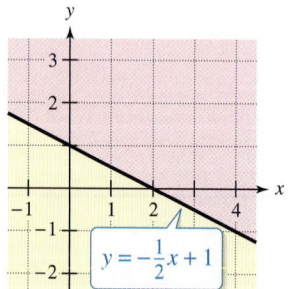

Graph of $x + 2y \geq 2$ is all points on and above $y = -\frac{1}{2}x + 1$.

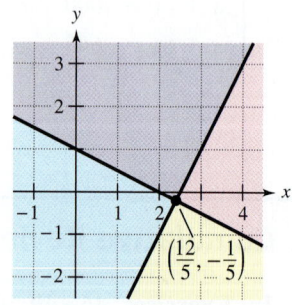

Graph of system is the purple wedge-shaped region.

Exercises Within Reach®   Solutions in English & Spanish and tutorial videos at AlgebraWithinReach.com

Graphing a System of Linear Inequalities In Exercises 1–6, sketch the graph of the system of linear inequalities.

1. $\begin{cases} y \geq 3x - 3 \\ y \leq -x + 1 \end{cases}$
2. $\begin{cases} y \geq 2x - 3 \\ y \leq 3x + 1 \end{cases}$
3. $\begin{cases} x + 2y > -4 \\ y < x + 5 \end{cases}$
4. $\begin{cases} x + y < -3 \\ y > 3x - 4 \end{cases}$
5. $\begin{cases} x + y \leq 3 \\ x - y \leq 1 \end{cases}$
6. $\begin{cases} x + y \geq 2 \\ x - y \leq 2 \end{cases}$

Section 8.5 Systems of Linear Inequalities

EXAMPLE 2 Graphing a System of Linear Inequalities

Sketch the graph of the system of linear inequalities.

$$\begin{cases} y < 4 \\ y > 1 \end{cases}$$

SOLUTION

The graph of the first inequality is the half-plane *below* the horizontal line

$y = 4.$ Upper boundary

The graph of the second inequality is the half-plane *above* the horizontal line

$y = 1.$ Lower boundary

The graph of the system is the horizontal band that lies *between* the two horizontal lines (where $y < 4$ and $y > 1$), as shown below.

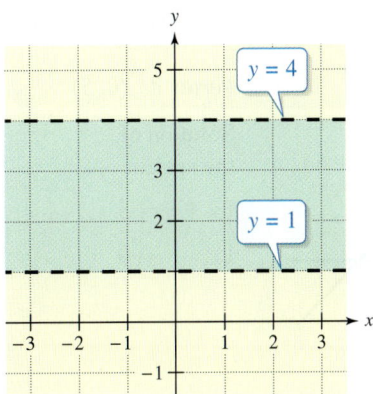

Exercises Within Reach®   Solutions in English & Spanish and tutorial videos at AlgebraWithinReach.com

Graphing a System of Linear Inequalities In Exercises 7–16, sketch the graph of the system of linear inequalities.

7. $\begin{cases} x < 3 \\ x > -2 \end{cases}$

8. $\begin{cases} y > -1 \\ y \leq 2 \end{cases}$

9. $\begin{cases} y < x - 2 \\ x < 5 \end{cases}$

10. $\begin{cases} y > x - 4 \\ x > -1 \end{cases}$

11. $\begin{cases} y \leq x - 5 \\ y > -7 \end{cases}$

12. $\begin{cases} y < -x + 3 \\ y \geq -1 \end{cases}$

13. $\begin{cases} y > 2x \\ y > -x + 4 \end{cases}$

14. $\begin{cases} y \leq -x \\ y \leq x + 1 \end{cases}$

15. $\begin{cases} x + y > -1 \\ x + y < 3 \end{cases}$

16. $\begin{cases} x - y > 2 \\ x - y < -4 \end{cases}$

EXAMPLE 3 Graphing a System of Linear Inequalities

Sketch the graph of the system of linear inequalities, and label the vertices.

$$\begin{cases} x + y \le 5 \\ 3x + 2y \le 12 \\ x \ge 0 \\ y \ge 0 \end{cases}$$

SOLUTION

Begin by sketching the half-planes represented by the four linear inequalities. The graph of $x + y \le 5$ is the half-plane lying on and below the line $y = -x + 5$. The graph of $3x + 2y \le 12$ is the half-plane lying on and below the line $y = -\frac{3}{2}x + 6$. The graph of $x \ge 0$ is the half-plane lying on and to the right of the y-axis, and the graph of $y \ge 0$ is the half-plane lying on and above the x-axis.

As shown in the figure, the region that is common to all four of these half-planes is a four-sided polygon. The vertices of the region are found as follows.

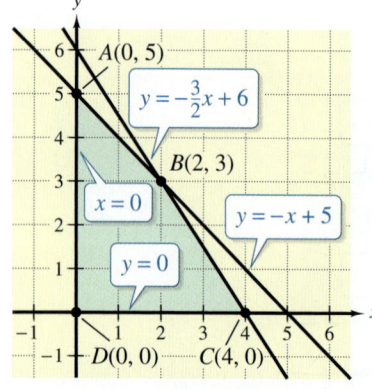

Vertex A: $(0, 5)$

Solution of the system

$$\begin{cases} x + y = 5 \\ x = 0 \end{cases}$$

Vertex B: $(2, 3)$

Solution of the system

$$\begin{cases} x + y = 5 \\ 3x + 2y = 12 \end{cases}$$

Vertex C: $(4, 0)$

Solution of the system

$$\begin{cases} 3x + 2y = 12 \\ y = 0 \end{cases}$$

Vertex D: $(0, 0)$

Solution of the system

$$\begin{cases} x = 0 \\ y = 0 \end{cases}$$

Exercises Within Reach®

Solutions in English & Spanish and tutorial videos at AlgebraWithinReach.com

Graphing a System of Linear Inequalities In Exercises 17–22, sketch the graph of the system of linear inequalities, and label the vertices.

17. $\begin{cases} y > -5 \\ x \le 2 \\ y \le x + 2 \end{cases}$

18. $\begin{cases} y \ge -1 \\ x \le 5 \\ y \le x + 4 \end{cases}$

19. $\begin{cases} x + y \le 5 \\ x \ge 2 \\ y \ge 0 \end{cases}$

20. $\begin{cases} 2x + y \ge 2 \\ x \le 2 \\ y \le 1 \end{cases}$

21. $\begin{cases} x + y \le 1 \\ -x + y \le 1 \\ y \ge 0 \end{cases}$

22. $\begin{cases} 3x + 2y < 6 \\ -3x + 2y \ge 6 \\ y \ge 0 \end{cases}$

Graphing a System of Linear Inequalities In Exercises 23–28, sketch the graph of the system of linear inequalities.

23. $\begin{cases} -3x + 2y < 6 \\ x - 4y > -2 \\ 2x + y < 3 \end{cases}$

24. $\begin{cases} x - 7y > -36 \\ 5x + 2y > 5 \\ 6x + 5y > 6 \end{cases}$

25. $\begin{cases} y \ge -6 \\ y \le -8x + 9 \\ x \ge 0 \\ y \le 0 \end{cases}$

26. $\begin{cases} y \le 10 \\ x + 3y \le 15 \\ x \ge 0 \\ y \ge 0 \end{cases}$

27. $\begin{cases} x - y \le 8 \\ 2x + 5y \le 25 \\ x \ge 0 \\ y \ge 0 \end{cases}$

28. $\begin{cases} 4x - y \le 13 \\ -x + 2y \le 22 \\ x \ge 0 \\ y \ge 0 \end{cases}$

Finding the Boundaries of a Region

EXAMPLE 4 Finding the Boundaries of a Region

Write a system of inequalities that describes the region shown in the figure.

SOLUTION

Three of the boundaries of the region are horizontal or vertical—they are easy to find. To find the diagonal boundary line, use the techniques from Section 4.4 to find the equation of the line passing through the points $(4, 4)$ and $(6, 0)$. You can use the formula for slope to find $m = -2$ and then use the point-slope form with point $(6, 0)$ and $m = -2$ to obtain

$$y - 0 = -2(x - 6).$$

So, the equation is

$$y = -2x + 12.$$

The system of linear inequalities that describes the region is as follows.

$$\begin{cases} y \leq 4 & \text{Region lies on and below line } y = 4. \\ y \geq 0 & \text{Region lies on and above } x\text{-axis.} \\ x \geq 0 & \text{Region lies on and to the right of } y\text{-axis.} \\ y \leq -2x + 12 & \text{Region lies on and below line } y = -2x + 12. \end{cases}$$

Exercises Within Reach®

Solutions in English & Spanish and tutorial videos at AlgebraWithinReach.com

Finding the Boundaries of a Region In Exercises 29–32, write a system of linear inequalities that describes the shaded region.

29.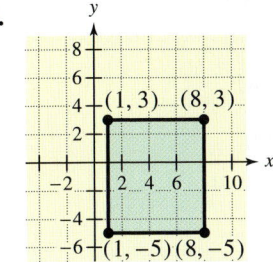

30.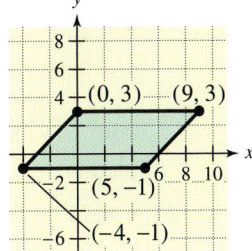

31.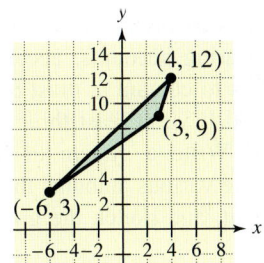

32.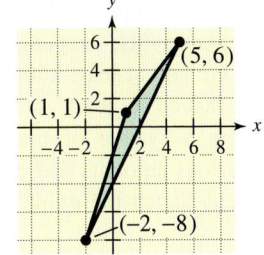

Applications

EXAMPLE 5 Target Heart Rate

A person's maximum heart rate is $220 - x$, where x is the person's age in years ($20 \leq x \leq 70$). It is recommended that during exercise a person strives for a heart rate that is at least 50% of the maximum and at most 85% of the maximum. Write a system of linear inequalities that describes the target heart rate zone, and sketch a graph of the system. (*Source:* American Heart Association)

SOLUTION

Let y represent the person's heart rate. From the given information, you can write the following system of inequalities.

$$\begin{cases} x \geq 20 \\ x \leq 70 \\ y \geq 0.50(220 - x) \\ y \leq 0.85(220 - x) \end{cases}$$

- Person's age must be at least 20.
- Person's age can be at most 70.
- Target rate is at least 50% of maximum rate.
- Target rate is at most 85% of maximum rate.

To graph this system, sketch the half-planes represented by the four linear inequalities. The graph of $x \geq 20$ is the half-plane lying on and to the right of the line $x = 20$. The graph of $x \leq 70$ is the half-plane lying on and to the left of the line $x = 70$. The graph of $y \geq 0.50(220 - x)$ is the half-plane lying on and above the line $y = 0.50(220 - x)$. The graph of $y \leq 0.85(220 - x)$ is the half-plane lying on and below the line $y = 0.85(220 - x)$.

As shown in the figure, the region that is common to all four of these half-planes is a four-sided polygon. The vertices of the region are found as follows.

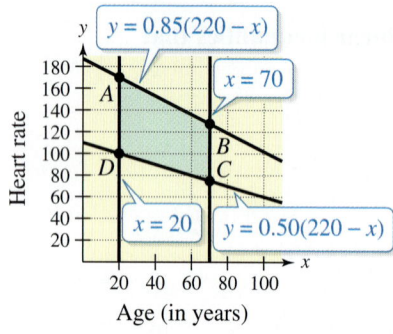

Target Heart Rate

Vertex A: (20, 170)
Solution of the system
$$\begin{cases} x = 20 \\ y = 0.85(220 - x) \end{cases}$$

Vertex B: (70, 127.5)
Solution of the system
$$\begin{cases} x = 70 \\ y = 0.85(220 - x) \end{cases}$$

Vertex C: (70, 75)
Solution of the system
$$\begin{cases} x = 70 \\ y = 0.50(220 - x) \end{cases}$$

Vertex D: (20, 100)
Solution of the system
$$\begin{cases} x = 20 \\ y = 0.50(220 - x) \end{cases}$$

Exercises Within Reach®
Solutions in English & Spanish and tutorial videos at AlgebraWithinReach.com

33. Investment A person plans to invest up to $20,000 in two different interest-bearing accounts, account X and account Y. Account X is to contain at least $5000. Moreover, account Y should have at least twice the amount in account X. Write a system of linear inequalities that describes the various amounts that can be deposited in each account, and sketch the graph of the system.

34. Investment A person plans to invest up to $10,000 in two different interest-bearing accounts, account X and account Y. Account Y is to contain at most $3000. Moreover, account X should have at least three times the amount in account Y. Write a system of linear inequalities that describes the various amounts that can be deposited in each account, and sketch the graph of the system.

Application

EXAMPLE 6 Geometry: Finding the Boundaries of a Region

The figure shows a chorus platform on a stage. Write a system of linear inequalities that describes the part of the audience that can see the full chorus. (Each unit in the coordinate system represents 1 meter.)

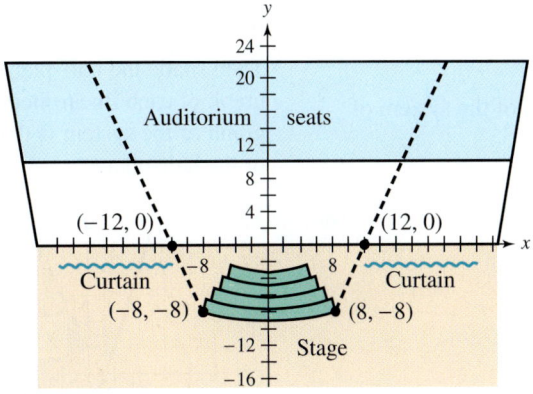

SOLUTION

Two of the boundary lines are horizontal. These boundaries can be described by the inequalities $y \leq 22$ and $y \geq 10$. To find the inequalities corresponding to the two diagonal boundary lines, use the two points given on each line.

$(-12, 0)$ and $(-8, -8)$: $y > -2x - 24$

$(12, 0)$ and $(8, -8)$: $y > 2x - 24$

So, the system of linear inequalities is:

$$\begin{cases} y \leq 22 \\ y \geq 10 \\ y > -2x - 24 \\ y > 2x - 24 \end{cases}$$

Exercises Within Reach®

Solutions in English & Spanish and tutorial videos at AlgebraWithinReach.com

35. Swimming The figure shows a cross section of a roped-off swimming area at a beach. Write a system of linear inequalities that describes the cross section. (Each unit in the coordinate system represents 1 foot.)

36. Geocaching You are using a GPS to find a container that is hidden in the section of land shown in the figure. Write a system of linear inequalities that describes the section of land. (Each unit in the coordinate system represents 1 foot.)

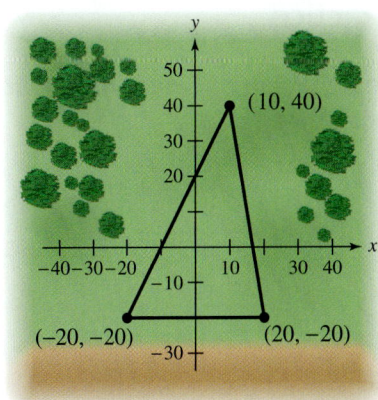

Concept Summary: Graphing Systems of Linear Inequalities

What
The graph of a **system of linear inequalities** shows *all* of the **solutions** of the system.

EXAMPLE
Sketch the graph of the system of linear inequalities.
$$\begin{cases} y < 2x + 1 \\ y \geq -3x - 2 \end{cases}$$

How
Sketch the line that corresponds to each inequality. Be sure to use dashed lines and solid lines appropriately.

Then shade the half-plane that is the graph of each linear inequality. The graph of the system is the intersection of the half-planes.

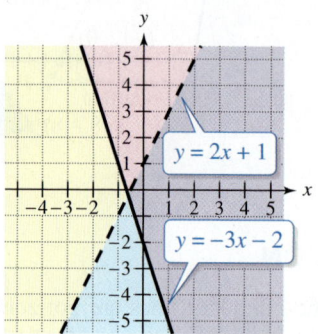

Why
Many practical problems in business, science, and engineering involve multiple constraints. These problems can be solved with systems of inequalities.

Exercises Within Reach®

Worked-out solutions to odd-numbered exercises at AlgebraWithinReach.com

Concept Summary Check

37. Reasoning When should the line that corresponds to an inequality be dashed? When should it be solid?

38. Logic How do you check a single point (x_1, y_1) to determine whether it is a solution of a system of inequalities?

39. Writing Explain how to sketch the graph of a system of inequalities in two variables.

40. Precision How do you determine the vertices of the solution region for a system of linear inequalities?

Extra Practice

Graphing a System of Linear Inequalities In Exercises 41–46, sketch the graph of the system of linear inequalities.

41. $\begin{cases} x + y \leq 4 \\ x \geq 0 \\ y \geq 0 \end{cases}$

42. $\begin{cases} 2x + y \leq 6 \\ x \geq 0 \\ y \geq 0 \end{cases}$

43. $\begin{cases} 4x - 2y > 8 \\ x \geq 0 \\ y \leq 0 \end{cases}$

44. $\begin{cases} 2x - 6y > 6 \\ x \leq 0 \\ y \leq 0 \end{cases}$

45. $\begin{cases} x \geq 1 \\ x - 2y \leq 3 \\ 3x + 2y \geq 9 \\ x + y \leq 6 \end{cases}$

46. $\begin{cases} x + y \leq 4 \\ x + y \geq -1 \\ x - y \geq -2 \\ x - y \leq 2 \end{cases}$

47. Ticket Sales Two types of tickets are to be sold for a concert. One type costs $15 per ticket and the other type costs $25 per ticket. The promoter of the concert must sell at least 15,000 tickets, including at least 8000 of the $15 tickets and at least 4000 of the $25 tickets. Moreover, the gross receipts must total at least $275,000 in order for the concert to be held. Write a system of linear inequalities that describes the different numbers of tickets that can be sold.

48. Nutrition A dietitian is asked to design a special dietary supplement using two different foods. Each ounce of food X contains 20 units of calcium, 15 units of iron, and 10 units of vitamin B. Each ounce of food Y contains 10 units of calcium, 10 units of iron, and 20 units of vitamin B. The minimum daily requirements in the diet are 280 units of calcium, 160 units of iron, and 180 units of vitamin B. Write a system of linear inequalities that describes the different amounts of food X and food Y that can be used in the diet.

49. Production A furniture company can sell all the tables and chairs it produces. Each table requires 1 hour in the assembly center and $1\frac{1}{2}$ hours in the finishing center. Each chair requires 2 hours in the assembly center and $\frac{3}{4}$ hour in the finishing center. The company's assembly center is available 12 hours per day, and its finishing center is available 18 hours per day. Write a system of linear inequalities that describes the different production levels. Graph the system.

50. Production An electronics company can sell all the HD TVs and DVD players it produces. Each HD TV requires 3 hours on the assembly line and $1\frac{1}{4}$ hours on the testing line. Each DVD player requires 2 hours on the assembly line and 1 hour on the testing line. The company's assembly line is available 20 hours per day, and its testing line is available 16 hours per day. Write a system of linear inequalities that describes the different production levels. Graph the system.

Explaining Concepts

51. Writing Explain the meaning of the term *half-plane*. Give an example of an inequality whose graph is a half-plane.

52. Reasoning Can the solution of a system of linear inequalities be a line? Explain.

53. Reasoning Can the solution of a system of linear inequalities be a single point? Explain.

54. Logic Is it possible for a system of linear inequalities to have no solution? If so, write an example.

Cumulative Review

In Exercises 55–58, rewrite the expression in exponential form.

55. $3 \cdot 3 \cdot 3 \cdot 3$

56. $(-6) \cdot (-6) \cdot (-6)$

57. $\frac{1}{2} \cdot \frac{1}{2} \cdot \frac{1}{2} \cdot \frac{1}{2} \cdot \frac{1}{2} \cdot \frac{1}{2}$

58. $(-1.3) \cdot (-1.3)$

In Exercises 59–62, rewrite the expression as a product.

59. $(-4)^4$

60. 7^5

61. $\left(-\frac{3}{4}\right)^2$

62. $(3.5)^8$

In Exercises 63–66, sketch the graph of the equation.

63. $2x + 4y = 8$

64. $y = 4(x - 1) + 3$

65. $0.3x - 0.2y = 0.8$

66. $x = 6$

8 Chapter Summary

	What did you learn?	Explanation and Examples	Review Exercises
8.1	Determine whether an ordered pair is a solution of a system of equations *(p. 354)*.	To determine whether an ordered pair is a solution, substitute the coordinates of the point into each equation and simplify each side of the equation.	1–4
	Use a coordinate system to solve systems of linear equations graphically *(p. 355)*.	1. The two lines can intersect in a single point (slopes are not equal). The corresponding system of linear equations has a single solution and is called *consistent*. 2. The two lines can coincide (slopes are equal) and have infinitely many points of intersection. The corresponding consistent system of linear equations has infinitely many solutions and is called *dependent*. 3. The two lines can be parallel (slopes are equal) and have no point of intersection. The corresponding system of linear equations has no solution and is called *inconsistent*.	5–14
	Use systems of equations to model real-life problems *(p. 359)*.	A common business application that involves systems of equations is break-even analysis. When enough units have been sold so that the total revenue equals the total cost, the sales have reached the break-even point.	15, 16
8.2	Use the method of substitution to solve systems of equations algebraically *(p. 362)*.	1. Solve one of the equations for one variable in terms of the other. 2. Substitute the expression obtained in Step 1 into the other equation to obtain an equation in one variable. 3. Solve the equation obtained in Step 2. 4. Back-substitute the solution from Step 3 into the expression obtained in Step 1 to find the value of the other variable. 5. Check the solution to see that it satisfies both of the original equations.	17–26
	Solve systems with no solution or infinitely many solutions *(p. 365)*.	Reduce the system of two linear equations to a single equation in one variable. 1. If the equation produces a false statement, the original system has no solution. 2. If the equation produces a true statement, the original system has infinitely many solutions.	27–30
	Use the method of substitution to solve application problems *(p. 366)*.	Using the algebraic method of substitution is an accurate way to solve applications involving more than one unknown.	31, 32

	What did you learn?	*Explanation and Examples*	*Review Exercises*
8.3	Solve systems of linear equations algebraically using the method of elimination *(p. 370)*.	1. Obtain opposite coefficients of x (or y) by multiplying all terms of one or both equations by suitable constants. 2. Add the equations to eliminate one variable and solve the resulting equation. 3. Back-substitute the value obtained in Step 2 into either of the original equations and solve for the other variable. 4. Check your solution in both of the original equations.	33–42
	Solve systems with no solution or infinitely many solutions *(p. 374)*.	As with the method of substitution, the key is to recognize the occurrence of a false or true statement.	43–46
	Use the method of elimination to solve application problems *(p. 375)*.	Using the algebraic method of elimination is an accurate way to solve applications involving more than one unknown.	47, 48
8.4	Construct a system of linear equations from an application problem *(p. 380)*.	Recognize applications of systems of linear equations. 1. Does the problem involve more than one unknown quantity? 2. Are there two (or more) equations or conditions to be satisfied? If one or both of these conditions occur, then the appropriate mathematical model for the problem may be a system of linear equations.	49–58
	Solve real-life applications using systems of linear equations *(p. 381)*.	You can use any of the following methods to solve real-life applications involving systems of linear equations. 1. Solving systems of equations by graphing 2. Solving systems of equations by substitution 3. Solving systems of equations by elimination	49–60
8.5	Solve systems of linear inequalities in two variables *(p. 388)*.	1. Sketch the line that corresponds to each inequality. (Use dashed lines for inequalities with < or > and use solid lines for inequalities with ≤ or ≥.) 2. Lightly shade the half-plane that is the graph of each linear inequality. 3. The graph of the system is the intersection of the half-planes.	61–70
	Find the boundaries of a region *(p. 391)*.	A system of linear inequalities can be used to describe the boundaries of a specified region.	71, 72
	Use systems of linear inequalities to model and solve real-life problems *(p. 392)*.	A system of linear inequalities can be used to find a number of acceptable solutions to real-life problems.	73–76

Review Exercises

Worked-out solutions to odd-numbered exercises at AlgebraWithinReach.com

8.1

Checking Solutions In Exercises 1−4, determine whether each ordered pair is a solution of the system.

System   Ordered Pairs

1. $\begin{cases} 3x - 5y = 11 \\ -x + 2y = -4 \end{cases}$ (a) $(2, -1)$ (b) $(3, -2)$

2. $\begin{cases} 10x + 8y = -2 \\ 2x - 5y = 26 \end{cases}$ (a) $(4, -4)$ (b) $(3, -4)$

3. $\begin{cases} 0.2x + 0.4y = 5 \\ x + 3y = 30 \end{cases}$ (a) $(0.5, -0.7)$ (b) $(15, 5)$

4. $\begin{cases} -\frac{1}{2}x - \frac{2}{3}y = \frac{1}{2} \\ x + y = 1 \end{cases}$ (a) $(-5, 6)$ (b) $(7, -3)$

Solving a System of Linear Equations In Exercises 5−14, solve the system by graphing.

5. $\begin{cases} y = x - 4 \\ y = 2x - 9 \end{cases}$

6. $\begin{cases} y = -\frac{5}{3}x + 6 \\ y = x - 10 \end{cases}$

7. $\begin{cases} x + y = 2 \\ x - y = 0 \end{cases}$

8. $\begin{cases} x - y = 9 \\ -x + y = 1 \end{cases}$

9. $\begin{cases} 2x + 3 = 3y \\ y = \frac{2}{3}x \end{cases}$

10. $\begin{cases} x + y = -1 \\ 3x + 2y = 0 \end{cases}$

11. $\begin{cases} -3x - 3y = -6 \\ x + y = 2 \end{cases}$

12. $\begin{cases} 2x - 3y = 11 \\ x + y = 3 \end{cases}$

13. $\begin{cases} x + 7y = 6 \\ -3x + y = 4 \end{cases}$

14. $\begin{cases} 5x - 4y = 12 \\ -x + 3y = 2 \end{cases}$

15. **Number Problem** The sum of two numbers x and y is 52, and the difference of the two numbers is 20. The system of equations that represents this problem is

$$\begin{cases} x + y = 52 \\ x - y = 20 \end{cases}.$$

Solve this system graphically to find the two numbers.

16. **Break-Even Analysis** A small company produces sunglasses that sell for $18 per unit. The cost of producing each unit is $10.25, and the company has fixed costs of $350.

(a) Use a verbal model to show that the cost C of producing x units is $C = 10.25x + 350$, and the revenue R from selling x units is $R = 18x$.

(b) Use a graphing calculator to graph the cost and revenue functions in the same viewing window. Approximate the point of intersection of the graphs and interpret the result.

8.2

Solving a System of Linear Equations In Exercises 17−30, solve the system by the method of substitution.

17. $\begin{cases} y = 2x \\ y = x + 4 \end{cases}$

18. $\begin{cases} x = -2y + 13 \\ x = \frac{y}{2} + 3 \end{cases}$

19. $\begin{cases} x = 3y - 2 \\ x = 6 - y \end{cases}$

20. $\begin{cases} y = -4x + 1 \\ y = x - 4 \end{cases}$

21. $\begin{cases} x - 2y = 6 \\ 3x + 2y = 10 \end{cases}$

22. $\begin{cases} 5x + y = 20 \\ 7x - 5y = -4 \end{cases}$

23. $\begin{cases} 2x - y = 2 \\ 6x + 8y = 39 \end{cases}$

24. $\begin{cases} 3x + 4y = 1 \\ x - 7y = -3 \end{cases}$

25. $\begin{cases} \frac{3}{5}x - y = 8 \\ 2x - 3y = 25 \end{cases}$

26. $\begin{cases} -x + 8y = -115 \\ 2x + \frac{2}{7}y = 2 \end{cases}$

27. $\begin{cases} x = y + 3 \\ x = y + 1 \end{cases}$

28. $\begin{cases} y = 3x + 4 \\ 9x = 3y - 12 \end{cases}$

29. $\begin{cases} -6x + y = -3 \\ 12x - 2y = 6 \end{cases}$

30. $\begin{cases} 3x + 4y = 7 \\ 6x + 8y = 10 \end{cases}$

31. **Investment** A total of $12,000 is invested in two funds paying 5% and 10% simple interest. (There is more risk in the 10% fund.) The combined annual interest for the two funds is $800. The system of equations that represents this situation is

$$\begin{cases} x + y = 12{,}000 \\ 0.05x + 0.10y = 800 \end{cases}$$

where x represents the amount invested in the 5% fund and y represents the amount invested in the 10% fund. Solve this system to determine how much of the $12,000 is invested at each rate.

32. **Comparing Costs** An MP3 player costs $200 plus $10 for every album purchased. A CD player costs $50 plus $15 for every album purchased. The system of equations that represents this situation is

$$\begin{cases} y = 200 + 10x \\ y = 50 + 15x \end{cases}$$

where y represents the total cost of each audio format and x is the number of albums purchased. Solve this system to determine after how many albums the total costs for the two music formats will be the same. What will be the cost?

8.3

Solving a System of Linear Equations In Exercises 33–46, solve the system by the method of elimination.

33. $\begin{cases} 2x + 4y = 2 \\ -2x - 7y = 4 \end{cases}$

34. $\begin{cases} 3x - y = 5 \\ 2x + y = 5 \end{cases}$

35. $\begin{cases} 3x - 2y = 9 \\ x + y = 3 \end{cases}$

36. $\begin{cases} 5x + 4y = 2 \\ -x + y = -22 \end{cases}$

37. $\begin{cases} x + 2y = 2 \\ x - 4y = 20 \end{cases}$

38. $\begin{cases} 2x + 6y = 16 \\ 2x + 3y = 7 \end{cases}$

39. $\begin{cases} \frac{1}{2}x - \frac{3}{5}y = \frac{1}{6} \\ -3x + 6y = 1 \end{cases}$

40. $\begin{cases} \frac{2}{3}x + \frac{1}{12}y = \frac{3}{4} \\ 3x - 4y = 2 \end{cases}$

41. $\begin{cases} 0.2x - 0.1y = 0.07 \\ 0.4x - 0.5y = -0.01 \end{cases}$

42. $\begin{cases} 0.2x + 0.1y = 0.03 \\ 0.3x - 0.1y = -0.13 \end{cases}$

43. $\begin{cases} 2x - 5y = 2 \\ 6x - 15y = 4 \end{cases}$

44. $\begin{cases} 8x - 6y = 4 \\ -4x + 3y = -2 \end{cases}$

45. $\begin{cases} 6x - 3y = 27 \\ -2x + y = -9 \end{cases}$

46. $\begin{cases} -\frac{1}{4}x + \frac{2}{3}y = 1 \\ 3x - 8y = 1 \end{cases}$

47. College Credits Each course at a college is worth either 3 or 4 credits. The members of the student council are taking a total of 47 courses that are worth a total of 156 credits. The system of equations that represents this situation is

$$\begin{cases} x + y = 47 \\ 3x + 4y = 156 \end{cases}$$

where x represents the number of 3-credit courses and y represents the number of 4-credit courses. Solve this system to determine the number of each type of course being taken.

48. Ticket Sales You are selling tickets to your school musical. Adult tickets cost $5 and children's tickets cost $3. You sell 1510 tickets and collect $6138. The system of equations that represents this situation is

$$\begin{cases} x + y = 1510 \\ 5x + 3y = 6138 \end{cases}$$

where x represents the number of adult tickets sold and y represents the number of children's tickets sold. Solve this system to determine how many of each type of ticket are sold.

8.4

Using a System of Linear Equations In Exercises 49–58, use a system of linear equations to solve the problem.

49. A cash register has 15 coins consisting of dimes and quarters. The total value of the coins is $2.85. Find the number of each type of coin.

50. You go to the video store to rent five movies for the weekend. Videos rent for $2 and $5. You spend $16. How many $2 videos do you rent?

51. A rectangular sign has a perimeter of 120 inches. The height of the sign is two-thirds of its width. Find the dimensions of the sign.

52. The perimeter of a table-tennis top is 28 feet. The difference between 4 times the length and 3 times the width is 21 feet. Find the dimensions of the table-tennis top.

53. The selling price of a DVD player is $109. The markup rate is 40% of the wholesale cost. Find the wholesale cost.

54. The sale price of a computer system is $952. The discount is 30% of the original price. Find the original price.

55. You buy 2 gallons of gasoline for your lawn mower and 5 gallons of diesel fuel for your garden tractor. The total bill is $28.45. Diesel fuel costs $0.30 more per gallon than gasoline. Find the price per gallon of each type of fuel.

56. Ten pounds of mixed bird seed sells for $6.97 per pound. The mixture is obtained from two kinds of bird seed, with one variety priced at $5.65 per pound and the other at $8.95 per pound. How many pounds of each variety of bird seed are used in the mixture?

57. A car travels for 3 hours at an average speed of 40 miles per hour. How much longer must the car travel at an average speed of 55 miles per hour so that the average speed for the entire trip will be 49 miles per hour?

58. A car travels for 4 hours at an average speed of 50 miles per hour. How much longer must the car travel at an average speed of 65 miles per hour so that the average speed for the entire trip will be 55 miles per hour?

Break-Even Analysis In Exercises 59 and 60, use a graphing calculator to graph the cost and revenue equations in the same viewing window. Find the sales x necessary to break even ($R = C$) and the corresponding revenue R obtained by selling x units. (Round x to the nearest whole unit.)

	Cost	Revenue
59.	$C = 650x + 12{,}500$	$R = 800x$
60.	$C = 3.30x + 1200$	$R = 4.75x$

8.5

Graphing a System of Linear Inequalities In Exercises 61−70, sketch the graph of the system of linear inequalities.

61. $\begin{cases} y < 2x - 2 \\ x \geq 3 \end{cases}$

62. $\begin{cases} 4x + 6y \leq 24 \\ y > 2 \end{cases}$

63. $\begin{cases} y > 2x - 5 \\ y \leq x + 1 \end{cases}$

64. $\begin{cases} y < \frac{1}{3}x + 2 \\ y \geq 3x - 8 \end{cases}$

65. $\begin{cases} 3x - y \leq 6 \\ x + y \geq 2 \end{cases}$

66. $\begin{cases} x + y < 4 \\ x - y < 4 \end{cases}$

67. $\begin{cases} x + y < 5 \\ x > 2 \\ y \geq 0 \end{cases}$

68. $\begin{cases} 2x + y > 2 \\ x < 2 \\ y < 1 \end{cases}$

69. $\begin{cases} x + 2y \leq 160 \\ 3x + y \leq 180 \\ x \geq 0 \\ y \geq 0 \end{cases}$

70. $\begin{cases} 2x + 3y \leq 24 \\ 2x + y \leq 16 \\ x \geq 0 \\ y \geq 0 \end{cases}$

Finding the Boundaries of a Region In Exercises 71 and 72, write a system of linear inequalities that describes the shaded region.

71.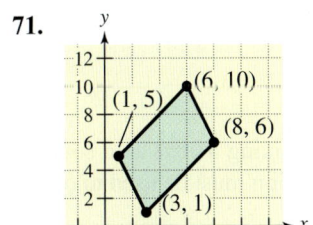

72.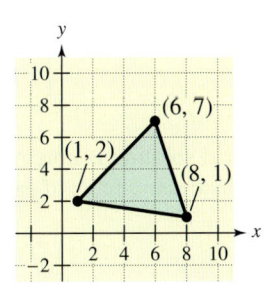

Graphing a System of Linear Inequalities In Exercises 73−76, write a system of linear inequalities that models the description, and then sketch the graph of the system.

73. A Pennsylvania fruit grower has up to 1500 bushels of apples that are to be divided between markets in Harrisburg and Philadelphia. These two markets need at least 400 bushels and 600 bushels, respectively.

74. A warehouse operator has up to 24,000 square feet of floor space in which to store two products. Each unit of product A requires 20 square feet of floor space and costs $12 per day to store. Each unit of product B requires 30 square feet of floor space and costs $8 per day to store. The total storage cost per day cannot exceed $12,400.

75. A company manufactures two types of hedge trimmers, a cordless model and a corded model. The corded trimmer requires 2 hours to assemble and the cordless trimmer requires 4 hours to assemble. The company has no more than 800 work hours to use in assembly each day, and the packing department can package no more than 300 trimmers per day.

76. A company manufactures two types of wood chippers, a standard model and a deluxe model. The deluxe model requires 3 hours to assemble and $\frac{1}{2}$ hour to paint. The standard model requires 2 hours to assemble and 1 hour to paint. The assembly line cannot be used for more than 24 hours each day. The painting line cannot be used for more than 8 hours each day.

Chapter Test

Solutions in English & Spanish and tutorial videos at AlgebraWithinReach.com

Take this test as you would take a test in class. After you are done, check your work against the answers in the back of the book.

1. Which is the solution of the system $x - 6y = -19$ and $4x - 5y = 0$: $(3, -2)$ or $(5, 4)$? Explain your reasoning.

In Exercises 2–4, determine the number of solutions of the system.

2. $\begin{cases} 3x + 4y = 16 \\ 3x - 4y = 8 \end{cases}$
3. $\begin{cases} x - 2y = -4 \\ x - 2y = 2 \end{cases}$
4. $\begin{cases} 2x - y = 5 \\ -4x + 2y = -10 \end{cases}$

In Exercises 5–8, solve the system of equations graphically.

5. $\begin{cases} -x + y = 1 \\ 5x - 2y = 4 \end{cases}$
6. $\begin{cases} 3x - y = 0 \\ -x + 2y = -10 \end{cases}$

7. $\begin{cases} 7x + 3y = 6 \\ 2x + y = 1 \end{cases}$
8. $\begin{cases} 2x = 3 \\ 2x + 3y = 9 \end{cases}$

In Exercises 9–11, solve the system of equations by the method of substitution.

9. $\begin{cases} x + 5y = 10 \\ -2x - 10y = 1 \end{cases}$
10. $\begin{cases} x + 3y = 15 \\ -2x + 5y = 14 \end{cases}$
11. $\begin{cases} y = 14 - 5x \\ x = y - 2 \end{cases}$

In Exercises 12–14, solve the system of equations by the method of elimination.

12. $\begin{cases} x + y = 8 \\ 2x - y = -2 \end{cases}$
13. $\begin{cases} 7x + 6y = 36 \\ 5x - 4y = 5 \end{cases}$
14. $\begin{cases} \frac{1}{2}x - \frac{1}{4}y = 1 \\ 4x + 5y = 22 \end{cases}$

In Exercises 15 and 16, sketch the graph of the system of linear inequalities.

15. $\begin{cases} 3x - y < 4 \\ x > 0 \\ y > 0 \end{cases}$
16. $\begin{cases} x + y < 6 \\ 2x + 3y > 9 \\ x \geq 0 \\ y \geq 0 \end{cases}$

17. Twenty liters of a 20% acid solution is obtained by mixing a 30% acid solution and a 5% acid solution. How many liters of each solution must be used to obtain the specified mixture?

18. You have decided to invest some money in certificates of deposit. You have up to $1000 to divide between two different programs. Program A requires a minimum deposit of $300 and program B requires a minimum deposit of $400. Write a system of linear inequalities that describes the various amounts that can be deposited in each certificate of deposit, and sketch the graph of the system.

9 Radical Expressions and Equations

- **9.1** Roots and Radicals
- **9.2** Simplifying Radicals
- **9.3** Operations with Radical Expressions
- **9.4** Radical Equations and Applications

MASTERY IS WITHIN REACH!

"When I was in my math courses, I had to get an early start on getting ready for my finals. I had to get a good grade in the last math course because I was competing with other students to get into the dental hygiene program. I studied with a friend from class a couple of weeks before the final. I stuck to studying for the final more since we did it together. We both did just fine on the final."

Mindy
Dental Hygiene

See page 423 for suggestions about preparing for the final exam.

9.1 Roots and Radicals

▶ Find the nth roots of real numbers.
▶ Use the radical symbol to denote the nth roots of numbers.
▶ Approximate the values of expressions involving square roots using a calculator.

Roots

> **Definition of nth Root of a Number**
>
> Let a and b be real numbers and let n be an integer such that $n \geq 2$. If
>
> $a = b^n$
>
> then b is an **nth root of a**. When $n = 2$, the root is a **square root**. When $n = 3$, the root is a **cube root**.

EXAMPLE 1 Finding the Square Roots of Numbers

a. The positive number 81 has two square roots, 9 and -9, because $9^2 = (9)(9) = 81$ and $(-9)^2 = (-9)(-9) = 81$.

b. The positive number $\frac{4}{9}$ has two square roots, $\frac{2}{3}$ and $-\frac{2}{3}$, because $\left(\frac{2}{3}\right)^2 = \left(\frac{2}{3}\right)\left(\frac{2}{3}\right) = \frac{4}{9}$ and $\left(-\frac{2}{3}\right)^2 = \left(-\frac{2}{3}\right)\left(-\frac{2}{3}\right) = \frac{4}{9}$.

c. The number 0 has only one square root: 0.

d. The negative number -4 has no square root because there is no real number that can be multiplied by itself to obtain -4.

Study Tip
Notice in Example 2 that there is only one (real) root associated with a cube root. In fact, when finding the nth root of a number, where n is odd, there is only one (real) root.

EXAMPLE 2 Finding the nth Roots of Numbers

a. The number 8 has one cube root, 2, because $2^3 = (2)(2)(2) = 8$.

b. The number -64 has one cube root, -4, because $(-4)^3 = (-4)(-4)(-4) = -64$.

c. The number 81 has two real fourth roots, 3 and -3, because $3^4 = (3)(3)(3)(3) = 81$ and $(-3)^4 = (-3)(-3)(-3)(-3) = 81$.

Exercises Within Reach®   Solutions in English & Spanish and tutorial videos at AlgebraWithinReach.com

Finding the Square Roots of a Number In Exercises 1–8, find the positive and negative square roots of the real number, if possible.

1. 16
2. 25
3. 36
4. 144
5. $\frac{9}{49}$
6. $\frac{4}{121}$
7. -16
8. -25

Finding the nth Root(s) of a Number In Exercises 9–16, find the nth root(s), if possible.

9. The cube root of 27
10. The cube root of 1
11. The cube root of -8
12. The cube root of -1000
13. The fourth root of 1
14. The fourth root of 625
15. The fourth root of -81
16. The fourth root of -625

Radicals

Principal nth Root of a Number

Let a be a real number that has at least one (real number) nth root. The **principal nth root of a** is the nth root that has the same sign as a, and it is denoted by the **radical**

$\sqrt[n]{a}.$ Principal nth root

The positive integer n is the **index** of the radical, and the number a is the **radicand**. When $n = 2$, omit the index and write $\sqrt{a}$ rather than $\sqrt[2]{a}$.

Study Tip

When used by itself, a radical *always* refers to the principal square root of the radicand. To denote a negative square root, place a negative sign in front of the radical. For instance, the principal square root of 36 is denoted by $\sqrt{36} = 6$, whereas the negative square root of 36 is denoted by $-\sqrt{36} = -6$.

EXAMPLE 3   Finding the Principal Square Root of a Number

a. The number 49 has two square roots, 7 and -7. The principal square root is the positive one.

$\sqrt{49} = 7$ Principal square root

b. The number $\frac{1}{4}$ has two square roots, $\frac{1}{2}$ and $-\frac{1}{2}$. The principal square root is the positive one.

$\sqrt{\frac{1}{4}} = \frac{1}{2}$ Principal square root

EXAMPLE 4   Finding the Principal nth Root of a Number

a. The number 125 has one cube root: 5.

$\sqrt[3]{125} = 5$ Cube root

b. The number -27 has one cube root: -3.

$\sqrt[3]{-27} = -3$ Cube root

c. The number 16 has two fourth roots: 2 and -2. The principal fourth root is the positive one.

$\sqrt[4]{16} = 2$ Principal fourth root

Exercises Within Reach®

Solutions in English & Spanish and tutorial videos at AlgebraWithinReach.com

Finding the Principal or Negative Square Root   In Exercises 17–28, **find** the indicated principal or negative square root, if possible.

17. $\sqrt{100}$
18. $\sqrt{64}$
19. $-\sqrt{100}$
20. $-\sqrt{64}$
21. $\sqrt{-100}$
22. $\sqrt{-64}$
23. $\sqrt{9}$
24. $-\sqrt{36}$
25. $-\sqrt{\frac{1}{9}}$
26. $\sqrt{\frac{36}{49}}$
27. $\sqrt{\frac{49}{64}}$
28. $-\sqrt{\frac{81}{400}}$

Finding the Principal nth Root   In Exercises 29–40, **find** the indicated principal nth root.

29. $\sqrt[3]{8}$
30. $\sqrt[3]{64}$
31. $\sqrt[3]{-125}$
32. $\sqrt[3]{-8}$
33. $\sqrt[3]{\frac{8}{27}}$
34. $\sqrt[3]{\frac{27}{64}}$
35. $\sqrt[4]{10,000}$
36. $\sqrt[4]{81}$
37. $\sqrt[4]{\frac{1}{16}}$
38. $\sqrt[4]{\frac{16}{81}}$
39. $\sqrt[5]{32}$
40. $\sqrt[5]{-243}$

Study Tip

Remember that a rational number is a real number that can be written as the ratio of two integers. So, when you say that $\sqrt{2}$ is irrational, you are saying that there is no fraction (with integer numerator and denominator) that can be multiplied by itself to obtain the number 2.

A real number is a **perfect square** if its square root is a rational number. Integers such as

1, 4, 9, 16, and 25   *Perfect squares*

are perfect squares because they have integer square roots. Rational numbers such as

$\frac{1}{4}$, $\frac{1}{9}$, and $\frac{9}{25}$ *Perfect squares*

are also perfect squares. The square roots of numbers that are not perfect squares are irrational numbers. For example, the numbers

$\sqrt{2}$, $\sqrt{3}$, $\sqrt{5}$, and $\sqrt{6}$ *Irrational numbers*

are all irrational.

EXAMPLE 5 Classifying Square Roots as Rational or Irrational

a. The negative square root of 100 is

$$-\sqrt{100} = -10$$ *Rational number*

which is a rational number. Remember that every integer is a rational number.

b. The principal square root of 11 is

$$\sqrt{11} \approx 3.3166\ldots$$ *Irrational number*

which is irrational, because 11 is not a perfect square.

c. The principal square root of $\frac{25}{16}$ is

$$\sqrt{\frac{25}{16}} = \frac{5}{4}$$ *Rational number*

which is a rational number.

Study Tip

You should note that

$-\sqrt{100} \neq \sqrt{-100}$.

The negative square root of 100 is -10 and the square root of -100 is not a real number.

d. Because $(0.6)(0.6) = 0.36$, the principal square root of 0.36 is

$$\sqrt{0.36} = 0.6$$ *Rational number*

which is the rational number $\frac{6}{10}$.

e. The principal square root of 3.2 is

$$\sqrt{3.2} = 1.788854\ldots$$ *Irrational number*

which is irrational, because 3.2 is not a perfect square.

Exercises Within Reach®   Solutions in English & Spanish and tutorial videos at AlgebraWithinReach.com

Classifying a Square Root In Exercises 41–54, classify the number as rational or irrational.

41. $\sqrt{15}$
42. $\sqrt{25}$
43. $-\sqrt{49}$
44. $-\sqrt{13}$
45. $-\sqrt{24}$
46. $\sqrt{42}$
47. $\sqrt{400}$
48. $-\sqrt{900}$
49. $-\sqrt{\frac{36}{25}}$
50. $\sqrt{\frac{4}{9}}$
51. $\sqrt{0.18}$
52. $-\sqrt{0.64}$
53. $\sqrt{1.21}$
54. $\sqrt{2.36}$

Radicals and Calculators

EXAMPLE 6 Decimal Approximations of Square Roots

	Square Root	Keystrokes	Calculator Display	Rounded Answer	
a.	$\sqrt{3}$	3 $\sqrt{x}$	1.732050808	1.7321	Scientific
	$\sqrt{3}$	$\sqrt{\ }$ 3) ENTER	1.732050808	1.7321	Graphing
b.	$-\sqrt{12}$	12 $\sqrt{x}$ +/−	−3.464101615	−3.464	Scientific
	$-\sqrt{12}$	(−) $\sqrt{\ }$ 12) ENTER	−3.464101615	−3.464	Graphing

When approximating a *negative square root* on a scientific calculator, be sure you press the square root key before pressing the change sign key. For instance, if you used the keystroke sequence 12 +/− $\sqrt{x}$, the calculator would think you were asking it to find $\sqrt{-12}$, and it would display an error message.

EXAMPLE 7 Approximating Expressions Involving Square Roots

a. The expression $2 + 3\sqrt{5}$ can be approximated using the following keystrokes.

Keystrokes	Display	
2 + 3 × 5 $\sqrt{x}$ =	8.708203933	Scientific
2 + 3 × $\sqrt{\ }$ 5) ENTER	8.708203932	Graphing

Rounded to two decimal places, the answer is $2 + 3\sqrt{5} \approx 8.71$.

b. The expression $(1 - 2\sqrt{3})/4$ can be approximated using the following keystrokes.

Keystrokes	Display	
1 − 2 × 3 $\sqrt{x}$ = ÷ 4 =	−0.616025404	Scientific
(1 − 2 × $\sqrt{\ }$ 3)) ÷ 4 ENTER	−.6160254038	Graphing

Rounded to two decimal places, the answer is

$$\frac{1 - 2\sqrt{3}}{4} \approx -0.62.$$

Exercises Within Reach®

Solutions in English & Spanish and tutorial videos at AlgebraWithinReach.com

Using a Calculator In Exercises 55−70, use a calculator to approximate the value of the expression, if possible. Round your answer to three decimal places.

55. $\sqrt{43}$ **56.** $\sqrt{38}$ **57.** $\sqrt{-12}$ **58.** $\sqrt{-8}$

59. $-\sqrt{137}$ **60.** $-\sqrt{150}$ **61.** $\sqrt{-632}$ **62.** $\sqrt{-326}$

63. $-\sqrt{517.8}$ **64.** $\sqrt{326.2}$ **65.** $\sqrt{\frac{95}{6}}$ **66.** $-\sqrt{\frac{43}{5}}$

67. $2 + 4\sqrt{7}$ **68.** $5 - 3\sqrt{6}$ **69.** $\dfrac{-4 - 3\sqrt{2}}{12}$ **70.** $\dfrac{-3 + 8\sqrt{24}}{2}$

EXAMPLE 8 Evaluating a Radical Expression

Evaluate each expression when $x = 2$, $y = 3$, and $z = 5$. Round your answer to two decimal places.

a. $\sqrt{x + y}$ **b.** $\sqrt{x^3 - z}$

SOLUTION

a. $\sqrt{x + y} = \sqrt{2 + 3}$
$= \sqrt{5} \approx 2.24$

b. $\sqrt{x^3 - z} = \sqrt{2^3 - 5}$
$= \sqrt{8 - 5}$
$= \sqrt{3} \approx 1.73$

EXAMPLE 9 Estimating Square Roots

Estimate each square root *without* using a calculator.

a. $\sqrt{200}$ **b.** $\sqrt{110}$

SOLUTION

a. The number $\sqrt{200}$ must lie between $\sqrt{196} = 14$ and $\sqrt{225} = 15$. Because 200 is closer to 196, you could estimate $\sqrt{200}$ to be about 14.1.

b. The number $\sqrt{110}$ must lie between $\sqrt{100} = 10$ and $\sqrt{121} = 11$. Because 110 is about halfway between 100 and 121, you could estimate $\sqrt{110}$ to be about 10.5.

Exercises Within Reach®

Solutions in English & Spanish and tutorial videos at AlgebraWithinReach.com

Evaluating a Radical Expression In Exercises 71–76, evaluate the expression for the given values of the variables. If it is not possible, state the reason. Round your answers to two decimal places, if necessary.

71. $\sqrt{5x - y}$

   (a) $x = 3, y = -1$
   (b) $x = -1, y = 4$

72. $\sqrt{x^2 + 3y}$

   (a) $x = -2, y = -1$
   (b) $x = 4, y = 0$

73. $\sqrt{z^2 - 4x}$

   (a) $x = 3, z = 1$
   (b) $x = -2, z = 4$

74. $\sqrt{-2xz}$

   (a) $x = 2, z = -5$
   (b) $x = 3, z = 8$

75. $\sqrt{b^2 - 4ac}$

   (a) $a = 4, b = 5, c = 1$
   (b) $a = -2, b = 7, c = 3$

76. $\sqrt{3a^3bc^2}$

   (a) $a = -1, b = 6, c = 2$
   (b) $a = 2, b = 2, c = -1$

Estimating a Square Root In Exercises 77–84, estimate the square root to one decimal place without using a calculator. Then check your estimate by using a calculator.

77. $\sqrt{55}$ **78.** $\sqrt{90}$ **79.** $\sqrt{70}$ **80.** $\sqrt{30}$

81. $\sqrt{130}$ **82.** $\sqrt{155}$ **83.** $\sqrt{300}$ **84.** $\sqrt{500}$

Application EXAMPLE 10 Estimating the Speed of a Car

In an emergency stop on dry pavement, the speed of a car v (in miles per hour) can be approximated by the equation $v = \sqrt{24l}$, where l is the length of the skid mark in feet. What is the estimated speed of a car that produces a skid mark of 110 feet?

SOLUTION

Because $l = 110$, you have

$$v = \sqrt{24(110)} = \sqrt{2640} \approx 51.4. \quad \text{Use a calculator.}$$

The estimated speed of the car is 51.4 miles per hour. Notice that the multiplication $24(110)$ is performed before you evaluate the square root.

Application EXAMPLE 11 Geometry: The Diagonal of a Rectangular Solid

The length of a diagonal of a rectangular solid of length l, width w, and height h is

$$\sqrt{l^2 + w^2 + h^2}.$$

Find the length of the diagonal of the solid shown in the figure. Round your answer to one decimal place.

SOLUTION

$$\sqrt{l^2 + w^2 + h^2} = \sqrt{12^2 + 9^2 + 3^2}$$
$$= \sqrt{144 + 81 + 9}$$
$$= \sqrt{234}$$
$$\approx 15.3$$

So, the length of the diagonal is about 15.3 inches.

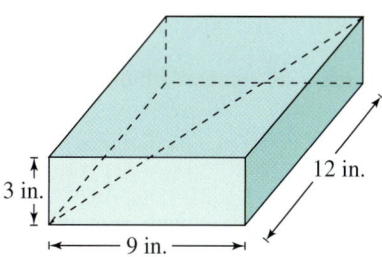

Exercises Within Reach®

Solutions in English & Spanish and tutorial videos at AlgebraWithinReach.com

85. Velocity of a Stream A stream of water moving at a rate of v feet per second can carry particles of size $0.03\sqrt{v}$ inches. Find the particle size that can be carried by a stream flowing at the rate of $\frac{3}{4}$ foot per second. Round your answer to three decimal places.

86. Geometry The length of a diagonal of a rectangular solid of length l, width w, and height h is $\sqrt{l^2 + w^2 + h^2}$. Find the length of the diagonal of the solid shown in the figure. Round your answer to one decimal place.

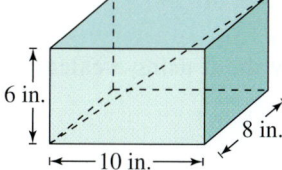

87. Geometry The square base of the Great Pyramid of Khufu in Egypt has an area of about 571,536 square feet. Find the dimensions of the base of the pyramid.

88. Walking Speed The maximum walking speed (in feet per second) for a person with leg length l (in feet) is $\sqrt{32l}$. Find the maximum walking speed for a person with legs 3 feet long. Round your answer to one decimal place.

Tumar/Shutterstock.com; Ziga Camernik/Shutterstock.com

Concept Summary: Using Radical Notation

What
An expression of the form $\sqrt[n]{a}$ represents the **principal nth root** of the number a.

EXAMPLE

Evaluate each expression.

(a) $\sqrt{\dfrac{1}{16}}$

(b) $\sqrt[3]{-1}$

How
To evaluate $\sqrt[n]{a}$, use the definition of the **nth root of a number** to find a number b with the same sign as a such that $b^n = a$.

EXAMPLE

(a) The positive number $\dfrac{1}{16}$ has two square roots because $\left(\dfrac{1}{4}\right)^2 = \dfrac{1}{16}$ and $\left(-\dfrac{1}{4}\right)^2 = \dfrac{1}{16}$, but the principal square root is positive: $\sqrt{\dfrac{1}{16}} = \dfrac{1}{4}$.

(b) There is only one **cube root** of -1. $(-1)^3 = -1$, so $\sqrt[3]{-1} = -1$.

Why
Although a number a may have more than one nth root, it is important to understand that the expression $\sqrt[n]{a}$ always represents the *principal nth root* of a.

Exercises Within Reach ®
Worked-out solutions to odd-numbered exercises at AlgebraWithinReach.com

Concept Summary Check

89. Vocabulary Identify the index and radicand in part (a) of the example above.

90. Finding the Principal Root What is the principal square root of $\dfrac{1}{16}$? What is the principal cube root of -1?

91. Understanding Roots Is $-\dfrac{1}{4}$ a square root of $\dfrac{1}{16}$? Explain.

92. Using Radical Notation Use radical notation to write an expression that represents the principal cube root of negative six.

Extra Practice

Square Root Notation In Exercises 93–98, **write** and **evaluate** the radical expressions that represent the positive and negative square roots of the number, if possible. (Do not use a calculator.)

93. $\dfrac{81}{121}$

94. $\dfrac{25}{196}$

95. 0.01

96. 0.36

97. -0.04

98. -0.49

Finding nth Roots In Exercises 99–104, **find** all of the nth roots. **Indicate** which root is the principal nth root.

99. The cube root of $\dfrac{8}{27}$

100. The cube root of $\dfrac{8}{125}$

101. The fifth root of -1

102. The fifth root of -32

103. The sixth root of 64

104. The sixth root of 729

Mental Math In Exercises 105–108, **find** the value of the expression without using a calculator.

105. $\sqrt[3]{27}\left(4 + \sqrt{25}\right)$

106. $\sqrt[4]{16}\left(10 - \sqrt{36}\right)$

107. $\dfrac{-5 + \sqrt{49}}{4}$

108. $\dfrac{12 - \sqrt{81}}{3}$

Geometry In Exercises 109 and 110, use the formula $h = \frac{\sqrt{3}}{2}s$ to approximate the height h of an equilateral triangle whose sides are of length s. Round your answer to two decimal places.

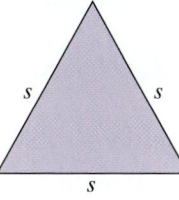

109. $s = 5$ inches   **110.** $s = 10$ centimeters

111. (a) Complete the table. Round your answers to two decimal places.

x	0	1	2	4	6	8
$\sqrt{x}$						

x	10	12	14	16	18	20
$\sqrt{x}$						

(b) Sketch a graph of the equation $y = \sqrt{x}$ by using the table generated in part (a).

112. (a) Complete the table. Round your answers to two decimal places.

x	0	$\frac{1}{2}$	1	$\frac{3}{2}$	2	$\frac{5}{2}$
$\sqrt{2x}$						

x	3	$\frac{7}{2}$	4	$\frac{9}{2}$	5
$\sqrt{2x}$					

(b) Sketch a graph of the equation $y = \sqrt{2x}$ by using the table generated in part (a).

113. *Exploration* Use a calculator to evaluate each expression.

(a) $(\sqrt{8.2})^2$ (b) $(\sqrt{142})^2$
(c) $(\sqrt{22})^2$ (d) $(\sqrt{850})^2$

114. *Think About It* Use the results of Exercise 113 to determine $(\sqrt{a})^2$, where a is a nonnegative real number.

Explaining Concepts

115. *Number Sense* Determine the values of x for which $\sqrt{x^2} \neq x$. Explain.

116. *Precision* Is it true that $\sqrt{2} = 1.414$? Explain.

117. *Number Sense* When does a real number have more than one real nth root?

118. (a) Find all possible last digits of integers that are perfect squares. (For instance, the last digit of 64 is 4.)

(b) Using the results of part (a), is it possible that 5,788,942,862 is a perfect square?

Cumulative Review

In Exercises 119–126, solve the rational equation.

119. $\frac{x}{4} + \frac{x}{4} = 2$

120. $\frac{4}{x-3} + \frac{6}{x-3} = -1$

121. $\frac{x+8}{3} = \frac{2x-5}{3}$

122. $\frac{x-1}{8} + \frac{x+5}{8} = \frac{x-6}{8}$

123. $\frac{x}{4} + \frac{x}{2} = 6$

124. $\frac{x}{2} = 10 - \frac{x}{3}$

125. $5 - \frac{1}{x} = 4$

126. $\frac{2}{3x} - \frac{1}{5} = \frac{4}{15x}$

In Exercises 127–130, sketch the graph of the system of linear inequalities.

127. $\begin{cases} x \leq 5 \\ y \geq 2 \end{cases}$ **128.** $\begin{cases} y \leq x \\ y > 3 \end{cases}$

129. $\begin{cases} x > 1 \\ y \leq 0 \\ x + y > 2 \end{cases}$ **130.** $\begin{cases} 3x - 2y \leq 6 \\ x \geq 0 \\ y \leq 0 \end{cases}$

9.2 Simplifying Radicals

▶ Simplify radical expressions involving constants.
▶ Simplify radical expressions involving variables.
▶ Simplify radical expressions by rationalizing the denominator.

Simplifying Radicals with Constant Factors

Product Rule for Radicals

Let a and b be real numbers, variables, or algebraic expressions. If the nth roots of a and b are real, then the following property is true.

$$\sqrt[n]{ab} = \sqrt[n]{a} \cdot \sqrt[n]{b} \qquad \text{Product Rule for Radicals}$$

EXAMPLE 1 Simplifying Radical Expressions

Simplify each radical expression.

a. $\sqrt{18} = \sqrt{9 \cdot 2}$ 9 is a perfect square factor of 18.
$= \sqrt{9} \cdot \sqrt{2}$ Product Rule for Radicals
$= 3\sqrt{2}$ Simplify.

b. $\sqrt{20} = \sqrt{4 \cdot 5}$ 4 is a perfect square factor of 20.
$= \sqrt{4} \cdot \sqrt{5}$ Product Rule for Radicals
$= 2\sqrt{5}$ Simplify.

c. $\sqrt[3]{54} = \sqrt[3]{27 \cdot 2}$ 27 is a perfect cube factor of 54.
$= \sqrt[3]{27} \cdot \sqrt[3]{2}$ Product Rule for Radicals
$= 3\sqrt[3]{2}$ Simplify.

d. $\sqrt[4]{80} = \sqrt[4]{16 \cdot 5}$ 16 is 2 raised to the fourth power.
$= \sqrt[4]{16} \cdot \sqrt[4]{5}$ Product Rule for Radicals
$= 2\sqrt[4]{5}$ Simplify.

Exercises Within Reach ®   Solutions in English & Spanish and tutorial videos at AlgebraWithinReach.com

Rewriting a Radical Expression In Exercises 1−4, write the expression as a product of two radicals and simplify.

1. $\sqrt{4 \cdot 15}$
2. $\sqrt{16 \cdot 3}$
3. $\sqrt{64 \cdot 11}$
4. $\sqrt{100 \cdot 7}$

Simplifying a Radical Expression In Exercises 5−12, simplify the radical expression.

5. $\sqrt{8}$
6. $\sqrt{12}$
7. $\sqrt{45}$
8. $\sqrt{28}$
9. $\sqrt[3]{48}$
10. $\sqrt[3]{40}$
11. $\sqrt[4]{48}$
12. $\sqrt[4]{32}$

EXAMPLE 2 Simplifying Radical Expressions

Simplify each radical expression.

a. $\sqrt{96}$

b. $\sqrt{108}$

SOLUTION

a. The greatest perfect square factor of 96 is 16.

$$\sqrt{96} = \sqrt{16 \cdot 6} \qquad \text{Factor 96.}$$
$$= \sqrt{16} \cdot \sqrt{6} \qquad \text{Product Rule for Radicals}$$
$$= 4\sqrt{6} \qquad \text{Simplify.}$$

b. The greatest perfect square factor of 108 is 36.

$$\sqrt{108} = \sqrt{36 \cdot 3} \qquad \text{Factor 108.}$$
$$= \sqrt{36} \cdot \sqrt{3} \qquad \text{Product Rule for Radicals}$$
$$= 6\sqrt{3} \qquad \text{Simplify.}$$

EXAMPLE 3 Simplifying Radical Expressions

Simplify each radical expression.

a. $\sqrt[3]{-192}$

b. $\sqrt[4]{512}$

SOLUTION

a. In the radical expression $\sqrt[3]{-192}$, the greatest perfect cube factor is -64.

$$\sqrt[3]{-192} = \sqrt[3]{-64 \cdot 3} \qquad \text{Factor } -192.$$
$$= \sqrt[3]{-64} \cdot \sqrt[3]{3} \qquad \text{Product Rule for Radicals}$$
$$= -4\sqrt[3]{3} \qquad \text{Simplify.}$$

b. In the radical expression $\sqrt[4]{512}$, the greatest perfect fourth-powered factor is 256.

$$\sqrt[4]{512} = \sqrt[4]{256 \cdot 2} \qquad \text{Factor 512.}$$
$$= \sqrt[4]{256} \cdot \sqrt[4]{2} \qquad \text{Product Rule for Radicals}$$
$$= 4\sqrt[4]{2} \qquad \text{Simplify.}$$

Exercises Within Reach ®

Solutions in English & Spanish and tutorial videos at AlgebraWithinReach.com

Simplifying a Radical Expression In Exercises 13–22, simplify the radical expression.

13. $\sqrt{72}$

14. $\sqrt{128}$

15. $\sqrt{180}$

16. $\sqrt{432}$

17. $\sqrt[3]{-24}$

18. $\sqrt[3]{-16}$

19. $\sqrt[3]{-128}$

20. $\sqrt[3]{-256}$

21. $\sqrt[4]{162}$

22. $\sqrt[4]{176}$

Simplifying Radicals with Variable Factors

Simplifying radicals that involve *variable* radicands is trickier than simplifying radicals involving only constant radicands. The reason for this can be seen by considering the radical $\sqrt{x^2}$. At first glance, it would appear that this radical simplifies as x. However, doing so overlooks the possibility that x might be negative. For instance, consider the following.

If $x = 2$, then $\sqrt{x^2} = \sqrt{2^2} = \sqrt{4} = 2 = x$.

If $x = -2$, then $\sqrt{x^2} = \sqrt{(-2)^2} = \sqrt{4} = 2 = |x|$.

In both of these cases, you can conclude that $\sqrt{x^2} = |x|$, but without knowing whether x is positive, zero, or negative, you *cannot* conclude that $\sqrt{x^2} = x$.

The Square Root of x^2

If x is a real number, then

$$\sqrt{x^2} = |x|. \qquad \text{Restricted by absolute value signs}$$

For the special case in which you know that x is a *nonnegative* real number, you can write $\sqrt{x^2} = x$.

EXAMPLE 4 Simplifying Radical Expressions

Simplify each radical expression.

a. $\sqrt{25x^2}$ **b.** $\sqrt{18a^2}, a \geq 0$ **c.** $\sqrt{x^4}$

SOLUTION

a. $\sqrt{25x^2} = \sqrt{5^2 \cdot x^2}$ Factor radicand.

$\phantom{\sqrt{25x^2}} = \sqrt{5^2} \cdot \sqrt{x^2}$ Product Rule for Radicals

$\phantom{\sqrt{25x^2}} = 5|x|$ $\sqrt{x^2} = |x|$.

b. $\sqrt{18a^2} = \sqrt{3^2 \cdot 2 \cdot a^2}$ Factor radicand.

$\phantom{\sqrt{18a^2}} = \sqrt{3^2} \cdot \sqrt{2} \cdot \sqrt{a^2}$ Product Rule for Radicals

$\phantom{\sqrt{18a^2}} = 3a\sqrt{2}$ $\sqrt{a^2} = a, a \geq 0$.

c. This problem is different. Note that the absolute value signs are not necessary in the final simplified version because you know that x^2 cannot be negative.

$\sqrt{x^4} = \sqrt{(x^2)^2}$

$\phantom{\sqrt{x^4}} = |x^2|$

$\phantom{\sqrt{x^4}} = x^2$

Study Tip

When writing a product involving a radical, write the radical last to avoid confusion. For instance, if you write the product of x and $\sqrt{2}$ as $\sqrt{2}x$, it might be read as $\sqrt{2x}$.

Exercises Within Reach®

Solutions in English & Spanish and tutorial videos at AlgebraWithinReach.com

Simplifying a Radical Expression In Exercises 23–30, simplify the radical expression. Use absolute value signs, if appropriate.

23. $\sqrt{4x^2}$ 24. $\sqrt{9x^4}$ 25. $\sqrt{84x^2}$ 26. $\sqrt{68y^2}$

27. $\sqrt{x^6}$ 28. $\sqrt{y^8}$ 29. $\sqrt{20x^6}$ 30. $\sqrt{28x^8}$

Section 9.2   Simplifying Radicals   415

In some cases, absolute value signs are not necessary. For instance, you do not need to use absolute value signs in simplifying $\sqrt{x^3} = x\sqrt{x}$ because the original radical is undefined when x is negative.

EXAMPLE 5 **Simplifying Square Roots Involving Odd Powers**

Simplify each radical expression.

a. $\sqrt{16x^3}$ b. $\sqrt{9a^5}$

SOLUTION

a. $\sqrt{16x^3} = \sqrt{4^2 \cdot x^2 \cdot x}$ Factor radicand.
   $= \sqrt{4^2} \cdot \sqrt{x^2} \cdot \sqrt{x}$ Product Rule for Radicals
   $= 4x\sqrt{x}$ Simplify.

b. $\sqrt{9a^5} = \sqrt{3^2 \cdot a^4 \cdot a}$ Factor radicand.
   $= \sqrt{3^2} \cdot \sqrt{(a^2)^2} \cdot \sqrt{a}$ Product Rule for Radicals
   $= 3a^2\sqrt{a}$ Simplify.

EXAMPLE 6 **Simplifying Radicals Involving nth Roots**

Simplify each radical expression.

a. $\sqrt[3]{54a^3}$ b. $\sqrt[5]{x^6}$ c. $\sqrt[4]{32x^7}$

SOLUTION

a. $\sqrt[3]{54a^3} = \sqrt[3]{27 \cdot 2 \cdot a^3}$ Factor radicand.
   $= \sqrt[3]{3^3} \cdot \sqrt[3]{2} \cdot \sqrt[3]{a^3}$ Product Rule for Radicals
   $= 3a\sqrt[3]{2}$ Simplify.

b. $\sqrt[5]{x^6} = \sqrt[5]{x^5 \cdot x}$ Factor radicand.
   $= \sqrt[5]{x^5} \cdot \sqrt[5]{x}$ Product Rule for Radicals
   $= x\sqrt[5]{x}$ Simplify.

c. $\sqrt[4]{32x^7} = \sqrt[4]{16 \cdot 2 \cdot x^4 \cdot x^3}$ Factor radicand.
   $= \sqrt[4]{2^4} \cdot \sqrt[4]{2} \cdot \sqrt[4]{x^4} \cdot \sqrt[4]{x^3}$ Product Rule for Radicals
   $= 2x\sqrt[4]{2x^3}$ Simplify.

Exercises Within Reach ®   Solutions in English & Spanish and tutorial videos at AlgebraWithinReach.com

Simplifying a Radical Expression In Exercises 31–42, simplify the radical expression.
Use absolute value signs, if appropriate.

31. $\sqrt{64x^3}$

32. $\sqrt{49z^5}$

33. $\sqrt{x^2y^3}$

34. $\sqrt{a^5b^4}$

35. $\sqrt{180x^5y^8}$

36. $\sqrt{128u^4v^7}$

37. $\sqrt[3]{27a^4}$

38. $\sqrt[3]{8x^5}$

39. $\sqrt[4]{t^7}$

40. $\sqrt[4]{x^5}$

41. $\sqrt[4]{16y^5}$

42. $\sqrt[4]{80t^7}$

Rationalizing Denominators

> **Quotient Rule for Radicals**
>
> Let a and b be real numbers, variables, or algebraic expressions. If the nth roots of a and b are real, then the following property is true.
>
> $$\sqrt[n]{\frac{a}{b}} = \frac{\sqrt[n]{a}}{\sqrt[n]{b}}, \quad b \neq 0 \qquad \text{Quotient Rule for Radicals}$$

EXAMPLE 7 Simplifying a Radical Expression Involving a Fraction

Simplify $\sqrt{\frac{21}{4}}$.

SOLUTION

$$\sqrt{\frac{21}{4}} = \frac{\sqrt{21}}{\sqrt{4}} \qquad \text{Quotient Rule for Radicals}$$

$$= \frac{\sqrt{21}}{2} \qquad \text{Simplify.}$$

EXAMPLE 8 Simplifying a Radical Expression Involving a Fraction

Simplify $\sqrt{\frac{3x^2}{12y^4}}$.

SOLUTION

$$\sqrt{\frac{3x^2}{12y^4}} = \sqrt{\frac{x^2}{4y^4}} \qquad \text{Simplify fraction.}$$

$$= \frac{\sqrt{x^2}}{\sqrt{4y^4}} \qquad \text{Quotient Rule for Radicals}$$

$$= \frac{|x|}{2y^2} \qquad \text{Simplify.}$$

Study Tip

In Examples 7 and 8, note that the denominators are free of radicals. This simplifying process is called **rationalizing the denominator**.

Exercises Within Reach®

Solutions in English & Spanish and tutorial videos at AlgebraWithinReach.com

Simplifying a Radical Expression In Exercises 43–54, simplify the radical expression.

43. $\frac{\sqrt{48}}{\sqrt{64}}$

44. $\frac{\sqrt{84}}{\sqrt{36}}$

45. $\sqrt{\frac{13}{49}}$

46. $\sqrt{\frac{7}{81}}$

47. $\sqrt{\frac{66}{88}}$

48. $\sqrt{\frac{24}{36}}$

49. $\sqrt{\frac{3x^2}{27}}$

50. $\sqrt{\frac{5x^4}{20}}$

51. $\sqrt{\frac{12x^2}{25}}$

52. $\sqrt{\frac{18u^4}{144}}$

53. $\sqrt{\frac{x^6}{16y^2}}$

54. $\sqrt{\frac{u^4}{36v^4}}$

Section 9.2   Simplifying Radicals   417

EXAMPLE 9 Rationalizing Denominators

Rationalize the denominator in each expression.

a. $\sqrt{\dfrac{13}{3}}$ b. $\sqrt{\dfrac{7}{20}}$ c. $\dfrac{12}{\sqrt{18}}$

SOLUTION

a. $\sqrt{\dfrac{13}{3}} = \dfrac{\sqrt{13}}{\sqrt{3}} \cdot \dfrac{\sqrt{3}}{\sqrt{3}} = \dfrac{\sqrt{39}}{\sqrt{3^2}} = \dfrac{\sqrt{39}}{3}$ Multiply by $\sqrt{3}/\sqrt{3}$ to create a perfect square in the denominator.

b. $\sqrt{\dfrac{7}{20}} = \dfrac{\sqrt{7}}{\sqrt{20}} \cdot \dfrac{\sqrt{5}}{\sqrt{5}} = \dfrac{\sqrt{35}}{\sqrt{10^2}} = \dfrac{\sqrt{35}}{10}$ Multiply by $\sqrt{5}/\sqrt{5}$ to create a perfect square in the denominator.

c. $\dfrac{12}{\sqrt{18}} = \dfrac{12}{\sqrt{18}} \cdot \dfrac{\sqrt{2}}{\sqrt{2}} = \dfrac{12\sqrt{2}}{\sqrt{6^2}} = \dfrac{12\sqrt{2}}{6} = 2\sqrt{2}$

Study Tip

When rationalizing a denominator, remember that for square roots, you want a perfect square in the denominator, for cube roots, you want a perfect cube, and so on. For instance, to find the radical factor needed to create a perfect square in the denominator in Example 9(b), you can write the prime factorization of 20.

$20 = 2 \cdot 2 \cdot 5$
$ = 2^2 \cdot 5$

From its prime factorization you can see that 2^2 is a square root factor of 20, and you need one more factor of 5 to create a perfect square in the denominator.

Simplifying Radical Expressions

A radical expression is said to be in simplest form when all three of the statements below are true.

1. All possible nth-powered factors have been removed from each radical.
2. No radical contains a fraction.
3. No denominator of a fraction contains a radical.

EXAMPLE 10 Rationalizing Denominators

a. $\sqrt{\dfrac{3}{a}} = \dfrac{\sqrt{3}}{\sqrt{a}} \cdot \dfrac{\sqrt{a}}{\sqrt{a}} = \dfrac{\sqrt{3a}}{\sqrt{a^2}} = \dfrac{\sqrt{3a}}{a}$

b. $\sqrt{\dfrac{1}{4x^3}} = \dfrac{\sqrt{1}}{\sqrt{4x^3}} \cdot \dfrac{\sqrt{x}}{\sqrt{x}} = \dfrac{\sqrt{x}}{\sqrt{4x^4}} = \dfrac{\sqrt{x}}{2x^2}$

c. $\sqrt{\dfrac{10x}{8y^5}} - \sqrt{\dfrac{5x}{4y^5}} - \dfrac{\sqrt{5x}}{\sqrt{4y^5}} \cdot \dfrac{\sqrt{y}}{\sqrt{y}} - \dfrac{\sqrt{5xy}}{\sqrt{4y^6}} - \dfrac{\sqrt{5xy}}{2y^3}$

Exercises Within Reach®   Solutions in English & Spanish and tutorial videos at AlgebraWithinReach.com

Rationalizing a Denominator In Exercises 55–64, rationalize the denominator of the expression and simplify. (Assume all variables are positive.)

55. $\sqrt{\dfrac{1}{3}}$ 56. $\sqrt{\dfrac{1}{5}}$ 57. $\dfrac{1}{\sqrt{7}}$ 58. $\dfrac{1}{\sqrt{10}}$

59. $\sqrt{\dfrac{11}{8}}$ 60. $\sqrt{\dfrac{7}{18}}$ 61. $\sqrt{\dfrac{5}{x}}$

62. $\sqrt{\dfrac{3}{a}}$ 63. $\sqrt{\dfrac{3}{16x^5}}$ 64. $\sqrt{\dfrac{6}{25u^3}}$

When the denominator of an expression contains a cube root, multiply the numerator and denominator by a factor that will create a perfect cube in the denominator.

EXAMPLE 11 Rationalizing Denominators with Cube Roots

Rationalize the denominator in each expression.

a. $\dfrac{5}{\sqrt[3]{9}}$

b. $\sqrt[3]{\dfrac{5}{x}}$

SOLUTION

a. $\dfrac{5}{\sqrt[3]{9}} = \dfrac{5}{\sqrt[3]{9}} \cdot \dfrac{\sqrt[3]{3}}{\sqrt[3]{3}}$ Multiply by $\sqrt[3]{3}/\sqrt[3]{3}$ to create a perfect cube in the denominator.

$= \dfrac{5\sqrt[3]{3}}{\sqrt[3]{27}}$ Multiply.

$= \dfrac{5\sqrt[3]{3}}{3}$ Simplify.

b. $\sqrt[3]{\dfrac{5}{x}} = \dfrac{\sqrt[3]{5}}{\sqrt[3]{x}}$ Quotient Rule for Radicals

$= \dfrac{\sqrt[3]{5}}{\sqrt[3]{x}} \cdot \dfrac{\sqrt[3]{x^2}}{\sqrt[3]{x^2}}$ Multiply by $\sqrt[3]{x^2}/\sqrt[3]{x^2}$ to create a perfect cube in the denominator.

$= \dfrac{\sqrt[3]{5x^2}}{\sqrt[3]{x^3}}$ Multiply.

$= \dfrac{\sqrt[3]{5x^2}}{x}$ Simplify.

Exercises Within Reach®

Solutions in English & Spanish and tutorial videos at AlgebraWithinReach.com

Rationalizing a Denominator In Exercises 65–74, rationalize the denominator of the expression and simplify. (Assume all variables are positive.)

65. $\dfrac{4}{\sqrt[3]{9}}$

66. $\dfrac{9}{\sqrt[3]{4}}$

67. $\dfrac{7}{\sqrt[3]{3}}$

68. $\dfrac{5}{\sqrt[3]{2}}$

69. $\dfrac{18}{\sqrt[3]{16}}$

70. $\dfrac{27}{\sqrt[3]{24}}$

71. $\sqrt[3]{\dfrac{1}{x^2}}$

72. $\sqrt[3]{\dfrac{3}{x}}$

73. $\sqrt[3]{\dfrac{1}{8y^2}}$

74. $\sqrt[3]{\dfrac{1}{27a}}$

Section 9.2 Simplifying Radicals 419

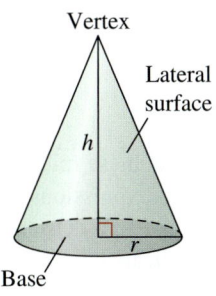

Radicals are also used in many geometric applications, such as applications involving right circular cones. The lateral surface of a cone consists of all segments that connect the vertex with points on the edge of the base, as shown in the figure. The lateral surface area S of a right circular cone is given by

$$S = \pi r \sqrt{r^2 + h^2}$$

where r is the radius of the base of the cone and h is the height.

Application **EXAMPLE 12** **Geometry: Lateral Surface Area**

The radius of the base of a traffic cone is 14 centimeters, and the height of the cone is 34 centimeters. What is the lateral surface area of the traffic cone?

SOLUTION

You can use the formula for the lateral surface area of a cone as follows.

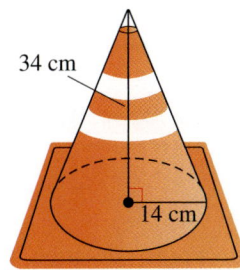

$S = \pi r \sqrt{r^2 + h^2}$	Formula for lateral surface area
$= \pi(14)\sqrt{14^2 + 34^2}$	Substitute 14 for r and 34 for h.
$= 14\pi\sqrt{1352}$	Simplify.
$= 14\pi\sqrt{676 \cdot 2}$	676 is a perfect square factor of 1352.
$= 14\pi \cdot 26\sqrt{2}$	Simplify.
$= 364\pi\sqrt{2}$	Simplify.
≈ 1617.2	Use a calculator.

So, the lateral surface area of the cone is about 1617.2 square centimeters.

Exercises Within Reach ®

Solutions in English & Spanish and tutorial videos at AlgebraWithinReach.com

Geometry In Exercises 75–78, find the lateral surface area of the cone. Round your answer to one decimal place.

75.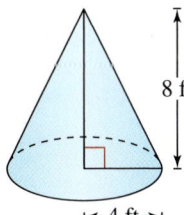

76.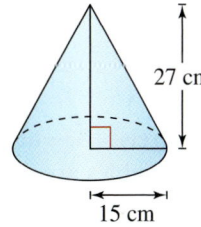

77.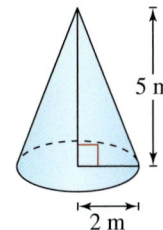

78.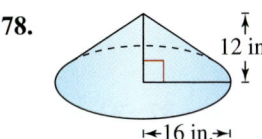

> **Concept Summary: Simplifying Radical Expressions**
>
> **What**
> You already knew how to simplify algebraic expressions. Now you can simplify **radical expressions** involving constants.
>
> **EXAMPLE**
> Simplify $\sqrt{\dfrac{54}{16}}$.
>
> **How**
> Use the Product and Quotient Rules for Radicals to simplify these types of expressions.
>
> **EXAMPLE**
> $$\sqrt{\dfrac{54}{16}} = \dfrac{\sqrt{54}}{\sqrt{16}} \quad \text{Quotient Rule for Radicals}$$
> $$= \dfrac{\sqrt{9} \cdot \sqrt{6}}{\sqrt{16}} \quad \text{Product Rule for Radicals}$$
> $$= \dfrac{3\sqrt{6}}{4} \quad \text{Simplify.}$$
>
> **Why**
> You can also use these rules to simplify radical expressions involving variables. It is trickier than simplifying radicals involving only constants. Just remember, $\sqrt{x^2} = |x|$.

Exercises Within Reach ® Worked-out solutions to odd-numbered exercises at AlgebraWithinReach.com

Concept Summary Check

79. Stating a Rule State the Product Rule for Radicals in words and give an example.

80. Stating a Rule State the Quotient Rule for Radicals in words and give an example.

81. Writing In your own words, describe the three conditions that must be true for a radical expression to be in simplest form.

82. Precision Explain why the Product Rule for Radicals cannot be applied to the expression $\sqrt{-8} \cdot \sqrt{-2}$.

Extra Practice

Rationalizing a Denominator In Exercises 83–86, **rationalize** the denominator of the expression and simplify. (Assume all variables are positive.)

83. $\dfrac{\sqrt{2t}}{\sqrt{8r}}$

84. $\dfrac{\sqrt{2x}}{\sqrt{50y}}$

85. $\dfrac{\sqrt{12x^3}}{\sqrt{3y}}$

86. $\dfrac{\sqrt{20x^2}}{\sqrt{5y^2}}$

Comparing Real Numbers In Exercises 87–92, **place** the correct inequality symbol (< or >) between the real numbers. Do not use a calculator.

87. $\sqrt{160}$ ☐ 12

88. $\sqrt{200}$ ☐ 15

89. $4\sqrt{2}$ ☐ 5

90. $5\sqrt{3}$ ☐ 64

91. $5\sqrt{6}$ ☐ $6\sqrt{5}$

92. $15\sqrt{7}$ ☐ $12\sqrt{10}$

Geometry In Exercises 93–96, find the area of the figure. Round your answer to two decimal places.

93.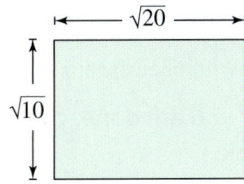
94.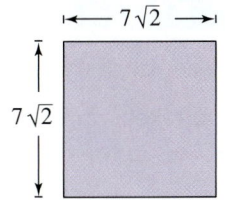
95.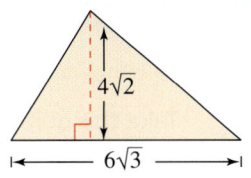
96.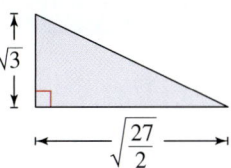

97. **Astronomy** The orbital period of a planet is the time that it takes the planet to travel around the Sun. You can find the orbital period P (in Earth years) using the formula $P = \sqrt{d^3}$, where d is the average distance (in astronomical units, abbreviated AU) of the planet from the Sun.

    (a) Simplify the formula.

    (b) Saturn's average distance from the Sun is about 9.54 AU. What is Saturn's orbital period? Round your answer to one decimal place.

    (c) Venus's average distance from the Sun is about 0.72 AU. What is Venus's orbital period? Round your answer to one decimal place.

98. **Distance to the Horizon** The distance d (in miles) from a person to the horizon is given by the formula
$$d = \sqrt{\frac{3h}{2}}$$
where h is the person's eye level (in feet) above sea level.

    (a) Rationalize the denominator and simplify the formula.

    (b) A person's eye level is 6 feet above sea level. How far is the person from the horizon?

    (c) A person's eye level is 96 feet above sea level. How far is the person from the horizon?

Explaining Concepts

99. **Precision** Determine which of the statements is true for any real number x. Give an example to show any false statements.

    (a) $\sqrt{x^2} = x$ (b) $\sqrt{x^2} = |x|$

    (c) $\sqrt[3]{x^3} = |x|$ (d) $\sqrt[3]{x^3} = x$

True or False? In Exercises 100–103, **decide** whether the statement is true or false. **Justify** your answer.

100. $\sqrt{3x^2} = x\sqrt{3}$

101. $\dfrac{\sqrt{50}}{\sqrt{2}} = 25$

102. $\sqrt{x^2 + 16} = x + 4$

103. $\sqrt[3]{72x^4} = 2|x|\sqrt[3]{9x}$

104. **Writing** Consider the expression $\sqrt{2^x}$, where x is a positive integer. For what values of x will the expression contain a radical when simplified? For what values of x will the expression contain no radical when simplified?

Cumulative Review

In Exercises 105 and 106, simplify the complex fraction.

105. $\dfrac{\left(\dfrac{x}{6}\right)}{\left(2 - \dfrac{5}{x}\right)}$

106. $\dfrac{\left(1 + \dfrac{7}{x}\right)}{\left(\dfrac{x}{3}\right)}$

In Exercises 107–114, find the indicated principal or negative nth root, if possible. (Do not use a calculator.)

107. $\sqrt{81}$

108. $-\sqrt{121}$

109. $\sqrt{-49}$

110. $\sqrt{\dfrac{36}{225}}$

111. $\sqrt[3]{-64}$

112. $\sqrt[3]{216}$

113. $-\sqrt[4]{16}$

114. $\sqrt[4]{256}$

Mid-Chapter Quiz: Sections 9.1–9.2

Solutions in English & Spanish and tutorial videos at AlgebraWithinReach.com

Take this quiz as you would take a quiz in class. After you are done, check your work against the answers in the back of the book.

In Exercises 1–6, find the indicated principal or negative nth root, if possible. (Do not use a calculator.)

1. $\sqrt{121}$
2. $-\sqrt{0.25}$
3. $\sqrt[3]{-8}$
4. $\sqrt[4]{-16}$
5. $\sqrt{-\frac{1}{49}}$
6. $-\sqrt{1.44}$

In Exercises 7–9, classify the number as rational or irrational.

7. $\sqrt{5}$
8. $\sqrt{\frac{3}{4}}$
9. $\sqrt{900}$

In Exercises 10–12, use a calculator to approximate the value of the expression. Round your answer to three decimal places.

10. $\sqrt{61}$
11. $13 + \sqrt{27.4}$
12. $\dfrac{7 - \sqrt{16}}{5}$

In Exercises 13–15, rewrite the expression using only one radical.

13. $\sqrt{5} \cdot \sqrt{19}$
14. $\sqrt{5x} \cdot \sqrt{6y}$
15. $\dfrac{\sqrt{42}}{\sqrt{6}}$

In Exercises 16–24, simplify the radical expression.

16. $\sqrt{50}$
17. $\sqrt{72x^2}$
18. $\sqrt{45b^8}$
19. $\sqrt[3]{64x^3}$
20. $\sqrt[4]{x^9}$
21. $\sqrt{18u^5v^2}$
22. $\dfrac{\sqrt{576}}{\sqrt{18}}$
23. $\dfrac{\sqrt{63x^2}}{\sqrt{64}}$
24. $\sqrt{\dfrac{90b^4}{2}}$

In Exercises 25–27, rationalize the denominator and simplify. (Assume all variables are positive.)

25. $\sqrt{\dfrac{3}{2}}$
26. $\dfrac{2}{\sqrt{12}}$
27. $\dfrac{\sqrt[3]{x}}{\sqrt[3]{27x^4}}$

Applications

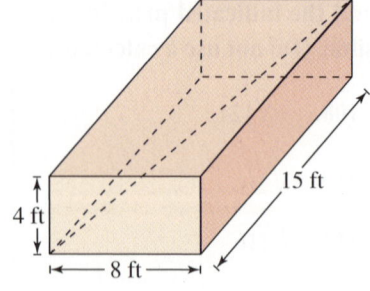

28. The length of a diagonal of a rectangular solid of length l, width w, and height h is $\sqrt{l^2 + w^2 + h^2}$. Find the length of the diagonal of the solid shown in the figure. Round your answer to one decimal place.

29. A square room has 361 square feet of floor space. Find the dimensions of the room.

Study Skills in Action

Preparing for the Final Exam

At the end of the semester, most students are inundated with projects, papers, and tests. Instructors may speed up the pace in lectures to get through all the material. If something unexpected is going to happen to a student, it often happens during this time.

Getting through the last couple of weeks of a math course can be challenging. This is why it is important to plan your review time for the final exam at least three weeks before the exam.

These students are planning how they will study for the final exam.

Smart Study Strategy

Form a Final Exam Study Group

1. ▶ **Form a study group of three or four students several weeks before the final exam.** The intent of this group is to review what you have already learned while continuing to learn new material.

2. ▶ **Find out what material you must know for the final, even if the instructor has not yet covered it.** As a group, meet with the instructor outside of class. A group is likely to receive more attention and can ask more questions.

3. ▶ **Ask for or create a practice final and have the instructor look at it.** Make sure the problems are at an appropriate level of difficulty. Look for sample problems in old tests and in cumulative tests in the textbook. Review what the textbook and your notes say as you look for problems. This will refresh your memory.

4. ▶ **Have each group member take the practice final exam.** Then have each member identify what he or she needs to study. Make sure you can complete the problems with the speed and accuracy that are necessary to complete the real final exam.

5. ▶ **Decide when the group is going to meet during the next couple of weeks and what you will cover during each session.** The tutoring or learning center on campus is an ideal setting in which to meet. Many libraries have small study rooms that study groups can reserve. Set up several study times for each week. If you live at home, make sure your family knows that this is a busy time.

6. ▶ **During the study group sessions, make sure you stay on track.** Prepare for each study session by knowing what material in the textbook you are going to cover and having the class notes for that material. When you have questions, assign a group member to go to the instructor for answers. Then this member can relay the correct information to the other group members. Save socializing for after the final exam.

9.3 Operations with Radical Expressions

- Add and subtract like radical expressions.
- Multiply radical expressions.
- Simplify quotients involving radicals by rationalizing the denominators.

Adding and Subtracting Radical Expressions

EXAMPLE 1 **Combining Radical Expressions**

a. $\sqrt{7} + 5\sqrt{7} = (1 + 5)\sqrt{7}$ Distributive Property
 $= 6\sqrt{7}$ Simplify.

b. $6\sqrt{6} - \sqrt{3} - 5\sqrt{6} + 2\sqrt{3}$
 $= (6\sqrt{6} - 5\sqrt{6}) + (-\sqrt{3} + 2\sqrt{3})$ Group like terms.
 $= (6 - 5)\sqrt{6} + (-1 + 2)\sqrt{3}$ Distributive Property
 $= \sqrt{6} + \sqrt{3}$ Simplify.

c. $3\sqrt[4]{5} + 7\sqrt[4]{5} = (3 + 7)\sqrt[4]{5}$ Distributive Property
 $= 10\sqrt[4]{5}$ Simplify.

d. $\sqrt[5]{2} + 6\sqrt[5]{2} - 2\sqrt[5]{2} = (1 + 6 - 2)\sqrt[5]{2}$ Distributive Property
 $= 5\sqrt[5]{2}$ Simplify.

e. $3 + 3\sqrt{x} - \sqrt{x} + 2$
 $= (3 + 2) + (3\sqrt{x} - \sqrt{x})$ Group like terms.
 $= (3 + 2) + (3 - 1)\sqrt{x}$ Distributive Property
 $= 5 + 2\sqrt{x}$ Simplify.

f. $2\sqrt[3]{2} + 2\sqrt[3]{y^2} - 4\sqrt[3]{2} + 5\sqrt[3]{y^2}$
 $= (2\sqrt[3]{2} - 4\sqrt[3]{2}) + (2\sqrt[3]{y^2} + 5\sqrt[3]{y^2})$ Group like terms.
 $= (2 - 4)\sqrt[3]{2} + (2 + 5)\sqrt[3]{y^2}$ Distributive Property
 $= -2\sqrt[3]{2} + 7\sqrt[3]{y^2}$ Simplify.

Study Tip

It is important to realize that the expression $\sqrt{a} + \sqrt{b}$ is not equal to $\sqrt{a + b}$. For instance, in Example 1(b), $\sqrt{6} + \sqrt{3}$ does not equal $\sqrt{9}$, or 3. But remember, because $\sqrt{6}$ and $\sqrt{3}$ are not like radicals, $\sqrt{6} + \sqrt{3}$ cannot be simplified further.

Exercises Within Reach®

Solutions in English & Spanish and tutorial videos at AlgebraWithinReach.com

Combining Radical Expressions In Exercises 1–18, simplify the expression.

1. $10\sqrt{11} + 8\sqrt{11}$
2. $\sqrt{15} + 7\sqrt{15}$
3. $3\sqrt{5} - \sqrt{5}$
4. $5\sqrt{6} - 10\sqrt{6}$
5. $\sqrt{3} - 5\sqrt{7} - 12\sqrt{3}$
6. $2\sqrt{14} - 3\sqrt{5} + 8\sqrt{14}$
7. $9\sqrt{17} + 7\sqrt{2} - 11\sqrt{17} + \sqrt{2}$
8. $3\sqrt{5} - 4\sqrt{11} + 5\sqrt{5} + 12\sqrt{11}$
9. $4\sqrt[3]{5} + 2\sqrt[3]{5}$
10. $6\sqrt[3]{3} - 5\sqrt[3]{3}$
11. $4\sqrt[4]{8} - 9\sqrt[4]{8}$
12. $7\sqrt[5]{4} + 3\sqrt[5]{4}$
13. $9\sqrt[3]{7} + 3\sqrt[3]{7} - 4\sqrt[3]{7}$
14. $5\sqrt[5]{3} + 3\sqrt[5]{3} - \sqrt[5]{3}$
15. $4\sqrt{u} - 3 + \sqrt{u} + 8$
16. $12\sqrt{v} + 6 - 5\sqrt{v} - 9$
17. $4\sqrt[3]{6} - 2\sqrt[3]{x^2} + 3\sqrt[3]{6} + 6\sqrt[3]{x^2}$
18. $5\sqrt[4]{9} + 2\sqrt[4]{a^3} - 3\sqrt[4]{9} + 3\sqrt[4]{a^3}$

Section 9.3   Operations with Radical Expressions   425

EXAMPLE 2   **Simplifying a Radical Expression**

Simplify the expression.

$2\sqrt{72} - 2\sqrt{32}$

SOLUTION

$2\sqrt{72} - 2\sqrt{32} = 2\sqrt{36 \cdot 2} - 2\sqrt{16 \cdot 2}$	Factor radicands.
$= 2 \cdot 6\sqrt{2} - 2 \cdot 4\sqrt{2}$	Product Rule for Radicals
$= 12\sqrt{2} - 8\sqrt{2}$	Simplify radicals.
$= 4\sqrt{2}$	Combine like radicals.

EXAMPLE 3   **Simplifying Radical Expressions**

Simplify each expression.

a. $3\sqrt{x} + \sqrt{4x}$ **b.** $\sqrt{45x} + 2\sqrt{20x}$ **c.** $5\sqrt{2x^3} - x\sqrt{8x}$

SOLUTION

a.
$3\sqrt{x} + \sqrt{4x} = 3\sqrt{x} + \sqrt{4 \cdot x}$	Factor radicand.
$= 3\sqrt{x} + 2\sqrt{x}$	Simplify radical.
$= 5\sqrt{x}$	Combine like radicals.

b.
$\sqrt{45x} + 2\sqrt{20x} = \sqrt{9 \cdot 5x} + 2\sqrt{4 \cdot 5x}$	Factor radicands.
$= 3\sqrt{5x} + 4\sqrt{5x}$	Simplify radicals.
$= 7\sqrt{5x}$	Combine like radicals.

c.
$5\sqrt{2x^3} - x\sqrt{8x} = 5\sqrt{2 \cdot x^2 \cdot x} - x\sqrt{4 \cdot 2 \cdot x}$	Factor radicands.
$= 5x\sqrt{2x} - 2x\sqrt{2x}$	Simplify radicals.
$= 3x\sqrt{2x}$	Combine like radicals.

Study Tip

Remember that the square root of a negative number is not a real number. Because of this, you can assume for $\sqrt{x^3}$ that the implied domain of the expression is the set of nonnegative real numbers. This assumption allows you to write $\sqrt{x^3} = x\sqrt{x}$.

Exercises Within Reach®   Solutions in English & Spanish and tutorial videos at AlgebraWithinReach.com

Simplifying a Radical Expression In Exercises 19–34, *simplify* the expression.

19. $\sqrt{50} + \sqrt{2}$

20. $3\sqrt{18} - \sqrt{8}$

21. $2\sqrt{45} + 12\sqrt{80}$

22. $\sqrt{75} - 4\sqrt{27}$

23. $\sqrt{81b} + \sqrt{b}$

24. $\sqrt{x} - \sqrt{25x}$

25. $\sqrt{9x} + \sqrt{36x}$

26. $\sqrt{64t} - \sqrt{16t}$

27. $\sqrt{45z} - \sqrt{125z}$

28. $\sqrt{32z} - \sqrt{98z}$

29. $\sqrt{18u} + 3\sqrt{8u}$

30. $\sqrt{12u} + 4\sqrt{27u}$

31. $2\sqrt{5x^3} + 6x\sqrt{20x}$

32. $4x\sqrt{12x} - 3\sqrt{3x^3}$

33. $\sqrt{48a^5} - 2a^2\sqrt{27a}$

34. $4a^2\sqrt{18a} + 7\sqrt{200a^5}$

Multiplying Radical Expressions

> **EXAMPLE 4** **Multiplying Radical Expressions**
>
> a. $\sqrt{6} \cdot \sqrt{3} = \sqrt{6 \cdot 3} = \sqrt{18} = \sqrt{9 \cdot 2} = 3\sqrt{2}$
> b. $\sqrt[3]{4} \cdot \sqrt[3]{12} = \sqrt[3]{4 \cdot 12} = \sqrt[3]{48} = \sqrt[3]{8 \cdot 6} = 2\sqrt[3]{6}$

> **EXAMPLE 5** **Multiplying Radical Expressions**
>
> Find each product and simplify.
>
> a. $\sqrt{5}(\sqrt{15} - \sqrt{5})$ b. $\sqrt[3]{4}(3 + \sqrt[3]{2})$
>
> **SOLUTION**
>
> a. $\sqrt{5}(\sqrt{15} - \sqrt{5}) = \sqrt{5} \cdot \sqrt{15} - \sqrt{5} \cdot \sqrt{5}$ Distributive Property
> $\phantom{\sqrt{5}(\sqrt{15} - \sqrt{5})} = \sqrt{75} - \sqrt{25}$ Product Rule for Radicals
> $\phantom{\sqrt{5}(\sqrt{15} - \sqrt{5})} = \sqrt{25 \cdot 3} - \sqrt{25}$ Factor radicand.
> $\phantom{\sqrt{5}(\sqrt{15} - \sqrt{5})} = 5\sqrt{3} - 5$ Simplify.
>
> b. $\sqrt[3]{4}(3 + \sqrt[3]{2}) = \sqrt[3]{4} \cdot 3 + \sqrt[3]{4} \cdot \sqrt[3]{2}$ Distributive Property
> $\phantom{\sqrt[3]{4}(3 + \sqrt[3]{2})} = 3\sqrt[3]{4} + \sqrt[3]{8}$ Product Rule for Radicals
> $\phantom{\sqrt[3]{4}(3 + \sqrt[3]{2})} = 3\sqrt[3]{4} + 2$ Simplify.

> **EXAMPLE 6** **Multiplying Radical Expressions**
>
> a. $(\sqrt{x} - 1)(\sqrt{x} + 3) = \overbrace{\sqrt{x \cdot x}}^{F} + \overbrace{3\sqrt{x}}^{O} - \overbrace{\sqrt{x}}^{I} - \overset{L}{3}$ FOIL Method
> $\phantom{(\sqrt{x} - 1)(\sqrt{x} + 3)} = x + (3 - 1)\sqrt{x} - 3$ Combine like radicals.
> $\phantom{(\sqrt{x} - 1)(\sqrt{x} + 3)} = x + 2\sqrt{x} - 3$ Simplify.
>
> b. $(2 - \sqrt{x})(2 + \sqrt{x}) = 2^2 - (\sqrt{x})^2$ Special product formula
> $\phantom{(2 - \sqrt{x})(2 + \sqrt{x})} = 4 - x$ Simplify.

Exercises Within Reach®

Solutions in English & Spanish and tutorial videos at AlgebraWithinReach.com

Multiplying Radical Expressions In Exercises 35−50, **multiply** and simplify.

35. $\sqrt{2} \cdot \sqrt{8}$
36. $\sqrt{3} \cdot \sqrt{12}$
37. $\sqrt{10} \cdot \sqrt{6}$
38. $\sqrt{7} \cdot \sqrt{21}$
39. $\sqrt[3]{4} \cdot \sqrt[3]{2}$
40. $\sqrt[3]{9} \cdot \sqrt[3]{3}$
41. $\sqrt{6}(\sqrt{12} + 8)$
42. $\sqrt{2}(\sqrt{14} + 3)$
43. $\sqrt[3]{2}(\sqrt[3]{4} + 5)$
44. $\sqrt[3]{3}(\sqrt[3]{9} - 7)$
45. $(\sqrt{2} - 1)(\sqrt{2} + 3)$
46. $(\sqrt{6} + \sqrt{5})(\sqrt{6} + \sqrt{3})$
47. $(\sqrt{x} + 7)(\sqrt{x} - 2)$
48. $(\sqrt{u} - 3)(\sqrt{u} - 4)$
49. $(1 + \sqrt{11})(1 - \sqrt{11})$
50. $(\sqrt{7} + 3)(\sqrt{7} - 3)$

Section 9.3 Operations with Radical Expressions

The expressions $3 + \sqrt{6}$ and $3 - \sqrt{6}$ are called **conjugates** of each other. Notice that they differ only in the sign between the terms. The product of two conjugates is the difference of two squares, which is given by the special product formula

$$(a + b)(a - b) = a^2 - b^2. \quad \text{Difference of two squares}$$

EXAMPLE 7 Multiplying Conjugates

Find the conjugate of each expression. Then find the product of each expression and its conjugate.

a. $\sqrt{3} + \sqrt{7}$

b. $\sqrt{t} - \sqrt{8}$

SOLUTION

a. The conjugate of $\sqrt{3} + \sqrt{7}$ is $\sqrt{3} - \sqrt{7}$. The product of $\sqrt{3} + \sqrt{7}$ and $\sqrt{3} - \sqrt{7}$ is

$$(\sqrt{3} + \sqrt{7})(\sqrt{3} - \sqrt{7}) = (\sqrt{3})^2 - (\sqrt{7})^2 \quad \text{Special product formula}$$
$$= 3 - 7 \quad \text{Simplify.}$$
$$= -4. \quad \text{Simplify.}$$

b. The conjugate of $\sqrt{t} - \sqrt{8}$ is $\sqrt{t} + \sqrt{8}$. The product of $\sqrt{t} - \sqrt{8}$ and $\sqrt{t} + \sqrt{8}$ is

$$(\sqrt{t} - \sqrt{8})(\sqrt{t} + \sqrt{8}) = (\sqrt{t})^2 - (\sqrt{8})^2 \quad \text{Special product formula}$$
$$= t - 8, \ t \geq 0. \quad \text{Simplify.}$$

Exercises Within Reach®

Solutions in English & Spanish and tutorial videos at AlgebraWithinReach.com

Multiplying Conjugates In Exercises 51–58, find the conjugate of the expression. Then find the product of the expression and its conjugate.

51. $4 + \sqrt{5}$

52. $\sqrt{7} - 3$

53. $\sqrt{t} - 5$

54. $6 + \sqrt{y}$

55. $\sqrt{15} - \sqrt{7}$

56. $\sqrt{11} + \sqrt{5}$

57. $\sqrt{u} - \sqrt{2}$

58. $\sqrt{a} + \sqrt{3}$

Geometry In Exercises 59 and 60, find the area of the rectangle.

59.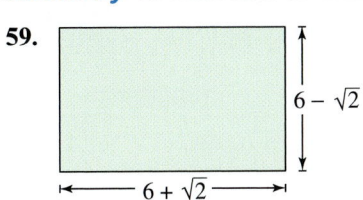

60.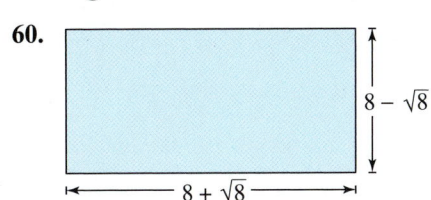

Application

EXAMPLE 8 Geometry: **Perimeter and Area of a Triangle**

Write and simplify expressions for the perimeter and area of the triangle shown in the figure.

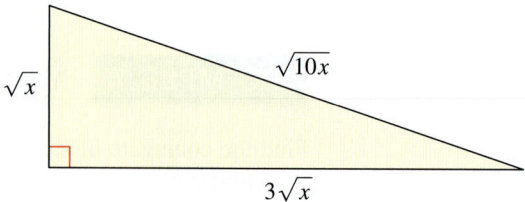

SOLUTION

$P = a + b + c$	Formula for perimeter of a triangle
$= \sqrt{x} + 3\sqrt{x} + \sqrt{10x}$	Substitute.
$= (1 + 3)\sqrt{x} + \sqrt{10x}$	Distributive Property
$= 4\sqrt{x} + \sqrt{10x}$	Simplify.

$A = \dfrac{1}{2}bh$	Formula for area of a triangle
$= \dfrac{1}{2}(3\sqrt{x})(\sqrt{x})$	Substitute.
$= \dfrac{3}{2}\sqrt{x^2}$	Product Rule for Radicals
$= \dfrac{3}{2}x$	Simplify.

Exercises Within Reach® Solutions in English & Spanish and tutorial videos at AlgebraWithinReach.com

Geometry In Exercises 61–64, write and simplify expressions for the perimeter and area of the rectangle.

61.

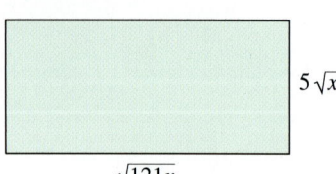

62.

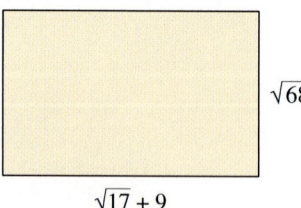

63.

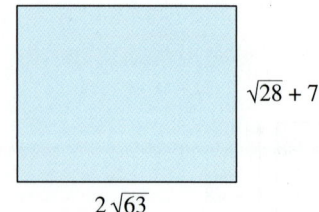

64.

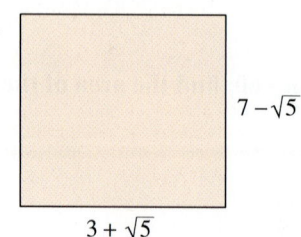

Section 9.3   Operations with Radical Expressions   429

Simplifying Quotients Involving Radicals

EXAMPLE 9   Simplifying a Quotient Involving a Radical

Study Tip
To simplify a quotient involving radicals, you can rationalize the denominator by multiplying both the numerator and denominator by the conjugate of the denominator.

$$\frac{2}{3-\sqrt{5}} = \frac{2}{3-\sqrt{5}} \cdot \frac{3+\sqrt{5}}{3+\sqrt{5}}$$ Multiply numerator and denominator by conjugate of denominator.

$$= \frac{2(3+\sqrt{5})}{(3)^2 - (\sqrt{5})^2}$$ Special product formula

$$= \frac{2(3+\sqrt{5})}{4} = \frac{3+\sqrt{5}}{2}$$ Simplify.

EXAMPLE 10   Simplifying Quotients Involving Radicals

a. $\dfrac{5\sqrt{2}}{\sqrt{3}+\sqrt{2}} = \dfrac{5\sqrt{2}}{\sqrt{3}+\sqrt{2}} \cdot \dfrac{\sqrt{3}-\sqrt{2}}{\sqrt{3}-\sqrt{2}}$ Multiply numerator and denominator by conjugate of denominator.

$= \dfrac{5\sqrt{6} - 5\sqrt{4}}{(\sqrt{3})^2 - (\sqrt{2})^2}$ Distributive Property and special product formula

$= \dfrac{5\sqrt{6} - 10}{3 - 2} = 5\sqrt{6} - 10$ Simplify.

b. $\dfrac{2-\sqrt{3}}{\sqrt{6}+\sqrt{2}} = \dfrac{2-\sqrt{3}}{\sqrt{6}+\sqrt{2}} \cdot \dfrac{\sqrt{6}-\sqrt{2}}{\sqrt{6}-\sqrt{2}}$ Multiply numerator and denominator by conjugate of denominator.

$= \dfrac{2\sqrt{6} - 2\sqrt{2} - \sqrt{18} + \sqrt{6}}{(\sqrt{6})^2 - (\sqrt{2})^2}$ FOIL Method and special product formula

$= \dfrac{3\sqrt{6} - 2\sqrt{2} - 3\sqrt{2}}{6 - 2}$ Combine like radicals.

$= \dfrac{3\sqrt{6} - 5\sqrt{2}}{4}$ Simplify.

c. $\dfrac{6}{\sqrt{x} - 2} = \dfrac{6}{\sqrt{x} - 2} \cdot \dfrac{\sqrt{x} + 2}{\sqrt{x} + 2}$ Multiply numerator and denominator by conjugate of denominator.

$= \dfrac{6(\sqrt{x} + 2)}{(\sqrt{x})^2 - (2)^2}$ Special product formula

$= \dfrac{6\sqrt{x} + 12}{x - 4}$ Simplify.

Exercises Within Reach®   Solutions in English & Spanish and tutorial videos at AlgebraWithinReach.com

Simplifying a Quotient Involving a Radical   In Exercises 65–72, **rationalize** the denominator of the expression and simplify.

65. $\dfrac{5}{\sqrt{14} - 2}$

66. $\dfrac{2}{\sqrt{10} - 4}$

67. $\dfrac{8\sqrt{6}}{\sqrt{6} + \sqrt{2}}$

68. $\dfrac{7\sqrt{5}}{\sqrt{3} - \sqrt{5}}$

69. $\dfrac{\sqrt{5} + 1}{\sqrt{13} + 7}$

70. $\dfrac{2 - \sqrt{7}}{1 + \sqrt{7}}$

71. $\dfrac{9}{\sqrt{x} + 2}$

72. $\dfrac{6}{\sqrt{x} - 1}$

Chapter 9 Radical Expressions and Equations

Concept Summary: Operations with Radical Expressions

What
To perform operations with radical expressions, use the guidelines for simplifying radical expressions along with the techniques for performing operations with polynomials.

EXAMPLE
Simplify each expression.
a. $2\sqrt{2} + 3\sqrt{2}$
b. $\sqrt{2}(\sqrt{8} - \sqrt{2})$
c. $\dfrac{1}{\sqrt{2} - 1}$

How
- Combine like terms to add and subtract radical expressions.
- Use the Distributive Property, FOIL Method, and special product formulas to multiply radical expressions.
- Rationalize the denominator to simplify quotients involving radicals.

EXAMPLE
a. $2\sqrt{2} + 3\sqrt{2} = (2+3)\sqrt{2} = 5\sqrt{2}$
b. $\sqrt{2}(\sqrt{8} - \sqrt{2}) = \sqrt{16} - \sqrt{4} = 2$
c. $\dfrac{1}{\sqrt{2} - 1} \cdot \dfrac{\sqrt{2} + 1}{\sqrt{2} + 1} = \dfrac{\sqrt{2} + 1}{2 - 1}$
$= \sqrt{2} + 1$

Why
As you continue your study of mathematics, you will use and build upon the concepts that you learned in this section.

Exercises Within Reach® Worked-out solutions to odd-numbered exercises at AlgebraWithinReach.com

Concept Summary Check

73. Like Terms Are $2\sqrt{2}$ and $3\sqrt{2}$ like terms? Explain.

74. Justifying Steps What properties or rules are used to rewrite $\sqrt{2}(\sqrt{8} - \sqrt{2})$ as $\sqrt{16} - \sqrt{4}$ in the solution above?

75. Describing a Relationship Describe the relationship between $\sqrt{2} + 1$ and $\sqrt{2} - 1$.

76. Identifying a Process What process is used to rewrite $\dfrac{1}{\sqrt{2} - 1}$ as $\sqrt{2} + 1$ in the solution above?

Extra Practice

Simplifying a Radical Expression In Exercises 77–88, simplify the expression.

77. $\dfrac{2}{5}\sqrt{3} - \dfrac{6}{5}\sqrt{3}$

78. $\dfrac{2}{3}\sqrt{13} - \dfrac{4}{3}\sqrt{13}$

79. $\sqrt{x^3 y} + 4\sqrt{xy}$

80. $3t\sqrt{st^3} - s\sqrt{s^3 t}$

81. $\sqrt{\dfrac{a}{4}} - \sqrt{\dfrac{a}{9}}$

82. $\sqrt{\dfrac{v}{144}} + \sqrt{\dfrac{v}{16}}$

83. $(\sqrt{13} + 2)^2$

84. $(\sqrt{7} + 3)^2$

85. $(\sqrt[3]{2} - 1)^2$

86. $(\sqrt[5]{6} + 3)^2$

87. $\dfrac{\sqrt{x} - 5}{\sqrt{x} - 1}$

88. $\dfrac{\sqrt{t} + 1}{\sqrt{t} - 4}$

Simplifying a Radical Expression In Exercises 89–92, rewrite the expression as a single fraction and simplify.

89. $3 - \dfrac{1}{\sqrt{3}}$

90. $\dfrac{2}{\sqrt{5}} - 9$

91. $\sqrt{50} - \dfrac{6}{\sqrt{2}}$

92. $\dfrac{7}{\sqrt{3}} + \sqrt{12}$

Comparing Expressions In Exercises 93–96, place the correct symbol (<, >, or =) between the expressions.

93. $\sqrt{5} + \sqrt{3}$ ▭ $\sqrt{5 + 3}$

94. $\sqrt{5} - \sqrt{3}$ ▭ $\sqrt{5 - 3}$

95. 5 ▭ $\sqrt{3^2 + 2^2}$

96. 5 ▭ $\sqrt{3^2 + 4^2}$

97. The Golden Section The ratio of the width of the Temple of Hephaestus to its height (see figure) is

$$\frac{w}{h} \approx \frac{2}{\sqrt{5} - 1}.$$

This number is called the **golden section**. Early Greeks believed that the most aesthetically pleasing rectangles were those whose sides had this ratio. Rationalize the denominator of this number. Approximate your answer, rounded to two decimal places.

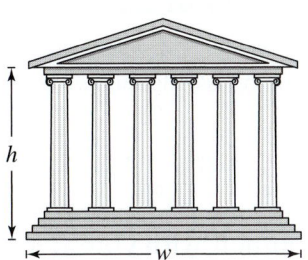

98. Geometry The surface area S of a right circular cone with a slant height of 1 unit (see figure) is given by $S = \pi r + \pi r^2$, where r is the radius of the cone. By solving for r, you obtain the equation

$$r = \frac{1}{\sqrt{\pi}} \sqrt{S + \frac{\pi}{4}} - \frac{1}{2}.$$

Find the radius of a right circular cone with a slant height of 1 unit and surface area of $\frac{3\pi}{8}$ square units.

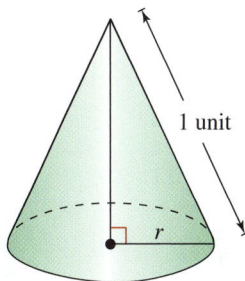

Explaining Concepts

99. Writing Explain how the Distributive Property can be used to add or subtract like radicals. Give an example.

100. Structure Are $\sqrt[3]{5}$ and $\sqrt[4]{5}$ like radical expressions? Explain.

101. Number Sense Is $\sqrt{2} + \sqrt{18}$ in simplest form? Explain.

102. Number Sense Is the number $3/(1 + \sqrt{5})$ in simplest form? If not, explain the steps for writing in simplest form.

103. Reasoning Square the real number $3/\sqrt{2}$. Is this equivalent to rationalizing the denominator? Explain.

104. Exploration Enter any positive real number in your calculator and repeatedly take the square root. What real number does the display appear to be approaching?

Cumulative Review

In Exercises 105–108, factor the polynomial completely.

105. $x^2 - 25$

106. $x^3 - 27$

107. $8x^3 + 1$

108. $(x + 5)^2 - 144$

In Exercises 109–112, solve the system of linear equations.

109. $\begin{cases} x - 3y = 2 \\ x + y = 6 \end{cases}$

110. $\begin{cases} 5x + y = 20 \\ 2x - 10y = 1 \end{cases}$

111. $\begin{cases} -x + 7y = -1 \\ 2x - 10y = 6 \end{cases}$

112. $\begin{cases} 9x - 7y = 39 \\ 4x + 3y = -1 \end{cases}$

In Exercises 113–116, rationalize the denominator of the expression and simplify. (Assume all variables are positive.)

113. $\sqrt{\dfrac{1}{6}}$

114. $\sqrt{\dfrac{2}{3}}$

115. $\sqrt{\dfrac{3}{8x}}$

116. $\sqrt{\dfrac{5}{4x^3}}$

9.4 Radical Equations and Applications

- Use the Squaring Property of Equality to solve radical equations.
- Solve application problems using radical equations.

Solving Radical Equations

Squaring Property of Equality

Let a and b be real numbers, variables, or algebraic expressions. If $a = b$, then it follows that $a^2 = b^2$. This operation is called *squaring each side of an equation*.

Study Tip

You will see in this section that squaring each side of an equation often introduces *extraneous solutions*. When you use this procedure, it is critical that you check each solution in the original equation.

EXAMPLE 1   A Radical Equation with One Radical

$\sqrt{2x+1} - 2 = 3$	Original equation
$\sqrt{2x+1} = 5$	Isolate the radical.
$(\sqrt{2x+1})^2 = (5)^2$	Square each side.
$2x + 1 = 25$	Simplify.
$2x = 24$	Subtract 1 from each side.
$x = 12$	Divide each side by 2.

By checking this in the original equation, you can see that it is a valid solution.

EXAMPLE 2   A Radical Equation with No Solution

$\sqrt{3x} = -9$	Original equation
$(\sqrt{3x})^2 = (-9)^2$	Square each side.
$3x = 81$	Simplify.
$x = 27$	Divide each side by 3.

Check

$\sqrt{3x} = -9$	Write original equation.
$\sqrt{3(27)} \stackrel{?}{=} -9$	Substitute 27 for x.
$9 \neq -9$	Solution does not check. ✗

So, $x = 27$ is *extraneous*. Thus, the original equation has no solution.

Exercises Within Reach ®   Solutions in English & Spanish and tutorial videos at AlgebraWithinReach.com

Solving a Radical Equation In Exercises 1−6, **solve** the equation. (Some of the equations have no solution.)

1. $\sqrt{x+4} = 3$
2. $\sqrt{x-2} = 5$
3. $\sqrt{3x+4} - 2 = 3$
4. $\sqrt{x-2} + 1 = 7$
5. $\sqrt{4x} = -6$
6. $\sqrt{2y} = -4$

EXAMPLE 3 A Radical Equation with Two Radicals

$\sqrt{5x+3} = \sqrt{x+11}$	Original equation
$(\sqrt{5x+3})^2 = (\sqrt{x+11})^2$	Square each side.
$5x + 3 = x + 11$	Simplify.
$5x = x + 8$	Subtract 3 from each side.
$4x = 8$	Subtract x from each side.
$x = 2$	Divide each side by 4.

Check

$\sqrt{5x+3} = \sqrt{x+11}$	Write original equation.
$\sqrt{5(2)+3} \stackrel{?}{=} \sqrt{2+11}$	Substitute 2 for x.
$\sqrt{13} = \sqrt{13}$	Solution checks. ✓

So, the solution is $x = 2$.

EXAMPLE 4 A Radical Equation with Two Radicals

$\sqrt{6x-4} - 2\sqrt{4-x} = 0$	Original equation
$\sqrt{6x-4} = 2\sqrt{4-x}$	Isolate radicals.
$6x - 4 = 2^2(4-x)$	Square each side.
$6x - 4 = 16 - 4x$	Distributive Property
$6x = 20 - 4x$	Add 4 to each side.
$10x = 20$	Add $4x$ to each side.
$x = 2$	Divide each side by 10.

Check

$\sqrt{6x-4} - 2\sqrt{4-x} = 0$	Write original equation.
$\sqrt{6(2)-4} - 2\sqrt{4-2} \stackrel{?}{=} 0$	Substitute 2 for x.
$\sqrt{8} - 2\sqrt{2} \stackrel{?}{=} 0$	Simplify.
$2\sqrt{2} - 2\sqrt{2} = 0$	Solution checks. ✓

So, the solution is $x = 2$.

Study Tip

In Example 4, it is necessary to isolate the radicals on each side of the equal sign. If both sides of the equation were squared before isolating the radicals, the resulting equation would still contain a radical, which would require isolating the radical and squaring each side a second time.

Exercises Within Reach®

Solutions in English & Spanish and tutorial videos at AlgebraWithinReach.com

Solving a Radical Equation In Exercises 7–16, solve the equation. (Some of the equations have no solution.)

7. $\sqrt{x+3} = \sqrt{6x-7}$

8. $\sqrt{6-7x} = \sqrt{4x+17}$

9. $\sqrt{x+3} = \sqrt{4x-3}$

10. $\sqrt{x+8} = \sqrt{5x-4}$

11. $\sqrt{3x+4} = -3\sqrt{x}$

12. $\sqrt{2u-9} = -\sqrt{u}$

13. $\sqrt{3t+11} - 5\sqrt{t} = 0$

14. $\sqrt{15-4u} - 4\sqrt{u} = 0$

15. $2\sqrt{y+1} - \sqrt{3y+6} = 0$

16. $2\sqrt{x+4} - 3\sqrt{x-1} = 0$

EXAMPLE 5 Solving a Radical Equation

$1 - 2x = \sqrt{x}$	Original equation
$(1 - 2x)^2 = (\sqrt{x})^2$	Square each side.
$1 - 4x + 4x^2 = x$	Square of a binomial
$4x^2 - 5x + 1 = 0$	Write in standard form.
$(4x - 1)(x - 1) = 0$	Factor.
$4x - 1 = 0 \implies x = \frac{1}{4}$	Set 1st factor equal to 0.
$x - 1 = 0 \implies x = 1$	Set 2nd factor equal to 0.

Check

First Solution
$$1 - 2x = \sqrt{x}$$
$$1 - 2\left(\tfrac{1}{4}\right) \stackrel{?}{=} \sqrt{\tfrac{1}{4}}$$
$$1 - \tfrac{1}{2} \stackrel{?}{=} \tfrac{1}{2}$$
$$\tfrac{1}{2} = \tfrac{1}{2} \checkmark$$

Second Solution
$$1 - 2x = \sqrt{x}$$
$$1 - 2(1) \stackrel{?}{=} \sqrt{1}$$
$$1 - 2 \stackrel{?}{=} 1$$
$$-1 \ne 1 \;\;✗$$

From the check, you can see that $x = 1$ is an extraneous solution. So, the only solution is $x = \frac{1}{4}$.

EXAMPLE 6 Solving a Radical Equation

$\sqrt{6x + 1} = x - 1$	Original equation
$(\sqrt{6x + 1})^2 = (x - 1)^2$	Square each side.
$6x + 1 = x^2 - 2x + 1$	Square of a binomial
$0 = x^2 - 8x$	Write in standard form.
$0 = x(x - 8)$	Factor.
$x = 0 \implies x = 0$	Set 1st factor equal to 0.
$x - 8 = 0 \implies x = 8$	Set 2nd factor equal to 0.

Check

First Solution
$$\sqrt{6x + 1} = x - 1$$
$$\sqrt{6(0) + 1} \stackrel{?}{=} 0 - 1$$
$$\sqrt{1} \stackrel{?}{=} -1$$
$$1 \ne -1 \;\;✗$$

Second Solution
$$\sqrt{6x + 1} = x - 1$$
$$\sqrt{6(8) + 1} \stackrel{?}{=} 8 - 1$$
$$\sqrt{49} \stackrel{?}{=} 7$$
$$7 \ne 7 \;\;\checkmark$$

From the check, you can see that $x = 0$ is an extraneous solution. So, the only solution is $x = 8$.

Exercises Within Reach® Solutions in English & Spanish and tutorial videos at AlgebraWithinReach.com

Solving a Radical Expression In Exercises 17–22, solve the equation.

17. $\sqrt{x} = 2 - x$
18. $6 - x = \sqrt{x}$
19. $2x + 1 = \sqrt{9x}$
20. $5\sqrt{x} = x + 4$
21. $\sqrt{6x + 7} = x + 2$
22. $\sqrt{5x + 11} = x + 3$

Section 9.4 Radical Equations and Applications 435

Applications of Radicals

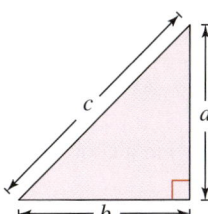

A common use of radicals occurs in applications involving right triangles. Recall that a right triangle is one that contains a right (or 90°) angle, as shown at left. The relationship among the three sides of a right triangle is described by the **Pythagorean Theorem**, which says that if a and b are the lengths of the legs (the two sides that form the right angle) and c is the length of the hypotenuse (the side across from the right angle), then

$c^2 = a^2 + b^2$ Pythagorean Theorem

$c = \sqrt{a^2 + b^2}$. Take square root of each side.

Application   **EXAMPLE 7**   Dimensions of a Softball Diamond

A softball diamond has the shape of a square with 60-foot sides, as shown at the left. The catcher is 4 feet behind home plate. How far does the catcher have to throw to reach second base?

SOLUTION

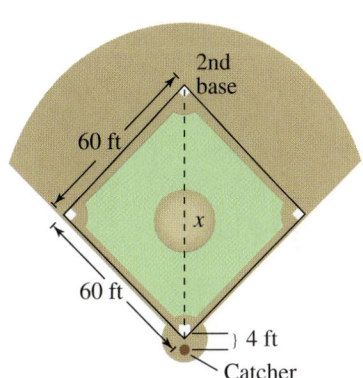

In the figure, let x be the hypotenuse of a right triangle with 60-foot legs. So, by the Pythagorean Theorem, you have the following.

$x = \sqrt{60^2 + 60^2}$ Pythagorean Theorem

$ = \sqrt{7200}$ Simplify.

$ \approx 84.9$ feet Use a calculator.

The distance from home plate to second base is approximately 84.9 feet. Because the catcher is 4 feet behind home plate, the catcher must make a throw of

$x + 4 \approx 84.9 + 4$

$ = 88.9$ feet.

Check this solution in the original statement of the problem.

Exercises Within Reach®   Solutions in English & Spanish and tutorial videos at AlgebraWithinReach.com

23. *Geometry* A volleyball court is a rectangle that is 30 feet wide and 60 feet long. Find the length of the diagonal of the court.

24. *Geometry* A baseball diamond is a square that is 90 feet on a side (see figure). Determine the distance between first base and third base.

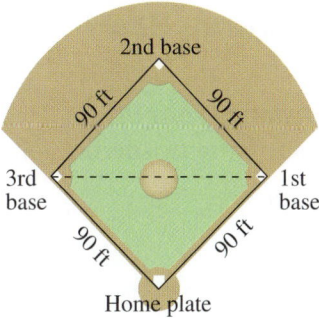

The Pythagorean Theorem can be used to establish the Distance Formula for finding the distance between two points in the coordinate plane.

> **The Distance Formula**
>
> The distance d between the two points (x_1, y_1) and (x_2, y_2) in a coordinate plane is
>
> $d = \sqrt{(x_2 - x_1)^2 + (y_2 - y_1)^2}.$

Application **EXAMPLE 8** **Using the Distance Formula**

Find the distance between the points $(-1, -5)$ and $(2, -2)$.

SOLUTION

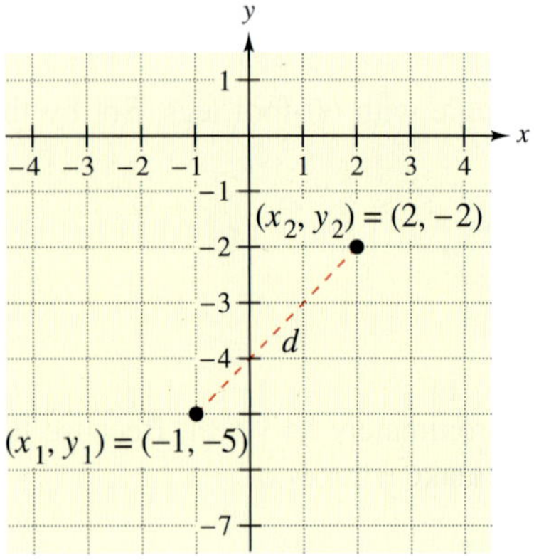

Let $(x_1, y_1) = (-1, -5)$ and $(x_2, y_2) = (2, -2)$, as shown in the figure. Then apply the Distance Formula as follows.

$d = \sqrt{(x_2 - x_1)^2 + (y_2 - y_1)^2}$ Distance formula

$= \sqrt{[2 - (-1)]^2 + [-2 - (-5)]^2}$ Substitute for x_1, y_1, x_2, and y_2.

$= \sqrt{3^2 + 3^2}$ Simplify.

$= \sqrt{18}$ Simplify.

≈ 4.24 Use a calculator.

So, the distance between the two points is about 4.24 units.

Notice in Example 8 that you could have chosen (x_1, y_1) to be the point $(2, -2)$ and (x_2, y_2) to be the point $(-1, -5)$. The results would have been the same.

$d = \sqrt{(-1 - 2)^2 + [-5 - (-2)]^2}$ Distance formula

$= \sqrt{(-3)^2 + (-3)^2}$ Simplify.

$= \sqrt{18}$ Simplify.

≈ 4.24 Use a calculator.

Exercises Within Reach®   Solutions in English & Spanish and tutorial videos at AlgebraWithinReach.com

Using the Distance Formula In Exercises 25–32, find the distance between the two points. Round your answer to two decimal places, if necessary.

25. $(1, 2), (5, 5)$

26. $(15, 7), (3, 5)$

27. $(-5, 4), (3, -2)$

28. $(1, -2), (-4, 8)$

29. $(3, -2), (4, 6)$

30. $(1, 5), (2, -6)$

31. $(-3, 2), (-2, 6)$

32. $(-1, 2), (7, -2)$

Application

EXAMPLE 9 Atmospheric Pressure

The atmospheric pressure decreases with increasing altitude. At sea level, the average air pressure is 1.033227 kilograms per square centimeter, which is called 1 atmosphere. Variations in weather conditions cause changes in the atmospheric pressure of up to ±5%. The table shows the pressures p (in atmospheres) at different altitudes h (in kilometers).

h	0	5	10	15	20	25
p	1	0.55	0.25	0.12	0.06	0.02

A model for this data is $p = 1 - 0.206\sqrt{h}$.

a. Use a spreadsheet to compare the data and the model. How good is the model?

b. Approximate the altitude when the atmospheric pressure is 0.4 atmosphere.

SOLUTION

a. The data from the table and from the model are shown below.

	A	B	C	D	E	F	G
1	h	0	5	10	15	20	25
2	p (table)	1	0.55	0.25	0.12	0.06	0.02
3	p (model)	1.00	0.54	0.35	0.20	0.08	-0.03

You can see that the square root model is not particularly good for representing atmospheric pressure.

b. The model is good between 0 and 5 kilometers. So, for an atmospheric pressure of 0.4 atmosphere, you can approximate the altitude as follows.

$p = 1 - 0.206\sqrt{h}$ Write model.

$0.4 = 1 - 0.206\sqrt{h}$ Substitute 0.4 for p.

$8.5 \approx h$ Solve for h.

So, the altitude is about 8.5 kilometers.

Exercises Within Reach®   Solutions in English & Spanish and tutorial videos at AlgebraWithinReach.com

33. *Analyzing Data* The vibration of a string on a guitar depends on the length of the string. The table shows the velocities v (in meters per second) at different lengths l (in meters).

l	0.40	0.50	0.60	0.70
v	196	219	240	259

A model for this data is $v = \sqrt{\dfrac{4800}{\left(\dfrac{0.05}{l}\right)}}$. Use a spreadsheet to compare the data and the model. How good is the model?

34. *Using a Model* Use the model in Exercise 33 to approximate the length of the string when the velocity is 200 meters per second.

Concept Summary: Solving Radical Equations

What
A **radical equation** is an equation that contains one or more radicals with variable radicands. Solving radical equations is somewhat like solving equations that contain fractions.

EXAMPLE
Solve $\sqrt{2x} + 1 = 5$.

How
First, try to get rid of the radicals and obtain a linear or quadratic equation. Then, solve the equation using the standard procedures.

EXAMPLE

$\sqrt{2x} + 1 = 5$	Original equation
$\sqrt{2x} = 4$	Isolate radical.
$(\sqrt{2x})^2 = (4)^2$	Square each side.
$2x = 16$	Simplify.
$x = 8$	Divide by 2.

Why
Knowing how to solve radical equations will help as you work through and solve problems involving the Distance Formula and the **Pythagorean Theorem**.

Exercises Within Reach®
Worked-out solutions to odd-numbered exercises at AlgebraWithinReach.com

Concept Summary Check

35. Vocabulary In your own words, describe a radical equation.

36. Logic Describe the steps used to solve a radical equation.

37. Vocabulary State the Pythagorean Theorem.

38. Reasoning Explain why the equation $\sqrt{x} = -4$ has no solution.

Extra Practice

Using the Pythagorean Theorem In Exercises 39–42, use the Pythagorean Theorem to **solve** for x. Round your answer to two decimal places, if necessary.

39.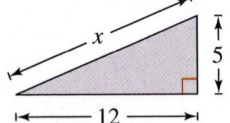

40.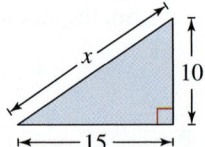

41.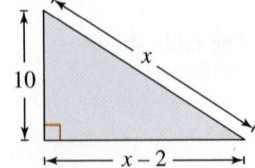

42.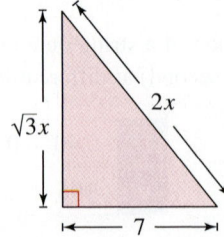

Free-Falling Object In Exercises 43 and 44, **use** the equation for the velocity of a free-falling object, $v = \sqrt{2gh}$, where v is measured in feet per second, $g = 32$ feet per second squared, and h is the height (in feet).

43. An object strikes the ground with a velocity of 45 feet per second. Estimate the height from which it was dropped.

44. An object strikes the ground with a velocity of 100 feet per second. Estimate the height from which it was dropped.

Section 9.4 Radical Equations and Applications 439

Pendulum Length In Exercises 45 and 46, the time t (in seconds) for a pendulum of length L (in feet) to go through one complete cycle, both forward and back (its period), is given by

$$t = 2\pi\sqrt{\dfrac{L}{32}}.$$

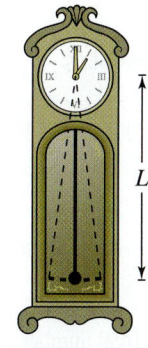

45. How long is the pendulum of a grandfather clock with a period of 2 seconds (see figure)?

46. How long is the pendulum of a mantel clock with a period of 0.8 second?

47. *Geometry* A 39-foot guy wire on a sailboat is attached to the top of the mast and to the deck 15 feet from the base of the mast (see figure). How tall is the mast?

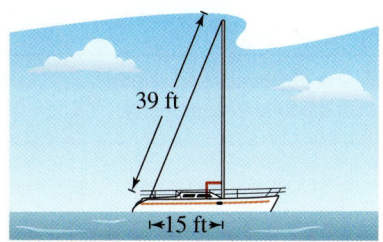

48. *Geometry* The distance between Memphis and New Orleans is 410 miles. The distance between Memphis and Chattanooga is 317 miles (see figure). Approximate the distance between Chattanooga and New Orleans.

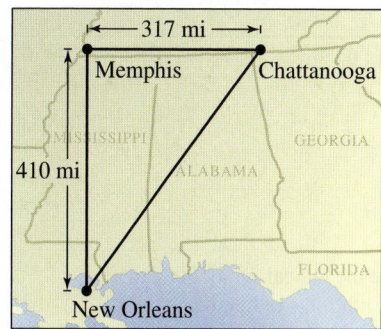

49. *Demand* The demand equation for a video game is

$$p = 40 - \sqrt{x - 1}$$

where x is the number of units demanded per day and p is the price per game. Find the demand when the price is set at $34.70.

50. *Airline Passengers* An airline offers daily flights between Chicago and Denver. The total monthly cost C (in millions of dollars) of these flights is modeled by

$$C = \sqrt{0.2x + 1},\ x \geq 0$$

where x is measured in thousands of passengers. The total cost of the flights for June is 2.5 million dollars. Approximately how many passengers flew in June?

Explaining Concepts

51. *Reasoning* Is $x = 25$ a solution of $\sqrt{x} = -5$? Explain.

52. *Writing* Give two reasons why it is important to check solutions of a radical equation.

53. *Think About It* Can a right triangle be isosceles (have two sides of the same length)? Explain.

Cumulative Review

In Exercises 54–57, determine whether the lines are parallel, perpendicular, or neither.

54. $y = 3x + 4$
$y = -3x + 4$

55. $y = -5x - 8$
$y = \dfrac{1}{5}x$

56. $y = \dfrac{1}{2}x + 7$
$y = 2x + 7$

57. $y = \dfrac{2}{3}x - \dfrac{3}{5}$
$y = \dfrac{2}{3}x + \dfrac{3}{5}$

In Exercises 58 and 59, perform the indicated operation and simplify.

58. $\dfrac{5}{x+1} \cdot \dfrac{x+1}{25}$

59. $\dfrac{r}{r-1} \div \dfrac{r^2}{r^2-1}$

In Exercises 60–63, rationalize the denominator of the expression and simplify.

60. $\dfrac{3}{\sqrt{5}}$

61. $\dfrac{2}{\sqrt{6} - 1}$

62. $\dfrac{\sqrt{3}}{\sqrt{3} + 2}$

63. $\dfrac{\sqrt{5} + 4}{\sqrt{7} - \sqrt{2}}$

9 Chapter Summary

What did you learn?

	What did you learn?	Explanation and Examples	Review Exercises
9.1	Find the *n*th roots of real numbers *(p. 404)*.	Let a and b be real numbers and let n be an integer such that $n \geq 2$. If $$a = b^n$$ then b is an nth root of a. When $n = 2$, the root is a square root. When $n = 3$, the root is a cube root.	1–8
	Use the radical symbol to denote the *n*th roots of numbers *(p. 405)*.	Let a be a real number that has at least one (real number) nth root. The principal nth root of a is the nth root that has the same sign as a, and it is denoted by the radical $$\sqrt[n]{a}.$$ The positive integer n is the index of the radical, and the number a is the radicand. When $n = 2$, omit the index and write $\sqrt{a}$ rather than $\sqrt[2]{a}$.	9–18
	Approximate the values of expressions involving square roots using a calculator *(p. 407)*.	The square roots of numbers that are not perfect squares are irrational numbers. You can use a calculator to find decimal approximations of square roots.	19–28
9.2	Simplify radical expressions involving constants *(p. 412)*.	Let a and b be real numbers, variables, or algebraic expressions. If the nth roots of a and b are real, then the following property is true. $$\sqrt[n]{ab} = \sqrt[n]{a} \cdot \sqrt[n]{b}$$	29–40
	Simplify radical expressions involving variables *(p. 414)*.	If x is a real number, then $$\sqrt{x^2} = |x|.$$ For the special case in which you know that x is a nonnegative real number, you can write $\sqrt{x^2} = x$.	41–52
	Simplify radical expressions by rationalizing the denominator *(p. 416)*.	Let a and b be real numbers, variables, or algebraic expressions. If the nth roots of a and b are real, then the following property is true. $$\sqrt[n]{\frac{a}{b}} = \frac{\sqrt[n]{a}}{\sqrt[n]{b}}, \quad b \neq 0$$ A radical expression is said to be in simplest form when all three of the statements below are true. **1.** All possible nth-powered factors have been removed from each radical. **2.** No radical contains a fraction. **3.** No denominator of a fraction contains a radical.	53–68

	What did you learn?	Explanation and Examples	Review Exercises
9.3	Add and subtract like radical expressions *(p. 424)*.	You can combine like radicals by using the Distributive Property. $2\sqrt{5} + 7\sqrt{5} = (2 + 7)\sqrt{5}$ $= 9\sqrt{5}$ You may have to simplify individual radicals before combining like radicals. $\sqrt{8} - \sqrt{2} = \sqrt{4 \cdot 2} - \sqrt{2}$ $= 2\sqrt{2} - \sqrt{2}$ $= \sqrt{2}$ It is important to realize that the expression $\sqrt{a} + \sqrt{b} \neq \sqrt{a + b}$.	69–84
	Multiply radical expressions *(p. 426)*.	You can multiply radicals by using the Distributive Property, the FOIL Method, or a special product formula.	85–100
	Simplify quotients involving radicals by rationalizing the denominators *(p. 429)*.	The expressions $3 + \sqrt{6}$ and $3 - \sqrt{6}$ are called conjugates of each other. They differ only in the sign between the terms. The product of two conjugates is the difference of two squares. $(a + b)(a - b) = a^2 - b^2$ To rationalize a denominator involving radicals, multiply both the numerator and the denominator by the conjugate of the denominator.	101–104
9.4	Use the Squaring Property of Equality to solve radical equations *(p. 432)*.	Let a and b be real numbers, variables, or algebraic expressions. If $a = b$, then it follows that $a^2 = b^2$. This operation is called squaring each side of an equation.	105–116
	Solve application problems using radical equations *(p. 435)*.	Use the Pythagorean Theorem: In a right triangle, if a and b are the lengths of the legs (the two sides that form the right angle) and c is the length of the hypotenuse (the side across from the right angle), then $c^2 = a^2 + b^2$ $c = \sqrt{a^2 + b^2}$. Use the Distance Formula: The distance d between the two points (x_1, y_1) and (x_2, y_2) in a coordinate plane is $d = \sqrt{(x_2 - x_1)^2 + (y_2 - y_1)^2}$.	117–134

Review Exercises

Worked-out solutions to odd-numbered exercises at AlgebraWithinReach.com

9.1

Finding the Square Roots of a Number In Exercises 1–4, find the positive and negative square roots of the real number, if possible. (Do not use a calculator.)

1. 49
2. 100
3. −4
4. −9

Finding the nth Root(s) of a Number In Exercises 5–8, find the nth root(s), if possible. (Do not use a calculator.)

5. The cube root of −125
6. The cube root of −1
7. The fourth root of 16
8. The fourth root of 256

Finding the Principal nth Root In Exercises 9–18, find the indicated principal nth root, if possible. (Do not use a calculator.)

9. $\sqrt{121}$
10. $\sqrt{-25}$
11. $\sqrt{1.44}$
12. $\sqrt{0.09}$
13. $\sqrt{-\frac{1}{16}}$
14. $\sqrt{\frac{64}{9}}$
15. $\sqrt[3]{-27}$
16. $\sqrt[4]{16}$
17. $\sqrt[3]{\frac{8}{125}}$
18. $\sqrt[4]{\frac{1}{625}}$

Using a Calculator In Exercises 19–26, use a calculator to approximate the value of the expression. Round your answer to three decimal places.

19. $\sqrt{53}$
20. $\sqrt{6142}$
21. $\sqrt{\frac{3}{20}}$
22. $-\sqrt{\frac{45}{8}}$
23. $3 + 2\sqrt{6}$
24. $-4 - 3\sqrt{10}$
25. $\frac{5 - 3\sqrt{3}}{2}$
26. $\frac{7 + 4\sqrt{2}}{4}$

Evaluating a Radical Expression In Exercises 27 and 28, evaluate the expression when $x = -2$ and $y = 3$. Round your answer to two decimal places.

27. $\sqrt{x^2 y}$
28. $\sqrt{y^2 - 5x}$

9.2

Simplifying a Radical Expression In Exercises 29–40, simplify the radical expression.

29. $\sqrt{48}$
30. $\sqrt{72}$
31. $\sqrt{160}$
32. $\sqrt{63}$
33. $\sqrt{\frac{23}{9}}$
34. $\sqrt{\frac{26}{16}}$
35. $\sqrt{\frac{20}{9}}$
36. $\sqrt{\frac{27}{16}}$
37. $\sqrt[3]{32}$
38. $\sqrt[4]{80}$
39. $\sqrt[4]{96}$
40. $\sqrt[3]{81}$

Simplifying a Radical Expression
In Exercises 41–52, simplify the radical expression. Use absolute value signs, if appropriate.

41. $\sqrt{36x^4}$
42. $\sqrt{81z^2}$
43. $\sqrt{4y^3}$
44. $\sqrt{100u^5}$
45. $\sqrt{x^3 y}$
46. $\sqrt{x^6 y^9}$
47. $\sqrt{32a^3 b}$
48. $\sqrt{75u^4 v^2}$
49. $\sqrt[3]{8x^6}$
50. $\sqrt[3]{x^7 y^{12}}$
51. $\sqrt[4]{81y^5}$
52. $\sqrt[4]{a^8 b^{11}}$

Rationalizing a Denominator
In Exercises 53–68, rationalize the denominator of the expression and simplify. (Assume all variables are positive.)

53. $\sqrt{\frac{3}{7}}$
54. $\sqrt{\frac{7}{10}}$
55. $\frac{6}{\sqrt{3}}$
56. $\frac{15}{\sqrt{5}}$
57. $\sqrt{\frac{5}{12}}$
58. $\sqrt{\frac{13}{32}}$
59. $\frac{4}{\sqrt{x}}$
60. $\frac{7}{\sqrt{t}}$
61. $\sqrt{\frac{11a}{b}}$
62. $\sqrt{\frac{4y}{z}}$
63. $\frac{\sqrt{6x^2}}{\sqrt{27y^3}}$
64. $\frac{\sqrt{10a^2}}{\sqrt{8b^2}}$
65. $\frac{3}{\sqrt[3]{2}}$
66. $\frac{5}{\sqrt[3]{4}}$
67. $\sqrt[3]{\frac{4}{x^2}}$
68. $\sqrt[3]{\frac{7}{y^5}}$

9.3

Combining Radical Expressions
In Exercises 69–84, simplify the expression.

69. $7\sqrt{2} + 5\sqrt{2}$
70. $15\sqrt{15} - 7\sqrt{15}$
71. $3\sqrt{5} - 7\sqrt{3} + 11\sqrt{3}$
72. $5\sqrt{11} + 6\sqrt{2} - 8\sqrt{11}$
73. $3\sqrt{20} - 10\sqrt{20}$
74. $25\sqrt{98} + 2\sqrt{98}$
75. $4\sqrt{48} + 2\sqrt{3} - 5\sqrt{12}$
76. $3\sqrt{8} - 12\sqrt{50} + \sqrt{32}$
77. $\sqrt[4]{4} + 5\sqrt[4]{4}$
78. $2\sqrt[3]{7} - 6\sqrt[3]{7} + 9\sqrt[3]{7}$
79. $5\sqrt[4]{x^3} - 3\sqrt[4]{y^3} + 4\sqrt[4]{x^3}$
80. $\sqrt[5]{x} - 8\sqrt[5]{x}$
81. $\sqrt{36y} - \sqrt{16y}$
82. $\sqrt{25x} + \sqrt{49x}$
83. $\sqrt{28y^5} + 4y\sqrt{7y^3}$
84. $\sqrt{18x^3} - 3x\sqrt{2x}$

Multiplying Radical Expressions
In Exercises 85–96, multiply and simplify.

85. $\sqrt{3}(\sqrt{6} + 1)$
86. $\sqrt{7}(10 - \sqrt{7})$

87. $\sqrt[4]{6}(\sqrt[4]{2} - 1)$

88. $\sqrt[3]{5}(\sqrt[3]{4} + 2)$

89. $(\sqrt{8} + 2)(3\sqrt{2} - 1)$

90. $(2\sqrt{3} + 10)(\sqrt{2} - 3)$

91. $(\sqrt{3} - \sqrt{5})(\sqrt{3} + \sqrt{5})$

92. $(\sqrt{7} - 2)(\sqrt{7} + 2)$

93. $(\sqrt{5} - 2)^2$

94. $(\sqrt{3} + 1)^2$

95. $(\sqrt[5]{2} + 3)^2$

96. $(\sqrt[4]{3} - 1)^2$

Multiplying Conjugates In Exercises 97–100, find the conjugate of the expression. Then find the product of the expression and its conjugate.

97. $\sqrt{x} + 9$

98. $\sqrt{y} - 15$

99. $12 - \sqrt{t}$

100. $\sqrt{2} + \sqrt{z}$

Simplifying a Quotient Involving a Radical In Exercises 101–104, rationalize the denominator of the expression and simplify.

101. $\dfrac{3}{\sqrt{12} - 3}$

102. $\dfrac{9}{\sqrt{7} - \sqrt{5}}$

103. $\dfrac{\sqrt{x} - 3}{\sqrt{x} + 3}$

104. $\dfrac{3\sqrt{s} + 2}{\sqrt{s} + 1}$

9.4

Solving a Radical Equation In Exercises 105–116, solve the equation. (Some of the equations have no solution.)

105. $\sqrt{y} = 13$

106. $\sqrt{z} = 25$

107. $\sqrt{x} + 2 = 0$

108. $\sqrt{y} + 5 = 0$

109. $\sqrt{x} - 10 = 0$

110. $\sqrt{y} - 6 = 0$

111. $\sqrt{2t + 8} = 5$

112. $\sqrt{2a - 7} = 15$

113. $\sqrt{4x - 3} = \sqrt{x + 6}$

114. $\sqrt{5x + 3} = \sqrt{x + 1}$

115. $\sqrt{x - 4} = x - 6$

116. $\sqrt{2x + 7} = x + 4$

Using the Pythagorean Theorem In Exercises 117 and 118, use the Pythagorean Theorem to solve for x. Round your answer to two decimal places, if necessary.

117.

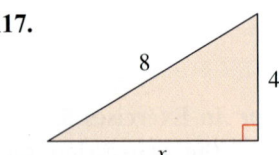

118.

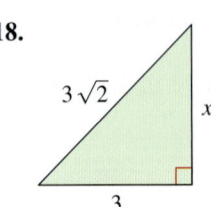

Using the Distance Formula In Exercises 119–126, find the distance between the two points. Round your answer to two decimal places, if necessary.

119. $(1, 4), (-2, 0)$

120. $(3, 8), (-3, 5)$

121. $(-5, -2), (-1, 9)$

122. $(6, 0), (1, -5)$

123. $(-4, -3), (2, -4)$

124. $(9, 20), (1, 5)$

125. $(7, 3), (2, -8)$

126. $(-6, 5), (9, -2)$

Free-Falling Object In Exercises 127 and 128, use the equation for the velocity of a free-falling object,

$$v = \sqrt{2gh}$$

where v is measured in feet per second, $g = 32$ feet per second squared, and h is the height (in feet).

127. An object strikes the ground with a velocity of 80 feet per second. Estimate the height from which it was dropped.

128. An object strikes the ground with a velocity of 50 feet per second. Estimate the height from which it was dropped.

129. *Pendulum Length* The time t (in seconds) for a pendulum of length L (in feet) to go through one complete cycle, both forward and back (its period), is given by

$$t = 2\pi\sqrt{\frac{L}{32}}.$$

How long is the pendulum of a grandfather clock with a period of 1.75 seconds (see figure)?

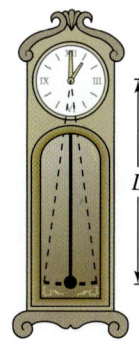

130. *Bridge* The time t (in seconds) for a free-falling object to fall d feet is given by

$$t = \sqrt{\frac{d}{16}}.$$

A child drops a rock from a bridge and sees it strike the water after approximately 1.5 seconds. Estimate the height of the bridge.

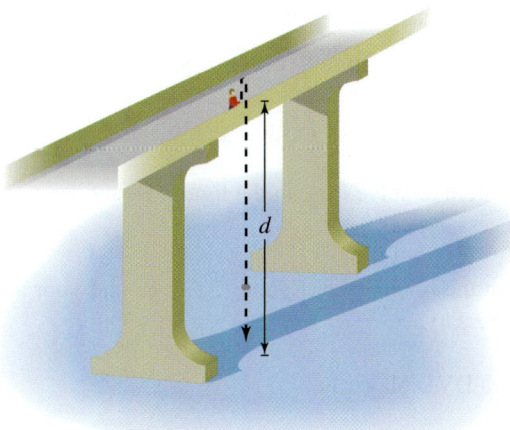

131. *Radio Tower* A guy wire on a radio tower is attached to the top of the tower and to an anchor 60 feet from the base of the tower (see figure). The tower is 100 feet high. How long is the wire?

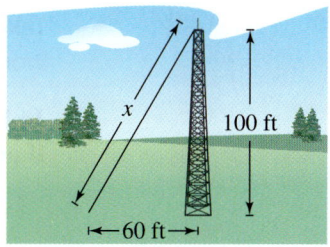

132. *Baseball* A baseball diamond is a square that is 90 feet on a side (see figure). The right fielder catches a fly ball on the first-base line approximately 70 feet beyond first base. He then throws a runner out at third base. Determine the distance d between the right fielder and third base.

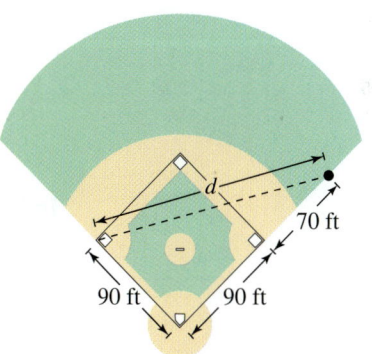

133. *Swimming Pool* A rectangular swimming pool is 20 feet wide and 42 feet long. Find the length of the diagonal of the pool.

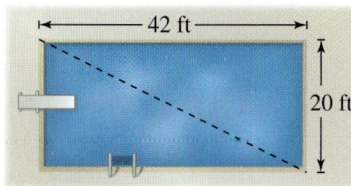

134. *Demand* The demand equation for a disposable camera is

$$p = 12 - \sqrt{0.75x - 9}$$

where x is the number of units demanded per day and p is the price per camera. Find the demand when the price is set at $6.

Chapter Test

Solutions in English & Spanish and tutorial videos at AlgebraWithinReach.com

Take this test as you would take a test in class. After you are done, check your work against the answers in the back of the book.

In Exercises 1 and 2, find the indicated principal or negative nth root, if possible. If it is not possible, explain why. (Do not use a calculator.)

1. (a) $\sqrt{121}$ (b) $\sqrt{-36}$

2. (a) $\sqrt[4]{81}$ (b) $\sqrt[3]{-64}$

In Exercises 3–8, simplify the radical expression. Use absolute value signs, if appropriate.

3. $\sqrt{28}$
4. $\sqrt[3]{54}$
5. $\sqrt{32x^2y^3}$
6. $\sqrt{\dfrac{3x^3}{y^4}}$
7. $\dfrac{5}{\sqrt{15}}$
8. $\dfrac{2}{\sqrt[3]{4}}$

In Exercises 9–16, perform the indicated operation(s).

9. $10\sqrt{2} - 7\sqrt{2}$
10. $5\sqrt{3x} + 3\sqrt{75x}$
11. $7\sqrt[3]{5} - 6\sqrt[3]{4} + \sqrt[3]{5}$
12. $4\sqrt{2x} - 6\sqrt{32x} + \sqrt{2x^2}$
13. $\sqrt{3}(2 - \sqrt{12})$
14. $\sqrt[3]{5}(\sqrt[3]{2} + 3)$
15. $(\sqrt{6} - 3)(\sqrt{6} + 5)$
16. $(1 - 2\sqrt{x})^2$

17. Find the conjugate of the expression $\sqrt{3} - 5$. Then find the product of the expression and its conjugate.

18. Rationalize the denominator: $\dfrac{10}{\sqrt{6} + 1}$.

In Exercises 19–22, solve the equation.

19. $\sqrt{y} = 5$
20. $2\sqrt{x+3} = 5$
21. $\sqrt{5x-4} = \sqrt{3x+6}$
22. $2\sqrt{6y} = y + 6$

23. Use the Pythagorean Theorem to solve for c in the figure. Round your answer to two decimal places, if necessary.

24. Find the distance between the points $(-3, 8)$ and $(5, 2)$. Round your answer to two decimal places, if necessary.

25. The demand equation for a DVD player is

$$p = 100 - \sqrt{x - 25}$$

where x is the number of units demanded per day and p is the price per player. Find the demand when the price is set at $90.

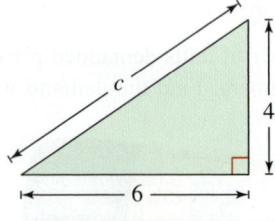

Cumulative Test: Chapters 7–9

Solutions in English & Spanish and tutorial videos at AlgebraWithinReach.com

Take this test as you would take a test in class. After you are done, check your work against the answers in the back of the book.

1. Find the domain of the rational expression $\dfrac{x-5}{x^2-4}$.

2. Fill in the missing factor: $\dfrac{7}{3x} = \dfrac{7\;\boxed{}}{18x^4}$.

In Exercises 3 and 4, simplify the expression.

3. $\dfrac{8-2x}{x^2-16}$

4. $\dfrac{x^2-3x-10}{x^2-4}$

In Exercises 5–8, perform the indicated operation and simplify.

5. $\dfrac{c}{c-1} \cdot \dfrac{c^2+9c-10}{c^3}$

6. $\dfrac{6}{(c-1)^2} \div \dfrac{8}{c^3-c^2}$

7. $\dfrac{3}{x-2} + \dfrac{x}{4-x^2}$

8. $\dfrac{5}{x-2} - \dfrac{2}{x^2}$

In Exercises 9 and 10, simplify the complex fraction.

9. $\dfrac{\left(\dfrac{9}{x}\right)}{\left(\dfrac{6}{x}+2\right)}$

10. $\dfrac{\left(a-\dfrac{1}{a}\right)}{\left(\dfrac{1}{2}+\dfrac{1}{a}\right)}$

In Exercises 11–14, solve the equation.

11. $\dfrac{5}{x} + \dfrac{3}{x} = 24$

12. $\dfrac{x}{5} - \dfrac{x}{2} = 3$

13. $\dfrac{1}{x} - \dfrac{2}{x-9} = 0$

14. $\dfrac{2x-1}{2x+1} = \dfrac{4}{5}$

In Exercises 15–20, solve the system of equations by the specified method.

15. Graphical method:
$\begin{cases} x + 5y = 0 \\ 7x + 5y - 30 \end{cases}$

16. Graphical method:
$\begin{cases} 3x + y = 2 \\ 5x \;\; y = 6 \end{cases}$

17. Substitution:
$\begin{cases} x - y = 0 \\ 5x - 3y = 10 \end{cases}$

18. Substitution:
$\begin{cases} x + 8y = 6 \\ 2x + 4y = -3 \end{cases}$

19. Elimination:
$\begin{cases} 4x + 3y = 15 \\ 2x - 5y = 1 \end{cases}$

20. Elimination:
$\begin{cases} 2x + y = 4 \\ 4x - 3y = 3 \end{cases}$

21. Give an example of a system of linear equations that has no solution.

22. Give an example of a system of linear equations that has infinitely many solutions.

In Exercises 23–26, simplify the radical expression, if possible.

23. $\sqrt{-\frac{4}{9}}$

24. $-\sqrt{\frac{4}{9}}$

25. $\sqrt{144}$

26. $\sqrt[3]{-125}$

In Exercises 27–30, simplify the radical expression. Use absolute value signs, if appropriate.

27. $-\sqrt{54}$

28. $\sqrt{50x^3}$

29. $\sqrt[3]{32u^4v^6}$

30. $\sqrt{\frac{32y}{9y^3}}$

In Exercises 31–34, perform the indicated operation and simplify.

31. $5\sqrt{x} - 3\sqrt{x}$

32. $\sqrt{7}(\sqrt{7} + 2)$

33. $(4 - \sqrt{8})^2$

34. $\dfrac{8y}{\sqrt{5} - 1}$

In Exercises 35–38, solve the equation.

35. $2\sqrt{y} = 14$

36. $\sqrt{a - 4} = 5$

37. $\sqrt{2x + 7} = 3\sqrt{x}$

38. $x(\sqrt{x} - 2) = 0$

In Exercises 39 and 40, find the distance between the two points. Round your answer to two decimal places, if necessary.

39. $(-2, -1), (1, 3)$

40. $(7, 4), (-3, 0)$

41. On the second half of a 200-mile trip, you average 10 more miles per hour than on the first half. What is your average speed on the second half of the trip if the total time for the trip is $4\frac{1}{2}$ hours?

42. A new employee takes twice as long as an experienced employee to complete a task. Together they can complete the task in 3 hours. Determine the time it takes each employee to complete the task individually.

43. The total cost of 10 gallons of regular gasoline and 12 gallons of premium gasoline is $89.30. Premium gasoline costs $0.20 more per gallon than regular gasoline. Find the price per gallon of each type of gasoline.

44. Use the Pythagorean Theorem to solve for x in the figure at the left. Round your answer to two decimal places, if necessary.

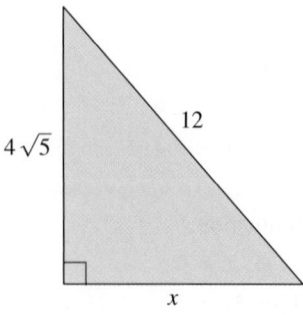

10 Quadratic Equations and Functions

- **10.1** Solving by the Square Root Property
- **10.2** Solving by Completing the Square
- **10.3** Solving by the Quadratic Formula
- **10.4** Graphing Quadratic Equations
- **10.5** Applications of Quadratic Equations
- **10.6** Complex Numbers
- **10.7** Relations, Functions, and Graphs

MASTERY IS WITHIN REACH!

"After I register for my classes, I go to the math department and ask for the syllabus for my next course. I look at how much work there will be and when I will have to do it. Then, I plan out what days I am going to need to study for the math class. It really helps when it comes to talking to my manager about work hours."

Joe
Engineering

See page 475 for suggestions about thinking ahead to the next course.

10.1 Solving by the Square Root Property

▶ Solve quadratic equations by factoring.
▶ Solve quadratic equations by the Square Root Property.

Solving Quadratic Equations by Factoring

Study Tip

In Section 6.5, you learned how to solve a quadratic equation by factoring. Example 1 reviews this procedure.

EXAMPLE 1 Solving Quadratic Equations by Factoring

Solve each quadratic equation.

a. $x^2 + 5x = 14$ b. $2x^2 = 32$

SOLUTION

a.
$x^2 + 5x = 14$	Write original equation.
$x^2 + 5x - 14 = 0$	Write in general form.
$(x + 7)(x - 2) = 0$	Factor.
$x + 7 = 0 \Rightarrow x = -7$	Set 1st factor equal to 0.
$x - 2 = 0 \Rightarrow x = 2$	Set 2nd factor equal to 0.

The solutions are $x = -7$ and $x = 2$. Check these in the original equation.

b.
$2x^2 = 32$	Write original equation.
$2x^2 - 32 = 0$	Write in general form.
$2(x^2 - 16) = 0$	Factor out common monomial factor.
$2(x + 4)(x - 4) = 0$	Factor as difference of squares.
$x + 4 = 0 \Rightarrow x = -4$	Set 1st factor equal to 0.
$x - 4 = 0 \Rightarrow x = 4$	Set 2nd factor equal to 0.

The solutions are $x = -4$ and $x = 4$. Check these in the original equation.

Exercises Within Reach® Solutions in English & Spanish and tutorial videos at AlgebraWithinReach.com

Solving a Quadratic Equation by Factoring In Exercises 1–14, solve the quadratic equation by factoring.

1. $y^2 - 3y = 0$
2. $t^2 + 5t = 0$
3. $4x^2 + 8x = 0$
4. $25y^2 - 100y = 0$
5. $a^2 - 25 = 0$
6. $16 - v^2 = 0$
7. $9m^2 = 64$
8. $16y^2 = 81$
9. $x^2 - 5x + 6 = 0$
10. $x^2 - 7x + 12 = 0$
11. $x^2 + 4x + 4 = 0$
12. $x^2 - 10x + 25 = 0$
13. $16x^2 - 40x + 25 = 0$
14. $9x^2 - 12x + 4 = 0$

Section 10.1 Solving by the Square Root Property

The Square Root Property

Study Tip

Not all quadratic equations have real number solutions. For instance, there is no real number that is the solution of the equation

$x^2 + 4 = 0$

because this would imply that there is a real number x such that $x^2 = -4$. Keep this possibility in mind as you do the exercises for this section.

> **Square Root Property**
>
> Let u be a real number, a variable, or an algebraic expression, and let d be a positive real number; then the equation $u^2 = d$ has exactly two solutions.
>
> If $u^2 = d$, then $u = \sqrt{d}$ and $u = -\sqrt{d}$.
>
> These solutions can also be written as $u = \pm\sqrt{d}$. This form of the solution is read as "u is equal to plus or minus the square root of d." Solving an equation of the form $u^2 = d$ by the Square Root Property is also called **extracting square roots**.

EXAMPLE 2 Using the Square Root Property

Solve $4x^2 = 12$ by the Square Root Property.

SOLUTION

$4x^2 = 12$	Write original equation.
$x^2 = 3$	Divide each side by 4.
$x = \pm\sqrt{3}$	Square Root Property

The solutions are $x = \sqrt{3}$ and $x = -\sqrt{3}$. Check these in the original equation.

EXAMPLE 3 Using the Square Root Property

Solve $(x - 3)^2 = 7$ by the Square Root Property.

SOLUTION

$(x - 3)^2 = 7$	Write original equation.
$x - 3 = \pm\sqrt{7}$	Square Root Property
$x = 3 \pm \sqrt{7}$	Add 3 to each side.

The solutions are $x = 3 + \sqrt{7}$ and $x = 3 - \sqrt{7}$. Check these in the original equation.

Exercises Within Reach®

Solutions in English & Spanish and tutorial videos at AlgebraWithinReach.com

Using the Square Root Property In Exercises 15–30, solve the quadratic equation by the Square Root Property. (Some equations have no real solutions.)

15. $6x^2 = 30$
16. $5x^2 = 35$
17. $7x^2 = 42$
18. $3x^2 = 33$

19. $9x^2 = 49$
20. $81z^2 = 121$
21. $u^2 - 100 = 0$
22. $v^2 - 25 = 0$

23. $x^2 + 1 = 0$
24. $a^2 + 9 = 0$
25. $(z + 4)^2 = 225$
26. $(y - 8)^2 = 121$

27. $(y - 7)^2 = 6$
28. $(t + 1)^2 = 10$
29. $(x + 6)^2 = 3$
30. $(x - 4)^2 = 5$

Chapter 10 Quadratic Equations and Functions

EXAMPLE 4 **Using the Square Root Property**

Solve the quadratic equation by the Square Root Property.

$$(3x + 5)^2 - 16 = 0$$

SOLUTION

$(3x + 5)^2 - 16 = 0$	Write original equation.
$(3x + 5)^2 = 16$	Add 16 to each side.
$3x + 5 = \pm 4$	Square Root Property
$3x = -5 \pm 4$	Subtract 5 from each side.
$x = -\frac{5}{3} \pm \frac{4}{3}$	Divide each side by 3.
$x = -\frac{5}{3} + \frac{4}{3} = -\frac{1}{3}$	First solution
$x = -\frac{5}{3} - \frac{4}{3} = -3$	Second solution

The solutions are $x = -\frac{1}{3}$ and $x = -3$. Check these in the original equation.

EXAMPLE 5 **Using the Square Root Property**

Solve the quadratic equation by the Square Root Property.

$$9(3x + 5)^2 - 16 = 0$$

SOLUTION

$9(3x + 5)^2 - 16 = 0$	Write original equation.
$9(3x + 5)^2 = 16$	Add 16 to each side.
$(3x + 5)^2 = \frac{16}{9}$	Divide each side by 9.
$3x + 5 = \pm\frac{4}{3}$	Square Root Property
$3x = -5 \pm \frac{4}{3}$	Subtract 5 from each side.
$x = -\frac{5}{3} \pm \frac{4}{9}$	Divide each side by 3.
$x = -\frac{5}{3} + \frac{4}{9} = -\frac{11}{9}$	First solution
$x = -\frac{5}{3} - \frac{4}{9} = -\frac{19}{9}$	Second solution

The solutions are $x = -\frac{11}{9}$ and $x = -\frac{19}{9}$. Check these in the original equation.

Exercises Within Reach®

Solutions in English & Spanish and tutorial videos at AlgebraWithinReach.com

Using the Square Root Property In Exercises 31–42, *solve* the quadratic equation by the Square Root Property. (Some equations have no real solutions.)

31. $(3x + 2)^2 = 9$
32. $(4x + 5)^2 = 49$
33. $(5x + 2)^2 = 5$
34. $(4x + 5)^2 = 10$
35. $(3x - 4)^2 - 27 = 0$
36. $(5x - 2)^2 - 20 = 0$

37. $(2x + 1)^2 + 4 = 0$
38. $(3x + 2)^2 + 7 = 0$
39. $4(x + 3)^2 - 25 = 0$
40. $9(x - 1)^2 - 16 = 0$
41. $8(4x + 3)^2 - 14 = 0$
42. $12(3x - 7)^2 - 15 = 0$

Application

EXAMPLE 6 Calculating a Compound Interest Rate

Five hundred dollars is deposited in an account. At the end of 2 years, the balance in the account is $561.80. The interest earned on the account is compounded annually. What is the interest rate?

SOLUTION

The mathematical model used to find the interest earned on this account is

$$A = P(1 + r)^t$$

where A is the balance in the account, P is the amount deposited, r is the annual interest rate (in decimal form), and t is the number of years.

$A = P(1 + r)^t$	Model for compound interest
$561.80 = 500(1 + r)^2$	Substitute 561.80 for A, 500 for P, and 2 for t.
$1.1236 = (1 + r)^2$	Divide each side by 500.
$1.06 = 1 + r$	Take positive square root.
$0.06 = r$	Subtract 1 from each side.

So, the interest rate is 6%. Check this in the original statement of the problem.

In Example 6, only the positive square root is used in the solution. The negative square root would have resulted in a negative value for r. In this real-life context, the value of r must be positive. Watch for this throughout the chapter.

Exercises Within Reach ®

Solutions in English & Spanish and tutorial videos at AlgebraWithinReach.com

Compound Interest The amount A in an account after 2 years when a principal of P dollars is invested at annual interest rate r (in decimal form) compounded annually is given by $A = P(1 + r)^2$. In Exercises 43 and 44, find the interest rate.

43. $P = \$1500$, $A = \$1685.40$

44. $P = \$5000$, $A = \$5724.50$

45. **Compound Interest** One thousand dollars is deposited in an account. At the end of 2 years, the balance in the account is $1166.40. The interest earned on the account is compounded annually. What is the interest rate?

46. **Compound Interest** Four hundred dollars is deposited in an account. At the end of 2 years, the balance in the account is $462.25. The interest earned on the account is compounded annually. What is the interest rate?

454   Chapter 10   Quadratic Equations and Functions

Application **EXAMPLE 7** Finding the Dimensions of an Oil Spill

An oil spill from an offshore drilling platform covers a circular region of about 10 square miles. Approximate the diameter of the region. (Use $\pi \approx 3.14$.)

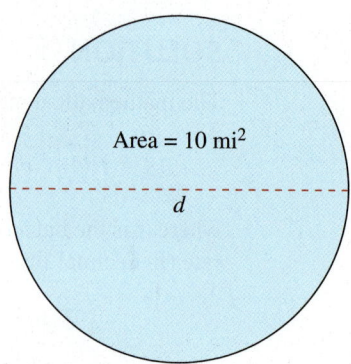

Area = 10 mi^2

d

SOLUTION

Verbal Model: Area = π · Square of Radius

Labels: Area = 10   (square miles)
Radius = r (miles)
Diameter = $2r$ (miles)

Equation:
$10 = \pi r^2$ Write equation.

$\dfrac{10}{\pi} = r^2$ Divide each side by π.

$\sqrt{\dfrac{10}{\pi}} = r$ Take positive square root.

The diameter of the oil spill is $2r = 2\sqrt{\dfrac{10}{\pi}} \approx 3.57$ miles.

Exercises Within Reach®   Solutions in English & Spanish and tutorial videos at AlgebraWithinReach.com

Geometry In Exercises 47–50, solve for x.

47. Area = 16 square centimeters

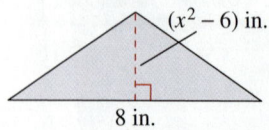

$(x + 6)$ cm

$(x + 6)$ cm

48. Area = 64 square feet

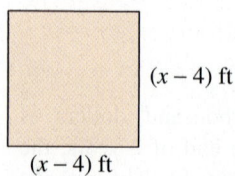

$(x - 4)$ ft

$(x - 4)$ ft

49. Area = 12 square inches

$(x^2 - 6)$ in.

8 in.

50. Area = 120 square feet

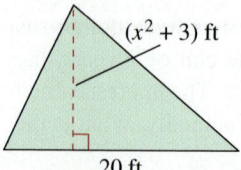

$(x^2 + 3)$ ft

20 ft

Application

EXAMPLE 8   A Falling-Object Problem

The height h (in feet) of an object dropped from the top of the Brooklyn Bridge, which is 272 feet high, is modeled by

$$h = 272 - 16t^2$$

where t is the time in seconds. How long does it take for the object to reach the water?

SOLUTION

The object hits the water when the height is 0.

$h = 272 - 16t^2$	Write equation.
$0 = 272 - 16t^2$	Let $h = 0$.
$16t^2 = 272$	Add $16t^2$ to each side.
$t^2 = 17$	Divide each side by 16.
$t = \sqrt{17} \approx 4.12$	Take positive square root.

It takes the object about 4.12 seconds to hit the water.

Exercises Within Reach®   Solutions in English & Spanish and tutorial videos at AlgebraWithinReach.com

51. **Free-Falling Object** The height h (in feet) of an object dropped from a tower 64 feet high is modeled by $h = 64 - 16t^2$, where t is the time in seconds. How long does it take for the object to reach the ground?

52. **Free-Falling Object** The height h (in feet) of a falling object on the moon is modeled by $h = -2.7t^2 + s$, where t is the time in seconds and s is the height from which the object is dropped. An astronaut drops a rock 5 feet from the surface of the moon. How long will it take for the rock to hit the surface?

> **Concept Summary:** Solving Quadratic Equations by the Square Root Property
>
> ### What
> Two ways to solve quadratic equations are (1) by factoring and (2) by the Square Root Property.
>
> **EXAMPLE**
>
> Solve $x^2 = 144$ by the Square Root Property.
>
> ### How
> To use the Square Root Property, take the square root of each side of the equation.
>
> **EXAMPLE**
>
> $x^2 = 144$ Original equation
>
> $x = \pm 12$ Square Root Property
>
> The solutions are $x = -12$ and $x = 12$.
>
> ### Why
> Many real-life situations can be modeled by quadratic equations. Knowing how to solve these types of equations will help you answer many real-life problems.

Exercises Within Reach®

Worked-out solutions to odd-numbered exercises at AlgebraWithinReach.com

Concept Summary Check

53. *Reasoning* How many solutions does $x^2 = 64$ have?

54. *Writing* In your own words, explain how to solve a quadratic equation by the Square Root Property.

55. *Reasoning* How can you recognize that a quadratic equation has no solution?

56. *Solving an Equation* Name two ways to solve the equation $2x^2 - 18 = 0$.

Extra Practice

Using the Square Root Property In Exercises 57–72, solve the quadratic equation by the Square Root Property. (Some equations have no real solutions.)

57. $2s^2 - 5 = 27$

58. $4x^2 - 7 = 93$

59. $9u^2 - 196 = 0$

60. $49v^2 - 144 = 0$

61. $\frac{1}{2}x^2 - 1 = 3$

62. $\frac{1}{5}x^2 + 4 = 5$

63. $\frac{1}{3}t^2 - 14 = 2$

64. $\frac{1}{6}z^2 + 2 = 4$

65. $\frac{1}{4}x^2 + 6 = 2$

66. $\frac{1}{3}y^2 - 5 = -6$

67. $(x - 3)^2 = 16$

68. $(x + 2)^2 = 25$

69. $3(2x + 3)^2 - 12 = 0$

70. $4(5x - 6)^2 + 36 = 0$

71. $9(7x + 4)^2 + 1 = 0$

72. $10(3x - 8)^2 - 4 = 0$

Solving a Quadratic Equation In Exercises 73–76, **factor** the left side of the equation and **solve** the resulting equation.

73. $x^2 - 12x + 36 = 49$

74. $x^2 + 6x + 9 = 36$

75. $9y^2 + 30y + 25 = 25$

76. $4x^2 - 12x + 9 = 16$

77. *Revenue* The revenue R (in dollars) when x units of a photo album are sold is modeled by

$$R = x\left(5 - \frac{1}{10}x\right), \quad 0 < x < 25.$$

Determine the number of photo albums that must be sold to produce a revenue of $60.

78. *Revenue* The revenue R (in dollars) when x units of a sewing machine are sold is modeled by

$$R = x\left(100 - \frac{1}{2}x\right), \quad 0 < x < 100.$$

Determine the number of sewing machines that must be sold to produce a revenue of $4200.

79. *Analyzing a Free-Falling Object* The height h (in feet) of an object dropped from JPMorgan Chase Tower in Houston, Texas, is modeled by $h = 1000 - 16t^2$, where t is the time in seconds.

(a) Complete the table. Round the times to two decimal places.

h	1000	950	900	850	800	750	700
t							

(b) Consecutive heights in the table differ by a constant amount of 50 feet. Do the corresponding times differ by a constant amount? Explain.

Explaining Concepts

80. *True or False?* The only solution of the equation $x^2 = 36$ is $x = 6$. Justify your answer.

81. *Writing* State whether you would use factoring or the Square Root Property to solve the quadratic equation $ax^2 + bx + c = 0$ when (a) $b = 0$ and (b) $c = 0$. Explain.

82. *Reasoning* Without attempting to solve the equations, state which of the equations do not have real solutions. Explain.

(a) $(x - 2)^2 = 0$

(b) $(x - 2)^2 = 36$

(c) $(x - 2)^2 + 36 = 0$

(d) $(x - 2)^2 - 5 = 0$

83. *Writing* Write a short paragraph explaining how to solve an equation of the form $ax^2 + c = 0$ by the Square Root Property. Use the described procedure to solve the equation $2x^2 + 6 = 0$. Can you suggest any revisions to make the procedure valid?

Cumulative Review

In Exercises 84–87, determine the quadrant in which the point is located without plotting it.

84. $(-4, 10)$

85. $(3, 55)$

86. $(2, -18)$

87. $(-1, -3)$

In Exercises 88–91, use a calculator to approximate the value of the expression. Round your answer to three decimal places.

88. $\sqrt{51}$

89. $\sqrt{88}$

90. $\sqrt{123}$

91. $\sqrt{782}$

In Exercises 92–95, find the distance between the two points. Round your answer to two decimal places, if necessary.

92. $(3, 2), (5, 8)$

93. $(-1, 0), (-4, -7)$

94. $(-8, 4), (3, 10)$

95. $(-5, -9), (-6, -10)$

10.2 Solving by Completing the Square

- Construct perfect square trinomials.
- Solve quadratic equations by completing the square.

Constructing Perfect Square Trinomials

Completing the Square

To **complete the square** for the expression

$$x^2 + bx$$

add $(b/2)^2$, which is the square of half the coefficient of x. Consequently,

$$x^2 + bx + \left(\frac{b}{2}\right)^2 = \left(x + \frac{b}{2}\right)^2.$$

EXAMPLE 1 Constructing a Perfect Square Trinomial

What term should be added to each expression to make it a perfect square trinomial? Write each new expression as the square of a binomial.

a. $x^2 + 12x$ **b.** $x^2 - 7x$

SOLUTION

a. For this expression, the coefficient of the x-term is 12. By taking half of this coefficient and squaring the result, you obtain $\left(\frac{12}{2}\right)^2 = 36$. This is the term that should be added to the expression to make it a perfect square trinomial.

$$x^2 + 12x + \left(\frac{12}{2}\right)^2 = x^2 + 12x + 36 \qquad \text{Add } \left(\frac{12}{2}\right)^2 = 36 \text{ to the expression.}$$

$$= (x + 6)^2 \qquad \text{Completed square form}$$

b. For this expression, the coefficient of the x-term is -7. By taking half of this coefficient and squaring the result, you obtain $\left(-\frac{7}{2}\right)^2 = \frac{49}{4}$. This is the term that should be added to the expression to make it a perfect square trinomial.

$$x^2 - 7x + \left(-\frac{7}{2}\right)^2 = x^2 - 7x + \frac{49}{4} \qquad \text{Add } \left(-\frac{7}{2}\right)^2 = \frac{49}{4} \text{ to the expression.}$$

$$= \left(x - \frac{7}{2}\right)^2 \qquad \text{Completed square form}$$

Exercises Within Reach® Solutions in English & Spanish and tutorial videos at AlgebraWithinReach.com

Constructing a Perfect Square Trinomial In Exercises 1–8, **determine** what term should be added to the expression to make it a perfect square trinomial. **Write** the new expression as the square of a binomial.

1. $x^2 + 8x$
2. $x^2 + 14x$
3. $y^2 - 24y$
4. $h^2 - 42h$
5. $t^2 + 3t$
6. $x^2 - x$
7. $t^2 - \frac{3}{4}t$
8. $u^2 + \frac{4}{5}u$

Completing the Square

EXAMPLE 2 Completing the Square: Leading Coefficient Is 1

Solve $x^2 + 10x = 0$ by completing the square.

$x^2 + 10x = 0$ Write original equation.

$x^2 + 10x + 5^2 = 25$ Add $5^2 = 25$ to each side.

(half of 10)^2

$(x + 5)^2 = 25$ Completed square form

$x + 5 = \pm\sqrt{25}$ Square Root Property

$x = -5 \pm 5$ Subtract 5 from each side.

$x = 0$ or $x = -10$ Solutions

The solutions are $x = 0$ and $x = -10$. Check these in the original equation.

Study Tip
In Example 2, completing the square is used for the sake of illustration. This particular equation would be easier to solve by factoring. Try reworking the problem by factoring to see that you obtain the same two solutions.

EXAMPLE 3 Completing the Square: Leading Coefficient Is 1

Solve $x^2 - 4x + 1 = 0$ by completing the square.

$x^2 - 4x + 1 = 0$ Write original equation.

$x^2 - 4x = -1$ Subtract 1 from each side.

$x^2 - 4x + (-2)^2 = -1 + 4$ Add $(-2)^2 = 4$ to each side.

(half of -4)^2

$(x - 2)^2 = 3$ Completed square form

$x - 2 = \pm\sqrt{3}$ Square Root Property

$x = 2 \pm \sqrt{3}$ Add 2 to each side.

The solutions are $x = 2 + \sqrt{3}$ and $x = 2 - \sqrt{3}$. Check these in the original equation.

Exercises Within Reach®

Solutions in English & Spanish and tutorial videos at AlgebraWithinReach.com

Completing the Square In Exercises 9–20, solve the quadratic equation by completing the square.

9. $x^2 - 8x = 0$

10. $x^2 + 14x = 0$

11. $y^2 + 20y = 0$

12. $u^2 - 16u = 0$

13. $y^2 + 4y - 1 = 0$

14. $a^2 - 10a + 15 = 0$

15. $x^2 - 8x - 2 = 0$

16. $x^2 + 6x - 3 = 0$

17. $x^2 + 2x - 35 = 0$

18. $x^2 - 6x - 27 = 0$

19. $x^2 + 3x - 4 = 0$

20. $x^2 + 5x + 6 = 0$

EXAMPLE 4 Completing the Square: Leading Coefficient Is Not 1

Solve $2x^2 + 5x = 3$ by completing the square.

SOLUTION

$$2x^2 + 5x = 3 \quad \text{Write original equation.}$$

$$x^2 + \frac{5}{2}x = \frac{3}{2} \quad \text{Divide each side by 2.}$$

$$x^2 + \frac{5}{2}x + \left(\frac{5}{4}\right)^2 = \frac{3}{2} + \frac{25}{16} \quad \text{Add } \left(\frac{5}{4}\right)^2 = \frac{25}{16} \text{ to each side.}$$

$$\left(\text{half of } \tfrac{5}{2}\right)^2$$

$$\left(x + \frac{5}{4}\right)^2 = \frac{49}{16} \quad \text{Completed square form}$$

$$x + \frac{5}{4} = \pm \frac{7}{4} \quad \text{Square Root Property}$$

$$x = -\frac{5}{4} \pm \frac{7}{4} \quad \text{Subtract } \tfrac{5}{4} \text{ from each side.}$$

The solutions are

$$x = -\frac{5}{4} + \frac{7}{4} = \frac{2}{4} = \frac{1}{2} \quad \text{and} \quad x = -\frac{5}{4} - \frac{7}{4} = -\frac{12}{4} = -3.$$

You can check these in the original equation as follows.

Check

Substitute $\frac{1}{2}$ into Equation

$$2\left(\tfrac{1}{2}\right)^2 + 5\left(\tfrac{1}{2}\right) \stackrel{?}{=} 3$$
$$2\left(\tfrac{1}{4}\right) + \tfrac{5}{2} \stackrel{?}{=} 3$$
$$\tfrac{1}{2} + \tfrac{5}{2} \stackrel{?}{=} 3$$
$$3 = 3 \checkmark$$

Substitute -3 into Equation

$$2(-3)^2 + 5(-3) \stackrel{?}{=} 3$$
$$2(9) - 15 \stackrel{?}{=} 3$$
$$18 - 15 \stackrel{?}{=} 3$$
$$3 = 3 \checkmark$$

Exercises Within Reach®

Solutions in English & Spanish and tutorial videos at AlgebraWithinReach.com

Completing the Square In Exercises 21–28, solve the quadratic equation by completing the square.

21. $2x^2 + x = 3$

22. $3x^2 + 10x = 8$

23. $3x^2 - 8x = 3$

24. $2x^2 + 15x = -25$

25. $2x^2 + 18x = -32$

26. $3x^2 - 12x = -7$

27. $2y^2 + 3y = 1$

28. $4z^2 - 3z = 2$

EXAMPLE 5 Completing the Square: Leading Coefficient Is Not 1

Solve $3x^2 - 2x - 4 = 0$ by completing the square. Use a calculator to approximate the solutions to two decimal places.

SOLUTION

$$3x^2 - 2x - 4 = 0 \qquad \text{Write original equation.}$$
$$3x^2 - 2x = 4 \qquad \text{Add 4 to each side.}$$
$$x^2 - \tfrac{2}{3}x = \tfrac{4}{3} \qquad \text{Divide each side by 3.}$$
$$x^2 - \tfrac{2}{3}x + \left(-\tfrac{1}{3}\right)^2 = \tfrac{4}{3} + \tfrac{1}{9} \qquad \text{Add } \left(-\tfrac{1}{3}\right)^2 = \tfrac{1}{9} \text{ to each side.}$$

$\left(\text{half of } -\tfrac{2}{3}\right)^2$

$$\left(x - \tfrac{1}{3}\right)^2 = \tfrac{13}{9} \qquad \text{Completed square form}$$
$$x - \tfrac{1}{3} = \pm \tfrac{\sqrt{13}}{3} \qquad \text{Square Root Property}$$
$$x = \tfrac{1}{3} \pm \tfrac{\sqrt{13}}{3} \qquad \text{Add } \tfrac{1}{3} \text{ to each side.}$$

The solutions are

$$x = \tfrac{1}{3} + \tfrac{\sqrt{13}}{3} \approx 1.54 \quad \text{and} \quad x = \tfrac{1}{3} - \tfrac{\sqrt{13}}{3} \approx -0.87.$$

Check these in the original equation.

EXAMPLE 6 A Quadratic Equation with No Real Solution

Show that $x^2 - 4x + 7 = 0$ has no real solution by completing the square.

SOLUTION

$$x^2 - 4x + 7 = 0 \qquad \text{Write original equation.}$$
$$x^2 - 4x = -7 \qquad \text{Subtract 7 from each side.}$$
$$x^2 - 4x + (-2)^2 = -7 + 4 \qquad \text{Add } (-2)^2 = 4 \text{ to each side.}$$
$$(x - 2)^2 = -3 \qquad \text{Completed square form}$$

Because the square of a real number cannot be negative, you can conclude that this equation has no real solution.

Exercises Within Reach®

Solutions in English & Spanish and tutorial videos at AlgebraWithinReach.com

Completing the Square In Exercises 29–36, solve the quadratic equation by completing the square, if possible. Use a calculator to approximate the solutions to two decimal places.

29. $2x^2 + 6x + 1 = 0$
30. $4x^2 - 12x + 7 = 0$
31. $3y^2 - y - 1 = 0$
32. $5x^2 + 3x - 4 = 0$
33. $x^2 - 2x + 3 = 0$
34. $x^2 - 6x + 14 = 0$
35. $3x^2 + 4x + 5 = 0$
36. $2z^2 - z + 1 = 0$

Application EXAMPLE 7 Using a Model

The revenue R (in dollars) from selling x units of a milk shake maker is modeled by

$$R = x\left(35 - \frac{1}{2}x\right), \quad 0 < x < 35.$$

Find the number of milk shake makers that must be sold to produce a revenue of $580.50.

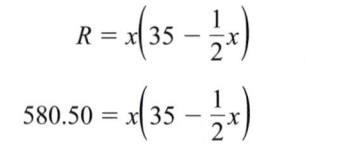

SOLUTION

$R = x\left(35 - \frac{1}{2}x\right)$	Write equation.
$580.50 = x\left(35 - \frac{1}{2}x\right)$	Let R equal 580.50.
$580.50 = 35x - \frac{1}{2}x^2$	Expand right side of equation.
$\frac{1}{2}x^2 - 35x + 580.50 = 0$	Write in standard form.
$x^2 - 70x + 1161 = 0$	Multiply each side by 2.
$x^2 - 70x = -1161$	Subtract 1161 from each side.
$x^2 - 70x + (-35)^2 = -1161 + 1225$	Add $(-35)^2 = 1225$ to each side.
(half of -70)^2	
$(x - 35)^2 = 64$	Completed square form
$x - 35 = \pm 8$	Square Root Property
$x = 35 \pm 8$	Add 35 to each side.

The solutions are $x = 35 + 8 = 43$ and $x = 35 - 8 = 27$. Because 43 is not in the interval $0 < x < 35$, the solution must be 27 units. Check this in the original statement of the problem.

Exercises Within Reach®

Solutions in English & Spanish and tutorial videos at AlgebraWithinReach.com

37. Revenue The revenue R (in dollars) from selling x units of an infant stroller is modeled by

$$R = x\left(90 - \frac{1}{2}x\right), \quad 0 < x < 90.$$

Find the number of strollers that must be sold to produce a revenue of $2800.

38. Revenue The revenue R (in dollars) from selling x units of an infant car seat is modeled by

$$R = x\left(60 - \frac{1}{4}x\right), \quad 0 < x < 120.$$

Find the number of car seats that must be sold to produce a revenue of $3200.

Application

EXAMPLE 8 Geometry: Dimensions of a Triangle

The sum of the base and the height of a triangle is 100 centimeters. The height is greater than the base. The area of the triangle is 1000 square centimeters.

a. Draw a figure that gives a visual representation of the problem.

b. The height of the triangle is h. Write an expression for the base in terms of h.

c. Write an expression for the area of the triangle in terms of h. Use the result to find the dimensions of the triangle. Round your answers to two decimal places.

SOLUTION

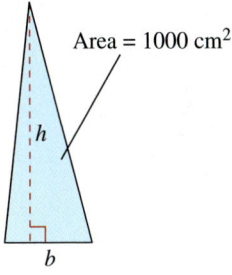

Area = 1000 cm²

a. One possible way to draw a figure showing the given information is shown at the left.

b. Because the sum of the base and the height is 100, the base must be $100 - h$.

c.

$A = \frac{1}{2}(\text{base})(\text{height})$	Write formula for area.
$1000 = \frac{1}{2}(100 - h)(h)$	Substitute given values.
$2000 = (100 - h)(h)$	Multiply each side by 2.
$2000 = 100h - h^2$	Expand right side of equation.
$h^2 - 100h + 2000 = 0$	Write in standard form.
$h^2 - 100h = -2000$	Subtract 2000 from each side.
$h^2 - 100h + (-50)^2 = -2000 + 2500$	Add $(-50)^2 = 2500$ to each side.
(half of -100)²	
$(h - 50)^2 = 500$	Completed square form
$h - 50 = \pm\sqrt{500}$	Square Root Property
$h = 50 \pm \sqrt{500}$	Add 50 to each side.

Because the height is greater than the base, the height is $50 + \sqrt{500} \approx 72.36$ centimeters and the base is $50 - \sqrt{500} \approx 27.64$ centimeters. Check this in the original statement of the problem.

Exercises Within Reach®

Solutions in English & Spanish and tutorial videos at AlgebraWithinReach.com

39. *Geometry* The sum of the base and height of a triangle is 120 centimeters. The height is greater than the base. The area of the triangle is 1785 square centimeters.

(a) Draw a figure that gives a visual representation of the problem.

(b) The height of the triangle is h. Write an expression for the base in terms of h.

(c) Write an expression for the area of the triangle in terms of h. Use the result to find the dimensions of the triangle. Round your answers to two decimal places.

40. *Geometry* The perimeter of a rectangle is 30 inches. The length is greater than the width. The area of the rectangle is 32 square inches.

(a) Draw a figure that gives a visual representation of the problem.

(b) The length of the rectangle is l. Write an expression for the width in terms of l.

(c) Write an expression for the area of the rectangle in terms of l. Use the result to find the dimensions of the rectangle. Round your answers to two decimal places.

Concept Summary: Solving Quadratic Equations by Completing the Square

What
Another way to solve a quadratic equation is by **completing the square**.

EXAMPLE
Solve $2x^2 + 12x + 10 = 0$ by completing the square.

How
1. Write the equation in the form $x^2 + bx = c$.
2. Add $(b/2)^2$ to each side.
3. Write the equation in completed square form and use the Square Root Property to solve.

EXAMPLE
$$2x^2 + 12x + 10 = 0$$
$$x^2 + 6x = -5 \quad \text{Step 1}$$
$$x^2 + 6x + 3^2 = -5 + 9 \quad \text{Step 2}$$
$$(x + 3)^2 = 4$$
$$x + 3 = \pm 2 \quad \text{Step 3}$$
$$x = -3 \pm 2$$
$$x = -1 \quad \text{and} \quad x = -5$$

Why
You can solve *any* quadratic equation by completing the square.

You can also use this method to identify quadratic equations that have no real solutions.

Exercises Within Reach ®
Worked-out solutions to odd-numbered exercises at AlgebraWithinReach.com

Concept Summary Check

41. Structure In the solution above, why is each term of the original equation divided by 2?

42. Identifying a Value In the solution above, what is the value of $(b/2)^2$?

43. Completed Square Form What is the completed square form of the equation in the solution above?

44. Identifying a Property In the solution above, what property is used to write the equation $x + 3 = \pm 2$?

Extra Practice

Solving an Equation In Exercises 45–48, solve the equation.

45. $\dfrac{x}{2} + \dfrac{1}{x} = 2$

46. $\dfrac{x}{3} + \dfrac{2}{x} = 4$

47. $\sqrt{2x + 3} = x - 2$

48. $\sqrt{4x + 5} = x - 6$

Geometry In Exercises 49 and 50, solve for x.

49. Area = 16 mm², $(x - 4)$ mm, x mm

50. Area = 115 ft², $\left(\dfrac{1}{4}x + 5\right)$ ft, x ft

51. Free-Falling Object An object is thrown upward with an initial velocity of 80 feet per second from a height of 4 feet. The height h (in feet) of the object t seconds after it is thrown is modeled by

$$h = -16t^2 + 80t + 4.$$

Find the two times when the object is at a height of 90 feet. Round your answers to two decimal places.

52. Free-Falling Object An object is thrown upward with an initial velocity of 60 feet per second from a height of 6 feet. The height h (in feet) of the object t seconds after it is thrown is modeled by

$$h = -16t^2 + 60t + 6.$$

Find the two times when the object is at a height of 30 feet. Round your answers to two decimal places.

53. Number Problem Find two consecutive positive integers such that the sum of their squares is 85.

54. Number Problem Find two consecutive positive integers such that the sum of their squares is 41.

Explaining Concepts

55. Think About It Add a term to the expression to make a perfect square trinomial.

$x^2 + \boxed{} + 144$

Explain how you obtained your answer.

56. Think About It Add a term to the expression to make a perfect square trinomial.

$x^2 - \boxed{} + 196$

Explain how you obtained your answer.

57. Error Analysis Describe and correct the error.

$9x^2 - 4x - 2 = 0$
$9x^2 - 4x = 2$
$9x^2 - 4x + 4 = 2 + 4$
$(3x - 2)^2 = 6$
$3x - 2 = \pm\sqrt{6}$
$x = \dfrac{2 \pm \sqrt{6}}{3}$

58. Reasoning Is it possible for a quadratic equation to have no real number solution? If so, give an example.

59. True or False? There exist quadratic equations with real solutions that cannot be solved by completing the square. Justify your answer.

60. Modeling The expression $x^2 + 2x$ represents the area of the figure on the left. The expression $(x + 1)^2$ represents the area of the figure on the right. Find the area of the part that was added to the left figure to "complete the square." Draw two such figures for the expression $x^2 + 5x$. What was added to complete the square?

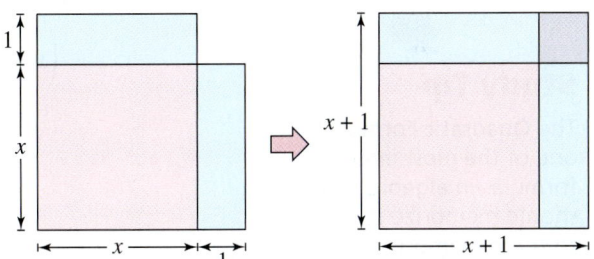

Cumulative Review

In Exercises 61–64, determine whether each pair is a solution of the system of linear equations.

61. $\begin{cases} x + 3y = 3 \\ 4x - y = -1 \end{cases}$ (a) $(0, 1)$ (b) $(3, 0)$

62. $\begin{cases} 5x - 6y = -2 \\ 7x + y = -31 \end{cases}$ (a) $(-4, -3)$ (b) $(-3, -4)$

63. $\begin{cases} -x - y = 6 \\ -5x - 2y = 3 \end{cases}$ (a) $(7, -13)$ (b) $(3, -9)$

64. $\begin{cases} 3x - 4y = -2 \\ -6x + 8y = 4 \end{cases}$ (a) $(2, 2)$ (b) $(10, 8)$

In Exercises 65–72, solve the quadratic equation by the Square Root Property. (Some equations have no real solutions.)

65. $5x^2 = 65$

66. $4y^2 = 81$

67. $u^2 + 8 = 0$

68. $16x^2 - 25 = 0$

69. $9t^2 - 72 = 28$

70. $81x^2 - 5 = 20$

71. $(x - 3)^2 = 24$

72. $(y + 4)^2 = 121$

10.3 Solving by the Quadratic Formula

▶ Use the discriminant to determine the number of real solutions of a quadratic equation.

▶ Solve quadratic equations using the Quadratic Formula.

The Quadratic Formula

Another technique for solving a quadratic equation involves the *Quadratic Formula*. This formula is obtained by completing the square for a general quadratic equation.

$$ax^2 + bx + c = 0 \quad \text{General form, } a \neq 0$$

$$ax^2 + bx = -c \quad \text{Subtract } c \text{ from each side.}$$

$$x^2 + \frac{b}{a}x = -\frac{c}{a} \quad \text{Divide each side by } a.$$

$$x^2 + \frac{b}{a}x + \left(\frac{b}{2a}\right)^2 = -\frac{c}{a} + \left(\frac{b}{2a}\right)^2 \quad \text{Add } \left(\frac{b}{2a}\right)^2 \text{ to each side.}$$

$$\left(x + \frac{b}{2a}\right)^2 = \frac{b^2 - 4ac}{4a^2} \quad \text{Simplify.}$$

$$x + \frac{b}{2a} = \pm\sqrt{\frac{b^2 - 4ac}{4a^2}} \quad \text{Square Root Property}$$

$$x = -\frac{b}{2a} \pm \frac{\sqrt{b^2 - 4ac}}{2|a|} \quad \text{Subtract } \frac{b}{2a} \text{ from each side.}$$

$$x = \frac{-b \pm \sqrt{b^2 - 4ac}}{2a} \quad \text{Simplify.}$$

Study Tip
The Quadratic Formula is one of the most important formulas in algebra, and you should memorize it. It helps to try to memorize a verbal statement of the rule. For instance, you might try to remember the following verbal statement of the Quadratic Formula: "The opposite of *b*, plus or minus the square root of *b* squared minus 4*ac*, all divided by 2*a*."

> **The Quadratic Formula**
>
> The solutions of $ax^2 + bx + c = 0$, $a \neq 0$, are given by the **Quadratic Formula**
>
> $$x = \frac{-b \pm \sqrt{b^2 - 4ac}}{2a}.$$
>
> The expression inside the radical, $b^2 - 4ac$, is called the **discriminant**.
>
> 1. If $b^2 - 4ac > 0$, the equation has two real solutions.
> 2. If $b^2 - 4ac = 0$, the equation has one (repeated) real solution.
> 3. If $b^2 - 4ac < 0$, the equation has no real solution.

Exercises Within Reach® Solutions in English & Spanish and tutorial videos at AlgebraWithinReach.com

Using the Discriminant In Exercises 1–6, use the discriminant to **determine** the number of real solutions of the quadratic equation.

1. $x^2 + 4x + 5 = 0$
2. $x^2 - 2x + 5 = 0$
3. $x^2 + 6x + 1 = 0$
4. $x^2 + 6x - 10 = 0$
5. $4x^2 + 4x + 1 = 0$
6. $9x^2 - 12x + 4 = 0$

Solving Equations by the Quadratic Formula

EXAMPLE 1 The Quadratic Formula: Two Distinct Solutions

Use the Quadratic Formula to solve $x^2 + 5x = 14$.

SOLUTION

To begin, write the equation in general form, $ax^2 + bx + c = 0$.

$x^2 + 5x = 14$ Write original equation.

$x^2 + 5x - 14 = 0$ Write in general form.

Use the values $a = 1$, $b = 5$, and $c = -14$ from the general form to substitute into the Quadratic Formula and obtain the solution.

$x = \dfrac{-b \pm \sqrt{b^2 - 4ac}}{2a}$ Quadratic formula

$x = \dfrac{-5 \pm \sqrt{5^2 - 4(1)(-14)}}{2(1)}$ Substitute 1 for a, 5 for b, and -14 for c.

$x = \dfrac{-5 \pm \sqrt{25 + 56}}{2}$ Simplify.

$x = \dfrac{-5 \pm \sqrt{81}}{2}$ Simplify.

$x = \dfrac{-5 \pm 9}{2}$ Simplify.

$x = \dfrac{-5 + 9}{2} = \dfrac{4}{2} = 2$ First solution

$x = \dfrac{-5 - 9}{2} = -\dfrac{14}{2} = -7$ Second solution

The solutions are $x = 2$ and $x = -7$. Check these in the original equation.

Study Tip

The equation in Example 1 could have been solved by factoring. Try doing so, and compare your results with those obtained in Example 1.

Exercises Within Reach®

Solutions in English & Spanish and tutorial videos at AlgebraWithinReach.com

Using the Quadratic Formula In Exercises 7–16, use the Quadratic Formula to solve the quadratic equation.

7. $x^2 - 3x - 18 = 0$

8. $x^2 - 3x - 10 = 0$

9. $x^2 + 8x + 15 = 0$

10. $x^2 + 8x + 12 = 0$

11. $t^2 - 5t = -6$

12. $t^2 + 3t = 10$

13. $x^2 - 4x = 5$

14. $x^2 - 12x = -20$

15. $-x^2 - 2x = -8$

16. $-x^2 = -3x - 4$

EXAMPLE 2 The Quadratic Formula: Two Distinct Solutions

Use the Quadratic Formula to solve $-x^2 - 2x = -4$.

SOLUTION

$-x^2 - 2x = -4$ — Write original equation.

$-x^2 - 2x + 4 = 0$ — Write in general form.

$x = \dfrac{-b \pm \sqrt{b^2 - 4ac}}{2a}$ — Quadratic Formula

$x = \dfrac{-(-2) \pm \sqrt{(-2)^2 - 4(-1)(4)}}{2(-1)}$ — Substitute -1 for a, -2 for b, and 4 for c.

$x = \dfrac{2 \pm \sqrt{4 + 16}}{-2}$ — Simplify.

$x = \dfrac{2 \pm 2\sqrt{5}}{-2}$ — Simplify radical.

$x = \dfrac{(\cancel{-2})(-1 \pm \sqrt{5})}{(\cancel{-2})}$ — Divide out common factor.

$x = -1 \pm \sqrt{5}$ — Simplify.

So, the solutions are $x = -1 + \sqrt{5} \approx 1.24$ and $x = -1 - \sqrt{5} \approx -3.24$. Check these in the original equation.

EXAMPLE 3 The Quadratic Formula: One Repeated Solution

Use the Quadratic Formula to solve $8x^2 - 24x + 18 = 0$.

Study Tip

Notice in Example 3 that 2 is factored out of the original equation. To use the Quadratic Formula, you do not have to factor out a common numerical factor; however, it makes the arithmetic easier.

SOLUTION

$8x^2 - 24x + 18 = 0$ — Write original equation.

$4x^2 - 12x + 9 = 0$ — Divide each side by 2.

$x = \dfrac{-b \pm \sqrt{b^2 - 4ac}}{2a}$ — Quadratic Formula

$x = \dfrac{-(-12) \pm \sqrt{(-12)^2 - 4(4)(9)}}{2(4)}$ — Substitute 4 for a, -12 for b, and 9 for c.

$x = \dfrac{2 \pm \sqrt{144 - 144}}{8}$ — Simplify.

$x = \dfrac{12 \pm \sqrt{0}}{8} = \dfrac{3}{2}$ — Simplify.

This equation has only one (repeated) solution, $x = \tfrac{3}{2}$. Check this in the original equation.

Exercises Within Reach ®   Solutions in English & Spanish and tutorial videos at AlgebraWithinReach.com

Using the Quadratic Formula In Exercises 17−22, use the Quadratic Formula to solve the quadratic equation.

17. $-x^2 + 6x = 7$

18. $-x^2 + 10x = 22$

19. $x^2 + 7x = 11$

20. $u^2 = -5u - 2$

21. $4x^2 - 20x + 25 = 0$

22. $9x^2 + 6x + 1 = 0$

EXAMPLE 4 The Quadratic Formula: No Real Solution

Use the Quadratic Formula to solve $2x^2 - 4x + 5 = 0$.

SOLUTION

$2x^2 - 4x + 5 = 0$ Write original equation.

$x = \dfrac{-b \pm \sqrt{b^2 - 4ac}}{2a}$ Quadratic Formula

$x = \dfrac{-(-4) \pm \sqrt{(-4)^2 - 4(2)(5)}}{2(2)}$ Substitute 2 for a, -4 for b, and 5 for c.

$x = \dfrac{4 \pm \sqrt{16 - 40}}{4}$ Simplify.

$x = \dfrac{4 \pm \sqrt{-24}}{4}$ Simplify.

Because $\sqrt{-24}$ is not a real number, you can conclude that the original equation has no real solution. Notice that -24 is the discriminant, $b^2 - 4ac$, which could have been calculated first to show that the original equation has no real solution.

Study Tip

You have now studied four ways to solve quadratic equations.
1. Square Root Property (10.1)
2. Factoring (6.5 & 10.1)
3. Completing the square (10.2)
4. The Quadratic Formula (10.3)

The guidelines at the right help you decide which of these to use.

Guidelines for Solving Quadratic Equations

1. First check to see whether you can solve the equation by the Square Root Property.
2. If you cannot use the Square Root Property, write the equation in general form and try factoring.
3. If you cannot factor the quadratic equation in general form, apply the Quadratic Formula.

Note that *completing the square* is not recommended as a practical technique—it is used more as a theoretical technique.

Remember that the Quadratic Formula can be used to solve *any* quadratic equation.

Exercises Within Reach®

Solutions in English & Spanish and tutorial videos at AlgebraWithinReach.com

Using the Quadratic Formula In Exercises 23–26, use the Quadratic Formula to solve the quadratic equation.

23. $y^2 + y + 1 = 0$

24. $3z^2 + 4z + 4 = 0$

25. $9z^2 + 10z + 4 = 0$

26. $\frac{3}{2}z^2 + 1 = 2z$

Solving a Quadratic Equation In Exercises 27–36, solve the quadratic equation by the most convenient method.

27. $x^2 = 20$

28. $t^2 = 27$

29. $y^2 + 8y = 0$

30. $7u^2 - 49u = 0$

31. $(x - 3)^2 - 75 = 0$

32. $(y - 8)^2 - 20 = 0$

33. $x^2 - 20x + 100 = 0$

34. $x^2 + 14x + 49 = 0$

35. $-2x^2 + 6x + 1 = 0$

36. $6x^2 + 20x + 5 = 0$

Application

EXAMPLE 5 Mountain Biker's Speed

A mountain biker spends a total of 5 hours going up a 25-mile mountain trail and coming back down. The biker's speed up the trail is 4 miles per hour less than the speed down the trail. What is the biker's speed coming down the trail?

SOLUTION

Verbal Model: Total time = Time up + Time down

Labels: Total time = 5 (hours)

Time up = $\dfrac{25}{x-4}$ (hours)

Time down = $\dfrac{25}{x}$ (hours)

Equation:

$5 = \dfrac{25}{x-4} + \dfrac{25}{x}$ Write equation.

$5 = \dfrac{25x}{x(x-4)} + \dfrac{25(x-4)}{x(x-4)}$ Rewrite using LCD of $x(x-4)$.

$x(x-4)(5) = x(x-4)\left(\dfrac{25x + 25x - 100}{x(x-4)}\right)$ Multiply each side by $x(x-4)$.

$5x^2 - 20x = 50x - 100$ Distribute and simplify.

$5x^2 - 70x + 100 = 0$ Write in general form.

$x^2 - 14x + 20 = 0$ Divide each side by 5.

This equation does not factor, so use the Quadratic Formula to solve the equation.

$x = \dfrac{-b \pm \sqrt{b^2 - 4ac}}{2a}$ Quadratic Formula

$x = \dfrac{-(-14) \pm \sqrt{(-14)^2 - 4(1)(20)}}{2(1)}$ Substitute 1 for a, -14 for b, and 20 for c.

$x = \dfrac{14 \pm \sqrt{116}}{2}$ Simplify.

$x = \dfrac{14 \pm 2\sqrt{29}}{2}$ Simplify radical.

$x = 7 \pm \sqrt{29}$ Solution of quadratic equation

The biker's speed coming down the trail is

$7 + \sqrt{29} \approx 12.4$ miles per hour.

The solution $7 - \sqrt{29} \approx 1.6$ is excluded because the uphill rate $x - 4$ would be negative. Check the solution in the original statement of the problem.

Study Tip

To solve Example 5, remember that

Distance = Rate × Time.

So,

Time = $\dfrac{\text{Distance}}{\text{Rate}}$.

Also, let x represent the rate coming down the trail and let $x - 4$ represent the rate going up the trail.

Exercises Within Reach® Solutions in English & Spanish and tutorial videos at AlgebraWithinReach.com

37. *Biker's Speed* A mountain biker spends a total of 4 hours going up a 20-mile mountain trail and coming back down. The biker's speed up the trail is 5 miles per hour less than the speed down the trail. What is the biker's speed coming down the trail?

38. *Hiker's Speed* A hiker spends a total of 4 hours going up a six-mile trail and coming back down. The hiker's speed up the trail is 1 mile per hour less than the speed down the trail. What is the hiker's speed coming down the trail?

Application

EXAMPLE 6 Geometry: **Dimensions of a Rectangle**

The area of the rectangle shown in the figure is 58.14 square inches. Use the Quadratic Formula to find its dimensions.

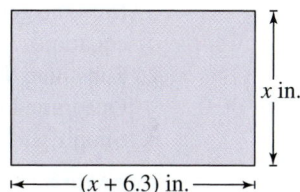

SOLUTION

Verbal Model: Area = Length × Width

Labels:
Area = 58.14 (square inches)
Width = x (inches)
Length = $x + 6.3$ (inches)

Equation:
$58.14 = x(x + 6.3)$	Write equation.
$58.14 = x^2 + 6.3x$	Distributive Property
$0 = x^2 + 6.3x - 58.14$	Write in general form.

Use the Quadratic Formula to solve the equation.

$x = \dfrac{-b \pm \sqrt{b^2 - 4ac}}{2a}$	Quadratic Formula
$x = \dfrac{-6.3 \pm \sqrt{6.3^2 - 4(1)(-58.14)}}{2(1)}$	Substitute 1 for a, 6.3 for b, and -58.14 for c.
$x = \dfrac{-6.3 \pm \sqrt{272.25}}{2}$	Simplify.
$x = \dfrac{-6.3 \pm 16.5}{2}$	Simplify radical.
$x = \dfrac{-6.3 + 16.5}{2} = 5.1$	First solution
$x = \dfrac{-6.3 - 16.5}{2} = 11.4$	Second solution

Choose the positive solution. The dimensions of the rectangle are 5.1 inches and $5.1 + 6.3 = 11.4$ inches. Check this in the original statement of the problem.

Exercises Within Reach®

Solutions in English & Spanish and tutorial videos at AlgebraWithinReach.com

39. *Geometry* The area of the rectangle shown below is 26.66 square inches. Use the Quadratic Formula to find its dimensions.

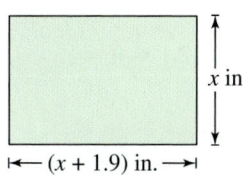

40. *Geometry* The area of the rectangle shown below is 38.25 square feet. Use the Quadratic Formula to find its dimensions.

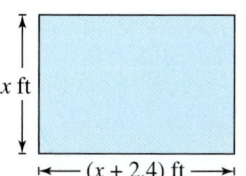

> **Concept Summary:** Solving Quadratic Equations by the Quadratic Formula
>
> ### What
> You can use the **Quadratic Formula** to solve quadratic equations.
>
> **EXAMPLE**
> Use the Quadratic Formula to solve $x^2 + 3x + 2 = 0$.
>
> ### How
> 1. Identify the values of a, b, and c from the general form of the equation.
> 2. Substitute these values into the Quadratic Formula.
> 3. Simplify to obtain the solution.
>
> **EXAMPLE**
>
> $x = \dfrac{-b \pm \sqrt{b^2 - 4ac}}{2a}$ Quadratic Formula
>
> $x = \dfrac{-3 \pm \sqrt{3^2 - 4(1)(2)}}{2(1)}$ Substitute.
>
> $x = \dfrac{-3 \pm \sqrt{1}}{2}$ Simplify.
>
> $x = \dfrac{-3 \pm 1}{2}$ Simplify.
>
> $x = -1$ and $x = -2$
>
> ### Why
> You now know four methods for solving quadratic equations.
>
> 1. Factoring
> 2. Square Root Property
> 3. Completing the square
> 4. The Quadratic Formula
>
> Remember that you can use the Quadratic Formula or completing the square to solve *any* quadratic equation.

Exercises Within Reach ®

Worked-out solutions to odd-numbered exercises at AlgebraWithinReach.com

Concept Summary Check

41. *Vocabulary* State the Quadratic Formula in words.

42. *Vocabulary* What can you determine by finding the discriminant of a quadratic equation?

43. *Four Methods* State the four methods used to solve quadratic equations.

44. *Reasoning* When is it best to use the Quadratic Formula to solve a quadratic equation?

Extra Practice

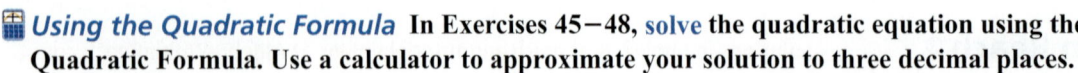
Using the Quadratic Formula In Exercises 45–48, solve the quadratic equation using the Quadratic Formula. Use a calculator to approximate your solution to three decimal places.

45. $3x^2 - 14x + 4 = 0$

46. $7x^2 + x - 35 = 0$

47. $-0.03x^2 + 2x - 0.5 = 0$

48. $1.7x^2 - 4.2x + 2.1 = 0$

Solving an Equation In Exercises 49–52, solve the equation.

49. $\dfrac{x+3}{2} - \dfrac{4}{x} = 2$

50. $\dfrac{2}{r+1} + \dfrac{2}{r} = 1$

51. $\sqrt{4x+3} = x - 1$

52. $\sqrt{3x-2} = x - 2$

53. *Free-Falling Object* An object is thrown upward with an initial velocity of 20 feet per second from the Sky Pod of the CN Tower 1465 feet above the ground. The height h (in feet) of the object t seconds after it is thrown is modeled by

   $h = -16t^2 + 20t + 1465.$

   (a) Find the two times when the object is 1465 feet above the ground.

   (b) Find the time when the object strikes the ground.

54. *Free-Falling Object* An object is thrown upward with an initial velocity of 24 feet per second from the Hearst Tower 597 feet above the ground. The height h (in feet) of the object t seconds after it is thrown is modeled by $h = -16t^2 + 24t + 597$.

   (a) Find the two times when the object is 605 feet above the ground.

   (b) Find the time when the object strikes the ground.

Explaining Concepts

55. *Reasoning* Can the Quadratic Formula be used to solve the equation $(x - 2)(x - 3) = 0$? If it can, would it be the simplest method? Explain.

56. *Writing* Explain why there is a $\pm$ symbol in the Quadratic Formula.

57. *Proof* Use the Quadratic Formula to show that the sum of the solutions of a quadratic equation is $-b/a$ and the product of the solutions is c/a.

58. *Logic* Explain how the discriminant of a quadratic equation can be used to determine the number of real solutions of the equation. Create a quadratic equation for each case.

Cumulative Review

In Exercises 59–64, find the greatest common factor of the expressions.

59. 3, 27

60. $-8y, 7y$

61. $10z^2, 5z^3$

62. $12x^5, 16x^3$

63. $4(x+1), 3(x+1)$

64. $-9(x+3), -12(x-9)$

In Exercises 65–68, solve the quadratic equation by completing the square.

65. $x^2 - 6x = 0$

66. $x^2 + 3x = 4$

67. $x^2 + 8x - 10 = 0$

68. $x^2 - 9x + 16 = 0$

Mid-Chapter Quiz: Sections 10.1–10.3

Solutions in English & Spanish and tutorial videos at AlgebraWithinReach.com

Take this quiz as you would take a quiz in class. After you are done, check your work against the answers in the back of the book.

In Exercises 1–12, solve the quadratic equation by the specified method.

1. Factoring:

    $x^2 - 7x + 10 = 0$

2. Square Root Property:

    $x^2 = 400$

3. Factoring:

    $2x^2 + 9x - 35 = 0$

4. Factoring:

    $8x(x - 4) + 3(x - 4) = 0$

5. Square Root Property:

    $x^2 - 2500 = 0$

6. Square Root Property:

    $9(z - 4)^2 - 81 = 0$

7. Completing the square:

    $y^2 + 6y - 11 = 0$

8. Completing the square:

    $4u^2 + 12u - 1 = 0$

9. Quadratic Formula:

    $x^2 + 3x + 1 = 0$

10. Quadratic Formula:

    $6x^2 - 8x - 20 = 0$

11. Quadratic Formula:

    $3x^2 - 48x = 96$

12. Quadratic Formula:

    $5x = 3x^2 + 1$

In Exercises 13–16, use the discriminant to determine the number of real solutions of the equation.

13. $x^2 + x + \frac{9}{4} = 0$

14. $y^2 - 7y - 1 = 0$

15. $3x^2 - 4x - 4 = 0$

16. $9x^2 + 6x + 1 = 0$

Applications

17. Fifteen hundred dollars is deposited in an account. At the end of 2 years, the balance in the account is $1669.54. The interest earned on the account is compounded annually. What is the interest rate?

18. On August 7, 1998, Stig Günther from Denmark dove 343 feet into a 39.4-by-49.2-by-14.8-foot air bag. His height h (in feet) after t seconds into the fall is modeled by $h = 343 - 16t^2$. How long was Günther in the air? (*Source:* Guinness Book of World Records)

19. The area of the rectangle shown below is 153.92 square inches. Find the dimensions of the rectangle.

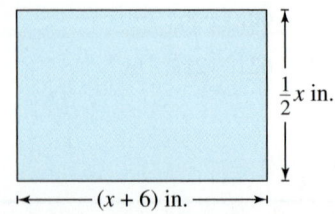

Study Skills in Action

Thinking Ahead to the Next Course

Most students do not want to think about their next math course. During the last couple of weeks of the semester, students just want to get done with math and everything else. In fact, some students want a break from math and intend to skip a semester. *Bad idea.* Students tend to forget almost all of what they just learned during the semester they take off, making the next math course more difficult when they take it.

Keep moving on with the math courses you must take. How can you make your next semester begin in a productive manner? Assess what you did well this semester and what you need to do better next semester.

Smart Study Strategy

Map Out Your Next Course

1. ▶ Sign up for your next math course if you have not already done so. If you have made friends in your class, explore signing up for the same section of the course.

2. ▶ Design your whole academic schedule in a way that allows plenty of time to study for math.
   - Select a class time when you are most alert.
   - Leave an hour free before your math class in order to prepare for it.
   - Leave an hour free after your math class so that you can review, work with a tutor, or complete homework.

3. ▶ Assess what worked well and what did not work well this semester.
   - What study strategies worked well for you?
   - What strategies did you ignore and now realize would probably help?
   - How much time do you need to study each week for math?

4. ▶ Write a plan for studying that you can use next semester. Include strategies for each of the learning tasks you must do well to be successful: class notes; textbook learning; homework; test preparation; test taking. Keep your plan in a place where you will remember to pull it out and use it.

5. ▶ If you know who your instructor is going to be, visit his or her office before the end of this semester and introduce yourself. Ask if a syllabus for the course is available. You could also go online and see if there is a website for the class.

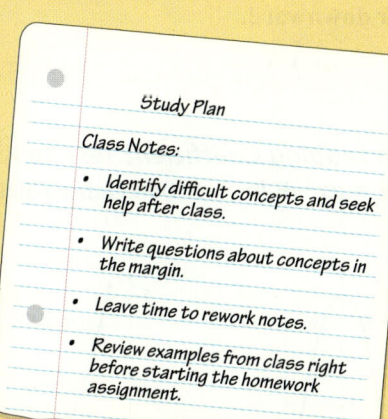

Study Plan

Class Notes:
- Identify difficult concepts and seek help after class.
- Write questions about concepts in the margin.
- Leave time to rework notes.
- Review examples from class right before starting the homework assignment.

10.4 Graphing Quadratic Equations

- Determine whether parabolas open upward or downward.
- Sketch graphs of quadratic equations using the point-plotting method.
- Sketch graphs of quadratic equations using the vertex of a parabola.

The Leading Coefficient Test

> **The Leading Coefficient Test for Parabolas**
>
> The graph of the quadratic equation $y = ax^2 + bx + c$ is a **parabola**.
>
> 1. If $a > 0$, the parabola opens upward.
> 2. If $a < 0$, the parabola opens downward.

Opens upward

Opens downward

EXAMPLE 1 Using the Leading Coefficient Test

Determine whether the parabola opens upward or downward.

$$y = -x^2 + 2x + 3$$

SOLUTION

In general form, you can see that the leading coefficient is negative.

$$y = ax^2 + bx + c$$
$$y = -x^2 + 2x + 3 \quad \Rightarrow \quad a = -1$$

The parabola opens downward, as shown.

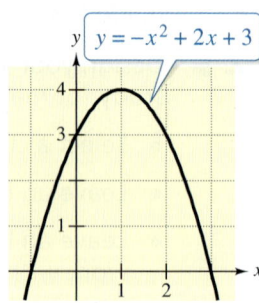

Exercises Within Reach®

Solutions in English & Spanish and tutorial videos at AlgebraWithinReach.com

Using the Leading Coefficient Test In Exercises 1–4, **determine** whether the parabola opens upward or downward.

1. $y = x^2 - 4x + 3$
2. $y = -x^2 + 2x + 2$
3. $y = -5x^2 - 4x + 8$
4. $y = 3x^2 - 5x - 1$

Using the Leading Coefficient Test In Exercises 5–8, the graph represents a quadratic equation of the form $y = ax^2 + bx + c$. **Determine** whether a is positive or negative.

5.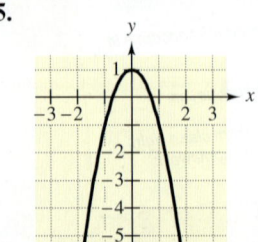
6.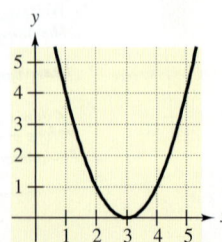
7.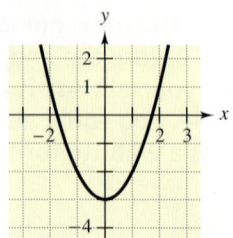
8.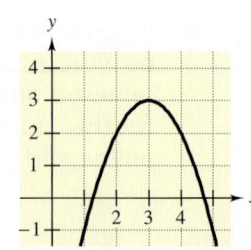

EXAMPLE 2 Using the Leading Coefficient Test

Determine whether the parabola opens upward or downward.

a. $y = 4x^2 - 1$ b. $y = 2 - 5x - 3x^2$ c. $y = 4 - 2x(3 - x)$

SOLUTION

a. In general form, you can see that the leading coefficient is positive.

$y = ax^2 + bx + c$ General form
$y = 4x^2 - 1$ ⟹ $a = 4$ Leading coefficient is positive.

The parabola opens upward, as shown below.

b. In general form, you can see that the leading coefficient is negative.

$y = ax^2 + bx + c$ General form
$y = -3x^2 - 5x + 2$ ⟹ $a = -3$ Leading coefficient is negative.

The parabola opens downward, as shown below.

c. In general form, you can see that the leading coefficient is positive.

$y = ax^2 + bx + c$ General form
$y = 4 - 6x + 2x^2$ Distributive Property
$y = 2x^2 - 6x + 4$ ⟹ $a = 2$ Leading coefficient is positive.

The parabola opens upward, as shown below.

a. 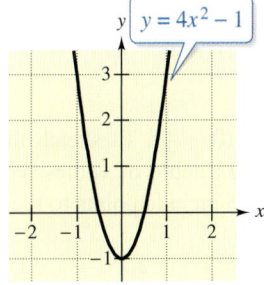 b. 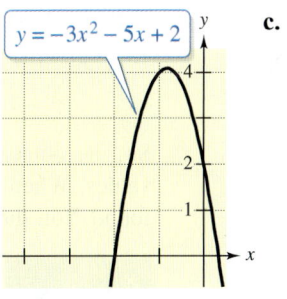 c.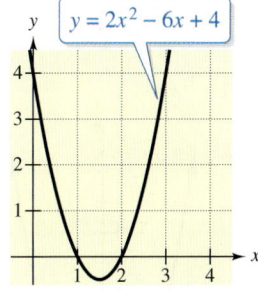

Exercises Within Reach®

Solutions in English & Spanish and tutorial videos at AlgebraWithinReach.com

Using the Leading Coefficient Test In Exercises 9–16, determine whether the parabola opens upward or downward.

9. $y = 6x^2 - 3$

10. $y = -3x^2 + 4$

11. $y = 5 - 6x + x^2$

12. $y = 6 + x - 2x^2$

13. $y = 3 + x(3 - x)$

14. $y = 6 - 2x(2 - x)$

15. $y = -(x + 1)^2 - 1$

16. $y = (x - 3)^2 - 2$

478 Chapter 10 Quadratic Equations and Functions

Sketching the Graph of a Quadratic Equation

There are three basic approaches to sketching the graph of a quadratic equation.

1. *Numerical Approach* You can create a table of values and use the point-plotting method.
2. *Analytic Approach* You can analyze the characteristics of the equation and use the results to draw the graph.
3. *Graphing Calculator* You can use a graphing calculator to graph the equation.

You will probably find that a combination of these approaches is most efficient.

EXAMPLE 3 Using the Point-Plotting Method

Find the intercepts of the graph of $y = x^2 - 4$. Then sketch the graph of the equation and label the intercepts.

SOLUTION

To find the *x*-intercepts, let *y* equal zero and solve the resulting equation for *x*.

$x^2 - 4 = 0$ Let $y = 0$ and solve for *x*.

$(x + 2)(x - 2) = 0$ Factor.

$x + 2 = 0 \implies x = -2$ Set 1st factor equal to 0.

$x - 2 = 0 \implies x = 2$ Set 2nd factor equal to 0.

From these two solutions, you can see that the graph has two *x*-intercepts: $(-2, 0)$ and $(2, 0)$. To find the *y*-intercept, let *x* equal zero in the original equation and solve for *y*, as follows.

$y = (0)^2 - 4$ Substitute 0 for *x*.

$= -4$ Simplify.

So, the *y*-intercept is $(0, -4)$. To sketch the graph of the equation, create a table of values. When creating a table of values, you should include any intercepts you have found. You should also include points to the left and right of the intercepts.

x	−3	−2	−1	0	1	2	3
$y = x^2 - 4$	5	0	−3	−4	−3	0	5
Solution point	(−3, 5)	(−2, 0)	(−1, −3)	(0, −4)	(1, −3)	(2, 0)	(3, 5)

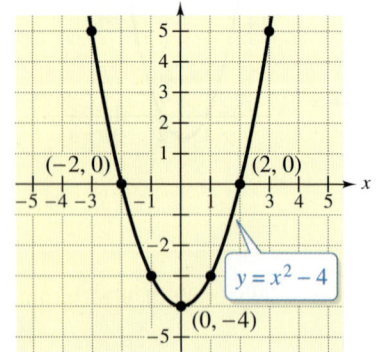

Plot the points and connect them with a smooth curve, as shown at the left. Note that the parabola opens upward because the leading coefficient is positive.

Exercises Within Reach ®

Solutions in English & Spanish and tutorial videos at AlgebraWithinReach.com

Using the Point-Plotting Method In Exercises 17−24, **find** the intercepts of the graph of the equation. Then **sketch** the graph of the equation and **label** the intercepts.

17. $y = 16 - x^2$ 18. $y = x^2 - 36$ 19. $y = x^2 - 2x$ 20. $y = 4x - x^2$

21. $y = x^2 - 4x + 3$ 22. $y = -x^2 + 8x - 12$ 23. $y = 3x^2 + 4x - 4$ 24. $y = 4x^2 - 6x + 7$

Section 10.4 Graphing Quadratic Equations 479

Finding the Vertex of a Parabola

Vertex of a Parabola

The **vertex** of a parabola given by $y = ax^2 + bx + c$ occurs at the point whose x-coordinate is

$$x = -\frac{b}{2a}.$$

To find the y-coordinate of the vertex, substitute the x-coordinate in the equation $y = ax^2 + bx + c$.

(a)
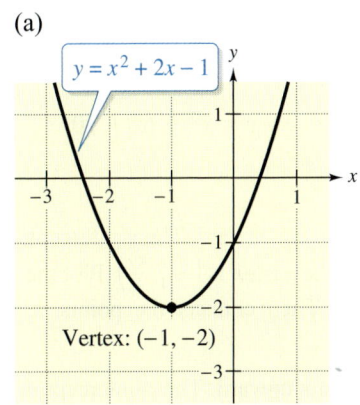

EXAMPLE 4 Finding the Vertex of a Parabola

Find the vertex of each parabola.

a. $y = x^2 + 2x - 1$ b. $y = -x^2 + 3x$

SOLUTION

a. For this equation, $a = 1$ and $b = 2$. So, the x-coordinate of the vertex is

$$x = -\frac{b}{2a} = -\frac{2}{2(1)} = -1.$$ Substitute 1 for *a* and 2 for *b*.

The y-coordinate of the vertex is

$$y = (-1)^2 + 2(-1) - 1 = -2.$$ Substitute −1 for *x*.

So, the vertex occurs at $(-1, -2)$, as shown at the left.

(b)
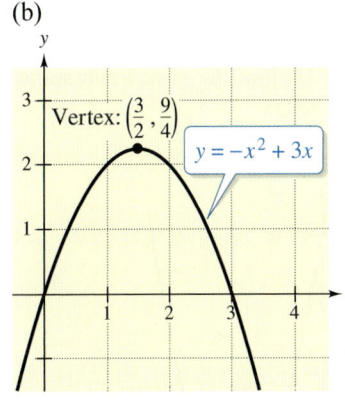

b. For this equation, $a = -1$ and $b = 3$. So, the x-coordinate of the vertex is

$$x = -\frac{b}{2a} = -\frac{3}{2(-1)} = \frac{3}{2}.$$ Substitute −1 for *a* and 3 for *b*.

The y-coordinate of the vertex is

$$y = -\left(\frac{3}{2}\right)^2 + 3\left(\frac{3}{2}\right) = \frac{9}{4}.$$ Substitute $\frac{3}{2}$ for *x*.

So, the vertex occurs at $\left(\frac{3}{2}, \frac{9}{4}\right)$, as shown at the left.

Exercises Within Reach®

Solutions in English & Spanish and tutorial videos at AlgebraWithinReach.com

Finding the Vertex of a Parabola In Exercises 25–36, find the vertex of the parabola.

25. $y = -x^2 + 2$

26. $y = 3x^2 - 3$

27. $y = x^2 - 4x + 7$

28. $y = 1 - 2x - x^2$

29. $y = 6 + 10x - x^2$

30. $y = x^2 - 12x + 9$

31. $y = x^2 + 5x - 3$

32. $y = x^2 + 3x + 4$

33. $y = 3x^2 + 6x - 8$

34. $y = -2x^2 - 4x - 1$

35. $y = 8 - 9x - 3x^2$

36. $y = 3x^2 + 4x - 1$

Guidelines for Sketching a Parabola

1. Use the Leading Coefficient Test to determine whether the parabola opens upward or downward.
2. Find and plot the x-intercepts (if any) and the y-intercept.
3. Find and plot the vertex.
4. Create a table of values that includes a few additional points.
5. Complete the graph with a smooth, U-shaped curve.

EXAMPLE 5 Sketching Parabolas

Sketch the graph of each parabola.

a. $y = -2x^2 - x + 6$ b. $y = x^2 - 4x + 4$ c. $y = x^2 - 6x + 10$

SOLUTION

a. From the Leading Coefficient Test, the parabola opens downward. The y-intercept is $(0, 6)$ and the x-intercepts are $\left(\frac{3}{2}, 0\right)$ and $(-2, 0)$. The vertex is $\left(-\frac{1}{4}, \frac{49}{8}\right)$. Plot the intercepts, the vertex, and some additional points. Then connect the points, as shown below.

b. From the Leading Coefficient Test, the parabola opens upward. The y-intercept is $(0, 4)$ and the x-intercept is $(2, 0)$. The vertex is $(2, 0)$. Plot the intercepts, the vertex, and some additional points. Then connect the points, as shown below.

c. From the Leading Coefficient Test, the parabola opens upward. The y-intercept is $(0, 10)$ and there is no x-intercept. The vertex is $(3, 1)$. Plot the vertex and some additional points. Then connect the points, as shown below.

(a) 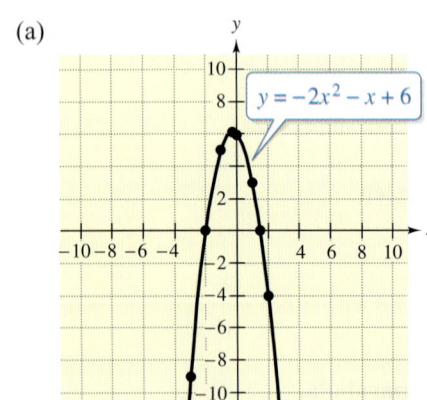 (b) 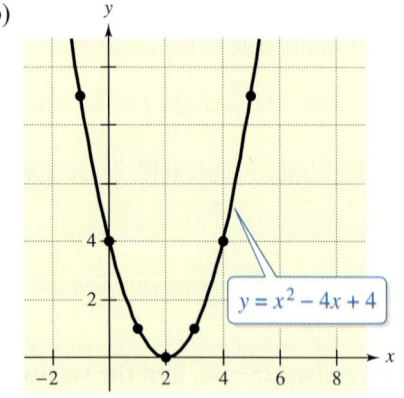 (c)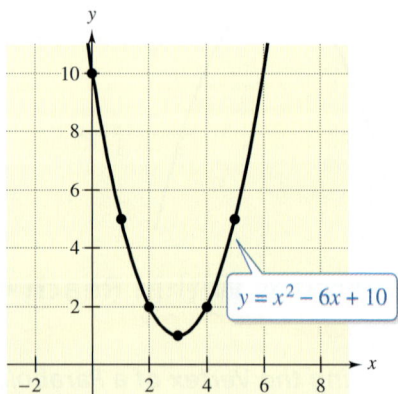

Exercises Within Reach®

Solutions in English & Spanish and tutorial videos at AlgebraWithinReach.com

Sketching a Parabola In Exercises 37–48, sketch the parabola. Label the vertex and any intercepts.

37. $y = x^2 - 1$
38. $y = x^2 - 9$
39. $y = -x^2 + 1$
40. $y = -x^2 + 9$
41. $y = x^2 - 4x$
42. $y = x^2 - 6x$
43. $y = -(x^2 + 4x + 4)$
44. $y = x^2 - 6x + 9$
45. $y = x^2 + 4x + 8$
46. $y = -(x^2 + 8x + 8)$
47. $y = x^2 - 4x + 1$
48. $y = x^2 - 6x + 12$

Application

EXAMPLE 6 Analyzing the Path of a Ball

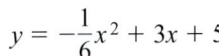

The height y (in feet) of a ball thrown on a parabolic path is modeled by

$$y = -\frac{1}{6}x^2 + 3x + 5$$

where x is the horizontal distance (in feet) from where the ball is thrown.

a. From what height is the ball thrown?
b. What is the maximum height reached by the ball?
c. How far does the ball travel horizontally through the air?

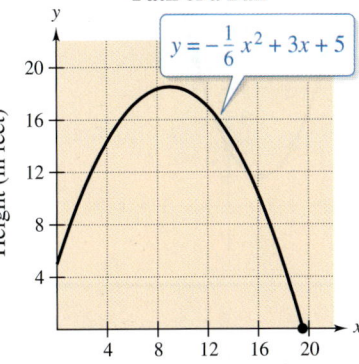

Path of a Ball

SOLUTION

a. The height from which the ball is thrown occurs when $x = 0$. When $x = 0$, the height is

$$y = -\frac{1}{6}(0)^2 + 3(0) + 5 = 5 \text{ feet.}$$

b. The maximum height occurs at the vertex of the parabolic path. The x-value of the vertex is

$$x = -\frac{b}{2a} = -\frac{3}{2\left(-\frac{1}{6}\right)} = 9. \qquad \text{Substitute } -\frac{1}{6} \text{ for } a \text{ and 3 for } b.$$

At this x-value, the height is

$$y = -\frac{1}{6}(9)^2 + 3(9) + 5 = 18.5 \text{ feet.}$$

c. The distance that the ball travels horizontally through the air corresponds to the x-intercept of the parabolic path. This intercept can be found by letting $y = 0$ and using the Quadratic Formula to solve the resulting equation.

$$x = \frac{-3 \pm \sqrt{3^2 - 4\left(-\frac{1}{6}\right)(5)}}{2\left(-\frac{1}{6}\right)} \qquad \text{Substitute } -\frac{1}{6} \text{ for } a, 3 \text{ for } b, \text{ and } 5 \text{ for } c \text{ in the Quadratic Formula.}$$

$$= \frac{-3 \pm \sqrt{\frac{37}{3}}}{-\frac{1}{3}} = 9 \pm \sqrt{111} \qquad \text{Simplify.}$$

Because you want the positive distance, the ball travels

$$x = 9 + \sqrt{111} \approx 9 + 10.5 = 19.5 \text{ feet}$$

horizontally through the air.

Exercises Within Reach®

Solutions in English & Spanish and tutorial videos at AlgebraWithinReach.com

49. *Path of a Ball* The height y (in feet) of a ball thrown on a parabolic path is modeled by $y = -0.1x^2 + 2x + 4$, where x is the horizontal distance (in feet) from where the ball is thrown.

(a) From what height is the ball thrown?

(b) What is the maximum height reached by the ball?

(c) How far does the ball travel horizontally through the air?

50. *Path of a Ball* The height y (in feet) of a ball thrown on a parabolic path is modeled by $y = -0.05x^2 + 2x + 4$, where x is the horizontal distance (in feet) from where the ball is thrown.

(a) From what height is the ball thrown?

(b) What is the maximum height reached by the ball?

(c) How far does the ball travel horizontally through the air?

Concept Summary: Graphing Quadratic Equations

What
There are three basic approaches you can use to sketch the graph of a quadratic equation.

EXAMPLE

Sketch the graph of the parabola $y = x^2 + 2x + 2$.

How
1. *Numerical Approach* Create a table of values and use the point-plotting method.
2. *Analytic Approach* Analyze the characteristics of the equation and use the results to draw the graph.
3. *Graphing Calculator* Use a graphing calculator to graph the equation.

Why
It is important to know all three approaches. You will probably find that a combination of these approaches is most efficient.

EXAMPLE

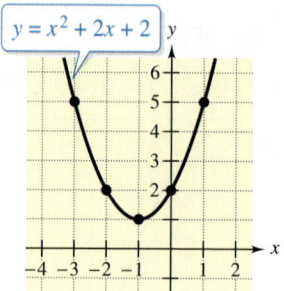

Exercises Within Reach ®

Worked-out solutions to odd-numbered exercises at AlgebraWithinReach.com

Concept Summary Check

51. Leading Coefficient Test What does the Leading Coefficient Test tell you about the parabola in the example above?

52. Identifying Intercepts How can you use a quadratic equation to find the x- and y-intercepts of its graph?

53. Identifying the Vertex How can you find the vertex of a parabola given by $y = ax^2 + bx + c$?

54. A Table of Values When creating a table of values to graph a quadratic equation, how do you choose the values to include?

Extra Practice

Think About It In Exercises 55–58, sketch the parabola and the horizontal line on the same set of coordinate axes. Find any points of intersection.

55. $y = -x^2 + 3$
    $y = -2$

56. $y = x^2 - 6x + 6$
    $y = 1$

57. $y = \frac{1}{2}x^2 - 4x + 10$
    $y = -4$

58. $y = -2x^2 + 12x - 14$
    $y = 5$

Exploration In Exercises 59–62, find two quadratic equations—one opening upward and one opening downward—whose graphs have the given x-intercepts. (There are many correct answers.)

59. $(-2, 0), (2, 0)$

60. $(-5, 0), (5, 0)$

61. $(-3, 0), (1, 0)$

62. $(2, 0), (5, 0)$

Exploration In Exercises 63 and 64, graph the equation. Then describe how the vertex can be determined from the completed square form of the equation.

63. $y = (x - 2)^2 + 3$

64. $y = -(x - 3)^2 + 1$

65. **Path of a Football** A football player kicks a punt. The path of the football is modeled by $y = -0.035x^2 + 1.4x + 1$, where y is the height (in yards) and x is the horizontal distance (in yards) from where the football is kicked.

   (a) What is the maximum height reached by the football?

   (b) The player kicks the football toward midfield from the 18-yard line. Over which yard line is the football at its maximum height?

66. **Weightless Flight** By flying in a parabolic path, a plane can allow passengers to experience weightlessness. A portion of the flight can be modeled by $y = -9.6t^2 + 624t + 22,360$, where y is the altitude (in feet) of the plane and t is the time (in seconds).

   (a) What is the maximum height of the plane?

   (b) A passenger experiences weightlessness when the altitude is at least 31,000 feet. Between what two times does weightlessness occur?

67. **Geometry** The perimeter of a rectangle is 36 meters.

   (a) The length of the rectangle is x. Write an expression for its width in terms of x.

   (b) Use the result of part (a) to write an expression for the area A of the rectangle in terms of x.

   (c) Use the result of part (b) to find the maximum area of the rectangle.

   (d) Find the dimensions of the rectangle of maximum area.

68. **Conjecture** Use the results of Exercise 67 to make a conjecture about the dimensions of the rectangle of maximum area for a given perimeter P.

Explaining Concepts

69. **Writing** Explain the relationship between the number of x-intercepts of the graph of $y = ax^2 + bx + c$ and the discriminant of $ax^2 + bx + c = 0$.

70. **Reasoning** What is the relationship between the x-coordinate of the vertex of a parabola and the x-intercepts?

71. **Structure** The parabola given by $y = (x - 6)(x - 2)$ has x-intercepts at $(2, 0)$ and $(6, 0)$.

   (a) Explain why the vertex of this parabola must be at $(4, -4)$.

   (b) Find equations of parabolas that have the given x-intercepts but that have vertices at the points $(4, 4)$, $(4, 8)$, and $(4, -8)$.

Cumulative Review

In Exercises 72–75, use the point-slope form to write an equation of the line that passes through the point and has the specified slope. Write the equation in slope-intercept form.

72. $(0, 3)$, $m = 2$

73. $(-4, 9)$, $m = -3$

74. $(2, -7)$, $m = \frac{1}{2}$

75. $(-5, -1)$, $m = -\frac{3}{5}$

In Exercises 76–79, find all integers b such that the trinomial can be factored.

76. $x^2 + bx + 4$

77. $2x^2 + bx - 9$

78. $3x^2 + bx - 5$

79. $5x^2 + bx + 8$

In Exercises 80–83, use the discriminant to determine the number of real solutions of the quadratic equation.

80. $x^2 + 3x - 4 = 0$

81. $3x^2 + 2x - 5 = 0$

82. $4x^2 - 4x + 2 = 0$

83. $2x^2 - 4x + 2 = 0$

10.5 Applications of Quadratic Equations

▶ Solve geometry problems.
▶ Use the Pythagorean Theorem to solve problems.
▶ Solve rate problems.

Solving Geometry Problems

Application

EXAMPLE 1   **Geometry: Dimensions of a Picture**

A picture is 3 inches taller than it is wide and has an area of 108 square inches. What are the dimensions of the picture?

SOLUTION

Begin by drawing a diagram, as shown at the left.

Verbal Model: Area of picture = Width · Height

Labels:
Picture width = w (inches)
Picture height = $w + 3$ (inches)
Area = 108   (square inches)

Equation:
$$108 = w(w + 3)$$
$$0 = w^2 + 3w - 108$$
$$0 = (w + 12)(w - 9)$$
$w + 12 = 0 \Rightarrow w = -12$
$w - 9 = 0 \Rightarrow w = 9$

Of the two possible solutions, choose the positive value of w and conclude that the width of the picture is 9 inches. Because the picture is 3 inches taller than it is wide, you can conclude that the height of the picture is $9 + 3 = 12$ inches. Check these dimensions in the original statement of the problem.

Exercises Within Reach®   Solutions in English & Spanish and tutorial videos at AlgebraWithinReach.com

Using a Quadratic Equation In Exercises 1 and 2, use a quadratic equation to **solve** the problem.

1. A picture is 6 inches longer than it is wide and has an area of 187 square inches. What are the dimensions of the picture?

2. A picture whose width is 7 inches less than its length has an area of 144 square inches. What are the dimensions of the picture?

Application

EXAMPLE 2 Geometry: Dimensions of a Region

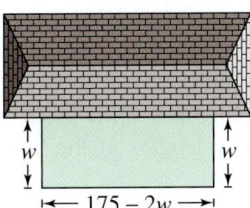

A rectangular region in a lumber yard is to be fenced for storage. The region will be fenced on three sides with 175 feet of fence, and the fourth side will be bounded by a building (see figure). The area of the fenced region is 3750 square feet. Find the dimensions of the region.

SOLUTION

Verbal Model: $\boxed{\text{Area of region}} = \boxed{\text{Width}} \cdot \boxed{\text{Height}}$

Labels:
Region width = w (feet)
Region length = $175 - 2w$ (feet)
Region area = 3750 (square feet)

Equation:
$$3750 = w(175 - 2w)$$
$$3750 = 175w - 2w^2$$
$$2w^2 - 175w + 3750 = 0$$
$$(w - 50)(2w - 75) = 0$$
$$w - 50 = 0 \implies w = 50$$
$$2w - 75 = 0$$
$$2w = 75 \implies w = 37.5$$

There are two possible solutions.

(1) The width is 50 feet, and the length is $175 - 2(50) = 75$ feet.

(2) The width is 37.5 feet, and the length is $175 - 2(37.5) = 100$ feet.

Exercises Within Reach®

Solutions in English & Spanish and tutorial videos at AlgebraWithinReach.com

Using a Quadratic Equation In Exercises 3−6, use a quadratic equation to solve the problem.

3. The original area of your yard is 9600 square feet. You plant a 4-foot-wide garden along the edge of your yard (see figure). What are the new dimensions of the yard?

4. The height of a triangle is one-third its base, and the area of the triangle is 24 square inches (see figure). Find the dimensions of the triangle.

5. A radio station advertises that its broadcasts are heard over a circular region covering approximately 10,000 square miles. Approximate the distance between the station and the listeners farthest from the station.

6. A lawn sprinkler waters a circular region of 1500 square feet. Approximate the diameter of the circular region that is watered by the sprinkler.

Using the Pythagorean Theorem

Application

EXAMPLE 3 **Geometry: Using the Pythagorean Theorem**

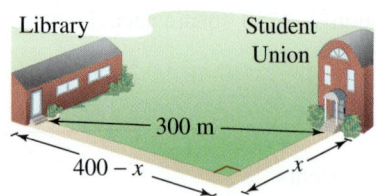

An L-shaped sidewalk from the library to the Student Union on a college campus is 400 meters long, as shown at the left. By cutting diagonally across the grass, students shorten the walking distance to 300 meters. What are the lengths of the two legs of the sidewalk?

SOLUTION

From the figure, you can see that the L-shaped sidewalk and the diagonal form a right triangle. So, to find the lengths of the two legs of the sidewalk, you can use the Pythagorean Theorem.

Common Formula: $a^2 + b^2 = c^2$ Pythagorean Theorem

Labels:
a = length of one leg = x (meters)
b = length of other leg = $400 - x$ (meters)
c = length of diagonal = 300 (meters)

Equation:
$$x^2 + (400 - x)^2 = 300^2$$
$$x^2 + 160{,}000 - 800x + x^2 = 90{,}000$$
$$2x^2 - 800x + 160{,}000 = 90{,}000$$
$$2x^2 - 800x + 70{,}000 = 0$$
$$x^2 - 400x + 35{,}000 = 0$$

Using the Quadratic Formula, you can find the solutions of this equation as follows.

$$x = \frac{-(-400) \pm \sqrt{(-400)^2 - 4(1)(35{,}000)}}{2(1)} = 200 \pm 50\sqrt{2}$$

Both solutions are positive, so it does not matter which you choose. If you let

$$x = 200 + 50\sqrt{2} \approx 270.7 \text{ meters}$$

the length of the other leg is

$$400 - x \approx 400 - 270.7 = 129.3 \text{ meters}.$$

Try choosing the other value of x to see that you obtain the same two lengths.

Exercises Within Reach®

Solutions in English & Spanish and tutorial videos at AlgebraWithinReach.com

Using the Pythagorean Theorem In Exercises 7 and 8, use the Pythagorean Theorem to **solve** the problem.

7. The perimeter of a rectangle is 68 inches and the length of the diagonal is 26 inches. Find the dimensions of the rectangle.

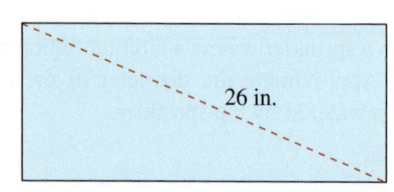

8. The perimeter of a rectangle is 84 centimeters and the length of the diagonal is 30 centimeters. Find the dimensions of the rectangle.

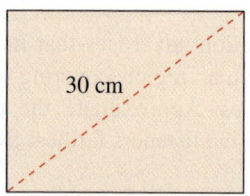

Application

EXAMPLE 4 **Geometry: Using the Pythagorean Theorem**

A windlass is used to pull a boat to a dock (see figure). The rope is attached to the boat at a point 15 feet below the level of the windlass. Find the distance from the boat to the dock when the length of the rope is 75 feet.

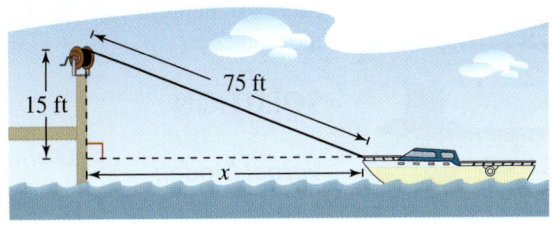

SOLUTION

Common Formula: $a^2 + b^2 = c^2$ Pythagorean Theorem

Labels:
a = length of one leg = x (feet)
b = length of other leg = 15 (feet)
c = length of diagonal = 75 (feet)

Equation:
$$x^2 + 15^2 = 75^2$$
$$x^2 + 225 = 5625$$
$$x^2 = 5400$$
$$x \approx 73.5$$

The boat is about 73.5 feet from the dock.

Exercises Within Reach®

Solutions in English & Spanish and tutorial videos at AlgebraWithinReach.com

Using the Pythagorean Theorem In Exercises 9 and 10, use the Pythagorean Theorem to solve the problem.

9. The distance between the end of a crane's arm and a beam is 100 feet (see figure). The crane's arm is 110 feet long. What is the distance between the beam and the crane?

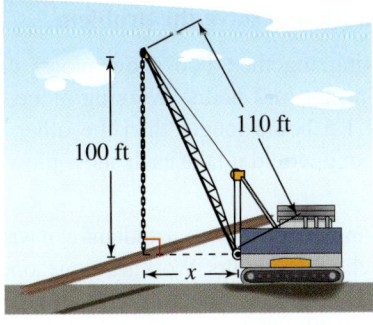

10. You are delivering pizzas to an apartment complex and a furniture store (see figure). You are required to keep a log of all mileages between stops. You forget to look at the odometer at the apartment complex, but after getting to the furniture store you record the total distance traveled from the pizza shop as 14 miles. The return distance on the beltway from the furniture store to the pizza shop is 10 miles. The route forms a right triangle. Find the possible distances from the pizza shop to the apartment complex.

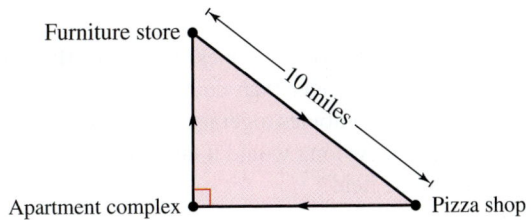

Solving Rate Problems

Application

EXAMPLE 5 Work-Rate Problem

An office has two copy machines. Machine B is known to take 12 minutes longer than machine A to copy the company's monthly report. Using both machines together, it takes 8 minutes to copy the report. How long would it take each machine alone to copy the report?

SOLUTION

Verbal Model: Rate for machine A + Rate for machine B = Rate for both machines

Labels:
Time for both machines = 8 (minutes)
Rate for both machines = $\frac{1}{8}$ (jobs per minute)
Time for machine A = t (minutes)
Rate for machine A = $1/t$ (jobs per minute)
Time for machine B = $t + 12$ (minutes)
Rate for machine B = $1/(t + 12)$ (jobs per minute)

Equation:

$$\frac{1}{t} + \frac{1}{t + 12} = \frac{1}{8}$$

$$\left(\frac{1}{t} + \frac{1}{t + 12}\right)(8t)(t + 12) = \frac{1}{8}(8t)(t + 12)$$

$$8(t + 12) + 8t = t(t + 12)$$

$$16t + 96 = t^2 + 12t$$

$$0 = t^2 - 4t - 96$$

$$0 = (t - 12)(t + 8)$$

$t - 12 = 0 \Rightarrow t = 12$

$t + 8 = 0 \Rightarrow t = -8$

By choosing the positive value of t, you can conclude that machine A would take 12 minutes to copy the report. Because machine B takes 12 minutes longer than machine A, you can conclude that machine B would take $12 + 12 = 24$ minutes to copy the report.

Exercises Within Reach®

Solutions in English & Spanish and tutorial videos at AlgebraWithinReach.com

Using a Rational Equation In Exercises 11–14, use a rational equation to solve the problem.

11. Working together, two people can complete a task in 4 hours. Working alone, one person takes 6 hours longer than the other. Working alone, how long would it take each person to complete the task?

12. Working together, two snow plows can clear a mall parking lot in 2 hours. Working alone, one snow plow takes 1.5 hours longer than the other. Working alone, how long would it take each snow plow to clear the lot?

13. A farmer has two combines. Combine B is known to take 2 hours longer than combine A to harvest a field. Using both combines together, it takes 4 hours to harvest the field. How long would it take each combine alone to harvest the field?

14. Working together, two custodians can wax all of the floors in a school in 3 hours. Working alone, one custodian takes 2 hours longer than the other. Working alone, how long would it take each custodian to wax the floors?

Application EXAMPLE 6 Reduced Rates

A ski club chartered a bus for a ski trip at a cost of $480. To lower the bus fare per skier, the club invited nonmembers to go along. When 5 nonmembers agreed to go on the trip, the fare per skier decreased by $4.80. How many club members are going on the trip?

SOLUTION

Verbal Model: Cost per skier · Number of skiers = 480

Labels:
Number of ski club members = x (people)
Number of skiers = $x + 5$ (people)
Original cost = $\dfrac{480}{x}$ (dollars per person)
New cost = $\dfrac{480}{x} - 4.80$ (dollars per person)

Equation:
$$\left(\dfrac{480}{x} - 4.80\right)(x + 5) = 480$$
$$\left(\dfrac{480 - 4.8x}{x}\right)(x + 5) = 480$$
$$(480 - 4.8x)(x + 5) = 480x, \quad x \neq 0$$
$$480x + 2400 - 4.8x^2 - 24x = 480x$$
$$-4.8x^2 - 24x + 2400 = 0$$
$$x^2 + 5x - 500 = 0$$
$$(x + 25)(x - 20) = 0$$
$$x + 25 = 0 \implies x = -25$$
$$x - 20 = 0 \implies x = 20$$

By choosing the positive value of x, you can conclude that 20 ski club members are going on the trip. Check this solution in the original statement of the problem.

Exercises Within Reach®

Solutions in English & Spanish and tutorial videos at AlgebraWithinReach.com

Using a Rational Equation In Exercises 15–18, use a rational equation to **solve** the problem.

15. A service organization paid $100 for a block of tickets to a baseball game. The block contained five more tickets than the organization needed for its members. By inviting 5 more people to attend (and share in the cost), the organization lowered the price per ticket by $1. How many people are going to the game?

16. A drama club paid $570 for a block of tickets to a musical performance. The block contained three more tickets than the club needed for its members. By inviting 3 more people to attend (and share in the cost), the club lowered the price per ticket by $9.50. How many people are going to the musical?

17. A science club chartered a bus for $360 to visit a science center. To lower the bus fare per member, the club invited nonmembers to go along. When 6 nonmembers agreed to go on the trip, the fare per person decreased by $5. How many people are going to the science center?

18. A literary club chartered a bus for $600 to attend a Shakespearean festival. To lower the bus fare per member, the club invited nonmembers to go along. When 10 nonmembers agreed to go on the trip, the fare per person decreased by $3. How many people are going to the festival?

Concept Summary: Using Quadratic Equations to Solve Problems

What
You can use quadratic equations to model many real-life problems.

EXAMPLE
A postcard is 2 inches wider than it is tall and has an area of 35 square inches. What are the dimensions of the postcard?

How
EXAMPLE
Here is one way to model this problem.

Create a Verbal Model:

$$\boxed{\text{Area of postcard}} = \boxed{\text{Height}} \cdot \boxed{\text{Width}}$$

Assign labels:

Postcard height $= t$

Postcard width $= t + 2$

Area $= 35$

Write an equation:

$35 = t(t + 2)$

Why
Notice that the equation that models the problem is a quadratic equation. You can now solve the problem by solving the quadratic equation.

EXAMPLE
$$35 = t(t + 2)$$
$$0 = t^2 + 2t - 35$$
$$0 = (t + 7)(t - 5)$$

So, $t = -7$ and $t = 5$.

Because the height of the postcard cannot be negative, the height is 5 inches and the width is 7 inches.

Exercises Within Reach®

Worked-out solutions to odd-numbered exercises at AlgebraWithinReach.com

Concept Summary Check

19. Another Method Name another method to solve the quadratic equation in the example above.

20. Identifying Operations What arithmetic operation does each word or phrase indicate?
   (a) Increased by   (b) Per
   (c) Product of   (d) Less than

21. Writing What are the four steps in solving a word problem?

22. Reasoning What formula can you use to find the length of the diagonal of the postcard in the example above?

Extra Practice

Number Problem In Exercises 23–26, find two positive integers satisfying the requirement.

23. The product of two consecutive integers is 132.

24. The product of two consecutive integers is 552.

25. The product of two consecutive odd integers is 323.

26. The product of two consecutive odd integers is 195.

Free-Falling Object In Exercises 27–30, the height h (in feet) of a falling object at any time t (in seconds) is modeled by $h = h_0 - 16t^2$, where h_0 is the initial height (in feet). Use the model to find the time it takes for an object to fall to the ground given h_0.

27. $h_0 = 1600$ **28.** $h_0 = 400$ **29.** $h_0 = 1122$ **30.** $h_0 = 2320$

Geometry In Exercises 31 and 32, use the area to find the length and width of the rectangle. Then find the perimeter.

	Width	Length	Perimeter	Area
31.	w	$2w$		50 ft²
32.	w	$1.2w$		1440 cm²

Section 10.5   Applications of Quadratic Equations

Using the Pythagorean Theorem In Exercises 33 and 34, use the Pythagorean Theorem to solve the problem.

33. A corner lot has sidewalks on two adjacent sides for a total length of 90 feet. The diagonal path across the lot is 64 feet. What are the lengths of the two sides of the sidewalk?

34. A public playground has sidewalks on two adjacent sides for a total length of 150 feet. The diagonal path across the playground is 115 feet. What are the lengths of the two sides of the sidewalk?

Using a Rational Equation In Exercises 35 and 36, use a rational equation to solve the problem.

35. A truck traveled the first 200 miles of a trip at one speed and the last 225 miles at an average speed of 5 miles per hour less. The entire trip took 10 hours. What were the two average speeds?

36. A truck traveled the first 150 miles of a trip at one speed and the last 200 miles at an average speed of 10 miles per hour less. The entire trip took 8 hours. What were the two average speeds?

Compound Interest The amount A in an account earning r percent (in decimal form) compounded annually for 2 years is given by $A = P(1 + r)^2$, where P is the original investment. In Exercises 37–40, find the interest rate r.

37. $P = \$1000$
$A = \$1123.60$

38. $P = \$2500$
$A = \$2862.25$

39. $P = \$200$
$A = \$235.44$

40. $P = \$10,000$
$A = \$11,990$

Explaining Concepts

41. *Logic* Explain why you must use a quadratic equation to solve Exercise 31.

42. *Reasoning* In what types of problems must you exclude negative solutions? Explain.

43. *Writing* Is it possible for a quadratic equation to have a rational solution and an irrational solution? Explain.

44. *Leading Coefficient Test* The height h (in feet) of an object thrown upward is modeled by $h = -16t^2 + 32t + 100$, where t is the time in seconds. Use the Leading Coefficient Test and the vertex of the parabola to explain why solving for t when $h = 125$ will result in no solution.

Cumulative Review

In Exercises 45–48, find the domain of the rational expression.

45. $\dfrac{3 + x}{2x}$

46. $\dfrac{2x}{x + 5}$

47. $\dfrac{6}{x^2 - 25}$

48. $\dfrac{3x + 8}{x^2 - 6x + 5}$

In Exercises 49–52, find the indicated principal or negative nth root.

49. $\sqrt{4}$

50. $-\sqrt{900}$

51. $\sqrt[3]{-64}$

52. $\sqrt[4]{81}$

In Exercises 53–56, find the vertex of the parabola.

53. $y = x^2 + 5$

54. $y = x^2 + 3x + 2$

55. $y = 2x^2 - 8x + 6$

56. $y = -3x^2 + 8x - 5$

10.6 Complex Numbers

- Write complex numbers in standard form.
- Perform operations with complex numbers.
- Solve quadratic equations with complex solutions.

Complex Numbers

In Section 9.1, you learned that a negative number has no *real* square root. For instance, $\sqrt{-1}$ is not real because there is no real number x such that $x^2 = -1$. So, as long as you are dealing only with real numbers, the equation $x^2 = -1$ has no solution. To overcome this deficiency, mathematicians have expanded the set of numbers by including the **imaginary unit i**, defined as

$i = \sqrt{-1}.$ Imaginary unit

This number has the property that $i^2 = -1$. So, the imaginary unit i is a solution of the equation $x^2 = -1$.

The Square Root of a Negative Number

Let c be a positive real number. Then the square root of $-c$ is given by

$$\sqrt{-c} = \sqrt{c(-1)} = \sqrt{c}\sqrt{-1} = \sqrt{c}\,i$$

When writing $\sqrt{-c}$ in the **i-form**, $\sqrt{c}\,i$, note that i is outside the radical.

EXAMPLE 1 Writing Numbers in i-Form

Write each number in i-form.

a. $\sqrt{-36}$ b. $\sqrt{-\dfrac{16}{25}}$ c. $\sqrt{-54}$ d. $\dfrac{\sqrt{-48}}{\sqrt{-3}}$

SOLUTION

a. $\sqrt{-36} = \sqrt{36(-1)} = \sqrt{36}\sqrt{-1} = 6i$

b. $\sqrt{-\dfrac{16}{25}} = \sqrt{\dfrac{16}{25}(-1)} = \sqrt{\dfrac{16}{25}}\sqrt{-1} = \dfrac{4}{5}i$

c. $\sqrt{-54} = \sqrt{54(-1)} = \sqrt{54}\sqrt{-1} = 3\sqrt{6}\,i$

d. $\dfrac{\sqrt{-48}}{\sqrt{-3}} = \dfrac{\sqrt{48}\sqrt{-1}}{\sqrt{3}\sqrt{-1}} = \dfrac{\sqrt{48}\,i}{\sqrt{3}\,i} = \sqrt{\dfrac{48}{3}} = \sqrt{16} = 4$

Exercises Within Reach®   Solutions in English & Spanish and tutorial videos at AlgebraWithinReach.com

Writing a Number in i-Form In Exercises 1−8, write the number in i-form.

1. $\sqrt{-4}$

2. $\sqrt{-9}$

3. $\sqrt{-\dfrac{4}{25}}$

4. $-\sqrt{-\dfrac{36}{121}}$

5. $-\sqrt{-8}$

6. $\sqrt{-75}$

7. $\dfrac{\sqrt{-12}}{\sqrt{-3}}$

8. $\dfrac{\sqrt{-45}}{\sqrt{-5}}$

Section 10.6 Complex Numbers

A number of the form $a + bi$, where a and b are real numbers, is called a **complex number**. The real number a is called the **real part** of the complex number $a + bi$, and the number bi is called the **imaginary part**.

> ### Definition of a Complex Number
> If a and b are real numbers, the number $a + bi$ is a **complex number**, and it is said to be written in **standard form**. If $b = 0$, the number $a + bi = a$ is a real number. If $b \neq 0$, the number $a + bi$ is called an **imaginary number**. A number of the form bi, where $b \neq 0$, is called a **pure imaginary number**.

EXAMPLE 2 Equality of Two Complex Numbers

To determine whether the complex numbers $\sqrt{9} + \sqrt{-48}$ and $3 - 4\sqrt{3}i$ are equal, begin by writing the first number in standard form.

$$\sqrt{9} + \sqrt{-48} = \sqrt{3^2} + \sqrt{4^2(3)(-1)} = 3 + 4\sqrt{3}i$$

From this form, you can see that the two numbers are not equal because they have imaginary parts that differ in sign.

EXAMPLE 3 Equality of Two Complex Numbers

Find values of x and y that satisfy the equation $3x - \sqrt{-25} = -6 + 3yi$.

SOLUTION

Begin by writing the left side of the equation in standard form.

$3x - 5i = -6 + 3yi$ Each side is in standard form.

For these two numbers to be equal, their real parts must be equal to each other and their imaginary parts must be equal to each other.

Real Parts	Imaginary Parts
$3x = -6$	$3yi = -5i$
$x = -2$	$y = -\frac{5}{3}$

So, $x = -2$ and $y = -\frac{5}{3}$.

Exercises Within Reach®

Solutions in English & Spanish and tutorial videos at AlgebraWithinReach.com

Equality of Two Complex Numbers In Exercises 9–12, **determine** whether the complex numbers are equal.

9. $\sqrt{64} + \sqrt{-25}$ and $8 + 5i$

10. $\sqrt{16} + \sqrt{-9}$ and $4 - 3i$

11. $\sqrt{27} - \sqrt{-8}$ and $3\sqrt{3} + 2\sqrt{2}i$

12. $\sqrt{12} - \sqrt{-18}$ and $2\sqrt{3} - 3\sqrt{2}i$

Equality of Two Complex Numbers In Exercises 13–18, **find** the values of a and b that satisfy the equation.

13. $-4 - \sqrt{-8} = a + bi$

14. $\sqrt{-36} - 3 = a + bi$

15. $3a + \sqrt{-81} = 15 + 3bi$

16. $5a + \sqrt{-100} = 20 + 4bi$

17. $8 + 2bi = 2a + \sqrt{-49}$

18. $-32 - 3bi = 4a + \sqrt{-144}$

Operations with Complex Numbers

EXAMPLE 4 **Adding and Subtracting Complex Numbers**

a. $(3 - i) + (-2 + 4i) = (3 - 2) + (-1 + 4)i = 1 + 3i$

b. $3i + (5 - 3i) = 5 + (3 - 3)i = 5$

c. $4 - (-1 + 5i) + (7 + 2i) = [4 - (-1) + 7] + (-5 + 2)i = 12 - 3i$

d. $(6 + 3i) + (2 - \sqrt{-8}) - \sqrt{-4} = (6 + 3i) + (2 - 2\sqrt{2}i) - 2i$
$= (6 + 2) + (3 - 2\sqrt{2} - 2)i$
$= 8 + (1 - 2\sqrt{2})i$

EXAMPLE 5 **Multiplying Complex Numbers**

a. $(7i)(-3i) = -21i^2$ — Multiply.
$= -21(-1) = 21$ — $i^2 = -1$

b. $(1 - i)(\sqrt{-9}) = (1 - i)(3i)$ — Write in i-form.
$= 3i - 3i^2$ — Distributive Property
$= 3i - 3(-1) = 3 + 3i$ — $i^2 = -1$

c. $(2 - i)(4 + 3i) = 8 + 6i - 4i - 3i^2$ — FOIL Method
$= 8 + 6i - 4i - 3(-1)$ — $i^2 = -1$
$= 11 + 2i$ — Combine like terms.

d. $(3 + 2i)(3 - 2i) = 3^2 - (2i)^2$ — Special product formula
$= 9 - 4i^2$ — Simplify.
$= 9 - 4(-1) = 13$ — $i^2 = -1$

Exercises Within Reach®

Solutions in English & Spanish and tutorial videos at AlgebraWithinReach.com

Adding and Subtracting Complex Numbers In Exercises 19–30, perform the operation(s) and write the result in standard form.

19. $(4 - 3i) + (6 + 7i)$
20. $(-10 + 2i) + (4 - 7i)$
21. $(-4 - 7i) + (-10 - 33i)$
22. $(15 + 10i) - (2 + 10i)$
23. $(3 + 4i) - (-2 - 5i)$
24. $(7 - 3i) - (6 - i)$
25. $13i - (14 - 7i)$
26. $20i + (9 - 26i)$
27. $6 - (3 - 4i) + (4 + 2i)$
28. $22 + (-5 + 8i) - (7 - 10i)$
29. $(15 - 3i) + (5 - \sqrt{-12}) - \sqrt{-81}$
30. $(9 + 2i) + (1 + \sqrt{-4}) - \sqrt{-2}$

Multiplying Complex Numbers In Exercises 31–42, perform the operation and write the result in standard form.

31. $(3i)(12i)$
32. $(-5i)(4i)$
33. $(3i)(-8i)$
34. $(-2i)(-10i)$
35. $(9 - 2i)(\sqrt{-4})$
36. $(11 + 3i)(\sqrt{-25})$
37. $(-4 - 5i)(4 - 2i)$
38. $(7 + 7i)(-7 + 4i)$
39. $(2 + 8i)(5 - 6i)$
40. $(9 - 12i)(15 + 18i)$
41. $(3 + 5i)(3 - 5i)$
42. $(2 + 15i)(2 - 15i)$

In Example 5(d), note that the product of two complex numbers can be a real number. This occurs with pairs of complex numbers of the form $a + bi$ and $a - bi$, called **complex conjugates**. In general, the product of complex conjugates has the following form.

$$(a + bi)(a - bi) = a^2 - (bi)^2 = a^2 - b^2i^2 = a^2 - b^2(-1) = a^2 + b^2$$

EXAMPLE 6 Dividing Complex Numbers

a.
$$\frac{2 - i}{4i} = \frac{2 - i}{4i} \cdot \frac{(-4i)}{(-4i)}$$ Multiply numerator and denominator by complex conjugate of denominator.

$$= \frac{-8i + 4i^2}{-16i^2}$$ Multiply fractions.

$$= \frac{-8i + 4(-1)}{-16(-1)}$$ $i^2 = -1$

$$= \frac{-8i - 4}{16}$$ Simplify.

$$= -\frac{1}{4} - \frac{1}{2}i$$ Write in standard form.

b.
$$\frac{5}{3 - 2i} = \frac{5}{3 - 2i} \cdot \frac{3 + 2i}{3 + 2i}$$ Multiply numerator and denominator by complex conjugate of denominator.

$$= \frac{5(3 + 2i)}{(3 - 2i)(3 + 2i)}$$ Multiply fractions.

$$= \frac{5(3 + 2i)}{3^2 + 2^2}$$ Product of complex conjugates

$$= \frac{15 + 10i}{13}$$ Simplify.

$$= \frac{15}{13} + \frac{10}{13}i$$ Write in standard form.

Exercises Within Reach®

Solutions in English & Spanish and tutorial videos at AlgebraWithinReach.com

Multiplying Complex Conjugates In Exercises 43–46, multiply the number by its complex conjugate and simplify.

43. $2 + i$

44. $3 + 4i$

45. $-2 - 8i$

46. $10 - 3i$

Dividing Complex Numbers In Exercises 47–54, write the quotient in standard form.

47. $\dfrac{2 + i}{-5i}$

48. $\dfrac{1 + i}{3i}$

49. $\dfrac{6 + 7i}{8i}$

50. $\dfrac{3 + 9i}{7i}$

51. $\dfrac{-12}{2 + 7i}$

52. $\dfrac{15}{5 - 2i}$

53. $\dfrac{3i}{5 + 2i}$

54. $\dfrac{4i}{5 - 3i}$

Quadratic Equations with Complex Solutions

> **Square Root Property (Complex Square Roots)**
>
> The equation $u^2 = d$, $d < 0$, has exactly two solutions:
>
> $u = \sqrt{|d|}\,i$ and $u = -\sqrt{|d|}\,i$.
>
> These solutions can also be written as $u = \pm\sqrt{|d|}\,i$.

EXAMPLE 7   **Quadratic Equations with Complex Solutions**

a. $x^2 + 8 = 0$ Original equation

$x^2 = -8$ Subtract 8 from each side.

$x = \pm\sqrt{8}\,i = \pm 2\sqrt{2}\,i$ Square Root Property

The solutions are $x = 2\sqrt{2}\,i$ and $x = -2\sqrt{2}\,i$. Check these in the original equation.

b. $(x - 4)^2 = -3$ Original equation

$x - 4 = \pm\sqrt{3}\,i$ Square Root Property

$x = 4 \pm \sqrt{3}\,i$ Add 4 to each side.

The solutions are $x = 4 + \sqrt{3}\,i$ and $x = 4 - \sqrt{3}\,i$. Check these in the original equation.

c. $2(3x - 5)^2 + 32 = 0$ Original equation

$2(3x - 5)^2 = -32$ Subtract 32 from each side.

$(3x - 5)^2 = -16$ Divide each side by 2.

$3x - 5 = \pm 4i$ Square Root Property

$3x = 5 \pm 4i$ Add 5 to each side.

$x = \dfrac{5 \pm 4i}{3}$ Divide each side by 3.

The solutions are $x = \dfrac{5 + 4i}{3}$ and $x = \dfrac{5 - 4i}{3}$. Check these in the original equation.

Exercises Within Reach®

Solutions in English & Spanish and tutorial videos at AlgebraWithinReach.com

Using the Square Root Property In Exercises 55–66, solve the equation by using the Square Root Property.

55. $z^2 = -36$

56. $x^2 = -9$

57. $x^2 + 32 = 0$

58. $x^2 + 80 = 0$

59. $(t - 3)^2 = -25$

60. $(x + 5)^2 = -81$

61. $(2x - 5)^2 = -54$

62. $(6y - 5)^2 = -8$

63. $9(x + 6)^2 + 121 = 0$

64. $4(x - 4)^2 + 169 = 0$

65. $2(9x - 4)^2 + 50 = 0$

66. $3(7x + 12)^2 + 192 = 0$

EXAMPLE 8 A Quadratic Equation with Complex Solutions

Solve $x^2 - 4x + 8 = 0$ by completing the square.

SOLUTION

$x^2 - 4x + 8 = 0$	Write original equation.
$x^2 - 4x = -8$	Subtract 8 from each side.
$x^2 - 4x + (-2)^2 = -8 + 4$	Add $(-2)^2 = 4$ to each side.
$(x - 2)^2 = -4$	Completed square form
$x - 2 = \pm 2i$	Square Root Property
$x = 2 \pm 2i$	Add 2 to each side.

The solutions are $x = 2 + 2i$ and $x = 2 - 2i$. Check these in the original equation.

EXAMPLE 9 A Quadratic Equation with Complex Solutions

Solve $x^2 - 4x + 5 = 0$ by using the Quadratic Formula.

SOLUTION

$x^2 - 4x + 5 = 0$	Write original equation.
$x = \dfrac{-b \pm \sqrt{b^2 - 4ac}}{2a}$	Quadratic Formula
$x = \dfrac{-(-4) \pm \sqrt{(-4)^2 - 4(1)(5)}}{2(1)}$	Substitute 1 for a, -4 for b, and 5 for c.
$x = \dfrac{4 \pm \sqrt{-4}}{2}$	Simplify.
$x = \dfrac{4 \pm 2i}{2}$	Write in i-form.
$x = \dfrac{2(2 \pm i)}{2}$	Factor numerator.
$x = \dfrac{2(2 \pm i)}{2}$	Divide out common factor.
$x = 2 \pm i$	Solutions

The solutions are $x = 2 + i$ and $x = 2 - i$. Check these in the original equation.

Exercises Within Reach ®
Solutions in English & Spanish and tutorial videos at AlgebraWithinReach.com

Completing the Square In Exercises 67–70, solve the equation by completing the square.

67. $x^2 + 10 = 6x$

68. $z^2 + 4z + 13 = 0$

69. $-x^2 + x - 1 = 0$

70. $-y^2 + 5y - 9 = 0$

Using the Quadratic Formula In Exercises 71–74, solve the equation by using the Quadratic Formula.

71. $x^2 - 8x + 19 = 0$

72. $2x^2 - x + 1 = 0$

73. $2x^2 + 3x + 3 = 0$

74. $8x^2 - 6x + 2 = 0$

Concept Summary: Complex Numbers

What
The **imaginary unit** *i* is used to define the square root of a negative number.
- $i = \sqrt{-1}$
- $\sqrt{-c} = \sqrt{c}\,i$, where $c > 0$.

A number of the form $a + bi$, where a and b are real numbers, is called a **complex number**.

Performing operations with complex numbers is similar to performing operations with polynomials.

EXAMPLE
Perform each operation.
a. $(2 - i) + (4 + 2i)$
b. $(1 + i)(3 - 2i)$
c. $\dfrac{3}{i}$

How
To perform operations with complex numbers:
- Combine like terms to add and subtract.
- Use the Commutative, Associative, and Distributive Properties along with the FOIL Method to multiply.
- Use **complex conjugates** to simplify quotients.

EXAMPLE
a. $(2 - i) + (4 + 2i)$
$= (2 + 4) + (-1 + 2)i = 6 + i$

b. $(1 + i)(3 - 2i)$
$= 3 - 2i + 3i - 2i^2$
$= 3 - 2i + 3i - 2(-1) = 5 + i$

c. $\dfrac{3}{i} = \dfrac{3}{i} \cdot \dfrac{-i}{-i} = \dfrac{-3i}{-(-1)} = -3i$

Why
The solutions of some quadratic equations are complex numbers. For instance, when a quadratic equation has no real solution, it has exactly two complex solutions. Knowing how to perform operations with complex numbers will help you identify such solutions.

Exercises Within Reach®
Worked-out solutions to odd-numbered exercises at AlgebraWithinReach.com

Concept Summary Check

75. *Writing a Number in i-Form* Write $\sqrt{-2}$ in *i*-form.

76. *Identifying Like Terms* Identify any like terms in the expression $(2 - i) + (4 + 2i)$.

77. *Multiplying Complex Numbers* What method is used to multiply $(1 + i)$ and $(3 - 2i)$ in the solution above?

78. *Complex Conjugate* What is the complex conjugate of *i*? What is the product of *i* and its complex conjugate?

Extra Practice

Writing a Number in i-Form In Exercises 79 and 80, write the number in *i*-form.

79. $\sqrt{-\dfrac{18}{64}}$

80. $\sqrt{-\dfrac{20}{49}}$

Operations with Complex Numbers In Exercises 81–86, perform the operation(s) and write the result in standard form.

81. $(30 - i) - (18 + 6i) + 3$

82. $(4 + 6i) + (15 + 24i) - 10$

83. $(-6i)(-i)(6i)$

84. $(10i)(12i)(-3i)$

85. $5i(13 + 2i)$

86. $-10i(8 - 6i)$

Multiplying Complex Conjugates In Exercises 87–90, multiply the number by its complex conjugate and simplify.

87. $5 - \sqrt{6}i$

88. $-4 + \sqrt{2}i$

89. $10i$

90. $20i$

Solving an Equation In Exercises 91–96, solve the equation.

91. $9u^2 + 17 = 0$

92. $4v^2 + 9 = 0$

93. $5x^2 - 5x + 7 = 0$

94. $6x^2 - 8x + 3 = 0$

95. $4z^2 - 3z + 2 = 0$

96. $7x^2 + 4x + 3 = 0$

Operations with Complex Conjugates In Exercises 97–100, perform the operation(s).

97. $(a + bi) + (a - bi)$

98. $(a + bi) - (a - bi)$

99. $(a + bi)(a - bi)$

100. $(a + bi)^2 + (a - bi)^2$

Explaining Concepts

101. **Writing** Explain why the equation $x^2 = -1$ does not have any real number solutions.

102. **Structure** Explain how adding two complex numbers is similar to adding two binomials.

103. **Structure** Explain how multiplying two complex numbers is similar to multiplying two binomials.

104. **Number Sense** Describe the values of a and b for which the complex number $a + bi$ is (a) a real number, (b) an imaginary number, and (c) a pure imaginary number.

105. **True or False?** Some numbers are both real and imaginary. Justify your answer.

106. **Complex Factors** The polynomial $x^2 + 1$ is prime with respect to the integers. It is not, however, prime with respect to the complex numbers. Show how $x^2 + 1$ can be factored using complex numbers.

107. **Repeated Reasoning** When performing operations with numbers in i-form, you sometimes need to evaluate powers of the imaginary unit i. The first eight powers of i are as follows.

$i^1 = i$
$i^2 = -1$
$i^3 = i(i^2) = i(-1) = -i$
$i^4 = (i^2)(i^2) = (-1)(-1) = 1$
$i^5 = i(i^4) = i(1) = i$
$i^6 = (i^2)(i^4) = (-1)(1) = -1$
$i^7 = (i^3)(i^4) = (-i)(1) = -i$
$i^8 = (i^4)(i^4) = (1)(1) = 1$

Describe the pattern of the powers of i.

108. **Number Sense** Explain how you can use the pattern of the powers of i to find the value of i^{86}. What is the value of i^{86}?

Cumulative Review

In Exercises 109–112, write the polynomial in standard form. Then identify its degree and leading coefficient.

109. $12 - 8x$

110. $7 - x + 13x^2$

111. 9

112. $7 + 3x + 5x^4 - 2x^3$

In Exercises 113–120, solve the equation.

113. $\sqrt{x} - 5 = 0$

114. $16 - \sqrt{t} = 7$

115. $\sqrt{4y} + 3 = 5$

116. $\sqrt{x - 4} = 6$

117. $\sqrt{3x + 2} = 5$

118. $\sqrt{2x - 15} - 2 = 3$

119. $4\sqrt{y + 1} = 5$

120. $\sqrt{x + 2} = \sqrt{5x - 7}$

10.7 Relations, Functions, and Graphs

- Identify the domain and range of a relation.
- Determine whether relations are functions.
- Use function notation and evaluate functions.
- Identify the domain and range of a function.

Relations

> **Definition of Relation**
>
> A **relation** is any set of ordered pairs. The set of first components in the ordered pairs is the **domain** of the relation. The set of second components is the **range** of the relation.

In mathematics, relations are commonly described by ordered pairs of *numbers*. The set of x-coordinates is the domain, and the set of y-coordinates is the range. In the relation $\{(3, 5), (1, 2), (4, 4), (0, 3)\}$, the domain D and range R are the sets $D = \{3, 1, 4, 0\}$ and $R = \{5, 2, 4, 3\}$.

EXAMPLE 1 Analyzing a Relation

Find the domain and range of the relation $\{(0, 1), (1, 3), (2, 5), (3, 5), (0, 3)\}$. Then sketch a graphical representation of the relation.

SOLUTION

The domain is the set of all first components of the relation, and the range is the set of all second components.

$$D = \{0, 1, 2, 3\}$$

and

$$R = \{1, 3, 5\}$$

A graphical representation of the relation is shown at the left.

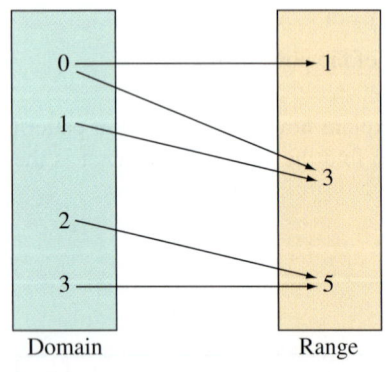

You should note that it is not necessary to list repeated components of the domain and range of a relation.

Exercises Within Reach® Solutions in English & Spanish and tutorial videos at AlgebraWithinReach.com

Analyzing a Relation In Exercises 1–6, find the domain and range of the relation.

1. $\{(-4, 3), (2, 5), (1, 2), (4, -3)\}$
2. $\{(-1, 5), (8, 3), (4, 6), (-5, -2)\}$
3. $\{(2, 16), (-9, -10), (\frac{1}{2}, 0)\}$
4. $\{(\frac{2}{3}, -4), (-6, \frac{1}{4}), (0, 0)\}$
5. $\{(-1, 3), (5, -7), (-1, 4), (8, -2), (1, -7)\}$
6. $\{(1, 1), (2, 4), (3, 9), (-2, 4), (-1, 1)\}$

Section 10.7 Relations, Functions, and Graphs 501

Functions

Study Tip

The ordered pairs of a relation can be thought of in the form (input, output). For a *function*, a given input cannot yield two different outputs. For instance, if the input is a person's name and the output is that person's month of birth, then your name as the input can yield only your month of birth as the output.

Definition of Function

A **function** is a relation in which no two ordered pairs have the same first component and different second components.

This definition means that a given first component cannot be paired with two different second components. For instance, the pairs $(1, 3)$ and $(1, -1)$ could not be ordered pairs of a function.

EXAMPLE 2 Testing Whether a Relation Is a Function

Decide whether each relation represents a function.

a. Input: a, b, c
   Output: $2, 3, 4$
   $\{(a, 2), (b, 3), (c, 4)\}$

b.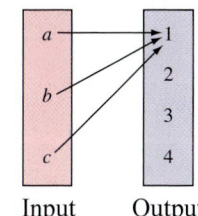

c.
Input x	Output y	(x, y)
3	1	(3, 1)
4	3	(4, 3)
5	4	(5, 4)
3	2	(3, 2)

SOLUTION

a. This set of ordered pairs *does* represent a function. No first component has two different second components.

b. This diagram *does* represent a function. No first component has two different second components.

c. This table *does not* represent a function. The first component 3 is paired with two different second components, 1 and 2.

Exercises Within Reach®

Solutions in English & Spanish and tutorial videos at AlgebraWithinReach.com

Testing Whether a Relation Is a Function In Exercises 7–12, determine whether the relation represents a function.

7. Domain   Range
   $-2 \longrightarrow 5$
   $-1 \longrightarrow 6$
   $0 \longrightarrow 7$
   $1 \longrightarrow 8$
   2

8. Domain   Range
   $-2 \longrightarrow 7$
   $-1 \longrightarrow 9$
   0
   1
   2

9. Input: a, b, c; Output: $0, 4, 9$
   $\{(a, 0), (b, 4), (c, 9)\}$

10. Input: $3, 5, 7$; Output: d, e, f
    $\{(3, d), (5, e), (7, f), (7, d)\}$

11.
Input, x	Output, y	(x, y)
0	2	(0, 2)
1	4	(1, 4)
2	6	(2, 6)
3	8	(3, 8)
4	10	(4, 10)

12.
Input, x	Output, y	(x, y)
0	2	(0, 2)
1	4	(1, 4)
2	6	(2, 6)
1	8	(1, 8)
0	10	(0, 10)

502 Chapter 10 Quadratic Equations and Functions

In algebra, it is common to represent functions by equations in two variables rather than by ordered pairs. For instance, the equation $y = x^2$ represents the variable y as a function of x. The variable x is the **independent variable** (the input) and y is the **dependent variable** (the output). In this context, the domain of the function is the set of all *allowable* values of x, and the range is the *resulting* set of all values taken on by the dependent variable y.

> ### Vertical Line Test
> A set of points on a rectangular coordinate system is the graph of y as a function of x if and only if no vertical line intersects the graph at more than one point.

EXAMPLE 3 Using the Vertical Line Test

Use the Vertical Line Test to determine whether y is a function of x.

a. 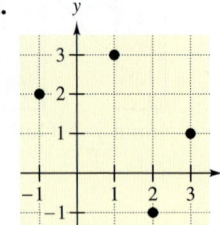 b. 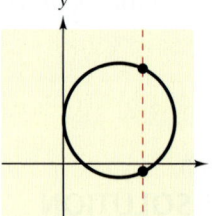 c.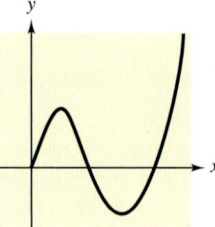

SOLUTION

a. From the graph, you can see that no vertical line intersects more than one point on the graph. So, the relation *does* represent y as a function of x.

b. From the graph, you can see that a vertical line intersects more than one point on the graph. So, the relation *does not* represent y as a function of x.

c. From the graph, you can see that no vertical line intersects more than one point on the graph. So, the relation *does* represent y as a function of x.

Exercises Within Reach®

Solutions in English & Spanish and tutorial videos at AlgebraWithinReach.com

Using the Vertical Line Test In Exercises 13–18, use the Vertical Line Test to determine whether y is a function of x.

13.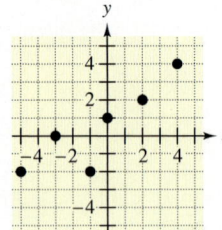

14.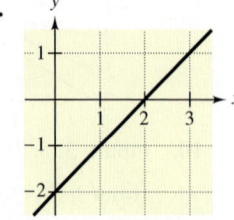

15.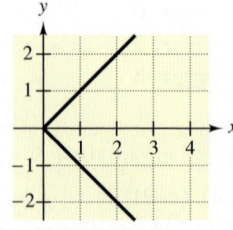

16.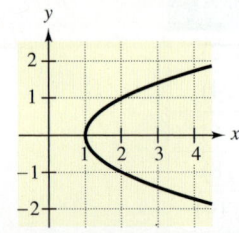

17.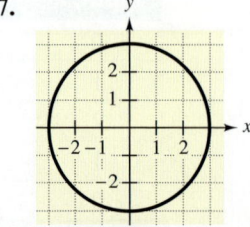

18.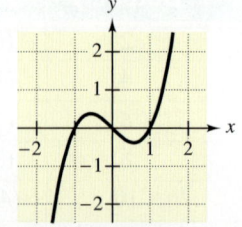

Section 10.7   Relations, Functions, and Graphs   503

Function Notation

Function Notation

In the notation $f(x)$:

f is the **name** of the function.

x is the **domain** (or input) value.

$f(x)$ is a **range** (or output) value y for a given x.

The symbol $f(x)$ is read as *the value of f at x* or simply *f of x*.

The process of finding the value of $f(x)$ for a given value of x is called **evaluating a function**. This is accomplished by substituting a given x-value (input) into the equation to obtain the value of $f(x)$ (output).

EXAMPLE 4   Evaluating a Function

Let $f(x) = x^2 + 1$. Find each value of the function.

a. $f(-2)$ **b.** $f(0)$

SOLUTION

a. $f(x) = x^2 + 1$ Write original function.

$f(-2) = (-2)^2 + 1$ Substitute -2 for x.

$= 4 + 1 = 5$ Simplify.

b. $f(x) = x^2 + 1$ Write original function.

$f(0) = (0)^2 + 1$ Substitute 0 for x.

$= 0 + 1 = 1$ Simplify.

EXAMPLE 5   Evaluating a Function

Let $g(x) = 3x - x^2$. Find each value of the function.

a. $g(2)$ **b.** $g(0)$

SOLUTION

a. Substituting 2 for x produces $g(2) = 3(2) - (2)^2 = 6 - 4 = 2$.

b. Substituting 0 for x produces $g(0) = 3(0) - (0)^2 = 0 - 0 = 0$.

Exercises Within Reach ®   Solutions in English & Spanish and tutorial videos at AlgebraWithinReach.com

Evaluating a Function   In Exercises 19–22, evaluate the function as indicated, and simplify.

19. $f(x) = 4x^2 + 2$ (a) $f(1)$ (b) $f(-1)$ **20.** $g(t) = 5 - 2t^2$ (a) $g\left(\frac{5}{2}\right)$ (b) $g(-10)$

(c) $f(-4)$ (d) $f\left(-\frac{3}{2}\right)$ (c) $g(0)$ (d) $g\left(\frac{3}{4}\right)$

21. $g(x) = 2x^2 - 3x + 1$ (a) $g(0)$ (b) $g(-2)$ **22.** $h(x) = 1 - 4x - x^2$ (a) $h(0)$ (b) $h(-4)$

(c) $g(1)$ (d) $g\left(\frac{1}{2}\right)$ (c) $h(10)$ (d) $h\left(\frac{3}{2}\right)$

The Domain and Range of a Function

The domain of a function may be explicitly described along with the function, or it may be *implied* by the context in which the function is used. For instance, if weekly pay is a function of hours worked (for a 40-hour work week), the implied domain is $0 \le x \le 40$. Certainly x cannot be negative in this context.

EXAMPLE 6 Finding the Domain and Range of a Function

Find the domain and range of each function.

a. $f: \{(-3, 0), (-1, 2), (0, 4), (2, 4), (4, -1)\}$

b. Area of a square: $A = s^2$

SOLUTION

a. The domain of f consists of all first components in the set of ordered pairs. So, the domain is

$$\{-3, -1, 0, 2, 4\}. \quad \text{Domain}$$

The range of f consists of all second components in the set of ordered pairs. So, the range is

$$\{0, 2, 4, -1\}. \quad \text{Range}$$

b. For the area of a square, you must choose positive values for the side s. So, the domain is the set of all real numbers s such that

$$s > 0. \quad \text{Domain}$$

The area of a square must be a positive number. So, the range is the set of all real numbers A such that

$$A > 0. \quad \text{Range}$$

Exercises Within Reach®

Solutions in English & Spanish and tutorial videos at AlgebraWithinReach.com

Finding Domain and Range In Exercises 23–30, **find** the domain and range of the function.

23. $f: \{(0, 4), (1, 3), (2, 2), (3, 1), (4, 0)\}$

24. $f: \{(-2, -1), (-1, 0), (0, 1), (1, 2), (2, 3)\}$

25. $g: \{(-8, -1), (-6, 0), (2, 7), (5, 0), (12, 10)\}$

26. $g: \{(-4, 4), (3, 8), (4, 5), (9, -2), (10, -7)\}$

27. $h: \{(-5, 2), (-4, 2), (-3, 2), (-2, 2), (-1, 2)\}$

28. $h: \{(10, 100), (20, 200), (30, 300), (40, 400)\}$

29. Area of a circle: $A = \pi r^2$

30. Perimeter of a square: $P = 4s$

Section 10.7   Relations, Functions, and Graphs   505

Application   **EXAMPLE 7**   **Estimating the Domain and Range of a Function**

The graph approximates the length of time L (in hours) between sunrise and sunset in Erie, Pennsylvania, for the year 2012. The variable t represents the day of the year.

Sunlight in Erie, PA

a. What is the domain of this function?

b. Estimate the range of this function.

SOLUTION

a. The year 2012 was a leap year with 366 days. So, the domain is

$$\{1, 2, 3, 4, \ldots, 365, 366\}. \quad \text{Domain}$$

Jan 1   Jan 2   Jan 3   Jan 4   Dec 30   Dec 31

b. The range is the set of different lengths of sunlight for the days of the year. These vary between about 9 hours and 15 hours. So, the range is

$$9 \leq L \leq 15. \quad \text{Range}$$

Exercises Within Reach ®

Solutions in English & Spanish and tutorial videos at AlgebraWithinReach.com

31. Estimating Domain The graph shows the SAT scores x and the grade-point averages y for 12 students.

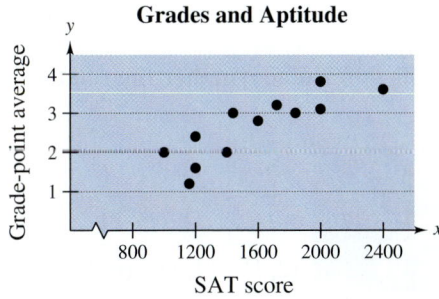

Grades and Aptitude

Estimate the greatest value in the domain of the relation.

32. Estimating Range Estimate the greatest value in the range of the relation in Exercise 31.

Concept Summary: Relations and Functions

What
A **relation** is any set of ordered pairs. A **function** is a special type of relation. How can you tell when a relation is a function?

EXAMPLE
Determine whether the relation represents a function.

$\{(-2, -1), (0, 1), (3, 1), (4, 4)\}$

How
One way to decide whether a relation represents a function is to use the Vertical Line Test. First, plot the ordered pairs on a rectangular coordinate system. Then decide whether any vertical line intersects more than one point on the graph.

EXAMPLE

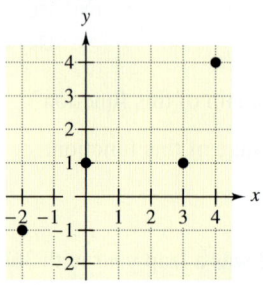

Because no vertical line intersects more than one point on the graph, the relation *does* represent a function.

Why
As you continue your study of mathematics, you will learn about many types of functions and how these functions model real-life situations.

Once you decide that a relation is a function, it is important that you (1) learn how to identify the **domain** and **range** of the function and (2) know how to **evaluate the function**.

Exercises Within Reach ®

Worked-out solutions to odd-numbered exercises at AlgebraWithinReach.com

Concept Summary Check

33. **Vocabulary** Explain the difference between a relation and a function.

34. **Vocabulary** Explain the meanings of the terms *domain* and *range* in the context of a function.

35. **Writing** In your own words, explain how to use the Vertical Line Test.

36. **Reasoning** Using the function in the example above, explain how to find $f(3)$.

Extra Practice

Testing Whether a Relation Is a Function In Exercises 37 and 38, determine whether the relation represents a function.

37.
Input, x	1	3	5	3	1
Output, y	1	2	3	4	5
(x, y)	(1, 1)	(3, 2)	(5, 3)	(3, 4)	(1, 5)

38.
Input, x	2	4	6	8	10
Output, y	1	1	1	1	1
(x, y)	(2, 1)	(4, 1)	(6, 1)	(8, 1)	(10, 1)

Evaluating a Function In Exercises 39–42, evaluate the function as indicated, and simplify.

39. $g(u) = |u + 2|$ (a) $g(2)$ (b) $g(-2)$ (c) $g(10)$ (d) $g\left(-\frac{5}{2}\right)$

40. $h(s) = |s| + 2$ (a) $h(4)$ (b) $h(-10)$ (c) $h(-2)$ (d) $h\left(\frac{3}{2}\right)$

41. $h(x) = x^3 - 1$ (a) $h(0)$ (b) $h(1)$ (c) $h(3)$ (d) $h\left(\frac{1}{2}\right)$

42. $f(x) = 16 - x^4$ (a) $f(-2)$ (b) $f(2)$ (c) $f(1)$ (d) $f(3)$

43. Demand The demand for a product is a function of its price. Consider the demand function

$f(p) = 20 - 0.5p$

where p is the price in dollars.

(a) Find $f(10)$ and $f(15)$.

(b) Describe the effect a price increase has on demand.

44. Maximum Load The maximum safe load L (in pounds) for a wooden beam 2 inches wide and d inches high is $L(d) = 100d^2$.

(a) Complete the table.

d	2	4	6	8
L(d)				

(b) Describe the effect of an increase in height on the maximum safe load.

45. Distance The function $d(t) = 50t$ gives the distance (in miles) that a car will travel in t hours at an average speed of 50 miles per hour. Find the distance traveled for (a) $t = 2$, (b) $t = 4$, and (c) $t = 10$.

46. Speed of Sound The function $S(h) = 1116 - 4.04h$ approximates the speed of sound (in feet per second) at altitude h (in thousands of feet). Use the function to approximate the speed of sound for (a) $h = 0$, (b) $h = 10$, and (c) $h = 30$.

47. Geometry Write the formula for the perimeter P of a square with sides of length s. Is P a function of s? Explain.

48. Geometry Write the formula for the volume V of a cube with sides of length t. Is V a function of t? Explain.

Explaining Concepts

49. Reasoning Is it possible to find a relation that is not a function? If it is, find one.

50. Reasoning Is it possible to find a function that is not a relation? If it is, find one.

51. Logic Is it possible for the number of elements in the domain of a relation to be greater than the number of elements in the range of the relation? Explain.

52. Writing Determine whether the statement uses the word *function* in a way that is mathematically correct. Explain your reasoning.

(a) The amount of money in your savings account is a function of your salary.

(b) The speed at which a free-falling baseball strikes the ground is a function of the height from which it is dropped.

Cumulative Review

In Exercises 53–56, solve the percent equation.

53. 5 is what percent of 35?

54. What number is 3% of 78?

55. 215 is 125% of what number?

56. 25 is what percent of 130?

In Exercises 57–62, perform the operation and write the result in standard form.

57. $(3 - 8i) + (8 - 4i)$

58. $7i - (10i + 2)$

59. $(9i)(i)$

60. $(6i)(11i)$

61. $2i(-5 + 8i)$

62. $(4 + 3i)(-1 + 9i)$

10 Chapter Summary

	What did you learn?	Explanation and Examples	Review Exercises
10.1	Solve quadratic equations by factoring *(p. 450)*.	To solve a quadratic equation by factoring, begin by factoring the equation. Then set each factor equal to zero and solve for the variable.	1–6
	Solve quadratic equations by the Square Root Property *(p. 451)*.	Let u be a real number, a variable, or an algebraic expression, and let d be a positive real number; then the equation $u^2 = d$ has exactly two solutions. If $u^2 = d$, then $u = \sqrt{d}$ and $u = -\sqrt{d}$. These solutions can also be written as $u = \pm\sqrt{d}$. This form of the solution is read as "u is equal to plus or minus the square root of d."	7–22
10.2	Construct perfect square trinomials *(p. 458)*.	To complete the square for the expression $$x^2 + bx$$ add $(b/2)^2$, which is the square of half the coefficient of x. Consequently, $$x^2 + bx + \left(\frac{b}{2}\right)^2 = \left(x + \frac{b}{2}\right)^2.$$	23–26
	Solve quadratic equations by completing the square *(p. 459)*.	1. Write the equation so that the leading coefficient is 1. 2. Complete the square and preserve the equality by adding the same constant to each side. 3. Use the Square Root Property to solve the quadratic equation.	27–34
10.3	Use the discriminant to determine the number of real solutions of a quadratic equation *(p. 466)*.	The solutions of $ax^2 + bx + c = 0$, $a \neq 0$, are given by the Quadratic Formula $$x = \frac{-b \pm \sqrt{b^2 - 4ac}}{2a}.$$ The expression inside the radical, $$b^2 - 4ac$$ is called the discriminant. 1. If $b^2 - 4ac > 0$, the equation has two real solutions. 2. If $b^2 - 4ac = 0$, the equation has one (repeated) real solution. 3. If $b^2 - 4ac < 0$, the equation has no real solution.	35–40
	Solve quadratic equations using the Quadratic Formula *(p. 467)*.	1. Write the equation in general form, $ax^2 + bx + c = 0$. 2. Identify the values of a, b, and c from the general form. 3. Substitute the values of a, b, and c into the Quadratic Formula, and simplify.	41–60

	What did you learn?	Explanation and Examples	Review Exercises
10.4	Determine whether parabolas open upward or downward *(p. 476)*.	The graph of the equation $y = ax^2 + bx + c$ is a parabola. 1. If $a > 0$, the parabola opens upward. 2. If $a < 0$, the parabola opens downward.	61–66
	Sketch graphs of quadratic equations using the point-plotting method *(p. 478)*.	1. Find the x-intercepts (if any) and the y-intercept. 2. Create a table of values that includes any intercepts and a few additional points to the left and right of the intercepts. 3. Plot the points and connect them with a smooth, U-shaped curve.	67–78
	Sketch graphs of quadratic equations using the vertex of a parabola *(p. 479)*.	The vertex of a parabola given by $y = ax^2 + bx + c$ occurs at the point whose x-coordinate is $x = -b/(2a)$. To find the y-coordinate of the vertex, substitute the x-coordinate in the equation $y = ax^2 + bx + c$.	67–78
10.5	Solve geometry problems *(p. 484)*.	You can solve area and perimeter applications using quadratic equations.	79–84
	Use the Pythagorean Theorem to solve problems *(p. 486)*.	You can solve applications involving right triangles using the Pythagorean Theorem and quadratic equations.	85, 86
	Solve rate problems *(p. 488)*.	You solved rate problems involving linear equations in Chapter 3. Now you can solve rate problems involving quadratic equations.	87–90
10.6	Write complex numbers in standard form *(p. 493)*.	If a and b are real numbers, the number $a + bi$ is a complex number, and it is said to be written in standard form.	91–100
	Perform operations with complex numbers *(p. 494)*.	When adding or subtracting complex numbers, add or subtract the real and imaginary parts separately. Multiplying complex numbers is similar to multiplying polynomials. Use complex conjugates to divide complex numbers.	101–114
	Solve quadratic equations with complex solutions *(p. 496)*.	The equation $u^2 = d$, where $d < 0$, has exactly two solutions: $u = \sqrt{\lvert d \rvert}\,i$ and $u = -\sqrt{\lvert d \rvert}\,i$.	115–128
10.7	Identify the domain and range of a relation *(p. 500)*.	A relation is any set of ordered pairs. The set of first components in the ordered pairs is the domain of the relation. The set of second components is the range of the relation.	129–132
	Determine whether relations are functions *(p. 501)*.	A function is a relation in which no two ordered pairs have the same first component and different second components.	133–144
	Use function notation and evaluate functions *(p. 503)*.	The function $y = 2x - 6$ can be given the name "f" and written in function notation as $f(x) = 2x - 6$.	145–152
	Identify the domain and range of a function *(p. 504)*.	The domain of a function may be explicitly described along with the function, or it may be implied by the context.	153–156

Review Exercises

Worked-out solutions to odd-numbered exercises at AlgebraWithinReach.com

10.1

Solving a Quadratic Equation by Factoring In Exercises 1–6, **solve** the quadratic equation by factoring.

1. $x^2 + 10x = 0$
2. $u^2 - 12u = 0$
3. $x^2 - 9x + 18 = 0$
4. $3y^2 + 7y - 6 = 0$
5. $8z^2 - 32 = 0$
6. $4y^2 - 25 = 0$

Using the Square Root Property In Exercises 7–22, **solve** the quadratic equation by the Square Root Property.

7. $x^2 = 49$
8. $a^2 = 81$
9. $y^2 - 18 = 0$
10. $y^2 - 12 = 0$
11. $x^2 - 48 = 0$
12. $x^2 - 72 = 0$
13. $(x - 5)^2 = 3$
14. $(x + 2)^2 = 5$
15. $(x - 2)^2 - 6 = 0$
16. $(x + 1)^2 - 3 = 0$
17. $8(x - 4)^2 - 32 = 0$
18. $7(y - 6)^2 - 63 = 0$
19. $2(x + 4)^2 - 16 = 0$
20. $3(y - 1)^2 - 72 = 0$
21. $9(x - 7)^2 - 25 = 25$
22. $4(x + 3)^2 - 9 = 9$

10.2

Constructing a Perfect Square Trinomial In Exercises 23–26, **determine** what term should be added to the expression to make it a perfect square trinomial. **Write** the new expression as the square of a binomial.

23. $x^2 + 12x$
24. $y^2 - 16y$
25. $t^2 - 15t$
26. $x^2 + 21x$

Completing the Square In Exercises 27–34, **solve** the quadratic equation by completing the square.

27. $x^2 - 6x - 1 = 0$
28. $x^2 + 10x + 12 = 0$
29. $x^2 - x - 1 = 0$
30. $t^2 + 3t + 1 = 0$
31. $2y^2 + 10y + 5 = 0$
32. $3x^2 - 2x - 1 = 0$
33. $4x^2 - 2x - 1 = 0$
34. $2y^2 + y - 3 = 0$

10.3

Using the Discriminant In Exercises 35–40, use the discriminant to **determine** the number of real solutions of the quadratic equation.

35. $x^2 - 4x - 5 = 0$
36. $x^2 + 3x + 5 = 0$
37. $4x^2 + 4x + 1 = 0$
38. $2x^2 + x + 1 = 0$
39. $9x^2 + 12x + 4 = 0$
40. $2x^2 - 5x + 3 = 0$

Using the Quadratic Formula In Exercises 41–56, use the Quadratic Formula to **solve** the quadratic equation.

41. $y^2 + y - 42 = 0$
42. $x^2 - x - 20 = 0$
43. $c^2 - 6c + 5 = 0$
44. $x^2 - 3x - 70 = 0$
45. $-c^2 + 6c - 6 = 0$
46. $y^2 + y - 1 = 0$
47. $2y^2 + y - 42 = 0$
48. $2x^2 - x - 20 = 0$
49. $9x^2 + 30x + 25 = 0$
50. $4x^2 + 4x + 1 = 0$
51. $v^2 = 250$
52. $x^2 - 45x = 0$
53. $0.3t^2 - 2t + 1 = 0$
54. $-u^2 + 3.1u + 5 = 0$
55. $0.7x^2 - 0.14x + 0.007 = 0$
56. $0.5y^2 + 0.75y - 2 = 0$

Solving an Equation In Exercises 57–60, solve the equation.

57. $\dfrac{1}{x} + \dfrac{1}{x+1} = \dfrac{1}{2}$

58. $\dfrac{3}{t-1} - \dfrac{2}{t^2+t-2} = 4$

59. $\sqrt{2x+5} = x - 3$

60. $x = \sqrt{4x+5}$

10.4

Using the Leading Coefficient Test In Exercises 61–66, determine whether the parabola opens upward or downward.

61. $y = x^2 - 9x + 3$

62. $y = -4x^2 + 2x + 10$

63. $y = 6 - 5x - 7x^2$

64. $y = 8 + 4x + 3x^2$

65. $y = 3 - (x+4)^2$

66. $y = 7 + (2x-1)^2$

Sketching a Parabola In Exercises 67–78, sketch the parabola. Label the vertex and any intercepts.

67. $y = x^2 - 2x + 1$
68. $y = -(x^2 - 2x + 1)$
69. $y = -x^2 + 4x - 3$
70. $y = x^2 - 6x + 9$
71. $y = -x^2 + 3x$
72. $y = x^2 - 10x$
73. $y = \frac{1}{4}(4x^2 - 4x + 3)$
74. $y = \frac{1}{3}(x^2 - 4x + 6)$
75. $y = 2x^2 + 4x + 5$
76. $y = -2x^2 + 8x - 5$
77. $y = -(3x^2 - 4x - 2)$
78. $y = 3x^2 + 2x + 3$

10.5

Geometry In Exercises 79–82, solve for x.

79. Area = 32 square centimeters

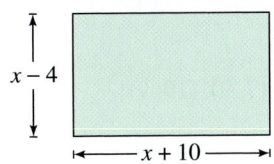

Figure for 79

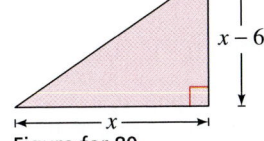

Figure for 80

80. Area = 20 square feet

81. Area = 1800 square meters

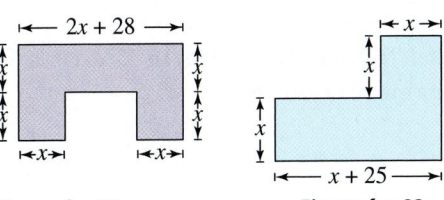

Figure for 81   Figure for 82

82. Area = 1300 square feet

83. *Geometry* The height of a triangle is one and one-half times its base, and its area is 60 square inches. Find the dimensions of the triangle.

84. *Geometry* The height of a triangle is three times its base, and its area is 24 square inches. Find the dimensions of the triangle.

85. *Geometry* The perimeter of a rectangular driveway is 94 feet and the length of the diagonal is 37 feet. Find the dimensions of the driveway.

86. *Geometry* The perimeter of a rectangular flag is 16 feet and the length of the diagonal is 5.8 feet. Find the dimensions of the flag.

87. *Work Rate* Working together, two people can complete a task in 10 hours. Working alone, one person takes 4 hours longer than the other. Working alone, how long would it take each person to complete the task?

88. *Work Rate* Working together, two people can complete a task in 15 hours. Working alone, one person takes 2 hours longer than the other. Working alone, how long would it take each person to complete the task?

89. *Reduced Rates* A Little League baseball team paid $72 for a block of tickets to a ball game. The block contained three more tickets than the team needed. By inviting 3 more people to attend (and share in the cost), the team lowered the price per ticket by $1.20. How many people are going to the game?

90. *Reduced Rates* A travel agency paid $780 for a block of tickets to a concert. The block contained four more tickets than the agency needed. By inviting 4 more people to attend (and share in the cost), the agency lowered the price per ticket by $9.75. How many people are going to the concert?

10.6

Writing a Number in *i*-Form In Exercises 91–96, write the number in *i*-form.

91. $\sqrt{-81}$
92. $\sqrt{-\dfrac{25}{196}}$
93. $\sqrt{-5}$
94. $\sqrt{-48}$
95. $\dfrac{\sqrt{-50}}{\sqrt{-2}}$
96. $\sqrt{-\dfrac{12}{49}}$

Equality of Two Complex Numbers In Exercises 97–100, find the values of *a* and *b* that satisfy the equation.

97. $12 - 5i = (a + 2) + (b - 1)i$
98. $-48 + 9i = (a - 5) + (b - 10)i$
99. $\sqrt{-49} + 4 = a + bi$
100. $-3 - \sqrt{-4} = a + bi$

Operations with Complex Numbers In Exercises 101–108, perform the operation and write the result in standard form.

101. $(-4 + 5i) - (-12 + 8i)$
102. $(-8 + 3i) - (6 + 7i)$
103. $(3 - 5i) + (7 + 12i)$
104. $(-6 + 3i) + (-1 + i)$
105. $(4 - 3i)(4 + 3i)$
106. $(12 - 5i)(2 + 7i)$
107. $(6 - 5i)^2$
108. $(2 - 9i)^2$

Dividing Complex Numbers In Exercises 109–114, write the quotient in standard form.

109. $\dfrac{7}{3i}$
110. $\dfrac{4}{5i}$
111. $\dfrac{5i}{6 + i}$
112. $\dfrac{i}{1 - 9i}$
113. $\dfrac{4i}{2 - 8i}$
114. $\dfrac{5i}{4 + 7i}$

Using the Square Root Property In Exercises 115–120, solve the equation by using the Square Root Property.

115. $z^2 = -121$
116. $u^2 = -25$
117. $y^2 + 50 = 0$
118. $x^2 + 48 = 0$
119. $(y + 4)^2 + 18 = 0$
120. $(x - 2)^2 + 24 = 0$

Completing the Square In Exercises 121–124, solve the equation by completing the square.

121. $x^2 - 2x + 26 = 0$
122. $t^2 - 16t + 208 = 0$
123. $x^2 - 3x + 3 = 0$
124. $y^2 - \dfrac{2}{3}y + 2 = 0$

Using the Quadratic Formula In Exercises 125–128, solve the equation by using the Quadratic Formula.

125. $x^2 + 6x + 13 = 0$
126. $a^2 + 4a + 29 = 0$
127. $3z^2 - 3z + \dfrac{49}{64} = 0$
128. $2y^2 + y + \dfrac{29}{32} = 0$

10.7

Analyzing a Relation In Exercises 129–132, find the domain and range of the relation.

129. $\{(8, 3), (-2, 7), (5, 1), (3, 8)\}$
130. $\{(0, 1), (-1, 3), (4, 6), (-7, 5)\}$
131. $\{(2, -3), (-2, 3), (2, 4), (4, 0)\}$
132. $\{(1, 7), (-3, 4), (6, 4), (-2, 4)\}$

Testing Whether a Relation Is a Function In Exercises 133–138, determine whether the relation represents a function.

133. Domain → Range: 1→2, 2→5, 3→7, 4→9, 5
134. Domain → Range: 5→5, 7→9, 9→13, 11→17, 13→19
135. $\{(-1, 3), (3, 3), (0, 3), (7, 9), (10, 9)\}$
136. Input: a, b, c
     Output: $4, 8, 9$
     $\{(a, 4), (b, 4), (b, 8), (c, 9)\}$

137.

Input x	Output y	(x, y)
0	0	(0, 0)
2	1	(2, 1)
4	1	(4, 1)
6	2	(6, 2)
2	3	(2, 3)

138.

Input x	Output y	(x, y)
−6	1	(−6, 1)
−3	0	(−3, 0)
0	1	(0, 1)
3	4	(3, 4)
6	2	(6, 2)

Using the Vertical Line Test In Exercises 139–144, use the Vertical Line Test to determine whether y is a function of x.

139.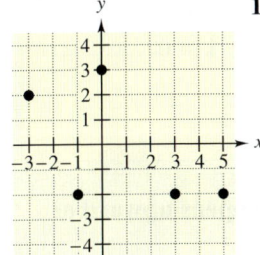

140.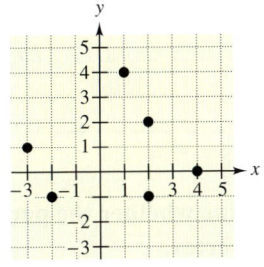

141.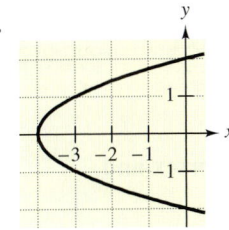

142.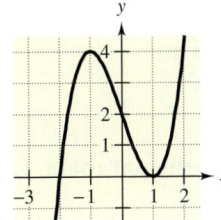

143.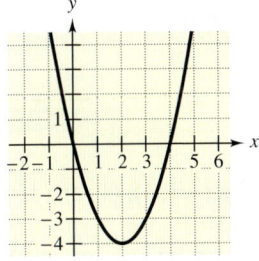

144.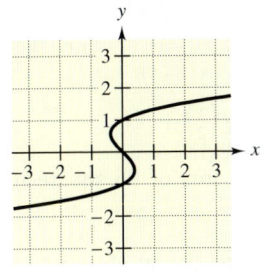

Evaluating a Function In Exercises 145–150, evaluate the function as indicated, and simplify.

145. $f(x) = \frac{3}{4}x$ (a) $f(-1)$ (b) $f(4)$
                  (c) $f(10)$ (d) $f\left(-\frac{4}{3}\right)$

146. $f(x) = 2x - 7$ (a) $f(-1)$ (b) $f(3)$
                     (c) $f\left(\frac{1}{2}\right)$ (d) $f(-4)$

147. $g(t) = -16t^2 + 64$ (a) $g(0)$ (b) $g\left(\frac{1}{4}\right)$
                             (c) $g(1)$ (d) $g(2)$

148. $h(u) = u^3 + 2u^2 - 4$ (a) $h(0)$ (b) $h(3)$
                                 (c) $h(-1)$ (d) $h(-2)$

149. $f(x) = |2x + 3|$ (a) $f(0)$ (b) $f(5)$
                     (c) $f(-4)$ (d) $f\left(-\frac{3}{2}\right)$

150. $f(x) = |x| - 4$ (a) $f(-1)$ (b) $f(1)$
                    (c) $f(-4)$ (d) $f(2)$

151. Demand The demand for a product is a function of its price. Consider the demand function

$$f(p) = 40 - 0.2p$$

where p is the price in dollars. Find the demand for (a) $p = 10$, (b) $p = 50$, and (c) $p = 100$.

152. Profit The profit for a product is a function of the amount spent on advertising for the product. In the profit function

$$f(x) = 8000 + 2000x - 50x^2$$

x is the amount (in hundreds of dollars) spent on advertising. Find the profit for (a) $x = 5$, (b) $x = 10$, and (c) $x = 20$.

Finding Domain and Range In Exercises 153–156, find the domain and range of the function.

153. $f: \{(1, 5), (2, 10), (3, 15), (4, -10), (5, -15)\}$

154. $g: \{(-3, 6), (-2, 4), (-1, 2), (0, 0), (1, -2)\}$

155. $f: \{(3, -1), (4, 6), (-2, -1), (0, -2), (7, 0)\}$

156. $g: \{(-8, 0), (3, -2), (10, 3), (-5, 1), (0, 0)\}$

Chapter Test

Solutions in English & Spanish and tutorial videos at AlgebraWithinReach.com

Take this test as you would take a test in class. After you are done, check your work against the answers in the back of the book.

In Exercises 1–8, solve the equation. If indicated, use the specified method.

1. Square Root Property:
   $x^2 - 400 = 0$

2. Square Root Property:
   $(x + 4)^2 + 100 = 0$

3. Completing the square:
   $t^2 - 6t + 11 = 0$

4. Completing the square:
   $3z^2 + 9z + 5 = 0$

5. Quadratic Formula:
   $x^2 - 2x + 3 = 0$

6. Quadratic Formula:
   $2u^2 + 4u + 1 = 0$

7. $\dfrac{x}{3} + \dfrac{1}{x} = 2$

8. $\sqrt{2x} = x - 1$

In Exercises 9–11, determine whether the parabola opens upward or downward. Then find the vertex.

9. $y = -2x^2 - 7$

10. $y = 5 - 2x - x^2$

11. $y = (x - 2)^2 + 3$

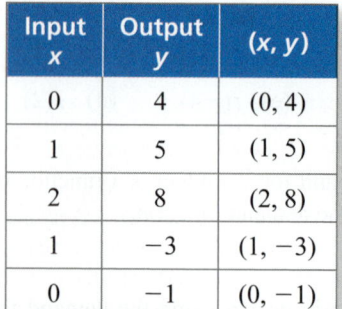

Table for 18

In Exercises 12 and 13, sketch the parabola. Label the vertex and any intercepts.

12. $y = x^2 - 8x + 12$

13. $y = -x^2 - 4x$

In Exercises 14 and 15, perform the operation and simplify.

14. $(2 + \sqrt{-9}) - (5 - \sqrt{-4})$

15. $(3 - 8i)(4 + i)$

16. Write $\dfrac{4 + 5i}{3i}$ in standard form.

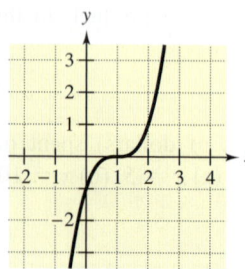

Figure for 19

17. Find the domain and range of the function $g: \{(-2, 3), (-1, 1), (1, 2), (4, -2)\}$.

18. Does the table at the left represent y as a function of x? Explain.

19. Does the graph at the left represent y as a function of x? Explain.

20. Evaluate $f(x) = x^3 - 2x^2$ as indicated, and simplify.
    (a) $f(0)$ (b) $f(2)$ (c) $f(-2)$ (d) $f\left(\dfrac{1}{2}\right)$

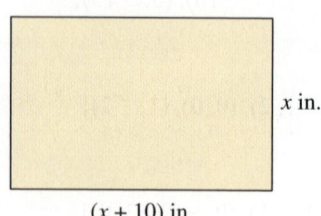

Figure for 21

21. The rectangle shown in the figure at the left has an area of 96 square inches. Find the dimensions of the rectangle.

22. Working together, two people can paint a house in 6 hours. Working alone, one person takes 5 hours longer than the other. Working alone, how long would it take each person to paint the house?

Answers to Odd-Numbered Exercises

CHAPTER 1

Section 1.1 (pp. 2–9)

1. (a) $20, \frac{9}{3}$ (b) $-3, 20, \frac{9}{3}$
   (c) $-3, 20, -\frac{3}{2}, \frac{9}{3}, 4.5$ (d) $\pi, -\sqrt{3}$
3. (a) none (b) $-\sqrt{25}, -\frac{5}{1}, 0, -12$
   (c) $-\sqrt{25}, -\frac{5}{1}, 9.4, 0, -12, \frac{7}{14}$ (d) $\sqrt{7}$
5. 3   7. 0   9. 1.5 or $\frac{3}{2}$
11.
13.
15.
17. >   19. <
21. <   23. >
25. <   27. >
29. Your scores decreased as you played more rounds.
31. 2   33. 8   35. 10   37. 3   39. 3.4   41. $\frac{7}{2}$
43. -23.6 45. 0   47. $=$ 49. >   51. <
53. 456 m
55. 49.12; As a change in the account balance, the amount is negative, as a payment, the amount is positive.
57. -2 is to the right of -2.5.
59. The fractions are converted to decimals and plotted on a number line to determine the order.
61.
63.
65. $-4, 20$ 67. $15.3, 27.3$ 69. $-5.5, 1.5$
71. Sample answer: $\sqrt{2}, \pi, -\sqrt{3}$
73. Sample answer: $-7, 1, 341$
75. Sample answer: $-\sqrt{7}, 0, -\frac{1}{3}$
77. $n \geq 0$
79. (a) Lo`ihi and Ruby   (b) Ruby   (c) Mauna Loa
81. The number 4 is plotted four units to the right of 0, and the number -4 is plotted four units to the left of 0.
83. The number $\frac{8}{4}$ is a natural number because it equals 2, and $\frac{7}{4}$ is not a natural number because it equals 1.75.
85. 3; $|3-(-4)| > |-10-(-4)|$
87. False. $|0| = 0$ 89. True. $\left|\frac{x}{y}\right| = \frac{x}{y}$
91. True. This is the definition of opposite.

Section 1.2 (pp. 10–17)

1.
3.
5.
7.

9.
11.
13. Yes   15. 23°C   17. 16   19. 0   21. 0
23. -27 25. -27 27. 6   29. 25   31. -5
33. 363   35. 726   37. 38   39. -10 41. -5
43. 300   45. -233 47. $109   49. 67 ft   51. 3
53. 26   55. 5   57. 25   59. 36   61. -30
63. -24 65. -109 67. -9 69. -11 71. -21
73. 0   75. -6 77. -103 79. -610 81. -80
83. -12 85. 17   87. -2 89. 7000 ft
91.

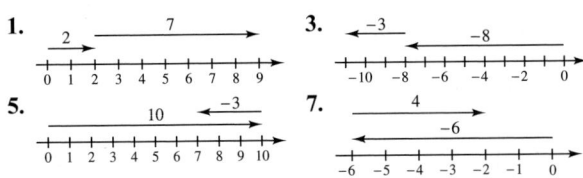

-10

93.

18

95.

-13

97. To add two integers with unlike signs, you subtract the smaller absolute value from the larger absolute value; You attach the sign of the integer with the larger absolute value.
99. To add two integers with like signs, add their absolute values and attach the common sign to the result.
101. -15 103. -36 105. 1371 m
107. (a)

Day	Daily Gain or Loss
Tuesday	$5
Wednesday	$8
Thursday	$-$5
Friday	$16

(b) $11
(c) $24; The stock gained $24 in value during the week; Find the difference between the first bar (Monday) and the last bar (Friday).
109. (a) $3 + 2 = 5$
(b) Adding two integers with like signs
111. To add two negative integers, add their absolute values and attach the negative sign.
113. No. To add two positive integers, add their absolute values and attach the common sign, which is always the positive sign.

Section 1.3 (pp. 18–27)

1. 35   3. 0   5. -32 7. -36 9. -690
11. 91   13. 1600   15. -90 17. 21   19. -30
21. 12   23. 90   25. 338   27. -4725 29. 260
31. 9009   33. 57,600 ft²   35. 180 in.³
37. 3   39. -6 41. -7 43. 7

45. Division by zero is undefined.   **47.** 0   **49.** 27
51. −7   **53.** 32   **55.** −160   **57.** −82   **59.** 331
61. 713   **63.** −1045   **65.** 5 mi/sec
67. (a) 82
(b) 73 77 82 87 91
   72 76 80 84 88 92
(c) 5, −9, −5, 9; Sum is 0; Explanations will vary.
69. Prime   **71.** Composite   **73.** Prime
75. Composite   **77.** Composite   **79.** Prime
81. 11 is prime.   **83.** 2 · 2   **85.** 2 · 2 · 2 · 2
87. 37 is prime.   **89.** 2 · 2 · 3   **91.** 3 · 11 · 17
93. 2 · 3 · 5 · 7   **95.** 2 · 2 · 2 · 2 · 2 · 2 · 3
97. 3 · 5 · 13 · 13
99. 1 row and 24 columns, 2 rows and 12 columns,
3 rows and 8 columns, 4 rows and 6 columns,
6 rows and 4 columns, 8 rows and 3 columns,
12 rows and 2 columns, 24 rows and 1 column
101. 3   **103.** 0
105. Example: $1 \cdot (-4) = -4$; Algebraic description: If a is a real number, then $1 \cdot a = a$.
107. $-1(0) = 0$
$-2(0) = 0$
The product of an integer and zero is 0.
109. $|0| = 0$
$|-1| = 1$
$|-2| = 2$
$|a| = \begin{cases} a, & \text{if } a \geq 0 \\ -a, & \text{if } a < 0 \end{cases}$
111. Unlike signs
113. Positive; Because −6 and −4 have like (negative) signs, the product is positive.
115. −120   **117.** 40   **119.** 840   **121.** −2520
123. −192   **125.** 0   **127.** Yes   **129.** No   **131.** Yes
133. Yes   **135.** No   **137.** Yes   **139.** −24°
141. −$0.84   **143.** 21 ft
145. 2; It is divisible only by 1 and itself. Any other even number is divisible by 1, itself, and 2.
147. $(2m)n = 2(mn)$. The product of two odd integers is odd.
149. The product is negative because there is an odd number of negative factors.
151. 5 and 7; 11 and 13; 17 and 19; 29 and 31; 41 and 43; 59 and 61; 71 and 73
153. 114, 115, 116, 117, 118, 119, 120, 121, 122, 123
155. (a) [Sieve of Eratosthenes table showing primes up to 100]
(b) Prime numbers; The multiples of 2, 3, 5, and 7, other than the numbers themselves, cannot be prime because they have 2, 3, 5, and 7 as factors.

Mid-Chapter Quiz   (p. 28)

1. <   **2.** >
3. <   **4.** >
5. −0.75   **6.** $\frac{17}{19}$ **7.** =   **8.** >   **9.** −9   **10.** 28
11. 99   **12.** −53   **13.** −27   **14.** −25   **15.** −50
16. −62   **17.** 8   **18.** −5   **19.** −60   **20.** 91
21. 15   **22.** −4   **23.** Prime
24. Composite; 7 · 13   **25.** Composite; 3 · 37
26. Composite; 2 · 2 · 2 · 2 · 3 · 3
27. $450,450   **28.** 128 ft³   **29.** 15 ft   **30.** $367

Section 1.4   (pp. 30–39)

1. 5   **3.** 5   **5.** 45   **7.** 6   **9.** $\frac{1}{2}$ **11.** $\frac{4}{5}$
13. $\frac{5}{16}$ **15.** $\frac{2}{25}$ **17.** 6   **19.** 10
21. $\frac{2}{3}$; 0.6666…; No   **23.** $\frac{3}{5}$; 0.6; Yes   **25.** $\frac{8}{15}$
27. 4   **29.** $\frac{3}{8}$ **31.** $-\frac{1}{2}$ **33.** $\frac{5}{6}$ **35.** $\frac{9}{16}$
37. $\frac{1}{2}$ **39.** $\frac{7}{20}$ **41.** $\frac{55}{6}$ **43.** $-\frac{17}{16}$ **45.** $-\frac{21}{4}$
47. $-\frac{39}{4}$ **49.** $\frac{17}{48}$ **51.** $\frac{3}{10}$ **53.** $\frac{3}{8}$ **55.** $-\frac{10}{21}$
57. $-\frac{3}{8}$ **59.** $\frac{1}{3}$ **61.** $\frac{3}{16}$ **63.** $\frac{12}{5}$ **65.** $\frac{1}{7}$; $7 \cdot \frac{1}{7} = 1$
67. $\frac{7}{4}$; $\frac{4}{7} \cdot \frac{7}{4} = 1$ **69.** $\frac{1}{2}$ **71.** $-\frac{8}{27}$ **73.** 1
75. $\frac{3}{7}$ **77.** $\frac{25}{24}$ **79.** −90
81. 2150 cal; $716\frac{2}{3}$ cal/day
83. $\frac{3}{5}$ hr or 36 min
85. 27.09   **87.** 4.1302   **89.** 106.65   **91.** 6.123
93. −2.128   **95.** 2.27   **97.** −57.02   **99.** 4.30
101. 39.08   **103.** −0.51   **105.** $4.71   **107.** $1872.11
109. Use the LCD to rewrite the two fractions so that they have like denominators. Then add the numerators and write the result over the LCD.
111. Use the decimal points to line up the digits of the decimals according to place value. Then add vertically down each column starting from the right and carrying when needed.
113. $\frac{13}{60}$ **115.** −1   **117.** $\frac{2}{5}$ **119.** $-\frac{7}{24}$ **121.** $\frac{7}{5}$
123. $\frac{121}{12}$ **125.** $-\frac{51}{2}$ **127.** $\frac{27}{40}$ **129.** 1
131. Division by zero is undefined.   **133.** $\frac{5}{2}$ **135.** −1.90
137. −63.22   **139.** $1.15   **141.** No; $-\frac{1}{2} + \left(-\frac{3}{4}\right) = -\frac{5}{4}$
143. The product of two fractions with like signs is positive. The product of two fractions with different signs is negative.
145. 12; $3 \div \frac{1}{4} = 12$
147. 43.6; 42.12
The first method produces the more accurate answer because you round only the answer, while in the second method, you round each dimension before you multiply to get the answer.
149. True. The reciprocal of a rational number can always be written as a ratio of two integers.
151. False. $\frac{1}{2} \cdot \frac{1}{3} = \frac{1}{6}$
153. False. If $u = 1$ and $v = 2$, then $u - v = 1 - 2 = -1 \not> 0$.
155. $\frac{4}{5} + \frac{3}{6} = \frac{13}{10}$

Section 1.5 (pp. 40–47)

1. 2^6 **3.** 9   **5.** 64   **7.** $\frac{1}{64}$ **9.** -125
11. -16 **13.** -1.728 **15.** 512 boxes
17. Stage 6

Stage	Emails sent, as a power	Emails sent
1	5^1	5
2	5^2	25
3	5^3	125
4	5^4	625
5	5^5	3125
6	5^6	15,625

19. 8   **21.** 12   **23.** 27   **25.** 17   **27.** 9
29. $-\frac{11}{2}$ **31.** 36   **33.** 8   **35.** 68   **37.** 17
39. 33   **41.** $\frac{7}{3}$ **43.** 21   **45.** $\frac{7}{80}$ **47.** $\frac{5}{6}$ **49.** $-\frac{1}{8}$
51. 4   **53.** Division by zero is undefined.   **55.** 0
57. 13   **59.** Associative Property of Addition
61. Commutative Property of Multiplication
63. Commutative Property of Addition
65. Distributive Property
67. Multiplicative Inverse Property
69. Additive Identity Property
71. $-3(10)$ **73.** $(18 + 12) + 9$
75. (a) $30(30 - 8)$ (b) $30(30) - 30(8)$ (c) 660 ft^2
77. (a) $2 \cdot 2 + 2 \cdot 3 = 4 + 6 = 10$
 (b) $2 \cdot 5 = 10$
 (c) Explanations will vary.
79. base: 2, exponent: 4
81. Added inside **P**arentheses, evaluated the **E**xponential expression, **M**ultiplied, and **A**dded.
83. -1 **85.** 10   **87.** 366.12   **89.** 10.69
91. (a) -50 (b) $\frac{1}{50}$
93. $8 + 2 + 6 \cdot 2 - 4 + 2 \cdot 3 = 24$
95. No. $-6^2 = -(6 \cdot 6) = -36$; $(-6)^2 = (-6)(-6) = 36$
97. $24^2 = (4 \cdot 6)^2 = 4^2 \cdot 6^2$ **99.** $4 - (6 - 2) = 4 - 6 + 2$
101. $100 \div 2 \times 50 = 50 \times 50 = 2500$
103. $5(7 + 3) = 5(7) + 5(3)$
105. Division by zero is undefined.
107. Fraction was simplified incorrectly.
$$-9 + \frac{9 + 20}{3(5)} - (-3) = -9 + \frac{29}{15} + 3$$
$$= -6 + \frac{29}{15}$$
$$= \frac{-90 + 29}{15}$$
$$= -\frac{61}{15}$$

109.
Expression	Value
$(6 + 2) \cdot (5 + 3)$	$= 64$
$(6 + 2) \cdot 5 + 3$	$= 43$
$6 + 2 \cdot 5 + 3$	$= 19$
$6 + 2 \cdot (5 + 3)$	$= 22$

111. $8 \cdot 15 - 8 \cdot 6 = 120 - 48 = 72$ **113.** No
$8(15 - 6) = 8(9) = 72$
Explanations will vary.

Review Exercises (pp. 50–53)

1. (a) $\sqrt{4}$ (b) $-1, \sqrt{4}$
 (c) $-1, 4.5, \frac{2}{5}, -\frac{1}{7}, \sqrt{4}$ (d) $\sqrt{5}$
3. (a) $\frac{30}{2}, 2$ (b) $\frac{30}{2}, 2$
 (c) $\frac{30}{2}, 2, 1.5, -\frac{10}{7}$ (d) $-\sqrt{3}, -\pi$
5. [number line]   **7.** [number line]
9. [number line]
11. $<$ **13.** $>$
[number lines]
15. $>$
[number line]
17. 152   **19.** $\frac{7}{3}$ **21.** 8.5   **23.** 3.4
25. -6.2 **27.** $-\frac{8}{5}$ **29.** $=$ **31.** $>$
33. $>$ **35.** $-2, 12$ **37.** $-2.4, 7.6$
39. 7   **41.** -5
[number lines]
43. 11   **45.** -95 **47.** -89 **49.** 5   **51.** -29
53. \$82,400   **55.** 21   **57.** -7 **59.** 33   **61.** -22
63. -9 **65.** \$765   **67.** 45   **69.** -72 **71.** -48
73. 45   **75.** -54 **77.** -40 **79.** \$3600   **81.** 9
83. -12 **85.** -15 **87.** 13   **89.** 0
91. Division by zero is undefined.   **93.** 65 mi/hr
95. Prime   **97.** Prime   **99.** Composite
101. $2 \cdot 2 \cdot 2 \cdot 3 \cdot 11$ **103.** $2 \cdot 3 \cdot 3 \cdot 3 \cdot 7$
105. $2 \cdot 2 \cdot 13 \cdot 31$ **107.** -36 **109.** 7   **111.** 18
113. 1   **115.** 21   **117.** $\frac{1}{4}$ **119.** $\frac{5}{8}$ **121.** 10
123. 15   **125.** $\frac{2}{5}$ **127.** $\frac{3}{4}$ **129.** $\frac{7}{8}$ **131.** $\frac{7}{8}$
133. $-\frac{103}{96}$ **135.** $\frac{5}{4}$ **137.** $\frac{17}{8}$ **139.** $2\frac{3}{4}$ in.
141. $-\frac{1}{12}$ **143.** 1   **145.** $-\frac{1}{36}$ **147.** $\frac{2}{3}$ **149.** $\frac{6}{7}$
151. Division by zero is undefined.   **153.** 0
155. $\frac{27}{32}$ in./hr   **157.** 5.65   **159.** -1.38
161. -0.75 **163.** 21   **165.** \$947.75   **167.** 6^5
169. $\left(\frac{6}{7}\right)^4$ **171.** 16   **173.** $-\frac{27}{64}$ **175.** -49
177. 6   **179.** 21   **181.** 52   **183.** 160   **185.** 81
187. $\frac{37}{8}$ **189.** 140   **191.** -3 **193.** 7   **195.** 0
197. 796.11   **199.** 1841.74
201. (a) \$10,546.88   (b) \$14,453.12
203. Additive Inverse Property
205. Commutative Property of Multiplication
207. Multiplicative Identity Property
209. Distributive Property
211. -16 **213.** $1 + 24$

215. $6 \cdot 18 - 6 \cdot 5 = 108 - 30 = 78$
$6(18 - 5) = 6(13) = 78$
Explanations will vary.

Chapter Test (p. 54)

1. (a) 4   (b) 4, −6, 0   (c) 4, −6, $\frac{1}{2}$, 0, $\frac{7}{9}$ (d) π
2. >   3. 13   4. −6.8   5. −4   6. 10   7. 10
8. 47   9. −160   10. 8   11. −30   12. 1
13. $\frac{17}{24}$ 14. $\frac{2}{15}$ 15. $\frac{7}{12}$ 16. −27   17. −0.64
18. 33   19. 235   20. −2   21. Distributive Property
22. Multiplicative Inverse Property
23. Associative Property of Addition
24. Commutative Property of Multiplication
25. $\frac{2}{9}$ 26. $2 \cdot 2 \cdot 2 \cdot 3 \cdot 3 \cdot 3$ 27. 58 ft/sec
28. $6.43

CHAPTER 2

Section 2.1 (pp. 56–63)

1. x 3. m, n 5. $4x, 3$ 7. $\frac{5}{3}, -3y^3$
9. $a^2, 4ab, b^2$ 11. 14   13. $-\frac{1}{3}$ 15. $\frac{2}{5}$
17. 2π 19. $y \cdot y \cdot y \cdot y \cdot y$ 21. $2 \cdot 2 \cdot x \cdot x \cdot x \cdot x$
23. $4 \cdot y \cdot y \cdot z \cdot z \cdot z$
25. $a^2 \cdot a^2 \cdot a^2 = a \cdot a \cdot a \cdot a \cdot a \cdot a$
27. $-4 \cdot x \cdot x \cdot x \cdot x \cdot x \cdot x \cdot x$ 29. $-9 \cdot a \cdot a \cdot a \cdot b \cdot b \cdot b$
31. $(x + y)(x + y)$ 33. $\left(\frac{a}{3s}\right)\left(\frac{a}{3s}\right)\left(\frac{a}{3s}\right)\left(\frac{a}{3s}\right)$
35. $2 \cdot 2 \cdot (a - b)(a - b)(a - b)(a - b)(a - b)$
37. (a) 0   (b) −9   39. (a) 3   (b) 13
41. (a) 6   (b) 4   43. (a) 3   (b) −20
45. (a) 33   (b) 112   47. (a) 5   (b) 14
49. (a) 0   (b) Division by zero is undefined.
51. (a) $-\frac{1}{5}$ (b) $\frac{3}{10}$ 53. (a) 0   (b) 11
55. $646   57. $7.55w$ 59. $3.79m$
61. $(n - 5)^2$; 9 square units   63. $a(a + b)$; 45 square units
65. No. $-3^2 = -9$ and $(-3)^2 = 9$. An exponent affects what is directly to its left. In the expression -3^2, the 3 is the only portion of the expression being squared.
67. addition, subtraction, multiplication, and division
69. k 71. $3(x + 5)$, 10   73. $-2u^4$ 75. $(-3)^3(x - y)^2$
77. (a) $\frac{15}{2}$ (b) 10   79. (a) 72   (b) 320
81. (a) $x + 6$ (b) 29 in.   (c) 26.5 in.
83. (a) $\frac{3(4)}{2} = 6 = 1 + 2 + 3$
    (b) $\frac{6(7)}{2} = 21 = 1 + 2 + 3 + 4 + 5 + 6$
    (c) $\frac{10(11)}{2} = 55 = 1 + 2 + 3 + 4 + 5 + 6 + 7 + 8 + 9 + 10$
85. No. The term includes the minus sign and is $-3x$.
87. The product of an even number and an odd number [$n(n + 1)$, where $n \geq 1$, and $n(n - 3)$, where $n \geq 4$] is even, so it divides evenly by 2. This will always yield a natural number.
89. 17   91. 10   93. 24   95. 12
97. Commutative Property of Multiplication
99. Distributive Property

Section 2.2 (pp. 64–73)

1. $2v$, Commutative Property of Multiplication
3. $t, -2$, Distributive Property   5. $32 + 16z$
7. $-10x + 5y$ 9. $8x + 8$ 11. $-36s^2 + 6s$
13. $ab; ac; a(b + c) = ab + ac$
15. $2a; 2(b - a); 2a + 2(b - a) = 2b$
17. $8(1) + 8(0.25) = 8 + 2 = 10$
19. $5(20) - 5(2) = 100 - 10 = 90$
21. $16t^3, 3t^3; 4t, -5t$ 23. $4rs^2, 12rs^2; -5, 1$
25. $-2y$ 27. $-2x + 5$ 29. $11x + 4$
31. $3r + 7$ 33. $12x$ 35. $-4x$ 37. $6x^2$
39. $-10z^3$ 41. $9a$ 43. $-\frac{x^3}{3}$ 45. $-24x^4y^4$
47. $2x$ 49. $13s - 2$ 51. $-2m + 21$ 53. $8x + 38$
55. $8x + 26$ 57. $2x - 17$ 59. $10x - 7x^2$
61. $3x^2 + 5x - 3$ 63. $4t^2 - 11t$ 65. $26t - 2t^2$
67. (a) $10x + 10$ (b) $4x^2 + 20x$ 69. $x^2 + 50x$
71. In an algebraic expression, two terms are said to be like terms if they are both constant terms or if they have the same variable factor(s).
Like terms: $-3x^2y, x^2y$ Unlike terms: x^2y, x^2y^2
73. Beginning with the innermost parentheses, use the Distributive Property to remove nested symbols of grouping.
75. $x^2 - xy + 4$ 77. $17z + 11$ 79. $-21y^3$
81. $\frac{x^2}{5}$ 83. $23x + 10$ 85. $3x - 5$
87. $5 + (3x - 1) + (2x + 5) = 5x + 9$
89. (a) $2(3x) + 2(x + 7); 8x + 14$ (b) $(3x)(x + 7); 3x^2 + 21x$
91. (a) Answers will vary.   (b) 56 square units
93. The exponents of y are not the same.
95. $(6x)^4 = (6x)(6x)(6x)(6x) = 6 \cdot 6 \cdot 6 \cdot 6 \cdot x \cdot x \cdot x \cdot x$;
$6x^4 = 6 \cdot x \cdot x \cdot x \cdot x$
97. 12   99. −11   101. $\frac{1}{80}$
103. (a) 4   (b) −5

Mid-Chapter Quiz (p. 74)

1. (a) 0   (b) 10
2. (a) Division by zero is undefined.   (b) 0
3. Terms: $4x^2, -2x$ 4. Terms: $5x, 3y, -z$
   Coefficients: 4, −2   Coefficients: 5, 3, −1
5. $(-3y)^4$ 6. $2^3(x - 3)^2$
7. Associative Property of Multiplication
8. Distributive Property   9. Multiplicative Inverse Property
10. Commutative Property of Addition   11. $6x^2 - 2x$
12. $-12y - 18y^2 + 36$ 13. $20y^2$ 14. $-\frac{x^2}{5}$
15. $9y^5$ 16. $\frac{10z^3}{21y}$ 17. $y^2 + 4xy + y$
18. $7x - 4$ 19. $8a - 7b$ 20. $-8x - 66$
21. $8 + (x + 6) + (3x + 1) = 4x + 15$
22. (a) $\frac{x}{6}$ (b) 5 students

Section 2.3 (pp. 76–85)

1. Pay per hour • Number of hours
3. Original number of coupons − Number of coupons used
5. Price per carton • Number of cartons
7. $0.10d$ 9. $0.10d + 0.25q$ 11. $x + 5$

13. $b - 25$ **15.** $g - 6$ **17.** $2h$ **19.** $\frac{w}{3}$
21. $\frac{x}{50}$ **23.** A number decreased by 10
25. The product of 3 and a number, increased by 2
27. One-half of a number, decreased by 6
29. Three times the difference of 2 and a number
31. The sum of a number and 1, all divided by 2
33. One-half decreased by the quotient of a number and 5
35. The square of a number, increased by 5
37. $0.06L$ **39.** $3m + 4v$ **41.** $12.50 + 0.75q$
43. $0.99a + 1.99r$ **45.** $t \approx 10.2$ yr
47. $t \approx 11.9$ yr **49.** $t \approx 11$ yr
51.

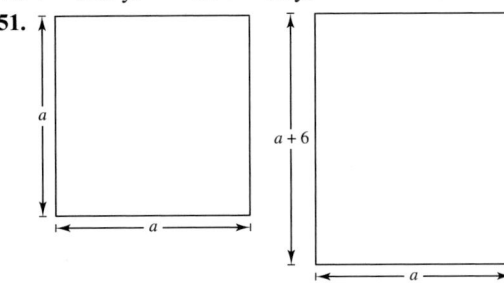

Perimeter of the square: $4a$ cm
Area of the square: a^2 cm^2
Perimeter of the rectangle: $4a + 12$ cm
Area of the rectangle: $a^2 + 6a$ cm^2
53. 5.75 ft
55. Guess, Check, and Revise; Make a Table; Look for a Pattern; Draw a Diagram; Solve a Simpler Problem
57. Answers will vary.
59. The product of a number and the sum of the same number and 16
61. 4 divided by the difference between a number and 2
63. $x(x + 3) = x^2 + 3x$ **65.** $x - (25 + x) = -25$
67. $x^2 - x(2x) = -x^2$ **69.** $0.25p + 0.1n$
71. One less than 2 times a number; 39
73. The start time is missing. **75.** (a), (b), and (e)
77. $\frac{5}{3n}, \frac{5}{n} \cdot 3 = \frac{15}{n}$; The expression $\frac{3n}{5}$ is not a possible interpretation because the phrase "the quotient of 5 and a number" indicates that the variable is in the denominator.
79. 78 **81.** $\frac{3}{4}$ **83.** $\frac{23}{9}$
85. Commutative Property of Addition
87. Distributive Property

Section 2.4 (pp. 86–93)

1. (a) Not a solution (b) Solution
3. (a) Not a solution (b) Solution
5. (a) Not a solution (b) Solution
7. (a) Not a solution (b) Not a solution
9. (a) Solution (b) Not a solution
11. Equation **13.** Expression **15.** Equation
17. Expression **19.** Equation **21.** Expression
23. Equation **25.** Expression **27.** $x = 13$
29. $x = 16$ **31.** $x = 10$
33. $x - 8 = 3$ Original equation
  $x - 8 + 8 = 3 + 8$ Add 8 to each side.
  $x = 11$ Solution
  Addition Property of Equality

35. $\frac{2}{3}x = 12$ Original equation
  $\frac{3}{2}\left(\frac{2}{3}x\right) = \frac{3}{2}(12)$ Multiply each side by $\frac{3}{2}$.
  $x = 18$ Solution
  Multiplication Property of Equality
37. $5x + 12 = 22$ Original equation
  $5x + 12 - 12 = 22 - 12$ Subtract 12 from each side.
  $5x = 10$ Combine like terms.
  $\frac{5x}{5} = \frac{10}{5}$ Divide each side by 5.
  $x = 2$ Solution
  Addition & Multiplication Properties of Equality
39. $x + 6 = 94$ **41.** $\frac{x}{18} = 4.5$ **43.** $7x = 150 - 72$
45. Yes. Answers will vary. **47.** Dividing by zero is undefined.
49.
  $\frac{x}{3} = x + 1$ Original equation
  $3\left(\frac{x}{3}\right) = 3(x + 1)$ Multiply each side by 3.
  $x = 3x + 3$ Distributive Property
  $x - 3x = 3x + 3 - 3x$ Subtract $3x$ from each side.
  $x - 3x = 3x - 3x + 3$ Group like terms.
  $-2x = 3$ Combine like terms.
  $\frac{-2x}{-2} = \frac{3}{-2}$ Divide each side by -2.
  $x = -\frac{3}{2}$ Solution
  Addition & Multiplication Properties of Equality
51. (a) Not a solution (b) Not a Solution
53. (a) Not a solution (b) Not a Solution
55. $0.25n + 7 = 8.75$
57. $10a + 6\left(\frac{3}{4}a\right) = 986$
  $\frac{29}{2}a = 986$
59. No, there is only one value of x, $\frac{b}{a}$, for which the equation is true.
61. *Sample answer:* The total cost of a shipment of bulbs is $840. Find the number of cases of bulbs when each case costs $35.
63. t^7 **65.** $15x$ **67.** $8b$ **69.** $x + 23$ **71.** $4y + 7$

Review Exercises (pp. 96–99)

1. x **3.** a, b
5. Terms: $12y, y^2$ **7.** Terms: $5x^2, -3xy, 10y^2$
  Coefficients: 12, 1   Coefficients: 5, -3, 10
9. Terms: $\frac{2y}{3}, -\frac{4x}{y}$
  Coefficients: $\frac{2}{3}, -4$
11. $(5z)^3$ **13.** $(-3x)^5$ **15.** $6^2(b - c)^2$
17. (a) 5 (b) 5 **19.** (a) 4 (b) -2
21. (a) 0 (b) -7 **23.** (a) -3 (b) 6
25. Multiplicative Inverse Property
27. Commutative Property of Multiplication
29. Associative Property of Addition **31.** $-2a$
33. $11p - 3q$ **35.** $\frac{15}{4}s - 5t$ **37.** $\frac{19}{15}a + \frac{1}{6}b$
39. $3x - 3y + 3xy$ **41.** $6n + 3$ **43.** $48t$ **45.** $45x^2$
47. $-12x^3$ **49.** $8x$ **51.** (a) $6x + 12$ (b) $2x^2 + 12x$
53. $(4x)(16x) - x(6x) = 58x^2$ **55.** $5u - 10$ **57.** $5s - r$
59. $10z - 1$ **61.** $2z - 2$ **63.** $8x - 32$ **65.** $-2x + 4y$

67. *Verbal model:*

Base pay per hour + Additional pay per unit · Number of units produced per hour

Algebraic expression: $8.25 + 0.60x$

69. $\frac{2}{3}x + 5$ **71.** $2y - 10$ **73.** $50 + 7z$ **75.** $\frac{s + 10}{8}$

77. $g^2 + 64$ **79.** $0.05x$ **81.** $625n$

83. Four more than 3 times a number; 64

85. (a) Not a solution (b) Solution
87. (a) Not a solution (b) Solution
89. (a) Solution (b) Not a solution
91. (a) Not a solution (b) Solution
93. (a) Solution (b) Solution

95.
$-7x + 20 = -1$	Original equation
$-7x + 20 - 20 = -1 - 20$	Subtract 20 from each side.
$-7x = -21$	Combine like terms.
$\frac{-7x}{-7} = \frac{-21}{-7}$	Divide each side by -7.
$x = 3$	Solution

Addition & Multiplication Properties of Equality

97.
$x = -(x - 14)$	Original equation
$x = -x + 14$	Distributive Property
$x + x = -x + 14 + x$	Add x to each side.
$x + x = -x + x + 14$	Group like terms.
$2x = 14$	Combine like terms.
$\frac{2x}{2} = \frac{14}{2}$	Divide each side by 2.
$x = 7$	Solution

Addition & Multiplication Properties of Equality

99. $x + \frac{1}{x} = \frac{37}{6}$ **101.** $6x - \frac{1}{2}(6x) = 24$

Chapter Test (p. 100)

1. Terms: $2x^2$, $-7xy$, $3y^3$; Coefficients: 2, -7, 3
2. $x^3(x + y)^2$ **3.** Associative Property of Multiplication
4. Commutative Property of Addition
5. Additive Identity Property
6. Multiplicative Inverse Property **7.** $3x + 24$
8. $20r - 5s$ **9.** $-3y + 2y^2$ **10.** $-36 + 18x - 9x^2$
11. $-a - 7b$ **12.** $8u - 8v$ **13.** $4z - 4$ **14.** $18 - 2t$
15. 6 **16.** -28 **17.** Division by zero is undefined.
18. $\frac{1}{3}n - 4$
19. (a) Perimeter: $2w + 2(2w - 4) = 6w - 8$; Area: $w(2w - 4) = 2w^2 - 4w$
(b) Perimeter: 34 units; Area: 70 square units
20. (a) $25m + 20n$ (b) $110
21. (a) Not a solution (b) Solution

CHAPTER 3

Section 3.1 (pp. 102–109)

1. 9 **3.** -6 **5.** Subtraction
7. Multiplication **9.** Subtraction **11.** Addition

13.
$5x + 15 = 0$	Original equation
$5x + 15 - 15 = 0 - 15$	Subtract 15 from each side.
$5x = -15$	Combine like terms.
$\frac{5x}{5} = \frac{-15}{5}$	Divide each side by 5.
$x = -3$	Simplify

15. -1 **17.** 2 **19.** 6 **21.** -26 **23.** $\frac{1}{4}$
25. -28 **27.** 9 **29.** -3 **31.** $\frac{1}{3}$ **33.** 2 **35.** $\frac{1}{3}$
37. -2 **39.** No solution **41.** 3
43. Infinitely many solutions **45.** $\frac{2}{5}$ **47.** $\frac{9}{2}$ **49.** 0
51. 75 cm **53.** 20 in. $\times$ 40 in. **55.** 150 seats
57. Yes. Subtract the cost of parts from the total to find the cost of labor. Then divide by 44 to find the number of hours spent on labor (2.25 hours).
59. $1430 **61.** 8 wk
63. Substitute the solution into the original equation and simplify each side.

$3x + 2 = 11$
$3(3) + 2 \stackrel{?}{=} 11$
$9 + 2 \stackrel{?}{=} 11$
$11 = 11$ ✓

65. You are trying to isolate the variable term on the left-hand side of the equation. To do this, you must eliminate the $+2$ by subtracting 2; add 2
67. 5 **69.** 30 **71.** 1 **73.** $\frac{2}{3}$
75. Infinitely many solutions **77.** No solution
79. $l = 80$ m, $w = 50$ m **81.** 12 units **83.** 12.5 hr
85. Yes; The Addition Property of Equality; 1 oz
87. False. Multiplying each side of the equation $3x = 9$ by 0 yields $0 = 0$. The equation $3x = 9$ has one solution, $x = 3$, and the result $0 = 0$ suggests that the equation has infinitely many solutions.
89. False. $(2m + 1) + 2n = 2m + 2n + 1 = 2(m + n) + 1$, which is odd.
91. [number line from -4 to 4 with points at -3 and 2]
93. [number line from -2 to 2 with points at -1 and $1\frac{3}{2}$]
95. (a) Solution (b) Not a solution
97. (a) Not a solution (b) Not a solution

Section 3.2 (pp. 110–117)

1. 4 **3.** 3 **5.** 2 **7.** -5 **9.** 2 **11.** -10
13. 9 **15.** 30 **17.** No solution **19.** -4
21. No solution **23.** No solution **25.** 1 **27.** $\frac{5}{6}$
29. 1 **31.** $\frac{5}{2}$ **33.** $-\frac{2}{5}$ **35.** $\frac{35}{2}$ **37.** $-\frac{10}{3}$ **39.** $\frac{3}{4}$
41. $\frac{1}{6}$ **43.** 4.8 hr **45.** 77 points **47.** 5
49. 6.18 **51.** 5 **53.** 7.71 **55.** 66.67
57. 0.42 **59.** 123 **61.** 3.51 **63.** 8.99 **65.** 2054
67. The least common multiple of the denominators of two or more fractions is the least number that is a common multiple of all of the denominators.
69. Multiplying each side of the equation by the least common multiple of the denominators clears the equation of fractions, making the equation easier to solve.
71. 6 **73.** 5 **75.** 6 **77.** No solution **79.** $\frac{4}{11}$
81. 10 **83.** 10 **85.** 25 qt
87. Because each brick is 8 inches long and there are n bricks, the width that is made up of bricks is represented by $8n$. Because there is $\frac{1}{2}$ inch of mortar between adjoining bricks and there are $n - 1$ widths of mortar, the width that is made up of mortar is represented by $\frac{1}{2}(n - 1)$. Because the width of the fireplace is 93 inches, the equation is $8n + \frac{1}{2}(n - 1) = 93$.

89. You could divide each side of the equation by 3.
91. Dividing by a variable assumes that is does not equal zero, which may yield a false solution.
93. $4x^6$ 95. $5z^5$ 97. $x - 4$ 99. $-y^4 + 2y^2$
101. $\frac{17}{3}$ 103. -9

Section 3.3 (pp. 118–125)

1. 62%   3. 7.5%   5. 80%   7. 125%   9. 0.12
11. 1.25   13. 0.085   15. 0.0075   17. $\frac{3}{10}$ 19. $\frac{13}{10}$
21. $\frac{7}{500}$ 23. $\frac{1}{200}$ 25. 45   27. 0.42   29. $37,380
31. 2100   33. 132   35. 430 points   37. 72%
39. 2.75%   41. 9.5%

	Cost	Selling Price	Markup	Markup Rate
43.	$26.97	$49.95	$22.98	85.2%
45.	$40.98	$74.38	$33.40	81.5%
47.	$69.29	$125.98	$56.69	81.8%
49.	$13,250.00	$15,900.00	$2650.00	20%

	Original Price	Sale Price	Discount	Discount Rate
51.	$39.95	$29.95	$5.00	9.8%
53.	$315.00	$18.95	$126.00	20%
55.	$189.99	$10.95	$30.00	42.2%
57.	$119.96	$29.73	$55.22	50%
59.	$394.97	$695.00	$300.00	34.2%

61. Percent means per hundred or parts of 100.
63. $x = 0.25y$

	Percent	Parts out of 100	Decimal	Fraction
65.	40%	40	0.40	$\frac{2}{5}$
67.	7.5%	7.5	0.075	$\frac{3}{40}$
69.	63%	63	0.63	$\frac{63}{100}$
71.	15.5%	15.5	0.155	$\frac{31}{200}$
73.	60%	60	0.60	$\frac{3}{5}$

75. $37\frac{1}{2}\%$ 77. $41\frac{2}{3}\%$ 79. 4%   81. 0.107%
83. If $a > b$, the percent is greater than 100%.
    If $a < b$, the percent is less than 100%.
    If $a = b$, the percent is equal to 100%.
85. False. $1\% = 0.01 \neq 1$
87. False. Because $68\% = 0.68$, $a = 0.68(50)$.
89. 0   91. (a) 7   (b) 16   93. $8x - 20$
95. -3 97. -12

Mid-Chapter Quiz (p. 126)

1. 6   2. 8   3. $\frac{19}{2}$ 4. 0   5. $-\frac{1}{3}$ 6. $\frac{35}{12}$
7. 36   8. $\frac{11}{5}$ 9. 5   10. -2 11. 2.06
12. 51.23   13. 15.5   14. 42   15. 200%   16. 455
17. 12 m × 18 m   18. 10 hr   19. 6 m², 12 m², 24 m²
20. 93   21. $11,550   22. $1017.59   23. 3 hr

Section 3.4 (pp. 128–135)

1. $\frac{4}{1}$ 3. $\frac{1}{2}$ 5. $\frac{17}{4}$ 7. $\frac{2}{3}$ 9. $\frac{9}{1}$ 11. $\frac{32}{53}$

13. $\frac{2}{1}$ 15. $\frac{2}{3}$ 17. $\frac{3}{50}$ 19. $\frac{7}{15}$ 21. $\frac{2}{1}$ 23. $\frac{3}{8}$
25. $0.049/oz   27. $0.073/oz   29. a   31. b
33. 12   35. $\frac{10}{3}$ 37. $\frac{175}{8}$ 39. $\frac{3}{16}$ 41. $\frac{1}{2}$
43. 27   45. $\frac{5}{2}$ 47. 6   49. 16 gal
51. 22,691 votes   53. 250 blocks
55. *Sample answer:* You want to calculate the average miles per hour of a trip.
57. Cross-multiply and then solve for the variable.
59. $\frac{1}{4}$ 61. $\frac{4}{5}$ 63. $\frac{3}{10}$ 65. a   67. b
69. 384 mi   71. $6\frac{2}{3}$ ft   73. $46\frac{2}{3}$ min
75. $12\frac{1}{2}$ lb   77. 20%
79. No. It is also necessary to know either the number of men in the class or the number of women in the class.
81. Answers will vary.   83. 13   85. 9,300,000
87. $\frac{77}{5}$ 89. 62.5   91. 60   93. 62.5

Section 3.5 (pp. 136–143)

1. $49.59   3. 6%   5. 4 in.   7. 2.5 m
9. $2000 at 7%, $4000 at 9%   11. 35 mL   13. $1\frac{1}{5}$ hr
15. Answers will vary.   17. 125 mL/hr
19. 16 dozen roses, 8 dozen carnations   21. $V = \pi r^2 h$
23. Perimeter: linear units—inches and meters
    Area: square units—square feet and square centimeters
    Volume: cubic units—cubic inches and cubic feet

	Distance, d	Rate, r	Time, t
25.	48 m	4 m/min	12 min
27.	210 mi	50 mi/hr	4.2 hr

29. $h = \dfrac{2A}{b}$
31. Solution 2: 75 gal; Final solution: 100 gal
33. Solution 2: 5 qt; Final solution: 10 qt
35. 8.6%   37. 15 m, 15 m, 53 m   39. 28 mi
41. 3 hours on the first part, 1 hour and 15 minutes on the last part
43. Divide by 2 to obtain 90 miles per 2 hours. Divide by 2 again to obtain 45 miles per hour.
    $d = rt \Rightarrow \dfrac{d}{t} = r \Rightarrow \dfrac{180 \text{ miles}}{4 \text{ hours}} = 45$ mi/hr
45. Use $h = \dfrac{2A}{x + y}$ to find the height of a trapezoid.
    Use $x = \dfrac{2A}{h} - y$ to find the base x of a trapezoid.
    Use $y = \dfrac{2A}{h} - x$ to find the base y of a trapezoid.
47. The circumference would double; the area would quadruple.
    Circumference: $C = 2\pi r$, Area: $A = \pi r^2$
    If r is doubled, $C = 2\pi(2r) = 2(2\pi r)$ and $A = \pi(2r)^2 = 4\pi r^2$
49. (a) 7, 1   (b) 7, 1, -3
    (c) 1.8, $\frac{1}{10}$, 7, -2.75, 1, -3 (d) None
51. 9   53. 6   55. 16

Section 3.6 (pp. 144–151)

1. x is greater than or equal to 3.
3. x is less than or equal to 10.
5. y is less than -9.
7. z is greater than or equal to 5 *and* less than or equal to 10.
9. y is greater than $-\frac{3}{2}$ *and* less than or equal to 5.

A8 Answers to Odd-Numbered Exercises

11. (a) Yes (b) No (c) Yes (d) No
13. (a) No (b) No (c) Yes (d) Yes
15. (a) No (b) Yes (c) No (d) No
17. d 19. b
21. $t \geq 5$ 23. $x \leq 2$

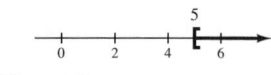

25. $x < 3$ 27. $x > -4$

26. $x \leq 18$ 31. $x \leq 4$

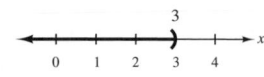

33. $x > \frac{1}{2}$ 35. $x < 8$

37. $x < -9$ 39. $t < 5$

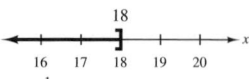

41. $x > -1$ 43. $x > -3$

45. $z \leq -1$ 47. $x > -2$

49. $x \leq -1$ 51. $0 < x \leq 6$ 53. $0 \leq x \leq 34$ bushels
55. Parenthesis; A parenthesis is used to exclude an endpoint from the solution interval. It is used when graphing inequalities containing the symbols < or >.
57. Solving linear inequalities is similar to solving linear equations in that you use properties to isolate the variable. However, when both sides of a linear inequality are multiplied or divided by a negative number, the direction of the inequality symbol must be reversed.
59. (a) Yes (b) Yes (c) No (d) Yes
61. (a) No (b) Yes (c) No (d) Yes
63. $x \geq -13$ 65. $x \geq 7$

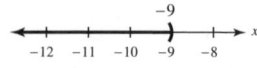

67. $y > \frac{1}{3}$ 69. $x \geq -\frac{10}{3}$

71. Mars is farther from the Sun than Mercury; Transitive Property
73. $x \geq 96$ points 75. $x > 360$ mi
77. True. Subtraction Property of Inequality
79. True. u is equal to 10 or greater than 10. 81. True. $0 < 4$
83. There should be parenthesis at 5, not a square bracket.

85. The sum of 7 and the product of 2 and a number
87. Negative five time the sum of a number and 7
89. One-third plus the quotient of a number and 2

	Distance, d	Rate, r	Time, t
91.	1824 ft	48 ft/sec	38 sec
93.	384 m	6 m/min	64 min
95.	240 km	12 km/hr	20 hr

Review Exercises (pp. 154–156)

1. 5 3. -4 5. $\frac{3}{5}$ 7. 3 9. 5 11. 4
13. $\frac{4}{3}$ 15. 20 17. 12 units 19. 103 m by 53 m
21. 20 23. 6 25. 1 27. 1 29. 20
31. 3 33. $-\frac{1}{7}$ 35. 23.26 37. 224.31 39. 3 hr

Percent	Parts out of 100	Decimal	Fraction
41. 60%	60	0.60	$\frac{3}{5}$
43. 80%	80	0.80	$\frac{4}{5}$
45. 20%	20	0.20	$\frac{1}{5}$
47. 12.5%	55	0.55	$\frac{11}{20}$

49. 20 51. 400 53. 60% 55. $85.44 57. $\frac{1}{8}$
59. $\frac{4}{3}$ 61. $\frac{4}{5}$ 63. (b) 24-ounce container 65. $\frac{7}{2}$
67. $-\frac{10}{3}$ 69. 9 71. $133.33

	Distance, d	Rate, r	Time, t
73.	520 mi	65 mi/hr	8 hr
75.	855 m	5 m/min	171 min
77.	3000 mi	60 mi/hr	50 hr

79. About 1108.3 mi 81. 30 ft × 26 ft 83. $475
85. 13 dimes, 17 quarters 87. $\frac{30}{11} \approx 2.7$ hr
89. x is less than 3.

91. x is greater than or equal to 1 and less than 4.

93. (a) Yes (b) No 95. (a) Yes (b) No
97. (a) No (b) Yes
99. $x \geq -2$; 101. $x < 3$;

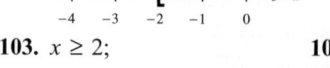 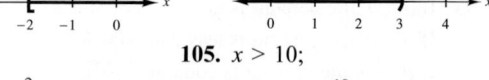

103. $x \geq 2$; 105. $x > 10$;

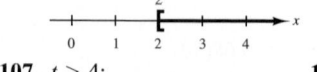

107. $t > 4$; 109. $y \leq -1$;

111. $z \geq 10$ 113. $0 < y \leq 100$ 115. $x \leq 88$ mi

Chapter Test (p. 157)

1. -13 2. $\frac{21}{4}$ 3. 7 4. 1 5. $-\frac{1}{3}$
6. -6 7. 11.03 8. $2\frac{1}{2}$ hr 9. $31\frac{1}{4}$%, 0.3125
10. 1200 11. 36% 12. 6
13. $\frac{5}{9}$; 2 yd = 6 ft = 72 in. 14. $\frac{12}{7}$ 15. 5
16. 66 mi/hr 17. $\frac{36}{7} \approx 5.1$ hr 18. $6250
19. $x \leq 4$ 20. $x < -6$ 21. $x \geq 5$
22. $x > -\frac{5}{3}$ 23. $x \leq \frac{1}{2}$ 24. $x > 1$

Cumulative Test   (p. 158)

1. < 2. 1200 3. $-\frac{11}{24}$ 4. $-\frac{25}{12}$ 5. 8
6. 14 7. 28 8. -30 9. $-\frac{11}{2}$ 10. $3^3(x+y)^2$
11. $-2x^2 + 6x$ 12. Associative Property of Addition
13. $15x^7$ 14. $7x^2 - 6x - 2$ 15. $-3x^2 + 18x$
16. 6 17. $\frac{52}{3}$ 18. -7
19. $x \geq -7$;

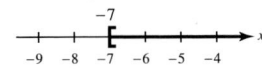

20. $\dfrac{15{,}000 \text{ miles}}{1 \text{ year}} \cdot \dfrac{1 \text{ gallon}}{30 \text{ miles}} \cdot \dfrac{\$4.00}{1 \text{ gallon}} \approx \$2000/\text{yr}$

21. $\frac{3}{4}$ 22. $920 23. $57,000

CHAPTER 4

Section 4.1   (pp. 160–167)

1. 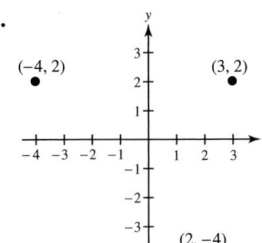 3.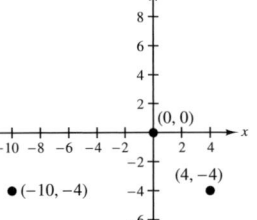

5. 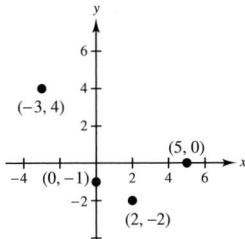 7.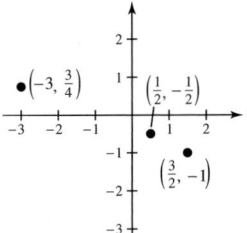

9. Quadrant II   11. Quadrant III   13. Quadrant III

15. (a)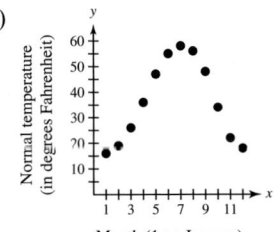

No, because there are only 12 months, but the temperature ranges from 16°F to 58°F.

(b) August

17.
x	-2	0	2	4	6
$y = 3x - 4$	-10	-4	2	8	14

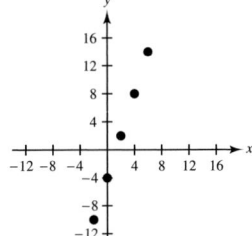

19.
x	-2	0	4	6	8
$y = -\frac{3}{2}x + 5$	8	5	-1	-4	-7

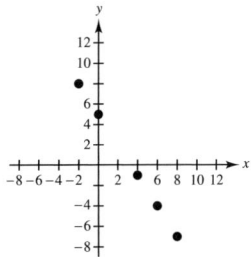

21.
x	-2	-1	0	1	2
$y = -4x - 5$	3	-1	-5	-9	-13

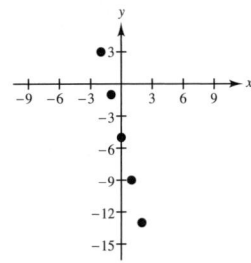

23. (a) Solution (b) Not a solution
    (c) Not a solution (d) Solution
25. (a) Solution (b) Solution
    (c) Not a solution (d) Solution
27. (a) Not solution (b) Solution
    (c) Solution (d) Not a solution
29. (a) Solution (b) Not a solution
    (c) Not a solution (d) Not a solution
31. (a) $-\frac{4}{3}$ (b) 16 (c) -2

33.
x	20	40	60	80	100
$y = 0.066x$	1.32	2.64	3.96	5.28	6.60

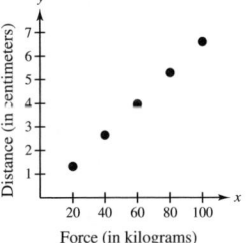

35. $y = 25x + 5000$;

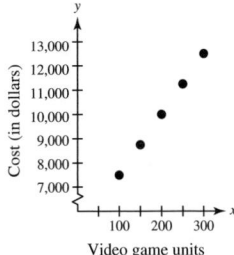

37. Yes. The scale on the vertical axis makes it appear that the changes in the surplus are dramatic.

A10 Answers to Odd-Numbered Exercises

39. (2, 2)   **41.** (−4, −1); (0, 1)
43. A: (5, 2), B: (−3, 4), C: (2, −5), D: (−2, −2)
45. A: (−1, 3), B: (5, 0), C: (2, 1), D: (−1, −2)

47.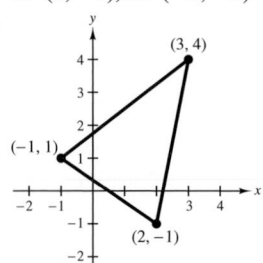

49.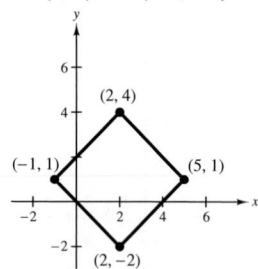

51.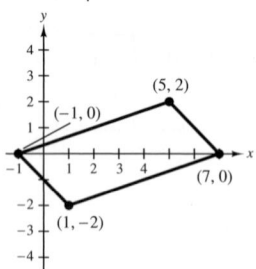

53.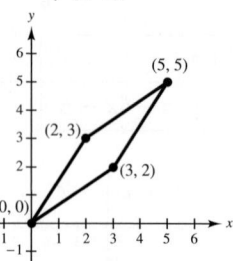

55. $y = -7x + 8$ **57.** $y = 10x - 2$ **59.** Quadrant I or IV

61. (a)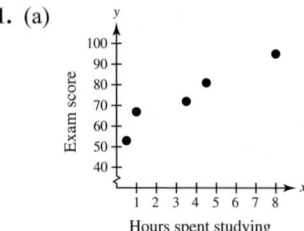
(b) Scores increase with increased study time.

63. (a) and (b)   (c) Reflection in the x-axis
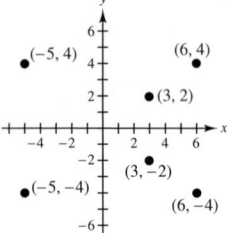

65. (6, 4)
67. No. *Sample answer:* If y is measuring revenue for a product and x is measuring time in years, then the scale on the y-axis may be in units of $100,000 and the scale on the x-axis may be in units of 1 year.
69. −10   **71.** 14   **73.** $x > -1$ **75.** $x < 4$

Section 4.2 (pp. 168–175)

1.

x	−2	−1	0	1	2
y	11	10	9	8	7

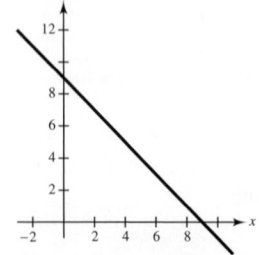

3.

x	−2	−1	0	1	2
y	7	4	3	4	7

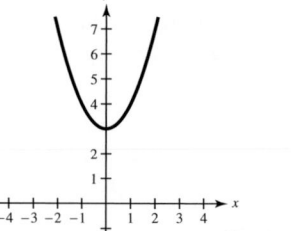

5.

x	−3	−2	−1	0	1
y	2	1	0	1	2

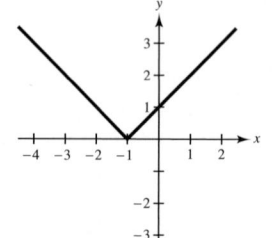

7.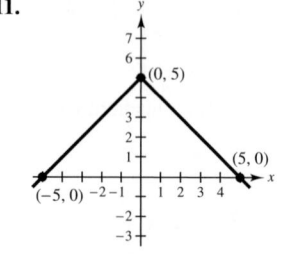

9.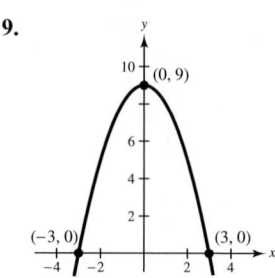

11.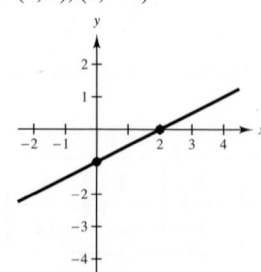

13. (−2, 0), (0, 4)

15. (−1, 0), (1, 0), (0, 1)
17. (2, 0), (0, −1)   **19.** (−1, 0), (0, −2)

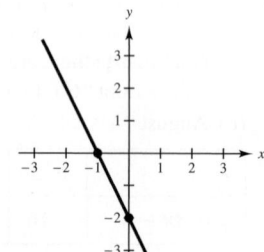

21. (0, 50)

Answers to Odd-Numbered Exercises **A11**

23. (a) $y = 1120 - 80x$
(b)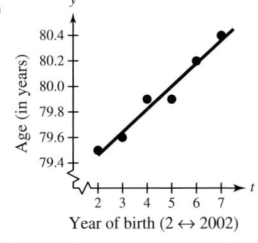
(c) $(0, 1120)$; the initial height of the hot-air balloon

25. (a) 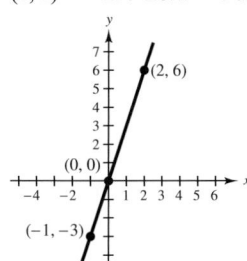 (b) 82.7 yr

27. $(2, 1)$ **29.** Let $x = 0$ and solve the equation for y.

31. 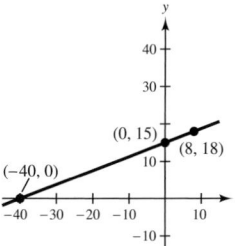   **33.**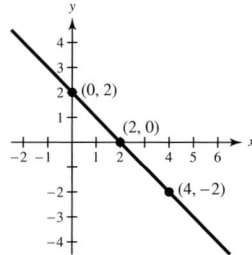

35. 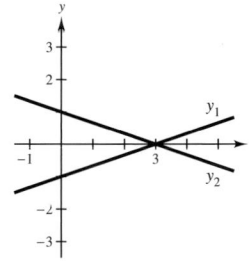   **37.**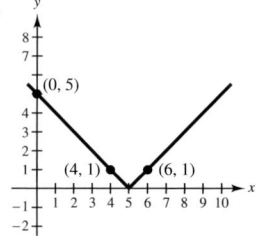

39. $(6, 0), (0, 2)$ **41.** $(6, 0), (0, 2)$
43. No   **45.** Yes; Distributive Property

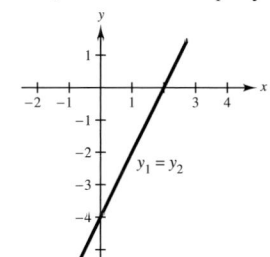

47. $y = 35t$;
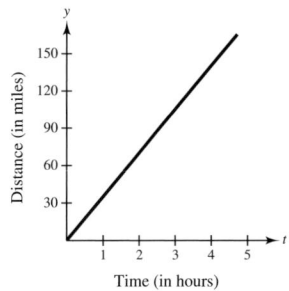

49. Yes. For any linear equation in two variables, x and y, there is a resulting value for y when $x = 0$. The corresponding point $(0, y)$ is the y-intercept of the graph of the equation.

51.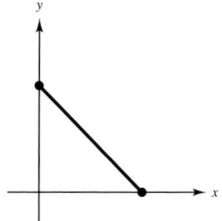

The distance between you and the tree decreases as you move from left to right on the graph. The x-intercept represents the number of seconds it takes you to reach the tree.

53. -13 **55.** 13 **57.** $-\frac{1}{6}$ **59.** $\frac{3}{10}$ **61.** $-\frac{36}{5}$

63. (a) Not a solution   (b) Not a solution
(c) Solution   (d) Solution

65. (a) Not a solution   (b) Not a solution
(c) Not a solution   (d) Not a solution

Section 4.3 *(pp. 176–183)*

1. 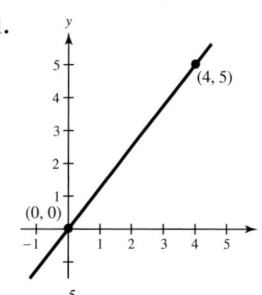   **3.**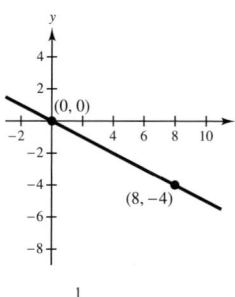

$m = \frac{5}{4}$ $m = -\frac{1}{2}$

5.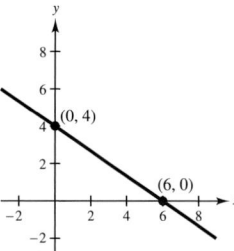

$m = -\frac{2}{3}$

7. $m = \frac{7}{6}$; The line rises

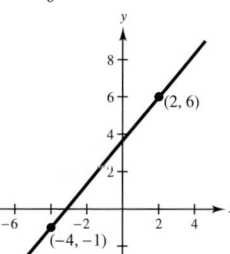

9. m is undefined; The line is vertical.

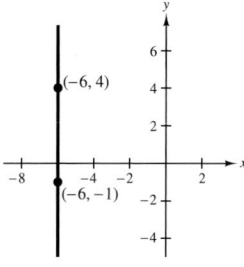

A12 Answers to Odd-Numbered Exercises

11. $m = 0$; The line is horizontal. **13.** $m = -\frac{18}{17}$; The line falls.

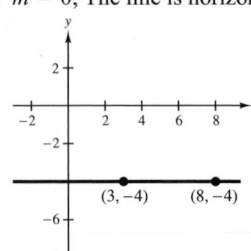 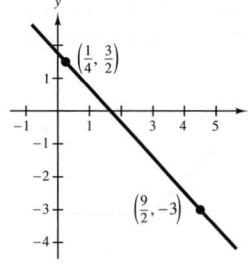

15. $-\frac{2}{3}$ **17.** $y = -\frac{1}{2}x + 2$; $m = -\frac{1}{2}$, $(0, 2)$
19. $y = -\frac{2}{3}x + 2$; $m = -\frac{2}{3}$, $(0, 2)$
21. $y = \frac{3}{4}x + \frac{1}{2}$; $m = \frac{3}{4}$, $\left(0, \frac{1}{2}\right)$
23. $y = 2x - 3$ **25.** $y = \frac{1}{2}x + 1$

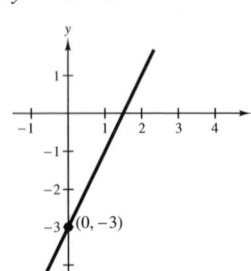 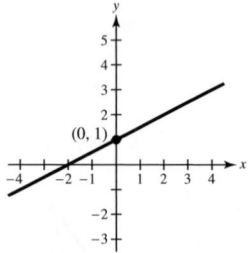

27. $y = \frac{1}{3}x - \frac{5}{2}$ **29.**

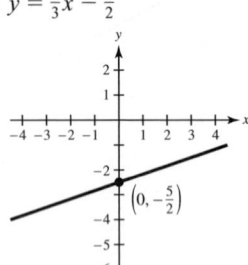 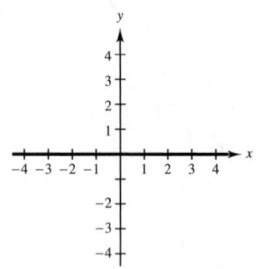

31. **33.**

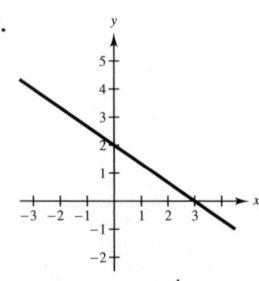

35. Perpendicular; The lines have slopes of -2 and $\frac{1}{2}$, which are negative reciprocals of each other.
37. Neither; The lines have slopes of $\frac{7}{8}$ and $\frac{5}{6}$, which are not the same and are not negative reciprocals of each other.
39. Parallel; The lines both have the same slope: 2.
41. Neither; The equations represent the same line, not 2 distinct lines.
43. Perpendicular; The lines have slopes of $-\frac{1}{3}$ and 3, which are negative reciprocals of each other.
45. (a) -2 (b) $\frac{1}{2}$ **47.** (a) 0 (b) Undefined
49. $-2; 3$ **51.** It falls from left to right.
53. (a) L_2 (b) L_3 (c) L_4 (d) L_1

55. **57.**

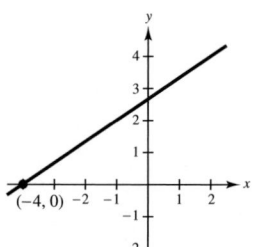 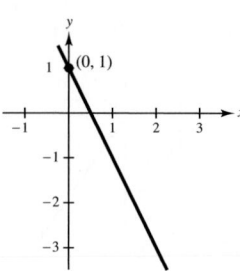

59. Perpendicular   **61.** Parallel
63. $\frac{2}{5}$ **65.** No; $\left|\frac{12}{60}\right| < \left|\frac{12}{50}\right|$
67. No. The slopes of nonvertical perpendicular lines have opposite signs. The slopes are the negative reciprocals of each other.
69. The slope
71. Yes. You are free to label either one of the points as (x_1, y_1) and the other as (x_2, y_2). However, once this is done, you must form the numerator and denominator using the same order of subtraction.
73. x^5 **75.** $-y^3$ **77.** $50x^5$ **79.** $x + 2$
81. $\left(\frac{1}{2}, 0\right), (0, -3)$ **83.** $\left(-\frac{3}{2}, 0\right), (0, -3)$

Mid-Chapter Quiz *(p. 184)*

1.

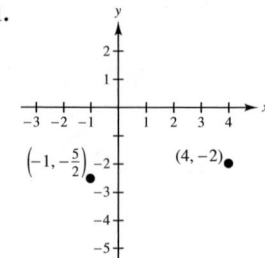

2. Quadrant I or IV
3. (a) Solution   (b) Not a solution
   (c) Solution   (d) Not a solution
4. $(12, 0), (0, -4)$ **5.** $\left(\frac{2}{7}, 0\right), (0, 2)$

6. **7.**

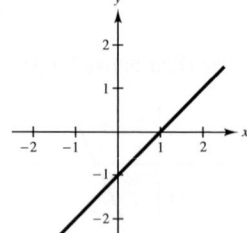 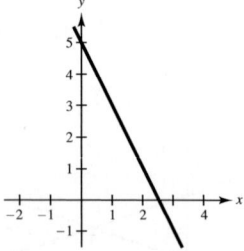

8. **9.**

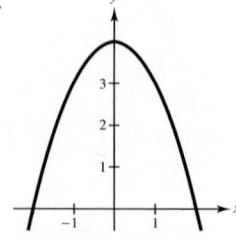 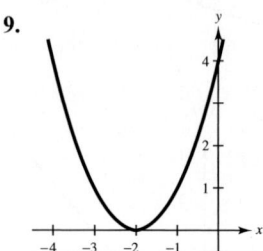

10. 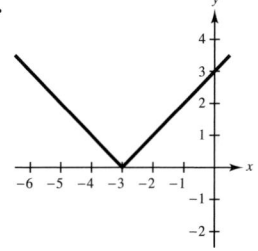   **11.**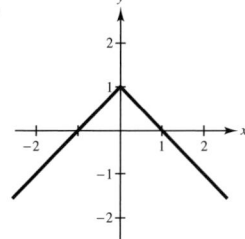

12. $m = -3$; The line falls.

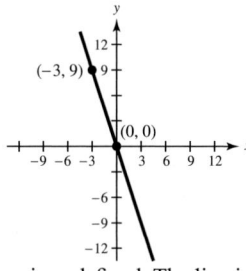

13. m is undefined. The line is vertical.

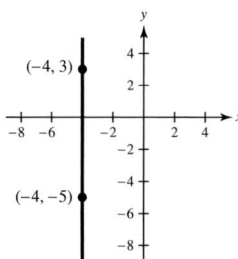

14. $m = \frac{3}{5}$; The line rises.   **15.** $y = x + 3$

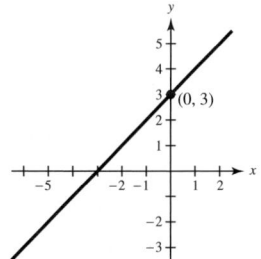

16. $y = \frac{2}{3}x + 2$

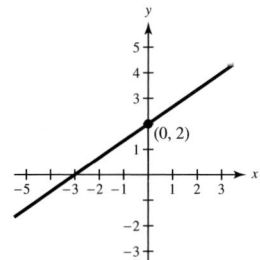

17. 2006: $120 million   2009: $122 million
    2007: $143 million   2010: $162 million
    2008: $133 million   2011: $162 million

18. (a) $y = 2000 - 500t$
(b)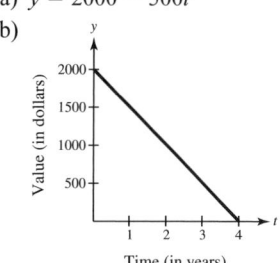
(c) $(0, 2000)$; the original value of the computer system

Section 4.4 *(pp. 186–193)*

1. $y = -\frac{1}{3}x + \frac{5}{3}$ **3.** $y = -\frac{1}{5}x - \frac{13}{5}$ **5.** $y = \frac{1}{3}x - \frac{1}{3}$
7. $y = -\frac{3}{5}x + \frac{8}{5}$ **9.** $y = -\frac{1}{2}x + \frac{13}{2}$
11. (a) $y = x - 1$ (b) $y = -x - 3$
13. (a) $y = -\frac{1}{2}x + \frac{5}{2}$ (b) $y = 2x - 5$
15. (a) $y = x - 1$ (b) $y = -x + 3$
17. (a) $y = -\frac{3}{4}x - 5$ (b) $y = \frac{4}{3}x + 20$ **19.** $x = -2$
21. $y = \frac{2}{3}$ **23.** $x = 4$ **25.** $y = -8$
27. $P = 146t + 735$; $1.027 billion
29. (a) $V = 25{,}000 - 2300t$ (b) $18,100
31. $-\frac{1}{2}$ mi/day; $y = -\frac{1}{2}x + 16$
33. The slope is needed to write an equation using the point-slope form.
35. The equation is written in slope-intercept form.
37. $y = \frac{2}{3}x + \frac{3}{2}$ **39.** $y = -0.8x + 5.6$ **41.** 5   **43.** $\frac{3}{2}$
45. $8x + 6y - 19 = 0$ **47.** $6x + 5y - 9 = 0$
49. $W = 2000 + 0.02S$
51. (a) f: $m = -10$; Loan decreases by $10 per week.
(b) e: $m = 1.50$; Pay increases by $1.50 per unit.
(c) g: $m = 0.32$; Amount increases by $0.32 per mile.
(d) h: $m = -100$; Annual depreciation is $100.
53. Let $y = 0$ and solve for x.
$$y = mx + b$$
$$0 = mx + b$$
$$-b = mx$$
$$-\frac{b}{m} = x$$
55. The lines are parallel or they coincide.   **57.** $12 - 8x$
59. $x + 10$ **61.** $y = -3x + 4$ **63.** $y = \frac{4}{5}x + \frac{2}{5}$
65. 2   **67.** $-\frac{1}{7}$

Section 4.5 *(pp. 194–201)*

1. (a) Not a solution   (b) Solution
(c) Not a solution   (d) Solution
3. (a) Solution   (b) Solution
(c) Solution   (d) Not a solution
5. Dashed   **7.** Solid   **9.** c   **11.** b

13.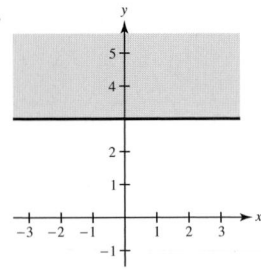
15.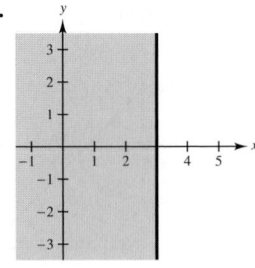
17. c **19.** b
21.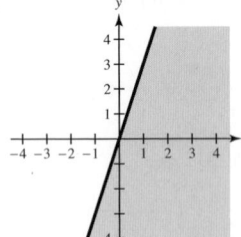
23.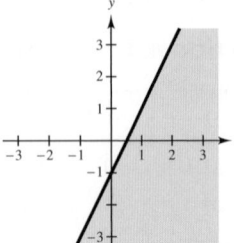
25.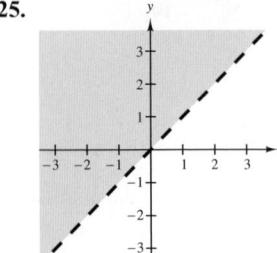
27.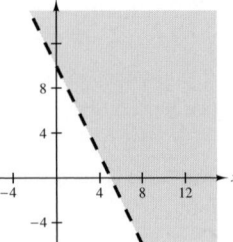
29. d **31.** c
33.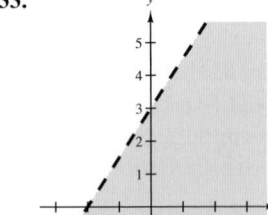
35.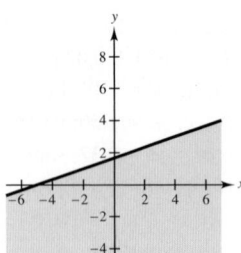
37.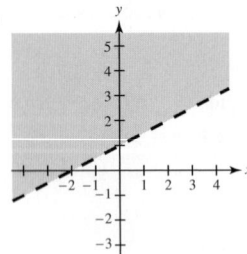

39. $9x + 12y \geq 210$, where x represents the number of hours at the grocery store and y represents the number of hours mowing lawns.

41.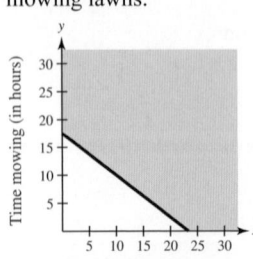
Sample answer: (2, 16), (15, 7), (20, 20)

43. The inequality symbol is $<$.
45. No. Only points in the shaded region below the dashed line represent solutions. (0, 1) is on the dashed line.
47. $y > 6x$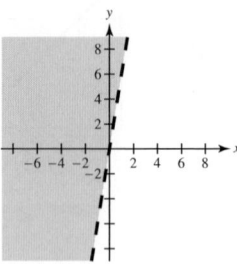
49. $x + y \geq 9$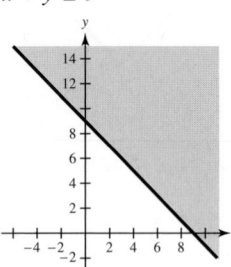
51. $y \leq x + 3$
53. $y > -\frac{2}{3}x + 2$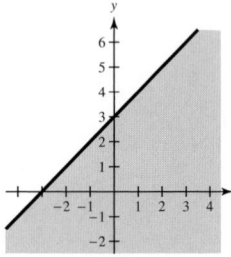
55. $y < \frac{1}{2}x + 1$
57. (a) $y \leq 0.35x$, where x represents the total calories consumed per day and y represents the fat calories consumed per day.
(b)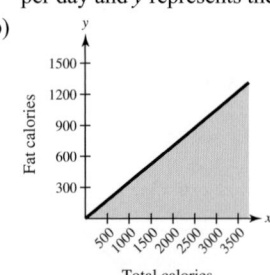
Sample answer: (1500, 500), (2000, 700), (2500, 800)

59. (a) $t + \frac{3}{2}c \leq 12$, where t represents the number of tables and c represents the number of chairs.
(b)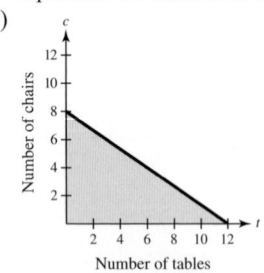

61. $y > 0$
63. Yes. When you divide each side of $2x < 2y$ by 2, you get $y > x$.
65. $\frac{9}{2}$ **67.** $\frac{16}{3}$ **69.** $\frac{18}{1}$ **71.** $y = -2x$ **73.** $y = \frac{2}{5}x - 2$
75. $y = \frac{2}{9}x - \frac{7}{3}$

Review Exercises *(pp. 204–207)*

1. 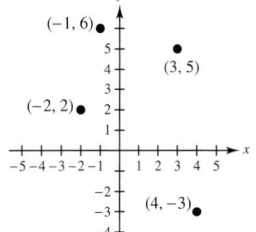 **3.**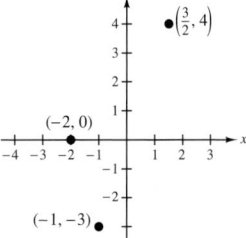

5. A: $(3, -2)$; B: $(0, 5)$; C: $(-1, 3)$; D: $(-5, -2)$ **7.** Quadrant II **9.** Quadrant III

11.

x	−1	0	1	2
y = 4x − 1	−5	−1	3	7

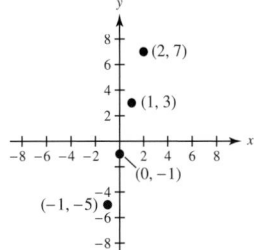

13.

x	−1	0	1	2
$y = -\frac{1}{2}x - 1$	$-\frac{1}{2}$	−1	$-\frac{3}{2}$	−2

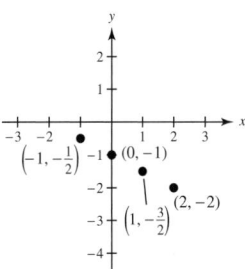

15. $y = -\frac{3}{4}x + 3$ **17.** $y = 3x - 4$

19. (a) Solution (b) Not a solution (c) Not a solution (d) Not a solution

21. (a) Solution (b) Solution (c) Not a solution (d) Solution

23. (a)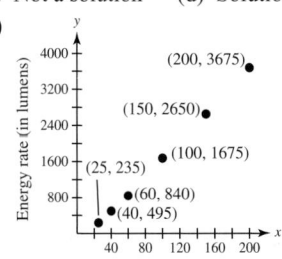

(b) The energy rate increases as the wattage increases.

25.

x	−2	−1	0	1	2
y	−7	−6	−5	−4	−3

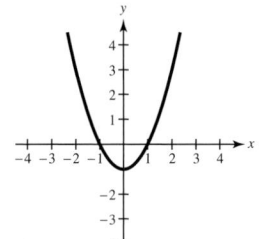

27.

x	−2	−1	0	1	2
y	3	0	−1	0	3

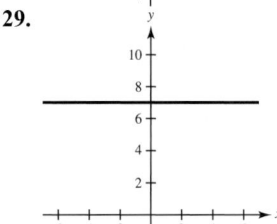

29. 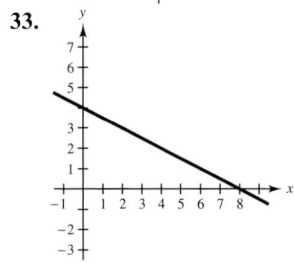 **31.**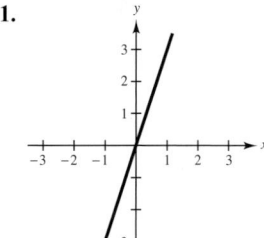

33. 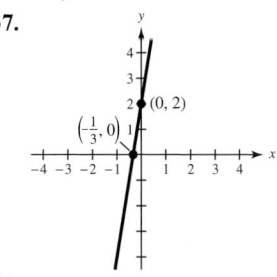 **35.**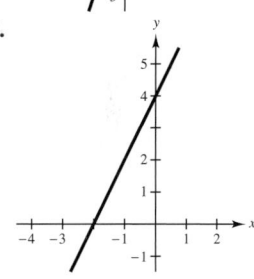

37. 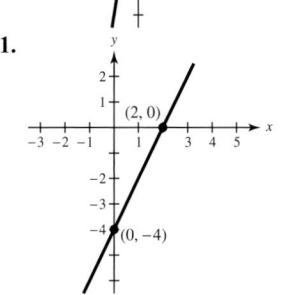 **39.**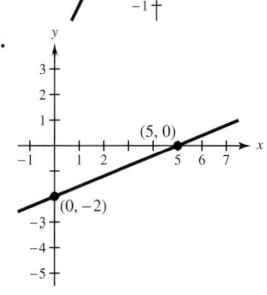

41. 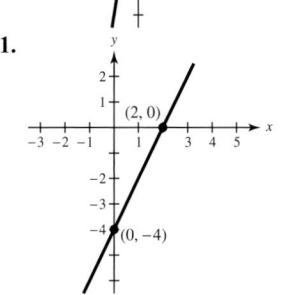 **43.**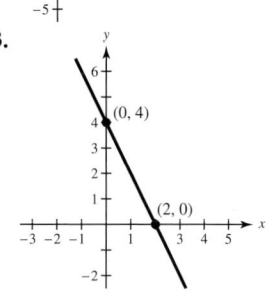

A16 Answers to Odd-Numbered Exercises

45. $C = 3x + 125;$

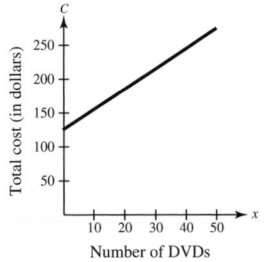

47. $\frac{1}{2}$ **49.** 3

51.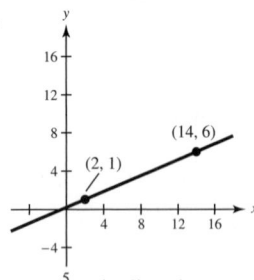

$m = \frac{5}{12};$ The line rises.

53.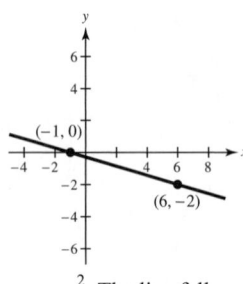

$m = -\frac{2}{7};$ The line falls.

55.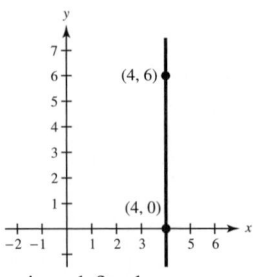

m is undefined.
The line is vertical.

57.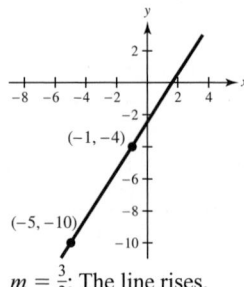

$m = \frac{3}{2};$ The line rises.

59.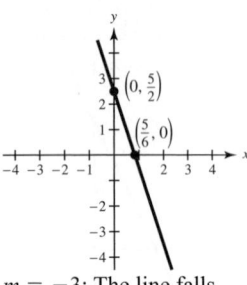

$m = -3;$ The line falls.

61. $\frac{2}{3}$

63. $y = -x + 6$

65. $y = 2x + 1$

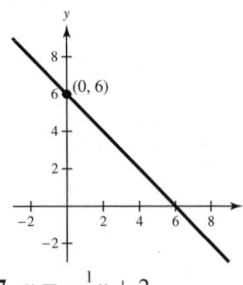

67. $y = -\frac{1}{2}x + 2$

69. $y = \frac{2}{5}x + 1$

71. Parallel   **73.** Neither   **75.** $y = 2x - 9$
77. $y = -4x + 6$ **79.** $y = -\frac{8}{3}x + \frac{1}{3}$ **81.** $x = 3$
83. $x + 2y + 4 = 0$ **85.** $y - 8 = 0$ **87.** $-5x + 3y + 1 = 0$
89. $25x - 20y + 6 = 0$ **91.** (a) $y = x + 9$ (b) $y = -x - 3$
93. (a) $y = 4$ (b) $x = \frac{3}{8}$ **95.** $y = 5$ **97.** $x = 5$
99. $y = 3$ **101.** $W = 5500 + 0.07S$
103. (a) Not a solution   (b) Not a solution
   (c) Solution   (d) Solution
105. (a) Not a solution   (b) Solution
   (c) Solution   (d) Not a solution
107. b   **108.** c   **109.** d   **110.** a
111. 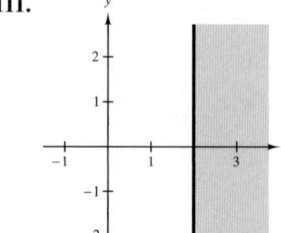   **113.**

115. 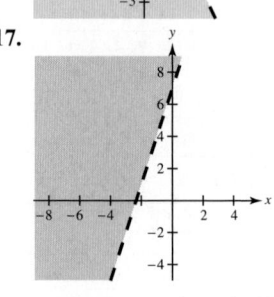   **117.**

119. (a) $2x + 3y \leq 120,$ where x represents the number of DVD players and y represents the number of camcorders.

(b)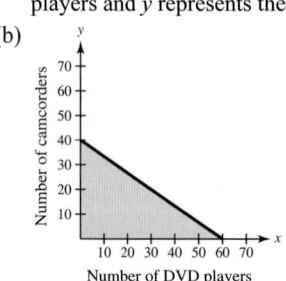

Sample answer: $(10, 15), (20, 20), (30, 20)$

Chapter Test (p. 208)

1.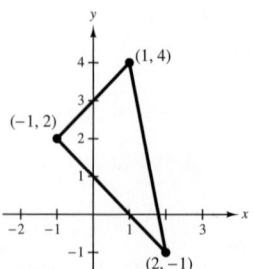

2. (a) Not a solution   (b) Solution
   (c) Solution   (d) Not a solution
3. 0   **4.** $(-2, 0), (0, 8)$

5.

x	−2	−1	0	1	2
y	2	−1	−4	−7	−10

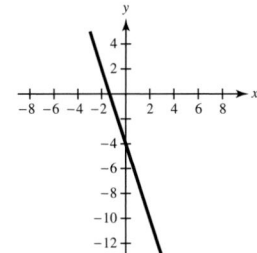

6. 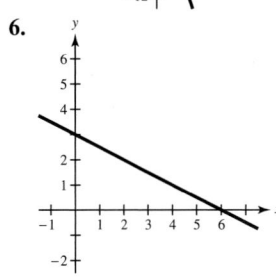 **7.**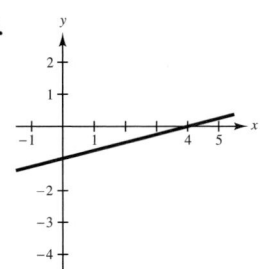

8. 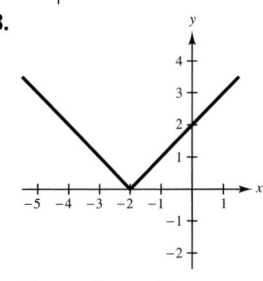 **9.**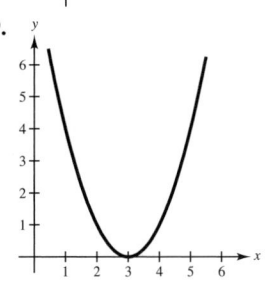

10. $\frac{3}{14}; y = \frac{3}{14}x + \frac{15}{14}$

11. 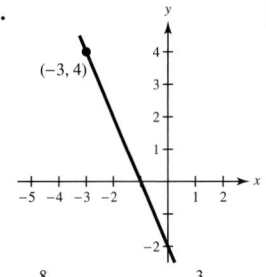 $(-2, 2), (-1, 0)$

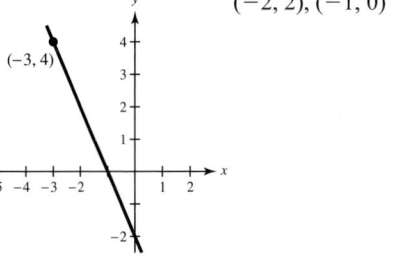

12. $-\frac{8}{7}$ **13.** $y = -\frac{3}{8}x + 6$ **14.** $x = 3$

15. (a) Solution (b) Solution
(c) Solution (d) Solution

16. 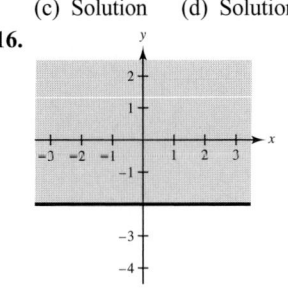 **17.**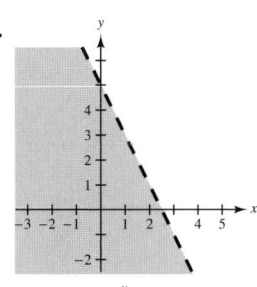

18. 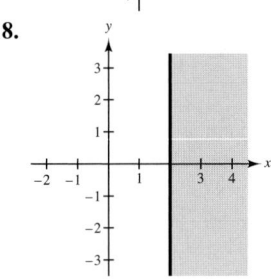 **19.**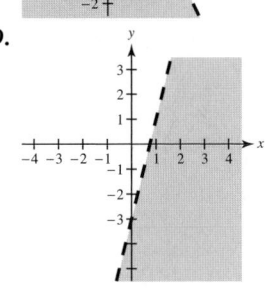

Answers to Odd-Numbered Exercises A17

20. Sales are increasing at a rate of 230 units per year.

CHAPTER 5

Section 5.1 (pp. 210–217)

1. 243   **3.** u^6 **5.** $5x^7$ **7.** t^6 **9.** t^8 **11.** $8v^3$
13. $-8s^3$ **15.** $a^{10}b^{11}$ **17.** x^3 **19.** x **21.** z^6
23. $3u$ **25.** $-4z^2$ **27.** $-\frac{8b}{3}$ **29.** $\frac{4v^2}{3}$ **31.** $\frac{1}{2^3}$
33. $\frac{1}{y^5}$ **35.** $\frac{8}{x^7}$ **37.** $\frac{z^4}{2}$ **39.** $\frac{7x^4}{y}$ **41.** $\frac{3}{4x}$
43. $\frac{1}{(-2)^5}$ **45.** $\frac{1}{3(2^4)}$ **47.** 4   **49.** x^2
51. $-\frac{8}{x}$ **53.** $\frac{x}{y}$ **55.** $\frac{1}{9x^4y^2}$ **57.** x^5 **59.** $\frac{1}{x^2}$
61. $\frac{x^5}{243z^{10}}$ **63.** $8x^3$ **65.** 9.3×10^7 **67.** 2.12×10^{-5}
69. 4.2×10^{-3} m   **71.** 867   **73.** 0.00852
75. 1,000,000,000 V   **77.** 3.27×10^{12}
79. 3.3554×10^{32} **81.** 2.2051×10^{-3} **83.** 5.2346×10^2
85. 1.119×10^{-2} mi²/person   **87.** 2.4
89. Write the number in the form $c \times 10^{-n}$, where $1 \le c \le 10$, by moving the decimal point n places to the right.
91. $\frac{81}{16}$ **93.** $\frac{9}{4}$ **95.** $-32u^6v^{13}$ **97.** $-\frac{15}{x^2}$
99. -2 **101.** -3 **103.** -2 **105.** -3
107. (a)

x	−1	−2	−3	−4	−5
$y = 2^x$	$\frac{1}{2}$	$\frac{1}{4}$	$\frac{1}{8}$	$\frac{1}{16}$	$\frac{1}{32}$

(b) 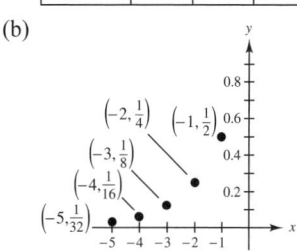 (c) When n is very large, 2^{-n} is very small; The value of 2^{-n} will never be negative.

109. 1.0772×10^8 km; 2.2740×10^8 km; 7.7795×10^8 km; 2.8694×10^9 km
111. False. Let $x = 5$ and $y = 2$.   **113.** True
115. Yes. $a^{-1}b^{-1} = a^{-1} \cdot b^{-1} = \frac{1}{a} \cdot \frac{1}{b} = \frac{1}{ab}$ **117.** $\frac{10^3}{4}$
119. $(3 \times 10^5)(4 \times 10^6)$ Commutative Property
                                                    of Multiplication
$= (3 \times 10^5)(10^6 \times 4)$
$= 3(10^5 \times 10^6)(4)$ Associative Property
                                                    of Multiplication
$= 4(10^{5+6})(4)$ Rule of exponents
$= (3 \cdot 4)10^{11}$ Commutative Property
                                                    of Multiplication
$= 12 \times 10^{11}$ Multiplication
$= 1.2 \times 10^{12}$ Scientific Notation

121. 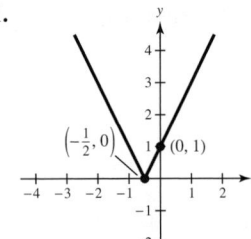 **123.**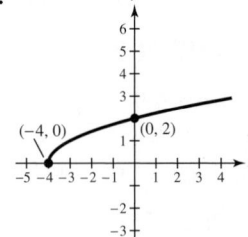

125. 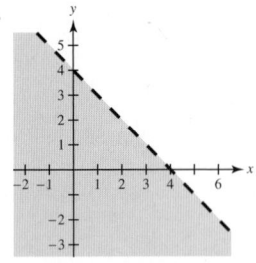 **127.**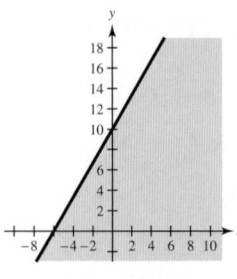

Section 5.2 (pp. 218–225)

1. Polynomial
3. Not a polynomial because the exponent in the first term is not an integer
5. Not a polynomial because the exponent is negative
7. Standard form: $-5x^2 + 7x + 10$; Degree: 2; Leading coefficient: -5
9. Standard form: $-3m^5 - m^2 + 6m + 12$; Degree: 5; Leading coefficient: -3
11. $20w - 4$ 13. $4z^2 - z - 2$ 15. $2b^3 - b^2$
17. $5x + 13$ 19. $-x - 28$ 21. $4x^2 + 2x + 2$
23. $10z + 4$
25. (a) $T = -0.026t^2 + 0.73t + 2.0$
    (b)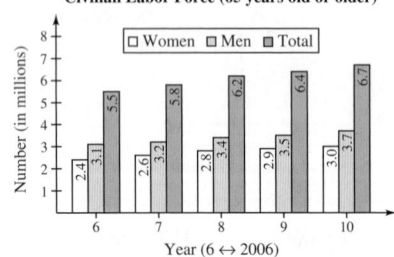
    Civilian Labor Force (65 years old or older)

27. $9x - 11$ 29. $x^2 - 2x + 2$ 31. $5z^3 - z - 4$
33. $x - 1$ 35. $-x^2 - 2x + 3$ 37. $-x^3 + 9x^2 - x - 10$
39. $-2x^4 - 5x^3 - 4x^2 + 6x - 10$ 41. $-2x - 20$
43. $3x^3 - 2x + 2$ 45. $2x^4 + 9x + 2$ 47. $12z + 8$
49. $y^3 - y^2 - 3y + 7$
51. $4t^2 + 20$ 53. $6v^2 + 90v + 30$ 55. $x^2 - 2x$
57. $2x^2 - 2x$ 59. $21x^2 - 8x$ 61. $3x^2$ and x^2, $2x$ and $-4x$
63. Lining up the like terms
65. $-3x^5 - 3x^4 + 2x^3 - 6x + 6$ 67. $\frac{3}{2}y^2 + \frac{5}{4}$
69. $1.6t^3 - 3.4t^2 - 7.3$ 71. $4b^2 - 3$ 73. $11x^2 - 3x$
75. (a) $P = -x^2 + 75x - 200$, $0 \le x \le 60$
    (b) $1200 (c) $-$150
77. No, two binomials may have no like terms or like terms that differ only in sign.
79. Two terms are like terms if they are both constant or if they have the same variable factor(s); Numerical coefficients

81. Yes. A polynomial is an algebraic expression whose terms are all of the form ax^k, where a is any real number and k is a nonnegative integer.
83. 6 85. $-\frac{3}{5}$
87. 89. -5 91. 3.3714×10^{-15}

Mid-Chapter Quiz (p. 226)

1. $81m^6$ 2. $-24x^7y$ 3. $-\dfrac{4}{3x^2y}$ 4. $\dfrac{t}{12}$
5. $\dfrac{5}{x^2y^3}$ 6. $\dfrac{3yz}{5x^2}$ 7. $\dfrac{a^6}{9b^4}$ 8. 1 9. 8.168×10^{12}
10. 0.005021
11. Because the exponent of the third term is negative.
12. Degree: 3; Leading coefficient: -4 13. $x^2 + 5x - 3$
14. $y^2 + 6y + 3$ 15. $-v^3 + v^2 + 6v - 5$ 16. $10s - 11$
17. $3x^2 + 5x - 4$ 18. $3x^3 - 2x + 2$
19. $x^3 - x^2 + 5x - 19$ 20. $x^2 - 2x + 6$
21. $5x^4 + 3x^3 - 4x^2 + 6$ 22. $10x + 36$

Section 5.3 (pp. 228–237)

1. $-x^3 + 4x$ 3. $-6x^3 - 15x$ 5. $24x^4 - 12x^3 - 12x$
7. $2x^3 - 4x^2 + 16x$ 9. $4x^4 - 3x^3 + x^2$
11. $4t^4 - 12t^3$ 13. $x^2 + 7x + 12$ 15. $3x^2 - 2x - 5$
17. x^2; $4x$; $2x$; 8 19. $2x^2 - 5xy + 2y^2$
21. $15x^2 + 23x + 6$ 23. $-8x^2 + 18x + 18$
25. $3x^2 - 5xy + 2y^2$ 27. $2x^2 + 4x = 2x(x + 2)$
29. $x^2 + 3x + 2 = (x + 1)(x + 2)$ 31. $4x^2 + 16x + 16$
33. $64x^2 + 32x + 4$ 35. $3x^3 + 6x^2 - 4x - 8$
37. $15s + 4$ 39. $-x^6 + 8x^3 + 32x^2 - 2x - 8$
41. $(x + 2)^2 = x^2 + 4x + 4$
43. $x^3 + 3x^2 + x - 1$ 45. $x^3 - 7x^2 + 11x - 5$
47. $5x^3 - 8x^2 - 8$ 49. $x^2 + x - 6$ 51. $8x^3 + 27$
53. $3x^5 + 4x^3 + 7x^2 + x + 7$ 55. $x^4 - x^2 + 4x - 4$
57. $-2x^4 - 9x^3 + 16x^2 + 11x - 12$
59. $x^3 - 6x^2 + 12x - 8$ 61. $x^2 - 9$ 63. $16t^2 - 36$
65. $16x^2 - y^2$ 67. $x^2 + 12x + 36$ 69. $t^2 - 6t + 9$
71. $64 - 48z + 9z^2$ 73. $16 + 56s^2 + 49s^4$
75. $4x^2 - 20xy + 25y^2$ 77. 4 ft × 4 ft
79. $x - y$ and $x + 2$
81. First: x, x; Outer: $2, x$; Inner: $x, -y$; Last: $2, -y$
83. $(x + 4)(x + 5) = x^2 + 9x + 20$
85. $8x$ 87. $x^3 - 12x - 16$
89. $x^2 + y^2 + 2xy + 2x + 2y + 1$
91. Yes; The right side is obtained by simplifying the left side.
93. $x^3 + 6x^2 + 12x + 8$ 95. $(x^2 + 10x)$ ft^2
97. False. $(3x)^2 = 3^2 \cdot x^2 = 9x^2 \ne 3x^2$
99. Multiplying a polynomial by a monomial is an application of the Distributive Property. Polynomial multiplication requires repeated use of this property. $4x(x - 2) = 4x^2 - 8x$

101. mn. Each term of the first polynomial must be multiplied by each term of the second polynomial.
103. $15x - 7$ **105.** $-4x + 17$ **107.** 11.25
109. $33\frac{1}{3}\%$ **111.** 4 **113.** $\frac{15}{4}$

Section 5.4 (pp. 238–245)

1. $z + 1$ **3.** $3x - \frac{5}{3}$ **5.** $b - 2$ **7.** $-5z^2 - 2z$
9. $4z^2 + \frac{3z}{2} - 1$ **11.** $3x - 1 + \frac{3}{2x} - \frac{1}{2x^2} + \frac{2}{x^3}$
13. $x - 2$ **15.** $x + 5$ **17.** $x + 4 - \frac{9}{x + 4}$ **19.** $z - 3$
21. $y + 1 - \frac{5}{3y + 1}$ **23.** $6t + 1$ **25.** $x^2 - 2x + 5 + \frac{3}{x - 2}$
27. $-3x^3 - x^2 - 4x - 10 + \frac{59}{-3x + 5}$ **29.** $x^2 + 2x + 4$
31. $3x^3 - 12x^2 + 8x - 4 - \frac{2}{x + 4}$ **33.** $3x - 1$
35. $x^3 + x^2 + x + 1$
37. (a) Verbal Model: Area = Length × Width
 (b) Area = $x^2 + 5x - 6$; Width = $x - 1$
 (c) Length = $\frac{x^2 + 5x - 6}{x - 1} = (x + 6)$ m
39. (a) $\frac{t + 12}{t + 8} = 1 + \frac{4}{t + 8}$
 (b)

t	0	10	20	30	40	50	60
$\frac{t + 12}{t + 3}$	1.5	1.22	1.14	1.11	1.08	1.07	1.06

 (c) The value of the ratio approaches 1, because the value of $\frac{4}{t + 8}$ approaches 0.
41. $x^2 - 4$ and $x + 2$ **43.** Yes; The remainder is zero.
45. $2x$ **47.** $4x$ **49.** $5uv$ **51.** $2xy$ **53.** $-2x + 5$
55. $5x - 5$
57. (a) $(x + 3)(x^2 + 2x - 1)$
 $= x(x^2) + x(2x) - x(1) + 3(x^2) + 3(2x) - 3(1)$
 $= x^3 + 5x^2 + 5x - 3$
 (b) $(x^3 + 5x^2 + 5x - 3) \div (x + 3) = x^2 + 2x - 1$
59. $(2h^2 + h)$ in.² **61.** 2
63. Use the reverse order of the rule for adding two fractions. $(a + b)/c = (a/c) + (b/c)$.
65. $-\frac{2}{3}$ **67.** $\frac{2}{5}$ **69.** $y = x + 1$ **71.** $y = \frac{13}{6}x + \frac{17}{6}$
73. $-10x^5$ **75.** $x^2 + 14x + 49$

Chapter Review (pp. 248–251)

1. x^8 **3.** x^{18} **5.** $-2t^{12}$ **7.** $-5x^6y^7$ **9.** $2x$
11. $2u^2v$ **13.** $\frac{1}{2^2}$ **15.** $\frac{6}{x^3}$ **17.** $4t^8$ **19.** $\frac{7}{2x}$
21. $\frac{1}{t}$ **23.** a^7 **25.** $4y^2$ **27.** $\frac{1}{9a^4}$ **29.** $\frac{4x^4}{y^6}$
31. $\frac{25}{y^2}$ **33.** $-\frac{y^2}{729}$ **35.** $\frac{2}{uv}$ **37.** 4.10×10^{-4} m
39. $180,900,000$
41. Standard form: $-5x^3 + 10x - 4$; Degree: 3
 Leading coefficient: -5
43. Standard form: $5x^4 + 4x^3 - 7x^2 - 2x$; Degree: 4
 Leading coefficient: 5
45. Standard form: $7x^4 + 11x^2 - 1$; Degree: 4
 Leading coefficient: 7
47. $9x$ **49.** $5y^3 + 5y^2 - 12y + 10$ **51.** $7u^2 + 8u + 5$
53. $x^3 + 6x + 2$ **55.** $2x^4 - 7x^2 + 3$ **57.** $(4x - 6)$ units
59. 4 **61.** $-6x^2 + 9x - 8$ **63.** $7y^2 - y + 6$
65. $3x^2 + 4x - 14$ **67.** $36x^2 - 10x$
69. $2x^2 + 8x$ **71.** $-12x^3 - 6x^2$ **73.** $x^2 + 12x + 20$
75. $14x^2 - 31x - 10$ **77.** $x^2 + 10x + 24$
79. $12x^2 + 7x - 12$ **81.** $-28x^2 + 45x - 18$
83. $x^2 + 4xy + 3y^2$ **85.** $x^3 - x^2 - 28x - 12$
87. $2t^3 - 7t^2 + 9t - 3$ **89.** $12x^3 - 11x^2 - 13x + 10$
91. $y^4 - 2y^3 - 6y^2 + 22y - 15$ **93.** $8x^3 + 12x^2 + 6x + 1$
95. $(6x^2 + 38x + 60)$ in.² **97.** $x^2 + 6x + 9$
99. $\frac{1}{4}x^2 - 4x + 16$ **101.** $u^2 - 36$
103. $4r^2 - 20rt + 25t^2$ **105.** $4x^2 - 16y^2$
107. $8x - \frac{4}{x}$ **109.** $\frac{3}{2x^2} - \frac{1}{4x} + \frac{3}{4}$
111. $x + 2$ **113.** $6x - 5 + \frac{7}{4x + 3}$
115. $2x^2 + 4x + 3 + \frac{5}{x - 1}$ **117.** $2x + 3$

Chapter Test (p. 252)

1. x^9 **2.** $25y^{14}$ **3.** $\frac{2a}{3}$ **4.** $\frac{9y^6}{x^4}$ **5.** $\frac{u^6}{16}$ **6.** $\frac{3y^6}{x}$
7. (a) $\frac{3}{8}$ (b) $40,000,000,000$ (c) 0.00003
8. 1.5×10^{-4} **9.** $80,000,000$
10. Degree: 4; Leading coefficient: -3
11. $2z^2 - 3z + 15$ **12.** $7u^3 - 1$ **13.** $y^2 + 8y - 3$
14. $-6x^2 + 12x$ **15.** $x^2 - 14x + 49$ **16.** $4x^2 - 9$
17. $2z^3 + z^2 - z + 10$ **18.** $y^3 + 3y^2 - 4y - 12$
19. $2z^2 + \frac{1}{2}$ **20.** $-2x^2 + \frac{3}{2}$ **21.** $x^2 + 2x + 3$
22. $2x^2 + 9x + 17 + \frac{63}{2x - 4}$ **23.** $4x^2 - x$
24. $2x^2 + 11x - 6$ **25.** $1500r^2 + 3000r + 1500$
26. $x - 3$

CHAPTER 6

Section 6.1 (pp. 254–261)

1. z^2 **3.** $2x$ **5.** u^2v **7.** $3y^5z^4$ **9.** 1
11. $14a^2$ **13.** $3(x + 1)$ **15.** $8(t - 2)$ **17.** $6(4y^2 - 3)$
19. $x(x + 1)$ **21.** $u(25u - 14)$ **23.** $2x^3(x + 3)$
25. No common factor **27.** $4(3x^2 + 4x - 2)$
29. $25(4 + 3z + 2z^2)$ **31.** $3x^2(3x^2 + 2x + 6)$
33. $5u(2u + 1)$ **35.** $10ab(1 + a)$
37. $4xy(1 - 2x + 6x^3y^4)$ **39.** $-5(2x - 1)$
41. $-5(3x^2 - x - 2)$ **43.** $-2(x^2 - 6x - 2)$
45. $(x - 3)(x + 5)$ **47.** $(q - 5)(y - 1)$
49. $(x - 4)(x^3 - 2)$ **51.** $(x + 10)(x + 1)$
53. $(x + 3)(x + 4)$ **55.** $(x + 3)(x - 5)$
57. $(2x - 7)(2x + 7)$ **59.** $(2x + 1)(3x - 1)$
61. $(x + 4)(8x + 1)$ **63.** $x + 1$ **65.** $3x^2 - 2$
67. The terms were grouped so that the terms in each group have a common monomial factor: x^3 and x^2 have a common factor of x^2, and $2x$ and 2 have a common factor of 2.
69. $x + 1$ **71.** $x + 3$ **73.** $10y - 1$ **75.** $14x + 5y$
77. $(a + b)(2a - b)$ **79.** $(y - 4)(ky + 2)$
81. $6x^2$ **83.** $2\pi r(r + h)$ **85.** $kx(Q - x)$

87. There are no more common monomials that can be factored out.
89. Sample answer: $4x^3 - 24x^2 = 4x^2(x - 6)$
91. Sample answer: $3x^3 + 2x + 7$
93. (a) Not a solution (b) Solution
95. (a) Solution (b) Not a solution
97. $\frac{9y^2}{4x^6}$ **99.** $\frac{1}{z^4}$ **101.** $3mn$ **103.** $x + 1$

Section 6.2 (pp. 262–269)

1. $x + 1$ **3.** $(x + 4)(x + 2)$ **5.** $(x + 5)(x - 3)$
7. $(x + 1)(x + 11); (x - 1)(x - 11)$
9. $(x + 14)(x + 1); (x - 14)(x - 1);$
$(x + 7)(x + 2); (x - 7)(x - 2)$
11. $(x - 11)(x + 2)$ **13.** $(x - 7)(x - 2)$
15. $-(x - 5)(x + 3)$ **17.** $(x + 10)(x - 7)$
19. 21 square units **21.** 9 units **23.** $(x - 9z)(x + 2z)$
25. $(x - 2y)(x - 3y)$ **27.** $(x + 5y)(x + 3y)$
29. $(a + 5b)(a - 3b)$ **31.** $4(x - 5)(x - 3)$
33. $9(x^2 + 2x - 2)$ **35.** $x(x - 10)(x - 3)$
37. $3x(x + 4)(x + 2)$ **39.** $x^2(x - 2)(x - 3)$
41. $2x^2(x - 3)(x - 7)$
43. (a) $4x(x - 4)(x - 3)$; The model was found by multiplying the length, width, and height of the box.
$V = lwh = (8 - 2x)(6 - 2x)(x) = 4x(x - 4)(x - 3)$
(b) About 1.1 ft
45. $c = -35$, which is negative.
47. A product of -35 and a sum of 2. **49.** Prime
51. $x(x + 2y)(x + 3y)$ **53.** Prime **55.** $2xy(x + 3y)(x - y)$
57. $\pm 9, \pm 11, \pm 19$ **59.** Sample answer: $2, -10$
61. $(x + 3)(x + 1)$

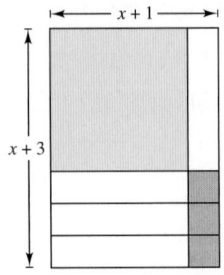

63. (a) $V = (y - 2x)(z - 2x)x$ (b) $4x(x - 4)(x - 2)$
(c) $y = 8$ ft, $z = 4$ ft
65. (a) and (d)
(a) Not completely factored; (d) Completely factored
67. Because c is positive, m and n have like signs that match the sign of b.
69. 7 **71.** $-\frac{1}{3}$ **73.** 3 **75.** $-\frac{7}{2}$ **77.** $2x^3$
79. $a^3 b^4$ **81.** $18r^3 s^3$ **83.** $2u^2 v^3$

Section 6.3 (pp. 270–277)

1. $x + 4$ **3.** $t + 3$ **5.** $(2x + 3)(x + 1)$
7. $(4y + 1)(y + 1)$ **9.** $(6y - 1)(y - 1)$
11. Prime **13.** $(x + 4)(4x - 3)$
15. $(3x - 2)(3x - 4)$ **17.** $2(v + 7)(v - 3)$
19. $3(z - 1)(3z - 5)$ **21.** $2s(4s - 1)(2s - 3)$
23. $-(2x - 9)(x + 1)$ **25.** $-(6x + 5)(x - 2)$
27. (a) $(2x - 1)$ ft (b) 168 ft^3
29. $(3x + 1)(x + 1)$ **31.** $(7x - 1)(x + 3)$
33. $(3x + 4)(2x - 1)$ **35.** $(5x - 2)(3x - 1)$
37. (a) $(x + 2)$ in. (b) 5865 in.^{3}
39. No
41. The middle term of the trinomial represented by $(3x - 1)(x + 16)$ would be $-13x$ rather than $47x$ if $(3x + 1)(x - 16)$ were the correct factorization.
43. $-x^2(5x + 4)(3x - 2)$ **45.** $6x(x - 4)(x + 8)$
47. $9u^2(2u^2 + 2u - 3)$ **49.** $\pm 11, \pm 13, \pm 17, \pm 31$
51. $\pm 1, \pm 4, \pm 11$
53. $\pm 22, \pm 23, \pm 26, \pm 29, \pm 34, \pm 43, \pm 62, \pm 121$
55. Sample answer: $-1, -7$ **57.** Sample answer: $-8, 3$
59. Sample answer: $-6, -1$ **61.** $2(x + 5) = 2x + 10$
63. Rewrite the middle term so that you can group the first two terms and the last two terms to factor the trinomial.
65. The product of the last terms of the binomials is 15, not -15.
67. Sample answer: $2x^3 + 2x^2 + 2x$
69. Four $(ax + 1)(x + c), (ax + c)(x + 1),$
$(ax - 1)(x - c), (ax - c)(x - 1)$
71. $2^2 \cdot 5^3$ **73.** $2^3 \cdot 3^2 \cdot 11$ **75.** $2x^2 + 9x - 35$
77. $49y^2 - 4$ **79.** $(x - 9)(x + 5)$ **81.** $(z + 2)(z + 20)$

Mid-Chapter Quiz (p. 278)

1. $2x - 1$ **2.** $x - y$ **3.** $y - 6$ **4.** $y + 3$
5. $3(3x^2 + 7)$ **6.** $5a^2(a - 5)$ **7.** $(x + 7)(x - 6)$
8. $(t - 3)(t^2 + 1)$ **9.** $(y + 6)(y + 5)$
10. $(u + 8)(u - 7)$ **11.** $x(x - 6)(x + 5)$
12. $2y(x + 8)(x - 4)$ **13.** $(2y - 9)(y + 3)$
14. $(3 + z)(2 - 5z)$ **15.** $(3x - 2)(4x + 1)$
16. $2s^2(5s^2 - 7s + 1)$
17. $\pm 7, \pm 8, \pm 13$; These integers are the sums of the factors of 12.
18. 16, 21; Find a number c that has a pair of factors with a sum of -10.
19. m and n are factors of 6.
$(3x + 1)(x + 6)$ $(3x - 1)(x - 6)$
$(3x + 6)(x + 1)$ $(3x - 6)(x - 1)$
$(3x + 2)(x + 3)$ $(3x - 2)(x - 3)$
$(3x + 3)(x + 2)$ $(3x - 3)(x - 2)$
20. $10(2x + 8) = 20x + 80$

Section 6.4 (pp. 280–287)

1. $(x + 3)(x - 3)$ **3.** $\left(u + \frac{1}{2}\right)\left(u - \frac{1}{2}\right)$
5. $(4y + 3)(4y - 3)$ **7.** $(x + 1)(x - 3)$
9. (a) $-(4t + 29)(4t - 29)$ (b) 1.5 sec
11. $2(x + 6)(x - 6)$ **13.** $x(2 + 5x)(2 - 5x)$
15. $2y(2y + 5)(2y - 5)$ **17.** $(y^2 + 9)(y + 3)(y - 3)$
19. $(1 + x^2)(1 + x)(1 - x)$ **21.** $2(x^2 + 9)(x + 3)(x - 3)$
23. Yes **25.** No **27.** Yes **29.** $x^2 + 2x + 1 = (x + 1)^2$
31. $(x - 4)^2$ **33.** $(x + 7)^2$ **35.** $(2t + 1)^2$
37. $(5y - 1)^2$ **39.** $(x - 3y)^2$ **41.** $(2y + 5z)^2$
43. x cm $\times (x + 8)$ cm $\times (x + 8)$ cm
45. $(x - 2)(x^2 + 2x + 4)$ **47.** $(y + 4)(y^2 - 4y + 16)$
49. $(1 + 2t)(1 - 2t + 4t^2)$ **51.** $(3u - 2)(9u^2 + 6u + 4)$
53. $(3x + 4y)(9x^2 - 12xy + 16y^2)$ **55.** No
57. $a = x, b = 2$ **59.** ± 2 **61.** ± 36 **63.** 9
65. 4 **67.** $y^2(y + 5)(y - 5)$ **69.** $(x - 1)^2$
71. $(t + 10)(t - 12)$ **73.** Prime

75. $(x^2 + 9)(x + 3)(x - 3)$ **77.** $(x + 1)(x - 1)(x - 4)$
79. 441 **81.** 3599 **83.** $(x + 3)$; 1; $(x + 4)(x + 2)$
85. Box 1: $(a - b)a^2$; Box 2: $(a - b)ab$; Box 3: $(a - b)b^2$
The sum of the volumes of boxes 1, 2, and 3 equals the volume of the large cube minus the volume of the small cube, which is the difference of two cubes.
87. False. $x(x + 2) - 2(x + 2) = (x + 2)(x - 2)$
89. False. $x^3 - 27 = (x - 3)(x^2 + 3x + 9)$, whereas $(x - 3)^3 = (x^2 - 6x + 9)(x - 3)$.
91. To identify the difference of two squares, look for coefficients that are squares of integers and for variables raised to even powers.
Let a and b be real numbers, variables, or algebraic expressions.
$a^2 - b^2 = (a + b)(a - b)$
93. 4 **95.** -1 **97.** $(2x + 1)(x + 3)$
99. $(3m - 4)(2m + 5)$

Section 6.5 (pp. 288–295)

1. 0, 5 **3.** 2, 3 **5.** $-\frac{1}{3}, \frac{5}{2}$ **7.** $-4, 4$ **9.** $-3, 3$
11. $-2, 4$ **13.** $-\frac{1}{2}, 3$ **15.** -7 **17.** $\frac{3}{2}$
19. $-8, 2$ **21.** $\frac{3}{2}, -1$ **23.** $-12, 2$ **25.** $\frac{2}{3}, -1$
27. 5, -3 **29.** $-10, 8$ **31.** 8, 9 **33.** 20, 22
35. 10 sec **37.** 9 in. × 12 in. **39.** 18 in. × 36 in.
41. Yes. It has the form $ax^2 + bx + c = 0$ with $a \neq 0$.
43. The factors $2x$ and $x + 9$ are set equal to zero to form two linear equations.
45. $-2, 0$ **47.** $-\frac{3}{5}, 8$ **49.** $-3, 1$ **51.** 5, 4
53. $(-3, 0), (1, 0)$; The solutions are the x-intercepts.
55. $(-3, 0), (4, 0)$; The solutions are the x-intercepts.
57. (a) 120 (b) 20
59. False.
$x^2 = 4x \Rightarrow x^2 - 4x = 0$
$x(x - 4) = 0$
$x = 0$
$x - 4 = 0 \Rightarrow x = 4$
$x = 0$ is also a solution.
61. True.
$ax^2 + bx = 0$
$x(ax + b) = 0$
$x = 0$
$ax + b = 0 \Rightarrow x = -\frac{b}{a}$
63. In order to use the Zero-Factor Property, the quadratic equation must be written in general form.
$x^2 + 4x = 12$
$x^2 + 4x - 12 = 0$
$(x + 6)(x - 2) = 0$
$x + 6 = 0 \Rightarrow x = -6$
$x - 2 = 0 \Rightarrow x = 2$
65. (a) Solution (b) Solution (c) Not a solution (d) Solution
67. $(x + 9)(x - 9)$ **69.** $3(x + 4)(x - 4)$ **71.** $(x + 2)^2$
73. $(2x - 5)^2$

Review Exercises (pp. 298–300)

1. t^2 **3.** $3x^2$ **5.** $7x^2y^3$ **7.** $4xy$ **9.** $3(x - 2)$
11. $t(3 - t)$ **13.** $5x^2(1 + 2x)$ **15.** $4a^2(2 - 3a)$
17. $5x(x^2 + x - 1)$ **19.** $4(2y^2 + y + 3)$ **21.** $x(3x + 4)$

23. $(x + 1)(x - 3)$ **25.** $(u - 2)(2u + 5)$
27. $(y + 3)(y^2 + 2)$ **29.** $(x^2 + 1)(x + 2)$
31. $(x + 3)(x - 4)$ **33.** $(x - 7)(x + 4)$
35. $(u - 4)(u + 9)$ **37.** $(x - 6)(x + 4)$
39. $(y + 7)(y + 3)$ **41.** $(b + 15)(b - 2)$
43. $(w + 8)(w - 5)$ **45.** $\pm 6, \pm 10$ **47.** ± 12
49. $(x - y)(x + 10y)$ **51.** $(y + 3x)(y - 9x)$
53. $(x + 2y)(x - 4y)$ **55.** $4(x - 2)(x - 4)$
57. $x(x + 3)(x + 6)$ **59.** $3(x + 9)(x - 3)$
61. $4x(x + 2)(x + 7)$ **63.** $(3x + 5)(x - 1)$
65. $(2x - 7)(x + 1)$ **67.** $(3x + 2)(2x + 1)$
69. $(4y + 1)(y - 1)$ **71.** $(3x - 2)(x + 3)$
73. $(3x - 1)(x + 2)$ **75.** $(2x - 1)(x - 1)$
77. $\pm 2, \pm 5, \pm 10, \pm 23$ **79.** Sample answer: 2, -6
81. $3(x + 6)(x + 5)$ **83.** $3(2y - 1)(y + 7)$
85. $3u(2u + 5)(u - 2)$ **87.** $4y(2y - 3)(y - 1)$
89. $2x(3x - 2)(x + 3)$ **91.** $(3x + 1)$ in.
93. $(2x - 7)(x - 3)$ **95.** $(4y - 3)(y + 1)$
97. $(3x - 2)(2x + 5)$ **99.** $(7x + 5)(2x + 1)$
101. $(a + 10)(a - 10)$ **103.** $(5 + 2y)(5 - 2y)$
105. $3(2x + 3)(2x - 3)$ **107.** $(u + 3)(u - 1)$
109. $-(z - 1)(z - 9)$ **111.** $3y(y + 5)(y - 5)$
113. $st(s + t)(s - t)$ **115.** $(x^2 + 9)(x + 3)(x - 3)$
117. $(x^2 + 4)(x - 2)$ **119.** No **121.** Yes
123. $(x - 4)^2$ **125.** $(3s + 2)^2$ **127.** $(y + 2z)^2$
129. $\left(x + \frac{1}{3}\right)^2$ **131.** $(a + 1)(a^2 - a + 1)$
133. $(3 - 2t)(9 + 6t + 4t^2)$ **135.** $(2x + y)(4x^2 - 2xy + y^2)$
137. $0, \frac{3}{2}$ **139.** $-3, 2$ **141.** $-\frac{1}{2}, \frac{7}{3}$ **143.** $-9, 9$
145. 6 **147.** $-\frac{1}{2}, 2$ **149.** 0, 4 **151.** 3, 4
153. $-4, 3$ **155.** 12, 14 **157.** 40 in. × 60 in.
159. 3 sec

Chapter Test (p. 301)

1. $9x^2(1 - 7x^3)$ **2.** $(z + 17)(z - 10)$
3. $(t - 10)(t + 8)$ **4.** $(3x - 4)(2x - 1)$
5. $3y(y - 1)(y + 25)$ **6.** $(2 + 5v)(2 - 5v)$
7. $(x + 2)(x^2 - 2x + 4)$ **8.** $-(z + 1)(z + 21)$
9. $(x + 2)(x + 3)(x - 3)$ **10.** $(4 + z^2)(2 + z)(2 - z)$
11. $2x - 3$ **12.** ± 6 **13.** 36
14. $3x^2 - 3x - 6 = 3(x + 1)(x - 2)$ **15.** $-4, \frac{3}{2}$
16. $-3, \frac{2}{3}$ **17.** $-\frac{3}{2}, 2$ **18.** $-1, 4$ **19.** $x + 4$
20. 7 in. × 12 in. **21.** 8.875 sec **22.** 24, 26

Cumulative Test (p. 302)

1. Quadrant II or Quadrant III.
2. (a) Not a solution (b) Solution
  (c) Solution (d) Not a solution
3. (0, 3) **4.** (6, 0), (0, 3)

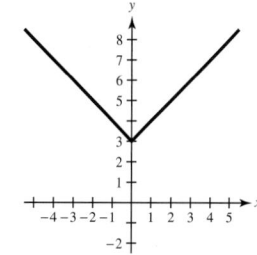 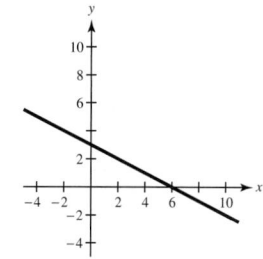

5. Sample answer: $(-2, 2)$; There are infinitely many points on a line.
6. $y = \frac{2}{5}x - \frac{3}{2}$
7. Perpendicular
8. Parallel

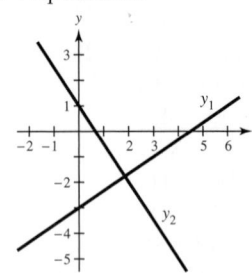
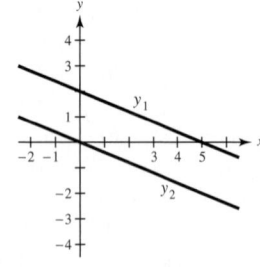

9. $-5x^2 + 5$ 10. $-42z^4$ 11. $3x^2 - 7x - 20$
12. $25x^2 - 9$ 13. $25x^2 + 60x + 36$ 14. $x + 12$
15. $x + 1 + \frac{2}{x-4}$ 16. $\frac{x}{54y^4}$ 17. $2u(u-3)$
18. $(x+2)(x-10)$ 19. $x(x+4)^2$ 20. $(x+2)^2(x-2)$
21. $0, 12$ 22. $-\frac{3}{5}, 3$ 23. $\frac{4}{x^2}$
24. $C = 150 + 0.45x$; $181.50
25. $x + y \le 10$; The second song can be up to 4 minutes long.

CHAPTER 7
Section 7.1 (pp. 304–311)
1. All real values of x such that $x \ne 4$
3. All real values of x such that $x \ne -2$
5. All real numbers   7. All real numbers
9. All real values of x such that $x \ne -1$ and $x \ne 2$
11. (a) $\frac{3000 + 8.50x}{x}$ (b) $\{1, 2, 3, 4, \ldots\}$ (c) $38.50
13. $\frac{7(x+3)}{x}$ 15. $\frac{5x}{x-4}$ 17. $\frac{1}{a+2}$ 19. $1 - y$
21. $\frac{y-4}{9}, y \ne 0$ 23. $\frac{x+8}{4}, x \ne -2$ 25. $\frac{x(x+3)}{x-2}, x \ne -2$
27. $x^2 + 1, x \ne 2$ 29. $\frac{5}{r+6s}, r \ne -s$ 31. $\frac{x+7y}{x+5y}, x \ne 3y$
33. $\frac{x-3}{x+5}$ 35. $\frac{1}{6}$
37. (a) $B = \frac{156.89 + 7.34x}{1 + 0.017x}, 10 \le x \le 100$ (b) $211.847°$ F
39. A rational expression is a fraction whose numerator and denominator are polynomials.
41. Factors; Factors are multiplied by everything in the numerator and denominator, but terms are not.
43. All real values of z such that $z \ne -\frac{2}{3}$ and $z \ne 1$
45. (a) 0 (b) Division by zero is undefined. (c) $\frac{10}{7}$ (d) $\frac{1}{2}$
47. $x + 6z, x \ne 6z$ 49. $-1, s^2 \ne 4t^2$ 51. $\frac{1-x}{1+x}, x \ne 1$
53. $-\frac{(x^2 + y^2)(x+y)}{(y-x)^3}$

55.

x	2	2.5	3	3.5	4
$\frac{x^3 - 3x^2}{x - 3}$	4	6.25	Undef.	12.25	16
x^2	4	6.25	9	12.25	16

The two expressions are equal for all values of x except 3. When $x = 3$, the first expression is undefined.

57. True; $\frac{ac}{bc} = \frac{a}{b}$ when a, b, and c are real numbers, variables, or algebraic expressions such that $b \ne 0$ and $c \ne 0$.
59. Yes. $\frac{4x}{2x} = \frac{\overset{2}{\cancel{4}} \cdot \cancel{x}}{\underset{1}{\cancel{2}} \cdot \cancel{x}} = 2$
61. Sample answer: $\frac{x}{x^2 + x} = \frac{x}{x(x+1)} = \frac{1}{x+1}$
63. 4 65. 9 67. 0, 8 69. $-\frac{6}{5}, -14$

Section 7.2 (pp. 312–319)
1. $\frac{3x}{2}, x \ne 0$ 3. $3, x \ne 0$ 5. $\frac{5a^2}{3}, a \ne 0$
7. $\frac{8xy}{3}, x \ne 0, y \ne 0$ 9. $1, y \ne 1$ 11. $2x, x \ne -1$
13. $\frac{1}{2x(x-4)}, x \ne -4, x \ne 3$ 15. $\frac{2y(y+5)}{y-2}$
17. $-\frac{3x}{5}, x \ne 3$ 19. $x - 5, x \ne -5, x \ne 5$
21. $-1, r \ne 1$ 23. $\frac{x-1}{x(x+1)^2}, x \ne -2$
25. $\frac{6}{7}, y \ne -2z, 3y \ne 5z$ 27. $\frac{4}{x+6}, x \ne 0, x \ne -2$
29. $3(a+1)^2(a-1), a \ne 0, a \ne 1$
31. $\frac{y}{y-6}, y \ne -8$ 33. $\frac{3}{4}, x \ne -4$
35. $x(x+2), x \ne -2, x \ne 0$ 37. $\frac{x-4}{(x+4)(x-3)}$
39. $\frac{x}{x+6}, x \ne 0, x \ne 3$ 41. $\frac{x+2}{x-3}, x \ne 7, x \ne 2$
43. $-\frac{6y+1}{2y+3}, y \ne -5, y \ne 5, y \ne -\frac{1}{6}$ 45. $\frac{2x^2}{3}$; increase
47. To multiply two rational expressions, let a, b, c, and d represent real numbers, variables, or algebraic expressions such that $b \ne 0$ and $d \ne 0$. Then the product of a/b and c/d is $\frac{a}{b} \cdot \frac{c}{d} = \frac{ac}{bd}$.
49. You change the division of two rational expressions into the multiplication of two rational expressions by multiplying the first expression by the reciprocal of the second expression.
51. $x - 3$ 53. $x - 4$ 55. $(x-4)(x+2)$
57. $\frac{8}{(z-3)^2}, z \ne -3$ 59. $\frac{(a-5)(a+3)}{(a+4)^2}, a \ne 5, a \ne -3$
61. $(x+2)(x+1), x \ne -2, x \ne -1$ 63. $1, t \ne -1, t \ne -2$
65. (a) $\frac{1}{5}$ min (b) $\frac{x}{5}$ min (c) 24 min
67. $\frac{x+2}{x} \cdot \frac{x^2}{(x+2)^3} \cdot \frac{(x+2)^2}{x}$

69. The divided was mistakenly inverted. Instead, the divisor should have been inverted.

$$\frac{x^2-4}{5x} \div \frac{x+2}{x-2} = \frac{x^2-4}{5x} \cdot \frac{x-2}{x+2}$$
$$= \frac{(x+2)(x-2)(x-2)}{5x(x+2)}$$
$$= \frac{(x-2)^2}{5x}, x \neq 2, x \neq -2$$

71. $x(3x+7)$ **73.** $(x+9)(x-2)$
75. All real numbers
77. All real values of x such that $x \neq -4$
79. All real values of x such that $x \neq 4$
81. All real values of x such that $x \neq -2$ and $x \neq 2$

Section 7.3 (pp. 320–327)

1. y **3.** $\frac{14}{3a}$ **5.** $\frac{1}{3}$ **7.** $\frac{3-7x}{x+2}$ **9.** $\frac{1}{x-1}, x \neq -1$
11. $\frac{7x+3}{(x+1)^2}$ **13.** $\frac{x+4}{x-7}, x \neq 0$ **15.** $-\frac{6}{x-5}$
17. $\frac{5x+9}{2x-3}$ **19.** $\frac{2x-1}{x-11}$ **21.** $\frac{2x+3}{x-15}$ **23.** $36y^2$
25. $6(y-3)$ **27.** $48x(x+2)$ **29.** $x^2(x+7)(x-7)$
31. $x(x+2)(x-2)$ **33.** $t(t+2)^2(t-4)$
35. $\frac{17}{10s}$ **37.** $\frac{1-3x}{5x}$ **39.** $\frac{5u+2}{u^2}$ **41.** $\frac{3b+5}{2b^2}$
43. $\frac{5x-1}{(x+3)(x-5)}$ **45.** $\frac{3}{(x+2)(x-1)}$
47. $\frac{-2x+11}{(x-3)(x-2)}$ **49.** $\frac{13x^2-24x+12}{6x(x-2)^2}$
51. $\frac{8-5x}{x-1}$ **53.** $\frac{2x^2+8x-9}{(x-3)(2x+5)}$ **55.** $\frac{4x-9}{(x+3)(x-3)}$
57. $\frac{7v+8}{v(v+4)}$ **59.** $0, x \neq -2, x \neq 2$ **61.** $\frac{24-3x^2}{(x+3)(x-2)}$

63. Let a, b, and c represent real numbers, variables or algebraic expressions such that $c \neq 0$.
$$\frac{a}{c} + \frac{b}{c} = \frac{a+b}{c}$$
$$\frac{a}{c} - \frac{b}{c} = \frac{a-b}{c}$$

65. To add or subtract rational expressions with unlike denominators, you must first rewrite the expressions so that they have like denominators. The like denominator is the least common multiple of the original denominators.

67. $\frac{x+5}{3(x-2)}, \frac{30}{3(x-2)}$ **69.** $\frac{2x}{x(x+3)^2}, \frac{5(x+3)}{x(x+3)^2}$
71. $\frac{(x-8)(x-4)}{(x+4)(x-4)^2}, \frac{9x(x+4)}{(x+4)(x-4)^2}$
73. $P = \frac{6(3y-2)}{y(y-2)}$ ft   **75.** $\frac{16r}{(r-2)(r+2)}$ hr
77. Yes. For example, for $2(x+2)$ and $x+2$, the least common multiple is $2(x+2)$.
79. The binomial is one of the factors of the trinomial. For example, the least common multiple of $x+3$ and x^2+5x+6 is x^2+5x+6 because $x^2+5x+6 = (x+3)(x+2)$.
81. (a) Not a solution   (b) Not a solution
(c) Not a solution   (d) Solution
83. $-\frac{12x}{7}, x \neq 5$ **85.** $\frac{9(x+3)}{x^2}, x \neq 3$

Mid-Chapter Quiz (p. 328)

1. The domain of a rational expression is the set of all real numbers for which it is defined.
(a) All real numbers   (b) All real values of x such that $x \neq \pm 2$
2. (a) $\frac{7}{12}$ (b) 0   (c) Division by zero is undefined.
3. $\frac{2z^3}{5}, z \neq 0$ **4.** $\frac{4a}{3b^2}, a \neq 0$ **5.** $-\frac{y+2}{4}, y \neq 2$
6. $\frac{3(u-3)}{5u}, u \neq 3$ **7.** $\frac{1}{b-1}, b \neq 0, b \neq -3$
8. $\frac{2x-3}{x+1}, x \neq \frac{3}{2}$ **9.** $\frac{5y^2}{3}, y \neq 0$ **10.** $-\frac{4s}{5(s+5)}, s \neq 5$
11. $\frac{5x^3}{x-2}, x \neq -4$ **12.** $r(r-4), r \neq 0, r \neq -4$
13. $\frac{2}{5(x+2)^2}, x \neq 0$ **14.** $\frac{20}{y^3}, x \neq 0$ **15.** $\frac{5x+24}{(x+6)(x-6)}$
16. $\frac{20x}{(x-5)(x+5)}$ **17.** $\frac{x^2+x-14}{(x+4)(x+3)}$ **18.** $\frac{1}{x-7}, x \neq -2$
19. (a) $\frac{10,000+225x}{x}$
(b) As the number of units increases, the cost per unit decreases.

x	2000	3000	4000	5000
Average cost	$230.00	$228.33	$227.50	$227.00

20. $\frac{x}{x+10}$

Section 7.4 (pp. 330–337)

1. $\frac{5}{3}$ **3.** $\frac{x}{2}, x \neq 0$ **5.** $\frac{2xy^2}{5}, x \neq 0, y \neq 0$
7. $\frac{x+14}{x+7}, x \neq -14, x \neq -4$ **9.** $2, x \neq -1, x \neq 5$
11. $\frac{x-2}{3(x+4)}, x \neq -5$ **13.** $\frac{23}{5}$ **15.** $\frac{x^2}{2(2x+3)}, x \neq 0$
17. $\frac{x(x-4)}{3(5x+1)}, x \neq 0$ **19.** $\frac{x+4}{4x}, x \neq 4$ **21.** $\frac{20}{7}, x \neq -1$
23. $\frac{1}{x}, x \neq -1$ **25.** $\frac{9x}{40}$ **27.** $\frac{19x}{30}$
29. $\frac{R_1 R_2 R_3}{R_1 R_2 + R_1 R_3 + R_2 R_3}$
31. A complex fraction is a fraction that has a fraction in its numerator or denominator, or both.
$$\frac{\frac{x-3}{x}}{\frac{3x-9}{4}}$$
33. Invert the denominator fraction and multiply.
35. $-\frac{1}{y}, y \neq 3$ **37.** $\frac{(2x-5)(3x+1)}{3x(x+1)}, x \neq \pm\frac{1}{3}$ **39.** $\frac{y+3}{y^2}$
41. $\frac{x^2+5x+8}{8x}$ **43.** $\frac{x}{8}, \frac{5x}{36}, \frac{11x}{72}$ **45.** $r = \frac{288(MN-P)}{N(12P+MN)}$
47. Method 1: Combine the numerator into a single fraction. Combine the denominator into a single fraction. Then divide by inverting the denominator and multiplying.
Method 2: Multiply the numerator and denominator of the complex fraction by the least common denominator for all fractions in the numerator and denominator of the original complex fraction. Then simplify the results. Answers about preference will vary.

49. $11x^2 - 2x - 3$ 51. $24x^3 - 16x^2 + 8x$
53. $8x + 4, x \neq 0$ 55. $\dfrac{4x}{5}$ 57. $\dfrac{8}{15x}$ 59. $\dfrac{12 - x}{x + 2}$

Section 7.5   (pp. 338–345)
1. 12   3. 50   5. $\dfrac{21}{2}$ 7. 5   9. $\dfrac{23}{10}$ 11. -4
13. $\dfrac{2}{9}$ 15. $\dfrac{3}{4}$ 17. No solution   19. 2   21. $\dfrac{4}{3}$
23. 2   25. 3   27. 1   29. $-5, 4$
31. $-2, \dfrac{3}{2}$ 33. $\dfrac{1}{2}, 2$
35. First car: 55 mi/hr; Second car: 52 mi/hr
37. First skater: 13.3 m/sec; Second skater: 12 m/sec
39. $1\dfrac{1}{3}$ hr   41. 75 min   43. 5 hits   45. 28 serves
47. $\dfrac{2}{x} - \dfrac{2}{5} = \dfrac{4}{x}$ is a rational equation and $\dfrac{2}{x} - \dfrac{2}{5} + \dfrac{4}{x}$ is a rational expression.
49. An extraneous solution is a "trial solution" that does not satisfy the original equation. When you substitute this "trial solution" into the original equation, the result is false–the solution does not check, or is undefined.
51. No solution   53. $\dfrac{1}{2}$ 55. $-2, 3$
57. Truck: 50 mi/hr; Car: 60 mi/hr
59. Only one side of the equation was multiplied by the LCD, $2(x + 5)$.
$$\dfrac{12}{x+5} + \dfrac{1}{2} = 2$$
$$2(x+5)\left(\dfrac{12}{x+5} + \dfrac{1}{2}\right) = 2(x+5)(2)$$
$$2(12) + x + 5 = 4(x+5)$$
$$x + 29 = 4x + 20$$
$$9 = 3x$$
$$3 = x$$
61. True. The equation has "trial solutions" $\dfrac{3}{2}$ and -3. By checking them, you can conclude that -3 is extraneous because it results in division by zero.
63. True
$$\dfrac{3 + x}{x} = \dfrac{7 + 2x}{2x}$$
$$\dfrac{3}{x} + \dfrac{x}{x} = \dfrac{7}{2x} + \dfrac{2x}{2x}$$
$$\dfrac{3}{x} + 1 = \dfrac{7}{2x} + 1$$
$$\dfrac{3}{x} = \dfrac{7}{2x}$$
65. When given $x = 1$, there is no reason to multiply each side by x. By doing so, you are introducing an extraneous solution 0. Because $x = 1, 0 \neq 1$ does not check. Also, when solving the equation
$$x^2 - x = 0$$
$$x(x - 1) = 0$$
the solutions are 0 and 1. You do not divide each side by $x - 1$ to solve the equation.
67. $(2t + 5)(t^2 - 3)$ 69. $(2y + 1)(3y^2 - 1)$
71. $(4 - x)(x^2 - 2)$ 73. $\dfrac{9x^3}{2}, x \neq 0$ 75. $\dfrac{12xy}{55}, x \neq 0, y \neq 0$
77. $\dfrac{x + 6}{2(x + 5)}, x \neq 3$

Review Exercises   (pp. 348–351)
1. All real values of x such that $x \neq 5$
3. All real values of t such that $t \neq 1$ and $t \neq 2$
5. $\dfrac{t}{3}$ 7. $4x^3, x \neq 0$ 9. $\dfrac{x}{3y}, x \neq 0$ 11. $\dfrac{3}{4}, b \neq 2$
13. $-4, x \neq y$ 15. $\dfrac{x + 3}{x + 2}, x \neq 3$ 17. $\dfrac{x^2 + x + 1}{x + 1}, x \neq 1$
19. $\dfrac{x - 6y}{x + y}, x \neq -3y$ 21. $\dfrac{x - 4}{x}, x \neq -7$
23. $\dfrac{x - 3}{2x + 1}$ 25. $\dfrac{1}{3}, x \neq 0$ 27. $\dfrac{3x^3}{4y^2}$ 29. $\dfrac{1}{2}, x \neq 5$
31. $-\dfrac{4}{x + 2}, x \neq -3, x \neq 2$ 33. $\dfrac{4x(x + 6)}{x - 8}$
35. $\dfrac{x + 6}{2(x - 6)}$ 37. $\dfrac{5v}{8u}, v \neq 0$ 39. $50y, y \neq 0$
41. $\dfrac{3}{28x(x + 7)}$ 43. $-\dfrac{3x - 2}{2x + 1}, x \neq -3, x \neq 3, x \neq \dfrac{2}{3}$
45. $6(x - 4), x \neq -2, x \neq -4$ 47. $\dfrac{x(x + 1)}{x - 8}, x \neq -1, x \neq 1$
49. $12(x - 7), x \neq 0$ 51. $\dfrac{x}{4}$ 53. $\dfrac{6x - 4}{x + 2}$ 55. $\dfrac{3t + 8}{2t - 3}$
57. $\dfrac{4x + 9}{x^2 + 2x - 3}$ 59. $120x^3$ 61. $12(x - 2)$
63. $x(x^2 - 16)$ 65. $\dfrac{5t}{48}$ 67. $\dfrac{29}{6x}$ 69. $-\dfrac{1}{(x + 1)(x + 2)}$
71. $\dfrac{4x + 3}{x - 1}$ 73. $\dfrac{5x + 1}{5 - 3x}$ 75. $\dfrac{x + 8}{(x + 4)(x + 2)^2}$
77. $\dfrac{4x + 5}{(x + 4)(x - 4)}$ 79. $\dfrac{x^3 - x + 3}{(x + 2)(x - 1)}$
81. $\dfrac{2x^3 - 4x^2 - 15x + 5}{(x - 4)(x + 2)}$ 83. $\dfrac{98}{27}$ 85. $5y, x \neq 0, y \neq 0$
87. $\dfrac{x(x - 3)}{4(x + 4)}, x \neq 3, x \neq -1$ 89. $\dfrac{2(x + 1)}{x - 1}, x \neq 0$
91. $\dfrac{-1}{4(x + 1)}, x \neq 3$ 93. $\dfrac{3}{8}, x \neq -3$
95. $\dfrac{3x}{8}$ 97. 24   99. $\dfrac{1}{2}$ 101. 8   103. $\dfrac{1}{2}$
105. $\dfrac{3}{2}$ 107. $-4, 3$ 109. -2 111. $\dfrac{5}{4}, \dfrac{4}{5}$
113. First car: 53 mi/hr; Second car: 58 mi/hr
115. $1\dfrac{1}{2}$ hr

Chapter Test   (p. 352)
1. All real values of x such that $x \neq \pm 9$ 2. $x^2 + x$
3. $\dfrac{1}{2y^2(2y - 1)}$ 4. $\dfrac{x + 8}{x + 5}, x \neq 8$ 5. $\dfrac{18}{x^2}$
6. $\dfrac{(x + 2)(x - 2)}{x^2}, x \neq -2$ 7. $\dfrac{5}{6x}$ 8. $\dfrac{x^3}{(x - 3)^8}, x \neq 0$
9. $\dfrac{6x + 15}{x + 2}$ 10. $\dfrac{2}{(x + 1)^2}$ 11. $\dfrac{8x}{4x + 1}, x \neq 0$
12. $-\dfrac{6}{5}, x \neq 1, x \neq -1$ 13. $-\dfrac{1}{t}, t \neq 5$
14. $-3x - 1, x \neq 0, x \neq \dfrac{1}{3}$
15. (a) Not a solution   (b) Solution
    (c) Not a solution   (d) Solution
16. $\dfrac{9}{2}$ 17. $-\dfrac{7}{2}$ 18. 4
19. Van: 36 mi/hr; Car: 42 mi/hr

CHAPTER 8

Section 8.1 (pp. 354–361)

1. (a) Solution   (b) Not a solution
3. (a) Not a solution   (b) Solution
5. (1, 2)   7. (2, 0)   9. (3, 0)   11. One solution
13. Infinitely many solutions   15. $(-1, -1)$
17. No solution   19. No solution   21. $(7, -2)$
23. (1, 1)   25. Infinitely many solutions   27. $(2, -1)$
29. (0, 6)   31. (5, 4)   33. Infinitely many solutions
35. No solution
37. (a)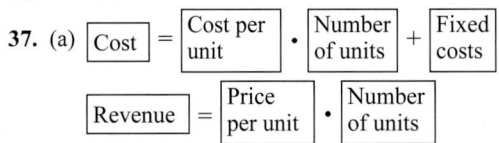

(b) 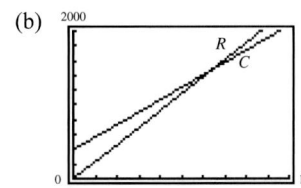 $x = 64$ units, $C = R = \$1472$; This means that the company must sell 64 feeders to cover its cost. Sales over 64 feeders will generate profit.

39. • The two lines intersect. They have a single point of intersection.
  • The two lines coincide (are identical). They have infinitely many points of intersection.
  • The two lines are parallel. They have no point of intersection.
41. An inconsistent system of linear equations has two lines that are parallel, no point of intersection, slopes that are equal, and no solution.
43. $y = \frac{2}{3}x + 4$, $y = \frac{2}{3}x - 1$; No solution
45. $y = \frac{2}{3}x + \frac{4}{3}$, $y = -\frac{2}{3}x + \frac{8}{3}$; One solution
47. $y = \frac{1}{4}x + \frac{7}{4}$, $y = \frac{1}{4}x + \frac{7}{4}$; Infinitely many solutions
49. (11, 9)
51. Because the slopes of the two lines are not equal, the lines intersect and the system has one solution: (79,400, 398).
53. False. It may have one solution or infinitely many solutions.
55. Sample answer:
$$\begin{cases} x + y = 0 \\ x + y = 1 \end{cases}$$
57. 1   59. $-\frac{1}{21}$ 61. $-\frac{3}{2}$ 63. $\frac{5}{11}$ 65. 5   67. -13

Section 8.2 (pp. 362–369)

1. (2, 3)   3. (0, 0)   5. (15, 5)   7. $(4, -4)$ 9. (2, 6)
11. $\left(\frac{1}{2}, 3\right)$ 13. $(-3, 2)$ 15. $\left(\frac{1}{4}, -1\right)$ 17. $(-4, 2)$
19. $(5, -2)$ 21. Infinitely many solutions
23. No solution   25. Infinitely many solutions
27. 5%: \$10,000; 8%: \$5000   29. 26 years; \$31,750
31. • Solve one of the equations for one variable in terms of the other.
  • Substitute the expression obtained in Step 1 into the other equation to obtain an equation in one variable.
  • Solve the equation obtained in Step 2.
  • Back-substitute the solution from Step 3 into the expression obtained in Step 1 to find the value of the other variable.
  • Check the solution to see that it satisfies both of the original equations.

33. When you obtain a true result such as $15 = 15$, then the system of linear equations has infinitely many solutions.
35. (5, 6)   37. $\left(\frac{18}{5}, \frac{3}{5}\right)$ 39. (3, 0)   41. (6, 0)
43. $\left(\frac{5}{2}, \frac{3}{4}\right)$ 45. (8, 4)
47. Sample answer: 49. (25, 15)   51. $2x - y - 9 = 0$
$$\begin{cases} x - 2y = 0 \\ x + y = 3 \end{cases}$$
53. The substitution method yields exact solutions.
55. $b = 2$ 57. $b = -\frac{1}{3}$ 59. $(3 - x)(x - 2)$
61. $(2y - 5)^2$ 63. 4   65. $-2, 6$ 67. No solution
69. $(0, -3)$

Section 8.3 (pp. 370–377)

1. (8, 4)   3. $(-4, 4)$ 5. (2, 1)   7. $\left(\frac{21}{4}, -\frac{3}{2}\right)$
9. $(-2, 2)$ 11. $\left(\frac{3}{4}, \frac{1}{2}\right)$ 13. $(-2, 3)$ 15. $(4, -1)$
17. $\left(\frac{50}{7}, \frac{4}{7}\right)$ 19. (8, 7)   21. $\left(\frac{15}{11}, \frac{15}{11}\right)$ 23. $\left(\frac{1}{2}, 0\right)$
25. Infinitely many solutions   27. No solution
29. Student ticket: \$3; General admission ticket: \$5   31. $(-1, 2)$
33. When you obtain a false result such as $0 = 11$, then the system of linear equations has no solution.
35. $\left(\frac{4}{3}, \frac{4}{3}\right)$ 37. $\left(1, -\frac{5}{4}\right)$ 39. \$4000   41. (48, 34)
43. $y = \frac{1}{3}x + \frac{2}{3}$
45. Sample answer:
$$\begin{cases} 0.02x - 0.03y = 0.12 \\ 0.5x + 0.3y = 0.9 \end{cases}$$
Multiply each side of the first equation by 100 and multiply each side of the second equation by 10.
$$\begin{cases} 2x - 3y = 12 \\ 5x + 3y = 9 \end{cases}$$
47. Sample answer:
$$\begin{cases} y = 3x + 4 \\ y = -x - 8 \end{cases}$$

49. 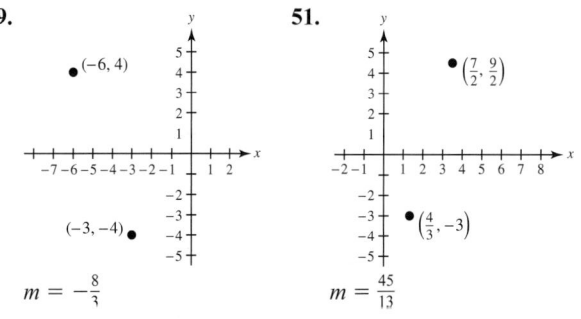 $m = -\frac{8}{3}$

51. $m = \frac{45}{13}$

53. 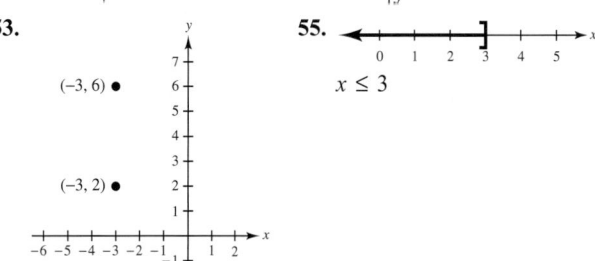 m is undefined.

55. $x \leq 3$

57. 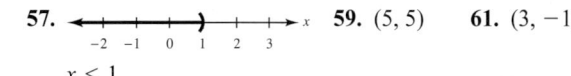 $x < 1$

59. (5, 5)   61. $(3, -1)$

Mid-Chapter Quiz (p. 378)

1. Not a solution, because substituting $x = 4$ and $y = 2$ into $3x + 4y = 4$ yields $20 = 4$, which is a contradiction.
2. Solution, because substituting $x = 2$ and $y = -1$ into the equations yields true equalities.
3. $(3, 2)$ 4. $(4, 1)$ 5. $(4, -1)$ 6. $(-3, 11)$
7. $(2, 2)$ 8. $(-1, 4)$ 9. $(6, 2)$ 10. $(3, 3)$
11. $\left(-\frac{1}{2}, 6\right)$ 12. $(1, -3)$ 13. $(-2, 5)$ 14. $\left(\frac{55}{23}, \frac{95}{23}\right)$
15. Sample answer:
$$\begin{cases} 2x + y = 7 \\ -5x - 2y = -13 \end{cases}$$
16. Sample answer:
$$\begin{cases} -x + y = -3 \\ 4x - 3y = 12 \end{cases}$$
17. $k = -2$ 18. $k = 0.8$ 19. $(36, 14)$
20. Book: $26, Calendar: $6

Section 8.4 (pp. 380–387)

1. (a) $\boxed{\text{Number of gallons of regular}} \cdot \boxed{\text{Price of regular}} + \boxed{\text{Number of gallons of premium}}$
$\cdot \boxed{\text{Price of premium}} = \boxed{\text{Total cost}}$

$\boxed{\text{Price of premium}} = \boxed{\text{Price of regular}} + \boxed{\text{Price difference of premium and regular}}$

(b) Number of gallons of regular = 15 (gallons)
Price of regular = x (dollars per gallon)
Number of gallons of premium = 10 (gallons)
Price of premium = y (dollars per gallon)
Price difference of premium and regular = 0.24 (dollars per gallon)
Total cost = 97.15 (dollars)

(c) $\begin{cases} 15x + 10y = 97.15 \\ y = x + 0.24 \end{cases}$

(d) Regular: $3.79/gal; Premium: $4.03/gal
3. Large truck: $32,000; Small truck: $21,500
5. 35% solution: 4 L; 60% solution: 6 L
7. $75   9. $137.50   11. 8 ft × 12 ft
13. 2 yd × 6 yd   15. 8 m × 9.6 m
17. 5 mi/hr   19. Graphing, substitution, elimination
21. A verbal model is a word equation that represents a real-life problem.
23. 8 dimes, 13 quarters   25. 15 nickels, 20 quarters
27. 28 nickels, 16 dimes   29. $276.15
31. Regular: $3.69/gal; Premium: $3.84/gal   33. 1 hr
35. [graph] 96 units; $R = \$859,200$
37. [graph] 133,333 units; $R = \$113,333.05$

39. $m = \frac{1}{2}, b = -2$ 41. $m = -\frac{1}{2}, b = \frac{9}{2}$
43. Answers will vary
45. The equation $2x - y = -3$ is equivalent to $2x + 3 = y$, which means that the larger number is 3 more than twice the smaller number.

$$\begin{cases} x + y = 42 \\ y = 2x - 3 \end{cases} \quad \text{or} \quad \begin{cases} x + y = 42 \\ 2x - y = 3 \end{cases}$$

47. Parallel   49. Perpendicular   51. $(3, 10)$ 53. $(5, 3)$

Section 8.5 (pp. 388–395)

1. 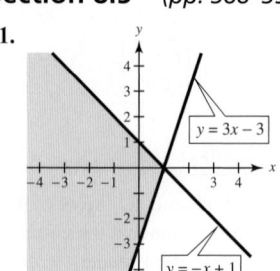   3.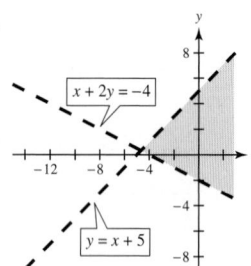

5. 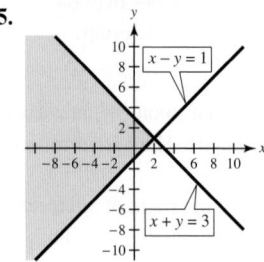   7.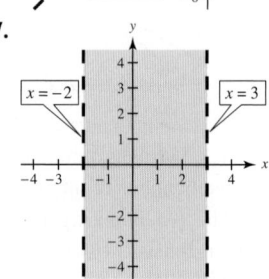

9. 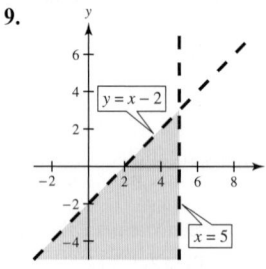   11.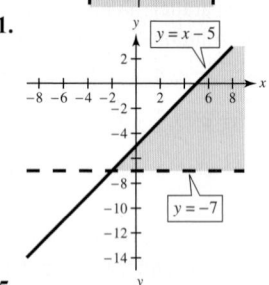

13. 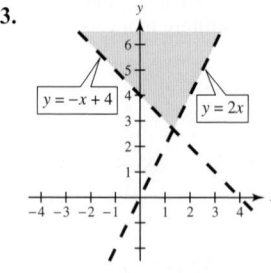   15.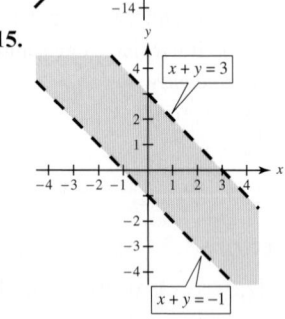

17. 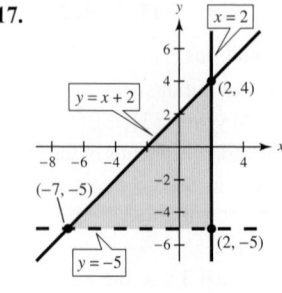   19.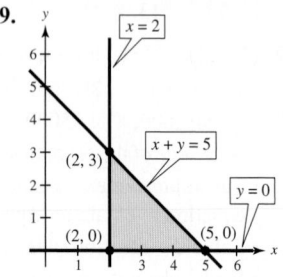

21.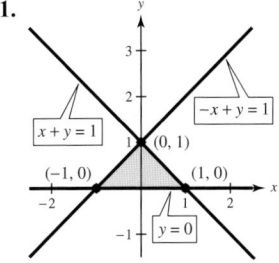

23.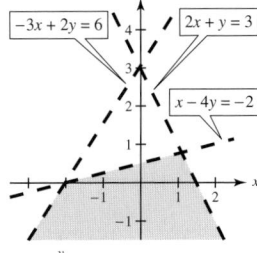

25.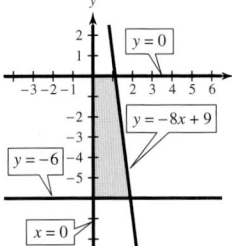

27.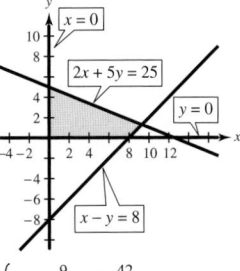

29. $\begin{cases} x \geq 1 \\ x \leq 8 \\ y \geq -5 \\ y \leq 3 \end{cases}$

31. $\begin{cases} y \leq \frac{9}{10}x + \frac{42}{5} \\ y \geq 3x \\ y \geq \frac{2}{3}x + 7 \end{cases}$

33.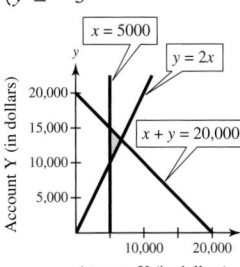

$\begin{cases} x + y \leq 20{,}000 \\ x \geq 5{,}000 \\ y \geq 2x \end{cases}$

35. $\begin{cases} x \leq 90 \\ y \leq 0 \\ y \geq -10 \\ y \geq -\frac{1}{7}x \end{cases}$

37. Use dashed lines for inequalities with $<$ or $>$, and use solid lines for equalities with $\leq$ or $\geq$.

39. • Sketch the line that corresponds to each inequality.
• Lightly shade the half-plane that is the graph of each linear inequality.
• The graph of the system is the intersection of the half-planes.

41.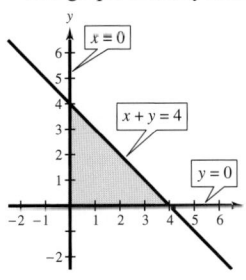

43.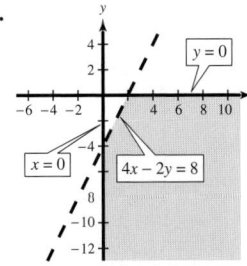

45.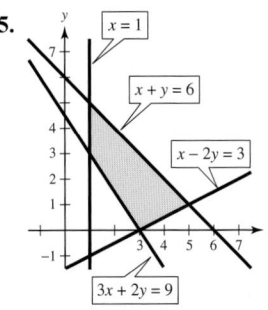

47. $\begin{cases} x + y \geq 15{,}000 \\ 15x + 25y \geq 275{,}000 \\ x \geq 8{,}000 \\ y \geq 4{,}000 \end{cases}$

49.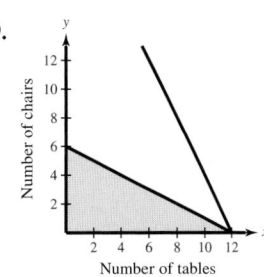

$\begin{cases} x + 2y \leq 12 \\ \frac{3}{2}x + \frac{3}{4}y \leq 18 \\ x \geq 0 \\ y \geq 0 \end{cases}$

51. The graph of a linear equation splits the xy-plane into two parts, each of which is a half-plane. $y < 5$ is a half-plane.

53. Yes; The solution of a system of linear inequalities is a single point if the system consists of two pairs of inequalities graphed as solid lines, each having different inequalities but the same corresponding equations. For example,
$\begin{cases} y \geq 3 \\ y \leq 3 \\ y \geq x \\ y \leq x \end{cases}$

55. 3^4 **57.** $\left(\frac{1}{2}\right)^6$ **59.** $(-4) \cdot (-4) \cdot (-4) \cdot (-4)$
61. $\left(-\frac{3}{4}\right) \cdot \left(-\frac{3}{4}\right)$

63.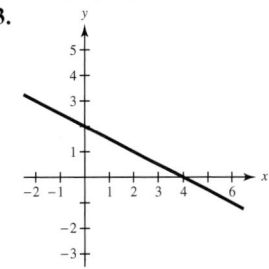

65.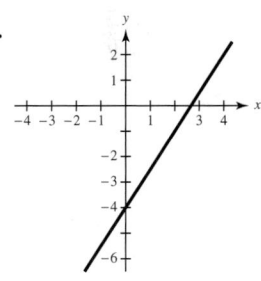

Review Exercises (pp. 398–401)

1. (a) Solution (b) Not a solution
3. (a) Not a solution (b) Solution
5. $(5, 1)$ **7.** $(1, 1)$ **9.** No solution
11. Infinitely many solutions **13.** $(-1, 1)$ **15.** $(36, 16)$
17. $(4, 8)$ **19.** $(4, 2)$ **21.** $(4, -1)$ **23.** $\left(\frac{5}{2}, 3\right)$
25. $(5, -5)$ **27.** No solution
29. Infinitely many solutions **31.** 5%: \$8000; 10%: \$4000
33. $(5, -2)$ **35.** $(3, 0)$ **37.** $(8, -3)$ **39.** $\left(\frac{4}{3}, \frac{5}{6}\right)$
41. $\left(\frac{3}{5}, \frac{1}{2}\right)$ **43.** No solution **45.** Infinitely many solutions
47. 3-credit courses: 32; 4-credit courses: 15
49. 6 dimes, 9 quarters **51.** 24 in. by 36 in. **53.** \$77.86
55. Gasoline: \$3.85/gal; Diesel fuel: \$4.15/gal **57.** $\frac{9}{2}$ hr

59.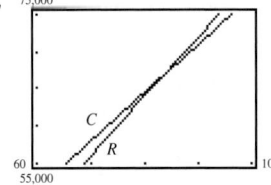

83 units; $R(83) = \$66{,}400$

61. 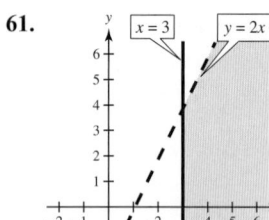   **63.**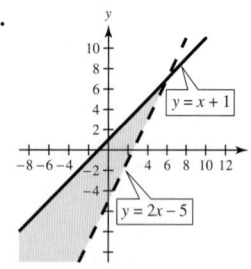

65. 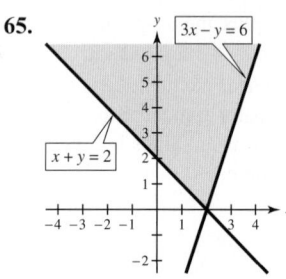   **67.**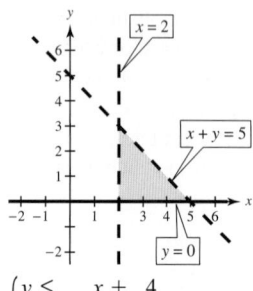

69. 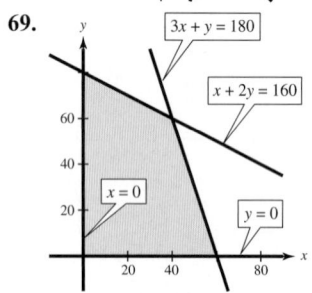   **71.** $\begin{cases} y \leq x + 4 \\ y \geq x - 2 \\ y \geq -2x + 7 \\ y \leq -2x + 22 \end{cases}$

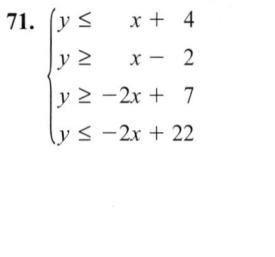

73. 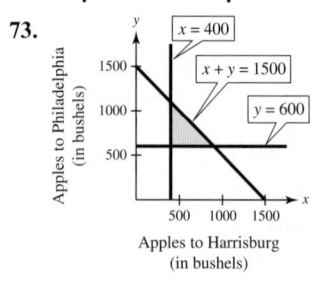   **75.**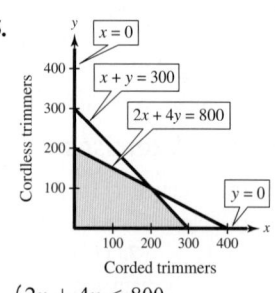

$\begin{cases} x + y \leq 1500 \\ x \geq 400 \\ y \geq 600 \end{cases}$ $\begin{cases} 2x + 4y < 800 \\ x + y \leq 300 \\ x \geq 0 \\ y \geq 0 \end{cases}$

Chapter Test (p. 402)

1. $(5, 4)$ because it satisfies both equations, whereas $(3, -2)$ does not.
2. One solution   **3.** No solution
4. Infinitely many solutions   **5.** $(2, 3)$ **6.** $(-2, -6)$
7. $(3, -5)$ **8.** $\left(\tfrac{3}{2}, 2\right)$ **9.** No solution   **10.** $(3, 4)$
11. $(2, 4)$ **12.** $(2, 6)$ **13.** $\left(3, \tfrac{5}{2}\right)$ **14.** $(3, 2)$
15. 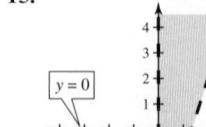   **16.**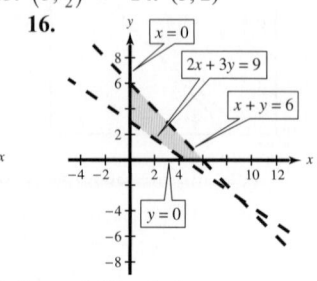

17. 12 liters of 30% solution; 8 liters of 5% solution

18.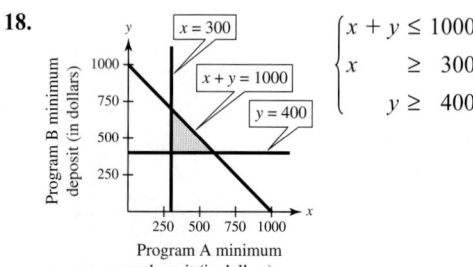

$\begin{cases} x + y \leq 1000 \\ x \geq 300 \\ y \geq 400 \end{cases}$

CHAPTER 9

Section 9.1 (pp. 404–411)

1. $4, -4$ **3.** $6, -6$ **5.** $\tfrac{3}{7}, -\tfrac{3}{7}$ **7.** Not possible
9. 3   **11.** -2 **13.** $1, -1$ **15.** Not possible
17. 10   **19.** -10 **21.** Not possible   **23.** 3
25. $-\tfrac{1}{3}$ **27.** $\tfrac{7}{8}$ **29.** 2   **31.** -5 **33.** $\tfrac{2}{3}$ **35.** 10
37. $\tfrac{1}{2}$ **39.** 2   **41.** Irrational   **43.** Rational
45. Irrational   **47.** Rational   **49.** Rational
51. Irrational   **53.** Rational   **55.** 6.557
57. Not possible   **59.** -11.705 **61.** Not possible
63. -22.755 **65.** 3.979   **67.** 12.583   **69.** -0.687
71. (a) 4   (b) Not possible   **73.** (a) Not possible   (b) 4.90
75. (a) 3   (b) 8.54   **77.** 7.4   **79.** 8.4   **81.** 11.4
83. 17.3   **85.** 0.026 in.   **87.** 756 ft × 756 ft
89. Index: 2, Radicand: $\tfrac{1}{16}$ **91.** Yes. $\left(-\tfrac{1}{4}\right)^2 = \tfrac{1}{16}$
93. $\sqrt{\tfrac{81}{121}} = \tfrac{9}{11}, -\sqrt{\tfrac{81}{121}} = -\tfrac{9}{11}$ **95.** $\sqrt{0.01} = 0.1, -\sqrt{0.01} = -0.1$
97. Not possible   **99.** $\tfrac{2}{3}$; Principal: $\tfrac{2}{3}$
101. -1; Principal: -1 **103.** $-2, 2$; Principal: 2
105. 27   **107.** 0.5   **109.** 4.33 in.
111. (a)

x	0	1	2	4	6	8
$\sqrt{x}$	0	1	1.41	2	2.45	2.83

x	10	12	14	16	18	20
$\sqrt{x}$	3.16	3.46	3.74	4	4.24	4.47

(b)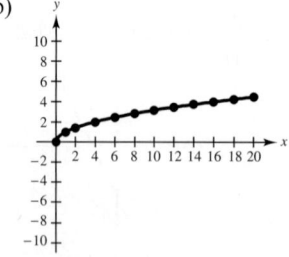

113. (a) 8.2   (b) 142   (c) 22   (d) 850
115. $x < 0$. $\sqrt{(-4)^2} = \sqrt{16} = 4$
117. A positive real number has two real nth roots when n is an even number.
119. 4   **121.** 13   **123.** 8   **125.** 1

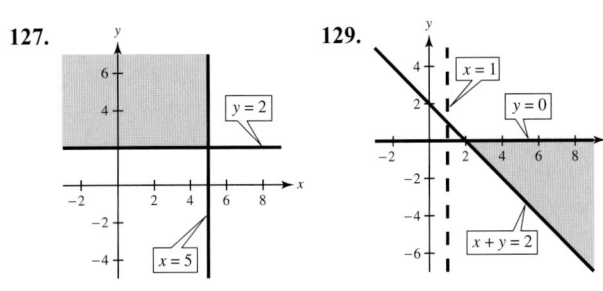

127.

129.

5. Not possible 6. -1.2 7. Irrational 8. Irrational
9. Rational 10. 7.810 11. 18.235 12. 0.6
13. $\sqrt{95}$ 14. $\sqrt{30xy}$ 15. $\sqrt{7}$ 16. $5\sqrt{2}$
17. $6|x|\sqrt{2}$ 18. $3b^4\sqrt{5}$ 19. $4x$ 20. $x^2\sqrt[4]{x}$
21. $3u^2|v|\sqrt{2u}$ 22. $4\sqrt{2}$ 23. $\dfrac{3|x|\sqrt{7}}{8}$ 24. $3b^2\sqrt{5}$
25. $\dfrac{\sqrt{6}}{2}$ 26. $\dfrac{\sqrt{3}}{3}$ 27. $\dfrac{1}{3x}$ 28. 17.5 ft
29. 19 ft by 19 ft

Section 9.2 (pp. 412–421)

1. $\sqrt{4} \cdot \sqrt{15} = 2\sqrt{15}$ 3. $\sqrt{64} \cdot \sqrt{11} = 8\sqrt{11}$ 5. $2\sqrt{2}$
7. $3\sqrt{5}$ 9. $2\sqrt[3]{6}$ 11. $2\sqrt[4]{3}$ 13. $6\sqrt{2}$ 15. $6\sqrt{5}$
17. $-2\sqrt[3]{3}$ 19. $-4\sqrt[3]{3}$ 21. $3\sqrt[4]{2}$ 23. $2|x|$
25. $2|x|\sqrt{21}$ 27. $|x^3|$ 29. $2|x^3|\sqrt{5}$ 31. $8x\sqrt{x}$
33. $|x|y\sqrt{y}$ 35. $6x^2y^4\sqrt{5x}$ 37. $3a^3\sqrt{a}$
39. $t\sqrt[4]{t^3}$ 41. $2y\sqrt[4]{y}$ 43. $\dfrac{\sqrt{3}}{2}$ 45. $\dfrac{\sqrt{13}}{7}$ 47. $\dfrac{\sqrt{3}}{2}$
49. $\dfrac{|x|}{3}$ 51. $\dfrac{2|x|\sqrt{3}}{5}$ 53. $\dfrac{|x^3|}{4|y|}$ 55. $\dfrac{\sqrt{3}}{3}$
57. $\dfrac{\sqrt{7}}{7}$ 59. $\dfrac{\sqrt{22}}{4}$ 61. $\dfrac{\sqrt{5x}}{x}$ 63. $\dfrac{\sqrt{3x}}{4x^3}$ 65. $\dfrac{4\sqrt[3]{3}}{3}$
67. $\dfrac{7\sqrt[3]{9}}{3}$ 69. $\dfrac{9\sqrt[4]{4}}{2}$ 71. $\dfrac{\sqrt[3]{x}}{x}$ 73. $\dfrac{\sqrt[3]{y}}{2y}$
75. $16\pi\sqrt{5} \approx 112.4$ ft² 77. $2\pi\sqrt{29} \approx 33.8$ m²
79. If the nth roots of a and b are real, then
$\sqrt[n]{ab} = \sqrt[n]{a} \cdot \sqrt[n]{b}$.
$\sqrt[3]{108} = \sqrt[3]{27} \cdot \sqrt[3]{4} = 3\sqrt[3]{4}$
81. • All possible nth-powered factors have been removed from each radical.
• No radical contains a fraction.
• No denominator of a fraction contains a radical.
83. $\dfrac{\sqrt{rt}}{2r}$ 85. $\dfrac{2x\sqrt{xy}}{y}$ 87. $>$ 89. $>$ 91. $<$
93. $10\sqrt{2} \approx 14.14$ square units
95. $12\sqrt{6} \approx 29.39$ square units
97. (a) $P = d\sqrt{d}$ (b) 29.5 yr (c) 0.6 yr
99. (a) False. (b) True
$\sqrt{(-3)^2} \stackrel{?}{=} -3$
$\sqrt{9} \stackrel{?}{=} -3$
$3 \neq -3$
(c) False. (d) True
$\sqrt[3]{(-2)^3} \stackrel{?}{=} |-2|$
$\sqrt[3]{-8} \stackrel{?}{=} |-2|$
$-2 \neq 2$
101. False. $\dfrac{\sqrt{50}}{\sqrt{2}} = \sqrt{\dfrac{50}{2}} = \sqrt{25} = 5$
103. False. $\sqrt[3]{72x^4} = \sqrt[3]{8 \cdot 9 \cdot x^3 \cdot x} = 2x\sqrt[3]{9x}$
105. $\dfrac{x^2}{6(2x-5)}, x \neq 0$ 107. 9 109. Not possible
111. -4 113. -2

Mid-Chapter Quiz (p. 422)

1. 11 2. -0.5 3. -2 4. Not possible

Section 9.3 (pp. 424–431)

1. $18\sqrt{11}$ 3. $2\sqrt{5}$ 5. $-11\sqrt{3} - 5\sqrt{7}$
7. $-2\sqrt{17} + 8\sqrt{2}$ 9. $6\sqrt[3]{5}$ 11. $-5\sqrt[4]{8}$
13. $8\sqrt[3]{7}$ 15. $5\sqrt{u} + 5$ 17. $7\sqrt[3]{6} + 4\sqrt[3]{x^2}$ 19. $6\sqrt{2}$
21. $54\sqrt{5}$ 23. $10\sqrt{b}$ 25. $9\sqrt{x}$ 27. $-2\sqrt{5z}$
29. $9\sqrt{2u}$ 31. $14x\sqrt{5x}$ 33. $-2a^2\sqrt{3a}$ 35. 4
37. $2\sqrt{15}$ 39. 2 41. $6\sqrt{2} + 8\sqrt{6}$ 43. $2 + 5\sqrt[3]{2}$
45. $-1 + 2\sqrt{2}$ 47. $x + 5\sqrt{x} - 14$ 49. -10
51. $4 - \sqrt{5}; 11$ 53. $\sqrt{t} + 5; t - 25, t \geq 0$
55. $\sqrt{15} + \sqrt{7}; 8$ 57. $\sqrt{u} + \sqrt{2}; u - 2, u \geq 0$ 59. 34
61. Perimeter: $32\sqrt{x}$; Area: $55x$
63. Perimeter: $16\sqrt{7} + 14$; Area: $42\sqrt{7} + 84$
65. $\dfrac{\sqrt{14} + 2}{2}$ 67. $12 - 4\sqrt{3}$
69. $-\dfrac{\sqrt{65} - 7\sqrt{5} + \sqrt{13} - 7}{36}$ 71. $\dfrac{9\sqrt{x} - 18}{x - 4}$
73. Yes. Both terms have the same radical factor, $\sqrt{2}$.
75. They are conjugates. 77. $-\dfrac{4}{5}\sqrt{3}$ 79. $(|x| + 4)\sqrt{xy}$
81. $\dfrac{\sqrt{a}}{6}$ 83. $4\sqrt{13} + 17$ 85. $\sqrt[3]{4} - 2\sqrt[3]{2} + 1$
87. $\dfrac{x - 4\sqrt{x} - 5}{x - 1}$ 89. $\dfrac{9 - \sqrt{3}}{3}$ 91. $2\sqrt{2}$ 93. $>$
95. $>$ 97. $\dfrac{\sqrt{5} + 1}{2} \approx 1.62$
99. The Distributive Property can be used to add or subtract like radicals by factoring out the like radicand, and then adding or subtracting.
101. No. $\sqrt{2} + \sqrt{18} = \sqrt{2} + 3\sqrt{2} = 4\sqrt{2}$
103. No. Rationalizing the denominator does not change the value of the expression as squaring the expression does.
105. $(x+5)(x-5)$ 107. $(2x+1)(4x^2 - 2x + 1)$
109. $(5, 1)$ 111. $(8, 1)$ 113. $\dfrac{\sqrt{6}}{6}$ 115. $\dfrac{\sqrt{6x}}{4x}$

Section 9.4 (pp. 432–439)

1. 5 3. 7 5. No solution 7. 2 9. 2
11. No solution 13. $\dfrac{1}{2}$ 15. 2 17. 1 19. $\dfrac{1}{4}$, 1
21. $-1, 3$ 23. $30\sqrt{5} \approx 67$ ft 25. 5 27. 10
29. $\sqrt{65} \approx 8.06$ 31. $\sqrt{17} \approx 4.12$
33. The model is a good fit for representing velocities of a string on a guitar.
35. A radical equation is an equation that contains one or more radicals with variable radicands.
37. If a and b are the lengths of the legs of a right triangle, and c is the length of the hypotenuse, then $a^2 + b^2 = c^2$.
39. 13 41. 26 43. 31.64 ft 45. 3.24 ft

47. 36 ft   **49.** 29 units
51. No.
$\sqrt{x} = -5$
$\sqrt{25} \stackrel{?}{=} -5$
$5 \neq -5$
53. Yes. The two legs can be of the same length l and the hypotenuse of length $\sqrt{2}l$.
55. Perpendicular   **57.** Parallel   **59.** $\dfrac{r+1}{r}, r \neq 1$
61. $\dfrac{2\sqrt{6}+2}{5}$ **63.** $\dfrac{\sqrt{35}+\sqrt{10}+4\sqrt{7}+4\sqrt{2}}{5}$

Review Exercises (pp. 442–445)
1. 7, −7   **3.** Not possible   **5.** −5   **7.** 2, −2
9. 11   **11.** 1.2   **13.** Not possible   **15.** −3   **17.** $\frac{2}{5}$
19. 7.280   **21.** 0.387   **23.** 7.899   **25.** −0.098
27. $\sqrt{12} \approx 3.46$ **29.** $4\sqrt{3}$ **31.** $4\sqrt{10}$ **33.** $\dfrac{\sqrt{23}}{3}$
35. $\dfrac{2\sqrt[3]{5}}{3}$ **37.** $2\sqrt[3]{4}$ **39.** $2\sqrt[4]{6}$ **41.** $6x^2$ **43.** $2y\sqrt{y}$
45. $x\sqrt{xy}$ **47.** $4a\sqrt{2ab}$ **49.** $2x^2$ **51.** $3y\sqrt[4]{y}$
53. $\dfrac{\sqrt{21}}{7}$ **55.** $2\sqrt{3}$ **57.** $\dfrac{\sqrt{15}}{6}$ **59.** $\dfrac{4\sqrt{x}}{x}$
61. $\dfrac{\sqrt{11ab}}{b}$ **63.** $\dfrac{x\sqrt{2y}}{3y^2}$ **65.** $\dfrac{3\sqrt[3]{4}}{2}$ **67.** $\dfrac{\sqrt[3]{4x}}{x}$
69. $12\sqrt{2}$ **71.** $3\sqrt{5}+4\sqrt{3}$ **73.** $-14\sqrt{5}$
75. $8\sqrt{3}$ **77.** $6\sqrt[4]{4}$ **79.** $9\sqrt[4]{x^3}-3\sqrt[4]{y^3}$ **81.** $2\sqrt{y}$
83. $6y^2\sqrt{7y}$ **85.** $3\sqrt{2}+\sqrt{3}$ **87.** $\sqrt[4]{12}-\sqrt[4]{6}$
89. $10+4\sqrt{2}$ **91.** −2   **93.** $9-4\sqrt{5}$
95. $\sqrt[5]{4}+6\sqrt[5]{2}+9$ **97.** $\sqrt{x}-9; x=81, x \geq 0$
99. $12+\sqrt{t}; 144-t, t \geq 0$ **101.** $2\sqrt{3}+3$
103. $\dfrac{x-6\sqrt{x}+9}{x-9}$ **105.** 169   **107.** No solution
109. 100   **111.** $\frac{17}{2}$ **113.** 3   **115.** 8
117. $4\sqrt{3} \approx 6.93$ **119.** 5   **121.** $\sqrt{137} \approx 11.70$
123. $\sqrt{37} \approx 6.08$ **125.** $\sqrt{146} \approx 12.08$ **127.** 100 ft
129. 2.48 ft   **131.** $20\sqrt{34} \approx 116.6$ ft
133. $2\sqrt{541} \approx 46.52$ ft

Chapter Test (p. 446)
1. (a) 11
  (b) Not possible; There is no real number that can be multiplied by itself to obtain −36.
2. (a) 3   (b) −4   **3.** $2\sqrt{7}$ **4.** $3\sqrt[3]{2}$ **5.** $4|x|y\sqrt{2y}$
6. $\dfrac{x\sqrt{3x}}{y^2}$ **7.** $\dfrac{\sqrt{15}}{3}$ **8.** $\sqrt[3]{2}$ **9.** $3\sqrt{2}$ **10.** $20\sqrt{3x}$
11. $8\sqrt[3]{5}-6\sqrt[3]{4}$ **12.** $\sqrt{2}|x|-20\sqrt{2x}$ **13.** $2\sqrt{3}-6$
14. $\sqrt[3]{10}+3\sqrt[3]{5}$ **15.** $2\sqrt{6}-9$ **16.** $1-4\sqrt{x}+4x$
17. $\sqrt{3}+5; -22$ **18.** $2\sqrt{6}-2$ **19.** 25   **20.** $\frac{13}{4}$
21. 5   **22.** 6   **23.** $2\sqrt{13} \approx 7.21$ **24.** 10
25. 125 units/day

Cumulative Test (pp. 447–448)
1. All real numbers x such that $x \neq 2$ and $x \neq -2$ **2.** $(6x^3)$
3. $-\dfrac{2}{x+4}, x \neq 4$ **4.** $\dfrac{x-5}{x-2}, x \neq -2$ **5.** $\dfrac{c+10}{c^2}, c \neq 1$
6. $\dfrac{3c^2}{4(c-1)}, c \neq 0$ **7.** $\dfrac{2(x+3)}{(x+2)(x-2)}$ **8.** $\dfrac{5x^2-2x+4}{x^2(x-2)}$
9. $\dfrac{9}{6+2x}, x \neq 0$ **10.** $\dfrac{2a^2-2}{a+2}, a \neq 0$ **11.** $\dfrac{1}{3}$
12. −10   **13.** −9   **14.** $\frac{9}{2}$ **15.** (5, −1)   **16.** (1, −1)
17. (5, 5)   **18.** $\left(-4, \frac{5}{4}\right)$ **19.** (3, 1)   **20.** $\left(\frac{3}{2}, 1\right)$
21. Answers will vary.   **22.** Answers will vary.
Sample answer:
$\begin{cases} x+y=1 \\ x+y=-1 \end{cases}$
Sample answer:
$\begin{cases} x+y=1 \\ 2x+2y=2 \end{cases}$
23. Not possible   **24.** $-\frac{2}{3}$ **25.** 12   **26.** −5
27. $-3\sqrt{6}$ **28.** $5x\sqrt{2x}$ **29.** $2uv^2\sqrt[3]{4u}$ **30.** $\dfrac{4\sqrt{2}}{3|y|}$
31. $2\sqrt{x}$ **32.** $7+2\sqrt{7}$ **33.** $24-16\sqrt{2}$
34. $2\sqrt{5y}+2y$ **35.** 49   **36.** 29   **37.** 1   **38.** 0, 4
39. 5   **40.** $2\sqrt{29} \approx 10.77$ **41.** 50 mi/hr
42. Experienced employee: $4\frac{1}{2}$ hr; New employee: 9 hr
43. Regular: $3.95; Premium: $4.15   **44.** 8

CHAPTER 10
Section 10.1 (pp. 450–457)
1. 0, 3   **3.** 0, −2   **5.** −5, 5   **7.** $-\frac{8}{3}, \frac{8}{3}$ **9.** 2, 3
11. −2   **13.** $\frac{5}{4}$ **15.** $-\sqrt{5}, \sqrt{5}$ **17.** $-\sqrt{6}, \sqrt{6}$
19. $-\frac{7}{3}, \frac{7}{3}$ **21.** −10, 10   **23.** No real solution
25. 11, −19   **27.** $7-\sqrt{6}, 7+\sqrt{6}$
29. $-6+\sqrt{3}, -6-\sqrt{3}$ **31.** $-\frac{5}{3}, \frac{1}{3}$
33. $\dfrac{-2-\sqrt{5}}{5}, \dfrac{-2+\sqrt{5}}{5}$ **35.** $\dfrac{4-3\sqrt{3}}{3}, \dfrac{4+3\sqrt{3}}{3}$
37. No real solution   **39.** $-\frac{11}{2}, -\frac{1}{2}$
41. $\dfrac{-6-\sqrt{7}}{8}, \dfrac{-6+\sqrt{7}}{8}$ **43.** 6%   **45.** $0.08 = 8\%$
47. −2   **49.** 3, −3   **51.** 2 sec   **53.** Two solutions
55. If $u^2 = d$, where d is a negative real number, the equation would have no real solution.
57. −4, 4   **59.** $\frac{14}{3}, -\frac{14}{3}$ **61.** $-2\sqrt{2}, 2\sqrt{2}$
63. $-4\sqrt{3}, 4\sqrt{3}$ **65.** No real solution   **67.** −1, 7
69. $-\frac{1}{2}, -\frac{5}{2}$ **71.** No real solution   **73.** 13, −1
75. 0, $-\frac{10}{3}$ **77.** 20 units
79. (a)

h	1000	950	900	850	800	750	700
t	0	1.77	2.50	3.06	3.54	3.95	4.33

(b) No. The corresponding times differ by less and less as the heights decrease.
81. (a) Square Root Property unless you have the difference of two squares then you should factor
(b) Factoring
$ax^2 + bx = 0$
$x(ax + b) = 0$

83. Isolate the variable and take the square root of each side.

$ax^2 + c = 0 \quad\quad 2x^2 + 6 = 0$
$ax^2 = -c \quad\quad 2x^2 = -6$
$x^2 = -\dfrac{c}{a} \quad\quad x^2 = -3$
$x = \pm\sqrt{-\dfrac{c}{a}} \quad\quad x = \pm\sqrt{-3}$

Suggested revision: If $a > 0$, restrict the values of c to $c \le 0$ so that the solutions will be real. If $a < 0$, restrict the values of c to $c \ge 0$ so that the solutions will be real.

85. Quadrant I **87.** Quadrant III **89.** 9.381
91. 27.964 **93.** $\sqrt{58} \approx 7.62$ **95.** $\sqrt{2} \approx 1.41$

Section 10.2 (pp. 458–465)

1. $16, (x+4)^2$ **3.** $144, (y-12)^2$ **5.** $\dfrac{9}{4}, \left(t + \dfrac{3}{2}\right)^2$
7. $\dfrac{9}{64}, \left(t - \dfrac{3}{8}\right)^2$ **9.** $0, 8$ **11.** $-20, 0$ **13.** $-2 \pm \sqrt{5}$
15. $4 \pm 3\sqrt{2}$ **17.** $-7, 5$ **19.** $-4, 1$ **21.** $-\dfrac{3}{2}, 1$
23. $-\dfrac{1}{3}, 3$ **25.** $\dfrac{-9 \pm \sqrt{17}}{2}$ **27.** $\dfrac{-3 \pm \sqrt{17}}{4}$
29. $\dfrac{-3 + \sqrt{7}}{2} \approx -0.18; \dfrac{-3 - \sqrt{7}}{2} \approx -2.82$
31. $\dfrac{1 + \sqrt{13}}{6} \approx 0.77; \dfrac{1 - \sqrt{13}}{6} \approx -0.43$
33. No real solution   **35.** No real solution   **37.** 40 strollers
39. (a)

(b) $120 - h$
(c) $\dfrac{1}{2}(120 - h)h$; Height: $60 + \sqrt{30} \approx 65.48$ cm; Base: $60 - \sqrt{30} \approx 54.52$ cm

41. The leading coefficient must be 1.   **43.** $(x + 3)^2 = 4$
45. $2 \pm \sqrt{2}$ **47.** $3 + 2\sqrt{2}$ **49.** $2 + 2\sqrt{5}$
51. $\dfrac{10 - \sqrt{14}}{4} \approx 1.56$ sec; $\dfrac{10 + \sqrt{14}}{4} \approx 3.44$ sec
53. $6, 7$ **55.** $24x, (x + 12)^2 = x^2 + 24x + 144$
57. The error occurred when attempting to complete the square. The expression $9x^2 - 4x$ needs to be in the form $x^2 + bx$ before completing the square.

$9x^2 - 4x - 2 = 0$
$9x^2 - 4x = 2$
$x^2 - \dfrac{4}{9}x = \dfrac{2}{9}$
$x^2 - \dfrac{4}{9}x + \left(-\dfrac{2}{9}\right)^2 = \dfrac{2}{9} + \left(-\dfrac{2}{9}\right)^2$
$\left(x - \dfrac{2}{9}\right)^2 = \dfrac{22}{81}$
$x - \dfrac{2}{9} = \pm\sqrt{\dfrac{22}{81}}$
$x = \dfrac{2 \pm \sqrt{22}}{9}$

59. True. Quadratic equations with real solutions that have no x-term cannot be solved by completing the square. For example, $x^2 - 9 = 0$.
61. (a) Solution   (b) Not a solution
63. (a) Not a solution   (b) Solution

65. $\sqrt{13}, -\sqrt{13}$ **67.** No real solution   **69.** $\dfrac{10}{3}, -\dfrac{10}{3}$
71. $3 \pm 2\sqrt{6}$

Section 10.3 (pp. 466–473)

1. No real solution   **3.** Two real solutions
5. One real solution   **7.** $-3, 6$ **9.** $-5, -3$
11. $2, 3$ **13.** $-1, 5$ **15.** $-4, 2$ **17.** $3 \pm \sqrt{2}$
19. $\dfrac{-7 \pm \sqrt{93}}{2}$ **21.** $\dfrac{5}{2}$ **23.** No real solution
25. No real solution   **27.** $\pm 2\sqrt{5}$ **29.** $-8, 0$
31. $3 \pm 5\sqrt{3}$ **33.** 10   **35.** $\dfrac{3 \pm \sqrt{11}}{2}$
37. 13.1 mi/hr   **39.** 6.2 in. × 4.3 in.
41. The opposite of b, plus or minus the square root of b squared minus $4ac$, all divided by $2a$
43. The four methods are factoring, the Square Root Property, completing the square, and the Quadratic Formula.
45. $\dfrac{7 + \sqrt{37}}{3} \approx 4.361; \dfrac{7 - \sqrt{37}}{3} \approx 0.306$
47. $\dfrac{100 + 5\sqrt{394}}{3} \approx 66.416; \dfrac{100 - 5\sqrt{394}}{3} \approx 0.251$
49. $\dfrac{1 \pm \sqrt{33}}{2}$ **51.** $3 + \sqrt{11}$
53. (a) 0 sec; $\dfrac{5}{4}$ sec   (b) 10.21 sec
55. After the binomial factors are multiplied, the Quadratic Formula could be used. This would not be the most efficient method, because the quadratic equation is already factored.
57. Proof   **59.** 3   **61.** $5z^2$ **63.** $x + 1$ **65.** $0, 6$
67. $-4 \pm \sqrt{26}$

Mid-Chapter Quiz (p. 474)

1. $2, 5$ **2.** ± 20 **3.** $-7, \dfrac{5}{2}$ **4.** $4, -\dfrac{3}{8}$ **5.** ± 50
6. $1, 7$ **7.** $-3 \pm 2\sqrt{5}$ **8.** $\dfrac{-3 \pm \sqrt{10}}{2}$ **9.** $\dfrac{-3 \pm \sqrt{5}}{2}$
10. $\dfrac{2 \pm \sqrt{34}}{3}$ **11.** $8 \pm 4\sqrt{6}$ **12.** $\dfrac{5 \pm \sqrt{13}}{6}$
13. No real solution   **14.** Two real solutions
15. Two real solutions   **16.** One real solution   **17.** 5.5%
18. $\dfrac{7\sqrt{7}}{4} \approx 4.63$ sec   **19.** 7.4 in. × 20.8 in.

Section 10.4 (pp. 476–483)

1. Upward   **3.** Downward   **5.** Negative   **7.** Positive
9. Upward   **11.** Upward   **13.** Downward
15. Downward
17.

19.

$(-4, 0), (4, 0), (0, 16)$

$(0, 0), (2, 0)$

21. 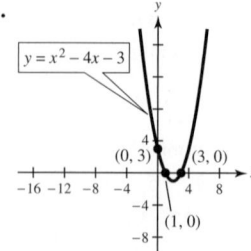 **23.**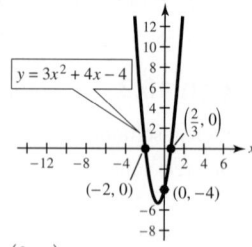

(1, 0), (3, 0), (0, 3) $\left(\frac{2}{3}, 0\right)$, (−2, 0), (0, −4)

25. (0, 2) **27.** (2, 3) **29.** (5, 31)
31. $\left(-\frac{5}{2}, -\frac{37}{4}\right)$ **33.** (−1, −11) **35.** $\left(-\frac{3}{2}, \frac{59}{4}\right)$
37. 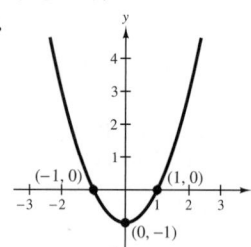   **39.**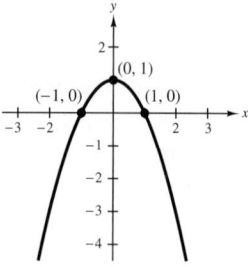

41. 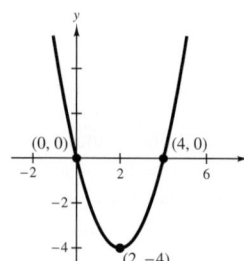   **43.**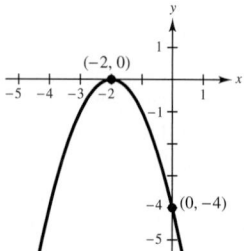

45. 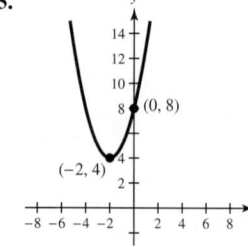   **47.**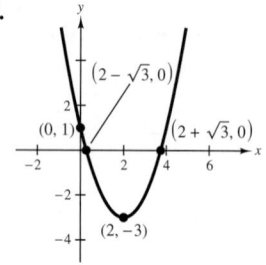

49. (a) 4 ft (b) 14 ft (c) $10 + 2\sqrt{35} \approx 21.8$ ft
51. The parabola opens upward.
53. To find the x-coordinate, use $x = -b/(2a)$.
To find the y-coordinate, substitute the x-coordinate into $y = ax^2 + bx + c$.

55. 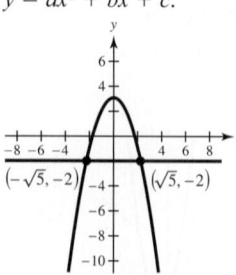   **57.**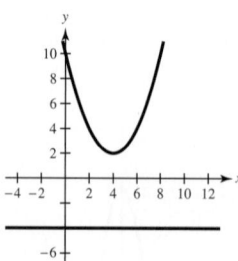

59. $y = x^2 − 4$; $y = 4 − x^2$
61. $y = x^2 + 2x − 3$; $y = -x^2 − 2x + 3$

63.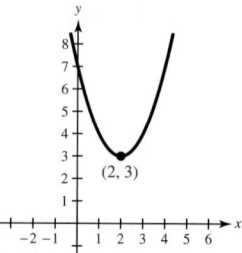

The vertex is (h, k) when the equation is in the form $y = (x − h)^2 + k$. So, the vertex is (2, 3).
65. (a) 15 yd (b) 38-yard line
67. (a) $18 − x$ (b) $A = x(18 − x)$
(c) 81 m² (d) 9 m × 9 m
69. When the discriminant is positive, there are two x-intercepts. When the discriminant is 0, there is one x-intercept. When the discriminant is negative, there are no x-intercepts.
71. (a) The x-coordinate of the vertex is halfway between 2 and 6 at $x = 4$. The y-coordinate is $(4 − 6)(4 − 2) = −4$. So, the vertex is (4, −4).
(b) $y = -x^2 + 8x − 12$; $y = -2x^2 + 16x − 24$
$y = 2x^2 + 16x + 24$
73. $y = -3x − 3$ **75.** $y = -\frac{3}{5}x − 4$ **77.** ±3, ±7, ±17
79. ±13, ±14, ±22, ±41 **81.** Two real solutions
83. One real solution

Section 10.5 (pp. 484–491)

1. Width: 11 in.; Length: 17 in. **3.** 96 ft × 96 ft
5. 56.4 mi **7.** 10 in. × 24 in. **9.** 45.83 ft
11. 6 hr; 12 hr
13. Combine A: 7.12 hr; Combine B: 9.12 hr
15. 25 people **17.** 24 people
19. *Sample answer*: Quadratic Formula
21. • Write a verbal model that will describe what you need to know.
• Assign labels to each part of the verbal model—numbers to the known quantities, and letters to the variable quantities.
• Use the labels to write an algebraic model based on the verbal model.
• Solve the resulting algebraic equation and check your solution.
23. 11, 12 **25.** 17, 19 **27.** 10 sec **29.** 8.37 sec
31. 30 ft, Width: 5 ft; Length: 10 ft
33. 49.80 ft, 40.20 ft **35.** 45.3 mi/hr; 40.3 mi/hr
37. 6% **39.** about 8.5%
41. To solve Exercise 31, you have to multiply the length and width, which will give you a squared term.
43. No. The square root in each solution of a quadratic equation will either simplify to a rational number, giving two rational solutions, or it will simplify to an irrational number, giving two irrational solutions.
45. All real values of x such that $x \neq 0$
47. All real values of x such that $x \neq -5$ and $x \neq 5$
49. 2 **51.** −4 **53.** (0, 5) **55.** (2, −2)

Section 10.6 (pp. 492–499)

1. $2i$ **3.** $\frac{2}{5}i$ **5.** $-2\sqrt{2}i$ **7.** 2 **9.** Equal
11. Not equal **13.** $a = -4, b = -2\sqrt{2}$ **15.** $a = 5, b = 3$

17. $a = 4$, $b = \frac{7}{2}$ 19. $10 + 4i$ 21. $-14 - 40i$
23. $5 + 9i$ 25. $-14 + 20i$ 27. $7 + 6i$
29. $20 - (12 + 2\sqrt{3})i$ 31. -36 33. 24
35. $4 + 18i$ 37. $-26 - 12i$ 39. $58 + 28i$ 41. 34
43. 5 45. 68 47. $-\frac{1}{5} + \frac{2}{5}i$ 49. $\frac{7}{8} - \frac{3}{4}i$
51. $-\frac{24}{53} + \frac{84}{53}i$ 53. $\frac{6}{29} + \frac{15}{29}i$ 55. $\pm 6i$ 57. $\pm 4\sqrt{2}i$
59. $3 \pm 5i$ 61. $\frac{5}{2} \pm \frac{3\sqrt{6}}{2}i$ 63. $-6 \pm \frac{11}{3}i$ 65. $\frac{4}{9} \pm \frac{5}{9}i$
67. $3 \pm i$ 69. $\frac{1}{2} \pm \frac{\sqrt{3}}{2}i$ 71. $4 \pm \sqrt{3}i$
73. $-\frac{3}{4} \pm \frac{\sqrt{15}}{4}i$ 75. $\sqrt{2}i$ 77. The FOIL Method
79. $\frac{3\sqrt{2}}{8}i$ 81. $15 - 7i$ 83. $-36i$ 85. $-10 + 65i$
87. 31 89. 100 91. $\pm\frac{\sqrt{17}}{3}i$ 93. $\frac{1}{2} \pm \frac{\sqrt{115}}{10}i$
95. $\frac{3}{8} \pm \frac{\sqrt{23}}{8}i$ 97. $2a$ 99. $a^2 + b^2$

101. The equation $x^2 = -1$ does not have any real number solutions because a negative number has no real square root.
103. You can use the FOIL Method to multiply two complex numbers such as $(2 - i)(4 + 3i)$ just as you can to multiply binomials such as $(x + 5)(x - 3)$.
105. False. All real and imaginary numbers can be written in the form $a + bi$, which is real when $b = 0$ and imaginary when $b \neq 0$.
107. Every four consecutive powers have the values i, -1, $-i$, and 1.
109. $-8x + 12$; Degree: 1; Leading coefficient: -8
111. 9; Degree: 0; Leading coefficient: 9
113. 25 115. 1 117. $\frac{23}{3}$ 119. $\frac{9}{16}$

Section 10.7 (pp. 500–507)

1. Domain: $\{-4, 1, 2, 4\}$; Range: $\{-3, 2, 3, 5\}$
3. Domain: $\{-9, \frac{1}{2}, 2\}$; Range: $\{-10, 0, 16\}$
5. Domain: $\{-1, 1, 5, 8\}$; Range: $\{-7, -2, 3, 4\}$
7. Function 9. Function 11. Function
13. Function 15. Not a function 17. Not a function
19. (a) 6 (b) 6 (c) 66 (d) 11
21. (a) 1 (b) 15 (c) 0 (d) 0
23. $D = \{0, 1, 2, 3, 4\}$
    $R = \{4, 3, 2, 1, 0\}$
25. $D = \{-8, -6, 2, 5, 12\}$
    $R = \{-1, 0, 7, 10\}$
27. $D = \{-5, -4, -3, -2, -1\}$
    $R = \{2\}$
29. Domain: The set of all real numbers r such that $r > 0$
    Range: $A > 0$
31. 2400
33. A relation is any set of ordered pairs. A function is a relation in which no two ordered pairs have the same first component and different second components.
35. Given the graph of a set of points on a rectangular coordinate system, if a vertical line intersects the graph at more than one point, the relation does not represent a function.
37. Not a function 39. (a) 4 (b) 0 (c) 12 (d) $\frac{1}{2}$

41. (a) -1 (b) 0 (c) 26 (d) $-\frac{7}{8}$
43. (a) $f(10) = 15$, $f(15) = 12.5$ (b) Demand decreases.
45. (a) 100 mi (b) 200 mi (c) 500 mi
47. $P = 4s$; P is a function of s. If you make a table of values where $s > 0$, no first component will have two different second components.

49. Yes; Domain Range 51. Yes; Domain Range
    1 → 4   1 → 6
    2 → 5   2 → 7
    3 → 6   3 → 8
                       4 → 9
                       5

53. 14.3% 55. 172 57. $11 - 12i$ 59. -9
61. $-16 - 10i$

Review Exercises (pp. 510–513)

1. $-10, 0$ 3. $3, 6$ 5. $-2, 2$ 7. ± 7
9. $\pm 3\sqrt{2}$ 11. $\pm 4\sqrt{3}$ 13. $5 \pm \sqrt{3}$ 15. $2 \pm \sqrt{6}$
17. $2, 6$ 19. $-4 \pm 2\sqrt{2}$ 21. $\frac{21 \pm 5\sqrt{2}}{3}$
23. 36; $(x + 6)^2$ 25. $\frac{225}{4}$; $\left(t - \frac{15}{2}\right)^2$ 27. $3 \pm \sqrt{10}$
29. $\frac{1 \pm \sqrt{5}}{2}$ 31. $\frac{-5 \pm \sqrt{15}}{2}$ 33. $\frac{1 \pm \sqrt{5}}{4}$
35. Two real solutions 37. One real solution
39. One real solution 41. $-7, 6$ 43. $1, 5$
45. $3 \pm \sqrt{3}$ 47. $\frac{-1 \pm \sqrt{337}}{4}$ 49. $-\frac{5}{3}$ 51. $\pm 5\sqrt{10}$
53. $\frac{10 \pm \sqrt{70}}{3}$ 55. 0.1 57. $\frac{3 \pm \sqrt{17}}{2}$ 59. $4 + 2\sqrt{3}$
61. Upward 63. Downward 65. Downward

67. 69.
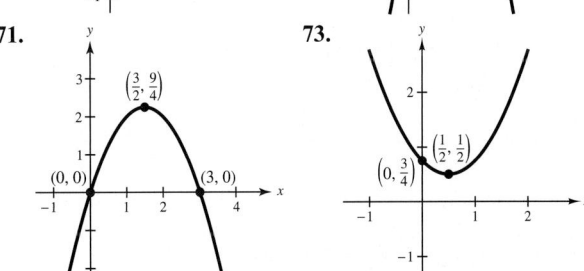

71. 73.
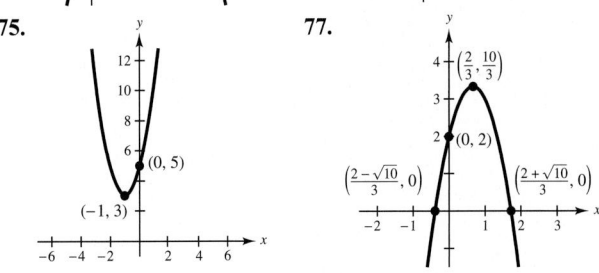

75. 77.

79. 6 cm 81. 18 m

83. Base: $4\sqrt{5} \approx 8.94$ in.; Height: $6\sqrt{5} \approx 13.42$ in.
85. 12 ft × 35 ft   87. 18.2 hr; 22.2 hr
89. 15 people   91. $9i$ 93. $\sqrt{5}i$ 95. 5
97. $a = 10, b = -4$ 99. $a = 4, b = 7$ 101. $8 - 3i$
103. $10 + 7i$ 105. 25   107. $11 - 60i$ 109. $-\frac{7}{3}i$
111. $\frac{5}{37} + \frac{30}{37}i$ 113. $-\frac{8}{17} + \frac{2}{17}i$ 115. $\pm 11i$
117. $\pm 5\sqrt{2}i$ 119. $-4 \pm 3\sqrt{2}i$ 121. $1 \pm 5i$
123. $\frac{3}{2} \pm \frac{\sqrt{3}}{2}i$ 125. $-3 \pm 2i$ 127. $\frac{1}{2} \pm \frac{\sqrt{3}}{24}i$
129. Domain: $\{-2, 3, 5, 8\}$; Range: $\{1, 3, 7, 8\}$
131. Domain: $\{-2, 2, 4\}$; Range: $\{-3, 0, 3, 4\}$
133. Function   135. Function   137. Not a function
139. Function   141. Not a function   143. Function
145. (a) $-\frac{3}{4}$ (b) 3   (c) $\frac{15}{2}$ (d) -1
147. (a) 64   (b) 63   (c) 48   (d) 0
149. (a) 3   (b) 13   (c) 5   (d) 0
151. (a) 38   (b) 30   (c) 20
153. $D = \{1, 2, 3, 4, 5\}$
    $R = \{5, 10, 15, -10, -15\}$
155. $D = \{-2, 0, 3, 4, 7\}$
    $R = \{-1, 6, -2, 0\}$

Chapter Test (p. 514)

1. $-20, 20$ 2. $-4 \pm 10i$ 3. $3 \pm \sqrt{2}i$ 4. $\dfrac{-9 \pm \sqrt{21}}{6}$
5. $1 \pm \sqrt{2}i$ 6. $\dfrac{-2 \pm \sqrt{2}}{2}$ 7. $3 \pm \sqrt{6}$ 8. $2 + \sqrt{3}$
9. Downward; Vertex: $(0, -7)$
10. Downward; Vertex: $(-1, 6)$
11. Upward; Vertex: $(2, 3)$
12. 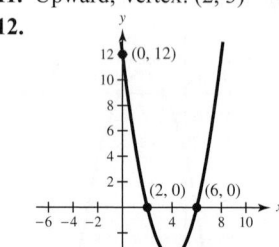   13.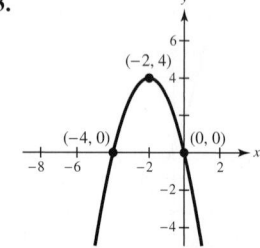
14. $-3 + 5i$ 15. $20 - 29i$ 16. $\frac{5}{3} - \frac{4}{3}i$
17. $D = \{-2, -1, 1, 4\}, R = \{-2, 1, 2, 3\}$
18. No, because some input values (0 and 1) have two different output values.
19. Yes, because it passes the Vertical Line Test.
20. (a) 0   (b) 0   (c) -16 (d) $-\frac{3}{8}$
21. 6 in. × 16 in.   22. 10 hr, 15 hr

Index of Applications

Biology and Life Sciences
Bamboo plant growth, 183
Calories burned
    playing basketball, 35
    playing tennis, 35
    racewalking, 35
Comparing ages, 243
Fitness trail, 172
Length of a dust mite, 248
Length of a leaf beetle, 214
Life expectancy, 173
Number of air sacs in human lungs, 214
Nutrition, 201, 395
Population, United States, 115
Smokestack emissions, 309
Target heart rate, 392
Walking speed and leg length, 409
Weight loss, 113
Weight of a mosquito, 214
Width of a human hair, 214

Business
Apartment rental, 207
Break-even analysis, 359, 387, 398, 400
Budget comparisons, 151
Cost
    of airline flights, 439
    of daily expenses, 193, 302
    per unit, 305, 328
    of printing a book, 175
Defective units, expected number, 133
Demand for a product, 439, 445, 446, 507, 513
Depreciation of equipment, 91, 164
Inventory of air conditioning units, 198, 199
Manufacturing, 201, 207, 395
    two products, 401
Petty cash, 201
Production cost, 164
Profit, 5, 15, 17, 27, 28, 50, 165, 225, 513
    of Coach, 190
    of Hewlett-Packard, 161
Real estate commission, 121
Retail discount, 123
Retail markup, 122
Revenue, 457, 462
Sales, 208
    of Aaron's, 190
    of AutoZone, 190
    of Coach, 115
Stock cost, 37
Stock value, 17, 21
Total salaries of the Boston Red Sox, 184
Warehouse utilization, 401

Chemistry and Physics
Analyzing the path of a ball, 481, 483
Astronomy, 151, 217, 421
Atmospheric pressure, 437
Boat and current, 327, 385
Boiling temperature of water, 309
Boltzmann's constant, 217
Car stopping distance, 167
Chemical reaction, 261
Electrical resistance, 335
Fluid rate problem, 135, 141, 191, 319
Force on a spring, 133, 164
Free-falling object, 281, 292, 300, 301, 438, 445, 455, 457, 464, 473, 474, 490
Frequencies of piano notes, 23
Hydraulic compression, 215
Light bulb wattage and lumens, 204
Pendulum length and period, 439, 445
Potential energy of a lightning bolt, 214
Relative density of hydrogen, 214
Safe load for a wooden beam, 507
Salt water mixture, 135, 243
Solution mixture, 117, 139, 142, 382, 402
Speed of light, 214
Sunlight by day of the year, 505
Temperature, 60
    Anchorage, Alaska, 161
    as measured by a weather balloon, 27
    of thawing meat, 5
Velocity of a stream, 409
Vibration velocity of a guitar string, 437
Weightless flight, 483

Construction
Bike path, 191
Cement block wall, 133
Dry mix mortar, 135
Exercise area for a Border Collie, 105
Fireplace, 117
Roof pitch, 183
Skateboarding ramp, 183
Tracking progress, 33

Consumer
Account, 7, 13, 28, 51
Amount owed on a loan, 193
Bird seed mixture, 156, 400
Calling card time available, 149
Car rental, 156, 171
Change due, 37, 54
Commissions
    production, 107, 108, 154, 164
    sales, 98, 107, 121, 206
Comparing costs
    car ownership, 367
    car rentals, 149, 151
    cell phone plans, 151
    home heating systems, 367
    music listening systems, 399
Comparing salaries, 243
Cost(s)
    of admission, 81, 90, 100, 489, 511
    of a band performance, 155
    of camping, 81, 98
    of car repair, 106
    of a cell phone, 37, 81
    of a chartered bus trip, 489
    of an engagement ring, 52
    of fruit, 77
    of fuel, 37, 133, 158, 380, 386
    of labor, 108, 154
    of living raise, 119
    of multiple units, 66, 85
    percent increase in, 125, 126
    of a plasma television, 52
    to produce a DVD, 205
    of renting movies and videos, 80
    of vehicles, 81, 381
Depreciation
    of a computer system, 184
    of a pizza oven, 191
    of a television, 193
    of a vehicle, 53, 172, 191, 309
Fuel consumption, 52
Housing budget, 126
Price(s)
    of a book and a calendar, 378
    of bottled soda, 39
    of electronics, 98, 155, 158
    of fuel, 37, 51, 400, 448
    of gold, 183
    of golf equipment, 155
    original, before discount, 383, 386, 400
Property assessment, 137
Rent for an apartment, 98
Salary, 107, 119
Savings, 27
Tax
    income, 80, 134
    sales, 45, 80
Unit price, 51, 130, 134, 155, 261
Wages, 66, 81, 90, 98, 193, 204
Wholesale cost, 383, 386, 400
Working two jobs, 109, 198, 199

Geometry
Area
    of an annulus, 281
    of an apartment, 83
    of the base of a box, 245
    circular, 124, 125, 504
    of the ground floor of a building, 245
    of a rectangle, 19, 45, 61, 65, 73, 83, 97, 124, 158, 251, 421, 427, 428

of a rectangle of fixed perimeter, 109, 483
of a region, 223, 224, 250, 252, 260, 264, 277, 278, 298, 299, 328
of a square, 504
trapezoidal, 71
of a triangle, 97, 237, 252, 317, 421, 428
Boundaries of a region, 401
Circle, doubling the radius, 143
Circumference of the Sun, 249
Comparing measurements, 129, 134, 155, 158
Diagonal of a rectangular solid, 409, 422
Diagonals of a polygon, 63
Dimension(s)
of a base drum, 137
of a baseball diamond, 435
of a box, 284, 295, 299, 300, 301
of a building, 284
of a circle, 485
of a circular broadcast region, 485
of a circular oil spill, 454
of a complex figure, 511
of a court, 105, 435
of a golf tee area, 235
of a Jamaican flag, 105
of a pyramid, 409
of a rectangle, 126, 154, 242, 251, 252, 259, 293, 384, 400, 463, 464, 471, 474, 484, 485, 486, 490, 511, 514
of a room, 235, 293, 422, 454
of a sandbox, 235, 273
of a softball diamond, 435
of a swimming pool, 137, 155
of a triangle, 137, 143, 155, 454, 463, 485, 511
of a window, 300
Geometric model, 465
for a difference of two cubes, 287
for factoring a trinomial, 268, 283
for a polynomial product, 230, 231, 236
Geometric probability, 308, 348
Golden section, 431
Height of an equilateral triangle, 411
Height of a prism, 27
Lateral surface area of a cone, 419
Length
of pieces of a board, 105, 108, 154
of a rectangle, 99
of rope segments, 28
of the sides of a sign, 105
Perimeter, 97, 219, 226
of a rectangle, 70, 73, 80, 97, 108, 116, 249, 326, 428
of a square, 507
of a triangle, 70, 72, 74, 116, 326, 428
Pythagorean Theorem, 438, 439, 444, 445, 446, 448, 486, 487, 491
Radius of a dime, 137
Resizing a rectangular picture, 135, 155

Segments of equal length on a number line, 337, 351
Stacking capacity, 23, 41
Surface area of a cone, 431
Surface area of a cylinder, 261
Triangle, doubling the height, 143
Using similar triangles, 132, 135, 157
Volume
of a block of ice, 275
of a box, 19, 267, 269, 273, 275
of a cord of wood, 28
of a cube, 213, 507
of Earth, 214
of a hot tub, 53
of a shower stall, 273
of a swimming pool, 275

Interest Rate

Compound interest, 237, 252, 453
Finding the interest rate, 142, 156, 453, 474, 491
Finding the time invested, 142
Interest rate in terms of monthly payment, 337
Investment mixture, 138, 366, 376, 380, 392, 399, 402
Simple interest, 60, 136, 156, 157, 261
Time to double an investment, 82

Miscellaneous

Auditorium seating, 393
Babylonian number system, 31
Batting average, 343
Book publishing, 305
Bowl-a-thon, 156
College credits, 400
Cooking, 39, 135
Course grade, 120
Election poll, 133
Exam scores, 21, 90, 113, 126, 151, 154, 167
Exponential communication, 41
Flowers, 141
Forensic archaeology, 13
Free throw percentage, 343
Fruit distribution, 401
Fund drive, 33
Game scores, 5, 11
Geocaching, 393
Jazz band audition, 302
Joint rate of flow, 342
Joint time to complete a task, 113, 126, 154
Legal cargo weight for a truck, 149
Number attending an event, 93
Number problem, 291, 300, 301, 351, 361, 369, 398, 465, 490
Oven repair, 157
Parachutist, 308
Photo printing, 319
Precipitation, 52, 90
Public TV station membership, 125

SAT score and grade-point average, 505
Sieve of Eratosthenes, 27
Slope of a ladder, 177
Slope of a loading ramp, 206
Super Bowl scores, 161
Swimming area, 393
Temperature change, 11, 15
Ticket sales, 91, 93, 106, 141, 367, 375, 395, 400
Time spent in class and studying, 134
Value of coins or bills, 77, 80, 93, 156, 386, 400
Video rentals, 400
Volleyball serve percentage, 343
Volunteer services, 91
Work-rate problem, 140, 156, 157, 342, 351, 448, 488, 511, 514

Time and Distance

Average speed, 341, 344, 351, 352, 386, 400, 448, 491
Distance
between two vehicles, 143
to the horizon, 421
jogging, 142
traveled by a bicycle, 80
traveled by a car, 60, 63, 83, 85, 98, 99, 175, 507
traveled by a plane, 155
traveled by a train, 83, 205
Elevation, 7, 9, 15, 171, 172, 191
at Death Valley, California, 7
at the Grand Canyon, 16
at the summits of volcanoes, 9
Fishing depth, 13
Flight path of an aircraft, 206
Jumping height with new shoes, 63
Map scale, 134
Speed
determined from skid marks, 409
jogging, 21
of sound, 507
up and down a trail, 470
of a vehicle, 11, 21, 51, 54, 91, 157
walking, 155
Time
to cross-country ski, 35
to overtake a slower jogger, 143
of travel, 81
traveled at each rate on a trip, 143
to walk to the subway, 35
Yards gained or lost in football, 11, 21, 90

U.S. Demographics

Civilian labor force, 220
College enrollments, 220
Government surplus, 165
Land per person in U.S., 215
Marital status, 173
Per capita cost to eliminate federal debt, 215
Postsecondary school enrollment, 115

Index

A

Absolute value, 6, 6*
Abundant number, 27
Add two integers, 12
Addition
  Associative Property of, 43, 64*
  Commutative Property of, 43, 64*
  of fractions, 32
    alternative rule, 32*
  of integers, 12
  Property of Inequality, 146
Additional problem-solving strategies, summary of, 82
Additive Identity Property, 43, 64*
Additive inverse, 11*
Additive Inverse Property, 43, 64*
Algebra, properties of, 64*
Algebraic
  equation, 86
  expression, 56, 56*, 87
    evaluating, 58
    expanding, 64*
    simplifying, 68, 68*
    terms of, 56
    translating phrases into, 78, 78*
  inequality, 144*
Algorithm
  borrowing, 14*
  carrying, 12*
  long division, 20
  vertical multiplication, 18
Alternative rule
  for adding two fractions, 32*
  for subtracting two fractions, 32*
Approximately equal to, 36*
Area formulas, 137
Arithmetic summary, 24
Associative Property
  of Addition, 43, 64*
  of Multiplication, 43, 64*
Average, 21*

B

Back-substitute, 362
Base, 40
Binomial, 218
  square of a, 234, 234*
Borrowing algorithm, 14*
Break-even point, 359*

C

Carrying algorithm, 12*
Cartesian plane, 160*
Check a solution
  of an equation, 86, 86*
  of an inequality, 145*
Clearing an equation of fractions, 112, 112*

Coefficient, 56, 218*
Combining rational expressions
  with like denominators, 320, 320*
  with unlike denominators, 323*
Common formulas, 137
  miscellaneous, 136
Commutative Property
  of Addition, 43, 64*
  of Multiplication, 43, 64*
Completing the square, 458, 458*
Complex conjugate, 495
Complex fraction, 330, 330*
Complex number, 493, 493*
  standard form of a, 493
Composite number, 22
Compound inequality, 144*
Conjugate, 427
  complex, 495
Consistent system, 356, 356*
Constant, 56, 56*
Constant term, 218
Constructing systems of linear equations, 380*
Coordinate, 160
Cost, 122*
Cross-multiplication, 131
Cube root, 404
Cubes
  difference of two, 285
  sum of two, 285

D

Decimal
  repeating, 36
  rounding a, 36, 36*
  terminating, 36
Decision digit, 36
Degree of a polynomial, 218, 218*
Degree of a term, 218*
Denominator, 20
Dependent system, 356, 356*
Dependent variable, 502
Difference, 14
  of two cubes, 285
  of two squares, 280, 280*
Discount, 123
Discount rate, 123
Discriminant, 466
Distance Formula, 436, 436*
Distance-Rate-Time Formula, 136
Distributive Property, 43, 64*
Divide evenly, 240*
Dividend, 20, 239
Dividing
  integers, rules for, 20
  a polynomial by a monomial, 238, 238*
  rational expressions, 315, 315*

Divisibility tests, 22*
Divisible, 22*
Division
  of fractions, 34
  of integers, 20
  Property of Inequality, 146
Divisor, 20, 24, 239
Domain
  of a function, 503
  of a rational expression, 304, 304*
  of a relation, 500

E

Equation(s), 86
  algebraic, 86
  clearing of fractions, 112, 112*
  equivalent, 88, 88*
  first-degree, 102*
  graph of an, 168, 168*
  of a line
    general form, 186, 189*
    point-slope form, 186
    slope-intercept form, 178
    summary, 191*
    two-point form, 187*
  linear
    in one variable, 102, 102*
    in two variables, 168*
  percent, 119, 119*
  quadratic, 289
    guidelines for solving, 289
    in two variables, 476*
  radical, 432, 432*
  second-degree polynomial, 288
  solution of, 162*
  solving, 86
  system of linear, 354, 354*
Equivalent
  equations, 88, 88*
  fractions, 30
  inequalities, 146
Evaluating an expression, 6*, 58
Evaluating a function, 503
Expanding an algebraic expression, 64*
Exponent(s), 40, 40*
  negative, 212, 212*
    rule, 212, 212*
  product and power rules of, 210, 210*
  quotient rules of, 211, 211*
  summary of rules of, 212
Exponential form, 40, 57, 57*
Expression, 6*
  algebraic, 56
  radical, 412*
  rational, 304, 304*
Extracting square roots, 451
Extraneous solutions, 339, 339*

*Terms that appear in the Math Help feature at AlgebraWithinReach.com

F

Factor, 22
  greatest common, 30*, 254*
    monomial, 255, 255*
  proper, 27
Factoring, 254
  $ax^2 + bx + c$, 270, 270*
    by grouping, guidelines for, 274
    guidelines for, 271
  by grouping, 257, 257*
  completely, 266
  out, 255*
  polynomials, guidelines for, 285
  $x^2 + bx + c$, guidelines for, 262, 262*
Factors, variable, 67, 67*
First-degree equation, 102*
FOIL Method, 229, 229*
Forming equivalent equations, 88, 88*
Formula, Distance, 436, 436*
Formula, Quadratic, 466, 466*
Formulas
  area, 137
  common, 137
  Distance-Rate-Time, 136
  miscellaneous, 136
  perimeter, 137
  simple interest, 136
  temperature, 136
  volume, 137
Fractions, 2*, 30*
  addition of, 32
    alternative rule, 32*
  clearing an equation of, 112, 112*
  complex, 330, 330*
  division of, 34
  equivalent, 30
  multiplication of, 34
  rules of signs, 30*
  subtraction of, 32
    alternative rule, 32*
  summary of rules of, 35*
  writing in simplest form, 30*
Function, 501
  domain, 503
  evaluating, 503
  name, 503
  notation, 503, 503*
  range, 503

G

General form
  of the equation of a line, 186, 189*
  of a quadratic equation, 289*
Golden section, 431
Graph
  of an equation, 168, 168*
  of an inequality, 144
  of a linear inequality in two variables, sketching, 195, 195*
  of a parabola, 169*
  of a quadratic equation, sketching, 478

Graphing
  point-plotting method, 168
  solution by, 355*
  a system of linear inequalities, 388
Greater than, 4
Greater than or equal to, 4*
Greatest common factor (GCF), 30*, 254*
Greatest common monomial factor, 255, 255*
Grouping, factoring by, 257, 257*
Guidelines
  for factoring
    $ax^2 + bx + c$, 271
    by grouping, 274
    polynomials, 285*
    $x^2 + bx + c$, 262
  for finding the least common multiple, 322
  for sketching a parabola, 480
  for solving
    a linear equation containing symbols of grouping, 110
    quadratic equations, 289, 469
    a system of linear equations, 375*
  for verifying solutions, 163

H

Half-planes, 195*

I

Identity
  additive, 43
  multiplicative, 43
If and only if, 180
i-form, 492
Imaginary number, 493, 493*
Imaginary part of a complex number, 493
Imaginary unit i, 492
Inconsistent system, 356, 356*
Independent variable, 502
Index of a radical, 405
Inequality (inequalities)
  algebraic, 144*
  checking a solution of, 145*
  compound, 144*
  equivalent, 146
  graph of, 144
  linear, 144*
    in two variables, 194
      graph of, 195, 195*
      solution of, 194
  properties of, 146
  satisfy, 144
  solution of, 144
  solution set of, 144
  solving, 144
  symbol, 4
  system of linear, 388, 388*
Integer(s), 2*
  addition of, 12

  negative, 2*
  positive, 2*
  rules for
    dividing, 20
    multiplying, 18
    subtraction of, 14
Intercepts, 170
Inverse
  additive, 11*
  multiplicative, 34*
Irrational number, 2, 2*

L

Leading coefficient of a polynomial, 218, 218*
Leading coefficient test for a parabola, 476
Least common denominator (LCD), 323*
Least common multiple (LCM), 32, 322*
  guidelines for finding, 322
Less than, 4
Less than or equal to, 4*
Like denominators, combining rational expressions, 32, 320, 320*
Like radicals, 424*
Like terms, 67, 67*
Linear
  equation(s), 102, 102*
    constructing systems of, 380*
    containing symbols of grouping, 110
    guidelines for solving a system of, 375*
    in one variable, 102, 102*
    solution of a system of, 354
    system of, 354, 354*
    in two variables, 168*
  extrapolation, 190*
  inequality (inequalities), 144, 194
    graph of, 195, 195*
    graphing a system of, 388
    solution of, 194
    solution of a system of, 388*
    system of, 388, 388*
    in two variables, 194
  interpolation, 190*
Lines
  parallel, 180
  perpendicular, 180
  summary of equations of, 191*
Long division algorithm, 20
Long division of polynomials, 239

M

Markup, 122
Markup rate, 122
Mathematical model, 90
  verbal, 76
Method of elimination, 370, 370*
Method of substitution, 362, 362*

*Terms that appear in the Math Help feature at AlgebraWithinReach.com

Miscellaneous common formulas, 136
Mixture problem, 138
Model
  mathematical, 90
  verbal, 90
  verbal mathematical, 76
Monomial, 218
Multiplication
  Associative Property of, 43, 64*
  Commutative Property of, 43, 64*
  of fractions, 34
  of integers, 18
  Property of Inequality, 146
Multiplicative Identity Property, 43, 64*
Multiplicative inverse, 34*
Multiplicative Inverse Property, 43, 64*
Multiplying
  integers, rules for, 18
  rational expressions, 312, 312*

N

Name of a function, 503
Natural number, 2, 2*
Negative, 3*
  exponent, 212, 212*
    rule, 212, 212*
  integer, 2*
  number, square root of, 492
Nonnegative, 3*
Notation, function, 503, 503*
nth root of a, 404
nth root of a number, 404
Number
  abundant, 27
  complex, 493, 493*
  composite, 22
  fraction, 2*
  imaginary, 493, 493*
  integer, 2, 2*
  irrational, 2, 2*
  natural, 2*
  negative, 3*
  nonnegative, 3*
  nth root of, 404
  perfect, 27
  positive, 3*
  prime, 22
  principal nth root of, 405, 405*
  pure imaginary, 493
  rational, 2, 2*
  real, 2
  whole, 2, 2*
Numerator, 20

O

Opposite of a number, 6, 6*, 11*
Order, 4*
  of operations, 42, 42*, 59*
Ordered pair, 160
Origin, 3*, 160

P

Parabola, 169*, 476, 476*
  graph of, 169*
  guidelines for sketching, 480
  leading coefficient test for, 476
  vertex of, 476*, 479, 479*
Parallel lines, 180
Percent, 118, 118*
  equation, 119, 119*
Perfect number, 27
Perfect square, 406
Perfect square trinomial, 283, 283*
Perimeter formulas, 137
Perpendicular lines, 180, 180*
Plotting, 3*
  points, 160, 160*
Point-plotting method of sketching a
    graph, 168
Point-slope form of the equation of a
    line, 186, 186*
Polynomial(s), 218, 218*
  constant term, 218
  degree, 218, 218*
  division by a monomial, 238, 238*
  equations, second-degree, 288
  guidelines for factoring, 285*
  leading coefficient, 218, 218*
  long division of, 239
  in one variable, 218*
  prime, 263
  standard form, 218*
  in x, 218
Positive, 3*
  integer, 2*
Power, 40, 40*
  rules of exponents, 210, 210*
Price, 122*
  unit, 130, 130*
Prime number, 22
Prime polynomials, 263
Principal nth root of a, 405
Principal nth root of a number,
    405, 405*
Problem-solving strategies, summary of,
    82
Product, 18
  rule of exponents, 210, 210*
  rule for radicals, 412
  of the sum and difference of two
    terms, 234, 234*
Proper factor, 27
Properties
  Additive Identity, 43, 64*
  Additive Inverse, 43, 64*
  of algebra, 64*
  Associative Property of Addition,
    43, 64*
  Associative Property of
    Multiplication, 43, 64*
  Commutative Property of Addition,
    43, 64*
  Commutative Property of
    Multiplication, 43, 64*
  Distributive, 43, 64*
  of equality, 88
  of inequalities, 146
  Multiplicative Identity, 43, 64*
  Multiplicative Inverse, 43, 64*
  of real numbers, 43
  Square Root, 451, 451*
    complex, 496
  Squaring Property of Equality, 432
  Zero-Factor, 288
Proportion, 131
  solving, 131
Pure imaginary number, 493
Pythagorean Theorem, 435

Q

Quadrant, 160
Quadratic equation(s), 289
  general form of, 289*
  guidelines for solving, 289, 469
  sketching the graph of, 478
  in two variables, 476*
Quadratic Formula, 466, 466*
Quotient, 20, 239
  rule for radicals, 416
  rules of exponents, 211, 211*

R

Radical(s), 405
  equation, 432, 432*
  expressions, 412*
    simplifying, 417
  index of, 405
  like, 424*
  product rule, 412
  quotient rule, 416
  symbol, 405*
Radicand, 405
Range of a function, 503
Range of a relation, 500
Rate
  discount, 123
  markup, 122
Ratio, 128
Rational expression(s), 304, 304*
  combining with like denominators,
    320, 320*
  combining with unlike denominators,
    323*
  dividing, 315, 315*
  domain of, 304, 304*
  multiplying, 312, 312*
  simplifying, 306, 306*
Rational number, 2, 2*
Rationalizing the denominator,
    416, 416*
Real number, 2
  line, 3
  properties of, 43
Real part of a complex number, 493

*Terms that appear in the Math Help feature at AlgebraWithinReach.com

Reciprocal, 34*
Rectangular coordinate system, 160, 160*
Reduced form of a rational expression, 306*
Relation, 500, 500*
  domain of, 500
  range of, 500
Remainder, 239
Repeated solution, 290
Repeating decimal, 36
Root
  cube, 404, 404*
  nth, 404
  principal nth, 405, 405*
  square, 404, 404*
Roots, extracting square, 451
Rounding a decimal, 36, 36*
Rounding digit, 36
Rules
  for dividing integers, 20
  of exponents
    product and power, 210, 210*
    quotient, 211, 211*
    summary, 212
  of fractions, summary of, 35*
  for multiplying integers, 18
  of signs for fractions, 30*

S

Satisfy
  an equation, 86, 163*
  an inequality, 144
Scientific notation, 214, 214*
Second-degree polynomial equation, 288
Set, 2*
Simple interest formula, 136
Simplest form, 30*
Simplified form of a rational expression, 306*
Simplify
  an algebraic expression, 68, 68*
  radical expressions, 417
  rational expressions, 306, 306*
Sketching
  the graph of a linear inequality in two variables, 195, 195*
  a graph, point-plotting method of, 168
  the graph of a quadratic equation, 478
  a parabola, guidelines for, 480
Slope of a line, 176, 177
Slope-intercept form of the equation of a line, 178
Solution(s)
  checking,
    of an equation, 86, 86*
    of an inequality, 145*
  of an equation, 86, 162*
  extraneous, 339, 339*
  guidelines for verifying, 163
  of a linear inequality, 144, 194
  point, 162*

repeated, 290
satisfy an equation, 86
set
  of an inequality, 144
  of a system of linear inequalities, 388*
steps, 88, 102*
of a system of linear equations, 354
  by graphing, 355*
of a system of linear inequalities, 388*
Solving
  an equation, 86
  an inequality, 144
  a linear equation containing symbols of grouping, 110
  a linear equation in nonstandard form, 104*
  a proportion, 131
  quadratic equations, guidelines for, 469
Special products, 234, 234*
Square of a binomial, 234, 234*
Square root(s), 404, 404*
  extracting, 451
  of a negative number, 492
  property, 451, 451*
    complex, 496
  of x^2, 414
Squaring each side of an equation, 432
Squaring Property of Equality, 432
Standard form
  of a complex number, 493
  of a polynomial, 218*
Steps of a solution, 88, 102*
Subset, 2*
Substitution, method of, 362, 362*
Subtract one integer from another, 14
Subtraction
  of fractions, 32
    alternative rule, 32*
  of integers, 14
  Property of Inequality, 146
Sum, 10
  of two cubes, 285
  of two squares, 282
Sum or difference of two cubes, 285
Summary
  of additional problem-solving strategies, 82
  arithmetic, 24
  of equations of lines, 191*
  of rules of exponents, 212
  of rules of fractions, 35
Symbols of grouping, 42, 42*
System
  consistent, 356, 356*
  dependent, 356, 356*
  inconsistent, 356, 356*
  of linear equations, 354, 354*
    constructing a, 380*
    guidelines for solving, 375*
    solution of, 354

of linear inequalities, 388, 388*
  graphing, 388
  solution of, 388*

T

Table of values, 162*
Temperature formula, 136
Terminating decimal, 36
Terms
  of an algebraic expression, 56
  like, 67, 67*
Tests, divisibility, 22*
Three approaches to problem solving, 162
Transitive Property of Inequality, 146
Translating phrases into algebraic expressions, 78, 78*, 90*
Trinomial, 218
  perfect square, 283, 283*
Two variables, quadratic equation in, 476*
Two-point form of the equation of a line, 187*

U

Undefined, 20*
Unit price, 130, 130*
Unlike denominators, combining rational expressions, 32, 323*

V

Variable, 56, 56*
  dependent, 502
  factors, 67
  independent, 502
Verbal mathematical model, 76
Verbal model, 76, 90
Verifying solutions, guidelines for, 163
Vertex of a parabola, 476*, 479, 479*
Vertex of a region, 388
Vertical Line Test, 502
Vertical multiplication algorithm, 18
Volume formulas, 137

W

Whole number, 2, 2*
Work-rate problem, 140
Writing a fraction in simplest form, 30*

X

x-axis, 160
x-coordinate, 160*
x-intercept, 170

Y

y-axis, 160
y-coordinate, 160*
y-intercept, 170

Z

Zero-Factor Property, 288